Lecture Notes in Computer Science 8469

Commenced Publication in 1973
Founding and Former Series Editors:
Gerhard Goos, Juris Hartmanis, and Jan van Leeuwen

David Pointcheval   Damien Vergnaud (Eds.)

Progress in Cryptology – AFRICACRYPT 2014

7th International Conference on Cryptology in Africa
Marrakesh, Morocco, May 28-30, 2014
Proceedings

 Springer

Volume Editors

David Pointcheval
Damien Vergnaud
Ecole Normale Supérieure
Computer Science Department
45, rue d'Ulm
75005 Paris, France
E-mail: {david.pointcheval, damien.vergnaud}@ens.fr

ISSN 0302-9743   e-ISSN 1611-3349
ISBN 978-3-319-06733-9   e-ISBN 978-3-319-06734-6
DOI 10.1007/978-3-319-06734-6
Springer Cham Heidelberg New York Dordrecht London

Library of Congress Control Number: Applied for

LNCS Sublibrary: SL 4 – Security and Cryptology

Typesetting: Camera-ready by author, data conversion by Scientific Publishing Services, Chennai, India

Printed on acid-free paper

Springer is part of Springer Science+Business Media (www.springer.com)

Francesco Regazzoni
Benedikt Schmidt
Hadi Soleimany
Katherine Stange
Valentin Suder
Xiao Tan
Joana Treger
Dominique Unruh

Joop van de Pol
Kerem Varici
Frederik Vercauteren
Nicolas Veyrat-Charvillon
Pengwei Wang
Gaven Watson
Liang Feng Zhang

Table of Contents

Hardware Implementation

Protocols

Lattice-Based Cryptography

Public-Key Cryptography

Secret-Key Cryptography

New Results for Rank-Based Cryptography

Philippe Gaborit[1], Olivier Ruatta[1], Julien Schrek[1], and Gilles Zémor[2]

[1] Université de Limoges, XLIM-DMI,
123, Av. Albert Thomas
87060 Limoges Cedex, France
{gaborit,ruatta,schrek}@unilim.fr
[2] Université Bordeaux I, A2X,
351 cours de la Libération
33400 Talence Cedex, France
gilles.zemor@math.u-bordeaux.fr

Abstract. In this paper we survey new results for rank-based cryptography: cryptosystems which are based on error-correcting codes embedded with the rank metric. These new results results first concern the LRPC cryptosystem, a cryptosystem based on a new class of decodable rank codes: the LRPC codes (for Low Rank Parity Check codes) which can be seen as an analog of the classical LDPC codes but for rank metric. The LRPC cryptosystem can benefit from very small public keys of less than 2,000 bits and is moreover very fast. We also present new optimized attacks for solving the general case of the rank syndrome decoding problem, together with a zero-knowledge authentication scheme and a new signature scheme based on a mixed errors-erasures decoding of LRPC codes, both these systems having public keys of a few thousand bits. These new recent results highlight that rank-based cryptography has many good features that can be used for practical cryptosystems.

Keywords: Public key cryptosystem, rank metric, error-correcting codes.

1 Introduction

In recent year there has been a burst of activities regarding post-quantum cryptography, the interest of such a field has become even more obvious since the recent attacks on the discrete logarithm problem in small characteristic [2]. These attacks show that finding new attacks on classical cryptographic systems is always a possibility and that it is important not have all its eggs in the same basket.

Among potential candidate for alternative cryptography, lattice-based and code-based cryptography are strong candidates. Rank-based cryptography relies on the difficulty of decoding error-correcting codes embedded with the rank metric (codes over extension fields of type $GF(q^m)$), when code-based cryptography relies on difficult problems related to error-correcting codes embedded with the

D. Pointcheval and D. Vergnaud (Eds.): AFRICACRYPT 2014, LNCS 8469, pp. 1–12, 2014.

Hamming metric (often over small fields $GF(q)$) and when lattice-based cryptography is mainly based on the study of q-ary lattices, which can be seen as codes over rings of type Z/qZ (for large q), embedded with the Euclidean distance.

The particular interest of rank metric is that the practical difficulty of the problems grows very fast with the size of parameters. In particular it is possible to reach a complexity of 2^{80} for random instances with size only a few thousand bits, when for lattices or codes, at least hundred thousand bits are needed. Of course with codes and lattices it is possible to decrease to a few thousand bits but with additional structure like quasi-cyclicity [4], but then the reduction properties to difficult problems is lost.

The rank metric was introduced by Gabidulin in 1985 in [12], together with the Gabidulin codes which are an equivalent of the Reed-Solomon codes for the rank metric. Since then, rank metric codes have been used in many applications: for coding theory and space-time codes and also for cryptography. Until now the main tool for rank based cryptography relied on masking the Gabidulin codes [13] in different ways and using the McEliece (or Niederreiter) setting with these codes. Meanwhile most of the systems were broken by using structural attacks which use the particular structure of the Gabidulin codes ([26], [11], [5], [21] - see also [11]). A similar situation exists in the Hamming case for which most of cryptosystems based on the Reed-Solomon have been broken for the same reason: the Reed-Solomon are so structured that their structure is difficult to mask and there is always structural information leaking.

Recently a new family of decodable codes in rank metric was proposed in [14], namely the LRPC codes. The basic idea to decode these codes is similar to the NTRU decryption method [19] or the LDPC approach and only uses the fact of knowing a dual matrix with low weight in order to decode (or decrypt) the code. The advantage of this approach is that the induced structure is very poor compared to very structured codes like Reed-Solomon or Gabidulin codes. For instance, the NTRU cryptosystem has never really been attacked for almost 20 years (nor has the more recent MDPC cryptosystem [23], an analog of the NTRU cryptosystem with LDPC codes for Hamming distance and with small public keys of $4,800$ bits).

In this paper we first review in Section 2, basic facts about rank metric and the difficulty of decoding rank metric codes, then we consider LRPC codes in Section 3 and their application to cryptography in Section 4, at last sections 5 and 6 deal with authentication and signature with rank metric.

2 Background on Rank Metric Codes and Cryptography

2.1 Definitions and Notation

Let q be a power of a prime p, m an integer and let V_n be a n dimensional vector space over the finite field $GF(q^m) = F_{q^m}$. Let $\beta = (\beta_1, \ldots, \beta_m)$ be a basis of $GF(q^m)$ over $GF(q)$. Let $\mathcal{F}_i$ be the map from $GF(q^m)$ to $GF(q)$ where $\mathcal{F}_i(x)$ is the i-th coordinate of x in the basis β. To any $v = (v_1, \ldots, v_n)$ in V_n we associate the matrix $\overline{v} \in \mathcal{M}_{m,n}(F_q)$ in which $\overline{v}_{i,j} = \mathcal{F}_i(v_j)$. The rank weight of a vector v can be defined as the rank of the associated matrix $\overline{v}$. If we name this value $\text{rank}(v)$ we can have a distance between two vectors x, y using the formula $\text{rd}(x, y) = \text{rank}(x - y)$. We refer to [22] for more details on codes for the rank distance.

A rank code C of length n and dimension k over $GF(q^m)$ is a subspace of dimension k of $GF(q^m)^n$ embedded with rank metric. The minimum rank distance of the code C is the minimum rank of non-zero vectors of the code. One also considers the usual inner product which allows to define the notion of dual code. An important notion which differs from the Hamming distance, is the notion of support. Let $x = (x_1, x_2, \cdots, x_n) \in GF(q^m)^n$ be a vector of rank r. We denote by $E :=< x_1, x_2, \cdots, x_n >$, the $GF(q)$-sub vector space of $GF(q^m)$ generated by the coordinates of x, $\{x_1, x_2, \cdots, x_n\}$. The vector space E is called the **support** of x. In the following, C is a rank metric code of length n and dimension k over $GF(q^m)$. The matrix G denotes a $k \times n$ generator matrix of C and H one of its parity check matrix. At last, the notion of **isometry** which in Hamming metric corresponds to the action on the code of $n \times n$ permutation matrices, is replaced for rank metric by the action of $n \times n$ invertible matrices over the base field $GF(q)$.

2.2 Difficult Problem for Rank-Based Cryptography

The main problem used for rank codes in the cryptographic context is the generalization of the classical syndrome decoding problem with Hamming distance in the case of rank metric:

Rank Syndrome Decoding Problem (RSD). Let H be a $((n - k) \times n)$ matrix over $GF(q^m)$ with $k \leq n$, $s \in GF(q^m)^k$ and r an integer. The problem is to find x such that $\text{rank}(x) = r$ and $Hx^t = s$.

The RSD problem is not proven NP-hard, but this problem is very close to the syndrome decoding problem in Hamming distance, which is NP-hard. In practice the problem has been studied for more than 20 years and is considered as difficult by the communauty.

2.3 Complexity of the Rank Decoding Problem

The complexity of practical attacks grows very fast with the size of parameters, there is a structural reason for this: for Hamming distance a key notion in the attacks is counting the number of words of length n and support size t, which corresponds to the notion of Newton binomial coefficient $\binom{n}{t}$, whose value is exponential and upper bounded by 2^n. In the case of rank metric, counting the number of possible supports of size r for a rank code of length n over $GF(q^m)$ corresponds to counting the number of subspaces of dimension r in $GF(q^m)$: **the Gaussian binomial coefficient** of size roughly q^{rm}, whose value is also exponential but with a quadratic term in the exponent.

There exist two types of generic attacks on the problem:

- **Combinatorial Attacks:** these attacks are usually the best ones for small values of q (typically $q = 2$) and when n and k are not too small, when q increases, the combinatorial aspect makes them less efficient. The first non-trivial attack on the problem was proposed by Chabaud and Stern [6] in 1996, then in 2002 Ourivski and Johannson [25] improved the previous attack and proposed a new attack, meanwhile these two attacks did not take account of the value of n in the exponent. Very recently the two previous attacks were generalized in [17] by Gaborit et al. in $(n-k)^3 m^3 q^{(r-1)\lfloor \frac{(k+1)m}{n} \rfloor})$) and took the value of n into account and were used to break some repaired versions of the GPT cryposystem. The idea of the latter new approach is to use the notion of support of a word in rank metric and then applying the classical Information Set Decoding [3] approach with this generalized notion of support.

- **Algebraic Attacks:** the particular nature of rank metric makes it a natural field for algebraic attacks and solving by Groebner basis, since these attacks are largely independent of the value of q and in some cases may also be largely independent on m. These attacks are usually the most efficient ones when q increases. There exist different type of algebraic equations settings to try to solve a multivariate system with Groebner basis. The algebraic context proposed by Levy and Perret [20] in 2006 considers a quadratic setting over $GF(q)$ by taking as unknowns the support E of the error and the error coordinates regarding E. It is also possible to consider the Kernel attack by [9] and the minor approach [10] which give multivariate equations of degree $r + 1$ over $GF(q)$ obtained from minors of matrices At last, more recently the annulator setting by Gaborit et al. in [17] (which is valid on certain type of parameters but may not be independent on m) give multivariate sparse equations of degree q^{r+1} but on the large field $GF(q^m)$ rather than on the base field $GF(q)$. The latter attack is based on the notion of q-polynomial [24] and is particularly efficient when r is small. Moreover all these attacks can be declined in an hybrid approach where some unknowns are guessed.

3   Low Rank Parity Check Codes and Their Decoding

3.1   Definition of Low Rank Parity Check Codes

The idea of these codes is to generalize the classical LDPC codes approach for Hamming distance to the rank metric. There is a natural analogy between low density matrices and matrices with low rank.

Definition 1. *A Low Rank Parity Check (LRPC) code of rank d, length n and dimension k over $GF(q^m)$ is a code such that the code has for parity check matrix, a $(n - k) \times n$ matrix $H(h_{ij})$ such that the sub-vector space of $GF(q^m)$ generated by its coefficients h_{ij} has dimension at most d. We call this dimension the weight of H. Denoting F the sub-vector space of $GF(q^m)$ generated by the coefficients h_{ij} of H, we denote by $\{F_1, F_2, \cdots, F_d\}$ one of its basis.*

In practice it means that for any $1 \leq i \leq n - k, 1 \leq j \leq n$, there exist $h_{ijl} \in GF(q)$ such that $h_{ij} = \sum_{l=1}^{d} h_{ijl} F_l$. Naturally the LRPC codes can also be considered in a quasi-cyclic context in which the matrix H is double circulant (a concatenation of two circulant matrices), which permits to dramatically decrease the size of the description of the matrix H.

3.2 Decoding Algorithm for LRPC Codes

The general idea of the algorithm is to use the fact that all coordinates of the parity check matrix H belong to the same vector space F of small dimension d. Given an error e with associated support E of dimension r and its associated syndrome $s = H.e^t$, we take advantage that when r and d are such that $rd \leq n - k$, the $GF(q)$-vector space S generated by the coordinates of the syndrome $S = < s_1, \ldots, s_{n_k} >$ permits to recover, with a strong probability, the whole product space $P = < E.F >$ (generated by the $E_i F_j$, $1 \leq i \leq r, 1 \leq i \leq d$), of the error support E and the LRPC small vector space F. Then knowing the *whole* product space $P = < E.F > = S$ and the space F, allows to recover E by a simple intersection of subspaces of the form: S times the inverse of the elements of a basis of F. Once the support E of the error e is recovered, it is easy to compute the exact value of each coordinate of e by solving a linear system.

Consider a $[n, k]$ LRPC code C of low weight d over $GF(q^m)$, with generator matrix G and dual $(n - k) \times n$ matrix H, such that all the coordinates h_{ij} of H belong to a space F of rank d with basis $\{F_1, \cdots, F_d\}$.

Suppose the received word to be $y = xG + e$ for x and e in $GF(q^m)^n$, and where $e(e_1, \cdots, e_n)$ is the error vector of rank r, which means that for any $1 \leq i \leq n$, $e_i \in E$, a vector space of dimension r with basis (say) $\{E_1, \ldots, E_r\}$. The Fig. 1 describes a general probabilistic algorithm which decodes an error e up to rank distance $r = (n - k)/d$.

Correctness, Probability of Failure and Complexity of the Decoding. The decoding algorithm is probabilistic since the probability to recover a set of maximal independant elements of the syndrome space is probabilistic and also since there is very small probability that Step 2 recovers a greater space than E, but all these probabilities can be easily evaluated. In term of complexity of decoding it is possible to use a formal description of the matrix H to compute the inversion of the coefficients of the error vector e with only a quadratic complexity (see [14] for details). Overall we have the following theorem:

Theorem 1 ([14]). *Let H be a $(n - k) \times n$ dual matrix of a LRPC codes with low rank $d \geq 2$ over $GF(q^m)$, then algorithm 1 decodes a random error e of low rank r such that $rd \leq n - k$, with failure probability $q^{-(n-k+1-rd)}$ and complexity $r^2(4d^2m + n^2)$.*

1. **Syndrome space computation**
   Compute the syndrome vector $H.y^t = s(s_1, \cdots, s_{n-k})$ and the syndrome space $S = <s_1, \cdots, s_{n-k}>$.

2. **Recovering the support E of the error**
   Define $S_i = F_i^{-1}S$, the subspace where all generators of S are multiplied by F_i^{-1}. Compute the support of the error $E = S_1 \cap S_2 \cap \cdots \cap S_d$, and compute a basis $\{E_1, E_2, \cdots, E_r\}$ of E.

3. **Recovering the error vector e**
   Write $e_i (1 \leq i \leq n)$ in the error support as $e_i = \sum_{i=1}^{n} e_{ij}E_j$, solve the system $H.e^t = s$, where the equations $H.e^t$ and the syndrome coordinates s_i are written as elements of the product space $P = <E.F>$ in the basis $\{F_1E_1, \cdots, F_1E_r, \cdots, F_dE_1, \cdots, F_dE_r\}$. The system has nr unknowns (the e_{ij}) in F_q and $(n-k).rd$ equations from the syndrome.

4. **Recovering the message x**
   Recover x from the system $xG = y - e$.

Fig. 1. Algorithm 1:a general decoding algorithm for LRPC codes

4   Application of LRPC Codes to Cryptography: The LRPC Cryptosystem

4.1   The LRPC Cryptosystem

The LRPC cryptosystem consists in applying a McEliece-like or a Niederreiter-like encryption setting to the LRPC family of decodable codes: Figure 2 presents the LRPC cryptosytem in a McEliece setting.

The system works for any LRPC code C and considering G in systematic form permits to decrease a little the size of the public key. The case of double circulant LRPC codes (DC-LRPC) is of particular interest since it permits to dramatically decrease the size of the public key: in that case the matrix H can be written $(A|B)$ where A and B are two circulant invertible LRPC matrices of low rank d and G can be written $G = ((A^{-1}B)^t|I)$ (for A^t the transposed matrix of A).

- General parameters of the LRPC cryptosystem:

Writing the matrices of the system in systematic form we obtain:

1. Size of public key (bits): LRPC: $(n-k)kmLog_2(q)$ / DC-LRPC: $\frac{nm}{2}Log_2(q)$
2. Size of secret key (bits): a seed can be used to recover the different parameters
3. Size of message: LRPC: $nmLog_2(q)$ / DC-LRPC: $nmLog_2(q)$
4. Encryption rate: LRPC: $\frac{k}{n}$ / DC-LRPC: $\frac{1}{2}$

For decryption and encryption, the computational cost is dominated by the matrix-vector multiplication and the cost of syndrome computing. In the case of DC-LRPC , one can use the double-circulant structure to improve computations. The cost of a multiplication in the extension field $GF(q^m)$, in binary

1. **Key creation** Choose a random $[n, k]$ LRPC code C over $GF(q^m)$, with low rank support F of weight d, which corrects errors of rank r and with parity check matrix, a $(n - k) \times n$ matrix H. Let G be a generator matrix of the LRPC code C.
   - **Secret key**: the LRPC $(n - k) \times n$ dual matrix H
   - **Public key**: a $k \times n$ generator matrix G of the LRPC code C

2. **Encryption**
   Translate the information vector M into a word x, choose a random error e of rank r on $GF(q^m)$. The encryption of M is $c = xG + e$.

3. **Decryption**
   Compute the syndrome $s = H.c^t$ and recover the error vector e by decoding s with the LRPC code, then compute $xG = c - e$ and recover x.

Fig. 2. The LRPC cryptosystem

operations, is $mLog_2(m)Log_2(Log_2(m))$ ([18]). The system in Fig. 2 is presented in a McEliece setting, in that case the size of the message is larger than for the Niederreiter setting but more can be proven regarding semantic security.

4.2 Security of the LRPC Cryptosystem

Attacks on the LRPC Cryptosystem. There are two type of attacks. The first type of attacks are **direct attacks on the message**, in which the attacker tries to recover directly the message by finding the error e of rank r with classical attacks described in Section 2.3. For the type of considered parameters, the recent combinatorial attacks or algebraic attacks of [14] are the most efficient ones.

It is also possible to consider **structural attacks** and try to attack directly the structure of the public key to recover the secret key. In particular one can use the fact that all the elements of the dual LRPC matrix H belong to the same subspace F of rank d. Let D be the dual code generated by H. Denote by $H_i (1 \le i \le n - k)$ the $n - k$ rows of H and consider a word x of D obtained from linear combinations in the small field $GF(q)$: $x = \sum_{i=1}^{n-k} a_i H_i$ for $a_i \in GF(q)$. All the coordinates of x belongs to F, now since F has dimension d, fixing d variables a_i in $GF(q)$ can allow to put to zero a coordinate of x, overall since there are $n - k$ variables a_i one can put to zero (with a good probability depending of the matrix H), $\lfloor (n - k)/d \rfloor$ coordinates positions of x. Therefore with a good probability the dual code D contains a word x with all coordinates in F and with $\lfloor (n - k)/d \rfloor$ coordinates to zero which can be the first $\lfloor (n - k)/d \rfloor$ coordinates without loss of generality. Hence the attacker can attack the subcode D' of D of all the words of D which are zero on the first $\lfloor (n - k)/d \rfloor$ coordinates. This code D' is a $[n - \lfloor (n - k)/d \rfloor, n - k - \lfloor (n - k)/d \rfloor]$ code which, by the previous discussion, contains a word of rank d.

The previous structural attack uses deeply the structure of the LRPC matrix so that the attacker has only to attack a smaller code which contains at least one word of rank d. This exponential attacks slightly reduces the computational cost of the attack on the system and can be easily handled. This attack has an equivalent attack for NTRU [19] and for MDPC codes [23] in which the attacker uses the cyclicity to decrease slightly the number of columns of the attacked matrix: by removing columns corresponding to zeros of a small weight vector of the secret key.

4.3   Examples of Parameters

We give three examples of parameters for the DC-LPRC case: an example with security 2^{80} operations which optimizes the size of the public key at 1680 bits with a decryption probability of 2^{-22}, an example with security 2^{128}, and at last an example with decryption failure probability of 2^{-80}.

These parameters update the parameters from [14] after the weak structural attack described in Section 4.4.

In the table 'failure' stands for probability of 'decryption failure', the size of the public key is in bits, the security is in bits. We give parameters for different level of security, but also for different decryption failure, in particular it is possible to reach a 2^{-80} easily at the cost of doubling the size of the key. Notice that the parameters are very versatile. Although no special attack is known for non prime number we choose to consider prime numbers in general. The complexity of decryption for the first set of parameters is 2^{20} bit operations. In particular in terms of computation cost the LRPC cryptosystem seems to compare very well with the MDPC cryptosystem.

n	k	m	q	d	r	failure	public key	security
82	41	41	2	5	4	-22	1681	80
106	53	53	2	6	5	-24	2809	128
74	37	23	2^4	4	4	-88	3404	110

5   Zero-Knowledge Authentication with Rank Metric

5.1   Previous Work and Definitions

In 1995 Chen proposed in [7] a 5-pass zero-knowledge protocol based on rank metric with cheating probability $1/2$, the protocol was in the spirit of the Stern SD protocol [27] meanwhile it turned out that the protocol was not correct and was subsequently broken in [16]. The main reason was that the zero-knowledge proof was false, especially since the author failed to construct an equivalent notion of permutation for Hamming distance which would associate any word of rank weight r to any particular given word of rank r. Indeed let x be a word of length n and rank r with support E, then for any $n \times n$ random invertible matrix P in the small field $GF(q)$, the word xP has also rank r but the support of x and xP are the same so that information leaks if one tries to hide x only by turning it into xP.

The definition of the product "$*$" allows to obtain such a property for rank metric. With the notation of Section 2.1: for a given basis β, we denote Φ_β the inverse of the function $V_n \to \mathcal{M}_{m,n}(\mathrm{GF}(q)) : x \to \overline{x}$ computed with the basis β.

Definition 2 (product). *Let Q be in $\mathcal{M}_{m,m}(\mathrm{GF}(q))$, $v \in V_n$ and β a basis. We define the product $Q * v$ such that $Q * v = \Phi_\beta(Q\overline{v})$, where $\overline{v}$ is constructed from the basis β.*

Then one can prove the following proposition which gives the equivalent notion of permutation for Hamming distance but in a rank metric context:

Proposition 1 ([16]). *For any $x, y \in V_n$ and $\mathrm{rank}(x) = \mathrm{rank}(y)$, it is possible to find $P \in \mathcal{M}_{n,n}(\mathrm{GF}(q))$ and $Q \in \mathcal{M}_{m,m}(\mathrm{GF}(q))$ such that $x = Q * yP$.*

5.2   Description of the Protocol

The previous definition of the "$*$" product permits to obtain a rank metric adaptation of the Stern protocol [27] which was presented in [16]: the masking of a codeword by a permutation is replaced by the masking $x \to Q * xP$ which has the same property in terms of rank distance as a permutation for a codeword with Hamming distance, since it can transform any given x with given rank to *any* element with the same rank. In the following the notation $(a|b)$ corresponds to the concatenation of a and b. The notation $hash(a)$ is the hash value of a. A given basis β is fixed and known in advance for the "$*$" product.

For the protocol a public $k \times n$ matrix over $GF(q^m)$ H is fixed. The **secret key** is a vector s of $V_n(= (GF(q^m)^n)$ with rank r. The **public key** consists of the matrix H, the syndrome $i = Hs^t$ and the rank r of s. The protocol is described in Fig. 3. For the protocol the small base field is $GF(2)$, (ie: $q = 2$). It is proven in [16] that the protocol described in Fig. 3, is a 3-pass zero-knowledge protocol with cheating probability $2/3$.

---

1.   [Commitment step] The prover $\mathcal{P}$ chooses $x \in V_n$, $P \in GL_n(\mathrm{GF}(q))$ and $Q \in GL_m(q)$. He sends c_1, c_2, c_3 such that :

$$c_1 = hash(Q|P|Hx^t), c_2 = hash(Q * xP), c_3 = hash(Q * (x + s)P)$$

2.   [Challenge step] The verifier $\mathcal{V}$ sends $b \in \{0, 1, 2\}$ to P.
3.   [Answer step] there are three possibilities :
   − if $b = 0$, $\mathcal{P}$ reveals x and $(Q|P)$
   − if $b = 1$, $\mathcal{P}$ reveals $x + s$ and $(Q|P)$
   − if $b = 2$, $\mathcal{P}$ reveals $Q * xP$ and $Q * sP$
4.   [Verification step] there are three possibilities :
   − if $b = 0$, $\mathcal{V}$ checks c_1 and c_2.
   − if $b = 1$, $\mathcal{V}$ checks c_1 and c_3.
   − if $b = 2$, $\mathcal{V}$ checks c_2 and c_3 and that $rank(Q * sP) = r$.

Fig. 3. Rank-SD protocol

Example of Parameters. If we consider, $q = 2, n = 22, m = 23$ and $k = 9$ one obtains a minimal distance of 8 by the rank Gilbert-Varshamov bound [22], hence we can take $r = 7$ for the rank weight of the secret. The security of the protocol relies then on the security of the hash function and on a general random instance of the RSD problem defined in Section 2. In that case with these parameters, the best practical known attacks lead to a complexity of at least 2^{80} operations (these parameters are updated from [16] after the recent improvements on generic attacks of [17]). The fact that one can take a rank weight r close to the rank Gilbert-Varshamov bound permits to greatly decrease the size of the parameters.

Public matrix H : $(n - k) \times k \times m = 2691$ bits

Public key i : $(n - k)m = 299$ bits

Secret key s : $r(m + n) = 360$ bits

Average Number of Bits Exchanged in One Round: 2 hash + one word of $GF(q^m) \sim 820$ bits.

Overall the protocol is more efficient than the Stern SD scheme and can probably be optimized as in [1]. For instance by cyclicity: if one considers a double-circulant matrix, the size of the public key decreases to only a few hundred bits: a $[22, 11]$ double circulant code, $q = 2$, $m = 29$, $r = 7$ leads to a public key of 319 bits. A security of 2^{100} can be reached with $k = 9, n = 27, q = 2, m = 24$ and $r = 10$.

6   Signature with Rank Metric

The signature is usually the most difficult primitive to obtain. In the case of rank metric, a first way to obtain a signature is to use the authentication scheme of the previous section in a Fiat-Shamir paradigm context (see [1]). It permits to build a signature scheme with small public key of a few thousand bits (even a few hundred if one uses quasi-cyclicity) but with large signature size of order a hundred thousand bits (although optimized rank-SD schemes should *a priori* probably do always better than optimized SD schemes [1]).

Another approach consists (like for the CFS scheme [8]) in trying to construct a hash and sign signature scheme. This approach is possible with rank metric codes, by using the notion of generalized erasure. The notion of generalized erasure means that the decoder knows not only the syndrome but has also information on the support of the error. In a rank metric context it means knowing a subspace of the support of the error e. The idea for the signature scheme is then to try to decode a random syndrome but not as usual with only errors, but with a mixed approach of errors and generalized erasures. It turns out that it is possible to modify the LRPC decoding algorithm so that it is possible to obtain a better decoding when a subspace of the error is known. Then for a fixed known part of the error support it is possible to decode beyond the rank Gilbert-Varshamov bound and then obtain a hash and sign signature algorithm.

Overall the RankSign signature scheme presented in [15] permits to obtain a fast signature scheme with public key of order $10,000$ bits and signature of size less than $2,000$ bits.

7 Conclusion

In this short survey paper we presented recent results for rank-based cryptography. These results show that rank-based cryptography has a strong potential in terms of size of keys because of the inherent difficulty of the RSD problem and its links with the Gaussian binomial which counts subspaces of given dimension. We highlighted the LRPC cryptosystem which benefits from a very low public key of less than $2,000$ bits and is moreover very fast. We also highlighted the analogy between the NTRU cryptosystem, the MDPC cryptosystem and the present LRPC cryptosystem which are based on similar ideas. We also presented the rank-SD authentication scheme and eventually gave the general ideas on which relies the RankSign signature scheme which has also relatively small public keys.

Overall even if more scrutiny is needed from the communauty, all these recent results propose new promising direction for rank-based cryptography and for obtaining fast asymmetric systems with small public keys.

References

1. Aguilar, C., Gaborit, P., Schrek, J.: A new zero-knowledge code based identification scheme with reduced communication. In: 2011 IEEE Information Theory Workshop (ITW), pp. 648–652 (2011)
2. Barbulescu, R., Gaudry, P., Joux, A., Thomé, E.: A quasi-polynomial algorithm for discrete logarithm in finite fields of small characteristic, eprint iacr 2013/400
3. Becker, A., Joux, A., May, A., Meurer, A.: Decoding Random Binary Linear Codes in 2 n/20: How $1 + 1 = 0$ Improves Information Set Decoding. In: Pointcheval, D., Johansson, T. (eds.) EUROCRYPT 2012. LNCS, vol. 7237, pp. 520–536. Springer, Heidelberg (2012)
4. Berger, T.P., Cayrel, P.-L., Gaborit, P., Otmani, A.: Reducing Key Length of the McEliece Cryptosystem. In: Preneel, B. (ed.) AFRICACRYPT 2009. LNCS, vol. 5580, pp. 77–97. Springer, Heidelberg (2009)
5. Berger, T., Loidreau, P.: Designing an Efficient and Secure Public-Key Cryptosystem Based on Reducible Rank Codes. In: Canteaut, A., Viswanathan, K. (eds.) INDOCRYPT 2004. LNCS, vol. 3348, pp. 218–229. Springer, Heidelberg (2004)
6. Chabaud, F., Stern, J.: The Cryptographic Security of the Syndrome in Decoding Problem for Rank Distance Codes. In: Kim, K.-C., Matsumoto, T. (eds.) ASIACRYPT 1996. LNCS, vol. 1163, pp. 368–381. Springer, Heidelberg (1996)
7. Chen, K.: A New Identification Algorithm. In: Dawson, E., Golić, J. (eds.) Cryptography: Policy and Algorithms 1995. LNCS, vol. 1029, pp. 244–249. Springer, Heidelberg (1996)
8. Courtois, N.T., Finiasz, M., Sendrier, N.: How to achieve a Mc-Eliece-based digital signature scheme. In: Boyd, C. (ed.) ASIACRYPT 2001. LNCS, vol. 2248, pp. 157–174. Springer, Heidelberg (2001)

9. Faugère, J.-C., Levy-dit-Vehel, F., Perret, L.: Cryptanalysis of MinRank. In: Wagner, D. (ed.) CRYPTO 2008. LNCS, vol. 5157, pp. 280–296. Springer, Heidelberg (2008)

10. Faugère, J.-C., El Din, M.S., Spaenlehauer, P.-J.: Computing loci of rank defects of linear matrices using Grbner bases and applications to cryptology. In: ISSAC 2010, pp. 257–264 (2010)

11. Faure, C., Loidreau, P.: A New Public-Key Cryptosystem Based on the Problem of Reconstructing p-Polynomials. In: Ytrehus, Ø. (ed.) WCC 2005. LNCS, vol. 3969, pp. 304–315. Springer, Heidelberg (2006)

12. Gabidulin, E.M.: Theory of Codes with Maximum Rank Distance. Probl. Peredachi Inf. (21), 3–16 (1985)

13. Gabidulin, E.M., Paramonov, A.V., Tretjakov, O.V.: Ideals over a Non-Commutative Ring and their Applications in Cryptology. In: Davies, D.W. (ed.) EUROCRYPT 1991. LNCS, vol. 547, pp. 482–489. Springer, Heidelberg (1991)

14. Gaborit, P., Murat, G., Ruatta, O., Zémor, G.: Low Rank Parity Check Codes and their application in cryptography. In: The Preproceedings of Workshop on Coding and Cryptography (WCC) 2013, Borgen, Norway, pp. 167–179 (2013)

15. Gaborit, P., Ruatta, O., Schrek, J., Zémor, G.: RankSign: An efficient signature algorithm based on the rank metric. eprint iacr (submitted)

16. Gaborit, P., Schrek, J., Zémor, G.: Full Cryptanalysis of the Chen Identification Protocol. In: Yang, B.-Y. (ed.) PQCrypto 2011. LNCS, vol. 7071, pp. 35–50. Springer, Heidelberg (2011)

17. Gaborit, P., Ruatta, O., Schrek, J.: On the complexity of the rank syndrome decoding problem. eprint. Submitted to IEEE Trans. Information Theory

18. von zur Gathen, J., Gerhard, J.: Modern computer algebra. Cambridge University Press (2003)

19. Hoffstein, J., Pipher, J., Silverman, J.H.: NTRU: A Ring-Based Public Key Cryptosystem. In: Buhler, J.P. (ed.) ANTS 1998. LNCS, vol. 1423, pp. 267–288. Springer, Heidelberg (1998)

20. Levy-dit-Vehel, F., Perret, L.: Algebraic decoding of rank metric codes. In: Proceedings of YACC 2006 (2006)

21. Loidreau, P.: Designing a Rank Metric Based McEliece Cryptosystem. In: Sendrier, N. (ed.) PQCrypto 2010. LNCS, vol. 6061, pp. 142–152. Springer, Heidelberg (2010)

22. Loidreau, P.: Properties of codes in rank metric, http://arxiv.org/abs/cs/0610057

23. Misoczki, R., Tillich, J.-P., Sendrier, N., Barreto, P.S.L.M.: MDPC-McEliece: New McEliece Variants from Moderate Density Parity-Check Codes. Cryptology ePrint Archive: Report 2012/409

24. Ore, O.: On a special class of polynomials. Trans. American Math. Soc. (1933)

25. Ourivski, A.V., Johansson, T.: New Technique for Decoding Codes in the Rank Metric and Its Cryptography Applications. Probl. Inf. Transm. (38), 237–246 (2002)

26. Overbeck, R.: Structural Attacks for Public Key Cryptosystems based on Gabidulin Codes. J. Cryptology 21(2), 280–301 (2008)

27. Stern, J.: A new paradigm for public key identification. IEEE Transactions on Information Theory 42(6), 2757–2768 (1996)

Proxy Re-Encryption Scheme Supporting a Selection of Delegatees

Julien Devigne[1], Eleonora Guerrini[2], and Fabien Laguillaumie[3]

[1] Orange Labs Applied Crypto Group, GREYC (CNRS, UCBN, ENSICAEN)
julien.devigne@orange.com
[2] Université Montpellier 2, LIRMM (CNRS UM2)
eleonora.guerrini@lirmm.fr
[3] Université Claude Bernard Lyon 1, LIP
(U. Lyon, CNRS, ENS Lyon, INRIA, UCBL)
fabien.laguillaumie@ens-lyon.fr

Abstract. Proxy re-encryption is a cryptographic primitive proposed by Blaze, Bleumer and Strauss in 1998. It allows a user, Alice, to decide that in case of unavailability, one (or several) particular user, the delegatee, Bob, will be able to read her confidential messages. This is made possible thanks to a semi-trusted third party, the proxy, which is given by Alice a re-encryption key, computed with Alice's secret key and Bob's public key. This information allows the proxy to transform a ciphertext intended to Alice into a ciphertext intended to Bob. Very few constructions of proxy re-encryption scheme actually handle the concern that the original sender may *not* want his message to be read by Bob instead of Alice. In this article, we adapt the primitive of proxy re-encryption to allow a sender to *choose* who among the potential delegatees will be able to decrypt his messages, and propose a simple and efficient scheme which is secure under chosen plaintext attack under standard algorithmic assumptions in a bilinear setting. We also add to our scheme a *traceability* of the proxy so that Alice can detect if it has leaked some re-encryption keys.

Keywords: proxy re-encryption, cloud storage, traceability.

1 Introduction

Proxy re-encryption (PRE), invented in 1998 by Blaze *et al.* [8], allows a user to delegate its decryption capability. To do so, this user, Alice, computes a piece of information with her secret key sk_A and her delegatee Bob's public key pk_B, called a re-encryption key and denoted by $R_{A \to B}$. This key is given to a semi-trusted proxy, accessible (at least) during Alice's unavailability. This re-encryption key allows the proxy to transform a ciphertext intended to Alice into a ciphertext intended to Bob. While doing this transformation, the proxy cannot learn *any* information about plaintext messages. We are here interested in *unidirectional* and *single-hop* schemes, which means (1) that with a re-encryption key $R_{A \to B}$, a proxy cannot translate a ciphertext intended to Bob, into a ciphertext intended to Alice[1] and (2) that once a message has been moved into a ciphertext

---

[1] Schemes which allow this symmetrical transformation are called *bi-directional*.

D. Pointcheval and D. Vergnaud (Eds.): AFRICACRYPT 2014, LNCS 8469, pp. 13–30, 2014.

intended to Bob, no more transformation on the new ciphertext to Bob is possible[2]. PRE schemes found applications in digital rights management, distributed file systems, privacy of medical records, outsourced filtering of encrypted spam, and encrypted email forwarding,...

We here address an issue which has not been intensively studied so far: the original sender, who encrypts a message to Alice, wants *a priori* his message to be read by Alice, and only Alice. This is the definition of the notion of confidentiality. In case of unavailability of Alice, she may authorize some users (who she trusts, we can imagine), to read some of her messages. Nevertheless, the sender may not want his message to Alice to be read by *all* these delegatees, i.e., he might prefer to decide *also* who among these delegatees can actually decrypt the message. Of course, we cannot prevent Alice to give afterward a message she decrypted to anyone she wants, as well as the delegatees.

To prevent this transformation of Alice's messages into messages intended to all of the delegatees, we propose a specific proxy re-encryption scheme which supports the selection of the delegatees by the original sender, who must have at his disposal a description of the identities of the delegatees. The idea is that the delegator assigns a temporary identity to each delegatee for the purpose we mentioned, using a structure of labeled tree (that we describe bellow). Corresponding re-encryption keys will be computed by Alice and given to a proxy which will re-encrypt her ciphertexts. The initial sender will then be able to decide that a "second-level" ciphertext (which can be re-encrypted) that he produced for Alice, will be decryptable only by the delegatees satisfying a certain pattern, in the spirit of identity-based encryption with wildcards schemes [2,1].

Application. Our (medical) use-case is the following: several hospitals want to cooperate to improve medical assistance and research, so they want to share data on a cloud storage involving their medical patients with different illness. Some parties from these hospitals decide to store their data on a common remote server. A cancer expert from hospital 1 (let us call him Dr. John Carter) wants to store some data about some patient behavior and he wants the advice of his colleagues from other hospitals. If the cancer expert of hospital 2 (Dr. Susan Lewis) is available, she can access these data on the server to work on, because Dr. Carter has encrypted this data for Dr. Lewis. When Dr. Lewis is on vacation, it might be important that other people from her hospital could access this information, like the nurses of her unit or colleagues in some other units. So Dr. Lewis chooses her delegatees, usually colleagues with whom she is used to work, among these trusted colleagues and assigns them their identities as delegatees. In this scenario the doctor Carter knows that his colleague is on vacation and he gets her list of delegatees. If he estimates that the data are too sensitive or private to be shared with all the delegatees of doctor Susan Lewis, he can choose in this list, for example, only the professor in charge of the maternity ward (assuming that his patient is also a pregnant woman).

---

[2] Schemes which allow several translation of ciphertexts are called *multi-hop*.

We want do design a proxy re-encryption scheme which supports this choice of delegatee's by the sender. We propose the following solution: any hospital is structured in different sectors and the delegator attributes an identity to each of his delegatees, according to his position in the hospital. An identity is chosen by the delegator, and is composed of different fields specifying the hospital where they work, the field, the position in the group and the name, structured in a labeled tree, as in Figure 1. For example a possible identity is "hospital_2.cancer-ward.headoftheunit.susan.lewis" or "hospital_1.maternity-ward.nurses.nurse$_2$". If professor John Carter of hospital 1 wants to share information with the professor in charge of the cancer unit of hospital 2, he encrypts data for his colleague, as well as for the identity of the delegatee he may accept, for instance "hospital_2.maternity-ward.headoftheunit.lucy.knight" if he wants this data to be read by the head of the maternity unit. Therefore, if this professor is in vacation, a proxy re-encrypts the data for someone chosen directly by the sender. More generally, these data might also be available for authorized entities all the time (for instance in cloud applications). We want also to allow a re-encryption for more delegatees in the following way: the proxy might re-encrypt data for *all* the nurses of the maternity ward of the hospital 2. In this case, we allow the sender who estimates that his message concern also all the nurses to encrypt directly for the address "hospital_2.maternity-ward.nurses.*" where the "*" means "any nurses from the maternity-ward of hospital 2".

In our scheme, a delegator chooses a set of delegatees structured according to a labeled tree and creates for each of them a re-encryption key. Then the original sender of the message chooses among these delegatees who can decrypt it, thanks to a target path specified through a sequence of fixed strings and wildcards, where any string can take the place of a wildcard in a matching identity. We propose in the last section other applications related to cloud storage.

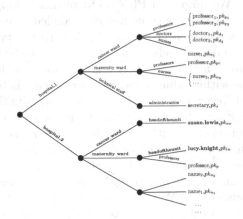

Fig. 1. Labeled tree of the delegatees chosen by the delegator

Related Work. After Blaze, *et al.*'s seminal work [8], PRE have been widely studied. The security model for proxy re-encryption scheme has been formalized in [13] (only for bidirectional schemes but easily adaptable for the unidirectional case, see [23]). Unidirectional schemes, built on bilinear maps, predominantly reach a chosen plaintext security [5,22,25,3]. The scheme in [23] is only *replayable* CCA secure in the sense of [14,13]. The one in [12] which is an amelioration of [15] was proved CCA secure in the random oracle model.

Some functionalities have been also added to PRE, such as *conditional proxy re-encryption* initiated by Tang [26] as *type-based proxy re-encryption*, and later developed in [29,30,16,17,27,17], which already addresses the problem of controlling the decryption rights. The motivation from [26] was that the *delegator* may want to decide whom of the delegatees might decrypt some ciphertexts. It has been extended by Chu *et al.* in [16]. Our point of view is different, since we want *the original sender* to choose who can eventually decrypt. To this end, Tang's and Chu *et al.*'s approach could also be used, but the number of re-encryption keys would be more important, since if a delegator which corresponds to different keywords will have as many corresponding re-encryption keys. Moreover Tang's scheme is proven in the random oracle model, whereas Chu *et al.*'s schemes are proven in the standard model but need a master authority. In our protocol, we want to exploit a structure among the delegatee's, which can be naturally described by a kind of hierarchical organization, as we have seen. Tang's approach is also related to the notion of proxy with keyword search from [20] (already mentioned in [24]) and attributed-based proxy [21]. We could also have thought about broadcast encryption to fit our cloud-storage applications, nevertheless, PRE is better adapted, since it offers a more flexible system for users. For instance, the proxy re-encryption primitive naturally offers dynamic right access to confidential messages as explain in [11].

Our Contributions. We propose a PRE protocol which supports a selection of the delegatees by the sender, once he knows a labeled tree organization of the delegatees. We define this concept, which is close to the type-based proxy re-encryption, then we propose a scheme which is proven to be semantically secure under chosen plaintext attacks in the standard model under reasonable algorithmic assumptions within bilinear groups. Our scheme is based on Libert and Vergnaud's traceable proxy re-encryption [22], and exploits the tag used to trace malicious proxies, by introducing the labeled tree organization, in the spirit of the identity-based encryption with wildcards [1,2]. It is therefore more efficient than any trivial scheme which would consist in encrypting (with a classical PRE, for instance) independently for any of the chosen delegatees. As in Libert and Vergnaud's paper, our work is related to Waters' IBE scheme [28]. The idea is that the original sender uses Waters' hash function to set the identity of the chosen delegatees. Nonetheless, for our security proof, we follow the alternative security proof proposed by Bellare and Ristenpart [6] which avoids some tricky step (namely the artificial abort) of the original proof. Moreover our scheme is still compatible with ideas from Libert and Vergnaud's traceable scheme and can therefore be adapted to reach also their same traceability functionalities.

2 Preliminaries

2.1 Definitions and Security Model

We define here a single-hop unidirectional PRE scheme which supports a selection of delegatees and its security requirements. In this context, the users who want to send some message M to Alice can decide, in case of Alice's unavailability whom of her delegatees may be able to decrypt. To this end, Alice firstly organized her delegatees according to a labeled tree denoted by $\mathcal{T}$ (as in Figure 1), and attributes to each of them a corresponding *identity path* (like "hospital_1.cancer-ward.professor.john.carter"). Then, the sender incorporates to the ciphertext intended to Alice a description of the identities of the delegatees he accepts, called the *target path*, which can contain wildcards (like "hospital_2.*.professor.*"). This ciphertext will be decryptable by all of delegatees whose identity matches this description. Formally, we say that an identity path $\mathbf{ID}^{(B)} = (\mathbf{id}_1^{(B)}, \dots, \mathbf{id}_\ell^{(B)}) \in \{0,1\}^{n_1} \times \cdots \times \{0,1\}^{n_\ell}$ (for a set of integers $\{n_1, \dots, n_\ell\}$) *matches* a target path $\mathbf{S} = (\mathbf{s}_1, \dots, \mathbf{s}_\ell) \in (\{0,1\}^{n_1} \cup \{*\}^{n_1}) \times \cdots \times (\{0,1\}^{n_\ell} \cup \{*\}^{n_\ell})$, if $\mathbf{s_i} = \mathbf{id}_i^{(B)}$ for all i such that $\mathbf{s_i} \neq (*, \dots, *)$ (and we note $\mathbf{ID}^{(B)} \in \mathbf{S}$). We point out here that we suppose Alice to be honest in the sense that she won't publish a tree which will hide some path to fool the senders. This is widely sufficient for most of the applications, but open an interesting cryptographic question, which seems hard to answer without the use of a trusted authority. We will discuss this issue in conclusion.

Syntactic Definition

Definition 1 (PRE). *Let κ be an integer and $\mathcal{T}$ be a labeled tree which describes all the identities $\mathbf{ID}^{(B)}$ of each delegatee B (including a sequence of ℓ integers n_i). A single-hop unidirectional selective proxy re-encryption scheme which supports a selection of delegatees consists of the following eight algorithms.*

- Setup$(\kappa, \mathcal{T}) \to \mathcal{P}$: *this setup algorithm takes a security parameter κ as input and produces the public parameters $\mathcal{P}$ (which includes the tree $\mathcal{T}$).*
- KeyGen$(\mathcal{P}) \to (sk, pk)$: *this key generation algorithm, whose inputs are the public parameters, outputs a pair of secret and public keys (sk, pk), and is executed by every user.*
- ReKeygen$(\mathcal{P}, sk_A, pk_A, pk_B, \mathbf{ID}^{(B)}) \to R_{A \to B}$: *given the public parameters, user A's pair of keys, the public key of the user B and its identity[3], this algorithm produces a re-encryption key $R_{A \to B}$ which allows to transform second level ciphertexts intended to A into first level ciphertexts for B.*
- Encrypt$_1(\mathcal{P}, pk, M) \to C$: *this first level encryption algorithm takes as inputs $\mathcal{P}$, a public key and a message. It outputs a first level ciphertext C that cannot be re-encrypted.*

---

[3] A user may have several identities ; for instance, in a Facebook-like application, a user may be in the group of friends, as well as in the group of colleagues.

- $\text{Encrypt}_2(\mathcal{P}, pk, M, \mathbf{S}) \rightarrow C$: *this second level encryption algorithm takes* $\mathcal{P}$, *a public key, a message and a target path* $\mathbf{S}$ *as inputs, and produces a second level ciphertext* C *that can be re-encrypted. It includes the target path.*
- $\text{ReEncrypt}(\mathcal{P}, R_{A \rightarrow B}, C) \rightarrow C' / \perp$: *this algorithm takes as input the public parameters, a re-encryption key* $R_{A \rightarrow B}$ *and a second level ciphertext intended to user* A, *to which is appended the target path. The output is a first level ciphertext* C' *re-encrypted for user* B *if his identity matches the target path or an invalid message* $\perp$.
- $\text{Decrypt}_1(\mathcal{P}, sk, C) \rightarrow m / \perp$: *this first level decryption algorithm takes as input* $\mathcal{P}$, *a secret key and a first level ciphertext and outputs a plaintext* M *or an invalid message* $\perp$.
- $\text{Decrypt}_2(\mathcal{P}, sk, C) \rightarrow m / \perp$: *this second level decryption algorithm takes as input the public parameters, a secret key and a second level ciphertext and outputs a plaintext* M *or* $\perp$.

For correctness, *these algorithms must satisfy the following properties: for all public parameters* $\mathcal{P}$ *generated by the* Setup *algorithm, for any message* M, *and any couple of valid secret/public key pair* (sk_A, pk_A), (sk_B, pk_B), $\text{Decrypt}_1(\mathcal{P}, sk_A, \text{Encrypt}_1(\mathcal{P}, pk_A, M)) \rightarrow M$, $\text{Decrypt}_2(\mathcal{P}, sk_A, \text{Encrypt}_2(\mathcal{P}, pk_A, M, \mathbf{S})) \rightarrow M$, *and* $\text{Decrypt}_1(\mathcal{P}, sk_B, \text{ReEncrypt}(\mathcal{P}, \text{ReKeygen}(\mathcal{P}, sk_A, pk_A, pk_B, \mathbf{ID}^{(\mathbf{B})}),$ $\text{Encrypt}_2(\mathcal{P}, pk_A, M, \mathbf{S}))) \rightarrow M$, *when* B's *identity* $\mathbf{ID}^{(\mathbf{B})}$ *matches the target path.*

Security Model. As for public-key encryption, the semantic security is the relevant security requirement for a PRE. We consider here chosen plaintext attacks, in a static model where the adversary does not choose which users he can corrupt. This is a classical approach: this security model is the one considered in [4,5,13,22]. The differences due to the modifications of our definition of a PRE is highlighted in the descriptions which follow. The security experiment is depicted in Fig. 3. We consider also in this case that Alice, the original recipient, is honest in the sense that she publishes a labeled tree which describes correctly her delegatees. For most of the applications, this notion is sufficient (for instance, in our preliminary example, Dr. Carter won't add a nurse in the group of doctors).

Our model implicitly makes the *knowledge of secret key (KOSK)* assumption, which means that all users know the secret key corresponding to their published public key. This implies a trusted key generation model, or a model where a user who wants its public key to be certified by a certification authority, to provide a proof of knowledge of his secret key. See [23] for a discussion of the stronger scenario of chosen-key model. Such model is also chosen in [4,5,13,22].

sIND-CPA SECURITY OF PRE.Encrypt₂. We define the static indistinguishability under a chosen plaintext attack of the Encrypt_2 algorithm of a single-hop unidirectional selective proxy re-encryption scheme by describing a three-stage attacker $\mathcal{A} = (\mathcal{A}_{init}, \mathcal{A}_f, \mathcal{A}_g)$. The challenge given to the adversary $\mathcal{A}$ is $C^\star = \text{Encrypt}_2(\mathcal{P}, pk^\star, M_\delta, \mathbf{S})$. As for standard proxy re-encryption protocols, this attacker has access to the secret keys of users which are corrupted (sk_c) and to all re-encryption keys except those from the target user to a corrupted one $(R_{\star \to c})$. This restriction prevents the adversary to re-encrypt the challenge ciphertext into a first-level ciphertext intended to a user whose secret key is known to the adversary, turning in a trivial success. Our model for PRE which supports a selection of delegatees differs a bit from the classical one, in the sense that during a first stage, the adversary $(\mathcal{A}_{init})$ chooses a target path **S**, based on the labeled tree that he knows. We stress out that our model allows the attacker to have *also* access to the re-encryption keys from the target user to a corrupted one when its identity does *not* match the target path **S** $(R_{\star \to c \notin \mathbf{S}})$. It captures the fact that only users whose identity matches a path can decrypt a ciphertext which has been re-encrypted by the proxy for an identity matching this path.

Definition 2 (sIND-CPA security of PRE.Encrypt$_2$). *Let κ be an integer and $\mathcal{T}$ be a labeled tree. Let $\mathcal{A} = (\mathcal{A}_{init}, \mathcal{A}_f, \mathcal{A}_g)$ be an adversary against the* CPA *indistinguishability of* PRE.Encrypt$_2$. *Let* $Adv_{\text{PRE.Encrypt}_2, \mathcal{A}}^{sind\text{-}cpa}(\kappa, \mathcal{T}) :=$
$2 \cdot \Pr \left[Exp_{\text{PRE.Encrypt}_2, \mathcal{A}}^{sind\text{-}cpa}(\kappa, \mathcal{T}) \to true \right] - 1$ *with* $Exp_{\text{PRE.Encrypt}_2, \mathcal{A}}^{sind\text{-}cpa}$ *as defined in Fig. 3 We say that* PRE *has* sIND-CPA *security of* PRE.Encrypt$_2$ *if for every p.p.t. adversary* $\mathcal{A} = (\mathcal{A}_{init}, \mathcal{A}_f, \mathcal{A}_g)$, $Adv_{\text{PRE.Encrypt}_2, \mathcal{A}}^{sind\text{-}cpa}(\kappa, \mathcal{T})$ *is negligible.*

sIND-CPA SECURITY OF PRE.Encrypt$_1$. We define the static indistinguishability under a chosen plaintext attack of the Encrypt$_1$ algorithm. The challenge given to the adversary $\mathcal{A}$ is $C^{\star} = \text{Encrypt}_1(\mathcal{P}, pk^{\star}, M_\delta)$ where $pk^{\star}$ is the public key of the target user. In this case, all re-encryption keys can be given to the adversary, since first level ciphertexts cannot be re-encrypted. A PRE which satisfies this definition is said to have sIND-CPA security for non-transformable ciphertexts.

sIND-CPA SECURITY OF PRE.ReEncrypt. Eventually, we define the static indistinguishability under CPA of the ReEncrypt algorithm. The challenge given to the adversary $\mathcal{A}$ is $C^{\star} = \text{ReEncrypt}(\mathcal{P}, R_{u \to \star}, \text{Encrypt}_2(\mathcal{P}, pk_u, M_\delta, \mathbf{S}))$ where pk_u is the public key of the user u that can be either corrupted or not and a target path **S** such that the target user's identity matches this path **S**. In this case also, all re-encryption keys can be given to the adversary, since first level ciphertexts cannot be re-encrypted again. A PRE which satisfies this definition is said to have sIND-CPA security for transformed ciphertexts.

Transferability. According to Hohenberger [19], a PRE is *transferable* if a coalition of malicious proxies and delegatees succeeds in delegating their decryption rights to non-authorized users. This means that given some re-encryption keys and some delegatee's secret key, it is possible to create a new re-encryption key for a fresh user. It is always possible for a delegatee to transfer its decryption capability, by giving his secret key. If this property is undesirable, most of the proxy re-encryption scheme are transferable. We will see that our proposed PRE is transferable as well, but fortunately, it is possible to incriminate a proxy that took part of this collusion. Following the work of Libert and Vergnaud [22], our scheme supports a white-box tracing, in the sense that if a proxy reveals a re-encryption key which was not produced by the delegator, then the proxy (but not the malicious delegatee), will be identified. Our scheme is therefore white-box *traceable* [22], and can be extended to a black-box traceable scheme at the expense of a loss of efficiency.

2.2 Algorithmic Assumption and Notations

Our scheme is built upon a bilinear group system, made up with an algorithm BGSGen which takes a security parameter κ as input, and outputs a tuple $(q, \mathbb{G}_1, \mathbb{G}_2, \mathbb{G}_T, e, \Phi)$, where q is a κ-bit prime, $\mathbb{G}_1$, $\mathbb{G}_2$ and $\mathbb{G}_T$ are three multiplicative groups of order q, $e : \mathbb{G}_1 \times \mathbb{G}_2 \to \mathbb{G}_T$ is an admissible bilinear map, and $\Phi : \mathbb{G}_2 \to \mathbb{G}_1$ is an isomorphism not publicly invertible. The semantic security relies on the hardness of the *decision bilinear Diffie-Hellman* problem:

$$\mathbf{Exp}^{\mathrm{dbdh}}_{\mathsf{BGSGen}, \mathcal{D}}(\kappa)$$

$\beta \xleftarrow{\$} \{0, 1\}$

$(q, \mathbb{G}_1, \mathbb{G}_2, \mathbb{G}_T, e, \Phi) \leftarrow \mathsf{BGSGen}(\kappa)$

$h \xleftarrow{\$} \mathbb{G}_2, g \leftarrow \Phi(h)$

$(a, b, c) \xleftarrow{\$} (\mathbb{Z}_q^*)^3$

if $\beta = 1$ then $d \leftarrow abc$ else $d \xleftarrow{\$} \mathbb{Z}_q^*$

$\beta' \leftarrow \mathcal{D}(g, h, h^a, h^b, h^c, e(g, h)^d)$

Return $(\beta' = \beta)$

$$\mathbf{Exp}^{\mathrm{sind\text{-}cpa}}_{\mathsf{PRE.Encrypt}_2, \mathcal{A}}(\kappa, \mathcal{T})$$

$\mathcal{P} \leftarrow \mathsf{Setup}(\kappa, \mathcal{T})$

$(\mathbf{S}^*, st) \leftarrow \mathcal{A}_{init}(\mathcal{P})$

$(sk^*, pk^*) \leftarrow \mathsf{KeyGen}(\mathcal{P})$

$\{(sk^h, pk^h) \leftarrow \mathsf{KeyGen}(\mathcal{P})\}$

$\{(sk^c, pk^c) \leftarrow \mathsf{KeyGen}(\mathcal{P})\}$

$\{R_{c \to *} \leftarrow \mathsf{ReKeyGen}(\mathcal{P}, sk_c, pk_c, pk^*, \mathbf{ID}^h)\}$,

$\{R_{* \to h} \leftarrow \mathsf{ReKeyGen}(\mathcal{P}, sk^*, sk^*, pk_h, \mathbf{ID}^h)\}$

$\{R_{h \to *} \leftarrow \mathsf{ReKeyGen}(\mathcal{P}, sk_h, pk_h, pk^*, \mathbf{ID}^*)\}$,

$\{R_{h \to c} \leftarrow \mathsf{ReKeyGen}(\mathcal{P}, sk_h, pk_h, pk_c, \mathbf{ID}^c)\}$

$\{R_{c \to h} \leftarrow \mathsf{ReKeyGen}(\mathcal{P}, sk_c, pk_c, pk_h, \mathbf{ID}^h)\}$,

$\{R_{h \to h'} \leftarrow \mathsf{ReKeyGen}(\mathcal{P}, sk_h, pk_h, pk'_h, \mathbf{ID}^{h'})\}$

$\{R_{c \to c'} \leftarrow \mathsf{ReKeyGen}(\mathcal{P}, sk_c, pk_c, pk'_c, \mathbf{ID}^{c'})\}$

$\{R_{* \to c \notin \mathbf{S}^*} \leftarrow \mathsf{ReKeyGen}(\mathcal{P}, sk^*, pk^*, pk_{c \notin \mathbf{S}^*}, \mathbf{ID}^c)\}$

$(M_0, M_1, st') \leftarrow \mathcal{A}_f(st, \mathcal{P}, pk^*, \{pk_c, sk_c\}, \{pk_h\}$,

$\{R_{c \to *}\}, \{R_{h \to *}\}, \{R_{* \to h}\}, \{R_{c \to h}\}, \{R_{h \to c}\}$,

$\{R_{h \to h'}\}, \{R_{c \to c'}\}, \{R_{* \to c \notin \mathbf{S}^*}\})$

$\delta \xleftarrow{\$} \{0, 1\}$

$C^* \leftarrow \mathsf{Encrypt}_2(\mathcal{P}, pk^*, M_\delta, \mathbf{S}^*)$

$\delta' \leftarrow \mathcal{A}_g(st', C^*)$

Return $(\delta' = \delta)$

Fig. 2. DBDH experiment   **Fig. 3.** sIND-CPA experiment

Definition 3 (DBDH). *Let $(q, \mathbb{G}_1, \mathbb{G}_2, \mathbb{G}_T, e, \Phi)$ be obtained from a bilinear group system generator* BGSGen *with a security parameter κ as input, the* decision bilinear Diffie-Hellman problem *(DBDH) is to distinguish elements $e(g, h)^{abc} \in \mathbb{G}_T$ from random elements from $\mathbb{G}_t$ given $(g, h, h^a, h^b, h^c) \in \mathbb{G}_1 \times \mathbb{G}_2^4$. A distinguisher $\mathcal{D}$ described in Fig. 2 (τ, ε)-breaks the problem if it has running time τ and* $\mathbf{Adv}^{\mathrm{dbdh}}_{\mathsf{BGSGen}, \mathcal{D}}(\kappa) = 2 \Pr[\mathbf{Exp}^{\mathrm{dbdh}}_{\mathsf{BGSGen}, \mathcal{D}}(\kappa) \Rightarrow \mathsf{true}] - 1 = \varepsilon$. *The* DBDH *problem is said to be hard if for any p.p.t. distinguisher $\mathcal{D}$, its advantage is negligible.*

Notations: For $n \in \mathbb{N}$ and $\mathbf{g} \in \mathbb{G}^n$, we define the function[4] $F_{\mathbf{g}} : \{0, 1\}^n \to \mathbb{G}$ as $F_{\mathbf{g}}(\omega) = \prod_{j=1}^n g_j^{w_j}$ (the group $\mathbb{G}$ will be either $\mathbb{G}_1$ or $\mathbb{G}_2$). Note that for any $s \in \mathbb{Z}_q^*$, $(F_{\mathbf{g}}(\omega))^s = F_{\mathbf{g}^s}(\omega)$. Moreover, if $\mathbf{g} \in (\mathbb{G}_2)^n$ then $\Phi(F_{\mathbf{g}}) = F_{\Phi(\mathbf{g})}$. In what follows, if $\mathbf{M}$ is a vector, $\mathbf{M}_i$ denotes its i-th element, so that if $\mathbf{M}$ is a vector of vectors, $\mathbf{M}_{i,j}$ denotes the j-th element of its i-th subvector.

---

[4] Note that this is essentially Waters' function $W(m) = \mathbf{g}_0 \prod_{i=1}^n \mathbf{g}_i^{m_i}$ without a $\mathbf{g}_0$.

3   The Scheme

3.1   Intuition

Our scheme is a modification of Libert and Vergnaud's scheme [22] (based on Waters' identity-based encryption scheme [28]), which exploits the string ω_{ij} contained in the re-encryption key from user i to j (originally used to trace malicious proxies) to incorporate information about the identity path of a delegatee. This information on the delegatee's identity is chosen and defined by the delegator, and will actually also help later in our protocol, in case of dispute, to trace back a malicious proxy. The technique to embed the authorized set of delegatees chosen by the sender is similar to the one used in identity-based encryption with wildcards [2,1] (which is already the core of the design of Libert and Vergnaud's scheme). The sender of the message chooses a target path $\mathbf{S} = \{\mathbf{s}_1, \ldots, \mathbf{s}_\ell\}$ such that either some block $\mathbf{s}_{i_1}, \ldots, \mathbf{s}_{i_k}$ are equal to a block of star pattern "*", either a label that must be fulfilled by the delegatee's identity. The wildcards corresponding to the different starred blocks, indicate the parts of the identity path that will be freely given in the second level ciphertext (using a random exponent). In this way, the proxy will be able to reform a complete (Water's) function of the identity of the authorized delegatees during the re-encryption process for *each* authorized delegatee, by completing these free blocks with their corresponding identity blocks, when they match the path. More precisely, to make the proxy to be able to reconstruct a whole Waters' hash function of the identity of an authorized user, a function with fixed predetermined part from $\mathbf{S}$ (i.e., not corresponding to wildcards) is pre-computed (as $(F_{\mathbf{U}_i^{(A)}}(\mathbf{s}_i))^s$), whereas the "wild part" is left free as a set $((\mathbf{U}_{i,j}^{(A)})^s)$ during the second level encryption. This last set will be raised at the power of the identity path of the delegatees which matches the target path $\mathbf{S} = (\mathbf{s}_1, \ldots, \mathbf{s}_\ell)$. In what follows, if ℓ is an integer, $\mathbf{n} = (n_1, \ldots, n_\ell) \in \mathbb{N}^\ell$ is the lengths of the different paths at each level: $\mathcal{T}$ is an authenticated (meaning that it has been signed by Alice) labeled tree, whose paths of level i are of size n_i.

3.2   Description of Our New Scheme

- Setup($\kappa, \mathcal{T}$): Let $(q, \mathbb{G}_1, \mathbb{G}_2, \mathbb{G}_T, e, \Phi)$ be a bilinear group system obtained from BGSGen with κ as input. The integer ℓ will be the number of blocks of an identity. Let $\mathbf{n} = (n_1, \ldots, n_\ell)$ a vector whose i-th coordinate n_i defines the size of the i-th block of an identity. Let $h \in \mathbb{G}_2$ be a random generator and $g = \Phi(h)$. The output of Setup is $(\mathcal{T}, (q, \mathbb{G}_1, \mathbb{G}_2, \mathbb{G}_T, e, \Phi), g, h, \mathbf{n}, \ell)$.

- KeyGen($\mathcal{P}$): On input the public parameters, this algorithm picks at random $(z, y, v') \xleftarrow{\$} (\mathbb{Z}_q^*)^3$, $v_{i,j} \xleftarrow{\$} \mathbb{Z}_q^*$, $\forall i \in [\![1, \ell]\!], \forall j \in [\![1, n_i]\!]$ and produces the vector (of vectors of different size) $\mathbf{V}$ such that $\mathbf{V}_{i,j} = h^{v_{i,j}} \in \mathbb{G}_2$, $\forall i \in [\![1, \ell]\!], \forall j \in [\![1, n_i]\!]$ and $V' = h^{v'}$, as well as $\mathbf{U} = \Phi(\mathbf{V})$ (which means that Φ

is applied coordinate-wise) and[5] $U' = \Phi(V')$. Eventually, $sk = (z, y, \mathbf{V}, V')$ and $pk = (Z, Y, \mathbf{U}, U')$ where $Z = e(g, h)^z \in \mathbb{G}_T$ and $Y = h^y \in \mathbb{G}_2$.

– ReKeyGen($\mathcal{P}, sk_A, pk_A, pk_B, \mathbf{ID}^{(B)}$): On input the public parameters, user A's private key $sk_A = (z_A, y_A, \mathbf{V}^{(A)}, V'^{(A)})$, user B's public key $pk_B = (Z_B, Y_B, \mathbf{U}^{(B)}, U'^{(B)})$ and identity $\mathbf{ID}^{(B)}$.

   1. Pick $r \overset{\$}{\leftarrow} \mathbb{Z}_q^*$ and parse user B's identity as $\mathbf{ID}^{(B)} = (\mathbf{id}_i^{(B)})_{i \in [\![1,\ell]\!]}$ such that $\mathbf{id}_i^{(B)}$ belongs to $\{0,1\}^{n_i}, \forall i \in [\![1, \ell]\!]$.

   2. $R_{A \to B} = \left(\mathbf{ID}^{(B)}, Y_B^{z_A} \cdot (V'^{(A)})^r \cdot \prod_{i=1}^{\ell} F_{\mathbf{V}_i^{(A)}} \left(\mathbf{id}_i^{(B)} \right)^r, h^r \right)$.

– Encrypt$_1$($\mathcal{P}, pk_B, M$): Given the public parameters, a plaintext $M \in \mathbb{G}_t$, user B's public key $pk_B = (Z_B, Y_B, \mathbf{U}^{(B)}, U'^{(B)})$, it picks $s \overset{\$}{\leftarrow} \mathbb{Z}_q^*$ and outputs the first level ciphertext $C' = (M.e(g, h)^s, e(g, Y_B)^s)$.

– Encrypt$_2$($\mathcal{P}, pk_A, M, \mathbf{S}$): On input the public parameters, a plaintext $M \in \mathbb{G}_t$, user A's public key $pk_A = (Z_A, Y_A, \mathbf{U}^{(A)}, U'^{(A)})$ and a target path $\mathbf{S} = (\mathbf{s}_1, \ldots, \mathbf{s}_\ell)$, where $\mathbf{s}_i \in \{0,1\}^{n_i} \cup \{*\}^{n_i}, \forall i \in [\![1, \ell]\!]$.

   1. Pick $s \overset{\$}{\leftarrow} \mathbb{Z}_q^*$ and set $I_* = \{i \in [\![1, \ell]\!] : \mathbf{s}_i = (*, \ldots, *)\}$.

   2. Output the second level ciphertext:
$$C = (M \cdot Z_A^s, g^s, \left(U'^{(A)} \cdot \prod_{i \notin I_*} F_{\mathbf{U}_i^{(A)}}(\mathbf{s}_i) \right)^s, \left((\mathbf{U}_{i,j}^{(A)})^s \right)_{i \in I_*, j \in [\![1, n_i]\!]}, \mathbf{S}).$$

– ReEncrypt($\mathcal{P}, R_{A \to B}, C$): On input public parameters, a re-encryption key $R_{A \to B} = (\mathbf{ID}^{(B)}, t, h_t)$ and a second level ciphertext $C = (C_1, C_2, C_3, \mathbf{c}^{(4)}, \mathbf{S})$.

   1. If $\mathbf{ID}_B \notin \mathbf{S}$, return $\bot$.

   2. Otherwise, let $I_* = \{i \in [\![1, \ell]\!] : \mathbf{s}_i = (*, \ldots, *)\}$.

   3. Compute $C_3' = C_3 \cdot \prod_{i \in I_*} F_{\mathbf{c}_i^{(4)}}(\mathbf{id}_i^{(B)})$ with the full identity of B.

   4. Output the first level ciphertext $C' = (C_1, e(C_2, t)/e(C_3', h_t))$.

– Decrypt$_1$($\mathcal{P}, sk_B, C'$): On input public parameters, user B's secret key $sk_B = (z_B, y_B, \mathbf{V}^{(B)}, V'^{(B)})$ and a first level ciphertext $C' = (C_1', C_2')$, return the plaintext $M = C_1'/C_2'^{1/y_B}$.

– Decrypt$_2$($\mathcal{P}, sk_A, C$): On input public parameters, user A's secret key $sk_A = (z_A, y_A, \mathbf{V}^{(A)}, V'^{(A)})$ and a second level ciphertext $C = (C_1, C_2, C_3, \mathbf{c}^{(4)}, \mathbf{S})$, return the plaintext $M = C_1/e(C_2, h)^{z_A}$.

The correctness of the scheme is easy to verify and is left to the reader.

Remark 1. Note that we could generalize the setting by allowing ℓ and $\mathbf{n}$ to be chosen independently for each delegator during the KeyGen algorithm. In this case if we denote these values by ℓ_A and $\mathbf{n}_A$ for a delegator A, each identity $\mathbf{ID}^{(B)}$ would be composed of ℓ_A blocks $\mathbf{id}_1^{(B)}, \ldots, \mathbf{id}_{\ell_A}^{(B)}$ of size respectively $n_{A,1}, \ldots, n_{A,\ell}$. Note that we could also allow wildcards in the $\mathbf{s}_i$'s themselves, but to match our target application we didn't choose these options.

---

[5] Note that U' will play the role of $\mathbf{g}_0$ in Waters' function $W(m) = \mathbf{g}_0 \prod_{i=1}^n \mathbf{g}_i^{m_i}$.

3.3   sIND − CPA Security of Our Scheme

In this section, we first prove the semantic security of the second level encryption algorithm. We actually consider a stronger model than the one presented in Section 2.1. The adversary, instead of choosing during his init part the whole challenge identity path $\mathbf{S}^\star = (\mathbf{s}_1^\star, \ldots, \mathbf{s}_\ell^\star)$, he only chooses the set of positions $I_*^\star = \{i \in [\![1, \ell]\!] : \mathbf{s}_i^\star = (*, \ldots, *)\}$ of the wildcards. He will precise the *whole* challenge during the challenge phase. This model is a bit stronger, but not necessarily more intuitive, so that's why we defined a slightly weaker model in Section 2.1. Our scheme relies on Waters' IBE scheme [28]. Contrary to Libert and Vergnaud, we have to adapt Waters' machinery to prove the security of our scheme. Instead of following Waters' original proof, we follow the alternative one of Bellare and Ristenpart [6]. Indeed, they propose a simpler proof (with better concrete security) where they avoid the *artificial abort* step.

Theorem 1. *The scheme has* sIND-CPA *security of* PRE.Encrypt$_2$ *under the* DBDH *assumption in the standard model.*

Proof. Keeping notations from Section 3.2, let $\mathcal{A}$ be an sIND-CPA attacker and $\varepsilon = \mathsf{Adv}_{\mathsf{PRE.Encrypt}_2, \mathcal{A}}^{\mathsf{sind\text{-}cpa}}(\kappa, \mathcal{T})$. Let $1 \leq Q \leq \frac{q\varepsilon}{9\nu}$ denote the number of re-encryption key queries where $\nu = \sum_{i \notin I_*^\star} n_i$ and let $m = \left\lceil \frac{9Q}{\varepsilon} \right\rceil$. We describe an algorithm $\mathcal{B}$ which will break a DBDH instance, using an $\mathcal{A}$ throughout a sequence of games.

Preliminaries: We define J as the set $\{(i, j) \mid i \in [\![1, \ell]\!] \setminus I_*^\star, j \in [\![1, n_i]\!]\}$, ω as a vector of vectors $(\omega_{i,j})_{(i,j) \in J}$ such that $\omega_{i,j} \in [\![0, (m-1)]\!]$ for all $(i, j) \in J$, ω' as an element belonging to $[\![-\nu(m-1), 0]\!]$, $\mathbf{Z}$ as a vector of vectors $(z_{i,j})_{i \in [\![1, \ell]\!], j \in [\![1, n_i]\!]}$ such that all $z_{i,j}$ belongs to $\mathbb{Z}_q^*$ and z' as an element of $\mathbb{Z}_q^*$. For an identity $\mathbf{ID} \in \{0,1\}^{n_1} \times \cdots \times \{0,1\}^{n_\ell}$, we denote by $\Omega_{\mathbf{ID}}$ the set of (i, j) such that $\mathbf{id}_{i,j} = 1$ *and* $i \notin I_*^\star$. We also denote by $\Lambda_{\mathbf{ID}}$ the set of (i, j) such that $\mathbf{id}_{i,j} = 1$. We define two functions $G(\omega, \omega', \mathbf{ID}) = \omega' + \sum_{(i,j) \in \Omega_{\mathbf{ID}}} \omega_{i,j}$ and $H(\mathbf{Z}, z', \mathbf{ID}) = z' + \sum_{(i,j) \in \Lambda_{\mathbf{ID}}} z_{i,j}$. Note that the sum in G is in $\mathbb{Z}$ whereas the sum in H is modulo q. It is important to note that, if $G(\omega, \omega', \mathbf{ID}) = 0 \pmod{q}$, then $G(\omega, \omega', \mathbf{ID}) = 0$. This is because $|G(\omega, \omega', \mathbf{ID})| \leq \nu(m-1) < q$ for our choice of m, q and of the range of the entries of ω. Finally, we denote by $\mathbf{ID}_{\mathbf{S}^\star}$ the identity which consists of $\mathbf{s}_i^\star$ (of the target path $\mathbf{S}^\star$) for all $i \notin I_*^\star$ and of 0^{n_i} otherwise.

Game 0. Let (g, h, h^a, h^b, h^c, T) be a DBDH instance, *i.e.*, $T = e(g, h)^{abc}$ if $\beta = 1$ and T is random if $\beta = 0$. $\mathcal{B}$ will output a guess β' of β. Let bad be a flag initialized at false, and set to true if $\mathcal{B}$ is not able to simulate all answers to $\mathcal{A}$'s queries. When $\mathcal{B}$ runs $\mathcal{A}$, it first receives its choice of wildcarded positions $I_*^\star$.

-   *Target-key generation*: $\mathcal{B}$ picks $y^\star \xleftarrow{\$} \mathbb{Z}_q^*$ and produces the vector of vectors $\mathbf{V}^\star$ composed of element in $\mathbb{G}_2$ where $\forall (i, j) \in J$, $\mathbf{V}_{i,j}^\star = (h^b)^{\omega_{i,j}} h^{z_{i,j}^\star}$ and the element $V'^\star = (h^b)^{\omega'} h^{z'^\star}$ of $\mathbb{G}_2$ where J is chosen as in the preliminaries and $(\omega_{i,j})_{(i,j) \in J}$, ω', $\mathbf{Z}^\star = (z_{i,j}^\star)_{i \in [\![1, \ell]\!], j \in [\![1, n_i]\!]}$ and $z'^\star$ are randomly picked in the set described in the preliminaries. This implicitly defines $v_{i,j}^\star = b \times \omega_{i,j} + z_{i,j}^\star$ and $v'^\star = b \times \omega' + z'^\star$. Moreover, $\forall (i, j) \notin J$, we have $\mathbf{V}_{i,j}^\star = h^{z_{i,j}^\star}$ which defines $v_{i,j}^{\star} = z_{i,j}^{\star}$, $\mathbf{U}^{\star} = \Phi(\mathbf{V}^{\star})$ and $U'^{\star} = \Phi(V'^{\star})$. It gives to $\mathcal{A}$ a $pk^{\star} = (Z^{\star}, Y^{\star}, \mathbf{U}^{\star}, U'^{\star})$, where $Z^{\star} = e(\Phi(h^a), h^b)$ and $Y^{\star} = h^{y^{\star}}$.

- *Other key generation:* $\mathcal{B}$ runs the KeyGen algorithm and gives (sk, pk) to $\mathcal{A}$ for a corrupted-key generation and only pk otherwise.
- **Re-Encryption key generation.** $\mathcal{B}$ generates re-encryption keys from a user A to a user B as follows, for the identity $\mathbf{ID}^{(B)} \in \{0,1\}^{n_1} \times \cdots \times \{0,1\}^{n_\ell}$:
  - If A *is not the target user:* $\mathcal{B}$ runs the ReKeyGen algorithm and gives $R_{A \to B}$ to $\mathcal{A}$.
  - If A *is the target user and B is corrupted*[6], *then $\mathcal{B}$ proceeds as follows:*
    1. Recover the secret key of B, $sk_B = (z_B, y_B, \mathbf{V}^{(B)}, V'^{(B)})$.
    2. Pick randomly $r \xleftarrow{\$} \mathbb{Z}_q^*$.
    3. If $G(\omega, \omega', \mathbf{ID}^{(B)}) = 0 \pmod{q}$, set bad $\leftarrow$ true and return $R_{A \to B} = \bot$.
    4. Compute $R_{A \to B} = (\mathbf{ID}^{(B)}, R_1, R_2)$ with

    $$R_1 = (h^b)^{y_B \cdot r \cdot G(\omega,\omega',\mathbf{ID}^{(B)})} \cdot h^{y_B \cdot r \cdot H(\mathbf{Z}^{\star}, z'^{\star}, \mathbf{ID}^{(B)})} \cdot (h^a)^{-y_B \cdot \frac{H(\mathbf{Z}^{\star}, z'^{\star}, \mathbf{ID}^{(B)})}{G(\omega,\omega',\mathbf{ID}^{(B)})}}$$

    and $R_2 = (h^a)^{\frac{-y_B}{G(\omega,\omega',\mathbf{ID}^{(B)})}} \cdot h^{r y_B}$.

    It implicitly defines $R_1 = h^{ab y_B} \cdot \left(V'^{\star} \cdot \prod_{i=1}^{\ell} F_{\mathbf{V}_i^{\star}}\left(\mathbf{id}_i^{(B)}\right) \right)^{\tilde{r}}$ and $R_2 = h^{\tilde{r}}$ with $\tilde{r} = r y_B - \frac{a y_B}{G(\omega,\omega',\mathbf{ID}^{(B)})}$.
  - If A *is the target user and B is honest, then $\mathcal{B}$ proceeds as follows. $\mathcal{B}$* returns a random re-encryption key: $R_{A \to B} = (\mathbf{ID}^{(B)}, R_1, R_2)$, where R_1 and R_2 are randomly picked from $\mathbb{G}_2$. The adversary will not detect that this re-encryption key is inconsistent since it does not have B's secret key and the considered attack is *chosen plaintext* and not chosen ciphertext. Note that all other cases are impossible due to the IND-CPA rules.
- **Challenge.** $\mathcal{A}$ outputs two messages M_0 and M_1, and a target path[7] $\mathbf{S}^{\star}$. We now consider the identity $\mathbf{ID_S}^{\star}$ defined during the preliminaries. If $G(\omega, \omega', \mathbf{ID_{S^\star}}) \neq 0$, then bad is set to true and $\mathcal{B}$ gives $C^{\star} = \bot$ to $\mathcal{A}$. Otherwise $\mathcal{B}$ picks $\delta \xleftarrow{\$} \{0,1\}$ and gives to $\mathcal{A}$:

$$C^{\star} = \left(M_\delta \cdot T, \Phi(h^c), \Phi(h^c)^{H(\mathbf{Z}^{\star}, z'^{\star}, \mathbf{ID_{S^\star}})}, \left(\Phi(h^c)^{z_{i,j}^{\star}} \right)_{i \in I_*, j \in \llbracket 1, n_i \rrbracket}, \mathbf{S}^{\star} \right).$$

An adaptation of [6, Lemma 3.2] shows that the distribution of the re-encryption keys simulated by $\mathcal{B}$ when $G(\omega, \omega', \mathbf{ID}^{(B)}) \neq 0$ is identically distributed as the genuine algorithm, and when $G(\omega, \omega', \mathbf{ID_{S^\star}}) = 0$, the third component of the challenge ciphertext is well-formed.
- **Guess.** If bad $=$ true, which means that $\mathcal{B}$ was not able to simulate correctly all answers to $\mathcal{A}$'s queries, then it returns a random guess β'. Otherwise $\mathcal{A}$ returns δ' and $\mathcal{B}$ returns 1 if $\delta = \delta'$ and 0 otherwise.

Game 1. In this game, we do not use the DBDH instance, so that $\mathcal{B}$ will be able to simulate all re-encryption key and challenge queries. Let $h \in \mathbb{G}_2$, $g = \Phi(h)$, $a, b, c \xleftarrow{\$} \mathbb{Z}_p^*$, $\beta \xleftarrow{\$} \{0,1\}$ and bad $\leftarrow$ false. If $\beta = 1$ then $T = e(g, h)^{abc}$, otherwise T is random. $\mathcal{B}$ runs $\mathcal{A}$ and receives its choice of wildcarded positions $I_*^{\star}$.

---

[6] We recall that this key is given to the adversary *only* if B's id. does not match $\mathbf{S}^{\star}$.
[7] *Cf.* the introduction of Section 3.3.

- *Target-key generation*: It picks $y^\star \xleftarrow{\$} \mathbb{Z}_q^\star$, sets $z^\star = ab$ and produces vector of vectors $\mathbf{V}^\star$ composed of element in $\mathbb{G}_2$ where $\forall (i,j) \in J$, $\mathbf{V}_{i,j}^\star = (h)^{b.\omega_{i,j}+z_{i,j}^\star}$ and the element $V'^\star = (h)^{b.\omega'+z'^\star}$ which defines $v_{i,j}^\star = b \times \omega_{i,j} + z_{i,j}^\star$ and $v'^\star = b \times \omega' + z'^\star$. Moreover, $\forall (i,j) \notin J$, we have $\mathbf{V}_{i,j}^\star = h^{z_{i,j}^\star}$ which defines $v_{i,j}^\star = z_{i,j}^\star$, $\mathbf{U}^\star = \Phi(\mathbf{V}^\star)$ and $U'^\star = \Phi(V'^\star)$. It gives to $\mathcal{A}$ $pk^\star = (Z^\star, Y^\star, \mathbf{U}^\star, U'^\star)$, where $Z^\star = e(g,h)^{z^\star}$ and $Y^\star = h^{y^\star}$.
- *Other key generation*: $\mathcal{B}$ proceeds as in **Game 0**.
- **Re-Encryption key generation.** $\mathcal{B}$ generates re-encryption keys from a user A to a user B for the path $\mathbf{ID}^{(B)} = (\mathbf{id}_1^{(B)}, \dots, \mathbf{id}_\ell^{(B)})$ as follows.
  - If A is the target user and B is corrupted [6] then $\mathcal{B}$ proceeds as follows:
    1. If $G(\omega, \omega', \mathbf{ID}^{(B)}) = 0 \pmod q$, then bad $\leftarrow$ true, define $R_{A \to B}$ as in the scheme and return it to $\mathcal{A}$.
    2. Otherwise define $R_{A \to B}$ as in the **Game 0**.
  - *Other re-encryption key generation*: $\mathcal{B}$ proceeds as in **Game 0**.
- **Challenge.** $\mathcal{A}$ outputs two messages M_0 and M_1, and a target path $\mathbf{S}^\star$. If $G(\omega, \omega', \mathbf{ID_{S^\star}}) \neq 0$ then bad $\leftarrow$ true and define the challenge as follows: Pick $\delta \xleftarrow{\$} \{0,1\}$, give $C^\star = (M_\delta \cdot T, g^c, \left(\prod_{i \notin I_\star} F_{\mathbf{U}_i^\star}(\mathbf{s}_i^\star)\right)^c, ((\mathbf{U}_{i,j}^\star)^c)_{i \in I_\star, j \in [\![1,n_i]\!]}, \mathbf{S}^\star)$ to $\mathcal{A}$, and, otherwise define it as in the **Game 0**.
- **Guess.** $\mathcal{A}$ returns δ' and eventually $\mathcal{B}$ returns 1 if $\delta = \delta'$ and 0 otherwise, even if bad $-$ true.

Game 2. We modify the simulation of the re-encryption keys and the challenge, in the case where the flag bad is set to true during **Game 0**. Let $h \in \mathbb{G}_2$, $g = \Phi(h)$, $a, b, c \xleftarrow{\$} \mathbb{Z}_p^\star$, $\beta \xleftarrow{\$} \{0,1\}$ and bad $\leftarrow$ false (bad will be set to true if $\mathcal{B}$ fails to simulate correctly during the **Game 0**.). If $\beta = 1$ then $T = e(g,h)^{abc}$, otherwise T is random. $\mathcal{B}$ runs $\mathcal{A}$ and receives its wildcarded positions $I_\star^\star$.

- *Target-key generation*: $\mathcal{B}$ proceeds as in the previous game
- *Other key generation*: $\mathcal{B}$ proceeds as in the previous game.
- **Re-Encryption key generation.** $\mathcal{B}$ generates the re-encryption keys from user A to user B as follows:
  - If A is the target user and B is corrupted [6], then $\mathcal{B}$ proceeds as follows:
    1. If $G(\omega, \omega', \mathbf{ID}^{(B)}) = 0 \pmod q$, then set bad $\leftarrow$ true.
    2. Define the re-encryption $R_{A \to B}$ as in the scheme.
  - *Other re-encryption key generation*: $\mathcal{B}$ proceeds as in the previous game.
- **Challenge.** $\mathcal{A}$ outputs two messages M_0 and M_1, and a target path $\mathbf{S}^\star = (\mathbf{s}_1^\star, \dots, \mathbf{s}_\ell^\star)$. If $G(\omega, \omega', \mathbf{ID_{S^\star}}) \neq 0$, $\mathcal{B}$ sets bad $\leftarrow$ true. Then $\mathcal{B}$ picks $\delta \xleftarrow{\$} \{0,1\}$ and gives to $\mathcal{A}$ the ciphertext $C^\star = (M_\delta \cdot T, g^c, \left(\prod_{i \notin I_\star} F_{\mathbf{U}_i^\star}(\mathbf{s}_i^\star)\right)^c, ((\mathbf{U}_{i,j}^\star)^c)_{i \in I_\star, j \in [\![0,n]\!]}, \mathbf{S}^\star)$.
- **Guess.** $\mathcal{A}$ returns δ' and $\mathcal{B}$ returns 1 if $\delta = \delta'$ and 0 otherwise.

Game 3. This game proceeds as the previous one, except that all checks done during the simulation to finally set bad to true are now done in the **Guess** phase.

Game 4. In this last game, we modify the definition of $v_{i,j}^*$. Indeed, $\forall (i,j) \in [\![1, \ell]\!] \times [\![1, n_i]\!]$, we have $v_{i,j}^* = z_{i,j}^*$ (which are all random), and *at the beginning of* the **Guess** phase, $\mathcal{B}$ picks $\omega' \in [\![-t(m-1), 0]\!]$ and $\omega_{i,j} \in [\![0, (m-1)]\!]$ for all $(i,j) \in J$ and proceeds all checks as in **Game 3**.

The idea behind game-playing techniques is to build a chain of *identical-until-bad* games. Two games are *identical-until*-bad if they are equivalent until the flag bad is not set (see [7] for a proper definition): here from **Game 0** to **Game 4**, two consecutive games are actually *identical-until*-bad (since we follow [6]). This is useful when we are not able to carry out the analysis for bounding directly the advantage of an adversary (this is the case in the **Game 0**), but that we can manage in some equivalent game (**Game 4**). In this way, we can switch the analysis of **Game 0** to a simpler analysis of the **Game 4**. Let now $\mathsf{GD_4}$ denote the event "**Game 4** does not set bad".

Lemma 1. *(Adapted from [6, Lemma 3.3])* $\mathbf{Adv}_{\mathsf{BGSGen}, \mathcal{B}}^{\mathsf{dbdh}}(\kappa) = 2 \Pr[\mathbf{Game\ 4} \Rightarrow \beta \wedge \mathsf{GD_4}] - \Pr[\mathsf{GD_4}]$.

Let $\mathcal{I} = (\mathbf{ID_{S^*}}, \mathbf{ID_1}, \ldots, \mathbf{ID_Q})$ be a sequence of identities that are involved in the same game (note that $\mathbf{ID_{S^*}} \neq \mathbf{ID_i}$ for all $1 \leq i \leq Q$). We need to compute the probability $\gamma(\mathcal{I})$ of $\mathsf{GD_4}$ according to such a sequence of identities. The idea is to obtain a fitting formula like in [6, Lemma 3.4 and 3.5]. Our function $G(\omega, \omega', \mathbf{ID})$ can be seen as the function $F(x, \mathbf{ID'})$ defined in Bellare and Ristenspart's paper [6] where $x = \{\omega', \omega_{1,1}, \ldots, \omega_{1,n_1}, \ldots \omega_{\ell,1}, \ldots, \omega_{\ell,n_\ell}\}$ and $\mathbf{ID'} = \{\mathrm{id}_{1,1}, \ldots, \mathrm{id}_{1,n_1}, \ldots, \mathrm{id}_{\ell,1}, \ldots, \mathrm{id}_{\ell,n_\ell}\}$. Let $\mathcal{I}_{sp} = \{(\mathbf{ID_{S^*}}, \mathbf{ID_1}, \ldots, \mathbf{ID_Q}) \in (\{0,1\}^{n_1} \times \cdots \times \{0,1\}^{n_\ell})^{Q+1} : \forall i \in [\![1, Q]\!] \ (\mathbf{ID_{S^*}} \neq \mathbf{ID_i})\}$. For $\mathcal{I} \in \mathcal{I}_{sp}$ we have $\gamma(\mathcal{I}) = Pr[G(\omega, \omega', \mathbf{ID_{S^*}}) = 0 \wedge G(\omega, \omega', \mathbf{ID_1}) \neq 0 \wedge \ldots \wedge G(\omega, \omega', \mathbf{ID_Q}) \neq 0]$ where the probability is taken over the ω. As pointed out by Bellare and Ristenpart, the problem here is that $\gamma(\mathcal{I})$ varies with any chosen sequence $\mathcal{I}$ in $\mathcal{I}_{sp}$. For that, let's introduce $\mathsf{Q}(\mathcal{I})$ the event "**Game 4** results in the identity sequence $\mathcal{I} = (\mathbf{ID_{S^*}}, \mathbf{ID_1}, \ldots, \mathbf{ID_Q})$ being queried by $\mathcal{A}$". As in [6, Lemma 3.4], we have

Lemma 2. *For* $\mathcal{I} \in \mathcal{I}_{sp}$, $\Pr[\mathbf{Game\ 4} \Rightarrow \beta \wedge \mathsf{GD_4} \wedge \mathsf{Q}(\mathcal{I})] = \gamma(\mathcal{I}) \Pr[\mathbf{Game\ 4} \Rightarrow \beta \wedge \mathsf{Q}(\mathcal{I})]$ *and* $\Pr[\mathsf{GD_4} \wedge \mathsf{Q}(\mathcal{I})] = \gamma(\mathcal{I}) \Pr[\mathsf{Q}(\mathcal{I})]$.

We now complete the analysis by computing an upper bound γ_{max} and a lower bound γ_{min} for $\gamma(\mathcal{I})$. We adapt [6, Lemma 3.4, Lemma 3.5] and obtain:

Lemma 3. $\frac{1}{\nu(m-1)+1}(1 - \frac{Q}{m}) \leq \gamma_{min} \leq \gamma_{max} \leq \frac{1}{\nu(m-1)+1}$.

We now get into the analysis of the security of our scheme. Thanks to Lemma 1 we have that $\mathbf{Adv}_{\mathsf{BGSGen}, \mathcal{B}}^{\mathsf{dbdh}}(\kappa) = \sum_{\mathcal{I}} 2 \Pr[\mathbf{Game\ 4} \Rightarrow \beta \wedge \mathsf{GD_4} \wedge \mathsf{Q}(\mathcal{I})] - \sum_{\mathcal{I}} \Pr[\mathbf{Game\ 4} \wedge \mathsf{Q}(\mathcal{I})]$ and using the "independence" of Lemma 2, we deduce that $\mathbf{Adv}_{\mathsf{BGSGen}, \mathcal{B}}^{\mathsf{dbdh}}(\kappa) \geq \gamma_{min} \sum_{\mathcal{I}} 2 \Pr[\mathbf{Game4} \Rightarrow \beta \wedge \mathsf{Q}(\mathcal{I})] - \gamma_{max} \sum_{\mathcal{I}} \Pr[\mathsf{Q}(\mathcal{I})]$.

Following [6] we have, $\Pr[\mathbf{Game4} \Rightarrow \beta] = \frac{1}{4}\mathbf{Adv}_{\mathsf{PRE.Encrypt_2}, \mathcal{A}}^{\mathsf{sind\text{-}cpa}}(\kappa, \mathcal{T}) + \frac{1}{2}$, which leads to $\mathbf{Adv}_{\mathsf{BGSGen}, \mathcal{B}}^{\mathsf{dbdh}}(\kappa) \geq 2\gamma_{min}(\frac{1}{4}\mathbf{Adv}_{\mathsf{PRE.Encrypt_2}, \mathcal{A}}^{\mathsf{sind\text{-}cpa}}(\kappa, \mathcal{T}) + \frac{1}{2}) - \gamma_{max}$. If we set $\alpha = \big(\nu(m-1) + 1\big)^{-1}$, and substitute γ_{min} and γ_{max} we get $\mathbf{Adv}_{\mathsf{BGSGen}, \mathcal{B}}^{\mathsf{dbdh}}(\kappa) \geq$

$\frac{\alpha}{2}(1 - \frac{Q}{m})\varepsilon + \alpha(1 - \frac{Q}{m}) - \alpha \geq \alpha(\frac{1}{2}(1 - \frac{\varepsilon}{9})\varepsilon - \frac{\varepsilon}{9}) \geq \frac{\alpha\varepsilon}{3}$. Putting all together and using $m = \left\lceil \frac{9Q}{\varepsilon} \right\rceil$ in α's definition, we finally get $\mathbf{Adv}_{\mathsf{BGSGen},\mathcal{B}}^{\mathsf{dbdh}}(\kappa) \geq \frac{\varepsilon^2}{27Q\nu + 3\varepsilon}$. $\quad\square$

The following theorem states the security of the encryption schemes under DBDH, and the classical proof can be found in the full version of this paper.

Theorem 2. *The scheme has* sIND-CPA *security of* PRE.Encrypt₁ *(respectively* PRE.ReEncrypt) *under the* DBDH *assumption in the standard model.*

3.4   Transferability Issues and Traceability

To deal with traceability properties, we need to consider the following CheckKey algorithm which checks the validity of a re-encryption key:

– CheckKey($\mathcal{P}, sk_A, pk_B, R_{A\to B}$): Given the public parameters, user A's private key $sk_A = (z_A, y_A, \mathbf{V}^{(A)}, V'^{(A)})$, user B's public key $pk_B = (Z_B, Y_B, \mathbf{U}^{(B)}, U'^{(B)})$ and a re-encryption key $R_{A\to B} = (\mathbf{ID}^{(B)}, t, h_t)$ as inputs, it returns 1 if $e(g, t) = e(g, Y_B)^{z_A} \cdot e(V'^{(A)} \cdot \prod_{i=1}^{\ell} F_{\mathbf{V}_i^{(A)}}(\mathsf{id}_i^{(B)}), h_t)$ and 0 otherwise. This algorithm returns 1 if and only if the re-encryption key $R_{A\to B}$ really allows to re-encrypt a ciphertext intended to a user A into a ciphertext intended to a user B.

As already mentioned in the introduction, our scheme suffers from transferability. Indeed, a corrupted proxy with a valid[8] re-encryption key from user A to user B, $R_{A\to B} = (\mathbf{ID}^{(B)}, Y_B^{z_A}.(V'^{(A)})^r. \prod_{i=1}^{\ell} F_{\mathbf{V}_i^{(A)}}(\mathsf{id}_i^{(B)})^r, h^r)$ can create, by colluding with B a re-encryption key from A to another corrupted user B'. To do so, given $R_{A\to B}$ and the two secret keys $sk_B = (z_B, y_B, \mathbf{V}^{(B)}, V'^{(B)})$ and $sk_{B'} = (z_{B'}, y_{B'}, \mathbf{V}^{(B')}, V'^{(B')})$, they can compute the key:
$R_{A\to B}^{\mathsf{fake}} = (\mathbf{ID}^{(B)}, Y_{B'}^{z_A}.(V'^{(A)})^{r.y_{B'}/y_B} \cdot \prod_{i=1}^{\ell} F_{\mathbf{V}_i^{(A)}}(\mathsf{id}_i^{(B)})^{r.y_{B'}/y_B}, h^{r.y_{B'}/y_B})$
which allows to re-encrypts ciphertexts from A to the user B' and such that[9] CheckKey($sk_A, pk_{B'}, R_{A\to B}^{\mathsf{fake}}$) = 1. But B' has to usurp the identity of B since the first component of the new re-encryption key has to stay user B's identity $\mathbf{ID}^{(B)}$ to pass the key check. It is therefore easily traceable in white box. On the other hand, this traceability does not permit to detect which proxy is potentially malicious as a re-encryption key is the same for all proxies, this can be useful in the case where there is a single proxy such as PRE-based file storage [4,5].

Improving the Traceability. As our encryption is based on Libert and Vergnaud's scheme [22], we can add a white box traceability (useful when there are several proxies) or a black box-traceability by concatenating to a re-encryption $R_{A\to B}$, a random string ω to B's identity and adding to the re-encryption some information corresponding to this string. If ω is chosen randomly, then we can have a white-box traceability, whereas if it is chosen as a random word of a collusion secure code [10], we obtain a scheme with a black-box traceability. We give in the long version, some details on the scheme supporting white-box

---

[8] i.e., CheckKey($sk_A, pk_B, R_{A\to B}$) = 1.
[9] If $sk_{B'} = 1$ then this re-encryption key allows to decrypt all second level ciphertext intended to A as in this case the secret key $sk_{B'}$ is not necessary anymore.

traceability. To obtain the black-box traceability we just have to modify the way the random string ω is chosen in this scheme as explain in the previous paragraph. The obtained protocol is still CPA-secure under the same assumptions and is traceable as the traceability of the system is the same as the one use in [22].

4 Concluding Remarks

We proposed a PRE which allows a user to choose among some predetermined delegatees who may decrypt a ciphertext originally intended to a delegator. It allows to access efficiently a large range of delegatees thanks to wildcards and is CPA secure under classical algorithmic assumptions in a bilinear setting.

Efficiency Considerations. A trivial use of a classical proxy re-encryption would lead to produce as many ciphertext as chosen delegatees. The use of a tree structure and wildcards gives with this respect a large improvement in terms of efficiency. In our schemes, the size of the re-encryption keys does not depend on the number of delegatees, except for $\mathbf{ID}^{(B)}$ which is included in the re-encryption key, and a re-encryption key from A to B has to be computed only once. On the contrary, in Tang's and Chu's *et al.* approaches [26,16], such a re-encryption key has to be computed for *each* group A decides that B belongs to. Our two schemes with white box traceability have a ciphertext which contains the path, so that its size has an additive factor of $\sum_i n_i$. By associating re-encryption keys to codewords of a collusion secure code, we obtain a scheme with a black-box tracing algorithm and with ciphertext with linear size in the length of this code.

Another Application: Cloud Storage. We present another application for our selective PRE: Alice has some data on a cloud storage and wants other users (friends, family, colleagues) to access this space to add or read these data. So, she organizes these users into groups, according to the type of relation they have with her (friends, family, colleagues) and she publishes this group information. Imagine now that someone wants to share some data with Alice and just her friends (not with her family or colleagues). In this case, he encrypts these data using our PRE scheme (here, the proxy is a service proposed by the cloud provider) which supports the choice of the delegatees (here friends), so that the ciphertext will be accessible to her and to Alice's group of friends, as the original sender chose. Alice can also securely store on the cloud some files and decide with whom to share them (holiday's photos with friends and family, working files with colleagues) by encrypting with our PRE to herself and her choice of authorized persons as delegatees. Now, suppose that a new user is added in the tree (if there is room): she will just have to update the re-encryption key, to make the re-encryption possible for this new user, without having to remove all encrypted files and re-encrypt them. Our scheme extends to trees with branches of different sizes when the description of the delegatees is more complex. Moreover, like in [18], our scheme has the following interesting feature: even though the files encrypted on the cloud have a size which depends on the path length, the files downloaded by the users after conversion by the proxy have *constant* size, since level-one ciphertext are Elgamal-like. This is crucial since it saves bandwidth, and since the ciphertext may to be decrypted on a low-resource devices.

Open Questions. We considered a model where Alice is honest in the sense that her tree is fairly designed: if she were not, she could add a hidden delegatee in the labeled tree, and give to the proxy the re-encryption key from her to him, so that he would be able to decrypt everything. Preventing such a behavior from the delegator is an interesting cryptographic question. This problem seems to be hard to solve without a trusted authority, which would linked the delegatee's identity path to their secret. From a positive side, in our setting, Alice can adaptatively add new delegatee in her tree. Another interesting problem we left open, is a manner for the proxy to re-encrypt for all the delegatees satisfying the target path with only one constant-size ciphertext. Reaching a stronger security level in the standard model (for instance CCA security) would be also important.

References

1. Abdalla, M., Birkett, J., Catalano, D., Dent, A.W., Malone-Lee, J., Neven, G., Schuldt, J.C.N., Smart, N.P.: Wildcarded identity-based encryption. Journal of Cryptology 24(1), 42–82 (2011)
2. Abdalla, M., Catalano, D., Dent, A.W., Malone-Lee, J., Neven, G., Smart, N.P.: Identity-based encryption gone wild. In: Bugliesi, M., Preneel, B., Sassone, V., Wegener, I. (eds.) ICALP 2006, Part II. LNCS, vol. 4052, pp. 300–311. Springer, Heidelberg (2006)
3. Ateniese, G., Benson, K., Hohenberger, S.: Key-private proxy re-encryption. In: Fischlin, M. (ed.) CT-RSA 2009. LNCS, vol. 5473, pp. 279–294. Springer, Heidelberg (2009)
4. Ateniese, G., Fu, K., Green, M., Hohenberger, S.: Improved proxy re-encryption schemes with applications to secure distributed storage. In: NDSS. The Internet Society (2005)
5. Ateniese, G., Fu, K., Green, M., Hohenberger, S.: Improved proxy re-encryption schemes with applications to secure distributed storage. ACM Trans. Inf. Syst. Secur. 9(1), 1–30 (2006)
6. Bellare, M., Ristenpart, T.: Simulation without the artificial abort: Simplified proof and improved concrete security for Waters' IBE scheme. In: Joux, A. (ed.) EUROCRYPT 2009. LNCS, vol. 5479, pp. 407–424. Springer, Heidelberg (2009)
7. Bellare, M., Rogaway, P.: The security of triple encryption and a framework for code-based game-playing proofs. In: Vaudenay, S. (ed.) EUROCRYPT 2006. LNCS, vol. 4004, pp. 409–426. Springer, Heidelberg (2006)
8. Blaze, M., Bleumer, G., Strauss, M.: Divertible protocols and atomic proxy cryptography. In: Nyberg, K. (ed.) EUROCRYPT 1998. LNCS, vol. 1403, pp. 127–144. Springer, Heidelberg (1998)
9. Boneh, D., Naor, M.: Traitor tracing with constant size ciphertext. In: Proceedings of the 15th ACM Conference on Computer and Communications Security, CCS 2008, pp. 501–510. ACM (2008)
10. Boneh, D., Shaw, J.: Collusion-secure fingerprinting for digital data (extended abstract). In: Coppersmith, D. (ed.) CRYPTO 1995. LNCS, vol. 963, pp. 452–465. Springer, Heidelberg (1995)
11. Canard, S., Devigne, J.: Combined proxy re-encryption. In: ICISC (2013) (to appear)
12. Canard, S., Devigne, J., Laguillaumie, F.: Improving the security of an efficient unidirectional proxy re-encryption scheme. Journal of Internet Services and Information Security 1(2/3), 140–160 (2011)

13. Canetti, R., Hohenberger, S.: Chosen-ciphertext secure proxy re-encryption. In: ACM Conference on Computer and Communications Security 2007, pp. 185–194. ACM (2007)
14. Canetti, R., Krawczyk, H., Nielsen, J.B.: Relaxing chosen-ciphertext security. In: Boneh, D. (ed.) CRYPTO 2003. LNCS, vol. 2729, pp. 565–582. Springer, Heidelberg (2003)
15. Chow, S.S.M., Weng, J., Yang, Y., Deng, R.H.: Efficient unidirectional proxy re-encryption. In: Bernstein, D.J., Lange, T. (eds.) AFRICACRYPT 2010. LNCS, vol. 6055, pp. 316–332. Springer, Heidelberg (2010)
16. Chu, C.-K., Weng, J., Chow, S.S.M., Zhou, J., Deng, R.H.: Conditional proxy broadcast re-encryption. In: Boyd, C., González Nieto, J. (eds.) ACISP 2009. LNCS, vol. 5594, pp. 327–342. Springer, Heidelberg (2009)
17. Fang, L., Susilo, W., Ge, C., Wang, J.: Interactive conditional proxy re-encryption with fine grain policy. Journal of Systems and Software 12, 2293–2302 (2011)
18. Green, M., Hohenberger, S., Waters, B.: Outsourcing the decryption of ABE ciphertexts. In: USENIX Security Symposium (2011)
19. Hohenberger, S.: Advances in Signatures, Encryption, and E-Cash from Bilinear Groups. PhD thesis, MIT (May 2006)
20. Shao, X.L.J., Cao, Z., Lin, H.: Proxy re-encryption with keyword search. Information Sciences 180(13), 2576–2587 (2010)
21. Liang, X., Cao, Z., Lin, H., Shao, J.: Attribute based proxy re-encryption with delegating capabilities. In: ASIACCS 2009, pp. 276–286 (2009)
22. Libert, B., Vergnaud, D.: Tracing malicious proxies in proxy re-encryption. In: Galbraith, S.D., Paterson, K.G. (eds.) Pairing 2008. LNCS, vol. 5209, pp. 332–353. Springer, Heidelberg (2008)
23. Libert, B., Vergnaud, D.: Unidirectional chosen-ciphertext secure proxy re-encryption. In: Cramer, R. (ed.) PKC 2008. LNCS, vol. 4939, pp. 360–379. Springer, Heidelberg (2008)
24. Libert, B., Vergnaud, D.: Unidirectional chosen-ciphertext secure proxy re-encryption. IEEE Transactions on Information Theory 57(3), 1786–1802 (2011)
25. Shao, J., Cao, Z.: CCA-secure proxy re-encryption without pairings. In: Jarecki, S., Tsudik, G. (eds.) PKC 2009. LNCS, vol. 5443, pp. 357–376. Springer, Heidelberg (2009)
26. Tang, Q.: Type-based proxy re-encryption and its construction. In: Chowdhury, D.R., Rijmen, V., Das, A. (eds.) INDOCRYPT 2008. LNCS, vol. 5365, pp. 130–144. Springer, Heidelberg (2008)
27. Vivek, S.S., Sharmila Deva Selvi, S., Radhakishan, V., Pandu Rangan, C.: Conditional proxy re-encryption — a more efficient construction. In: Wyld, D.C., Wozniak, M., Chaki, N., Meghanathan, N., Nagamalai, D. (eds.) CNSA 2011. CCIS, vol. 196, pp. 502–512. Springer, Heidelberg (2011)
28. Waters, B.: Efficient identity-based encryption without random oracles. In: Cramer, R. (ed.) EUROCRYPT 2005. LNCS, vol. 3494, pp. 114–127. Springer, Heidelberg (2005)
29. Weng, J., Deng, R.H., Ding, X., Chu, C.K., Lai, J.: Conditional proxy re-encryption secure against chosen-ciphertext attack. In: ASIACCS 2009, pp. 322–332. ACM (2009)
30. Weng, J., Yang, Y., Tang, Q., Deng, R.H., Bao, F.: Efficient conditional proxy re-encryption with chosen-ciphertext security. In: Samarati, P., Yung, M., Martinelli, F., Ardagna, C.A. (eds.) ISC 2009. LNCS, vol. 5735, pp. 151–166. Springer, Heidelberg (2009)

Trapdoor Privacy in Asymmetric Searchable Encryption Schemes

Afonso Arriaga, Qiang Tang, and Peter Ryan

SnT, University of Luxembourg
{afonso.delerue,qiang.tang,peter.ryan}@uni.lu

Abstract. Asymmetric searchable encryption allows searches to be carried over ciphertexts, through delegation, and by means of trapdoors issued by the owner of the data. Public Key Encryption with Keyword Search (PEKS) is a primitive with such functionality that provides delegation of exact-match searches. As it is important that ciphertexts preserve data privacy, it is also important that trapdoors do not expose the user's search criteria. The difficulty of formalizing a security model for trapdoor privacy lies in the verification functionality, which gives the adversary the power of verifying if a trapdoor encodes a particular keyword. In this paper, we provide a broader view on what can be achieved regarding trapdoor privacy in asymmetric searchable encryption schemes, and bridge the gap between previous definitions, which give limited privacy guarantees in practice against search patterns. Since it is well-known that PEKS schemes can be trivially constructed from any Anonymous IBE scheme, we propose the security notion of Key Unlinkability for IBE, which leads to strong guarantees of trapdoor privacy in PEKS, and we construct a scheme that achieves this security notion.

Keywords: Asymmetric Searchable Encryption, PEKS, Trapdoor Privacy, Function Privacy, Search Pattern Privacy, Key Unlinkability.

1 Introduction

As cloud services become increasingly popular, security concerns arise from exposing the user's data to third-party service providers. Encryption can be used to protect the user's data privacy, but usability is sacrificed if not even the most basic operations, such as searching over the user's data, can be delegated to the service provider. In the public key setting, Boneh et al. [6] were the first to propose a primitive to tackle this problem. They called it Public Key Encryption with Keyword Search (PEKS), a primitive that provides delegation of *exact-match* searches over ciphertexts. A typical scenario where this primitive can bring great benefits to users (and consequently to service providers wishing to increase their customer base as well) is that of any email system.

Suppose user Alice stores her emails in the servers of some email service provider, so that she can access them from either her laptop or her smartphone. Alice does not trust the service provider or fears that government agencies may

D. Pointcheval and D. Vergnaud (Eds.): AFRICACRYPT 2014, LNCS 8469, pp. 31–50, 2014.
© Springer International Publishing Switzerland 2014

require the service provider to hand over all her data. Using standard public key encryption, any user with Alice's public key can send her encrypted emails that only she can decrypt. For Alice to find a particular email later on, the sender could also attach to the email some *searchable ciphertexts*, produced from a PEKS scheme, with keywords that Alice might use when searching for this email. These ciphertexts are searchable upon delegation, meaning that only Alice can authorize the email service provider to search on her behalf by issuing a trapdoor that encodes Alice's search criteria (e.g. ciphertexts that encrypt the keyword "project xx123 - meeting"), generated from her own secret key. The service provider searches through all Alice's emails for those containing searchable ciphertexts that match the issued trapdoor, and returns to her only those with a positive match.

Many efforts have been put in asymmetric searchable encryption in general, as surveyed in [16], most towards more efficient PEKS schemes (or relying on weaker assumptions) or towards primitives with more flexible search queries, such as conjunctive, disjunctive, subset and inner product types of queries. Until recently [14,9,10], the concern was always to preserve data privacy in the ciphertexts and no attention was paid to possible information leakage from the trapdoors. In fact, some schemes, as the statistically consistent scheme proposed in [1], include the keyword itself in the trapdoor. In this paper we focus on defining trapdoor privacy for PEKS and constructing a scheme that provably stands up to the definition. Nevertheless, the definition can easily be extended to asymmetric searchable encryption in general [13].

The difficulty of formalizing a security model for trapdoor privacy lies in the verification functionality of PEKS, which in the public key setting depends on the trapdoor itself and ciphertexts created from publicly known parameters. This provides to any adversary the power to verify if a trapdoor encodes a particular keyword. (The adversary encrypts the chosen keyword under the public key associated with the trapdoor; if the ciphertext matches the trapdoor then the trapdoor encodes the chosen keyword.) Therefore, an offline dictionary attack can always be launched, putting aside the possibility of formalizing the security notion of trapdoor privacy as a traditional choose-then-guess indistinguishability game in the public-key setting - although possible in the symmetric setting [15]. In many cases, the keywords encoded in trapdoors are *sufficiently unpredictable* for a dictionary attack to be infeasible. So, defining the right notion of trapdoor privacy is crucial to guarantee that the user's privacy is fully protected.

RELATED WORK. Abdalla et al. [1], by extending the results left implicit in [6], proposed a general black-box transformation from *Anonymous* Identity-Based Encryption (IBE) to PEKS, where the resulting PEKS scheme is secure in the traditional ciphertext indistinguishability sense. Identities and their secret keys in the original IBE scheme map to keywords and trapdoors in resulting PEKS scheme, respectively. The anonymity requirement informally states that ciphertexts leak no information regarding the identity of the recipient, leading to the commonly desired keyword-privacy guarantees over ciphertexts in PEKS. The standard notion of ciphertext indistinguishability in IBE leads to *Computational*

Consistency in the resulting PEKS scheme, which informally means that it is hard for computationally bounded adversaries to find two distinct keywords such that the trapdoors for the first keyword positively match the ciphertexts of the second keyword. (Note that if keywords are hashed before used in the scheme, *inconsistency* happens at least every time $H(w_1) = H(w_2)$, where $w_1 \neq w_2$.) We refer the reader to Section 2 for precise details on this transformation and to [1] for formal proofs.

This (black-box) transformation allows us to define a "dual" security notion for IBE that will lead to the desired trapdoor privacy notion in PEKS, and motivates the construction of IBE schemes that can provably satisfy it. This approach was also followed by [14,9,10], which, to the best of our knowledge, are the only works to address the concerns on trapdoor privacy in asymmetric searchable encryption.

Two distinct scenarios have to be considered to model trapdoor privacy. One in the presence of ciphertexts that positively match the trapdoors, and the other in the absence of such ciphertexts. Consider a toy example where the service provider possesses one ciphertext that belongs to Alice and two trapdoors that Alice issued for searches to be performed on her behalf. The service provider executes the test-search and one of the following cases occurs:

(a) Both trapdoors positively match the stored ciphertext, in which case the trapdoors encode the same keyword.
(b) Only one of the trapdoors match the ciphertext, in which case the trapdoors encode different keywords.
(c) None of the trapdoors positively match the stored ciphertext.

From cases (a) and (b), we can see that, in the presence of ciphertexts that match the trapdoors, an equality relation between the keywords encoded under the trapdoors can be determined trivially. In such cases, the notion of trapdoor privacy focus on revealing as little information as possible on the keywords themselves[1]. Recently, Boneh et al. [9] put forward two formal definitions of different strengths for IBE, inspired by the security definition given for Deterministic Encryption in [3]: *Function Privacy* and *Enhanced Function Privacy*. The latter leads to a security notion in PEKS (after the black-box transformation in [1]), which addresses this scenario.

Case (c) covers the scenario where trapdoors do not match any ciphertext. It is in Alice's best interest to hide her search pattern from the service provider. If the search pattern is revealed, the attacker could concentrate its resources on breaking the privacy of trapdoors encoding the most frequent keywords, which a priori are the most relevant to Alice. This issue is particularly important for PEKS due to the possibility of launching dictionary attacks. Nishioka [14] proposed a model denoted *Search Pattern Privacy*, which partially addresses this scenario. However, the model limits the distinguishing game to two trapdoors,

---

[1] Note that some information is inevitably leaked because of the verification functionality, e.g. if the trapdoor does not match a ciphertext which encrypts a particular known keyword, it means that the trapdoor does not encode this keyword.

which provides insufficient privacy guarantees in practice, considering that an actual attacker may have access to a much larger number of (possibly related) trapdoors. As we show in Section 4, and contrarily to intuition, the so-called hybrid argument does not apply here, unless trapdoors can be efficiently re-randomized. Also, after the transformation from IBE to PEKS, Function Privacy leads to a security definition that provides limited privacy guarantees against search patterns, since the resulting model prevents the adversary from being challenged with trapdoors encoding the same keyword.

OUR CONTRIBUTIONS. We formulate the "dual" notion of Search Pattern Privacy [14], which we call *Weak Key Unlinkability* for IBE. We then show that Weak Key Unlinkability is insufficient in practice. We do so by constructing a new Anonymous IBE scheme with Weak Key Unlinkability, based on the Anonymous IBE scheme by Boyen and Waters [12], and show that the resulting PEKS scheme (by applying black-box transformation in [1]) fails to hide search patterns when more than two trapdoors have been issued. We then propose a new security model, strictly stronger than Weak Key Unlinkability, which we call *Strong Key Unlinkability*. We compare the different notions of security and show that Key Unlinkability and Function Privacy [9] are orthogonal notions. Finally, we extend our IBE scheme to groups of composite order, and prove its security in the Strong Key Unlinkability model.

2 Preliminaries

NOTATION. We write $a \leftarrow b$ to denote the algorithmic action of assigning the value of b to the variable a. We use $\perp \notin \{0, 1\}^*$ to denote a special failure symbol. If S is a set, we write $a \leftarrow_\$ S$ for sampling a from S uniformly at random. If X is a joint probability distribution with L random variables, we write $(x_1, ..., x_L) \leftarrow_\$ X$ for sampling $(x_1, ..., x_L)$ from X. If A is a probabilistic algorithm we write $a \leftarrow_\$ A(i_1, i_2, \ldots, i_n)$ for the action of running A on inputs $i_1, i_2, \ldots, i_n$ with random coins, and assigning the result to a. If a is a variable, $|a|$ denotes the length in bits of its representation. We denote by $a \| b$ the concatenation of variables a and b, represented as bit-strings.

GAMES. In this paper we use the code-based game-playing language [4]. Each game has an Initialize and a Finalize procedure. It also has specifications of procedures to respond to an adversary's various queries. A game is run with an adversary A as follows. First Initialize runs and its outputs are passed to A. Then A runs and its oracle queries are answered by the procedures of the game. When A terminates, its output is passed to Finalize, which returns the outcome of the game. In each game, we restrict attention to legitimate adversaries, which is defined specifically for each game. We use lists as data structures to keep relevant state in the games. The empty list is represented by empty square brackets [ ]. We denote by list $\leftarrow$ a : list the action of appending element a to the head of list. To access the value stored in index i of list and assign it to a, we write $a \leftarrow$ list[i]. To denote the number of elements in list, we use |list|. Unless stated otherwise, lists are initialized empty and variables are first assigned with $\perp$.

2.1 Bilinear Groups

We first revise pairings over prime-order groups and the associated *Decision Bilinear Diffie-Hellman* (DBDH) and *Decision Linear* (DLIN) assumptions [7,5]. We then revise pairings over composite-order groups [8], introduce the new *Composite Decision Diffie-Hellman* (CDDH) assumption, and show that this assumption is weaker than the well-established *Composite 3-party Diffie-Hellman* (C3DH) assumption made in [11].

Bilinear Groups of Prime Order

Definition 1. *A prime-order bilinear group generator is an algorithm $\mathcal{G}_P$ that takes as input a security parameter λ and outputs a description $\Gamma = (p, \mathbb{G}, \mathbb{G}_T, e, g)$ where $\mathbb{G}$ and $\mathbb{G}_T$ are groups of order p with efficiently-computable group laws, where p is a λ-bit prime, g is a generator of $\mathbb{G}$ and e is an efficiently-computable bilinear pairing $e : \mathbb{G} \times \mathbb{G} \to \mathbb{G}_T$.*

Definition 2. *Let $\Gamma = (p, \mathbb{G}, \mathbb{G}_T, e, g)$ be the description output by $\mathcal{G}_P(\lambda)$. We say the DBDH assumption holds for description Γ if, for every PPT adversary $\mathcal{A}$, the following definition of advantage is negligible in λ.*

$$\mathbf{Adv}^{\mathsf{DBDH}}_{\Gamma,\mathcal{A}} := 2 \cdot \Pr[\mathsf{DBDH} \Rightarrow \mathsf{True}] - 1,$$

where game DBDH is described in Fig. 1.

```
procedure Initialize(λ):
Γ ←$ 𝒢ₚ(λ)
(p, 𝔾, 𝔾_T, e, g) ← Γ
z₁ ←$ ℤ_p
z₂ ←$ ℤ_p
z₃ ←$ ℤ_p
Z ←$ 𝔾_T
bit ←$ {0,1}
if bit = 0 return (Γ, g^z₁, g^z₂, g^z₃, e(g,g)^z₁z₂z₃)
else return (Γ, g^z₁, g^z₂, g^z₃, Z)

procedure Finalize(bit'):
if bit = bit' return True
else return False
```

Fig. 1. Game DBDH

```
procedure Initialize(λ):
Γ ←$ 𝒢ₚ(λ)
(p, 𝔾, 𝔾_T, e, g) ← Γ
z₁ ←$ ℤ_p
z₂ ←$ ℤ_p
z₃ ←$ ℤ_p
z₄ ←$ ℤ_p
Z ←$ 𝔾_T
bit ←$ {0,1}
if bit = 0 return (Γ, g^z₁, g^z₂, g^z₁z₃, g^z₂z₄, g^z₃+z₄)
else return (Γ, g^z₁, g^z₂, g^z₁z₃, g^z₂z₄, Z)

procedure Finalize(bit'):
if bit = bit' return True
else return False
```

Fig. 2. Game DLIN

Definition 3. *Let $\Gamma = (p, \mathbb{G}, \mathbb{G}_T, e, g)$ be the description output by $\mathcal{G}_P(\lambda)$. We say the DLIN assumption holds for description Γ if, for every PPT adversary $\mathcal{A}$, the following definition of advantage is negligible in λ.*

$$\mathbf{Adv}^{\mathsf{DLIN}}_{\Gamma,\mathcal{A}} := 2 \cdot \Pr[\mathsf{DLIN} \Rightarrow \mathsf{True}] - 1,$$

where game DLIN is described in Fig. 2.

Bilinear Groups of Composite Order

Definition 4. *A composite-order bilinear group generator is an algorithm $\mathcal{G}_C$ that takes as input a security parameter λ and outputs a description $\Gamma = (\mathsf{p}, \mathsf{q}, \mathbb{G}, \mathbb{G}_T, \mathsf{e}, \mathsf{g})$ where $\mathbb{G}$ and $\mathbb{G}_T$ are groups of order $\mathsf{n} = \mathsf{pq}$, where p and q are independent λ-bit primes, with efficiently computable group laws, g is a generator of $\mathbb{G}$ and e is an efficiently-computable bilinear pairing $\mathsf{e} : \mathbb{G} \times \mathbb{G} \to \mathbb{G}_T$.*

Subgroups $\mathbb{G}_\mathsf{p} \subset \mathbb{G}$ and $\mathbb{G}_\mathsf{q} \subset \mathbb{G}$ of order p and order q can be generated respectively by $\mathsf{g}_\mathsf{p} = \mathsf{g}^\mathsf{q}$ and $\mathsf{g}_\mathsf{q} = \mathsf{g}^\mathsf{p}$. We recall some important facts regarding these groups:

- $\mathbb{G} = \mathbb{G}_\mathsf{p} \times \mathbb{G}_\mathsf{q}$
- $\mathsf{e}(\mathsf{g}_\mathsf{p}, \mathsf{g}_\mathsf{q}) = \mathsf{e}(\mathsf{g}^\mathsf{q}, \mathsf{g}^\mathsf{p}) = \mathsf{e}(\mathsf{g}, \mathsf{g})^\mathsf{n} = 1$
- $\mathsf{e}(\mathsf{g}_\mathsf{p}, (\mathsf{g}_\mathsf{p})^\mathsf{a} \cdot (\mathsf{g}_\mathsf{q})^\mathsf{b}) = \mathsf{e}(\mathsf{g}_\mathsf{p}, (\mathsf{g}_\mathsf{p})^\mathsf{a}) \cdot \mathsf{e}(\mathsf{g}_\mathsf{p}, (\mathsf{g}_\mathsf{q})^\mathsf{b}) = \mathsf{e}(\mathsf{g}_\mathsf{p}, \mathsf{g}_\mathsf{p})^\mathsf{a}$

Definition 5. *Let $\Gamma = (\mathsf{p}, \mathsf{q}, \mathbb{G}, \mathbb{G}_T, \mathsf{e}, \mathsf{g})$ be the description output by $\mathcal{G}_C(\lambda)$ and $\Gamma' = (\mathsf{n}, \mathbb{G}, \mathbb{G}_T, \mathsf{e}, \mathsf{g})$, where $\mathsf{n} \leftarrow \mathsf{pq}$. We say the C3DH assumption holds for description Γ' if, for every PPT adversary $\mathcal{A}$, the following definition of advantage is negligible in λ.*

$$\mathbf{Adv}^{\mathsf{C3DH}}_{\Gamma', \mathcal{A}} := 2 \cdot \Pr[\mathsf{C3DH} \Rightarrow \mathsf{True}] - 1,$$

where game C3DH is described in Fig. 3.

Definition 6. *Let $\Gamma = (\mathsf{p}, \mathsf{q}, \mathbb{G}, \mathbb{G}_T, \mathsf{e}, \mathsf{g})$ be the description output by $\mathcal{G}_C(\lambda)$ and $\Gamma' = (\mathsf{n}, \mathbb{G}, \mathbb{G}_T, \mathsf{e}, \mathsf{g})$, where $\mathsf{n} \leftarrow \mathsf{pq}$. We say the CDDH assumption holds for description Γ' if, for every PPT adversary $\mathcal{A}$, the following definition of advantage is negligible in λ.*

$$\mathbf{Adv}^{\mathsf{CDDH}}_{\Gamma', \mathcal{A}} := 2 \cdot \Pr[\mathsf{CDDH} \Rightarrow \mathsf{True}] - 1,$$

where game CDDH is described in Fig. 4.

procedure Initialize(λ):	**procedure Initialize**(λ):
$(\mathsf{p}, \mathsf{q}, \mathbb{G}, \mathbb{G}_T, \mathsf{e}, \mathsf{g}) \leftarrow_\$ \mathcal{G}_C(\lambda)$	$(\mathsf{p}, \mathsf{q}, \mathbb{G}, \mathbb{G}_T, \mathsf{e}, \mathsf{g}) \leftarrow_\$ \mathcal{G}_C(\lambda)$
$\mathsf{n} \leftarrow \mathsf{pq}; \quad \mathsf{g}_\mathsf{p} \leftarrow \mathsf{g}^\mathsf{q}; \quad \mathsf{g}_\mathsf{q} \leftarrow \mathsf{g}^\mathsf{p}$	$\mathsf{n} \leftarrow \mathsf{pq}; \quad \mathsf{g}_\mathsf{p} \leftarrow \mathsf{g}^\mathsf{q}; \quad \mathsf{g}_\mathsf{q} \leftarrow \mathsf{g}^\mathsf{p}$
$\Gamma' \leftarrow (\mathsf{n}, \mathbb{G}, \mathbb{G}_T, \mathsf{e}, \mathsf{g})$	$\Gamma' \leftarrow (\mathsf{n}, \mathbb{G}, \mathbb{G}_T, \mathsf{e}, \mathsf{g})$
$\mathsf{X}_1 \leftarrow_\$ \mathbb{G}_\mathsf{q}; \quad \mathsf{X}_2 \leftarrow_\$ \mathbb{G}_\mathsf{q}; \quad \mathsf{X}_3 \leftarrow_\$ \mathbb{G}_\mathsf{q}$	$\mathsf{X}_1 \leftarrow_\$ \mathbb{G}_\mathsf{q}; \quad \mathsf{X}_2 \leftarrow_\$ \mathbb{G}_\mathsf{q}; \quad \mathsf{X}_3 \leftarrow_\$ \mathbb{G}_\mathsf{q}$
$\mathsf{a} \leftarrow_\$ \mathbb{Z}_\mathsf{n}; \quad \mathsf{b} \leftarrow_\$ \mathbb{Z}_\mathsf{n}; \quad \mathsf{c} \leftarrow_\$ \mathbb{Z}_\mathsf{n}; \quad \mathsf{R} \leftarrow_\$ \mathbb{G}$	$\mathsf{a} \leftarrow_\$ \mathbb{Z}_\mathsf{n}; \quad \mathsf{b} \leftarrow_\$ \mathbb{Z}_\mathsf{n}; \quad \mathsf{R} \leftarrow_\$ \mathbb{G}$
$\mathsf{bit} \leftarrow_\$ \{0, 1\}$	$\mathsf{bit} \leftarrow_\$ \{0, 1\}$
if $\mathsf{bit} = 0$ return	if $\mathsf{bit} = 0$ return
$\dots (\Gamma', \mathsf{g}_\mathsf{p}, \mathsf{g}_\mathsf{q}, (\mathsf{g}_\mathsf{p})^\mathsf{a}, (\mathsf{g}_\mathsf{p})^\mathsf{b}, \mathsf{X}_1(\mathsf{g}_\mathsf{p})^{\mathsf{ab}}, \mathsf{X}_2(\mathsf{g}_\mathsf{p})^{\mathsf{abc}}, \mathsf{X}_3(\mathsf{g}_\mathsf{p})^\mathsf{c})$	$\dots (\Gamma', \mathsf{g}_\mathsf{p}, \mathsf{g}_\mathsf{q}, \mathsf{X}_1(\mathsf{g}_\mathsf{p})^\mathsf{a}, \mathsf{X}_2(\mathsf{g}_\mathsf{p})^\mathsf{b}, \mathsf{X}_3(\mathsf{g}_\mathsf{p})^{\mathsf{ab}})$
else return	else return
$\dots (\Gamma', \mathsf{g}_\mathsf{p}, \mathsf{g}_\mathsf{q}, (\mathsf{g}_\mathsf{p})^\mathsf{a}, (\mathsf{g}_\mathsf{p})^\mathsf{b}, \mathsf{X}_1(\mathsf{g}_\mathsf{p})^{\mathsf{ab}}, \mathsf{X}_2(\mathsf{g}_\mathsf{p})^{\mathsf{abc}}, \mathsf{R})$	$\dots (\Gamma', \mathsf{g}_\mathsf{p}, \mathsf{g}_\mathsf{q}, \mathsf{X}_1(\mathsf{g}_\mathsf{p})^\mathsf{a}, \mathsf{X}_2(\mathsf{g}_\mathsf{p})^\mathsf{b}, \mathsf{R})$
procedure Finalize(bit'):	**procedure Finalize**(bit'):
if $\mathsf{bit} = \mathsf{bit}'$ return True	if $\mathsf{bit} = \mathsf{bit}'$ return True
else return False	else return False

Fig. 3. Game C3DH   **Fig. 4.** Game CDDH

In game C3DH, adversary is given a tuple $(\Gamma', g_p, g_q, (g_p)^a, (g_p)^b, X_1(g_p)^{ab}, X_2(g_p)^{abc}, Z)$ and has to decide whether $Z = X_3(g_p)^c$, for some $X_3 \in \mathbb{G}_q$. For convenience, we rewrite this as $(\Gamma', g_p, g_q, (g_p)^a, (g_p)^b, X_1(g_p)^{ab}, Y, X_3(g_p)^c)$, where Y is either $X_2(g_p)^{abc}$ or random in $\mathbb{G}$. Now, notice that $(\Gamma', g_p, g_q, X_1(g_p)^{ab}, X_3(g_p)^c, Y)$ is a CDDH tuple. Therefore, CDDH is a weaker assumption than C3DH.

2.2 Anonymous Identity-Based Encryption

An IBE scheme $\Pi = (\mathsf{Setup}, \mathsf{Extract}, \mathsf{Enc}, \mathsf{Dec})$ is specified by four polynomial-time algorithms associated with a message space $\mathcal{M}$ and an identity space $\mathcal{I}$.

- $\mathsf{Setup}(\lambda)$: On input the security parameter λ, this algorithm returns a master secret key msk and public parameters pp.
- $\mathsf{Extract}(\mathsf{pp}, \mathsf{msk}, \mathsf{id})$: On input public parameters pp, a master secret key msk and an identity $\mathsf{id} \in \mathcal{I}$, this algorithm outputs a secret key sk.
- $\mathsf{Enc}(\mathsf{pp}, \mathsf{m}, \mathsf{id})$: On input public parameters pp, a message $\mathsf{m} \in \mathcal{M}$ and an identity $\mathsf{id} \in \mathcal{I}$, this algorithm outputs a ciphertext c.
- $\mathsf{Dec}(\mathsf{pp}, \mathsf{c}, \mathsf{sk})$: On input public parameters pp, a ciphertext c and a secret key sk, this algorithm outputs either a message m or a failure symbol $\perp$.

The correctness of an IBE scheme requires that decryption reverses encryption, i.e., for any $\lambda \in \mathbb{N}$, any $(\mathsf{msk}, \mathsf{pp}) \leftarrow_\$ \mathsf{Setup}(\lambda)$, any $\mathsf{id} \in \mathcal{I}$, any $\mathsf{m} \in \mathcal{M}$, we have that $\mathsf{Dec}(\mathsf{pp}, \mathsf{Enc}(\mathsf{pp}, \mathsf{m}, \mathsf{id}), \mathsf{Extract}(\mathsf{pp}, \mathsf{msk}, \mathsf{id})) = \mathsf{m}$.

The standard notions of security for IBE are *anonymity* and *semantic security*. Intuitively, *anonymity* requires that ciphertexts conceal the identity and *semantic security* requires that ciphertexts conceal the message. We omit the formal definitions in this version due to space limitations. These properties lead to *semantic security* and *computational consistency*, respectively, in PEKS, after applying the black-box transformation described in [1].

3   Security Definitions

In this section, we formulate the notion of *Weak Key Unlinkability* for IBE, which leads to Nishioka's *Search Pattern Privacy* model for PEKS [14], after the black-box transformation from IBE to PEKS [1]. We then strengthen the model by allowing the adversary to be challenged with multiple secret keys, instead of just two. We refer to this new model as *Strong Key Unlinkability*. The resulting "dual" property for PEKS allows the adversary to be challenged with multiple trapdoors, which better reflects real-world scenarios. We then compare the new notions of security introduced here with those introduced by Boneh et al. in [9], and show that the two are orthogonal. Finally, we show that an easy and natural transformation from Strong Key Unlinkability to a more generalized definition, where the adversary is allowed to choose a joint probability distribution from which identities are sampled - instead of being sampled uniformly at random from the identity space - exists, as long as the adversary's choice does not depend on the public parameters of the scheme.

3.1   Key Unlinkability for IBE

Key Unlinkability models for IBE require that the size of the identity space is at least $\omega(\log \lambda)$, where λ is the security parameter of the scheme.

Definition 7. *An IBE scheme Π, associated with a non-polynomial size identity space $\mathcal{I}$, has Weak Key Unlinkability if, for every legitimate PPT adversary $\mathcal{A}$, the following definition of advantage is negligible in λ*

$$\mathbf{Adv}_{\Pi,\mathcal{A}}^{\mathsf{WEAK\text{-}KEY\text{-}UNLINK}}(\lambda) := 2 \cdot \Pr[\mathsf{WEAK\text{-}KEY\text{-}UNLINK}(\lambda) \Rightarrow \mathsf{True}] - 1,$$

where game WEAK-KEY-UNLINK *is described in Fig. 5.*

Definition 8. *An IBE scheme Π, associated with a non-polynomial size identity space $\mathcal{I}$, has Strong Key Unlinkability if, for every legitimate PPT adversary $\mathcal{A}$, the following definition of advantage is negligible in λ*

$$\mathbf{Adv}_{\Pi,\mathcal{A}}^{\mathsf{STRONG\text{-}KEY\text{-}UNLINK}}(\lambda) := 2 \cdot \Pr[\mathsf{STRONG\text{-}KEY\text{-}UNLINK}(\lambda) \Rightarrow \mathsf{True}] - 1,$$

where game STRONG-KEY-UNLINK *is described in Fig. 6.*

---

procedure Initialize(λ):
$(\mathsf{msk}, \mathsf{pp}) \leftarrow_\$ \mathsf{Setup}(\lambda)$
$\mathsf{bit} \leftarrow_\$ \{0, 1\}$
$\mathsf{id}_0 \leftarrow_\$ \mathcal{I}$
$\mathsf{id}_1 \leftarrow_\$ \mathcal{I}$
$\mathsf{sk}_0 \leftarrow_\$ \mathsf{Extract}(\mathsf{pp}, \mathsf{msk}, \mathsf{id}_0)$
$\mathsf{sk}_1 \leftarrow_\$ \mathsf{Extract}(\mathsf{pp}, \mathsf{msk}, \mathsf{id}_{\mathsf{bit}})$
return $(\mathsf{pp}, \mathsf{sk}_0, \mathsf{sk}_1)$

procedure Extract(id):
$\mathsf{sk}_{\mathsf{id}} \leftarrow_\$ \mathsf{Extract}(\mathsf{pp}, \mathsf{msk}, \mathsf{id})$
return $\mathsf{sk}_{\mathsf{id}}$

procedure Finalize(bit$'$):
return $(\mathsf{bit} = \mathsf{bit}')$

---

procedure Initialize(λ):
$(\mathsf{msk}, \mathsf{pp}) \leftarrow_\$ \mathsf{Setup}(\lambda)$
$\mathsf{bit} \leftarrow_\$ \{0, 1\}$
$\mathsf{list}_{\mathsf{id}} \leftarrow []$
$\mathsf{list}_{\mathsf{sk}} \leftarrow []$
return pp

procedure Extract(id):
$\mathsf{sk} \leftarrow_\$ \mathsf{Extract}(\mathsf{pp}, \mathsf{msk}, \mathsf{id})$
return tp

procedure Finalize(bit$'$):
return $(\mathsf{bit} = \mathsf{bit}')$

---

procedure Challenge($\mathsf{list}_0, \mathsf{list}_1$):
$L \leftarrow |\mathsf{list}_0|$
for i in $\{1..L\}$
    ... get id for $\mathsf{list}_{\mathsf{bit}}[i]$ from $\mathsf{list}_{\mathsf{id}}$
    ... if id $= \perp$
    ... ... id $\leftarrow_\$ \mathcal{I}$
    ... ... $\mathsf{list}_{\mathsf{id}} \leftarrow (\mathsf{list}_{\mathsf{bit}}[i], \mathsf{id}) : \mathsf{list}_{\mathsf{id}}$
    ... $\mathsf{list}_{\mathsf{sk}}[i] \leftarrow_\$ \mathsf{Extract}(\mathsf{pp}, \mathsf{msk}, \mathsf{id})$
return $\mathsf{list}_{\mathsf{sk}}$

Fig. 5. Game WEAK-KEY-UNLINK

Fig. 6. Game STRONG-KEY-UNLINK. Adversary is legitimate if it only calls Challenge once with $|\mathsf{list}_0| = |\mathsf{list}_1|$.

3.2   Function Privacy for IBE: An Independent Security Notion

Recently, Boneh, Raghunathan and Segev [9] put forward two security notions, of different strength, for IBE, inspired by the security definition given for deterministic encryption in [3]: *Function Privacy* and *Enhanced Function Privacy*. These notions ask that "decryption keys reveal essentially no information on their corresponding identities, beyond the absolute minimum necessary". In both definitions, the adversary is first given the public parameters and then interacts with a Real-or-Random function privacy oracle, which takes as input an adversarially-chosen joint probability distribution – represented as a circuit – for random variables $\mathsf{X}_1, \mathsf{X}_2, ..., \mathsf{X}_\mathsf{L}$ defined over the identity space $\mathcal{I}$, and outputs L secret keys either for identities sampled from the given joint probability distribution or for independent and uniformly distributed identities over $\mathcal{I}$.

An adversary is legitimate if, for every $i \in \{1..L\}$ and every $x_1, ..., x_i \in \mathcal{I}$, it holds that: $\mathbf{H}_\infty(X_i | X_1=x_1, ..., X_{i-1}=x_{i-1}) = -\log(\max \Pr[X_i = x_i | X_1=x_1,...,X_{i-1}=x_{i-1}]) \geq \omega(\log \lambda)$. Put differently, the chosen joint probability distribution for $(X_1, ..., X_L)$ has to be such that every random variable X_i is *sufficiently unpredictable*, even if every random variable $X_{j<i}$ has been fixed. To discard exhaustive searches, a *conditional min-entropy* $\mathbf{H}_\infty(X_i | X_1=x_1,...,X_{i-1})$ of at least $\omega(\log \lambda)$ bits is required[2]. The *Enhanced* model provides the adversary with an extra function-privacy encryption oracle capable of encrypting adversarially-chosen messages under the identities sampled by Real-or-Random oracle. Formal definitions can be found in [9].

We first remark that Key Unlinkability and Function Privacy security models are essentially different in the way the challenger samples ids: in the former ids are sampled uniformly from the id space, whereas in the latter model ids may be sampled from an adversarial-chosen joint probability distribution, with (possibly) non-uniform random variables, but also high min-entropy requirements. In the following subsections we provide counterexamples to show that Function Privacy (both *Non-enhanced* and *Enhanced*) and Key Unlinkability (both *Weak* and *Strong*) are independent security notions. Meaningful counterexamples follow. For a quick overview, Figure 7 states the relations between Weak Key Unlinkability, Strong Key Unlinkability, Function Privacy and Enhanced Function Privacy security notions.

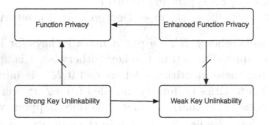

Fig. 7. Relations between Key Unlinkability and Function Privacy security notions

We stress that even Enhanced Function Privacy fails to capture the security guarantees of Weak Key Unlinkability. In practice, transforming an anonymous IBE with Enhanced Function Privacy to PEKS (according to the transformation described in Section 2) results in no guarantee that the service provider will not be able to find search patterns in the users' trapdoors.

ENHANCED FUNCTION PRIVACY $\not\Rightarrow$ WEAK KEY UNLINKABILITY. Consider $F : \{0,1\}^\lambda \times \mathcal{I} \to \{0,1\}^\lambda$ to be a secure PRF. We denote by $f \leftarrow_\$ F$ the operation: $k \leftarrow_\$ \{0,1\}^\lambda; f \leftarrow F(k, \cdot)$. Let $\Pi = (\mathsf{Setup}, \mathsf{Extract}, \mathsf{Enc}, \mathsf{Dec})$ be an enhanced function-private IBE. From Π we can construct Π', where Π' is still enhanced function-private but definitely not weak key-unlinkable. We do so by simply modifying the extraction algorithm and appending to each secret key

---

[2] The *minimal* unpredictability requirement of $\omega(\log \lambda)$ bits has only been achieve later in [10]. Schemes in [9] have only been proven secure for highly unpredictable identities with min-entropy of $\lambda + \omega(\log \lambda)$.

the result of a PRF on id. More precisely, $\Pi' = (\mathsf{Setup'}, \mathsf{Extract'}, \mathsf{Enc'}, \mathsf{Dec'})$ is constructed as follows:

- $\mathsf{Setup'}(\lambda) : (\mathsf{msk}, \mathsf{pp}) \leftarrow_\$ \mathsf{Setup}(\lambda); f \leftarrow_\$ F; \text{return } ((\mathsf{msk}, f), \mathsf{pp}).$
- $\mathsf{Extract'}(\mathsf{msk}, \mathsf{id}) : \mathsf{sk} \leftarrow_\$ \mathsf{Extract}(\mathsf{msk}, \mathsf{id}); \mathsf{sk'} \leftarrow (\mathsf{sk}, f(\mathsf{id})); \text{return } \mathsf{sk'}.$
- $\mathsf{Enc'}(\mathsf{pp}, \mathsf{m}, \mathsf{id}) : \mathsf{c} \leftarrow_\$ \mathsf{Enc}(\mathsf{pp}, \mathsf{m}, \mathsf{id}); \text{return } \mathsf{c}.$
- $\mathsf{Dec'}(\mathsf{pp}, \mathsf{c}, \mathsf{id}, \mathsf{sk'}) : (\mathsf{sk}, y) \leftarrow \mathsf{sk'}; \mathsf{m} \leftarrow \mathsf{Dec}(\mathsf{pp}, \mathsf{c}, \mathsf{id}, \mathsf{sk}); \text{return } \mathsf{m}.$

Informally, since f is unknown to the adversary, the adversary cannot choose distributions depending on f. Furthermore, F is a secure PRF, so no information on id can be acquired. Therefore, Π' is still an enhanced function-private IBE. But, because f is deterministic, it is trivial to identify with overwhelming probability if two keys have been extracted from the same identity.

STRONG KEY UNLINKABILITY $\not\Rightarrow$ FUNCTION PRIVACY. Again, we show this by counterexample. Let $\Pi = (\mathsf{Setup}, \mathsf{Extract}, \mathsf{Enc}, \mathsf{Dec})$ be a strong key-unlinkable IBE associated with id space $\mathcal{I} = \{0, 1\}^{2\lambda}$. We build $\Pi' = (\mathsf{Setup'}, \mathsf{Extract'}, \mathsf{Enc'}, \mathsf{Dec'})$ based on Π as follows:

- $\mathsf{Setup'}(\lambda) : (\mathsf{msk}, \mathsf{pp}) \leftarrow_\$ \mathsf{Setup}(\lambda); \text{return } (\mathsf{msk}, \mathsf{pp}).$
- $\mathsf{Extract'}(\mathsf{msk}, \mathsf{id}) : \mathsf{sk} \leftarrow_\$ \mathsf{Extract}(\mathsf{msk}, \mathsf{id}); \text{if } \mathsf{id} \in \{0, 1\}^\lambda 0^\lambda \text{ then } \mathsf{sk'} \leftarrow (\mathsf{sk}||0)$
  else $\mathsf{sk'} \leftarrow (\mathsf{sk}||1); \text{return } \mathsf{sk'}.$
- $\mathsf{Enc'}(\mathsf{pp}, \mathsf{m}, \mathsf{id}) : \mathsf{c} \leftarrow_\$ \mathsf{Enc}(\mathsf{pp}, \mathsf{m}, \mathsf{id}) \text{ return } \mathsf{c}.$
- $\mathsf{Dec'}(\mathsf{pp}, \mathsf{c}, \mathsf{id}, \mathsf{sk'}) : (\mathsf{sk}||b) \leftarrow \mathsf{sk'}; \mathsf{m} \leftarrow \mathsf{Dec}(\mathsf{pp}, \mathsf{c}, \mathsf{id}, \mathsf{sk}); \text{return } \mathsf{m}.$

In our counterexample scheme Π', we put a mark in keys for identities whose last λ bits are 0, by appending a 0 to the key (otherwise, 1 is appended). Since the subset containing these identities – let us call it $\mathcal{U}$ – is much smaller than the identity space $\mathcal{I}$, identities uniformly sampled from $\mathcal{I}$ are very unlikely to be in $\mathcal{U}$, and thus to possess the mark. In fact, this only happens with probability $\Pr = \frac{2^\lambda}{2^{2\lambda}} = \frac{1}{2^\lambda}$, which is a negligible function in the security parameter λ. Strong Key Unlinkability is therefore preserved in Π'. However, $\mathcal{U}$ is big enough so that the unpredictability of an id uniformly sampled from $\mathcal{U}$ is high. By choosing to be challenged on a random variable X that selects any element in $\mathcal{U}$ with probability $\frac{1}{2^\lambda}$ and any element in $\{x \in \mathcal{I} : x \notin \mathcal{U}\}$ with zero probability, an adversary could trivially win the function-privacy game, with overwhelming probability, just by looking into the key's mark. Also notice that the condition $\mathbf{H}_\infty(\mathsf{X}) > \omega(\log \lambda)$ is satisfied. Generically, we can conclude that a strong key-unlinkable scheme is not necessarily function-private secure.

3.3 Adversarially-Chosen Joint Probability Distributions of Keywords

In security game Strong Key Unlinkability [Fig. 6], identities are sampled uniformly at random from the identity space, as opposed to from a (possibly non-uniform) adversarially-chosen joint probability distribution. The latter approach was used by Boneh et al. [9] to form the challenge in Function Privacy security models. In most real-world applications of PEKS, keywords are not chosen uniformly from the keyword space. Therefore, it is important to discuss the choice of our model and the impact of generalizing it to deal with adversarially-chosen distributions.

The full version of [9] proposes a generic method for transforming any IBE scheme into an IBE scheme which achieves a *weaker* form of Enhanced Function Privacy, where the adversary is not allowed to choose a joint probability distribution (from which identities are sampled for the challenge) that depends on the public parameters of the scheme. In fact, the challenger only provides the public parameters *after* the joint probability distribution is fixed by the adversary. This relaxation results in a definition denoted *Non-Adaptive* Enhanced Function Privacy.

Adopting the same strategy as [9,3], we strengthen our model by allowing the adversary to choose a joint probability distribution from which identities are sampled, instead of lists defining equality relations between identities. The environment of the game becomes exactly that of Non-Enhanced Function Privacy defined in [9] (and described here, in Subsection 3.2), but the unpredictability requirements on what constitutes a legitimate joint probability distribution $\mathbb{X} = \{X_1, ..., X_L\}$ are relaxed to $\mathbf{H}_\infty(X_i) \geq \omega(\log \lambda)$, for every $i \in \{1..L\}$. Public parameters can be provided before or after the adversary fixes $\mathbb{X}$, resulting in two models of different strengths. We refer to the model where the adversary fixes a joint probability distribution (with possibly non-uniform random variables) from which the challenger samples the identities *after* (resp. *before*) receiving the public parameters as *Adaptive* (resp. *Non-Adaptive*) *Key Unlinkability*.

Definition 9. *An IBE scheme Π, associated with a non-polynomial size identity space $\mathcal{I}$, has Non-Adaptive Key Unlinkability if, for every legitimate PPT adversary $\mathcal{A}$, the following definition of advantage is negligible in λ*

$$\mathbf{Adv}_{\Pi, \mathcal{A}_{\text{nonadaptive}}}^{\text{KEY-UNLINK}}(\lambda) := 2 \cdot \Pr[\text{KEY-UNLINK}(\lambda, \text{mode} = \text{``non-adaptive''}) \Rightarrow \text{True}] - 1,$$

where game KEY-UNLINK *is described in Fig. 8.*

procedure **Initialize**(λ, mode):	procedure **Challenge**($\mathbb{X} = \{X_1, ..., X_L\}$):
(msk, pp) $\leftarrow_\$$ Setup(λ)	if bit = 0
bit $\leftarrow_\$$ $\{0, 1\}$	... $(id_1, ..., id_L) \leftarrow_\$$ $\mathbb{X}$
list $\leftarrow$ []	if bit = 1
if mode = "adaptive" return pp	... $(id_1, ..., id_L) \leftarrow_\$$ $\mathcal{I}^L$
	for $i \in \{1..L\}$
	... list[i] $\leftarrow_\$$ Extract(pp, msk, id_i)
	return (list, pp)
procedure **Extract**(id):	procedure **Finalize**(bit'):
sk $\leftarrow_\$$ Extract(pp, msk, id)	return (bit = bit')
return tp	

Fig. 8. Game KEY-UNLINK. $\mathbb{X} = \{X_1, ..., X_L\}$ is a joint probability distribution with L random variables defined over the identity space $\mathcal{I}$. Adversary is legitimate if $\mathbf{H}_\infty(X_i) \geq \omega(\log \lambda)$, for every $i \in \{1..L\}$.

Definition 10. *An IBE scheme Π, associated with a non-polynomial size identity space $\mathcal{I}$, has Adaptive Key Unlinkability if, for every legitimate PPT adversary $\mathcal{A}$, the following definition of advantage is negligible in λ*

$$\mathbf{Adv}_{\Pi,\mathcal{A}_{adaptive}}^{\text{KEY-UNLINK}}(\lambda) := 2 \cdot \Pr[\text{KEY-UNLINK}(\lambda, \text{mode} = \text{``adaptive''}) \Rightarrow \text{True}] - 1,$$

where game KEY-UNLINK *is described in Fig. 8.*

REMARK. The joint probability distribution $\mathbb{X} = \{X_1, X_2\}$ such that $\Pr[X_2 = x_1] = 1$ is legitimate for Adaptive (and Non-Adaptive) Key Unlinkability, as long as $\mathbf{H}_\infty(X_1) \geq \omega(\log \lambda)$. In particular, if X_1 is a uniform random variable in $\mathcal{I}$, then the game becomes that of Weak Key Unlinkability [Fig. 5]. However, as expected, $\mathbb{X}$ is not a legitimate joint probability distribution for Function Privacy (Enhanced or Non-Enhanced, Adaptive or Non-Adaptive).

TOWARDS NON-ADAPTIVE KEY UNLINKABILITY. We now show that there is an easy and natural transformation from Strong Key Unlinkability to Non-Adaptive Key Unlinkability. Let $\Pi = (\text{Setup}, \text{Extract}, \text{Enc}, \text{Dec})$ be an IBE scheme, associated with message space $\mathcal{M}$ and identity space $\mathcal{I}$, and let $\mathcal{H} : \mathcal{I}' \to \mathcal{I}$ be a family of hash functions. We construct an IBE scheme $\Pi' = (\text{Setup}', \text{Extract}', \text{Enc}', \text{Dec}')$, associated with message space $\mathcal{M}$ and identity space $\mathcal{I}'$, as follows:

- Setup$'(\lambda)$: (msk, pp) $\leftarrow_\$$ Setup(λ); H $\leftarrow_\$$ $\mathcal{H}$; return (msk, (pp, H)).
- Extract$'(\text{msk}, \text{id}')$: id $\leftarrow$ H(id'); sk $\leftarrow_\$$ Extract(msk, id); return sk.
- Enc$'((\text{pp}, H), m, \text{id}')$: id $\leftarrow$ H(id'); c $\leftarrow_\$$ Enc(pp, m, id); return c.
- Dec$'((\text{pp}, H), c, \text{id}, \text{sk}')$: id $\leftarrow$ H(id'); m $\leftarrow$ Dec(pp, c, id, sk); return m.

Lemma 1. *If $|\mathcal{I}'| \geq |\mathcal{I}| \geq 2^{\omega(\log \lambda)}$ and IBE scheme Π has Strong Key Unlinkability, then IBE scheme Π' has Non-Adaptive Key Unlinkability, in the random oracle model.*

Proof. Let $\mathcal{A}$ be a legitimate adversary against Non-Adaptive Key Unlinkability, and let $\mathbb{X} = \{X_1, ..., X_L\}$ be the joint probability distribution that $\mathcal{A}$ chooses for the challenge. We recall that a legitimate adversary is required to choose $\mathbb{X}$ such that $\forall X_i \in \mathbb{X}, \mathbf{H}_\infty(X_i) \geq \omega(\log \lambda)$, where λ is the security parameter of Π. Game$_0$ is the original Non-Adaptive Key Unlinkability game described above, instantiated with IBE scheme Π'. In Game$_1$, H is modeled as a random oracle. $(\text{id}'_1, ..., \text{id}'_L) \leftarrow_\$ \{X_1, ..., X_L\}$ forms a list of bit-strings. A simulator $\mathcal{S}$ could construct the challenge of Game$_1$ by setting list$_0 = (\text{id}'_1, ..., \text{id}'_L) \leftarrow_\$ \{X_1, ..., X_L\}$ and list$_1$ with L different bit-strings, and querying the challenge procedure of STRONG-KEY-UNLINK$_{\Pi,\mathcal{S}}$ with (list$_0$, list$_1$). The result is a well-formed tuple of L secret keys, and $\mathcal{A}$'s final guess can be forward to STRONG-KEY-UNLINK$_{\Pi,\mathcal{S}}$. Simulator $\mathcal{S}$ perfectly mimics the environment of Game$_1$, unless $\mathcal{A}$ queries the hash value of any id$'_i$, in which case the simulation aborts. However, this event only happens negligible probability. Therefore, we have that $\mathbf{Adv}_{\Pi',\mathcal{A}_{nonadaptive}}^{\text{KEY-UNLINK}}(\lambda) \leq$ $\mathbf{Adv}_{\Pi,\mathcal{S}}^{\text{STRONG-KEY-UNLINK}}(\lambda) + \frac{q \cdot L}{2^{\omega(\log \lambda)}}$, where q is the number of queries $\mathcal{A}$ asks to the random oracle. $\qquad\square$

Most IBE schemes, including the one introduced in this paper later on, only make use of the hash value of identities (instead of the identities themselves). Thus, the simplicity of Strong Key Unlinkability does not come at the expense of the model's security meaning. From a theoretical point of view, it seems interesting (but we leave it as future work) to investigate the construction of IBE schemes that achieve Key Unlinkability against *adaptive* adversaries. In practice, for what concerns PEKS, it seems reasonable to assume that keywords will not depend on the public parameters of the scheme, and, in particular, on the values output by the hash function.

4   From Weak to Strong Key Unlinkability

A SCHEME WITH WEAK KEY UNLINKABILITY. We construct a new anonymous IBE scheme with Weak Key Unlinkability, based on the anonymous IBE scheme of Boyen and Waters [12]. Our scheme relies on a bilinear group description Γ of prime order. To eliminate the selective-ID constraint, we replace identities with their hash values and model the hash function as a random oracle. Furthermore, we simplify the resulted scheme by removing two group elements from the public parameters and from private keys, and obtain the final scheme in Fig. 9. Compared with the original scheme, our scheme also saves two exponentiations in the key-extraction and encryption algorithms, and saves two pairing computations in the decryption algorithm. Our scheme preserves *anonymity* and *semantic security* properties, provided that the hash function H, selected from a family of hash functions $\mathcal{H} : \mathcal{I} \to \mathbb{G}$, is modeled as a random oracle. Added to this, the scheme also has the Weak Key Unlinkability property.

Setup(λ):	Extract(pp, msk, id):	Enc(pp, m, id):	Dec(pp, c, id, sk_{id}):
$\Gamma \leftarrow_\$ \mathcal{G}_\mathcal{P}(\lambda)$	$r \leftarrow_\$ \mathbb{Z}_p$	$s, s_1 \leftarrow_\$ \mathbb{Z}_p^2$	$(\Gamma, \Omega, v_1, v_2, H) \leftarrow pp$
$(p, \mathbb{G}, \mathbb{G}_T, e, g) \leftarrow \Gamma$	$(w, t_1, t_2) \leftarrow msk$	$(\Gamma, \Omega, v_1, v_2, H) \leftarrow pp$	$(p, \mathbb{G}, \mathbb{G}_T, e, g) \leftarrow \Gamma$
$w, t_1, t_2 \leftarrow_\$ \mathbb{Z}_p^3$	$(\Gamma, \Omega, v_1, v_2, H) \leftarrow pp$	$(p, \mathbb{G}, \mathbb{G}_T, e, g) \leftarrow \Gamma$	$(d_0, d_1, d_2) \leftarrow sk_{id}$
$\Omega \leftarrow e(g, g)^{t_1 t_2 w}$	$(p, \mathbb{G}, \mathbb{G}_T, e, g) \leftarrow \Gamma$	$h \leftarrow H(id)$	$(\hat{c}, c_0, c_1, c_2) \leftarrow c$
$v_1 \leftarrow g^{t_1}$	$h \leftarrow H(id)$	$\hat{c} \leftarrow \Omega^s m$	$e_0 \leftarrow e(c_0, d_0)$
$v_2 \leftarrow g^{t_2}$	$d_0 \leftarrow g^{r t_1 t_2}$	$c_0 \leftarrow h^s$	$e_1 \leftarrow e(c_1, d_1)$
$H \leftarrow_\$ \mathcal{H} : \mathcal{I} \to \mathbb{G}$	$d_1 \leftarrow g^{-w t_2} \cdot h^{-r t_2}$	$c_1 \leftarrow v_1^{s-s_1}$	$e_2 \leftarrow e(c_2, d_2)$
$pp \leftarrow (\Gamma, \Omega, v_1, v_2, H)$	$d_2 \leftarrow g^{-w t_1} \cdot h^{-r t_1}$	$c_2 \leftarrow v_2^{s_1}$	$m \leftarrow \hat{c} \cdot e_0 \cdot e_1 \cdot e_2$
$msk \leftarrow (w, t_1, t_2)$	$sk_{id} \leftarrow (d_0, d_1, d_2)$	$c \leftarrow (\hat{c}, c_0, c_1, c_2)$	return m
return (msk, pp)	return sk_{id}	return c	

Fig. 9. Anonymous IBE scheme Π with Weak Key Unlinkability

Theorem 1. *IBE scheme Π [Fig. 9] is semantically secure, in the random oracle model, assuming DBDH is intractable [Definition 2].*

Theorem 2. *IBE scheme Π [Fig. 9] is anonymous, in the random oracle model, assuming DBDH and DLIN are intractable [Definitions 2 and 3].*

We omit the proofs of Theorem 1 and Theorem 2 in this version due to space limitations.

Theorem 3 (Appendix A). *IBE scheme Π [Fig. 9] has the Weak Key Unlinkability property [Definition 7], in the random oracle model, assuming DLIN is intractable [Definition 3].*

WEAK KEY UNLINKABILITY $\not\Rightarrow$ STRONG KEY UNLINKABILITY. Standard real-or-random definitions for public-key encryption model the encryption of a single plaintext. These definitions are equivalent (with some loss in tightness) to those allowing an adversary to acquire multiple encryptions, which can be shown by applying the hybrid argument from [2]. One might be tempted to think that the same hybrid argument also applies to Weak Key Unlinkability model. However, this argument does *not* apply, since we can show that an adversary can still easily distinguish patterns when more than two keys are issued with scheme Π [Fig. 9].

Suppose that an adversary is asked to distinguished between tuples of the form $(\mathsf{Extract}(\mathsf{id}_0), \mathsf{Extract}(\mathsf{id}_0), \mathsf{Extract}(\mathsf{id}_0))$, where the three secret keys are extracted from the same id, from those of the form $(\mathsf{Extract}(\mathsf{id}_0), \mathsf{Extract}(\mathsf{id}_0), \mathsf{Extract}(\mathsf{id}_1))$, where the third key is extracted from an independent id, for uniformly sampled id_0 and $\mathsf{id}_1 \in \mathcal{I}$. Let $(\mathsf{sk}_0, \mathsf{sk}_1, \mathsf{sk}_2)$ be the tuple the adversary receives, and for which it has to decide its form. We further expand sk_i to $(\mathsf{d}_{i0}, \mathsf{d}_{i1}, \mathsf{d}_{i2})$ according to our scheme. If the keys were generated honestly, i.e. by following the algorithm Extract as described in Fig. 9, the adversary simply has to check if

$$e(\frac{\mathsf{d}_{10}}{\mathsf{d}_{00}}, \frac{\mathsf{d}_{21}}{\mathsf{d}_{01}}) \stackrel{?}{=} e(\frac{\mathsf{d}_{00}}{\mathsf{d}_{20}}, \frac{\mathsf{d}_{01}}{\mathsf{d}_{11}})$$

to determine the form of the tuple with overwhelming probability. If the result from the equality is true, then the three secret keys are very likely to have been extracted for the same id[3]. If the result is false, then the tuple is definitely of the form $(\mathsf{Extract}(\mathsf{id}_0), \mathsf{Extract}(\mathsf{id}_0), \mathsf{Extract}(\mathsf{id}_1))$. For completeness, we show this by expanding and simplifying the above expression.

$$e(\frac{\mathsf{d}_{10}}{\mathsf{d}_{00}}, \frac{\mathsf{d}_{21}}{\mathsf{d}_{01}}) = e(\frac{\mathsf{d}_{00}}{\mathsf{d}_{20}}, \frac{\mathsf{d}_{01}}{\mathsf{d}_{11}}) \Leftrightarrow$$

$$e(\frac{\mathsf{g}^{r_1 t_1 t_2}}{\mathsf{g}^{r_0 t_1 t_2}}, \frac{\mathsf{g}^{-w t_2} \cdot \mathsf{h}_2^{-r_2 t_2}}{\mathsf{g}^{-w t_2} \cdot \mathsf{h}_0^{-r_0 t_2}}) = e(\frac{\mathsf{g}^{r_0 t_1 t_2}}{\mathsf{g}^{r_2 t_1 t_2}}, \frac{\mathsf{g}^{-w t_2} \cdot \mathsf{h}_0^{-r_0 t_2}}{\mathsf{g}^{-w t_2} \cdot \mathsf{h}_1^{-r_1 t_2}}) \Leftrightarrow$$

$$e(\frac{\mathsf{g}^{r_1 t_1 t_2}}{\mathsf{g}^{r_0 t_1 t_2}}, \frac{\mathsf{h}_2^{-r_2 t_2}}{\mathsf{h}_0^{-r_0 t_2}}) = e(\frac{\mathsf{g}^{r_0 t_1 t_2}}{\mathsf{g}^{r_2 t_1 t_2}}, \frac{\mathsf{h}_0^{-r_0 t_2}}{\mathsf{h}_0^{-r_1 t_2}}) \Leftrightarrow$$

$$e(\mathsf{g}^{(r_1 - r_0)}, \mathsf{h}_0^{r_0} \cdot \mathsf{h}_2^{-r_2})^{t_1 (t_2)^2} = e(\mathsf{g}^{(r_0 - r_2)}, \mathsf{h}_0^{(r_1 - r_0)})^{t_1 (t_2)^2} \Leftrightarrow$$

$$e(\mathsf{g}, \mathsf{h}_0^{r_0} \cdot \mathsf{h}_2^{-r_2}) = e(\mathsf{g}, \mathsf{h}_0^{(r_0 - r_2)}) \Leftrightarrow$$

$$\mathsf{h}_2 = \mathsf{h}_0$$

---

[3] Collisions in the hash function H may lead to misleading results but only occur with negligible probability.

It is now clear that IBE scheme Π [Fig. 9] fails to achieve the Strong Key Unlinkability property.

A SCHEME WITH STRONG KEY UNLINKABILITY. We extend Π to groups of composite order and obtain Π' [Fig. 10]. The extension is very simple: let all the parameters in the original scheme be from the subgroup $\mathbb{G}_p$ (generated by g_p) and randomize each element of the extracted secret key by a random element from the subgroup $\mathbb{G}_q$ (generated by g_q). Note that the message space is $\mathbb{G}_T$.

Setup(1^λ):	Extract(pp, msk, id):	Enc(pp, m, id):
$(p, q, \mathbb{G}, \mathbb{G}_T, e, g) \leftarrow_\$ \mathcal{G}_C(\lambda)$	$(w, t_1, t_2) \leftarrow \text{msk}$	$(\Gamma, \Omega, v_1, v_2, H) \leftarrow \text{pp}$
$n \leftarrow pq;\ \ g_p \leftarrow g^q;\ \ g_q \leftarrow g^p$	$(\Gamma, \Omega, v_1, v_2, H) \leftarrow \text{pp}$	$(n, \mathbb{G}, \mathbb{G}_T, e, g, g_p, g_q) \leftarrow \Gamma$
$\Gamma \leftarrow (n, \mathbb{G}, \mathbb{G}_T, e, g, g_p, g_q)$	$(n, \mathbb{G}, \mathbb{G}_T, e, g, g_p, g_q) \leftarrow \Gamma$	$s, s_1 \leftarrow_\$ \mathbb{Z}_n;\ h \leftarrow H(\text{id})$
$w, t_1, t_2 \leftarrow_\$ \mathbb{Z}_n$	$r \leftarrow_\$ \mathbb{Z}_n$	$\hat{c} \leftarrow \Omega^s m;\ c_0 \leftarrow h^s;\ c_1 \leftarrow v_1^{s-s_1};\ c_2 \leftarrow v_2^{s_1}$
$\Omega \leftarrow e(g_p, g_p)^{t_1 t_2 w}$	$x_0, x_1, x_2 \leftarrow_\$ \mathbb{G}_q$	$c \leftarrow (\hat{c}, c_0, c_1, c_2);$ return c
$v_1 \leftarrow g_p^{t_1}$	$h \leftarrow H(\text{id})$	
$v_2 \leftarrow g_p^{t_2}$	$d_0 \leftarrow x_0 \cdot g_p^{rt_1 t_2}$	Dec(pp, c, id, sk_{id}):
$H \leftarrow_\$ \mathcal{H} : \mathcal{I} \to \mathbb{G}_p$	$d_1 \leftarrow x_1 \cdot g_p^{-wt_2} \cdot h^{-rt_2}$	$(\Gamma, \Omega, v_1, v_2, H) \leftarrow \text{pp}$
$\text{pp} \leftarrow (\Gamma, \Omega, v_1, v_2, H)$	$d_2 \leftarrow x_2 \cdot g_p^{-wt_1} \cdot h^{-rt_1}$	$(n, \mathbb{G}, \mathbb{G}_T, e, g, g_p, g_q) \leftarrow \Gamma$
$\text{msk} \leftarrow (w, t_1, t_2)$	$\text{sk} \leftarrow (d_0, d_1, d_2)$	$(d_0, d_1, d_2) \leftarrow \text{sk}_{\text{id}};\ (\hat{c}, c_0, c_1, c_2) \leftarrow c$
return (msk, pp)	return sk	$e_0 \leftarrow e(c_0, d_0);\ e_1 \leftarrow e(c_1, d_1)$
		$e_2 \leftarrow e(c_2, d_2);\ m \leftarrow \hat{c} \cdot e_0 \cdot e_1 \cdot e_2$
		return m

Fig. 10. Anonymous IBE scheme Π' with Strong Key Unlinkability

The decryption algorithm remains correct, since

$$e_0 = e(h^s, x_0 \cdot g_p^{rt_1 t_2}) = e(h^s, g_p^{rt_1 t_2})$$
$$e_1 = e(v_1^{s-s_1}, x_1 \cdot g_p^{-wt_2} \cdot h^{-rt_2}) = e(v_1^{s-s_1}, g_p^{-wt_2} \cdot h^{-rt_2})$$
$$e_2 = e(v_2^{s_1}, x_2 \cdot g_p^{-wt_1} \cdot h^{-rt_1}) = e(v_2^{s_1}, g_p^{-wt_1} \cdot h^{-rt_1})$$

Also, *semantic security* and *anonymity* properties are not affected, assuming DBDH and DLIN hold in $\mathbb{G}_p$. We only need to prove that Π' possesses the *Strong Key Unlinkability* property.

Theorem 4 (Appendix B). *IBE scheme Π' [Fig. 10] has Strong Key Unlinkability [Definition 8], assuming CDDH is intractable [Definition 5].*

5   Conclusions and Future Directions

Our work shows that two distinct scenarios have to be considered to model trapdoor privacy: one in the presence of ciphertexts that match trapdoors, and the other in the absence of such ciphertexts. The notion of Strong Search Pattern Privacy we introduced here addresses privacy concerns up to the point where ciphertexts matching the issued trapdoors become available, after which, search patterns can no longer be hidden from an attacker. Previous models provide limited privacy guarantees against search patterns. Of theoretical interest, it remains an open problem to prove if our scheme Π' [Fig. 10] (or any other) can achieve security according to the generalized definition of *Adaptive* Key

Unlinkability. The overarching goal would be to construct an Anonymous IBE scheme which satisfies both Adaptive Key Unlinkability and Enhanced Function Privacy, simultaneously.

Acknowledgements. The present project is supported by the National Research Fund, Luxembourg.

References

1. Abdalla, M., Bellare, M., Catalano, D., Kiltz, E., Kohno, T., Lange, T., Malone-Lee, J., Neven, G., Paillier, P., Shi, H.: Searchable encryption revisited: Consistency properties, relation to anonymous ibe, and extensions. Journal of Cryptology 21(3), 350–391 (2008)
2. Bellare, M., Boldyreva, A., Micali, S.: Public-key encryption in a multi-user setting: Security proofs and improvements. In: Preneel, B. (ed.) EUROCRYPT 2000. LNCS, vol. 1807, pp. 259–274. Springer, Heidelberg (2000)
3. Bellare, M., Boldyreva, A., O'Neill, A.: Deterministic and efficiently searchable encryption. In: Menezes, A. (ed.) CRYPTO 2007. LNCS, vol. 4622, pp. 535–552. Springer, Heidelberg (2007)
4. Bellare, M., Rogaway, P.: The security of triple encryption and a framework for code-based game-playing proofs. In: Vaudenay, S. (ed.) EUROCRYPT 2006. LNCS, vol. 4004, pp. 409–426. Springer, Heidelberg (2006)
5. Boneh, D., Boyen, X., Shacham, H.: Short group signatures. In: Franklin, M. (ed.) CRYPTO 2004. LNCS, vol. 3152, pp. 41–55. Springer, Heidelberg (2004)
6. Boneh, D., Di Crescenzo, G., Ostrovsky, R., Persiano, G.: Public key encryption with keyword search. In: Cachin, C., Camenisch, J.L. (eds.) EUROCRYPT 2004. LNCS, vol. 3027, pp. 506–522. Springer, Heidelberg (2004)
7. Boneh, D., Franklin, M.: Identity-based encryption from the weil pairing. In: Kilian, J. (ed.) CRYPTO 2001. LNCS, vol. 2139, pp. 213–229. Springer, Heidelberg (2001)
8. Boneh, D., Goh, E.-J., Nissim, K.: Evaluating 2-dnf formulas on ciphertexts. In: Kilian, J. (ed.) TCC 2005. LNCS, vol. 3378, pp. 325–341. Springer, Heidelberg (2005)
9. Boneh, D., Raghunathan, A., Segev, G.: Function-private identity-based encryption: Hiding the function in functional encryption. In: Canetti, R., Garay, J.A. (eds.) CRYPTO 2013, Part II. LNCS, vol. 8043, pp. 461–478. Springer, Heidelberg (2013)
10. Boneh, D., Raghunathan, A., Segev, G.: Function-private subspace-membership encryption and its applications. In: Sako, K., Sarkar, P. (eds.) ASIACRYPT 2013, Part I. LNCS, vol. 8269, pp. 255–275. Springer, Heidelberg (2013)
11. Boneh, D., Waters, B.: Conjunctive, subset, and range queries on encrypted data. In: Vadhan, S.P. (ed.) TCC 2007. LNCS, vol. 4392, pp. 535–554. Springer, Heidelberg (2007)
12. Boyen, X., Waters, B.: Anonymous hierarchical identity-based encryption (Without random oracles). In: Dwork, C. (ed.) CRYPTO 2006. LNCS, vol. 4117, pp. 290–307. Springer, Heidelberg (2006)
13. Katz, J., Sahai, A., Waters, B.: Predicate encryption supporting disjunctions, polynomial equations, and inner products. In: Smart, N.P. (ed.) EUROCRYPT 2008. LNCS, vol. 4965, pp. 146–162. Springer, Heidelberg (2008)

14. Nishioka, M.: Perfect keyword privacy in PEKS systems. In: Takagi, T., Wang, G., Qin, Z., Jiang, S., Yu, Y. (eds.) ProvSec 2012. LNCS, vol. 7496, pp. 175–192. Springer, Heidelberg (2012)
15. Shen, E., Shi, E., Waters, B.: Predicate privacy in encryption systems. In: Reingold, O. (ed.) TCC 2009. LNCS, vol. 5444, pp. 457–473. Springer, Heidelberg (2009)
16. Tang, Q.: Theory and Practice of Cryptography Solutions for Secure Information Systems. In: Search in Encrypted Data: Theoretical Models and Practical Applications, pp. 84–108. IGI (2013)

A   Proof of Theorem 3

Let $\mathcal{A}$ be any legitimate PPT adversary in game WEAK-KEY-UNLINK$_{\Pi,\mathcal{A}}$ [Fig. 5]. By building a simulator $\mathcal{S}_2$ [Fig. 11] that plays game DLIN$_{\Gamma,\mathcal{S}_2}$ [Fig. 2] and simulates game WEAK-KEY-UNLINK$_{\Pi,\mathcal{A}}$ in such a way that $\mathcal{A}$'s guess can be forward to game DLIN$_{\Gamma,\mathcal{S}_2}$, we upper-bound the adversary's advantage to the hardness of deciding on an instance of this problem.

The master secret key is set as following: $t_1 = z_1$, $t_2 = z_1 \cdot a$ for random $a \in \mathbb{Z}_p$, and $w = \frac{z_3 \cdot b}{z_1}$ for random $b \in \mathbb{Z}_p$. Although the values of t_1, t_2 and w are unknown to $\mathcal{S}_2$, the corresponding public parameters can still be consistently computed:

$$\Omega = e(g,g)^{t_1 t_2 w} = e(g,g)^{z_1 z_1 a \frac{z_3 \cdot b}{z_1}} = e(Z_{13},g)^{ab}$$
$$v_1 = g^{t_1} = Z_1$$
$$v_2 = g^{t_2} = (Z_1)^a$$

The hash function H is modeled as a random oracle and set to $(g^{z_1})^x \cdot g^{-\frac{1}{y}}$, for random $x, y \in \mathbb{Z}_p^2$. We assume, without loss of generality, that $\mathcal{A}$ always asks for the hash value of id before querying id to oracle Extract. Whenever asked to extract a private key on some id, we set $r = w \cdot y$, where y is the value used to compute the hash of that particular id. Note that this still makes r uniformly distributed over $\mathbb{Z}_p$ and independent of h and w. Given this, private keys can be extracted as follows:

$$d_0 = g^{rt_1 t_2} = g^{wyt_1 t_2} = g^{\frac{z_3 \cdot b}{z_1} yz_1 z_1 a} = (Z_{13})^{aby}$$
$$d_1 = g^{-wt_2} \cdot h^{-rt_2} = g^{-wt_2} \cdot [(g^{z_1})^x \cdot g^{-\frac{1}{y}}]^{-wyt_2} = g^{-z_1 xwyt_2} = g^{-z_1 x \frac{z_3 \cdot b}{z_1} yz_1 a} = (Z_{13})^{-abxy}$$
$$d_2 = g^{-wt_1} \cdot h^{-rt_1} = g^{-wt_1} \cdot [(g^{z_1})^x \cdot g^{-\frac{1}{y}}]^{-wyt_1} = g^{-z_1 xwyt_1} = g^{-z_1 x \frac{z_3 \cdot b}{z_1} yz_1} = (Z_{13})^{-bxy}$$

Finally, to complete the simulation, we extract two private keys to challenge $\mathcal{A}$, such that these private keys are for the same id if $\mathcal{S}_2$ received a valid DLIN tuple, and for different ids otherwise. Let $sk^\star = (d_0^\star, d_1^\star, d_2^\star)$ and $sk^\circ = (d_0^\circ, d_1^\circ, d_2^\circ)$ be the challenge keys. We set $h = g^{z_1 z_4}$, $r^\star = \frac{b}{(z_1)^2}$ and $r^\circ = \frac{z_2 + b}{(z_1)^2}$. Note that h is uniformly distributed over $\mathbb{G}$, and $r^\star$ and r° are uniformly distributed over $\mathbb{Z}_p$, independent of each other and of w. For completeness, we present the equalities between the original expressions and those computed by the simulator.

$$d_0^\star = g^{r^\star t_1 t_2} = g^{\frac{b}{(z_1)^2} z_1 z_1 a} = g^{ab}$$

$$d_1^\star = g^{-wt_2} \cdot h^{-r^\star t_2} = g^{-\frac{z_3 b}{z_1} z_1 a} \cdot (g^{z_1 z_4})^{-\frac{b}{(z_1)^2} z_1 a} = (g^{-ab})^{z_3} \cdot (g^{-ab})^{z_4} = Z^{-ab}$$

$$d_1^\star = g^{-wt_1} \cdot h^{-r^\star t_1} = g^{-\frac{z_3 b}{z_1} z_1} \cdot (g^{z_1 z_4})^{-\frac{b}{(z_1)^2} z_1} = (g^{-b})^{z_3} \cdot (g^{-b})^{z_4} = Z^{-b}$$

$$d_0^\circ = g^{r^\circ t_1 t_2} = g^{\frac{z_2+b}{(z_1)^2} z_1 z_1 \cdot a} = g^{z_2 \cdot a + ab} = (Z_2)^a \cdot g^{ab}$$

$$d_1^\circ = g^{-wt_2} \cdot h^{-r^\circ t_2} = g^{-\frac{z_3 b}{z_1} z_1 a} \cdot (g^{z_1 z_4})^{-\frac{z_2+b}{(z_1)^2} z_1 a} = (g^{-ab})^{(z_3+z_4)} \cdot (g^{z_2 z_4})^{-a} = Z^{-ab} \cdot (Z_{24})^{-a}$$

$$d_2^\circ = g^{-wt_1} \cdot h^{-r^\circ t_1} = g^{-\frac{z_3 b}{z_1} z_1} \cdot (g^{z_1 z_4})^{-\frac{z_2+b}{(z_1)^2} z_1} = (g^{-b})^{(z_3+z_4)} \cdot (g^{z_2 z_4})^{-1} = Z^{-b} \cdot (Z_{24})^{-1}$$

Therefore, we have that $\mathbf{Adv}_{\Pi,\mathcal{A}}^{\mathsf{WEAK\text{-}KEY\text{-}UNLINK}}(\lambda) = \mathbf{Adv}_{\Gamma,\mathcal{S}_2}^{\mathsf{DLIN}}$, which concludes our proof. $\square$

---

procedure Initialize(λ):
$(Z_1, Z_2, Z_{13}, Z_{24}, Z) \leftarrow \mathsf{DLIN.Initialize}$
$a \leftarrow_\$ \mathbb{Z}_p, b \leftarrow_\$ \mathbb{Z}_p$
$\mathsf{list}_H \leftarrow []$

$\Omega \leftarrow e(Z_{13}, g)^{ab}$
$v_1 \leftarrow Z_1$
$v_2 \leftarrow (Z_1)^a$

$d_0^\star \leftarrow g^{ab}, \quad d_0^\circ \leftarrow (Z_2)^a \cdot g^{ab}$
$d_1^\star \leftarrow Z^{-ab}, \quad d_1^\circ \leftarrow Z^{-ab} \cdot (Z_{24})^{-a}$
$d_1^\star \leftarrow Z^{-b}, \quad d_2^\circ \leftarrow Z^{-b} \cdot (Z_{24})^{-1}$

$\mathsf{sk}_0 \leftarrow (d_0^\star, d_2^\star, d_2^\star)$
$\mathsf{sk}_1 \leftarrow (d_0^\circ, d_2^\circ, d_2^\circ)$
$\mathsf{pp} \leftarrow (\Omega, v_1, v_2)$

return $(\mathsf{pp}, \mathsf{sk}_0, \mathsf{sk}_1)$

procedure H(id) :
get (x, y) for id from list_H
if $(x, y) == \bot$
$\quad$ ..... $x \leftarrow_\$ \mathbb{Z}_p$
$\quad$ ..... $y \leftarrow_\$ \mathbb{Z}_p$
$\quad$ ..... $\mathsf{list}_H \leftarrow (\mathsf{id}, x, y) : \mathsf{list}_H$
$h \leftarrow (g^{z_1})^x \cdot g^{-\frac{1}{y}}$
return h

procedure Extract(id):
get (x, y) for id from list_H

$d_0 \leftarrow (Z_{13})^{aby}$
$d_1 \leftarrow (Z_{13})^{-abxy}$
$d_2 \leftarrow (Z_{13})^{-bxy}$

$\mathsf{sk}_{\mathsf{id}} \leftarrow (d_0, d_1, d_2)$
return $\mathsf{sk}_{\mathsf{id}}$

procedure Finalize(bit):
$\mathsf{DLIN.Finalize(bit)}$

Fig. 11. Simulator $\mathcal{S}_2$ forwards $\mathcal{A}$'s guess from game $\mathsf{WEAK\text{-}KEY\text{-}UNLINK}_{\Pi,\mathcal{A}}$ to game $\mathsf{DLIN}_{\Gamma,\mathcal{S}_2}$

B   Proof of Theorem 4

First, let us show an important re-randomization property that scheme Π' possess and that is relevant for the completion of this proof. From two keys honestly extracted from the same identity, say $\mathsf{sk}_0 = (d_{00}, d_{01}, d_{02})$ and $\mathsf{sk}_1 = (d_{10}, d_{11}, d_{12})$, one can generate new valid keys for that identity with fresh random coins, without the knowledge of any secret parameter. Concretely, $\mathsf{sk}_2 = (d_{20}, d_{21}, d_{22})$ can be generated as follows, with a random $y \in \mathbb{Z}_n$ and random $R_0, R_1, R_2 \in \mathbb{G}_q$:

$$d_{20} = R_0 \cdot \left(\frac{d_{10}}{d_{00}}\right)^y \cdot d_{00} = \left[R_0 \cdot \frac{(x_{10})^y}{(x_{00})^{(y-1)}}\right] \cdot g^{[yr_1 - (y-1)r_0]t_1 t_2}$$

$$d_{21} = R_1 \cdot \left(\frac{d_{11}}{d_{01}}\right)^y \cdot d_{01} = \left[R_1 \cdot \frac{(x_{11})^y}{(x_{01})^{(y-1)}}\right] \cdot g^{-wt_2} \cdot h^{-[yr_1 - (y-1)r_0]t_2}$$

$$d_{22} = R_2 \cdot \left(\frac{d_{12}}{d_{02}}\right)^y \cdot d_{02} = \left[R_2 \cdot \frac{(x_{12})^y}{(x_{02})^{(y-1)}}\right] \cdot g^{-wt_1} \cdot h^{-[yr_1 - (y-1)r_0]t_1}$$

Let $\mathcal{A}$ be any PPT adversary against $\mathsf{STRONG\text{-}KEY\text{-}UNLINK}_{\Pi',\mathcal{A}}$ [Fig. 6]. We now drastically simplify the security model, so that it looks like the one presented in Fig. 12, which we call 5-KEY-UNLINK. Using a hybrid argument and taking advantage of the re-randomization property previously described, we show that the advantage of $\mathcal{A}$ against STRONG-KEY-UNLINK$_{\Pi',\mathcal{A}}$ is polynomially-bounded by the advantage of $\mathcal{A}$ against 5-KEY-UNLINK.

procedure Initialize(λ):	procedure Extract(id):
$(\mathsf{msk}, \mathsf{pp}) \leftarrow_\$ \mathsf{Setup}(1^\lambda)$	$\mathsf{sk}_{\mathsf{id}} \leftarrow_\$ \mathsf{Extract}(\mathsf{pp}, \mathsf{msk}, \mathsf{id})$
$\mathsf{bit} \leftarrow_\$ \{0, 1\}$	return $\mathsf{sk}_{\mathsf{id}}$
$\mathsf{id}_0 \leftarrow_\$ \mathcal{I}$	
$\mathsf{id}_1 \leftarrow_\$ \mathcal{I}$	procedure Finalize(bit'):
$\mathsf{sk}_0 \leftarrow_\$ \mathsf{Extract}(\mathsf{pp}, \mathsf{msk}, \mathsf{id}_0)$	return $(\mathsf{bit} = \mathsf{bit}')$
$\mathsf{sk}_1 \leftarrow_\$ \mathsf{Extract}(\mathsf{pp}, \mathsf{msk}, \mathsf{id}_0)$	
$\mathsf{sk}_2 \leftarrow_\$ \mathsf{Extract}(\mathsf{pp}, \mathsf{msk}, \mathsf{id}_{\mathsf{bit}})$	
$\mathsf{sk}_3 \leftarrow_\$ \mathsf{Extract}(\mathsf{pp}, \mathsf{msk}, \mathsf{id}_1)$	
$\mathsf{sk}_4 \leftarrow_\$ \mathsf{Extract}(\mathsf{pp}, \mathsf{msk}, \mathsf{id}_1)$	
return $(\mathsf{pp}, \mathsf{sk}_0, \mathsf{sk}_1, \mathsf{sk}_2, \mathsf{sk}_3, \mathsf{sk}_4)$	

Fig. 12. 5-KEY-UNLINK$_{\Pi,\mathcal{A}}$ Game

In STRONG-KEY-UNLINK$_{\Pi',\mathcal{A}}$, $\mathcal{A}$ submits two lists list$_0$ and list$_1$ of the same length, say L, for the challenge. For this argument, we construct $L + 1$ lists. The first list is list$_0$ and the last list is list$_1$. In between, we have L-1 intermediate lists that transition from list$_0$ to list$_1$, one element at the time. The L-1 intermediate lists are constructed such that the first list is list$_0$, and for every $i \in \{1..L\text{-}1\}$, list^i $-$ list^{i-1}, except for the element list^i[i] which is taken from list$_1$[i]. Again, the last list is list$_1$. The advantage $\mathcal{A}$ has in distinguishing list$_0$ from list$_1$ cannot be more than the sum of the advantages of distinguishing list^{i-1} from list^i, for every $i \in \{1..(L+1)\}$. The probability of distinguishing list^{i-1} from list^i cannot be more than that of identifying the form of the tuple in model 5-KEY-UNLINK. More precisely, one can expand the 5-tuple $(\mathsf{sk}_0^\circ, \mathsf{sk}_1^\circ, \mathsf{sk}_2^\circ, \mathsf{sk}_3^\circ, \mathsf{sk}_4^\circ)$ from 5-KEY-UNLINK into a L-tuple of keys that corresponds to the requirements of either list^{i-1} or list^i. Since the lists only (possibly) differ in position i, we set sk$_i$ of the L-tuple to sk$_2^\circ$. Every other key is extracted from the extraction oracle of model 5-KEY-UNLINK or generated from $(\mathsf{sk}_0^\circ, \mathsf{sk}_1^\circ)$ or $(\mathsf{sk}_3^\circ, \mathsf{sk}_4^\circ)$ if the key is required to be extracted from the identity in list^{i-1}[i] or list^i[i], respectively.

The model can be further simplified to that of Fig. 13, which we call 4-KEY--UNLINK. Again, we make use of the so-called hybrid argument and the re-randomization property introduced in the beginning of this proof that Π' possesses[4], the difficulty of distinguishing a 5-tuple of keys extracted from $(\mathsf{id}_0, \mathsf{id}_0, \mathsf{id}_0, \mathsf{id}_1, \mathsf{id}_1)$ from those extracted from $(\mathsf{id}_0, \mathsf{id}_0, \mathsf{id}_0, \mathsf{id}_0, \mathsf{id}_0)$, where id_0 and id_1 are sampled from $\mathcal{I}$, is equivalent to that of distinguishing a 4-tuple of keys that were extracted from $(\mathsf{id}_0, \mathsf{id}_0, \mathsf{id}_1, \mathsf{id}_1)$ from those extracted from $(\mathsf{id}_0, \mathsf{id}_0, \mathsf{id}_0, \mathsf{id}_0)$, since the fifth key the adversary could generate himself. This difficulty of distinguishing the 5-tuple of keys extracted from $(\mathsf{id}_0, \mathsf{id}_0, \mathsf{id}_1, \mathsf{id}_1, \mathsf{id}_1)$ from those extracted from $(\mathsf{id}_0, \mathsf{id}_0, \mathsf{id}_0, \mathsf{id}_0, \mathsf{id}_0)$ is also the same as distinguishing the key tuple in 4-KEY-UNLINK model. So, the advantage $\mathcal{A}$ has in distinguishing the

---

[4] From two keys honestly extracted for the same identity, we can generate a third one with random coins.

tuples in 5-KEY-UNLINK game cannot be more than *twice* the advantage $\mathcal{A}$ has in distinguishing the tuples in 4-KEY-UNLINK.

---

procedure Initialize(λ):
$(\mathsf{msk}, \mathsf{pp}) \leftarrow_\$ \mathsf{Setup}(1^\lambda)$
$\mathsf{bit} \leftarrow_\$ \{0, 1\}$
$\mathsf{id}_0 \leftarrow_\$ \mathcal{I}$
$\mathsf{id}_1 \leftarrow_\$ \mathcal{I}$
$\mathsf{sk}_0 \leftarrow_\$ \mathsf{Extract}(\mathsf{pp}, \mathsf{msk}, \mathsf{id}_0)$
$\mathsf{sk}_1 \leftarrow_\$ \mathsf{Extract}(\mathsf{pp}, \mathsf{msk}, \mathsf{id}_0)$
$\mathsf{sk}_2 \leftarrow_\$ \mathsf{Extract}(\mathsf{pp}, \mathsf{msk}, \mathsf{id}_{\mathsf{bit}})$
$\mathsf{sk}_3 \leftarrow_\$ \mathsf{Extract}(\mathsf{pp}, \mathsf{msk}, \mathsf{id}_{\mathsf{bit}})$
return $(\mathsf{pp}, \mathsf{sk}_0, \mathsf{sk}_1, \mathsf{sk}_2, \mathsf{sk}_3)$

procedure Extract(id):
$\mathsf{sk}_{\mathsf{id}} \leftarrow_\$ \mathsf{Extract}(\mathsf{pp}, \mathsf{msk}, \mathsf{id})$
return $\mathsf{sk}_{\mathsf{id}}$

procedure Finalize(bit'):
return $(\mathsf{bit} = \mathsf{bit}')$

Fig. 13. 4-KEY-UNLINK$_{\Pi, \mathcal{A}}$ Game

---

procedure Initialize(λ):
$(\Gamma, Z_a, Z_b, Z_{ab}) \leftarrow \mathsf{CDDH.Initialize}(\lambda)$
$(n, \mathbb{G}, \mathbb{G}_T, \mathbf{e}, g, g_p, g_q) \leftarrow \Gamma$
$w, t_1, t_2 \leftarrow_\$ \mathbb{Z}_n$
$\Omega \leftarrow \mathbf{e}(g_p, g_p)^{t_1 t_2 w}$
$v_1 \leftarrow g_p^{t_1}$
$v_2 \leftarrow g_p^{t_2}$
$\mathsf{msk} \leftarrow (w, t_1, t_2)$
$\mathsf{pp} \leftarrow (\Gamma, \Omega, v_1, v_2)$

$r_0^\star \leftarrow_\$ \mathbb{Z}_n$
$x_{00}', x_{01}', x_{02}' \leftarrow_\$ \mathbb{Z}_n; \quad x_{00} \leftarrow g_q^{x_{00}'}; \quad x_{01} \leftarrow g_q^{x_{01}'}; \quad x_{02} \leftarrow g_q^{x_{02}'}$
$\mathsf{sk}_0^\star \leftarrow (x_{00} \cdot (g_p)^{r_0^\star t_1 t_2}, x_{01} \cdot Z_a^{-r_0^\star t_2} \cdot (g_p)^{-w t_2}, x_{02} \cdot Z_a^{-r_0^\star t_1} \cdot (g_p)^{-w t_1})$

$r_1^\star \leftarrow_\$ \mathbb{Z}_n$
$x_{10}', x_{11}', x_{12}' \leftarrow_\$ \mathbb{Z}_n; \quad x_{10} \leftarrow g_q^{x_{10}'}; \quad x_{11} \leftarrow g_q^{x_{11}'}; \quad x_{12} \leftarrow g_q^{x_{12}'}$
$\mathsf{sk}_1^\star \leftarrow (x_{10} \cdot (g_p)^{r_1^\star t_1 t_2}, x_{11} \cdot Z_a^{-r_1^\star t_2} \cdot (g_p)^{-w t_2}, x_{12} \cdot Z_a^{-r_1^\star t_1} \cdot (g_p)^{-w t_1})$

$x_{20}', x_{21}', x_{22}' \leftarrow_\$ \mathbb{Z}_n; \quad x_{20} \leftarrow g_q^{x_{20}'}; \quad x_{21} \leftarrow g_q^{x_{21}'}; \quad x_{22} \leftarrow g_q^{x_{22}'}$
$\mathsf{sk}_2^\star \leftarrow (x_{20} \cdot Z_b^{t_1 t_2}, x_{21} \cdot Z_{ab}^{-t_2} \cdot (g_p)^{-w t_2}, x_{22} \cdot Z_{ab}^{-t_1} \cdot (g_p)^{-w t_1})$

$u \leftarrow_\$ \mathbb{Z}_n$
$x_{30}', x_{31}', x_{32}' \leftarrow_\$ \mathbb{Z}_n; \quad x_{30} \leftarrow g_q^{x_{30}'}; \quad x_{31} \leftarrow g_q^{x_{31}'}; \quad x_{32} \leftarrow g_q^{x_{32}'}$
$\mathsf{sk}_3^\star \leftarrow (x_{30} \cdot Z_b^{u t_1 t_2}, x_{31} \cdot Z_{ab}^{-u t_2} \cdot (g_p)^{-w t_2}, x_{32} \cdot Z_{ab}^{-u t_1} \cdot (g_p)^{-w t_1})$

return $(\mathsf{pp}, \mathsf{sk}_0^\star, \mathsf{sk}_1^\star, \mathsf{sk}_2^\star, \mathsf{sk}_3^\star)$

procedure Extract(id):
$\mathsf{sk}_{\mathsf{id}} \leftarrow_\$ \mathsf{Extract}(\mathsf{pp}, \mathsf{msk}, \mathsf{id})$
return $\mathsf{sk}_{\mathsf{id}}$

procedure Finalize(bit):
return $\mathsf{CDDH.Finalize(bit)}$

Fig. 14. Simulator $\mathcal{S}_3$ forwards $\mathcal{A}$'s guess from 4-KEY-ANO$_{\Pi', \mathcal{A}}$ to game CDDH

To complete the proof, we build a simulator $\mathcal{S}_3$ [Fig. 14] that by playing game CDDH$_{\Gamma', \mathcal{S}_3}$ outputs four keys $(\mathsf{sk}_0^\star, \mathsf{sk}_1^\star, \mathsf{sk}_2^\star, \mathsf{sk}_3^\star)$ such that the adversary's guess in 4-KEY-UNLINK$_{\Pi', \mathcal{A}}$ can be forward to game CDDH$_{\Gamma', \mathcal{S}_3}$. We refer to key $\mathsf{sk}_i^\star$ as the tuple $(\mathsf{d}_{i0}^\star, \mathsf{d}_{i1}^\star, \mathsf{d}_{i2}^\star)$, associated with $\mathsf{h}_i^\star$, the hashed-identity from which $\mathsf{sk}_i^\star$ was extracted. If the simulator receives a well-formed CDDH tuple, $\mathsf{h}_0^\star = \mathsf{h}_1^\star = \mathsf{h}_2^\star = \mathsf{h}_3^\star$ is set to g^a. Otherwise, $\mathsf{h}_0^\star = \mathsf{h}_1^\star = g^a$ and $\mathsf{h}_2^\star = \mathsf{h}_3^\star$ with an independent random value in $\mathbb{G}_p$. We also set $r_2^\star = b$ and $r_3^\star = b \cdot u$, for a random $u \in \mathbb{Z}_n$. Finally, we have that $\mathbf{Adv}_{\Pi', \mathcal{A}}^{\text{STRONG-KEY-UNLINK}}(\lambda) \leq 2\mathsf{L} \cdot \mathbf{Adv}_{\Gamma', \mathcal{S}_3}^{\text{CDDH}}$, which concludes our proof. $\qquad\square$

Kurosawa-Desmedt Key Encapsulation Mechanism, Revisited

Kaoru Kurosawa[1] and Le Trieu Phong[2]

[1] Ibaraki University, Japan
kurosawa@mx.ibaraki.ac.jp
[2] NICT, Japan
phong@nict.go.jp

Abstract. While the hybrid public key encryption scheme of Kurosawa and Desmedt (CRYPTO 2004) is provably secure against chosen ciphertext attacks (namely, IND-CCA-secure), its associated key encapsulation mechanism (KEM) is not IND-CCA-secure (Herranz et al. 2006, Choi et al. 2009). In this paper, we show a simple twist on the Kurosawa-Desmedt KEM turning it into a scheme with IND-CCA security under the decisional Diffie-Hellman assumption. Our KEM beats the standardized version of Cramer-Shoup KEM in ISO/IEC 18033-2 by margins of at least 20% in encapsulation speed, and 20% ~ 60% in decapsulation speed. Moreover, the public and secret key sizes in our schemes are at least 160-bit smaller than those of the Cramer-Shoup KEM. We then generalize the technique into hash proof systems, proposing several KEM schemes with IND-CCA security under decision linear and decisional composite residuosity assumptions respectively. All the KEMs are in the standard model, and use standard, computationally secure symmetric building blocks.

Keywords: Kurosawa-Desmedt KEM, IND-CCA security, hash proof systems, standard model.

1 Introduction

1.1 Background

Key Encapsulation Mechanism (KEM) is an asymmetric encryption technique allows generating *simultaneously* a random key K_s together with its encryption C, termed encapsulation. The key K_s then will be used for long data encryption, while the encapsulation C is used for sharing K_s. In other words, KEM serves as a delivery of secret keys used in symmetric encryption.

KEM implies public-key encryption (PKE). Indeed, it can be used to construct *hybrid* PKE, namely PKE with unrestricted message space, when combining with a data encapsulation mechanism (DEM) [11]. In practice, since the DEM part is already highly efficient, one usually concerns about the performance of the KEM part. Specific constructions of KEM are incorporated in the standards ISO/IEC 18033-2 [1], ANSI X9.44 [4], and can be considered for e-Government

D. Pointcheval and D. Vergnaud (Eds.): AFRICACRYPT 2014, LNCS 8469, pp. 51–68, 2014.

usage in the future [2]. KEM is widely yet implicitly used in the TLS Handshake Protocol [18].

In 2004, Kurosawa and Desmedt [19], improved upon the seminal work of Cramer and Shoup [10], published an efficient hybrid PKE, whose security proof was refined in [13], resisting chosen ciphetext attacks (IND-CCA) under the decisional Diffie-Hellman (DDH) assumption. Unlike Cramer-Shoup scheme, the KEM part of the Kurosawa-Desmedt scheme is not IND-CCA secure, as shown in 2006 in [9, 15]. In 2007, by creatively switching elements in the Kurosawa-Desmedt KEM, Kiltz [17] presented an IND-CCA-secure KEM, and yet under the less standard Gap Hashed Diffie-Hellman (GHDH) assumption. On the other hand, sticking to the DDH assumption, Abe, Gennaro, Kurosawa [3], and Hofheinz, Kiltz [16] showed the Kurosawa-Desmedt KEM only meets weakened notions of CCA security.

While weakened IND-CCA security as defined in [3, 16] can be converted into IND-CCA security (see Section 1.4), there is still no direct security proof for any variant of the Kurosawa-Desmedt KEM. A summarization of these discussions is in Table 1.

Table 1. Classification of Kurosawa-Desmedt (KD) KEM and its variants

Security (↓) Assumption (→)	GHDH	DDH
Weakened IND-CCA	–	[3], [16] (KD KEM)
IND-CCA	[17] (dual KD KEM)	This paper (with direct proof)

1.2 Our Contributions

Our results can be categorized as follows.

Theoretical Contribution. We show a slight twist on the insecure Kurosawa-Desmedt KEM turning it into an IND-CCA-*secure* one. Formally, we propose a variant of the Kurosawa-Desmedt KEM which can be proved IND-CCA-secure under the DDH assumption. That is, we fulfill Table 1 with the most "*desirable*" KEM in terms of security assumption (namely, DDH) and security notion (namely, IND-CCA).

The twist is simple. Details are discussed at length at the beginning of Section 3.1, but a high view is as follows. In the original Kurosawa-Desmedt KEM, the encapsulation of a symmetric key v consists of group elements (u_1, u_2). In our proposal, we do not return the whole v as the shared symmetric key, but split it into two independent keys k_s and k_a. The key k_s is then returned as the shared key, while the key k_a is internally used to authenticate the encapsulation (u_1, u_2). This authentication step is important as it protects the KEM against adversarial decapsulation queries, and is novel to this work in the sense that, with the twist, previous security proof for hybrid PKE in [13] can be *as is* reused for the KEM case, without any loss factor to the main complexity assumption.

Practical Impact. The result is not only of theoretical interest. Indeed, compared to the existing practice [1], namely the standardized ACE-KEM basing on the same assumption in the standard model, we achieve

- more than 20% improvement over encapsulation speed, and at least 20% improvement over decapsulation speed in general, and
- for specific choices of the base group such as prime-field NIST elliptic curves, the speed improvement on decapsulation can go up to 60%.

These theoretical estimations are checked by experimental results in Section 3.2. These improvements are significant, as frequently there are large amounts of asymmetric encryption and decryption works, e.g., in SSL/TLS servers.

In sizes, the public and secret keys in our schemes are one group element, or at least 160-bit, smaller than those of the ACE-KEM. The encapsulation length is also slightly shorter. See Table 2 in Section 3.2 for details.

DLIN-Based and DCR-Based Extensions. Our method can be extended to hash proofs systems. When coupling with known constructions of hash proof systems in the literature, we obtain KEMs under the decision linear (DLIN) and decisional composite residuosity (DCR) assumptions, respectively.

1.3 Other Usage of KEM Beyond Hybrid Encryption

While original application of KEM is hybrid PKE, the ability to output a shared symmetric key allows KEM to have other applications as well. For example, KEM can be used to build schemes for identification [5] and authenticated key exchange (AKE) [8,14,23]. In particular, Boyd et al. [8] showed that a one-round AKE protocol can be constructed from IND-CCA secure KEM, and Fujioka et al. [14] showed that a two-pass AKE protocol with weak perfect forward secrecy can be constructed from IND-CCA secure KEM. This additionally illustrates why KEM is preferable over PKE alone.

1.4 More Related Works

The proof given in [19] depends on some information theoretically secure components, which affects the efficiency of the hybrid PKE scheme. The refined proof in [13] weakens the components to computationally secure ones.

Already in [9, 15], it was remarked that, if one models the key derivation function as a random oracle and is content with a much stronger assumption than DDH, the Kurosawa-Desmedt KEM can be proved IND-CCA-secure.

Okamoto [22] presented a KEM derived from the Kurosawa-Desmedt hybrid PKE. The KEM is IND-CCA-secure under the DDH assumption, and yet theoretically relies on an arguably non-standard primitive called pseudo-random function with pairwise independent random sources.

We are informed by Takahiro Matsuda that constrained IND-CCA (CCCA) security [16] can be converted into standard IND-CCA security as done in [6] using essentially the same idea with this work. The transformation, while generic and applied to the original Kurosawa-Desmedt KEM, however has a loss factor of 4 in the security reduction. Our approach in this paper puts aside constrained IND-CCA definition, giving a direct proof for the KEM and related schemes from hash proof systems and yielding a theoretically better loss factor of 1 to the main complexity assumptions (namely DDH, DLIN, and DCR).

In the same vein, LCCA-secure KEM as defined in [3] can be converted to IND-CCA-secure Tag-KEM [3, Theorem 3] which in turn yields hybrid PKE. The conversion again has a loss factor of 2 to the main complexity assumption. The application of Tag-KEM beyond hybrid PKE is arguably less clear than KEM.

The conversions from CCCA or LCCA security to CCA security, while being generic, are of theoretical interests, since proving that a concrete scheme is CCCA-secure or LCCA-secure is apparently not easier than directly showing that scheme is IND-CCA-secure.

2   Preliminaries

KEM. A KEM consists of key generation KG, encapsulation Encap, and decapsulation Decap algorithms. $\mathsf{KG}(1^\kappa)$ with security parameter κ outputs public key pk and secret key sk. The algorithm $\mathsf{Encap}(pk)$ returns a pair (C, K). Correctness holds if $\mathsf{Decap}(sk, C) = K$.

IND-CCA Security of KEM. To define the security, consider the following game with adversary $\mathcal{A}$. First, $(pk, sk) \leftarrow \mathsf{KG}(1^\kappa)$ and pk is given to $\mathcal{A}$. In the so-called find stage, $\mathcal{A}$ can query any C of its choice to oracle $\mathsf{Decap}(sk, \cdot)$.

Then $\mathcal{A}$ invokes a challenge oracle who computes $(C^*, K^*) \leftarrow \mathsf{Encap}(pk)$, then takes K_* randomly satisfying $|K^*| = |K_*|$, and chooses $b \xleftarrow{\$} \{0, 1\}$. The oracle returns challenge pair $(C^*, K(b))$ in which $K(0) = K^*$ and $K(1) = K_*$.

After that, in the guess stage, $\mathcal{A}$ can again access to the oracle $\mathsf{Decap}(sk, \cdot)$, but is not allowed to query C^* to the decapsulation oracle. Finally, $\mathcal{A}$ returns b' as a guess of the hidden b.

The KEM is IND-CCA-secure if the advantage

$$\mathbf{Adv}_{\mathcal{A}}^{\mathrm{ind-cca}}(\kappa) = \left| \Pr[b' = b] - \frac{1}{2} \right|$$

is negligible in κ for all poly-time adversary $\mathcal{A}$.

Taking an element a randomly from a set A is notationally expressed by $a \xleftarrow{\$} A$. Let κ be the security parameter. We requires following building blocks. Concrete schemes can be found in [1, Section 6].

TCR. A target collision resistant hash function $\mathsf{TCR} : \mathcal{E}(\kappa) \to \mathcal{R}(\kappa)$ is defined as follows. Given a target $x^* \xleftarrow{\$} \mathcal{E}(\kappa)$, it is hard for all poly-time adversary $\mathcal{A}$ to find $x \in \mathcal{E}(\kappa)$ satisfying $\mathsf{TCR}(x) = \mathsf{TCR}(x^*)$. Formally, the advantage

$$\mathbf{Adv}_{\mathcal{A}}^{\mathsf{TCR}}(\kappa) = \Pr[x \leftarrow \mathcal{A}(x^*) : x \neq x^* \wedge \mathsf{TCR}(x) = \mathsf{TCR}(x^*)]$$

is negligible for all poly-time adversary $\mathcal{A}$.

KDF. We assume that there exists a key derivation function $\mathsf{KDF} : \mathcal{K}(\kappa) \to \{0,1\}^{2n(\kappa)}$ such that $\mathsf{KDF}(v)$ for random $v \in \mathcal{K}(\kappa)$ is computationally random over $\{0,1\}^{2n(\kappa)}$. Formally, the advantage

$$\mathbf{Adv}_{\mathcal{D}}^{\mathsf{KDF}}(\kappa) = \left| \Pr_{v \xleftarrow{\$} \mathcal{K}(\kappa)} [\mathcal{D}(\mathsf{KDF}(v)) = 1] - \Pr_{(k,k') \xleftarrow{\$} \{0,1\}^{2n(\kappa)}} [\mathcal{D}(k, k') = 1] \right|$$

is negligible for all poly-time distinguishers $\mathcal{D}$.

MAC. A message authentication code $\mathsf{MAC} : \{0,1\}^{n(\kappa)} \times \mathcal{E}(\kappa) \to \{0,1\}^{\tau(\kappa)}$ takes inputs $k \in \{0,1\}^{n(\kappa)}$ and $x \in \mathcal{E}(\kappa)$ to compute tag $t = \mathsf{MAC}_k(x)$. For random key $k \xleftarrow{\$} \{0,1\}^{n(\kappa)}$, the adversary $\mathcal{A}$ is given at most one pair $(x^*, t^* = \mathsf{MAC}_k(x^*))$ where x^* is of $\mathcal{A}$'s own choice. The adversary $\mathcal{A}$ then returns a pair (x, t). It is required that the following advantage

$$\mathbf{Adv}_{\mathcal{A}}^{\mathsf{MAC}}(\kappa) = \Pr[x \neq x^* \wedge t = \mathsf{MAC}_k(x)]$$

is negligible for all poly-time distinguishers $\mathcal{A}$. Note that the definition treats MAC as a function where $\mathcal{E}(\kappa)$ contains both messages and randomness (if any), the security notion already captures *strong* unforgeability against chosen-message attacks.

3 Kurosawa-Desmedt KEM, Revisited

Let $\mathbb{G} = \langle g \rangle$ be a group, generated by g, of prime public order $2^\kappa < q < 2^{\kappa+1}$ for security parameter κ.

The DDH assumption on $\mathbb{G}$ asserts that, for all poly-time distinguishers $\mathcal{D}$, non-unit random elements $g_1, g_2 \xleftarrow{\$} \mathbb{G}$, and $r \neq s \xleftarrow{\$} \mathbb{Z}_q$, the advantage

$$\mathbf{Adv}_{\mathcal{D}}^{\mathrm{ddh}}(\kappa) = \left| \Pr[\mathcal{D}(g_1, g_2, g_1^r, g_2^r) = 1] - \Pr[\mathcal{D}(g_1, g_2, g_1^r, g_2^s) = 1] \right|$$

is negligible on parameter κ.

3.1 Our Proposed KEM under DDH

The construction is depicted in Figure 1. In the construction, keys k_s and k_a are of n-bit length. In Decap, if $u_1 \notin \mathbb{G}$ or $u_2 \notin \mathbb{G}$ then $\perp$ is returned immediately at the beginning. The description of symmetric building blocks TCR, KDF, and MAC are in Section 2.

The main difference with the Kurosawa-Desmedt KEM is, in $\mathsf{Encap}(pk)$, the element v is spitted in two keys (k_s, k_a) by KDF. Then, the key k_a is used to authenticate elements (u_1, u_2) inside $\mathsf{Encap}(pk)$, while the key k_s is returned as the shared symmetric key. The crucial point here is the authentication of (u_1, u_2) by the MAC, which helps proving IND-CCA security of our proposal. This technique, while simple, has been neglected in the literature.

KG(1^κ) :	Encap(pk) :	Decap(sk, C) :
$g_1, g_2 \xleftarrow{\$} \mathbb{G}$	$r \xleftarrow{\$} \mathbb{Z}_q$	Parse $C = (u_1, u_2, t)$
$(x_1, x_2, y_1, y_2) \xleftarrow{\$} \mathbb{Z}_q^4$	$u_1 \leftarrow g_1^r, u_2 \leftarrow g_2^r$	$\alpha \leftarrow \mathsf{TCR}(u_1, u_2)$
$c \leftarrow g_1^{x_1} g_2^{x_2}$	$\alpha \leftarrow \mathsf{TCR}(u_1, u_2)$	$v \leftarrow u_1^{x_1 + \alpha y_1} u_2^{x_2 + \alpha y_2}$
$d \leftarrow g_1^{y_1} g_2^{y_2}$	$v \leftarrow c^r d^{r\alpha}$	$(k_s, k_a) \leftarrow \mathsf{KDF}(v)$
$pk \leftarrow (g_1, g_2, c, d)$	$(k_s, k_a) \leftarrow \mathsf{KDF}(v)$	If $t = \mathsf{MAC}_{k_a}(u_1, u_2)$
$sk \leftarrow (x_1, x_2, y_1, y_2)$	$t \leftarrow \mathsf{MAC}_{k_a}(u_1, u_2)$	return k_s
Return (pk, sk)	Return $C = (u_1, u_2, t)$ and $K = k_s$	Else return $\perp$

Fig. 1. Our IND-CCA-secure KEM under the DDH assumption

Perhaps it is illustrative to see how our KEM resists against the chosen ciphertext attack in [9,15] that breaks the Kurosawa-Desmedt KEM. Recall that, in the attack, the adversary first obtains the challenge encapsulation consisting of (u_1^*, u_2^*). The adversary then queries the decapsulation oracle with query of form $((u_1^*)^r, (u_2^*)^r)$ where $r \in \mathbb{Z}_q$ is random of its own choice. In [9,15], it is showed that, by only two such queries, the encapsulated symmetric key can be computed with overwhelming probability. In comparison, in our KEM, the tag t is effective as a hedge against such malformed queries. When the adversary submits $(u_1, u_2, t) = ((u_1^*)^r, (u_2^*)^r, t)$, the corresponding v can be proved randomly distributed under the DDH assumption (in the proof, see **Game₄**). This means corresponding keys $(k_s, k_a) = \mathsf{KDF}(v)$ are randomly distributed. For the decapsulation not returning $\perp$, the adversary had to come up with the tag t satisfying $t = \mathsf{MAC}_{k_a}((u_1^*)^r, (u_2^*)^r)$, which is computationally hard since k_a is random and MAC is assumed secure.

Our use of MAC is different from the counterpart in the hybrid PKE [13] in its input. In [13], MAC is used to authenticate a symmetrically encrypted plaintext e. Namely, using our notations, in [13], $e \leftarrow \mathsf{SymmetricEncryption}_{k_s}(\texttt{plaintext})$ and then $t \leftarrow \mathsf{MAC}_{k_a}(e)$. In contrast, in Figure 1, we take "early" MAC on (u_1, u_2). Nevertheless, the resemblance between our KEM and the hybrid PKE allows us to re-utilize the proof in the hybrid encryption case.

3.2 Comparison and Implementation

Base Group. There are primarily two choices for the group $\mathbb{G}$ so that DDH assumption is believed holds true. The first choice is to take $\mathbb{G}$ as the order q, multiplicative subgroup of $\mathbb{Z}_p^*$ in which $p = 1 \pmod{q}$ is a prime. The elements in $\mathbb{G}$ are thus represented modulo p, and hence of $|p| = 1024$ bits (for 80-bit security) or $|p| = 3072$ bits (for 128-bit security). See [11] for more details.

The second choice of $\mathbb{G}$ is to take elliptic curve groups of order q. This choice reduces the length of element representation, since the length of q in bits can be $|q| = 160$ (for 80-bit security), or $|q| = 256$ (for 128-bit security). See [21] for specific curves.

Theoretical Comparison. In Table 2, we compare our KEMs with the ACE-KEM in ISO/IEC 18033-2 [1], which refined the schemes in [10,11]. Both enjoys

Table 2. Comparison of KEMs in standard model based on the DDH assumption. Abbreviations in the table: **me** = multi-exponentiation, **se** = single-exponentiation, **gmc** = group membership check, **el** = group element.

Scheme	Assumption	Encap length	[Encap]; [Decap] main costs of computation	$[pk, sk]$ size
ACE-KEM [1]	DDH	$3\lvert q\rvert$	[1 me, 3 se]; [0 me, 3 se, 1 gmc]	[5 el, 4 el]
Ours, Figure 1	DDH	$2\lvert q\rvert + \lvert t\rvert$	[1 me, 2 se]; [1 me, 0 se, 2 gmc]	[4 el, 4 el]

a tight security reduction to the DDH assumption. Since the tag size $\lvert t\rvert$ can be 128 in our KEMs, our encapsulation size is slightly shorter than ACE-KEM. The public key in our KEMs is one group element shorter.

To compare computation costs, we consider ACE-KEM implemented a group of prime order q. We use the result that one multi-exponentiation in that group can be carried out in $(1 + 2/\log_2 \log_2 q) \log_2 q$ multiplications [7], therefore can be counted as approximately 1.2 single exponentiation, which also is supported by experimental results in [20].

First, in groups where group membership checks are trivial, our KEM in Figure 1 needs just one multi-exponentiation, thus beating the ACE-KEM at dramatic margin of 60% in decapsulation speed. Examples of the groups include NIST elliptic curves [21] defined over prime fields (P-192, P-224, P-256, P-384, P-521) and binary fields (B-163, B-233, B-283, B-409, B-571).

Now assume that a group membership check is costly as one single exponentiation, while more efficient methods (e.g., using the Legendre symbol) may be available depending on the base group [11, Section 4.2]. Using abbreviations in Table 2, we count: 1 **me** = 1.2 **se**, 1 **gmc** = 1 **se**.

Thus our encapsulation needs 3.2 (se), while that for ACE-KEM is 4.2 (se), meaning more than 20% improvement in speed. For decapsulation, our schemes in Figure 1 require 3.2 (se), while that of ACE-KEM is 4 (se), yielding at least 20% improvement.

Our KEM decapsulation speed is even either faster or comparable with standardized PSEC-KEM and ECIES-KEM schemes whose security proofs are not in the standard model. Interested readers can find more details in the full version of this paper [20].

Experimental Comparison. ISO/IEC 18033-2 comes with a reference implementation, written by Anshuman Rawat and Victor Shoup (see website of [1]). The implementation, among others, includes ACE-KEM, PSEC-KEM, and ECIES-KEM. We add an implementation of our proposed KEM based on that library. Timings of encryption and decryption are reported in Figure 2, in which our scheme in Figure 1 is named "newkd". The codes in [1] neither speed up multi-exponentiation nor use Legendre symbol for group membership check. Our code elaborates on these aspects by

– employing a square-and-multiply algorithm for multi-exponentiation (see [20] for details), and

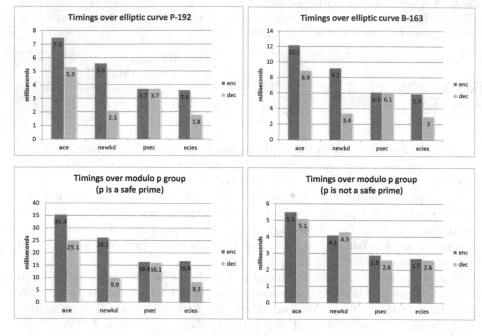

Fig. 2. Average timings, taken over 10000 executions, over different base groups. Experiment is done over a laptop (Intel 2.0GHz CPU, 8GB RAM) running Ubuntu 12.04 LTS. The C compiler is **g++** 4.6.3 using NTL 6.0.0 and GMP 5.1.1 libraries.

- using Legendre symbol for group membership check in $\mathbb{G} \subset \mathbb{Z}_p^*$ where p is a "safe" prime, namely $p = 2q + 1$ for a prime q (Sophie Germain prime).

Over all groups, one can confirm by Figure 2 that our proposed "newkd" is more efficient than ACE-KEM in both encapsulation and decapsulation. The bar charts also fit above theoretical comparisons.

Whenever above speedup tricks are applicable, namely over NIST's elliptic curves or over $\mathbb{G} \subset \mathbb{Z}_p^*$ with safe prime p, one can confirm that our proposal's decapsulation is faster than PSEC-KEM, and is even comparable to ECIES-KEM.

Over a subgroup $\mathbb{G} \subset \mathbb{Z}_p^*$ where p is not a safe prime, the decapsulation speed of "newkd" decreases. Here, two group membership checks, performed by two exponentiations, must be done since the Legendre symbol trick cannot be applied.

3.3   Security Proof

This subsection is devoted to prove the following theorem.

Theorem 1. *The KEM in Figure 1 is IND-CCA-secure under the DDH assumption.*

The following proof is similar to [13], adjusted for our KEM.

Proof. We will proceed in games, each of which is a modification of the previous one. Below, $\Pr[X_i] = \Pr[b' = b$ in **Game**$_i]$.

Game$_0$: This game is the IND-CCA attack game with an adversary $\mathcal{A}$. Recall that κ is the security parameter, and $\mathbf{Adv}_{\mathcal{A}}^{\text{ind-cca}}(\kappa) = |\Pr[b' = b] - \frac{1}{2}|$.

The challenge is $(C^*, K(b))$ where $C^* = (u_1^*, u_2^*, t^*)$. We denote by r^*, α^*, v^*, k_s^*, k_a^* the corresponding intermediate quantities. The key $K(b)$ is (k_s^*, k_a^*) or random depending on the bit b.

Game$_1$: The challenge oracle uses secrets (x_1, y_1, x_2, y_2) to compute v^*. Namely,

$$v^* = (u_1^*)^{x_1 + \alpha^* y_1} (u_2^*)^{x_2 + \alpha^* y_2}$$

where $u_1^* = g_1^{r^*}, u_2^* = g_2^{r^*}$ and $\alpha^* = \mathsf{TCR}(u_1^*, u_2^*)$.

Moreover, for any query (u_1, u_2, t) with $(u_1, u_2) \neq (u_1^*, u_2^*)$ and $\mathsf{TCR}(u_1, u_2) = \mathsf{TCR}(u_1^*, u_2^*)$, the decapsulation oracle returns $\perp$.

Then there exists a poly-time adversary $\mathcal{A}_1$ such that

$$|\Pr[X_0] - \Pr[X_1]| \leq \mathbf{Adv}_{\mathcal{A}_1}^{\mathsf{TCR}}(\kappa) \tag{1}$$

since the first change is notational, and the second one is based on the security of TCR. More formally, $\mathcal{A}_1$ gets inputs (u_1^*, u_2^*), and simulates the environment for $\mathcal{A}$ by generating the public and secret keys. $\mathcal{A}_1$ gives $\mathcal{A}$ the public key, and answers $\mathcal{A}$'s decapsulation queries using the secret key. In any decapsulation query (u_1, u_2, t), if $(u_1, u_2) \neq (u_1^*, u_2^*)$ and $\mathsf{TCR}(u_1, u_2) = \mathsf{TCR}(u_1^*, u_2^*)$, then $\mathcal{A}_1$ stops the simulation and returns the pair (u_1, u_2) as its output. The running time of $\mathcal{A}_1$ in the worst case is that of $\mathcal{A}$ plus time for doing arithmetic computations in $\mathbb{G}$ and time for some symmetric operations, so is of polynomial time.

Game$_2$: In this game, elements u_1^* and u_2^* are computed as follows: $r_1^* \xleftarrow{\$} \mathbb{Z}_q$, $u_1^* \leftarrow g_1^{r_1^*}$, and $r_2^* \xleftarrow{\$} \mathbb{Z}_q \setminus \{r_1^*\}, u_2^* \leftarrow g_2^{r_2^*}$. Then there is a poly-time adversary $\mathcal{A}_2$ such that

$$|\Pr[X_1] - \Pr[X_2]| = \mathbf{Adv}_{\mathcal{A}_2}^{\text{ddh}}(\kappa). \tag{2}$$

The description of $\mathcal{A}_2$ is as follows. Its input is a tuple (g_1, g_2, u_1^*, u_2^*). $\mathcal{A}_2$ itself generates the secret key, and then coupling with generators g_1, g_2 of $\mathbb{G}$, it computes the public key. Since $\mathcal{A}_2$ holds the secret key, it can answer all decapsulation queries from $\mathcal{A}$. The adversary $\mathcal{A}_2$ controls the hidden bit b, so that it can compare that bit with $\mathcal{A}$'s output bit b. In case $b' = b$, $\mathcal{A}_2$ returns 1; otherwise it returns 0. Any difference on the output b' of $\mathcal{A}$ depending on tuple (g_1, g_2, u_1^*, u_2^*) directly yields a difference on the probability $\mathcal{A}_2$ outputting 1, so that above equation claim is justified. The running time of $\mathcal{A}_2$ in the worst case is that of $\mathcal{A}$ plus time for doing arithmetic computations in $\mathbb{G}$ and time for some symmetric operations, so is of polynomial time.

Game$_3$: This game makes use of $\omega \in \mathbb{Z}_q^*$ satisfying $g_2 = g_1^\omega$. With ω, we can check in poly-time whether $\log_{g_1} u_1 = \log_{g_2} u_2$ by simply verifying $u_1^\omega = u_2$.

Initialization of the game	Decapsulation of adversarial query $C = (u_1, u_2, t)$
I1: $\omega \xleftarrow{\$} \mathbb{Z}_q^*,\ g_2 \leftarrow g_1^\omega$ **I2:** $(x_1, x_2, y_1, y_2) \xleftarrow{\$} \mathbb{Z}_q^4$ $\quad c \leftarrow g_1^{x_1} g_2^{x_2},\ d \leftarrow g_1^{y_1} g_2^{y_2}$ **I3:** $r_1^* \xleftarrow{\$} \mathbb{Z}_q,\ u_1^* \leftarrow g_1^{r_1^*}$ $\quad r_2^* \xleftarrow{\$} \mathbb{Z}_q \setminus \{r_1^*\},\ u_2^* \leftarrow g_2^{r_2^*}$ **I4:** $\alpha^* \leftarrow \mathsf{TCR}(u_1^*, u_2^*)$ $\quad v^* \leftarrow (u_1^*)^{x_1 + \alpha^* y_1} (u_2^*)^{x_2 + \alpha^* y_2}$ **I5:** $(k_s^*, k_a^*) \leftarrow \mathsf{KDF}(v^*)$	1. $\alpha = \mathsf{TCR}(u_1, u_2)$ 2. **if** $(u_1, u_2) \neq (u_1^*, u_2^*)$ and $\alpha = \alpha^*$ **then** 3. $\quad$ **return** $\perp$ 4. **end if** 5. **if** $(u_1, u_2) = (u_1^*, u_2^*)$ **then** 6. $\quad$ **if** $t \neq \mathsf{MAC}_{k_a^*}(u_1^*, u_2^*)$ **then return** $\perp$ 7. $\quad$ **else return** k_s^* 8. **else if** $(u_1, u_2) \notin \mathcal{V}$ **then** 9. $\quad \alpha \leftarrow \mathsf{TCR}(u_1, u_2)$ 10. $\quad v \leftarrow u_1^{x_1 + \alpha y_1} u_2^{x_2 + \alpha y_2}$ 11. $\quad (k_s, k_a) \leftarrow \mathsf{KDF}(v)$ 12. $\quad$ **if** $t \neq \mathsf{MAC}_{k_a}(u_1, u_2)$ **then return** $\perp$ 13. $\quad$ **else return** $\perp$ {Rejection rule in **Game$_3$**} 14. **else** 15. $\quad \alpha \leftarrow \mathsf{TCR}(u_1, u_2)$ 16. $\quad v \leftarrow u_1^{x_1 + \alpha y_1} u_2^{x_2 + \alpha y_2}$ 17. $\quad (k_s, k_a) \leftarrow \mathsf{KDF}(v)$ 18. $\quad$ **if** $t \neq \mathsf{MAC}_{k_a}(u_1, u_2)$ **then return** $\perp$ 19. $\quad$ **else return** k_s 20. **end if**

Fig. 3. Oracles in **Game$_3$** for the proof of Theorem 1

Denote $\mathcal{V} = \{(u_1, u_2) \in \mathbb{G}^2 : u_1^\omega = u_2\}$. In this game, any decapsulation query (u_1, u_2, t) with $(u_1, u_2) \notin \mathcal{V}$ is rejected. The initialization and decapsulation oracle in this game are depicted in Figure 3.

Let F_i $(i \geq 3)$ be the event that a query is rejected at line 13 of the decapsulation oracle in Game$_i$. Let Q be the bound on the total number of decapsulation queries $\mathcal{A}$ makes, we have

$$|\Pr[X_2] - \Pr[X_3]| \leq Q \Pr[F_3]. \tag{3}$$

Game$_4$: In this game, take $v^* \xleftarrow{\$} \mathbb{G}$ (at line **I4**) and $v \xleftarrow{\$} \mathbb{G}$ (at line 10 in the decapsulation). This is because

$$
\begin{bmatrix} \log_{g_1} c \\ \log_{g_1} d \\ \log_{g_1} v^* \\ \log_{g_1} v \end{bmatrix}
=
\underbrace{\begin{bmatrix} 1 & 0 & \omega & 0 \\ 0 & 1 & 0 & \omega \\ r_1^* & r_1^* \alpha^* & r_2^* \omega & r_2^* \omega \alpha^* \\ r_1 & r_1 \alpha & r_2 \omega & r_2 \omega \alpha \end{bmatrix}}_{M}
\begin{bmatrix} x_1 \\ y_1 \\ x_2 \\ y_2 \end{bmatrix}
$$

and determinant $\det(M) = \omega^2 (r_2^* - r_1^*)(r_2 - r_1)(\alpha - \alpha^*) \neq 0$ shows that (c, d, v^*, v) are uniformly distributed as (x_1, y_1, x_2, y_2) are. We have

$$\Pr[X_3] = \Pr[X_4] \tag{4}$$

$$\Pr[F_3] = \Pr[F_4]. \tag{5}$$

Game₅: At line **I5**, take $(k_s^*, k_a^*) \xleftarrow{\$} \{0,1\}^{2n}$. This is because v^* is taken randomly in the previous game. Then there exists an adversary $\mathcal{A}_5$ against KDF such that

$$|\Pr[X_4] - \Pr[X_5]| \leq \mathbf{Adv}_{\mathcal{A}_5}^{\mathsf{KDF}}(\kappa). \tag{6}$$

The description of $\mathcal{A}_5$ is as follows. Its input is a string in $\{0,1\}^{2n}$. It uses the input for the keys (k_s^*, k_a^*) at line **I5**, while generating the secret key and public key and others as in lines **I1** to **I4**. Since $\mathcal{A}_2$ holds the trapdoor for membership testing ω and the secret key, it can handle decapsulation queries as in Figure 3. When $\mathcal{A}$ returns b', the adversary $\mathcal{A}_5$ checks whether b' equals its chosen bit b. If $b' = b$, $\mathcal{A}_5$ returns 1. The running time of $\mathcal{A}_5$ in the worst case is that of $\mathcal{A}$ plus time for doing arithmetic computations in $\mathbb{G}$ and time for some symmetric operations, so is of polynomial time.

Game₆: At line 7 in the decapsulation, return $\bot$. This is because $(u_1, u_2) = (u_1^*, u_2^*)$ with probability $\frac{1}{q^2}$ before the challenge phase. Moreover, after the challenge phase when (u_1^*, u_2^*, t^*) was already announced, querying (u_1^*, u_2^*, t) with $t = \mathsf{MAC}_{k_a^*}(u_1^*, u_2^*)$ and $t \neq t^*$ to the oracle means the adversary can break the MAC. We have

$$|\Pr[X_5] - \Pr[X_6]| \leq Q\left(\frac{1}{q^2} + \mathbf{Adv}_{\mathcal{A}_6}^{\mathsf{MAC}}(\kappa)\right) \text{ and } \Pr[X_6] = \frac{1}{2} \tag{7}$$

since (k_s^*, k_a^*) are perfectly random in this game.

The description of $\mathcal{A}_6$ is as follows. Its input is (u_1^*, u_2^*, t^*) where $t^* = \mathsf{MAC}_{k_a^*}(u_1^*, u_2^*)$ for random key k_a^*. It generates the secret key and then simulates the environment for $\mathcal{A}$. Whenever $\mathcal{A}$ queries (u_1, u_2, t) for decapsulation in which $t \neq t^*$ and $t = \mathsf{MAC}_{k_a^*}(u_1^*, u_2^*)$, the adversary $\mathcal{A}_6$ halts the simulation and returns (u_1^*, u_2^*, t). The running time of $\mathcal{A}_6$ in the worst case is that of $\mathcal{A}$ plus time for doing arithmetic computations in $\mathbb{G}$ and time for some symmetric operations, so is of polynomial time.

Game₅′: Now we move back to consider **Game₄** again. This game is the same as **Game₄**, except that, $(k_s, k_a) \xleftarrow{\$} \{0,1\}^{2n}$ at line 11. We have

$$|\Pr[F_4] - \Pr[F_{5'}]| \leq \mathbf{Adv}_{\mathcal{A}_5'}^{\mathsf{KDF}}(\kappa). \tag{8}$$

Since the MAC key has been turned random,

$$\Pr[F_{5'}] \leq \mathbf{Adv}_{\mathcal{A}_5''}^{\mathsf{MAC}}(\kappa) \tag{9}$$

in which, as a recall, $F_{5'}$ is the event that a query is rejected at line 13 of the decapsulation oracle in this game. The descriptions of adversaries $\mathcal{A}_5'$ against KDF and $\mathcal{A}_5''$ against MAC are similar to those in **Game₅** and **Game₆**.

By (5), (8), (9), we have

$$\Pr[F_3] = \Pr[F_4] \leq \Pr[F_{5'}] + \mathbf{Adv}_{\mathcal{A}_5'}^{\mathsf{KDF}}(\kappa) \leq \mathbf{Adv}_{\mathcal{A}_5''}^{\mathsf{MAC}}(\kappa) + \mathbf{Adv}_{\mathcal{A}_5'}^{\mathsf{KDF}}(\kappa) \quad (10)$$

and by (1), (2), (3), (4), (6), (7), and the bound (10),

$$\mathbf{Adv}_{\mathcal{A}}^{\mathrm{ind-cca}}(\kappa) \leq \mathbf{Adv}_{\mathcal{A}_1}^{\mathsf{TCR}}(\kappa) + \mathbf{Adv}_{\mathcal{A}_2}^{\mathrm{ddh}}(\kappa) + Q\left(\mathbf{Adv}_{\mathcal{A}_5''}^{\mathsf{MAC}}(\kappa) + \mathbf{Adv}_{\mathcal{A}_5'}^{\mathsf{KDF}}(\kappa)\right)$$
$$+ \mathbf{Adv}_{\mathcal{A}_5}^{\mathsf{KDF}}(\kappa) + Q\left(\frac{1}{q^2} + \mathbf{Adv}_{\mathcal{A}_6}^{\mathsf{MAC}}(\kappa)\right)$$

ending the proof.   □

4 Generalization to Universal Hash Proof System

4.1 Hash Proof System

The notion of hash proof systems was introduced by Cramer and Shoup [12]. Let $\mathcal{SK}, \mathcal{PK}$, and $\mathcal{K}$ be sets of secret keys, public keys, and encapsulated symmetric keys. Let $\mathcal{E}$ be the set of all "valid" and "invalid" encapsulation, and $\mathcal{V} \subset \mathcal{E}$ be the set of all "valid" ones. To illustrate the above notation, in the DDH-based scheme, $\mathcal{SK} = \mathbb{G}^4$, $\mathcal{PK} = \mathbb{G}^2$, $\mathcal{E} = \mathbb{G}^2$, $\mathcal{K} = \mathbb{G}$, $\mathcal{V} = \{(g_1^r, g_2^r) : r \in \mathbb{Z}_q\}$.

The subset membership assumption says that $\mathcal{V}$ is indistinguishable from $\mathcal{E}$. If $\mathcal{V} = \{(g_1^r, g_2^r) : r \in \mathbb{Z}_q\}$ and $\mathcal{E} = \mathbb{G}^2$ as above, this is exactly the DDH assumption. Formally, the advantage

$$\mathbf{Adv}_{\mathcal{D}}^{\mathrm{sm}}(\kappa) = \left| \Pr_{U \xleftarrow{\$} \mathcal{E}} [\mathcal{D}(U) = 1] - \Pr_{U \xleftarrow{\$} \mathcal{V}} [\mathcal{D}(U) = 1] \right|$$

is negligible for all poly-time distinguishers $\mathcal{D}$.

A function $\Lambda_{sk} : \mathcal{E} \to \mathcal{K}$ is *projective* if there exists a projection $\mu : \mathcal{SK} \to \mathcal{PK}$ such that $pk = \mu(sk)$ defines $\Lambda_{sk} : \mathcal{V} \to \mathcal{K}$. Namely, for every $E \in \mathcal{V}$, the value $K = \Lambda_{sk}(E)$ is uniquely determined by $pk = \mu(sk)$ and E. As an example, in our scheme of Sect.3, $\Lambda_{sk}(E = (u_1, u_2)) = u_1^{x_1+\alpha y_1} u_2^{x_2+\alpha y_2}$ where $\alpha = \mathsf{TCR}(E)$.

A projective function Λ_{sk} is called computationally universal-2 [16] if for all $E, E' \notin \mathcal{V}$ with $E \neq E'$,

$$\left(pk, \Lambda_{sk}(E'), \Lambda_{sk}(E)\right) \text{ and } \left(pk, \Lambda_{sk}(E'), K\right)$$

are computationally indistinguishable, where $sk \xleftarrow{\$} \mathcal{SK}$ and $K \xleftarrow{\$} \mathcal{K}$. Formally, consider an adversary $\mathcal{A} = (\mathcal{A}_{\mathrm{find}}, \mathcal{A}_{\mathrm{guess}})$ in the following experiment $\mathbf{Exp}_{\mathcal{A}}^{\mathrm{cu2}}(\kappa)$.

KG(1^κ) :	Encap(pk) :	Decap(sk, C) :
Run Param to define	Take random witness r	Parse $C = (E, t)$
$(group, \mathcal{SK}, \mathcal{PK}, \mathcal{K},$	$E = E(r) \xleftarrow{\$} \mathcal{V}$	$v \leftarrow \mathsf{Priv}(sk, E)$
$\mathcal{E}, \mathcal{V}, \Lambda_{(\cdot)}(\cdot), \mu)$	$v \leftarrow \mathsf{Pub}(pk, E, r)$	$(k_s, k_a) \leftarrow \mathsf{KDF}(v)$
$sk \xleftarrow{\$} \mathcal{SK}$	$(k_s, k_a) \leftarrow \mathsf{KDF}(v)$	If $t = \mathsf{MAC}_{k_a}(E)$
$pk \leftarrow \mu(sk)$	$t \leftarrow \mathsf{MAC}_{k_a}(E)$	return k_s
Return (pk, sk)	Return $C = (E, t)$ and $K = k_s$	Else return $\perp$

Fig. 4. Our generic KEM from hash proof system (Param, Pub, Priv)

In $\mathbf{Exp}_{\mathcal{A}}^{cu2}(\kappa)$ on the right, the oracle $\mathsf{Eval}_{sk}(F)$ returns $\Lambda_{sk}(F)$ if $F \in \mathcal{V}$ and $\perp$ otherwise. Computational universality requires that

$$\mathbf{Adv}_{\mathcal{A}}^{cu2}(\kappa) = \Pr[\mathbf{Exp}_{\mathcal{A}}^{cu2}(\kappa) = 1]$$

is negligible for all poly-time $\mathcal{A}$.

Hash Proof System. A hash proof system $\mathcal{HPS}$ consists of algorithms (Param, Pub, Priv) described as follows. Algorithm Param(1^κ) first generates the description of $group$, $\mathcal{SK}$, $\mathcal{PK}$, $\mathcal{K}$, $\mathcal{E}$, $\mathcal{V}$, $\Lambda_{(\cdot)}(\cdot)$, and $\mu : \mathcal{SK} \to \mathcal{PK}$. Algorithm Pub($pk, E, r$) returns $K = \Lambda_{sk}(E)$ for $E \in \mathcal{V}$,

Experiment $\mathbf{Exp}_{\mathcal{A}}^{cu2}(\kappa)$:

Run Param(1^κ) to generate
$\quad (group, \mathcal{SK}, \mathcal{PK}, \mathcal{K}, \mathcal{E}, \mathcal{V}, \Lambda_{(\cdot)}(\cdot), \mu)$
$sk \xleftarrow{\$} \mathcal{SK}, pk \leftarrow \mu(sk), E' \xleftarrow{\$} \mathcal{E} \setminus \mathcal{V}$
$K' \leftarrow \Lambda_{sk}(E')$
$(E \in \mathcal{E} \setminus \mathcal{V}, \mathsf{st}) \leftarrow \mathcal{A}_{\mathsf{find}}^{\mathsf{Eval}_{sk}(\cdot)}(pk, E', K')$
$b \xleftarrow{\$} \{0,1\}, K(0) \leftarrow \Lambda_{sk}(E), K(1) \xleftarrow{\$} \mathcal{SK}$
$b' \leftarrow \mathcal{A}_{\mathsf{guess}}(\mathsf{st}, K(b))$
If $b' = b$ then return 1 else return 0

where the computation does not use sk but makes use of r, a witness of the fact that $E \in \mathcal{V}$. Algorithm Priv(sk, E) returns $\Lambda_{sk}(E)$.

4.2 IND-CCA-Secure KEM from Hash Proof Systems

The KEM is depicted in Figure 4. The descriptions of symmetric building blocks KDF and MAC are in Section 2.

Theorem 2. *The generic construction of KEM in Figure 4 is IND-CCA-secure.*

Proof. We proceed in games as follows.

Game$_0$: This game is the IND-CCA attack game with leakage. Without loss of generality, assume that E^*, r^* are generated at the beginning of the game.

Game$_1$: Compute Pub(pk, E^*, r^*) in the challenge encapsulation as Priv(sk, E^*). This change is only notational since Priv(sk, E^*) = Pub(pk, E^*, r^*) = $\Lambda_{sk}(E^*)$ so that $\Pr[X_0] = \Pr[X_1]$.

Game$_2$: Take $E^* \xleftarrow{\$} \mathcal{C} \setminus \mathcal{V}$. We have

$$|\Pr[X_1] - \Pr[X_2]| \leq \mathbf{Adv}_{\mathcal{A}_2}^{sm}(\kappa) \tag{11}$$

thanked to the subset membership problem. The running time of $\mathcal{A}_2$ in the worst case is that of $\mathcal{A}$ plus time for doing some computations in the hash proof systems and time for some symmetric operations, so is of polynomial time.

Game$_3$: Any decapsulation query (E,t) with $E \neq E^*$ and $E \notin \mathcal{V}$ is answered by $\perp$. Let Q be the total number of decapsulation queries, we have

$$|\Pr[X_2] - \Pr[X_3]| \leq Q \Pr[F_3] \tag{12}$$

Initialization of the game	Decapsulation of adversarial query $C = (E, t)$
I1: $\omega \overset{\$}{\leftarrow}$ Trapdoors **I2:** $sk \overset{\$}{\leftarrow} \mathcal{SK}$, $pk \leftarrow \mu(sk)$ **I3:** $E^* \overset{\$}{\leftarrow} \mathcal{C} \setminus \mathcal{V}$ **I4:** $v^* \leftarrow \mathsf{Priv}(sk, E^*)$ **I5:** $(k_s^*, k_a^*) \leftarrow \mathsf{KDF}(v^*)$	1. **if** $E = E^*$ **then** 2. **if** $t \neq \mathsf{MAC}_{k_a^*}(E^*)$ **then return** $\bot$ 3. **else return** k_s^* 4. **else if** $E \notin \mathcal{V}$ **then** 5. $v \leftarrow \mathsf{Priv}(sk, E)$ 6. $(k_s, k_a) \leftarrow \mathsf{KDF}(v)$ 7. **if** $t \neq \mathsf{MAC}_{k_a}(E)$ **then return** $\bot$ 8. **else return** $\bot$ 9. **else** 10. $v \leftarrow \mathsf{Priv}(sk, E)$ 11. $(k_s, k_a) \leftarrow \mathsf{KDF}(v)$ 12. **if** $t \neq \mathsf{MAC}_{k_a}(E)$ **then return** $\bot$ 13. **else return** k_s 14. **end if**

Fig. 5. Oracles in **Game**$_3$ for the proof of Theorem 2

where F_3 is the event that a query is rejected by the above rule. The initialization and the decapsulation oracle are depicted in Figure 5, in which F_3 happens whenever line 8 of decapsulation is reached.

Game$_4$: In this game, take $v^* \overset{\$}{\leftarrow} \mathcal{K}$ (at line **I4**) and $v \overset{\$}{\leftarrow} \mathcal{K}$ (at line 5 in the decapsulation). We have

$$|\Pr[X_3] - \Pr[X_4]| \leq \mathbf{Adv}_{\mathcal{A}_4}^{\mathrm{cu2}}(\kappa) \tag{13}$$

$$|\Pr[F_3] - \Pr[F_4]| \leq \mathbf{Adv}_{\mathcal{A}_4'}^{\mathrm{cu2}}(\kappa) \tag{14}$$

where event F_4 happens whenever line 8 of decapsulation is reached in this game. The reasons are that $v = \Lambda_{sk}(E)$ is computationally random conditioned on $pk, v^* = \Lambda_{sk}(E^*)$; and that $v^* = \Lambda_{sk}(E^*)$ is computationally random conditioned on pk, v thanks to the computational universality of the hash proof system.

Game$_5$: At line **I5**, take $(k_s^*, k_a^*) \overset{\$}{\leftarrow} \{0, 1\}^{2n}$. This is because v^* is taken randomly in the previous game. Then there exists an adversary $\mathcal{A}_5$ against KDF such that

$$|\Pr[X_4] - \Pr[X_5]| \leq \mathbf{Adv}_{\mathcal{A}_5}^{\mathsf{KDF}}(\kappa). \tag{15}$$

The description of $\mathcal{A}_5$ is the same as its counterpart in the proof of Theorem 1.

Game$_6$: At line 3 in the decapsulation, return $\bot$. This is because $E = E^*$ with probability $\frac{1}{|\mathcal{E}|}$ before the challenge phase. Moreover, after the challenge phase when (E^*, t^*) was already announced, querying (E^*, t) with $t = \mathsf{MAC}_{k_a^*}(E^*)$ and $t \neq t^*$ to the oracle means the adversary can break the MAC. We have

$$|\Pr[X_5] - \Pr[X_6]| \leq Q\left(\frac{1}{|\mathcal{E}|} + \mathbf{Adv}_{\mathcal{A}_6}^{\mathsf{MAC}}(\kappa)\right) \text{ and } \Pr[X_6] = \frac{1}{2} \qquad (16)$$

since (k_s^*, k_a^*) are perfectly random in this game.

The description of $\mathcal{A}_6$ is as follows. Its input is (E^*, t^*) where $t^* = \mathsf{MAC}_{k_a^*}(E^*)$ for random key k_a^*. It generates the secret key and then simulates the environment for $\mathcal{A}$. Whenever $\mathcal{A}$ queries (E, t) for decapsulation in which $t \neq t^*$ and $t = \mathsf{MAC}_{k_a^*}(E^*)$, the adversary $\mathcal{A}_6$ halts the simulation and returns (E^*, t).

Game$_{5'}$: Now we move back to consider **Game$_4$** again. This game is the same as **Game$_4$**, except that, $(k_s, k_a) \xleftarrow{\$} \{0,1\}^{2n}$ at line 6. We have

$$|\Pr[F_4] - \Pr[F_{5'}]| \leq \mathbf{Adv}_{\mathcal{A}_5'}^{\mathsf{KDF}}(\kappa). \qquad (17)$$

The description of $\mathcal{A}_5$ is the same as its counterpart in the proof of Theorem 1. Since the MAC key k_a has been turned random,

$$\Pr[F_{5'}] \leq \mathbf{Adv}_{\mathcal{A}_5''}^{\mathsf{MAC}}(\kappa) \qquad (18)$$

in which, as a recall, $F_{5'}$ is the event that a query is rejected at line 8 of the decapsulation oracle in this game. The descriptions of adversaries $\mathcal{A}_5'$ against KDF and $\mathcal{A}_5''$ against MAC are similar to those in **Game$_5$** and **Game$_6$**.

By (14), (17), and (18),

$$\Pr[F_3] \leq \mathbf{Adv}_{\mathcal{A}_4'}^{\mathsf{cu2}}(\kappa) + \mathbf{Adv}_{\mathcal{A}_5'}^{\mathsf{KDF}}(\kappa) + \mathbf{Adv}_{\mathcal{A}_5''}^{\mathsf{MAC}}(\kappa) \qquad (19)$$

Summing up (11), (12), (13), (15), (16), and (19),

$$\mathbf{Adv}_{\mathcal{A}}^{\mathsf{ind-cca}}(\kappa) \leq \mathbf{Adv}_{\mathcal{A}_2}^{\mathsf{sm}}(\kappa) + Q\left(\mathbf{Adv}_{\mathcal{A}_4'}^{\mathsf{cu2}}(\kappa) + \mathbf{Adv}_{\mathcal{A}_5'}^{\mathsf{KDF}}(\kappa) + \mathbf{Adv}_{\mathcal{A}_5''}^{\mathsf{MAC}}(\kappa)\right)$$

$$+ \mathbf{Adv}_{\mathcal{A}_4}^{\mathsf{cu2}}(\kappa) + \mathbf{Adv}_{\mathcal{A}_5}^{\mathsf{KDF}}(\kappa) + Q\left(\frac{1}{|\mathcal{E}|} + \mathbf{Adv}_{\mathcal{A}_6}^{\mathsf{MAC}}(\kappa)\right)$$

ending the proof. $\qquad\qquad\qquad\qquad\qquad\qquad\qquad\qquad\qquad\qquad\qquad\qquad\square$

4.3   Instantiation under the DLIN Assumption

We use the HPS based on the decisional linear assumption (DLIN) given by [16]. In this HPS, $\mathcal{SK} = \mathbb{Z}_q^6$, $\mathcal{PK} = \mathbb{G}^4$, $\mathcal{K} = \mathbb{G}$. Also $\mathcal{E} = \mathbb{G}^3$ and $\mathcal{V} = \{(g_1^{r_1}, g_2^{r_2}, h^{r_1+r_2}) : r_1, r_2 \in \mathbb{Z}_q\}$, where $g_1, g_2, h \in \mathbb{G}$. The DLIN assumption asserts that $\mathcal{E}$ and $\mathcal{V}$ are indistinguishable. The projective function is

$$\Lambda_{sk}(u_1, u_2, u_3) = u_1^{x_1+\alpha y_1} u_2^{x_2+\alpha y_2} u_3^{z+\alpha z'} \iff \Lambda_{sk}(u_1, u_2, u_3) = (c_1 d_1^\alpha)^{r_1} (c_2 d_2^\alpha)^{r_2}$$

using the same notations as in Figure 6. To check $E \in \mathcal{E} \setminus \mathcal{V}$ in Figure 5, use trapdoors $\log_{g_1} h \in \mathbb{Z}_q$ and $\log_{g_2} h \in \mathbb{Z}_q$.

Lemma 1 (Lemma 6.3 in [16]). *The above hash proof system is computationally universal-2 if* TCR *is target collision resistant.*

KG(1^κ) :	Encap(pk) :	Decap(sk, C) :
$g_1, g_2, h \xleftarrow{\$} \mathbb{G}$	$r_1, r_2 \xleftarrow{\$} \mathbb{Z}_q$	Parse $C = (u_1, u_2, u_3, t)$
$(x_1, x_2, y_1, y_2) \xleftarrow{\$} \mathbb{Z}_q^4$	$u_1 \leftarrow g_1^{r_1}, u_2 \leftarrow g_2^{r_2}$	$\alpha \leftarrow \mathsf{TCR}(u_1, u_2, u_3)$
$(z, z') \xleftarrow{\$} \mathbb{Z}_q^2$	$u_3 \leftarrow h^{r_1 + r_2}$	$v \leftarrow u_1^{x_1 + \alpha y_1} u_2^{x_2 + \alpha y_2} u_3^{z + \alpha z'}$
$c_1 \leftarrow g_1^{x_1} h^z, c_2 \leftarrow g_2^{x_2} h^z$	$\alpha \leftarrow \mathsf{TCR}(u_1, u_2, u_3)$	$(k_s, k_a) \leftarrow \mathsf{KDF}(v)$
$d_1 \leftarrow g_1^{y_1} h^{z'}, d_2 \leftarrow g_2^{y_2} h^{z'}$	$v \leftarrow (c_1 d_1^\alpha)^{r_1} (c_2 d_2^\alpha)^{r_2}$	If $t = \mathsf{MAC}_{k_a}(u_1, u_2, u_3)$
$pk \leftarrow (g_1, g_2, h,$	$(k_s, k_a) \leftarrow \mathsf{KDF}(v)$	return k_s
$\quad c_1, d_1, c_2, d_2)$	$t \leftarrow \mathsf{MAC}_{k_a}(u_1, u_2, u_3)$	Else
$sk \leftarrow (x_1, x_2, y_1, y_2, z, z')$	Return $C = (u_1, u_2, u_3, t)$	return $\perp$
Return (pk, sk)	and $K = k_s$	
KG(1^κ) :	Encap(pk) :	Decap(sk, C) :
$g \xleftarrow{\$} \mathbb{G}, g_2 \leftarrow g^{N_1}$	$r \xleftarrow{\$} \{0, \dots, N_1/4\}$	Parse $C = (u, t)$
$(x, y) \xleftarrow{\$} \mathcal{SK}$	$u \leftarrow g_2^r \bmod N_1^2$	$\alpha \leftarrow \mathsf{TCR}(u)$
$c \leftarrow g_2^x \bmod N_1^2$	$\alpha \leftarrow \mathsf{TCR}(u)$	$v \leftarrow u^{x + y\alpha} \bmod N_1$
$d = g_2^y \bmod N_1^2$	$v \leftarrow (cd^\alpha)^r \bmod N_1$	$(k_s, k_a) \leftarrow \mathsf{KDF}(v)$
$pk = (N_1, g_2, c, d)$	$(k_s, k_a) \leftarrow \mathsf{KDF}(v)$	If $t = \mathsf{MAC}_{k_a}(u)$
$sk \leftarrow (x, y)$	$t \leftarrow \mathsf{MAC}_{k_a}(u)$	return k_s
Return (pk, sk)	Return $C = (u, t)$ and $K = k_s$	Else return $\perp$

Fig. 6. Our DLIN-based KEM (above) and DCR-based KEM (below)

Our DLIN-based KEM appears in Figure 6. The symmetric building blocks are $\mathsf{TCR} : \mathbb{G}^3 \to \mathbb{Z}_q$, $\mathsf{KDF} : \mathbb{G} \to \{0, 1\}^{2n}$, and $\mathsf{MAC} : \{0, 1\}^n \times \mathbb{G}^3 \to \{0, 1\}^\tau$. Security requirements are given in Section 2.

Theorem 3. *The construction of KEM in Figure 6 is IND-CCA-secure under the DLIN assumption.*

Proof. Directly from Lemma 1 and Theorem 2. □

4.4 Instantiation under the DCR Assumption

We use the HPS based on the decisional composite residuosity assumption (DCR) given in [16]. Let $p_1 = 2p_2 + 1$ and $q_1 = 2q_2 + 1$ be primes, where p_2 and q_2 are also primes. Let $N_1 = p_1 q_1$ and $N_2 = p_2 q_2$. Let $\mathbb{G}$ be the subgroup of $Z_{N_1^2}^*$ with order $N_1 N_2$. Note that $\mathbb{G}$ is written as $\mathbb{G} = \mathbb{G}_{N_1} \cdot \mathbb{G}_{N_2}$ where $\mathbb{G}_{N_i}$ denotes a cyclic group of order N_i. Let g be a generator of $\mathbb{G}$, so that $g_1 = g^{N_2}$ is a generator of $\mathbb{G}_{N_1}$ and $g_2 = g^{N_1}$ is a generator of $\mathbb{G}_{N_2}$.

In this HPS, $\mathcal{SK} = \{0, \dots, \lfloor N_1^2/2 \rfloor\}^2$, $\mathcal{PK} = \mathbb{G}_{N_2}^2$, $\mathcal{K} = \mathbb{Z}_{N_1}$. Also $\mathcal{E} = \mathbb{G}$ and $\mathcal{V} = \{g_2^r \bmod N_1^2 : r \in \{0, \dots, N_1/4\}\}$. The DCR assumption says that $\mathcal{E}$ and $\mathcal{V}$ are indistinguishable. To check $E \in \mathcal{E} \setminus \mathcal{V}$ in Figure 5, use trapdoor N_2.

The projection function is, using the same notation as in Figure 6,

$$\Lambda_{sk}(u) = u^{x + y\alpha} \bmod N_1 \iff \Lambda_{sk}(u = g_2^r \bmod N_1^2) = (cd^\alpha)^r \bmod N_1.$$

Lemma 2 (By [12, 16]). *The above hash proof system is computationally universal 2 if TCR is target collision resistant.*

Our DLIN-based KEM appears in Figure 6, which uses symmetric building blocks TCR : $\mathbb{Z}_{N_1^2} \to \mathbb{Z}_{\lfloor N_1^2/2 \rfloor}$, and KDF : $\mathbb{Z}_{N_1} \to \{0,1\}^{2n}$, and MAC : $\{0,1\}^n \times \mathbb{Z}_{N_1^2} \to \{0,1\}^\tau$.

Theorem 4. *The construction of KEM in Figure 6 is IND-CCA-secure under the DCR assumption.*

Proof. Directly from Lemma 2 and Theorem 2.

Acknowledgment. We are grateful to Takahiro Matsuda for informing us about reference [6]. We also thank Qiong Huang and the anonymous reviewers for comments that help refining this manuscript.

References

1. International Organization for Standardization, Genève, Switzerland. ISO/IEC 18033-2:2006, Information technology — Security techniques — Encryption Algorithms — Part 2: Asymmetric Ciphers, Final Committee Draft (2006), http://shoup.net/iso/
2. Cryptography Research and Evaluation Committees (CRYPTREC). Specifications of ciphers in the Candidate Recommended Ciphers List (March, 2013), http://www.cryptrec.go.jp/english/method.html
3. Abe, M., Gennaro, R., Kurosawa, K.: Tag-KEM/DEM: A new framework for hybrid encryption. J. Cryptology 21(1), 97–130 (2008)
4. American National Standards Institute. ANSI X9.44-2007: Key Establishment Using Integer Factorization Cryptography (2007)
5. Anada, H., Arita, S.: Identification schemes from key encapsulation mechanisms. IEICE Transactions 95-A(7), 1136–1155 (2012)
6. Baek, J., Galindo, D., Susilo, W., Zhou, J.: Constructing strong KEM from weak KEM (or how to revive the KEM/DEM framework). In: Ostrovsky, R., De Prisco, R., Visconti, I. (eds.) SCN 2008. LNCS, vol. 5229, pp. 358–374. Springer, Heidelberg (2008)
7. Bernstein, D.J.: Pippenger's exponentiation algorithm (2002), http://cr.yp.to/papers/pippenger.pdf
8. Boyd, C., Cliff, Y., Nieto, J.M.G., Paterson, K.G.: One-round key exchange in the standard model. IJACT 1(3), 181–199 (2009)
9. Choi, S.G., Herranz, J., Hofheinz, D., Hwang, J.Y., Kiltz, E., Lee, D.H., Yung, M.: The Kurosawa-Desmedt key encapsulation is not chosen-ciphertext secure. Inf. Process. Lett. 109(16), 897–901 (2009)
10. Cramer, R., Shoup, V.: A practical public key cryptosystem provably secure against adaptive chosen ciphertext attack. In: Krawczyk, H. (ed.) CRYPTO 1998. LNCS, vol. 1462, pp. 13–25. Springer, Heidelberg (1998)
11. Cramer, R., Shoup, V.: Design and analysis of practical public-key encryption schemes secure against adaptive chosen ciphertext attack. SIAM Journal on Computing 33, 167–226 (2001)
12. Cramer, R., Shoup, V.: Universal hash proofs and a paradigm for adaptive chosen ciphertext secure public-key encryption. In: Knudsen, L.R. (ed.) EUROCRYPT 2002. LNCS, vol. 2332, pp. 45–64. Springer, Heidelberg (2002)

13. Desmedt, Y., Gennaro, R., Kurosawa, K., Shoup, V.: A new and improved paradigm for hybrid encryption secure against chosen-ciphertext attack. J. Cryptology 23(1), 91–120 (2010)
14. Fujioka, A., Suzuki, K., Xagawa, K., Yoneyama, K.: Strongly secure authenticated key exchange from factoring, codes, and lattices. In: Fischlin, M., Buchmann, J., Manulis, M. (eds.) PKC 2012. LNCS, vol. 7293, pp. 467–484. Springer, Heidelberg (2012)
15. Herranz, J., Hofheinz, D., Kiltz, E.: The Kurosawa-Desmedt key encapsulation is not chosen-ciphertext secure. IACR Cryptology ePrint Archive 2006, 207 (2006)
16. Hofheinz, D., Kiltz, E.: Secure hybrid encryption from weakened key encapsulation. Cryptology ePrint Archive, Report 2007/288 (2007), http://eprint.iacr.org/. Full version of a paper at Menezes, A. (ed.): CRYPTO 2007. LNCS, vol. 4622. Springer, Heidelberg (2007)
17. Kiltz, E.: Chosen-ciphertext secure key-encapsulation based on gap hashed Diffie-Hellman. In: Okamoto, T., Wang, X. (eds.) PKC 2007. LNCS, vol. 4450, pp. 282–297. Springer, Heidelberg (2007)
18. Krawczyk, H., Paterson, K.G., Wee, H.: On the security of the TLS protocol: A systematic analysis. In: Canetti, R., Garay, J.A. (eds.) CRYPTO 2013, Part I. LNCS, vol. 8042, pp. 429–448. Springer, Heidelberg (2013)
19. Kurosawa, K., Desmedt, Y.G.: A new paradigm of hybrid encryption scheme. In: Franklin, M. (ed.) CRYPTO 2004. LNCS, vol. 3152, pp. 426–442. Springer, Heidelberg (2004)
20. Kurosawa, K., Phong, L.T.: Kurosawa-Desmedt key encapsulation mechanism, revisited. Cryptology ePrint Archive, Report 2013/765 (2013), http://eprint.iacr.org/; Full version of this manuscript
21. National Institute of Standards and Technology. Recommended elliptic curves for federal government use (1999), http://csrc.nist.gov/groups/ST/toolkit/documents/dss/NISTReCur.pdf
22. Okamoto, T.: Authenticated key exchange and key encapsulation in the standard model. In: Kurosawa, K. (ed.) ASIACRYPT 2007. LNCS, vol. 4833, pp. 474–484. Springer, Heidelberg (2007) Revised version available at http://eprint.iacr.org/2007/473
23. Yoneyama, K.: Compact authenticated key exchange from bounded CCA-secure KEM. In: Paul, G., Vaudenay, S. (eds.) INDOCRYPT 2013. LNCS, vol. 8250, pp. 161–178. Springer, Heidelberg (2013)

Differential Biases in Reduced-Round Keccak

Sourav Das and Willi Meier

Alcatel-Lucent India Ltd. and FHNW, Windisch, Switzerland
sourav10101976@gmail.com

Abstract. The Keccak hash function is the winner of the SHA-3 competition. In this paper, we examine differential propagation properties of Keccak constituent functions. We discover that low-weight differentials produce a number of biased and fixed difference bits in the state after two rounds and provide a theoretical explanation for the existence of such a bias. We also describe several other propagation properties of Keccak with respect to differential cryptanalysis. Combining our propagation analysis with results from the existing literature we find distinguishers on six rounds of the Keccak hash function with complexity 2^{52} for the first time in this paper.

Keywords: SHA-3, Propagation Analysis, Double-kernel, TDA.

1 Introduction

Cryptographic hash functions are important tools in cryptography that can be used for multiple purposes including authentication, integrity check of executables, digital signatures, etc. These are deterministic functions, H, that given an input or message of an arbitrary length, M, return a short pseudo-random value of fixed length, n. A hash function should be easy to compute and is usually defined by an iterative construction and a compression function.

The classical security requirements of a hash function are:

1. Collision resistance: finding two message $M1$ and $M2$ so that $H(M1) = H(M2)$ must be "hard". The minimum complexity of such an attack is given by birthday bound which is $2^{n/2}$ calls to the compression function.
2. Second preimage resistance: Given a message $M1$ and its hash value $h = H(M1)$, finding another message $M2$ so that $H(M2) = h$ must be hard. The generic second preimage attack has a complexity of 2^n calls to the compression function.
3. Preimage resistance: Given a hash value h, finding a message $M1$ so that $H(M1) = h$ must be hard. The generic preimage attack requires 2^n calls to the compression function.

In a practical sense, finding a collision or a (second) preimage of the hash function should require not significantly less calls to the compression function than the generic attacks. Besides these properties, a hash function is expected to

D. Pointcheval and D. Vergnaud (Eds.): AFRICACRYPT 2014, LNCS 8469, pp. 69–87, 2014.

satisfy a few other conditions, e.g, the hash output should not be distinguishable from random output.

In the last decade, many of the standardized hash functions e.g. MD5 [19], SHA1 [16] have suffered from serious collision attacks [21], [20]. The confidence in the standard SHA-2 has then been put into question due to its resemblance with SHA-1. As a consequence, the American National Institute of Standards and Technology (NIST) has launched in 2008 a competition to find a new hash function standard, SHA-3. From the 64 initial submissions, two rounds and four years later, Keccak emerged as the winner in 2012.

Keccak is a sponge based hash function. One of the stated reasons for the selection of Keccak by NIST as the new SHA-3 hash standard was its exceptional resistance to cryptanalytic attacks [7]. Even though it was a prime target for several years and a lot of cryptanalytic effort went into it [1], [5], [3], [8], [12], [4], [10], [11], [9], [13], [17], there was limited progress so far in mounting an attack even in greatly simplified versions of its various flavors. In particular, the best known results so far have been successful only up to five rounds of Keccak. One of the main reasons for this lack of progress is that the probabilities of the standard differential characteristics of Keccak's internal permutation are quite small, as was rigorously shown in [8].

This paper investigates on the propagation properties of certain low-weight differentials in Keccak. It discovers that there exist biased differential bits in the output part of Keccak after two rounds. It traces back the existence of these biased bits to the quadratic non-linear function, χ, of Keccak. There also exist state bits with fixed difference with probability 1 after two rounds. The existence of fixed difference bits with probability 1 after two rounds was reported in [14], [15], without giving further explanation. As a consequence of the propagation analysis as described in this paper we not only provide a clear explanation for the existence of such bits, but also discover many bits with strongly biased differences. Further, this paper extends the propagation properties to four rounds of Keccak by using a *double-kernel*, a concept studied in [17], extending from the concept of a *kernel* of differences in [4]. We exhaustively search all possible double-kernels upto weight six in Keccak permutation and discover some interesting properties of double-kernels.

As a main application, we show that some of our four-round distinguishers enable an extension to six-round distinguishers for Keccak-224. The complexity of the resulting distinguishers amounts to the evaluation of the hash function for 2^{52} message pairs. This uses computation backwards for two rounds and is based on results published in [9] and [10], where a heuristic algorithm is designed (called Target Difference Algorithm - TDA), that allows to find message pairs which satisfy a given target difference after one Keccak permutation round.

In [17] a very efficient distinguisher on four rounds of the Keccak hash function is described. This uses the concept of free bits combined with well chosen messages, and is not available for backwards computation over two rounds, to provide a distinguisher over six rounds. Also, there are previous results known on distinguishers of the Keccak permutation, that enter more than six rounds:

Zero-sum distinguishers [1], can distinguish the full 24-round internal permutation from a random permutation [5], [6], [12], and in [13], a differential distinguisher on eight rounds of the internal permutation is described. However these distinguishers are not easily amenable to provide distinguishers for the hash function. This paper provides a six round round distinguisher on Keccak hash function i.e. it considers the initial capacity bits set to zero. The Table 1 provides the best results of cryptanalysis of Keccak hash function.

Table 1. The best cryptanalytic results on Keccak on the *hash function* settings

Variant	No of Rounds	Complexity	Type of Attack	Reference
512, 384, 256, 224	4	Improvement by 2^3 - 2^8	preimage	[15]
224,256	4	practical	collision	[10]
224,256	5	practical	near-collision	[10]
256	5	2^{115}	collision	[11]
384, 512	3	practical	collision	[11]
224, 256	4	2^{24}	distinguisher	[17]
224	6	2^{52}	distinguisher	This Paper

As a final application of our propagation analysis, we provide an explanation why the near collision on five rounds of Keccak as reported in [9], [10], does not match with the expected near collision predicted by their differential trail.

This paper is organized as follows. Section 2 describes the Keccak hash function. The propagation properties of Keccak sponge constituent functions are given in Section 3. Section 4 shows how to get 2 rounds distinguishers using the propagation properties. Section 5 gives the application of the propagation properties to get biased difference bits upto six rounds of Keccak and thereby get a six round distinguisher. Finally, Section 6 shows another application of these propagation properties.

2   Description of Keccak

Keccak is a family of sponge hash functions. A sponge hash function absorbs a message block of r bits into its internal state and subsequently applies an internal permutation to the state. This step is repeated until all the blocks of the message to hash have been treated. Next, in the squeezing phase, r bits are generated from the state before each new permutation application, until the number of desired output bits has been generated. In the following we recall the recommended Keccak versions for SHA-3. All versions use the same internal permutation: $Keccak f[1600]$.

The $Keccak f[1600]$ state consists of 1600 bits, organized in 64 *slices* of 5×5 bits. The position of a bit in a slice can be given by its x and y value. The z coordinate gives the number of the slice $0 \leq z \leq 63$. Most of the steps in the round function of Keccak are invariant to a translation in z direction. The only part non-invariant is the round constant addition ι.

The permutation of the full Keccak hash function consists of 24 iterations of the round function. The round function itself is composed of five steps:

1. θ: Xor to each bit the XOR of two *columns* (column = same x value, y from 0 to 4). The first column is in the same slice as the bit and the second column is in the slice before the bit.
2. ρ: Translate a bit in z direction.
3. π: Permute the bits within a slice.
4. χ: Apply a 5 × 5 S-box on one *row* (row = same y value, x from 0 to 4).
5. ι: Addition of round constant.

Each of the versions (224, 256, 384 and 512 output bits) has a different block message size r. The capacity in a sponge construction is the size of the internal state minus the size of a message block. Consequently, they all have a different capacity c:

- For an output of 224 bits, r = 1152 and c = 448.
- For an output of 256 bits, r = 1088 and c = 512.
- For an output of 384 bits, r = 832 and c = 768.
- For an output of 512 bits, r = 576 and c = 1024.

In this paper, we have represented the state as a 5 × 5 matrix where each element of the matrix is a 64-bit *lane* (lane= same x and y value, z ranging from 0 to 63). In the lane, the LSB is at the right side i.e. it is in Little Endian notation.

3   Propagation Properties of the Keccak Constituent Functions

In this section, we consider some propagation properties of the constituent functions θ, ρ, π, χ and ι. In propagation analysis, we perform a large number of experiments and observe the state output differences. If the state difference is always 1, then we call that state difference bit as *active*. If it is always 0, then we call that difference bit as *inactive*. If the state difference is 0 for half the times (and 1 for half the times), then we call that state difference bit as *balanced*.

3.1   Propagation Properties of θ, ρ, π and ι

The θ transformation on each state bit depends linearly on 10 other state bits and the self bit.

Property 1. If an even number of bits of the input of θ are active (i.e. change their value with probability 1), and the remaining input bits are inactive (i.e. stay constant), then the output state bit is inactive.

Similarly, if an odd number of bits of the input of θ are active, and the remaining input bits are inactive, then the output state bit is active.

Property 2. If any bit of the input state of θ is balanced then the output state bit is also balanced.

Property 3. The functions ρ, π and ι, being simple bitwise permutations, do not change any propagation properties of the input.

3.2   Propagation Properties of χ

The S-box of the χ layer is a 5-bit S-box, where every output bit y_i depends on only three input bits. Out of these, it is dependent linearly on one bit and non-linearly on another two bits (operations are in $GF(2)$):

$$y_i = x_i + (x_{i+1} + 1) \cdot x_{i+2} \tag{1}$$

Assume a set of input differences causes a set of output differences. We describe how the input set affects the output set for a particular bit.

Property 4. If x_i is active and there is no difference in any of x_{i+1} and x_{i+2}, then the output bit, y_i, is active.

Property 5. If there is a difference either in x_{i+1} or x_{i+2} or in both, then the output is balanced.

If there is a difference in one of these inputs, the output difference will depend on the value of the other non-linear bit. If the value of the other bit is zero, then there is no difference in the output, else if it is 1, then there is a difference. Assume a difference in both inputs. If the input values for $\sim (x_{i+1})$ and x_{i+2} are 00 (11), the other input values are 11 (00), hence the product difference is 1 for these two cases. If the input values of these two bits are 01 (10), then the other values are 10 (01), hence the product difference is 0.

Next we consider the properties when some input differences are balanced.

Property 6. If the difference in x_i is balanced, then the output difference is also balanced (independently of the propagation properties of the other input variables).

Now consider the following two cases with fixed difference in x_i.

Property 7. If either the difference in x_{i+1} or x_{i+2} is balanced, and the other difference is fixed, then the output difference is biased with probability $3/4$.

Property 8. If differences in both x_{i+1} and x_{i+2} are balanced, then the output difference is biased with probability $5/8$.

If differences in both x_{i+1} and x_{i+2} are balanced, the following are the possibilities about their input differences:
00 implies difference zero for 25 percent cases.
01, 10, 11 implies the output is balanced for $25 \times 3 = 75$ percent cases.

So the output difference is zero for $(25+75/2)/100=125/200=5/8$ cases.

Next, we consider some properties from the difference distribution table (DDT) of the S-box of Keccak. The DDT of the Keccac S-box can be found in Appendix B, Table 7 of [13]. We can see an important property as below.

Property 9. In the DDT of the Keccac S-box, every bit is either fixed or balanced.

This follows immediately from the fact that χ is a quadratic function. The difference function for each output bit is linear for any given input difference in χ and is thus either fixed or balanced. For example, when the input difference is $0x01$ (which is used maximally in our analysis), the possible output differences are $0x01$, $0x09$, $0x11$ and $0x19$ each having occurrences of exactly eight times. One can observe that the output difference is always 1 in the 0th bit; it is always 0 on 1st and 2nd bits; and, always balanced on the 3rd and the 4th bits. This property can be easily proven for every case using the boolean equation of the S-box i.e. Equation 1.

4 Application of Propagation Properties: Two Rounds Distinguishers

In this section, we show the first application of the propagation properties as described in the previous section. First we present how to calculate the distinguishers using the active, inactive and the biased bits. The balanced bits seem of no assistance, as they don't remain exactly balanced bits after few rounds.

4.1 Method for Finding Distinguishers

The above properties suggest that there will be fixed and balanced difference bits after a round when we start with a low-weight differential (refer Properties 4, 5, 6). After another round, the balanced difference bits and the fixed difference bits will give rise to biased difference bits (refer Properties 7 and 8). It is evident from all the properties of χ functions that the there are five possibilities of the differential state bits after performing a large number of experiments, namely, always active (m), always inactive (n), biased towards zero or one in 75 percent trials (q), biased towards zero or one in 62.5 percent trials (r) and unbiased (s). We show below how we construct a favorable event for a distinguisher. A suitable favorable event for our distinguisher might be as follows:

1. We have $m + n$ deterministic bits. We require these to have the right value.
2. We have a number q of bits with probability about 0.75 to be either 0 or 1. Then the expected value of bits which take the value as predicted by the bias is $q \cdot p = 0.75 \cdot q$.
3. We have r bits to be either 0 or 1 with prob. about 0.625. Then the expected value of correct bits is $0.625 \cdot r$.

Thus our test first asks if the $(m + n)$ deterministic bits are correct. If no, ignore the rest. If yes, count the numbers of correct bits under the previous conditions 2. and 3. If one or both of these numbers are at least $k_1 = 0.75 \cdot q$ or $k_2 = 0.625 \cdot r$, respectively, we have a desirable event.

According to Binomial distribution, the probability for at least k successes in n trials having a probability p is:

$$P(X \geq k) = \Sigma_{i=k}^{n} C_i^n p^i (1 - p)^{n-i} \tag{2}$$

This can be approximated by a normal distribution. For this we make use of the following formulae for n trials with probability p:
Mean=$\mu = np$, Standard deviation=$\sigma = \sqrt{np(1 - p)}$.
Convert the binomial variable (X) to normal variable by: $Z = (X - \mu)/\sigma$.

Finally, calculate the probability of the event, $P((np - \mu)/\sigma < Z \leq (n - \mu)/\sigma) = P(0 < Z \leq (n - \mu)/\sigma)$ (since, $np = \mu$), from the normal distribution table.

Clearly, if we require all the bits are correct, the probability becomes p^n. We look for a favourable event that is able to distinguish later the hash output from random with a good tradeoff between distinguishing property and number of samples. An example of such a favourable event is given below.

Example 1. Assume that in a large number of experiments, we have got 16 deterministic difference bits (i.e. totally active or inactive bits), 35 bits with a differential bias 0.75 and 31 bits with a differential bias 0.625. Consider now an event to be favorable if all 16 deterministic bits are correct, and at the same time all 35 bits with bias 0.75 are correct. Moreover require that for the bits with bias 0.625, at least $\mu = np = 31 \times 0.625 = 19.375$ bits are correct.

Then, for the 16 deterministic bits and the 35 bits with the bias 0.75, the probability of success for the random case is, $2^{-16-35} = 2^{-51}$, whereas for the biased case is $1 \times 0.75^{35} \approx 2^{-15}$.

According to normal approximation, the probability that at least $\mu = 19.375$ of the 31 bits with a differential bias of 0.625 are correct is $1/2$. If the bits were random, then $\mu = 31 \times .5 = 15.5$ and $\sigma = 2.78$. In that case, the probability of the same number of bits to be correct were: $P(((19.375 - 15.5)/2.78) < Z \leq ((31 - 15.5)/2.78)) = P(1.39 < Z \leq 5.57)$. From a table, $P = 1 - 0.91774 = 0.08226 = 2^{-3.6}$.

Hence, the total probability for the biased case is $2^{-15-1} = 2^{-16}$ whereas for the random case it is $2^{-51-3.6} = 2^{-54.6}$. So, the total probability for the event in the biased case is almost 2^{39} times higher than the random case. This gives a strong distinguishing property for this event.

$\square$

4.2   Two Rounds Distinguishers for Low Weight Differentials

In this section, it will be shown that low-weight differentials can provide distinguishers for two rounds. Let us consider some low weight differentials of weight

w on the message part of the hash function. Now let us see how the differences propagate through different constituent functions of Keccak.

θ **Function of Round 1:** On the θ function of the first round, the differences of weight w, will spread the differences to roughly $11w$ bits if the differences are sparse enough. Hence, on a large number of experiments, the state difference remains fixed after the θ function of the first round.

ρ, π **and ι Functions of Round 1:** These functions keep the weight of the differences.

χ **Function of Round 1:** The χ function will keep the difference fixed (either 0 or 1) for some bits, and make it balanced for some other bits, as follows from Properties 4, 5, 6 and 9.

θ **Function of Round 2:** The balanced bit differences produced in the χ layer in the first round will make 11 bits per balanced bits of the state as balanced. On the other hand, the fixed differences (of value 1) will make 11 bits per state bit (excluding the balanced bits) as fixed differences (of value 1). If all the 11 bits involved in the θ transformation of a particular state bit have no differences (i.e. fixed difference 0) then, that particular state bit difference will be always 0. Since, at the end of round 1, the number of balanced state bits is still small, there are still a lot of fixed difference bits at the end of θ transformation of Round 2.

ρ, π **and ι Functions of Round 2:** These functions keep the weight of the differences.

χ **Function of Round 2:** At this point, the χ function has three types of input state differences. The first one is always 0, the second one is always 1 and the third one is balanced in a large number of experiments. By virtue of the Properties 4 to 9, it produces fixed output differences, balanced output differences and biased output differences in a large number of experiments.

Thus, at the end of Round 2, we still get fixed output differences and biased output differences in the state. If we get some of these conditions in the output part of the hash function, then we will have distinguishers. Now there are special techniques already developed in literature to have low-weight differences up to a certain number of rounds. Using those techniques combined with the propagation analysis shown so far, we will show how to get distinguishers for 3, 4, 5 and 6 rounds in the rest of the paper.

Note that we can expect biased bits starting with a low weight differential only after two rounds but not after one round. Further as we shall see, we cannot expect biased bits after three rounds.

Now, we are able to get biased difference bits after two rounds of Keccak, provided the input difference has relatively small weight. Let us try to quantify how much small i.e. what is the maximum weight that can give rise to the fixed/biased bits after the second round. The χ layer is the one that creates the balanced bits at round 1 and θ layer is the one which spreads the balanced bits at round 2. Let w_{max} be the maximum weight that will produce fixed bits and/or biased bits after round 2. After θ layer, this weight will spread to $11w_{max}$ state bits. In χ layer, every fixed difference bit makes two more bits balanced (by extension of Property 5). Hence after the first χ layer, $22w_{max}$ bits become balanced. After the second θ layer, the total number of bits that becomes balanced are $11 \times 22w_{max} = 242w_{max}$. Again in χ layer, if an input difference bit is balanced, the corresponding output difference bit is also balanced (by property 6). In order that all the 1600 state bits do not become balanced after θ layer of round 2, the maximum value of w_{max} can be only $1600/242 \approx 6$.

In our case, we have always started with a differential of weight 6, and indeed we have got some fixed or biased bits in the output after two rounds. The above heuristics assume that the weight is evenly distributed in the state. However, this is not the case always and some difference bits are clustered (θ and χ, tend to cluster but ρ and π attempt to distribute them evenly). This clustering gives us some more fixed/biased difference bits after round 2. But, most of the bits after round 2 are balanced, hence the θ operation and the χ operation in round 3 spreads the balanced bits to all the state bits. Thus we don't get any more fixed or biased bits after round 3.

4.3   The Effect of ι in Differential Bias Propagation

The round constant, ι, of the last round will not have any effect on the differentials. This is because the round constant is simply XORed at the end and the differences will cancel it out. However, in our biased differential propagation analysis, we have passed through one extra round by value before calculating the final differential for our propagation analysis. Hence, in our method of calculating differential bias propagation, the ι constant will matter for the last but one round. It will not matter for the last round. However, the round constants are different for different rounds and our differential bias propagation analysis are performed in the last two rounds. So, when we consider the differential bias propagation analysis for Keccak permutation for two, three and four rounds, we should consider the ι constants for first, second and third round, respectively. But, when we consider the distinguishers for Keccak hash functions later, we will prepend two more rounds. In that case, we have to consider the right ι constant accordingly (i.e. ι of round 5, for a six round distinguisher) in the subsequent sections.

5   Distinguishers of Six Rounds of Keccak

In this section, we proceed step by step to extend the distinguishers of two rounds using the propagation properties, to three, four and six rounds distinguishers. We extend the two rounds distinguishers for three rounds using the concept of *kernels*. A *kernel* is, as defined by Keccak authors, a state difference with even number of differences in each column. When the state difference is in a kernel, the θ transformation on that state differences will not change it. We further extend this for four rounds by using a *double-kernel*. A *double kernel*, as considered in [17], is a differential path where the state differences of two successive rounds are in a kernel. As a main application, a few of the distinguishers as found for four rounds will be extended to the hash function when reduced to six rounds by using *target-difference-algorithm (TDA)*. *TDA*, as defined in [10], allows to find message pairs which satisfy a given target difference after one Keccak permutation round.

5.1   Extension to 3 Rounds - Starting with a Kernel

We can easily extend the above distinguishers to 3 rounds by starting with a Kernel. If we start with a low difference kernel with the differences in the message part of the hash function, the θ, ρ, π will maintain the same difference; then it is already well-known that the χ layer will maintain this sparse difference with a probability 2^{-2} per difference bit [17]. If there are r difference bits, then the total probability of maintaining the same difference is 2^{-2r}. Thus, at the end of round 1, we get a low-weight difference with a probability 2^{-2r}. This can be now the starting point for the distinguisher of two rounds using our method. Hence, we can easily get three rounds distinguishers using this method with an additional complexity of 2^{-2r} ($r \leq 6$).

5.2   Extension to 4 Rounds - Starting with a Double Kernel

Most of the successful differential paths on various rounds have made use of the double kernels to increase the number of rounds. We also immediately increase the number of rounds for the double kernel. We have used some of the existing two round trails to find distinguishers up to four rounds.

We start with the double kernel that was found in [17]. The double kernel defined by them is shown in Table 2. Starting with this differential, we have performed 2^{20} experiments[1] with two more rounds and checked the output part after four rounds for Keccak-224, Keccak-256, Keccak-384 and Keccak-512 to see if there are biased bits. We have found many bits which are fixed (i.e. always active or always inactive) and many bits which are biased. The results are shown in Table 3 where the entries with 0.75 means that for those many bits, the number of zeros (or ones) occurred 75 or 25 percent times and the entries with 0.625 means that for those many bits, the number of zeros (or ones) occurred 62.5 or 37.5 percent times. These cases represent the biased bits. We can easily employ the methods given in Section 4 (see Example 1) to determine the distinguishers. Note that in [17] already a very efficient distinguisher on the Keccak hash function reduced to four rounds is described. This is based however on a trick using free bits and selecting appropriate messages that is not available when we later prepend two more rounds.

---

[1] We started our analysis with 2^{20} experiments; but later on we found that we can clearly distinguish the biased and fixed bits with 10000 or even 1000 experiments. As computational time difference between 10000 and 1000 experiments was not significant enough, in the subsequent sections we have performed experimentation with 10000 samples.

Table 2. The differential path of [17]

```
        --------------- 1|8 -------------------|-------------------|-------------------|-------------------
        --------------- 1|-------------------- |--------- 4 ------ |-------------------|-------------------
   δ0   --------------- |8 ------------------- |--------- 4 ------ |-------------------|-------------------
        ------------- |------------------------|-------------------|-------------------|-------------------

        --------------- 1|--------------------- |------------ 2 -- |-------------------|-------------------
   δ1   --------------- 1|------- 1----------- |-------------------|-------------------|-------------------
        ------------ |------- 1-------------- |------------ 2 -- |-------------------|-------------------

        --------------- 1|---------------------|-------------------|-------------------|-------------------
        ------------ |-------------------------|--------------- 8 |------------ 2 --- |-----------------
   δ2   ------------ |------------------------- |--- 4 ---------- |--------- 1 ----- |-----------------
        ------------ 8 -|-------------------------|-------------------|-------------------|-------------------
```

Table 3. The biased difference bits after 4-rounds using the differential path of [17]

Variant	Active	Inactive	0.75	0.625
224	7	9	35	31
256	7	11	45	34
384	7	17	77	59
512	10	20	100	73

5.3 Differential Distinguishers with All Possible Double Kernels Up to Weight Six

After finding deterministic and biased bits, we investigated towards finding the best possible results by looking into all possible double kernels upto weight six. As shown in the previous section, we get two rounds distinguishers using the propagation properties only when we start with a differential of weight upto six. Since we start with a kernel with three non-zero columns, each with weight 2, the computational complexity to find all the double kernels is reasonable. The computational complexity to find all the double kernels is: *(all possible combinations of two bits in three different columns)* × *(all possible combinations of three columns in 320 columns)* = $(C_2^5)^3 \times (C_3^{320}) = 5.4 \times 10^9$ ρ and π operations. We found that there exists an "equivalent class" of double kernels. Note that a kernel becomes a double kernel only when a specific permutation of ρ and π rearranges them to another kernel. Now if the difference bits in the original kernel are shifted along the z-direction equally, then ρ will shift those difference bits equally. The lane rearrangement done by the π function will keep the state again in a kernel. Hence, the following property of double kernel can easily be observed.

Property 10. For a low-weight kernel which results into a double-kernel, if the difference bits are equally shifted along the z-axis, then the new kernel also results in a double kernel.

Since there are 64 bits in a lane, the following property follows:

Property 11. The total number of double kernels must be divisible by 64.

Finally, each of these shifted double kernels will provide almost same but shifted propagation properties for the state bits. That means there will be similar number of active, inactive and biased bits in the state after four rounds for each of these double kernels. However, it is not exactly same because of the effect of ι in the last but one round (see Section 4.3). Hence, we give the following propagation property of a double kernel:

Property 12. There exists an equivalent class, with respect to the propagation properties of two additional rounds, of double kernels consisting of 64 members which will give rise to the almost same number of biased differential bits (active, inactive, biased) in the state after four rounds.

We have done an exhaustive search of double kernels and found a total of 512 double kernels in Keccak. Hence, there are a total of eight equivalent classes of double kernels. The double kernels for all the eight cases are given in the Table 4. For each of the double kernels of Table 3, the number of active bits, inactive bits, bits with bias 0.25 (towards 0), 0.75, 0.625, 0.375 and their corresponding μ, σ and Z (refer to Section 4) are given in Table 5. Using Table 4 and the normal distribution table, the distinguishers can be easily calculated as outlined in Section 4.

5.4   A Concrete Six Round Distinguisher of Keccak-224

In this section, we extend our work in the previous section to a six round distinguisher by using the results of [9] and [10]. It can be noted that it is easy to go one round backwards in Keccak *permutation* starting from a low-weight differential. The inverse χ layer maps 1-bit difference to a 1-bit difference with a probability 2^{-2}. Hence, going backwards by one round has a complexity of 2^{-12} for a differential with weight 6, as in our case. But, going back one round in the *hash function* is a challenging task. The main challenge is to keep the capacity bits to zero at the beginning of the hash function. The $\theta - inverse$ function of Keccak propagates too many bits even when we start with a low-weight difference. This problem was solved for the first time (heuristically) in [10] and [9]. They called this algorithm to solve this problem as Target Difference Algorithm (TDA). Prepending one more round, on page 14, Section 5.2 of [9], they say that *"For Keccak-224, the algorithm typically returned an affine subspace of message pairs with dimension of about 100 within one minute"* when they go backwards from a differential of double kernel i.e. with weight six. As done by them, we go back one more round directly with a probability of 2^{-12} and then appeal to their findings on TDA to get the hash inputs (and gain one more round). The two rounds given by the double kernels (which occurs with a probability of 2^{-24}) and additional two rounds to get biased differential bits enable us to get a six round distinguisher. This six rounds distinguisher is the result of a combination of TDA [9], double-kernel concept of [17] and the differential bias analysis of this paper.

Table 4. The equivalent classes of the double-kernels

Sl No	δ_i	Differentials				
1	δ_0	1 8	4			
		1 8	4			
1	δ_1	1	2			
		1 1				
		1	2			
2	δ_0	1			2	
		1	4			
			4		2	
2	δ_1	1				
		1		2		
		1 1		2		
3	δ_0	1	4			
			4	1		
		1		1		
3	δ_1	1	2	2		
			2			
		1		2		
4	δ_0	1		1		
			2	1		
		1	2			
4	δ_1	1			4	
		8			4	
		8				
5	δ_0	1		2		
		1		2		
		1 1				
5	δ_1			2		
			4			
		1	4			
		1		2		
6	δ_0	1			1	
			2		1	
		1	2			
6	δ_1		8	1	4	
				1	4	
			8			
7	δ_0	1			4	
		1	8			
			8		4	
7	δ_1		4			
		2		4		
		2	4	4		
8	δ_0		1	4		
				4	8	
		1			8	
8	δ_1	4	8	2		
			8	2		
		4				

Table 5. The differential biased bits after 4-rounds for all double kernels

Sl.	Variant	Active	Inactive	0.75	$\mu_{.75}$	$\sigma_{.75}$	$Z_{.75}$	.25	$\mu_{.25}$	$\sigma_{.25}$	$Z_{.25}$	0.625	$\mu_{.625}$	$\sigma_{.625}$	$Z_{.625}$	.375	$\mu_{.375}$	$\sigma_{.375}$	$Z_{.375}$
1	224	7	9	9	6.75	1.30	1.73	26	6.50	2.21	8.83	7	4.38	1.28	2.05	24	9.00	2.37	6.32
1	256	7	11	10	7.50	1.37	1.83	35	2.50	2.56	12.69	7	4.38	1.28	2.05	27	10.13	2.52	6.71
1	384	7	17	19	14.25	1.89	2.52	58	4.75	3.30	16.15	13	8.13	1.75	2.79	46	17.25	3.28	8.76
1	512	10	20	20	15.00	1.94	2.58	80	5.00	3.87	19.36	16	10.00	1.94	3.10	57	21.38	3.66	9.75
2	224	4	11	11	8.25	1.44	1.91	24	6.00	2.12	8.49	11	6.88	1.61	2.57	22	8.25	2.27	6.06
2	256	4	12	11	8.25	1.44	1.91	28	2.75	2.29	11.02	13	8.13	1.75	2.79	25	9.38	2.42	6.45
2	384	4	16	15	11.25	1.68	2.24	45	3.75	2.90	14.20	20	12.50	2.17	3.46	39	14.63	3.02	8.06
2	512	4	19	19	14.25	1.89	2.52	58	4.75	3.30	16.15	22	13.75	2.27	3.63	55	20.63	3.59	9.57
3	224	1	7	8	6.00	1.22	1.63	22	5.50	2.03	8.12	7	4.38	1.28	2.05	22	8.25	2.27	6.06
3	256	1	7	9	6.75	1.30	1.73	28	2.25	2.29	11.24	9	5.63	1.45	2.32	25	9.38	2.42	6.45
3	384	4	13	15	11.25	1.68	2.24	41	3.75	2.77	13.43	21	13.13	2.22	3.55	34	12.75	2.82	7.53
3	512	6	15	24	18.00	2.12	2.83	57	6.00	3.27	15.60	24	15.00	2.37	3.79	51	19.13	3.46	9.22
4	224	0	4	11	8.25	1.44	1.91	20	5.00	1.94	7.75	12	7.50	1.68	2.68	24	9.00	2.37	6.32
4	256	0	5	13	9.75	1.56	2.08	24	3.25	2.12	9.78	14	8.75	1.81	2.90	26	9.75	2.47	6.58
4	384	4	8	19	14.25	1.89	2.52	36	4.75	2.60	12.03	19	11.88	2.11	3.38	44	16.50	3.21	8.56
4	512	5	13	25	18.75	2.17	2.89	49	6.25	3.03	14.10	22	13.75	2.27	3.63	59	22.13	3.72	9.92
5	224	4	4	8	6.00	1.22	1.63	32	8.00	2.45	9.80	5	3.13	1.08	1.73	29	10.88	2.61	6.95
5	256	4	4	11	8.25	1.44	1.91	36	2.75	2.60	12.80	6	3.75	1.19	1.90	33	12.38	2.78	7.42
5	384	8	8	19	14.25	1.89	2.52	53	4.75	3.15	15.31	12	7.50	1.68	2.68	45	16.88	3.25	8.66
5	512	9	13	23	17.25	2.08	2.77	69	5.75	3.60	17.58	22	13.75	2.27	3.63	58	21.75	3.69	9.83
6	224	0	2	12	9.00	1.50	2.00	18	4.50	1.84	7.35	14	8.75	1.81	2.90	27	10.13	2.52	6.71
6	256	0	2	14	10.50	1.62	2.16	20	3.50	1.94	8.52	16	10.00	1.94	3.10	29	10.88	2.61	6.95
6	384	2	4	18	13.50	1.84	2.45	34	4.50	2.52	11.68	27	16.88	2.52	4.02	40	15.00	3.06	8.16
6	512	3	4	23	17.25	2.08	2.77	44	5.75	2.87	13.32	34	21.25	2.82	4.52	54	20.25	3.56	9.49
7	224	3	11	11	8.25	1.44	1.91	26	6.50	2.21	8.83	8	5.00	1.37	2.19	28	10.50	2.56	6.83
7	256	3	11	13	9.75	1.56	2.08	31	3.25	2.41	11.51	10	6.25	1.53	2.45	28	10.50	2.56	6.83
7	384	6	16	21	15.75	1.98	2.65	49	5.25	3.03	14.43	16	10.00	1.94	3.10	39	14.63	3.02	8.06
7	512	8	24	27	20.25	2.25	3.00	66	6.75	3.52	16.84	20	12.50	2.17	3.46	51	19.13	3.46	9.22
8	224	4	1	8	6.00	1.22	1.63	18	4.50	1.84	7.35	10	6.25	1.53	2.45	20	7.50	2.17	5.77
8	256	5	1	9	6.75	1.30	1.73	19	2.25	1.89	8.87	13	8.13	1.75	2.79	24	9.00	2.37	6.32
8	384	7	4	12	9.00	1.50	2.00	27	3.00	2.25	10.67	20	12.50	2.17	3.46	36	13.50	2.90	7.75
8	512	8	4	21	15.75	1.98	2.65	37	5.25	2.63	12.05	24	15.00	2.37	3.79	50	18.75	3.42	9.13

Now let us calculate the complexity of these distinguishers. The double-kernel has a probability of 2^{-24}. Going backwards by one round will have a complexity of 2^{-12}. Hence, the total complexity of the differential path for this distinguisher is 2^{-36}. To have a distinguisher of six rounds using our propagation analysis, the total probability of a suitably chosen favorable event should be larger than the probability of the event if it were purely random. For the double kernel of [17] which corresponds to the entry with Sl. No. 1 in Table 3 and 4, such a favorable event is shown in Example 1. We found that the probability of the event using our propagation analysis is 2^{-16}. Multiplying by the probability of the characteristic, we get the total probability of this event as $2^{-16-36} = 2^{-52}$, whereas in the random case, the probability is $2^{-54.6}$. Since, the probability of the event is at least four times higher than the random case, we have a distinguisher for this event. The complexity of the distinguisher is about 2^{52}.

Clearly, we can have stronger distinguishers with an additional complexity by repeating the experiment or by considering an event with more bits with probability 0.625 to be correct.

Since the differential path for the six round distinguisher itself has a complexity of 2^{36}, we cannot get any favorable event for some cases in Table 4 and Table 5. But for 3 out of 8 equivalent cases, we can get a favorable event leading to a distinguisher. The cases for which we have a distinguisher are with Sl. No. 1, 2 and 7 in Table 4 and 5.

This distinguisher does not carry over to a six round distinguisher of Keccak-256 without any further refinements (e.g., message modification), as the space of message pairs delivered by TDA as reported in [9] in this case may not always be large enough.

The methods as developed for ordinary differentials carry over to generalized internal differentials as brought up in [11]. However regarding distinguishers they appear to lead to no better results. This can also be extended for special differential characteristics as in [8]. Here again it doesn't appear to lead to better results regarding distinguishers.

6   A Second Application of Propagation Properties

As a final note, please observe that the difference bits in the near collision found in [10],[9] do not follow the difference bits predicted by the trail (see Appendix B of [10],[9]). Table 5 gives the 4-round characteristic that has been used to get near collisions. The near collision they have got are (the difference nibbles are shown in underline): For Keccak-256:

Output1=
407D4466 FEA8B231 EC9$\underline{6}$8181 $\underline{D}$F902165 23C$\underline{2}$19FF $\underline{54}$571D7$\underline{0}$ 2800F5$\underline{0}$6 E81$\underline{8}$6$\underline{44}$B
Output2−
407D4466 FEA8B231 EC9$\underline{2}$8181 $\underline{F}$F902165 23C$\underline{0}$19FF $\underline{1C}$571D7$\underline{4}$ 2800F5$\underline{1}$6 E81$\underline{0}$6$\underline{56}$B
For Keccak-224:
Output1=
85373497 97D871C2 FBD0A823 $\underline{5}$84C0ED4 $\underline{C}$1B3BF$\underline{4}$F BC4087$\underline{6}$6 0584B08D
Output2=
85373497 97D871C2 FBD0A823 $\underline{7}$84C0ED4 $\underline{E}$1B$\underline{1}$BF$\underline{5}$F BC4087$\underline{7}$6 0584B08D

Note that the output strings given here are with 32 bits whereas the differentials given in Table 5 are 64 bits; both with little-endian format. For Keccak-256, the bits that are different in the near-collision are (starting the bit numbering from 0), 82(0x52), 125(0x7D), 145(0x91), 162(0xA2), 187(0xBB), 190(0xBE), 196(0xC4), 229(0xE5), 232(0xE8), 243(0xF3). For Keccak-224, the positions are, 125 (0x7D), 132(0x84), 145(0x91), 157(0x9D), 164(0xA4). Clearly, the near collision found does not coincide with the difference bits predicted by the characteristic (as given in δ_4 of Table 5). The authors did not give an explanation for that. Here, we give an explanation.

We started with δ_3 of Table 6 and checked the number of difference bits in the output part of Keccak-256 (automatically implying Keccak-224 as well). Table 7 gives the number of times the bits were different in the first 320 bits of the state which includes the output part i.e. the first 256 bits of the state (the rows indicate the first nibble and the columns indicate the second nibble of the output state bit position). Note that the first 320 bits are the first row of the state along the 64-bit lanes. We can observe that even with the differential biases, we can't explain the near collision completely. For example, the state bit at position 0x52 is always zero, but this bit shows up in the near collision of [9] as a difference

Table 6. The 4-Round Characteristic Leading to Near Collision in [9]

δ_0	BD135E2FA6BD1346	12D789A92F12D78F	D7E26BC344D7E224	E69AF134B5E69AD5	98BC4D6BF898BC58
	BD135E2FA6BD1346	12D789A82F12D78F	D7E26BC344D7E264	E69AF134B5E69AD5	98BC4D6BF898BC58
	BD135E2FA6BD1346	12D789AB2F12D78F	D7E26BC344D7E224	E69AF134B5E29AD5	98BC4D6BF898BC58
	BD135E2FA6BD1346	12D789A92F12D78F	D7E26BC344D7E224	E69AF134B5E69AD5	98BC4D6BF898BC58
	BD135E2FA6BD1346	12D789A92F12D78F	D7E26BC344D7E224	E29AF134B5E69AD5	98BC4D6BF898BC58
δ_1	--------------	----------1---	--------------	--------------	---4----------
	--------------	----------1---	----8---------	--------------	--------------
	--------------	--------------	----8---------	--------------	---4----------
δ_2	----------2---	--------------	------4-------	--------------	--------------
	----------2---	--------------	--------------	-4-----------	--------------
	----------2---	--------------	------4-------	-4-----------	--------------
δ_3	--------------	--------------	--------------	--------8---	--------------
	--------------	--------------	-------1----	--------------	--------------
	--------------	--------------	-------8----	--------------	--------------
	--------------	--------------	--------------	-------2----	--------------
	-------1-----	--------------	--------------	--4---------	--------------
δ_4	--------------	----------2---	4 8----4---2----	4---12--------	---8---8 2-----1-
	----98------	-2---2-8-----4--	--------------	4------------	--1----8----2---
	--------------	2---1--------	---1 2--------	4---2----2-8----	-------4------
	---1-4----2---1	--------------	------8------	--2----8-----4--	----4--------9--
	--2---1-----4---	------------4 8-	1-4-----2---1---	--------------	----------8----

Table 7. The Difference Distribution in the first 320 bits after Extending One Round from δ_3 of the Table 6

	0	1	2	3	4	5	6	7	8	9	A	B	C	D	E	F
0	0	0	0	0	0	0	0	0	0	0	0	0	0	0	0	0
1	0	5039	0	0	0	0	0	0	0	0	0	0	0	0	0	0
2	0	0	5096	0	0	0	0	0	0	0	0	0	0	0	0	0
3	0	0	0	0	0	0	0	0	0	0	0	5019	0	4984	4982	0
4	0	0	0	0	0	0	0	0	0	0	0	0	0	0	0	0
5	0	4982	0	0	0	0	0	0	0	0	0	0	0	0	0	0
6	0	0	5005	0	0	4953	0	0	4993	0	0	0	0	0	0	0
7	0	0	0	0	0	0	0	0	0	0	4942	4978	0	10000	4996	0
8	0	0	0	0	5089	0	0	0	0	0	0	0	0	0	0	0
9	0	10000	0	0	0	0	0	0	0	0	0	0	0	5011	0	0
A	0	0	10000	4925	0	5030	0	0	4971	0	0	0	0	0	0	0
B	0	0	0	5006	0	0	0	0	0	0	5068	10000	0	0	10000	0
C	0	0	0	0	5055	0	0	0	0	0	0	0	0	0	0	0
D	0	0	0	0	0	0	0	0	0	0	0	0	0	4977	0	0
E	0	0	0	4984	0	10000	0	0	10000	0	0	0	0	0	0	0
F	0	0	0	4935	0	0	0	0	0	0	10000	0	0	0	0	0
10	0	0	0	0	10000	0	0	0	0	0	0	0	0	0	0	0
11	0	0	0	0	0	0	0	0	0	0	0	0	0	10000	0	0
12	0	0	10000	0	0	0	0	0	0	0	0	0	0	0	0	0
13	0	0	10000	0	0	0	0	0	0	0	0	0	0	5024	0	0

Table 8. The δ_3 of Table 6 with the difference bit in row 3, column 4 mapping to 11 in χ layer

```
      ------------------------------------------------------- 8 ---------------
      ------------------------------------------ 1 ----|----------------------
δ3 -- ------------------------------------- 8 ----|----------------------------
      ---------------------------------- 2 --------------- 2 -----------------
      ----------- 1 ----|------------------------- 4 -----------------------
```

Table 9. The Difference Distribution in the first 320 bits after Extending One Round from δ_3 of the Table 8

	0	1	2	3	4	5	6	7	8	9	A	B	C	D	E	F
0	0	0	0	0	0	0	0	0	0	0	0	0	0	0	0	0
1	0	5039	5027	0	0	0	0	0	0	0	0	0	0	0	0	0
2	0	0	5096	0	0	0	0	0	0	0	0	0	0	0	0	0
3	0	0	0	0	0	0	0	0	0	0	0	5019	0	4984	4982	0
4	0	0	0	0	0	0	0	0	0	0	0	0	0	0	0	0
5	0	4982	10000	0	0	0	0	0	0	0	0	0	0	0	0	0
6	0	0	5005	0	0	4953	0	0	4993	0	0	0	0	0	0	0
7	0	0	0	0	0	0	0	0	0	0	0	4978	0	10000	4996	0
8	0	0	0	0	5089	0	0	0	0	0	0	0	0	0	0	0
9	0	10000	0	0	0	0	0	0	0	0	0	0	0	5011	0	0
A	0	0	10000	4925	0	5030	0	0	4971	0	0	0	0	0	0	0
B	0	0	0	5006	0	0	0	0	0	0	0	10000	0	0	10000	0
C	0	0	0	0	5055	0	0	0	0	0	0	0	0	0	0	0
D	0	0	0	0	0	0	0	0	0	0	0	0	0	4977	0	0
E	0	0	0	4984	0	10000	0	0	10000	0	0	0	0	0	0	0
F	0	0	0	4935	0	0	0	0	0	0	0	0	0	0	0	0
10	0	0	0	0	10000	0	0	0	0	0	0	0	0	0	0	0
11	0	0	5111	0	0	0	0	0	0	0	0	0	0	10000	0	0
12	0	0	0	10000	0	0	0	0	0	0	0	0	0	0	0	0
13	0	0	0	10000	0	0	0	0	0	0	0	0	0	5024	0	0

bit. Now, take a look into the DDT of the S-box as given in [13]. Notice that for a single bit difference, there are only four possible outputs; each happens with the probability 2^{-2}. We found that the actual path taken in δ_3 of Table 6 is not a single bit difference mapping to a single bit difference in all the cases. The difference bit in row 3, column 3 (starting count with 0) of δ_3 of Table 6 actually mapped to a two bit difference in the output. The corresponding δ_3 is shown in Table 7. With this δ_3, we have performed experimentation with 10000 samples with one more round and Table 9 gives the number of output bits that were different in the first 320 state bits. Now we see that state bit position 0x52 is always active. This is the case for all the state bits mentioned above as the difference bits in near collision except the bit positions 0xC4 and 0xF3 where bits were active for 50 percent times. Recall that in Property 5, we have stated that if x_{i+1} is always different, then x_i is balanced in the χ layer. If we look for the next bit along the row for these two bit positions, we can see that both the bit positions (0xC4+0x40=)0x104 and (0xF3+0x40=)0x133 are always active;

hence bit positions 0xC4 and 0xF3 are balanced. Hence we can explain the near collision found in [9].

7   Conclusion

We have analysed the propagation properties of Keccak constituent functions. For low weight input differences this enables to derive a number of fixed or strongly biased difference bits after two rounds. Combined with the concept of double kernel [17] this leads to several differential distinguishers over four rounds of Keccak. Some of these distinguishers are flexible enough to be extended via the TDA algorithm in [10] to efficient differential distinguishers of the Keccak hash function when reduced to six rounds, despite the quite low probabilities of individual characteristics. We have discovered a few properties of Keccak that contribute to a better understanding of this hash function. The results found in this paper pose no threat to the security of full round Keccak.

Acknowledgement. We are grateful to the anonymous reviewers for their comments that have helped to improve the presentation of this paper.

References

1. Aumasson, J.P., Meier, W.: Zero-sum distinguishers for reduced Keccak-f and for the core functions of Luffa and Hamsi. NIST Mailing List (2009)
2. Bellare, M., Rogaway, P.: Random Oracles are Practical: A Paradigm for Designing Efficient Protocols. In: CCS, Proceedings of the 1st ACM Conference on Computer and Communications Security, pp. 62–73. ACM (1993)
3. Bernstein, D.J.: Second preimages for 6 (7?(8??)) rounds of keccak? NIST Mailing List (2010)
4. Bertoni, G., Daemen, J., Peeters, M., Van Assche, G.: The Keccak SHA-3 submission. Submission to NIST, Round 3 (2011)
5. Boura, C., Canteaut, A.: Zero-Sum Distinguishers for Iterated Permutations and Application to Keccak-f and Hamsi-256. In: Biryukov, A., Gong, G., Stinson, D.R. (eds.) SAC 2010. LNCS, vol. 6544, pp. 1–17. Springer, Heidelberg (2011)
6. Boura, C., Canteaut, A., De Cannière, C.: Higher Order Differential Properties of Keccak and Luffa. In: Joux, A. (ed.) FSE 2011. LNCS, vol. 6733, pp. 252–269. Springer, Heidelberg (2011)
7. Chang, S., Perlner, R., Burr, W.E., Turan, M.S., Kelsey, J.M., Paul, S., Bassham, L.E.: Third-Round Report of the SHA-3 Cryptographic Hash Algorithm Competition (2012),
http://csrc.nist.gov/groups/ST/hash/sha-3/Round3/documents/
Round3ReportNISTIR7896.pdf
8. Daemen, J., Van Assche, G.: Differential Propagation Analysis of Keccak. In: Canteaut, A. (ed.) FSE 2012. LNCS, vol. 7549, pp. 422–441. Springer, Heidelberg (2012)
9. Dinur, I., Dunkelman, O., Shamir, A.: Improved Practical Attacks on Round-Reduced Keccak. To appear in Journal of Cryptology

10. Dinur, I., Dunkelman, O., Shamir, A.: New Attacks on Keccak-224 and Keccak-256. In: Canteaut, A. (ed.) FSE 2012. LNCS, vol. 7549, pp. 442–461. Springer, Heidelberg (2012)
11. Dinur, I., Dunkelman, O., Shamir, A.: Collision Attacks on Up to 5 Rounds of SHA-3 Using Generalized Internal Differentials. In: FSE 2013. LNCS (2013)
12. Duan, M., Lai, X.: Improved Zero-Sum Distinguisher for Full Round Keccak-f Permutation. Cryptology ePrint Archive, Report 2011/023 (2011)
13. Duc, A., et al.: Unaligned Rebound Attack – Application to Keccak, http://eprint.iacr.org/2011/420
14. Morawiecki, P., Pieprzyk, J., Srebrny, M., Straus, M.: Preimage attacks on the round-reduced Keccak with the aid of differential cryptanalysis. Cryptology ePrint Archive, http://eprint.iacr.org/2013/561.pdf
15. Morawiecki, P., Pieprzyk, J., Srebrny, M.: Rotational cryptanalysis of round-reduced Keccak. In: FSE (2013), http://eprint.iacr.org/2012/546.pdf
16. National Institute of Standards and Technology. FIPS 180-1: Secure Hash Standard (April 1995), http://csrc.nist.gov
17. Naya-Plasencia, M., Röck, A., Meier, W.: Practical Analysis of Reduced-Round Keccak. In: Bernstein, D.J., Chatterjee, S. (eds.) INDOCRYPT 2011. LNCS, vol. 7107, pp. 236–254. Springer, Heidelberg (2011)
18. Peyrin, T.: Improved Differential Attacks for ECHO and Grøstl. In: Rabin, T. (ed.) CRYPTO 2010. LNCS, vol. 6223, pp. 370–392. Springer, Heidelberg (2010)
19. Rivest, R.L.: The MD5 message-digest algorithm. Request for Comments (RFC) 1320, Internet Activities Board, Internet Privacy Task Force (April 1992)
20. Wang, X., Yin, Y.L., Yu, H.: Finding Collisions in the Full SHA-1. In: Shoup, V. (ed.) CRYPTO 2005. LNCS, vol. 3621, pp. 17–36. Springer, Heidelberg (2005)
21. Wang, X., Yu, H.: How to Break MD5 and Other Hash Functions. In: Cramer, R. (ed.) EUROCRYPT 2005. LNCS, vol. 3494, pp. 19–35. Springer, Heidelberg (2005)

Practical Distinguishers against 6-Round Keccak-f Exploiting Self-Symmetry

Sukhendu Kuila[1], Dhiman Saha[2],
Madhumangal Pal[1], and Dipanwita Roy Chowdhury[2]

[1] Department of Mathematics, Vidyasagar University, India
{babu.sukhendu,mmpalvu}@gmail.com
[2] Department of Computer Science and Engineering, IIT Kharagpur, India
{dhimans,drc}@cse.iitkgp.ernet.in

Abstract. This paper presents new distinguishers against Keccak-f[1600] permutation reaching up to 6-rounds. The main intuition is to exploit the self-symmetry of the internal state of Keccak. Formal analysis reveals that the proposed distinguisher can penetrate up to 3 rounds and the penetration depends only on the hamming weight of the round-constant of the initial round. New strategies developed in this work, when combined, are shown to distinguish up to 5-rounds with a probability of 1 using a *single* query. Finally, the extension to 6-rounds with a complexity of 2^{11} gives us the most efficient 6-round distinguisher reported in literature. All claims and formal arguments conform to the results obtained by extensive experimentation.

Keywords: distinguisher, keccak, hash function analysis, internal differentials, self-symmetry.

1 Introduction

In the last 5 years, the cryptographic community has seen remarkable progress in the design and analysis of hash functions and the credit mainly goes to the introduction of the SHA-3 contest by NIST following the concerns over the security flaws in SHA-1 and SHA-2. The contest declared 5 finalists and in October 2012 announced **KECCAK** as the next SHA-3 standard. The hash function Keccak [3], has a permutation based internal function and employs the sponge-construction [2] as the mode of operation. Keccak has shown great strength against all classical and state-of-the-art cryptanalytic techniques. So far we have seen several directions on the cryptanalysis of Keccak. By exploiting low algebraic degrees of Keccak, attacks such as [9,1] have been reported. While [9] requires very high time complexity, [1] requires heavy memory overhead along with marginally less than brute-force time complexity. Classical differential propagation in Keccak internal permutation has been done by the designers in [5]. Another direction to cryptanalyze Keccak is based on some heuristic methods. Here the attack is conducted on experimental basis. In such an attack [13], SAT-SOLVER was used to obtain second-preimages on Keccak. Similarly another

D. Pointcheval and D. Vergnaud (Eds.): AFRICACRYPT 2014, LNCS 8469, pp. 88–108, 2014.

heuristic approach showing collision attack which exploits differential as well as algebraic properties was given by Dinur et al. using the Target Difference Algorithm [7]. There have been attempts to analyze Keccak by also exploiting structural properties such as Cube attack in [11]. Rotational cryptanalysis [12] and internal differential cryptanalysis [6] have also emerged to be powerful tools in the analysis of Keccak. The attacks reported on Keccak can be broadly classified into three types, namely, pre-image, collision and distinguishing attacks. In this paper, we deal with the third type of attack that aims at distinguishing the Keccak internal permutation function from a truly random permutation, thereby showing non-random behavior.

Most of the previous cryptanalysis results on Keccak that try to distinguish it are based on finding high probability differential paths [14,10]. In this respect, internal differential attack [15] is a relatively new type of cryptanalytic method extensively used in [6] for Keccak. Here also the main focus is to obtain *subset characteristic* which holds with high probability. On the other hand, in [12], rotational cryptanalysis is applied on Keccak which does not require any differential or rotational path. The only deterministic cryptanalysis which does not require any differential path/subset characteristic is zero-sum distinguisher [8]. However, the complexity of the distinguishing attack is very high. In this paper we first develop a distinguisher that requires a single query to the permutation oracle to succeed. Later, we augment it with some other strategies to reach a higher number of rounds. This work is primarily based on the evolution of self-symmetry through each round of the Keccak permutation. The idea of exploiting the internal relations of a state was introduced by Peyrin [15] in Crypto 2010 and generalized by Dinur [6] et al. in FSE 2013. However, in [6], the authors exploit the property to find characteristic which lead to collision attacks. To the best of our knowledge, our work is the first attempt to use the self-symmetry property to devise distinguishers against Keccak. A comparison with other contemporary distinguishing attacks on Keccak is given in Table 1. Our main contribution can be summarized as follows:

- Formalize the existence of self-symmetric properties through any 3-rounds of Keccak and use it to devise the basic distinguisher which works with a single query.
- Extend the self-symmetric property probabilistically to augment the basic distinguisher by using a special property of the θ-operation in the Keccak round function.
- Exploit the structural properties of Keccak permutation, to find an elegant state construction technique which extends the distinguisher one more round with the same complexity.
- Finally, we simplify the Target Internal Difference Algorithm [6] proposed by Dinur et. al., to further extend the attack to 6-rounds.
- Distinguishers up to 5-rounds succeed with a probability 1 and the 6-round distinguisher works with a theoretical complexity of 2^{11}.

The rest of the paper is organized as follows: The basic notations and definitions are given in Section 2. The different distinguishing strategies developed

Table 1. Summary of distinguishing attacks on Keccak

Reference	Exploiting property	Rounds	Complexity
[8]	Zero sum	24	2^{1579}
[14]	Differential path	4	$2^{25} + 2N, N \geq 1$
[12]	Rotational symmetry	4	$2^{8.6}$
[10]	Differential path	4	2^2
Our	Self-symmetry	4	**Single Query**
[12]	Rotational symmetry	5	2^{12}
[10]	Differential path	5	2^8
Our	Self-symmetry	5	**Single Query**
[10]	Differential path	6	2^{32}
Our	Self-symmetry	6	2^{11}

are illustrated individually in Section 3. The experimental results showcasing the attacks are detailed in Section 4. Section 5 gives the concluding remarks.

2   Notations

In this work we consider the largest variant of the Keccak-f permutation i.e., Keccak-f[1600] and denote it by $\mathcal{K}$, $\mathcal{K}^n$ signifying n rounds of Keccak where $1 \leq n \leq 24$. The entire state, denoted by $\mathcal{S}$, is visualized as a collection of 64 ordered *slices*[1], 64 being the lane-size. We next introduce concept of a substate followed by the definition of an index which is vital in modeling the type of symmetry that a given state $\mathcal{S}$ exhibits.

Definition 1. *A **Substate**[2] (σ) of the Keccak-f[1600] state $\mathcal{S}$ is a collection of n consecutive slices such that $n|64$.*

Definition 2. *A 1600-bit Keccak state $\mathcal{S}$ is called **Self-Symmetric** ($\mathcal{S}^\eta$), if all its substates are equal. This precludes that the substates must be equal-sized. The size of each-substate (i.e, the dimension in z-axis) is denoted by η. It is clear that $\eta \mid 64$ and so $\eta \in \{1, 2, \cdots, 32\}$. For a self-symmetric state, for each substate $(\sigma_i), i \in \{0, 1, \cdots, \frac{64}{\eta} - 1\}$, we have $(\sigma_0 = \sigma_1 = \cdots = \sigma_{\frac{64}{\eta}-1})$. The parameter η is called the **Self-Symmetry Index**. For arbitrary self-symmetric state $\mathcal{S}^\eta$*

$$\mathcal{S}^\eta = \overset{64/\eta}{\underset{i=1}{\bigparallel}} \sigma_i, \text{ where } \begin{cases} \sigma_0 = \sigma_1 = \cdots = \sigma_{64/\eta-1} \\ \sigma_i \xleftarrow{R} U_{25 \times \eta} \end{cases}$$

$\parallel$ *is the concatenation operator. U_n represents a **Uniform Distribution** of all possible 2^n strings over $\{0,1\}^n$. The co-ordinate relation between the state and the substates is as follows:*

---

[1] The definition of a slice is consistent with the Keccak submission document and is defined as a 5×5 matrix of bits.

[2] The concept of a substate is similar to the concept of *Consecutive Slice Sets* given by Dinur et al. in [6].

$$\mathcal{S}(x, y, z) = \sigma_{\lfloor \frac{z}{\eta} \rfloor}(x, y, z \bmod \eta)$$

It follows from Definition 2 that for a fixed η, we can have $2^{25 \times \eta}$ possible self-symmetric states. The set $\{\mathcal{S}^\eta\}$ represents the set of all self-symmetric states with Self-Symmetry Index $= \eta$. One can note that η actually signifies the *granularity* of the self-symmetry available in the entire state of Keccak-f permutation.

Definition 3. *Self-Symmetry Class* $(\omega_{x,y,z})$ *of a state* $\mathcal{S}^\eta$ *refers to the set of points in the same lane* (x, y), $\{(x, y, z_1), (x, y, z_2), \dots, (x, y, z_{\frac{64}{\eta}})\} \in \mathcal{S}^\eta$ *whose z-coordinates form the residue class* **z** *modulo* η *i.e,* $(z_1 \equiv z_2 \equiv \cdots \equiv z_{\frac{64}{\eta}} \equiv z \bmod \eta)$. *It is clear that each point in a fixed* $\omega_{x,y,z}$ *comes from a different substate and consequently* $|\omega_{x,y,z}| = \frac{64}{\eta}$.

$$\omega_{x,y,z} = \{(x, y, z') \in \mathcal{S}^\eta | z' \equiv z \bmod \eta\}, \ where \ \begin{cases} x, y \in \{0, 1, \cdots, 4\}, \\ z \in \{0, 1, \cdots, \eta - 1\} \end{cases}$$

In this work we study the behavior of these Self-Symmetry Classes under the influence of Keccak round operations. Moreover, we are interested in all values of η, as it magnifies our degrees of freedom. We can now bring into the picture, the well-known idea of *internal difference* which is the basis of the distinguisher proposed in this paper.

Definition 4. *The* **Internal Difference** $(\Delta_{i,j})$ *of a state is the point-wise XOR of its substates* σ_i *and* σ_j :

$$\Delta_{i,j} = \Delta_{j,i} = \sigma_i \oplus \sigma_j, \ where \ \begin{cases} i \neq j, \\ i, j \in \{0, 1, \cdots, \frac{64}{\eta} - 1\} \end{cases}$$

For a Self-Symmetric State, $\Delta_{i,j} = \mathbf{0}, \forall (i, j)$

Definition 5. *The quantity* $p_{x,y}^\eta(z, z')$ *is the probability that for the pair of substates* $\left(\sigma_{\lfloor \frac{z}{\eta} \rfloor}, \sigma_{\lfloor \frac{z'}{\eta} \rfloor}\right)$ *where* $z \neq z'$, *we have* $\sigma_{\lfloor \frac{z}{\eta} \rfloor}(x, y, z \bmod \eta) = \sigma_{\lfloor \frac{z'}{\eta} \rfloor}(x, y, z' \bmod \eta)$. *i.e., the internal difference* $\Delta_{\lfloor \frac{z}{\eta} \rfloor, \lfloor \frac{z'}{\eta} \rfloor}(x, y, z \bmod \eta) = 0$.

$$p_{x,y}^\eta(z, z') = \Pr[\sigma_{\lfloor \frac{z}{\eta} \rfloor}(x, y, z \bmod \eta) = \sigma_{\lfloor \frac{z'}{\eta} \rfloor}(x, y, z' \bmod \eta)], \ where \ \begin{cases} x, y \in \{0, 1, \cdots, 4\}, \\ z, z' \in \{0, 1, \cdots, 63\} \end{cases}$$

Definition 6. *A* **Symmetric-Pair** $(x, y, z, z')^\eta$ *is a pair of co-ordinates* $\{(x, y, z), (x, y, z')\}$ *at which the internal symmetry of the initial state* $(\mathcal{S}^\eta)$ *is still* [3] *preserved. For all symmetric-pairs* $(x, y, z, z')^\eta$ *of a state,* $p_{x,y}^\eta(z, z') = 1$. *Each point of a symmetric-pair is called a* **Symmetric-Point** *and is denoted by* $(x, y, z)^s$.

---

[3] It could have been that due to the cumulative effect of the Keccak internal operations, the point had lost its symmetry.

For a self-symmetric state, any pair of points belonging to the same self-symmetry class will form a symmetric-pair i.e., for $\mathcal{S}^{\eta}$, $p_{x,y}^{\eta}(z, z') = 1 \; \forall \{(x, y, z), (x, y, z')\} \in \omega_{x,y,\text{zmod } \eta}$.

Definition 7. *The **Kernel** $(\mathcal{C}_n^{\eta})$ of a state is the set of all symmetric-pairs of the state after n rounds of Keccak permutation i.e., $\forall \{(x, y, z), (x, y, z')\} \in \mathcal{C}_n^{\eta}$, $p_{x,y}^{\eta}(z, z') = 1$.*

Later in the paper we furnish theoretical analysis on the existence of such Kernels and then devise ways to find them for different rounds of Keccak. We next give a measure of the asymmetry exhibited by a state. This helps us to capture the asymmetry induced in a state by the various operations in the Keccak permutation. In the next section we develop a formal argument to quantify the dispersion of the asymmetry in a single Keccak round.

Definition 8. *The **Asymmetry** of a state $\mathcal{S}$, denoted by $\mathcal{N}_a^{\eta}$, is the cardinality of the set of co-ordinates which do **not** belong to any symmetric-pair. For such a point (x, y, z) of the state, $p_{x,y}^{\eta}(z, z') = 0$ w.r.t to any symmetric-point $(x, y, z')^s$ which belongs to the same self-symmetry class i.e,. $\{(x, y, z), (x, y, z') \in \omega_{x,y,\text{zmod } \eta}\}$.*

$$\mathcal{N}_a^{\eta}(\mathcal{S}) = |\mathcal{A}|, \text{ where } \mathcal{A} = \{(x, y, z) : p_{x,y}^{\eta}(z, z') = 0, \forall (x, y, z')^s \in \omega_{x,y,z \text{ mod } \eta}\}$$

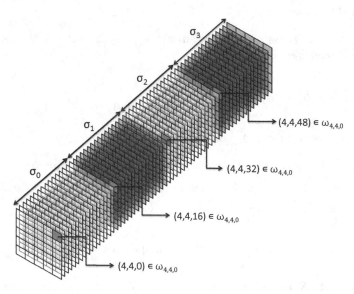

Fig. 1. A self-symmetric state with $\eta = 16$

The relation between a self-symmetric state and its substates and the concept of the self-symmetry classes is illustrated in Fig. 1. Here we see a self-symmetric state with $\eta = 16$. There are $(\frac{64}{\eta}) = 4$ substates σ_i. Four points are highlighted to convey that they belong to the same self-symmetry class $\omega_{4,4,0}$.

3 Distinguishing Strategies Exploiting Self-Symmetry

Distinguishing attacks try to characterize the output distribution of a crypto-graphic construction. In the ideal setting, the output of a cryptographic primitive ($\mathcal{F}$) is said to be *pseudorandom* if it is impossible to distinguish in polynomial time between interactions with $\mathcal{F}$ and interactions with a truly random permu-tation. Thus, the primary aim of distinguishing attacks is to nullify this claim by showing that the output of $\mathcal{F}$ is non-random. In this work, we try to devise dis-tinguishers against various round-reduced versions of the Keccak permutation. In order to do so we exploit various properties of the Keccak round functions to develop three new strategies and modify an existing one. Combining these strategies together we are able to distinguish up to 6-rounds of Keccak. Interest-ingly most of the distinguishers succeed with a probability 1 using only a single query. In the next subsections we illustrate all the strategies one by one.

3.1 The Kernel Strategy

The Kernel strategy is the most important weapon that helps us to penetrate any three rounds of Keccak. It exploits the self-symmetry available in an initial state. In particular, it tries to track how the symmetry of a state changes as it passes through different rounds of the Keccak permutation. We, first, give formal arguments on quantifying the diffusion of asymmetry in one round of Keccak.

Analysis of Diffusion of Asymmetry in One Round of Keccak. We now analyze how different operations of the Keccak-f contribute in dispersing the asymmetry at the input to the output of a single round.
- *theta* (θ) : The operation θ has a maximum diffusion of 11, so it disperses the asymmetry at its input to a maximum of 11 times.
- *rho and pie* (ρ, π) : The mappings ρ, π only translate the bit positions and hence the number of asymmetric bits remains the same.
- *chi* (χ) : The maximum diffusion of χ is 3. We measured the average diffusion of χ using computer experiments by observing the hamming weight of output difference of χ for 1-bit change in the input. The average diffusion of χ can be computed from Table 2 as : $\frac{1 \times 40 + 2 \times 80 + 3 \times 40}{40 + 80 + 40} = \frac{320}{160} = 2$. So, after χ the number of asymmetric bits can increase by a factor of 2 on an average.
- *iota* (ι) : The final contribution to the asymmetry comes from ι which in-creases it by the hamming-weight of the round constant for a given round.

The above analysis reveals the following property that gives a recurrence relation to model the spread of asymmetry between two consecutive rounds of Keccak.

Table 2. Diffusion table for χ for 1-bit input difference

Diffusion(# of bits)	1	2	3
Count	40	80	40

Property 1. *The asymmetry of a self-symmetric state after the ith round of Keccak can be expressed in terms of the asymmetry after the $(i-1)^{th}$ round by the following recurrence relation:*

$$\mathcal{N}_a^{\eta}(\mathcal{K}^i(\mathcal{S}^{\eta})) \leq \mathcal{N}_a^{\eta}(\mathcal{K}^{i-1}(\mathcal{S}^{\eta})) \times 22 + h_i, \tag{1}$$

where h_i is the hamming-weight($\mathcal{HW}$) of the round-constant for the ith round and initial condition is $\mathcal{N}_a^{\eta}(\mathcal{S}^{\eta}) = 0$.

Analysis: First, let us verify the initial condition of the recursion.

$$\mathcal{N}_a^{\eta}(\mathcal{S}^{\eta}) = 0 \ [\mathcal{S}^{\eta} \text{ is a self-symmetric state}]$$

The rest of the formalization follows from the analysis furnished in subsection 3.1 which leads us to the following derivation:

$$\begin{aligned}
\mathcal{N}_a^{\eta}(\mathcal{K}^i(\mathcal{S}^{\eta})) &= \mathcal{N}_a^{\eta}(\iota \circ \chi \circ \rho \circ \pi \circ \theta(\mathcal{K}^{i-1}(\mathcal{S}^{\eta})) \\
&\leq \mathcal{N}_a^{\eta}(\iota \circ \chi \circ \rho \circ \pi(\mathcal{K}^{i-1}(\mathcal{S}^{\eta})) \times 11 \ [\because \text{ Max. diffusion of } \theta \text{ is } 11] \\
&= \mathcal{N}_a^{\eta}(\iota \circ \chi(\mathcal{K}^{i-1}(\mathcal{S}^{\eta})) \times 11 \ [\because \text{ Diffusion of } \rho, \pi \text{ is } 0] \\
&\approx \mathcal{N}_a^{\eta}(\iota(\mathcal{K}^{i-1}(\mathcal{S}^{\eta})) \times 11 \times 2 \ [\because \text{ Avg. diffusion of } \chi \text{ is } 2] \\
&= \mathcal{N}_a^{\eta}((\mathcal{K}^{i-1}(\mathcal{S}^{\eta})) \times 22 + h_i \qquad \blacksquare
\end{aligned}$$

Remark. The positions of non zero bits in round constants are fixed. So all the $\mathcal{N}_a^{\eta}(\mathcal{K}^i(\mathcal{S}))$ symmetry disturbing positions remain fixed and consequently symmetry preserving positions also remain fixed. These symmetry preserving positions constitute the Kernel of the state defined earlier and play a central role in devising distinguishers that succeed with a single query. It is important to note that the above bound is **not a strict** bound as we are considering the average diffusion in case of χ. However, we later show that it suffices to consider the average case and is also supported well by experimental evidence.

Property 2. *The self-symmetric property of a state is destroyed[4] after 4 rounds of Keccak permutation.*

Analysis: This property becomes evident by unraveling the recurrence in (1).

$$\begin{aligned}
\mathcal{N}_a^{\eta}(\mathcal{K}^i(\mathcal{S})) &\leq \mathcal{N}_a^{\eta}(\mathcal{K}^{i-1}(\mathcal{S})) \times 22 + h_i \\
&= \mathcal{N}_a^{\eta}[\mathcal{N}_a^{\eta}(\mathcal{K}^{i-2}(\mathcal{S})) \times 22 + h_{i-1}] \times 22 + h_i \\
&= \mathcal{N}_a^{\eta}[\mathcal{N}_a^{\eta}(\mathcal{K}^{i-2}(\mathcal{S})) \times 22^2] + \mathcal{N}_a^{\eta}(h_{i-1}) \times 22 + h_i \\
&= \mathcal{N}_a^{\eta}(\mathcal{K}^{i-2}(\mathcal{S})) \times 22^2 + h_{i-1} \times 22 + h_i
\end{aligned}$$

The first term is the most dominant term of the recurrence and grows by 22 times in every round. It is clear that further unraveling of the recursion would imply that more than 1600 bits of the state have been rendered asymmetric due to the dominant term alone. Consequently, the state looses its entire self-symmetry and hence, the self-symmetry will no longer hold *deterministically*. $\qquad \blacksquare$

---

[4] It can be noted that this property holds strictly even while considering the average case.

Remark. The converse of Property (2) implies that in the *average* case the self-symmetry of an initial state holds deterministically for at least 3 rounds of the Keccak permutation. This property is used in the construction of 3-round distinguishers against Keccak that hold with a probability of 1.

With all necessary tools in place we are now ready to introduce the basic distinguisher that utilizes the Kernel. We first describe the basic technique.

The Distinguishing Algorithm Using The Kernel.

1: **procedure** DISTINGUISHER($\mathcal{P}(\mathcal{S})$, η, n) $\triangleright \mathcal{S} \xleftarrow{R} \{\mathcal{S}^\eta\}$
2:     Load the Kernel $\mathcal{C}_n^\eta$
3:     Select any symmetric-pair $(x, y, z, z')^\eta \in \mathcal{C}_n^\eta$
4:     Compute $p_{x,y}^\eta(z, z')$
5:     **if** $p_{x,y}^\eta(z, z') = 1$ **then**
6:        **return** 1
7:     **end if**
8:     Repeat steps 3-7 to increase probability of success
9: **end procedure**

Based on the above algorithm we formally define the distinguisher as follows:

Definition 9. *Let $\mathcal{C}_n^\eta$ be the Kernel after n-rounds of Keccak and $\mathcal{D}$ be a polynomial-time distinguisher which takes as input the output of a permutation. $\mathcal{D}$ returns 1 if and only if for any randomly selected symmetric-pair $(x, y, z, z')^\eta \in \mathcal{C}_n^\eta$, we have $p_{x,y}^\eta(z, z') = 1$. Now if $\mathcal{R}$ be a random permutation, then the distinguishing probability of $\mathcal{D}$ for n-rounds of Keccak permutation is given by*

$$\left| \Pr[\mathcal{D}(\mathcal{K}^n(\mathcal{S}^\eta)) = 1] - \Pr[\mathcal{D}(\mathcal{R}(\mathcal{S}^\eta)) = 1] \right| \tag{2}$$

For the random-permutation $\mathcal{R}$, the probability of $\mathcal{D}$ returning 1 is

$$\Pr[\mathcal{D}(\mathcal{R}(\mathcal{S}^\eta)) = 1] = \frac{1}{2} \tag{3}$$

In the previous section we furnished theoretical claims about the existence of the Kernel for 3 rounds of Keccak. In the next subsection we experimentally show that for $n = 3, 4, 5$ $|\mathcal{C}_n^\eta| \gg 0$ i.e.,

$$\Pr[\mathcal{D}(\mathcal{K}^n(\mathcal{S}^\eta)) = 1] = 1 \tag{4}$$

Hence for $n = 3, 4, 5$, the advantage of the distinguisher $\mathcal{D}$ can be obtained by substituting values from Equations (3) and (4) in Equation (2).

$$\mathcal{A}dv(\mathcal{D}) = 1 - \frac{1}{2} \tag{5}$$

It is interesting to note that Equation (5) gives the least value of the adversarial advantage. Since, $|\mathcal{C}_n^\eta| \gg 0$, the adversary could verify the hypothesis for a different symmetric-pair from the Kernel, thereby, increasing its probability of success. Thus, actual advantage of the distinguisher in terms of the number of symmetric-pairs it checks for is given by

$$\mathcal{A}dv^r(\mathcal{D}) = 1 - \left(\frac{1}{2}\right)^r, \text{where } r \text{ is the number of distinct symmetric-pairs} \quad (6)$$

The adversarial advantage $\mathcal{A}dv^r(\mathcal{D}) \to 1$, as r increases. This justifies why the proposed distinguishers succeed with a probability very close to 1 using a single query only. We now illustrate the technique of finding any Kernel $\mathcal{C}_n^\eta$.

Finding the Kernel. Here we show a procedure to ascertain the symmetric-pairs that maintain the self-symmetry of an initial state($\mathcal{S}^\eta$) after n rounds of Keccak permutation. This is an one-time overhead on the part of the adversary. The procedure is illustrated below. We randomly select around m self-symmetric states $\mathcal{S}^\eta$. For each state we apply the Keccak permutation for n rounds. We compute internal-differences for all possible combinations of substates. We then find pairs of points $\{(x, y, z), (x, y, z')\}$ belonging to the same self-symmetry class $\omega_{x,y,z\bmod\eta}$ such that the probability $p_{x,y}^\eta(z, z') = 1$. We store these points for each iteration of the algorithm in a set. At the end we take an intersection among all the sets. The resulting set constitutes the symmetric-pairs and is the desired Kernel for the self-symmetry index η after n rounds. It is important to take the intersection over m-sets because it eliminates the noisy pairs of points i.e., $\{(x, y, z), (x, y, z') \mid p_{x,y}^\eta(z, z') \xleftarrow{R} \{0, 1\}\}$. It has been experimentally verified that a value of $m = 30$ suffices to extract the actual Kernel.

1: **procedure** FINDKERNEL(η, n)
2:    **for** $k = 1 : m$ **do**   ▷ $m \approx 30$ (verified experimentally)
3:       $\mathcal{S}^\eta \xleftarrow{R} \{\mathcal{S}^\eta\}$
4:       Obtain $\mathcal{K}^n(\mathcal{S}^\eta)$
5:       Compute $\Delta_{i,j} \forall (\sigma_i, \sigma_j) \in \mathcal{K}^n(\mathcal{S}^\eta)$.
6:       $\mathcal{I}(k) = \{\{(x, y, z), (x, y, z')\} \in \omega_{x,y,z\bmod\eta} \mid p_{x,y}^\eta(z, z') = 1\}$
7:    **end for**
8:    **return** $\mathcal{C}_n^\eta = \bigcap_{k=1}^{m} \mathcal{I}(k)$
9: **end procedure**

We now enlist some interesting properties of Kernels inferred experimentally. These properties are consistent with the theoretical analysis covered in the previous section of this work.

- The size of a Kernel is fixed for a particular value of η.
- Kernel size decreases as one increases the number of rounds. In particular after 3-rounds Kernel-size reduces to zero which conforms to Property (2).
  Remark. To find the Kernels for $n > 3$, we run Keccak permutation starting from the $(n-2)^{th}$ round to the nth round.
- Kernels for lower values of η are supersets of Kernels for higher values i.e., $\mathcal{C}_n^{\eta_1} \subset \mathcal{C}_n^{\eta_2}$, if $\eta_1 > \eta_2$

Table 3. Kernel-sizes for different rounds and self-symmetry indices

	Kernel-Size : $|\mathcal{C}_n^\eta|$		
η	n = 3	n = 4	n = 5
1	12855	1017	109
2	6242	631	95
4	3002	266	69
8	1355	147	44
16	532	20	12
32	161	3	0

Table 3 gives the sizes of the Kernel for each value of η and for round $n = 3, 4, 5$. It is important to note that by Property (2) an n-round Kernel works only if the input to $(n-2)^{th}$ round is self-symmetric. This implies that for a 5-round Kernel, the input to the 3rd-round must be self-symmetric. Later in this work, we develop strategies that help to achieve this requirement. The Kernel along with these methods helps in devising distinguishers against $\mathcal{K}^4$ and $\mathcal{K}^5$ that require a single query to succeed. Interestingly, the Kernel can also be extended in the forward direction using a property that is detailed in the next subsection.

3.2 The Quartet Strategy

The Quartet strategy is an interesting technique that helps to probabilistically extend any n-round Kernel distinguisher to $(n + 1)$ rounds. The extension to the $(n + 1)^{th}$ round exploits the Kernel $\mathcal{C}_n^\eta$ and a particular property of the θ operation. The strategy is detailed below. We consider two symmetric-pairs $\{(p_1, p_2), (p_3, p_4)\} \in \mathcal{C}_n^\eta$ which satisfy the condition that (p_1, p_3) and (p_2, p_4) belong to the same columns respectively. Thus the points p_1, p_2, p_3, p_4 are of the following form :

$$\left. \begin{array}{l} \{p_1 = (x, y_1, z_1), p_2 = (x, y_1, z_2)\} \in \omega_{x,y_1,z} \\ \{p_3 = (x, y_2, z_1), p_4 = (x, y_2, z_2)\} \in \omega_{x,y_2,z} \end{array} \right\} \text{ where } z_1 \equiv z_2 \equiv z \bmod \eta$$

A pictorial representation of a quartet is given in Fig. 2. Our aim is to track the behavior of the quartet (p_1, p_2, p_3, p_4) through the $(n + 1)^{th}$ round of the Keccak permutation. In particular, we are interested in finding out the value of $p_1 \oplus p_2 \oplus p_3 \oplus p_4$ after the round. Our primary intuition is that the distribution of the variable $X = \mathcal{K}(p_1) \oplus \mathcal{K}(p_2) \oplus \mathcal{K}(p_3) \oplus \mathcal{K}(p_4)$ is non-random. Here, $\mathcal{K}(p_i)$ refers the value of $\pi \circ \rho(p_i)$ after the $(n + 1)^{th}$ round. In order to prove this we use the following property of the θ-transformation:

Property 3. *The symmetric-difference of the values of the points belonging to the same column is not affected by the θ-transformation. If $p_i = (x, y_i, z), p_j = (x, y_j, z), i \neq j$, then*

$$\theta(p_i \oplus p_j) = p_i \oplus p_j$$

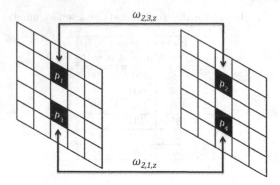

Fig. 2. A typical quartet $(p_1, p_2, p_3, p_4) \in \mathcal{C}_n^{\eta}$

This property is evident from the fact that θ affects all points in the same column in the same way i.e., it will either invert all of them or none of them. Thus the difference between points in the same column is preserved. Now, let us look at our quartet (p_1, p_2, p_3, p_4). Based on Property (3) we have :

$$\theta(p_1 \oplus p_3) = p_1 \oplus p_3$$
$$\theta(p_2 \oplus p_4) = p_2 \oplus p_4$$

Also, since (p_1, p_2) and (p_3, p_4) are symmetric-pairs from $\mathcal{C}_n^{\eta}$, by virtue of the n-round Kernel distinguisher the following will hold for $\mathcal{K}^n$ with a probability 1.

$$\Pr[p_1 \oplus p_2 = 0] = 1$$
$$\Pr[p_3 \oplus p_4 = 0] = 1 \tag{7}$$

We next analyze the effect of the $(n+1)^{th}$ round on the value of the variable X by looking at all the internal transformations individually:

1. **θ-transformation :** The θ-transformation preserves the value of the expression $p_1 \oplus p_2 \oplus p_3 \oplus p_4$.

$$\theta(p_1) \oplus \theta(p_2) \oplus \theta(p_3) \oplus \theta(p_4) = \theta(p_1) \oplus \theta(p_3) \oplus \theta(p_2) \oplus \theta(p_4)$$
$$= \theta(p_1 \oplus p_3) \oplus \theta(p_3 \oplus p_4)$$
$$= p_1 \oplus p_2 \oplus p_3 \oplus p_4 \text{ [By Property (3)]}$$

2. **π and ρ-transformations :** π and ρ have no effect on the value of $p_1 \oplus p_2 \oplus p_3 \oplus p_4$. They just permute the position of each p_i. So now we look at the value of the positions $p_i' = \pi \circ \rho(p_i)$.

$$\pi \circ \rho(p_1) \oplus \pi \circ \rho(p_2) \oplus \pi \circ \rho(p_3) \oplus \pi \circ \rho(p_4) = p_1' \oplus p_2' \oplus p_3' \oplus p_4'$$
$$\text{[Value unchanged } |p_i'| = |p_i|]$$

3. **χ-transformation :** χ which is the only non-linear transformation will probabilistically preserve the value of each p_i' i.e., each p_i. Let the probability of

$\mathcal{X}$ to preserve the value of bit be t. Then we have :

$$\Pr[\mathcal{X}(p_1') = p_1' \ \& \ \mathcal{X}(p_2') = p_2' \ \& \ \mathcal{X}(p_3') = p_3' \ \& \ \mathcal{X}(p_4') = p_4'] = t^4$$

From the Difference Distribution Table (DDT) of $\mathcal{X}$, we computed the value of t in the following way: For all values of a bit variable $x = b_0b_1b_2b_3b_4$ we computed $\mathcal{X}(x) = \mathcal{X}(b_0b_1b_2b_3b_4) = b_0'b_1'b_2'b_3'b_4'$. Then we computed the probability[5] $\Pr[b_i = b_i'] = \frac{3}{4}$.

4. **ι-transformation :** ι can be safely ignored since $\iota \circ \iota$ gives the *identity* transformation. Thus applying ι on $\mathcal{K}$ would nullify its effect.

Now let us look at the probability that the random-variable $\left[\bigoplus_{i=1}^{4} \mathcal{K}(p_i) \right] = 0$

$$\Pr\left[\bigoplus_{i=1}^{4} \mathcal{K}(p_i) = 0 \right] = \Pr\left[\mathcal{K}(p_i) = p_i, \forall p_i \right] + \Pr\left[\mathcal{K}(p_i) = p_i, \text{ for exactly a single pair } (p_i, p_j) \right]$$

$$+ \Pr\left[\mathcal{K}(p_i) \neq p_i, \forall p_i \right]$$

$$= t^4 + \binom{4}{2} t^2 (1-t)^2 + (1-t)^4$$

$$= 0.53125 \text{ [Substituting value of } t]$$

Now, if we would be dealing with a random permutation $\mathcal{R}$, then we would have $\Pr[\bigoplus_{i=1}^{4} \mathcal{R}(p_i) = 0] = \frac{1}{2}$. We now get the theoretical value of the adversarial advantage as follows:

$$Adv(\mathcal{D}) = \left| \Pr[\mathcal{D}(\mathcal{K}^{n+1}(\mathcal{S}^n)) = 1] - \Pr[\mathcal{D}(\mathcal{R}(\mathcal{S}^n)) = 1] \right| = 0.53125 - \frac{1}{2} \quad (8)$$

Equation (8) verifies our intuition that the distribution of the variable $X = \mathcal{K}(p_1) \oplus \mathcal{K}(p_2) \oplus \mathcal{K}(p_3) \oplus \mathcal{K}(p_4)$ is non-random. Thus given an n-round Kernel, we can search for a quartet $\{p_i \in \mathcal{C}_n^{\eta}, i = 1, 2, 3, 4\}$. If found, we can devise an $(n + 1)$-round distinguisher by looking at the distribution of the variable $(p_1' \oplus p_2' \oplus p_3' \oplus p_4')$ where $p_i' = \pi \circ \rho(p_i)$ over a reasonable number of queries. A concrete lower bound for the number of queries can be expressed in the form of the following inequality given by Chernoff bound [4] :

$$q \geq \frac{1}{Adv(\mathcal{D})^2} \ln \frac{1}{\sqrt{\epsilon}} \quad \text{where } \epsilon \to \text{Error bound} \quad (9)$$

The typical value for $\epsilon = 0.05$. From Equation (8) and Chernoff inequality (9), we get $q \geq 1534 \approx 2^{11}$. Thus the theoretical complexity of any quartet based distinguisher is around 2^{11}.

3.3   Self-Symmetric State Construction (SSC)

The SSC strategy when used in conjunction with the Kernel strategy can penetrate **any** four rounds of Keccak. As stated earlier, the Kernel strategy can

---

[5] We mean to say that $\Pr[b_0 = b_0'] = \Pr[b_1 = b_1'] = \Pr[b_2 = b_2'] = \Pr[b_3 = b_3'] = \Pr[b_4 = b_4'] = \frac{3}{4}$.

deterministically distinguish any three rounds of Keccak and solely depends on the self-symmetry of the initial state. The SSC technique generates states that become self-symmetric after one round i.e., it produces states X such that $\mathcal{K}(X) = \mathcal{S}^\eta$, and is applicable for any round of Keccak. Hence, if we prepend SSC before the Kernel strategy, we get $(n+1)$-round distinguisher from an n-round distinguisher. Before describing the technique, we define a function (ϕ), that modifies a state at specific points to get another state which is self-symmetric, following certain constraints.

Definition 10. *Let us consider a self-symmetric state $\mathcal{S}_1^\eta$ and let $\mathcal{S}_1^\eta(c_{i,j})$, where $(0 \le i \le 4)$ and $(0 \le j \le 63)$ denote the ith column in the jth slice of $\mathcal{S}_1^\eta$. Also let $par(\mathcal{S}_1^\eta(c_{i,j}))$ denote the parity of the column $\mathcal{S}_1^\eta(c_{i,j})$. The function ϕ is defined as follows:*

$$\phi(\mathcal{S}_1^\eta) = \mathcal{S}_2^\eta, \; where \begin{cases} \mathcal{S}_2^\eta(x,y,z) = s, s \xleftarrow{R} \{0,1\} \; if \; \mathcal{S}_1^\eta(x,y,z) = 0, \forall(x,y) \; for \; a \; particular \; z \\ \begin{cases} \mathcal{S}_2^\eta(x,y,z) = \mathcal{S}_1^\eta(x,y,z) \; if \; \mathcal{S}_1^\eta(x,y,z) = 1 \vee x = y, \\ otherwise \; \mathcal{S}_2^\eta(x,y,z) = s, s \xleftarrow{R} \{0,1\}, \\ such \; that \\ \begin{cases} par(\mathcal{S}_2^\eta(c_{i,j})) - par(\mathcal{S}_1^\eta(c_{i,j})) \equiv 0 \quad (mod\,2) \forall(i,j) \\ or \\ par(\mathcal{S}_2^\eta(c_{i,j})) - par(\mathcal{S}_1^\eta(c_{i,j})) \equiv 1 \quad (mod\,2) \forall(i,j) \end{cases} \end{cases} \end{cases}$$

Our basic aim is to construct states $\mathcal{S}$ for which $\mathcal{K}(\mathcal{S})$ becomes self-symmetric. Since all the operations θ, ρ, π, χ preserve self-symmetry except ι, the problem reduces to finding states which when applied to θ, ρ, π, χ give states which are symmetric except the bit positions where ι is supposed to act. We split the problem into smaller and relatively simpler sub-problems by exploiting the properties of θ and χ. An abstract form of the algorithm is given below:

1: **procedure** SSC(η)
2:     Compute fixed states A, B each of which is 1600-bit
3:     Generate state C, such that C $= \phi(B)$
4:     Compute $(A \oplus C)$ is the desired state i.e., $\mathcal{K}(A \oplus C) = \mathcal{S}^\eta$
5:     **return** State $(A \oplus C)$
6: **end procedure**

The computation of the fixed states A, B is illustrated next. Though this method can be generalized for any $\eta \in \{1, 2, 4, 8, 16, 32\}$ and any intermediate round, for simplicity, we limit our discussion to the first round and find states with $\eta = 1$. The primary intuition to get self-symmetric state after first round is to analyze what happens as we go backward one round from a self-symmetric state. So we first look at the five operations of Keccak permutation $\iota, \chi, \rho, \pi, \theta$ *reversely*. Here we recall that the only symmetry disturbing operation is ι and the Hamming-weight of the round-constant in first round is 1.

Since ι in first round acts only on bit-position $(0,0,63)$, any self-symmetric state remains self symmetric after χ^{-1} except for the row $(*, 0, 63)$. This is because χ acts independently on rows. The row $(*, 0, 63)$ and the corresponding symmetric positions (for $\eta = 1$, all the rows $(i, 0, j); i = 0, 1, 2, 3, 4; j = 0, 1, \cdots, 62$) at the input of χ are assigned specific values represented by five-bit vectors a_j $(j = 0, 1, \cdots, 63)$ such that $a_0 = a_1 = \cdots = a_{62} \ne a_{63}$.

Let $X(a_j) = b_j$, then it follows that $b_0 = b_1 = \cdots = b_{62} \neq b_{63}$. Now if a_k and a_{63} ($k \neq 63$) differ in i ($i \leq 5$) positions (strictly including MSB position) these i positions are called **asymmetric** bits while remaining $(5 - i)$ fixed bits of the row are called **fixed symmetric** bits. Note here that we gain many degrees of freedom as we can choose any a_k's satisfying the above conditions. However, for the sake of compact representation, we look more into X function and observe that there exist four value pairs (a_0, a_{63}) for which (a_0, a_{63}) differ in 0th position and corresponding (b_0, b_{63}) also differ in 0th position. The values are given in Table 4. When going backward, if we denote the state just after X^{-1} as S then S has the property that it is any self-symmetric state except for row $(*, 0, 63)$ which is assigned the value of the fixed five-bit vector a. π^{-1} and ρ^{-1} just permute the positions of bits of S with their values unaltered. When going through π^{-1}, ρ^{-1} we track the positions of asymmetric i bits as well as $(5 - i)$ fixed symmetric bits which were situated in row $(*, 0, 63)$ of the state S. Let $\pi^{-1} \circ \rho^{-1}(S) = S'$. We now split S' into two states : **fixed asymmetric state** (A_1) and fixed symmetric state (B_1) such that $S' = A_1 \oplus B_1$. The state A_1 is formed by taking an *all-zero* state and then overwriting it with the i asymmetric bits in their respective positions. State B_1 is the concatenation of 64 identical slices and each slice consists of $(5 - i)$ fixed symmetric bits maintaining their positions and remaining bits are filled with $0's$. By inspecting the operation π^{-1}, we observe that 0th row of any slice in S comes into the diagonal positions of the corresponding slice of the state S'. Finally, we see the effect of θ^{-1}:

$$\therefore \theta^{-1}(S') = \theta^{-1}(A_1 \oplus B_1)$$
$$= \theta^{-1}(A_1) \oplus \theta^{-1}(B_1)$$
$$= A \oplus B$$

Since A_1 is a fixed state we can easily compute $A = \theta^{-1}(A_1)$ and store the states A which will be utilized later. Since θ (consequently θ^{-1}) is translation-invariant so self-symmetry of B_1 is preserved in B. We next try to construct symmetric states C_j from B such that $B \subset \bigcap_j C_j$ and $B_1 \subset \bigcap_j \theta(C_j)$. Here our main intention is to construct any self-symmetric state which consists of fixed rows whose Hamming weight is non zero. Consequently, we have to construct states such that when θ is applied to the state, it does not disturb the values of the fixed row in B_1. This property will hold if the parity of **all** the columns of $(C_j \oplus B)\forall j$ is either even or odd. Herein, comes the application of function ϕ defined earlier which ensures this property of states C_j. Since in this example,

Table 4. Input and output pairs for χ that differ only in MSB

Input pair	Output pair
00001 & 10001	00101 & 10101
00011 & 10011	01011 & 11011
00101 & 10101	10001 & 00001
00111 & 10111	10111 & 00111

$\eta = 1$, all slices of C becomes identical and hence degrees of freedom lies in how many distinct states C_j can be obtained applying the concept of ϕ on a *single slice* of B. For larger values of η, larger number of distinct states C_j can be obtained from fixed state B because ϕ can then be applied on η *number of consecutive slices*. So by precomputing A and B, one can easily construct C and consequently get $(A \oplus C)$ which becomes self-symmetric after the first round of Keccak. The following instance illustrates the construction of such a state.

Table 5. Generation of one-round self-symmetric state (Here, $C = \phi(B)$)

(a)

B(Slice)   (b) State A

0 1 0 1 0	EBC69AF135E26BC4 7134BC4D789AF135 CD79135E26BC4D78 DE26D789AF135E26 09ABE26BC4D789AF
0 0 0 1 0	EBC69AF135E26BC5 7134BC4D789AF135 CD79135E26BC4D78 DE26D789AF135E26 09ABE26BC4D789AF
0 1 1 1 0	EBC69AF135E26BC5 7134BC4D789AF135 CD79135E26BC4D78 DE26D789AF135E26 09ABE26BC4D789AF
0 1 0 0 0	EBC69AF135E26BC5 7134BC4D789AF135 CD79135E26BC4D78 DE26D789AF135E26 09ABE26BC4D789AF
0 1 0 1 0	EBC69AF135E26BC5 7134BC4D789AF135 CD79135E26BC4D78 DE26D789AF135E26 09AFE26BC4D789AF

(c)

C(Slice)   (d) State $(C \oplus A)$

0 1 0 1 0	EBC69AF135E26BC4 8ECB43B287650ECA CD79135E26BC4D78 21D9287650ECA1D9 09ABE26BC4D789AF
1 0 0 1 0	1439650ECA1D943A 7134BC4D789AF135 CD79135E26BC4D78 21D9287650ECA1D9 09ABE26BC4D789AF
1 1 1 1 0	1439650ECA1D943A 8ECB43B287650ECA 3286ECA1D943B287 21D9287650ECA1D9 09ABE26BC4D789AF
1 1 1 0 0	1439650ECA1D943A 8ECB43B287650ECA 3286ECA1D943B287 DE26D789AF135E26 09ABE26BC4D789AF
1 1 1 1 0	1439650ECA1D943A 8ECB43B287650ECA 3286ECA1D943B287 21D9287650ECA1D9 09AFE26BC4D789AF

(e) $\mathcal{K}(C \oplus A) \in \{\mathcal{S}^1\}$

0000000000000000	FFFFFFFFFFFFFFFF	FFFFFFFFFFFFFFFF	FFFFFFFFFFFFFFFF	FFFFFFFFFFFFFFFF
FFFFFFFFFFFFFFFF	0000000000000000	0000000000000000	0000000000000000	FFFFFFFFFFFFFFFF
0000000000000000	0000000000000000	FFFFFFFFFFFFFFFF	0000000000000000	FFFFFFFFFFFFFFFF
0000000000000000	0000000000000000	0000000000000000	FFFFFFFFFFFFFFFF	FFFFFFFFFFFFFFFF
0000000000000000	FFFFFFFFFFFFFFFF	0000000000000000	FFFFFFFFFFFFFFFF	0000000000000000

Table 5 depicts a typical example of the message construction. Here, states A, B are pre-computed. Since state B is self-symmetric with $\eta = 1$, it is represented using a single slice. State C is randomly generated from State B following function ϕ. After applying the first round of Keccak on $C \oplus A$ we get a self-symmetric state with $\eta = 1$. Based on SSC, we give the following definition:

Definition 11. *An **One-Symmetric State** $(\mathbb{S}^\eta)$ is a state that becomes self-symmetric after one round of Keccak permutation.*

$$\mathbb{S}^\eta = \{\mathcal{S} : \mathcal{K}(\mathcal{S}) \in \{\mathcal{S}^\eta\}\}$$

Thus $\{\mathbb{S}^\eta\}$ is the set of all one-symmetric states. Using SSC we can generate subsets of $\{\mathbb{S}^\eta\}$ for all values of η. In the example above $(C \oplus A) \in \{\mathbb{S}^\eta\}$.

3.4   Simplified Target Internal Difference Algorithm (sTIDA)

Here, our primary aim is to produce one-symmetric (Definition 11) states. This when combined with SSC gives a way to make states that are two-round self-symmetric. Let us start by the following definition:

Definition 12. *A **Two-Symmetric State** ($\mathscr{S}^\eta$) is a state that becomes one-symmetric after one round of the Keccak permutation or alternatively becomes self-symmetric after two rounds of the Keccak permutation.*

$$\mathscr{S}^\eta = \mathcal{S} : \left\{ \begin{array}{l} \mathcal{K}(\mathcal{S}) \in \{\mathbb{S}^\eta\} \\ \mathcal{K}^2(\mathcal{S}) \in \{\mathcal{S}^\eta\} \end{array} \right\}$$

Here, we intent to generate states that belong to $\{\mathscr{S}^\eta\}$. For this purpose we tweak the Target Internal Difference Algorithm(TIDA) algorithm proposed by Dinur et al. in [6]. The authors of TIDA have claimed that despite being a heuristic algorithm, it has high rate of success. Given a target internal difference, TIDA can produce states that yield that difference after one round of Keccak. In our case the requirement is similar but has some additional constraints. We want TIDA to generate arbitrary states $\mathcal{S}$ satisfying the following constraints :

$$\text{Input} \rightarrow \left\{ \begin{array}{l} \Delta_T(\text{target internal difference}) \\ \mathcal{S}_T(\text{specific target state}) \end{array} \right\} \text{Output} \rightarrow \mathcal{S} : \left\{ \begin{array}{l} \Delta_T = (\mathcal{K}(\mathcal{S})) \text{ and} \\ \mathcal{S}_T \subset (\mathcal{K}(\mathcal{S}) \cap \mathcal{S}_T) \end{array} \right\} \forall \mathcal{S}$$

The basic TIDA consists of two phases : difference phase and value phase. Let us first look at the difference phase and handle the parts that are not relevant to us. The goal of difference phase is achieved by two constraints:
- Some specified positions[6] of initial internal difference(Δ_I) are bound to 0.
- Δ_I be such that $\iota \circ \chi \circ \rho \circ \pi \circ \theta(\Delta_I) = \Delta_T$.

However, we are only interested in analyzing Keccak permutation rather than Keccak hash algorithm. So we discard[7] the first constraint of the difference phase. In this case we have full $(25 \times \eta)$ degrees of freedom[8] for any self-symmetric state S^η. In difference phase, the basic TIDA exploits the property that for a particular output difference of a Keccak Sbox, the input difference consists of at least 5 two-dimensional affine subspaces [7]. In our simplified TIDA, we also go with the same strategy for all the active Sboxes except some specific *fixed* Sboxes which are determined by the SSC technique. The Sboxes of fixed slices in SSC method are regarded here as fixed Sboxes. If u be the number of specific target slices, we preserve their values by assigning appropriate values at the inputs of the corresponding Sboxes. In doing, so we loose $(25 \times u)$ degrees of freedom. We next move our focus to the value phase of the basic TIDA which also imposes two constraints. The first constraint simply equates the capacity part to $p||0^*$, where p represents padding bits. We discard this value constraint. However, we preserve the second value constraint.

It is to be noted that the success of sTIDA depends heavily on the chosen value of the parameter η. The largest value of η for which a quartet is found in the 5th round Kernel is 8. As we have already lost $25 \times u$ degrees of freedom in the difference phase, we are left with $(8 \times 25 - 25 \times u)$ degrees of freedom. To make sTIDA successful, the quantity $(8 \times 25 - 25 \times u)$ must be positive. Moreover, the

---

[6] Capacity part of Sponge construction.

[7] This is why we call it as Simplified Target Internal Difference Algorithm (sTIDA).

[8] On the contrary, for the hash function there are only (r-8) degrees of freedom; where r is the bit-rate of sponge function.

theoretical lower bound on the number of queries to verify the Quartet strategy is 2^{11}. Thus for the whole attack to succeed we need to have $(8 \times 25 - 25 \times u) > 11$. The above expression directly implies that to maximize the degrees of freedom one has to minimize the value of u. During our experimentation, in creating an one-symmetric state using SSC with $\eta = 8$ for round-2 we found cases where $u = 4$. Thus $(8 \times 25 - 4 \times 25)$ degrees of freedom are left which are sufficient for our algorithm. In general, as TIDA survives with more constraints [6], sTIDA will also successfully produce two-symmetric states for lower values of u.

4   Experimental Results

In this section we show how we can combine the various strategies developed so far to generate different types of distinguishers against Keccak. We start with an overview and then look at each in detail. Figure 3 gives an idea about all the distinguishers developed in this work and the relevant strategies adopted. It is also important to note the sequence in which the methods are applied. The first level of the distinguishers require a single query to succeed and all use the Kernel strategy at the end. We already mentioned, that the Quartet technique can extend the Kernel probabilistically to one more round. As a result, the second level distinguishers use the Quartet as the extension strategy and we are able to reach up to 6-rounds. It is also worth noting that SSC and sTIDA are used before applying the Kernel method in order to make the appropriate changes as per the requirement of the Kernel mentioned in subsection 3.1. We now look at these results individually, starting with the basic 3-round distinguisher.

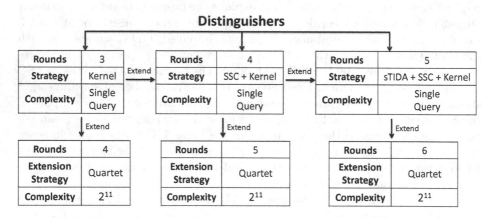

Fig. 3. An overview of all the distinguishers with corresponding strategies

4.1   The Basic 3-Round Distinguisher Using Only the Kernel

Once the Kernels ($\mathcal{C}_3^{\eta}$) have been extracted, it is easy to mount the 3-round distinguisher. Let us detail an instance to make things clear. For this, we take

Table 6. Input state : $\mathcal{S}^{16}$ (hexadecimal)

```
B443B443B443B443  745F745F745F745F  C38CC38CC38CC38C  348B348B348B348B  4678467846784678
5C825C825C825C82  74F074F074F074F0  D0F9D0F9D0F9D0F9  85F085F085F085F0  772B772B772B772B
CD96CD96CD96CD96  7CA87CA87CA87CA8  429B429B429B429B  0D430D430D430D43  C8A9C8A9C8A9C8A9
229C229C229C229C  76D376D376D376D3  5840584058405840  B172B172B172B172  8521852185218521
35D135D135D135D1  C6B6C6B6C6B6C6B6  81BE81BE81BE81BE  E389E389E389E389  9E879E879E879E87
```

$\eta = 16$. From Table (3) we have $|\mathcal{C}_3^{16}| = 532$. We choose any five[9] symmetric-pairs from $\mathcal{C}_3^{16}$. The pairs chosen here are as follows:

$$(3, 3, 9, 57)^{16} \rightarrow \{(3, 3, 9), (3, 3, 57) \in \omega_{3,3,9}\}$$
$$(0, 0, 5, 21)^{16} \rightarrow \{(0, 0, 5), (0, 0, 21) \in \omega_{0,0,5}\}$$
$$(0, 2, 6, 38)^{16} \rightarrow \{(0, 2, 6), (0, 2, 38) \in \omega_{0,2,6}\}$$
$$(2, 1, 30, 46)^{16} \rightarrow \{(2, 1, 30), (2, 1, 46) \in \omega_{2,1,14}\}$$
$$(2, 4, 31, 63)^{16} \rightarrow \{(2, 4, 31), (2, 4, 63) \in \omega_{2,4,15}\}$$

These pairs should preserve the self-symmetry of an initial state $(\mathcal{S}^\eta)$ after 3 rounds of Keccak. The state $\mathcal{S}^{16}$ used in this example is shown in hexadecimal lane-wise format in Table 6. Figure 4(a) shows the same state pictorially in the form of a two-dimensional (2D) matrix. Each row represents a lane and each column is a slice. The output $\mathcal{K}^3(\mathcal{S}^{16})$ is depicted in Fig. 4(b). The values of the five symmetric-pairs $\in \mathcal{C}_3^{16}$ chosen in this example are shown as ▨ and ▧ squares. Interestingly, here one can also verify the distinguishing hypothesis visually. Picking any symmetric-pair given above one can see that in Fig. 4(b) they are either both ▨ or both ▧ which means that they are equal. This implies that the self-symmetry of $\mathcal{S}^{16}$ where points belonging to the same symmetry class are either all ▨ or ▧ squares as shown in Fig. 4(a) is preserved by the symmetric-pairs chosen from $\mathcal{C}_3^{16}$ as shown in Fig. 4(b). For a random permutation, the probability that chosen pairs would be symmetric $\left(\frac{1}{2}\right)^5$. Thus the adversarial advantage $\mathcal{A}dv^5(\mathcal{D}) = 1 - \left(\frac{1}{2}\right)^5$ is close to 1. Recall that we need only one self-symmetric state to verify the hypothesis. Thus the proposed distinguisher enables an adversary to distinguish $\mathcal{K}^3$ from $\mathcal{R}$ using only a single query.

4.2 The 4-Round Distinguisher Employing Multiple Strategies

Due to Property (2), we **cannot** directly extend the 3-round distinguisher to 4 rounds. However, we can reach 4-rounds in two ways. One employs SSC prior to using the Kernel while the other extends the Kernel using the Quartet.

Using SSC before the 4-Round Kernel ($\mathcal{C}_4^\eta$). The main idea here is that if we can have a self-symmetric state at the input of 2nd round the distinguisher

---

[9] An adversary can choose any number of pairs.

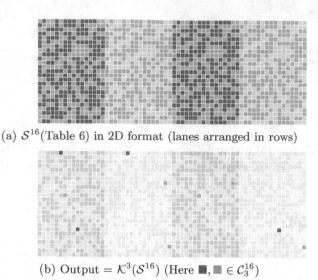

(a) $\mathcal{S}^{16}$(Table 6) in 2D format (lanes arranged in rows)

(b) Output = $\mathcal{K}^3(\mathcal{S}^{16})$ (Here ■, ■ ∈ $\mathcal{C}_3^{16}$)

Fig. 4. Example showing that symmetric-pairs ∈ $\mathcal{C}_3^{16}$ in $\mathcal{K}^3(\mathcal{S}^{16})$ preserve the self-symmetry of the initial state $\mathcal{S}^{16}$ (■ = 0, ■ = 1)

will hold for 4-rounds too. The only assumption here is that the input to 2nd round is self-symmetric. Herein, comes the role of SSC. It helps, with high degree of freedom, to generate states that become self-symmetric after the first round of Keccak. Once this is done the attacks proceeds as before. We choose symmetric-pairs from $\mathcal{C}_4^{\eta}$ and the rest of the attack follows like the 3-round distinguisher. Thus, we get a single-query distinguisher for $\mathcal{K}^4$.

Using a Quartet from the 3-Round Kernel ($\mathcal{C}_3^{\eta}$). Here, we extract a quartet from $\mathcal{C}_3^{\eta}$, then employ the Quartet strategy developed earlier to probabilistically extend the distinguisher to the next round. Let us look at the following example where we take $\eta = 32$ and choose the quartet $(p_1, p_2, p_3, p_4) \in \mathcal{C}_3^{32}$ as follows:

$$\left.\begin{array}{l} \{p_1 = (0,0,24), p_2 = (0,0,56)\} \in \omega_{0,0,24} \\ \{p_3 = (0,1,24), p_4 = (0,1,56)\} \in \omega_{0,1,24} \end{array}\right\} \text{ where } p_i \in \mathcal{C}_3^{32}$$

We generated a random self-symmetric state with $\eta = 32$ and computed $\mathcal{K}^4(\mathcal{S}^{32})$ and looked at the value of $[\bigoplus_{i=1}^4 p_i']$ where[10] $p_i' = \rho \circ \pi(p_i) \in \mathcal{K}^4(\mathcal{S}^{32})$. This process was repeated for around 2^{11} times as per the lower bound derived from Chernoff inequality (9) in subsection 3.2. The probability was computed to be $\Pr[\bigoplus_{i=1}^4 p_i' = 0] = 0.5322$, very close to the theoretical probability (0.53215) of a Quartet distinguisher. This verifies the distinguisher against $\mathcal{K}^4$ that succeed with non-negligible probability and have a practical complexity of 2^{11}.

---

[10] $p_1' = (0,0,24), p_2' = (0,0,56), p_3' = (1,3,20), p_4' = (1,3,52)$.

4.3   The 5-Round Distinguisher Employing Multiple Strategies

The 5-round distinguisher also has two variants : one working with a single query while the other is probabilistic.

Using sTIDA and SSC before the 5-Round Kernel ($\mathcal{C}_5^\eta$). The concept is similar to the one mentioned in subsection 4.2, the only difference being the fact that we now need states that become self-symmetric after two rounds. Consequently, we use both sTIDA and SSC. Using these strategies together, we are able to generate states S such that $\mathcal{K}^2(S) \in \mathcal{S}^\eta$ with reasonable degrees of freedom. This satisfies the requirement of the 5-round Kernel ($\mathcal{C}_5^\eta$) that the input to $5 - 2 = 3^{rd}$ round must be self-symmetric. Next we choose symmetric-pairs from ($\mathcal{C}_5^\eta$) and proceed as before getting a single-query distinguisher against $\mathcal{K}^5$.

Using a Quartet from the 4-Round Kernel ($\mathcal{C}_4^\eta$). This attack exploits the quartets available in $\mathcal{C}_4^\eta$ and extends the single-query 4-round distinguisher described in subsection 4.2 to the 5th round. As before it requires around 2^{10} queries to conclude that the quartet does behave randomly.

4.4   The 6-Round Distinguisher Deploying All the Strategies

All the properties studied in this work are used together to reach the 6th round of the Keccak permutation. The main idea is to use the single-query 5-round attack developed in subsection 4.3 and use the Quartet strategy on the 5-round Kernel ($\mathcal{C}_5^\eta$). Thus if we unfold the distinguisher we look at the following sequence : sTIDA $\rightarrow$ SSC $\rightarrow$ Kernel $\rightarrow$ Quartet $\rightarrow$ $\mathcal{D}(\mathcal{K}^6)$. For verifying the distinguisher, we take $\eta = 8$ and choose the quartet $(p_1, p_2, p_3, p_4) \in \mathcal{C}_5^8$ as follows:

$$\left. \begin{array}{l} \{p_1 = (0,3,30), p_2 = (0,3,38)\} \in \omega_{0,3,6} \\ \{p_3 = (0,4,30), p_4 = (0,4,38)\} \in \omega_{0,4,6} \end{array} \right\} \text{ where } p_i \in \mathcal{C}_5^8$$

So at the output of $\mathcal{K}^6$ we look at the points $p_1' = (3,4,53), p_2' = (3,4,51), p_3' = (4,2,12), p_4' = (4,2,20)$ where $(p_i' = \rho \circ \pi(p_i))$. Conforming to theoretical estimates, we find that 2^{11} queries are sufficient to conclude that the variable $\left[\bigoplus_{i=1}^4 p_i' \right]$ does not follow a random distribution.

5   Conclusion

In this work we have used the concept of internal-differentials to find self-symmetric properties of the Keccak permutation that hold with a probability of 1 for any three rounds. We have studied new properties of Keccak and developed strategies based on our observations. Combining these strategies 3-round distinguisher can easily be extended to 4 and 5 rounds with a probability 1. It is further extended to 6-rounds with a complexity of 2^{11}. As a future work it would be interesting to study whether the proposed strategies can be further exploited to reach a higher number of rounds.

References

1. Bernstein, D.: Second preimages for 6 (7?(8??)) rounds of Keccak? (November 2010),
   http://ehash.iaik.tugraz.at/uploads/6/65/
   NIST-mailing-list_Bernstein-Daemen.txt
2. Bertoni, G., Daemen, J., Peeters, M., Assche, G.V.: Sponge functions. In: Ecrypt Hash Workshop 2007 (May 2007)
3. Bertoni, G., Daemen, J., Peeters, M., Assche, G.V.: The Keccak SHA-3 submission. Submission to NIST, Round 3 (2011),
   http://keccak.noekeon.org/Keccak-submission-3.pdf
4. Chernoff, H.: A Note on an Inequality Involving the Normal Distribution. The Annals of Probability 9(3), 533–535 (1981),
   http://dx.doi.org/10.1214/aop/1176994428
5. Daemen, J., Van Assche, G.: Differential propagation analysis of keccak. In: Canteaut, A. (ed.) FSE 2012. LNCS, vol. 7549, pp. 422–441. Springer, Heidelberg (2012), http://dx.doi.org/10.1007/978-3-642-34047-5_24
6. Dinur, I., Dunkelman, O., Shamir, A.: Collision Attacks on Up to 5 Rounds of SHA-3 Using Generalized Internal Differentials. Cryptology ePrint Archive, Report 2012/672 (2012), http://eprint.iacr.org/
7. Dinur, I., Dunkelman, O., Shamir, A.: New attacks on keccak-224 and keccak-256. In: Canteaut, A. (ed.) FSE 2012. LNCS, vol. 7549, pp. 442–461. Springer, Heidelberg (2012), http://dx.doi.org/10.1007/978-3-642-34047-5_25
8. Duan, M., Lai, X.: Improved zero-sum distinguisher for full round Keccak-f permutation. Cryptology ePrint Archive, Report 2011/023 (2011),
   http://eprint.iacr.org/2011/023.pdf
9. Duan, M., Lai, X.: Improved zero-sum distinguisher for full round keccak-f permutation. Chinese Science Bulletin 57(6), 694–697 (2012),
   http://dx.doi.org/10.1007/s11434-011-4909-x
10. Duc, A., Guo, J., Peyrin, T., Wei, L.: Unaligned rebound attack: Application to Keccak. In: Canteaut, A. (ed.) FSE 2012. LNCS, vol. 7549, pp. 402–421. Springer, Heidelberg (2012)
11. Lathrop, J.: Cube attacks on cryptographic hash functions. Master's thesis (2009),
    http://www.cs.rit.edu/~jal6806/thesis/
12. Morawiecki, P., Pieprzyk, J., Srebrny, M.: Rotational cryptanalysis of round-reduced Keccak. Cryptology ePrint Archive, Report 2012/546 (2012),
    http://eprint.iacr.org/
13. Morawiecki, P., Srebrny, M.: A sat-based preimage analysis of reduced keccak hash functions. IACR Cryptology ePrint Archive 2010, 285 (2010),
    http://dblp.uni-trier.de/db/journals/iacr/iacr2010.html#MorawieckiS10
14. Naya-Plasencia, M., Röck, A., Meier, W.: Practical analysis of reduced-round Keccak. In: Bernstein, D.J., Chatterjee, S. (eds.) INDOCRYPT 2011. LNCS, vol. 7107, pp. 236–254. Springer, Heidelberg (2011)
15. Peyrin, T.: Improved Differential Attacks for ECHO and Grøstl. In: Rabin, T. (ed.) CRYPTO 2010. LNCS, vol. 6223, pp. 370–392. Springer, Heidelberg (2010)

Preimage Attacks on Reduced-Round Stribog

Riham AlTawy and Amr M. Youssef

Concordia Institute for Information Systems Engineering,
Concordia University, Montréal, Québec, Canada

Abstract. In August 2012, the Stribog hash function was selected as
the new Russian cryptographic hash standard (GOST R 34.11-2012).
Stribog employs twelve rounds of an AES-based compression function
operating in Miyaguchi-Preneel mode. In this paper, we investigate the
preimage resistance of the Stribog hash function. In particular, we apply
a meet in the middle preimage attack on the compression function which
allows us to obtain a 5-round pseudo preimage for a given compression
function output with time complexity of 2^{448} and memory complexity of
2^{64}. Additionally, we adopt a guess and determine approach to obtain a
6-round chunk separation that balances the available degrees of freedom
and the guess size. The proposed chunk separation allows us to attack 6
out of 12 rounds with time and memory complexities of 2^{496} and 2^{112},
respectively. Finally, by employing a multicollision attack, we show that
preimages of the 5 and 6-round reduced hash function can be generated
with time complexity of 2^{481} and 2^{505}, respectively. The two preimage
attacks have equal memory complexity of 2^{256}.

Keywords: Cryptanalysis, Hash functions, Meet in the middle, Preimage attack, GOST R 34.11-2012, Stribog.

1 Introduction

The attacks by Wang *et al.* on MD5 [23] and SHA-1 [22] followed by the SHA-3
competition [18] have led to a flurry in the area of hash function cryptanalysis.
The primary targets of these attacks are the Add-Rotate-Xor (ARX) based hash
functions where one can find differential patterns that propagate with acceptable
probabilities. Additionally, using message modification techniques, significant
complexity reduction is achieved. Consequently, during the SHA-3 competition,
different design concepts were introduced, out of which are the Advanced Encryption Standard (AES) based designs that are known for their resistance to
standard differential attacks due to the wide trail strategy. The ISO standard
Whirlpool [19], the SHA-3 finalist Grøstl [7], and the new Russian hash standard
Stribog [1] are among the proposed AES-based hash functions.

Stribog was proposed in 2010 [13]. It has an output length of 512/256-bit.
The compression function employs a 12-round AES-like cipher with 8×8-byte
internal state preceded with one round of nonlinear whitening of the chaining
value. The compression function operates in Miyaguchi-Preneel (MP) mode and

D. Pointcheval and D. Vergnaud (Eds.): AFRICACRYPT 2014, LNCS 8469, pp. 109–125, 2014.

is plugged in Merkle-Damgård domain extender with a finalization step [1]. Stribog officially replaces the previous standard GOST R 34.11-94 which has been theoretically broken in [16,15] and recently analyzed in [14]. Early works related to the cryptanalysis of Stribog have been introduced in [2,3] and [11].

Following the work of Lai and Massey [12], the meet in the middle (MitM) preimage attack [6] was proposed by Aoki and Sasaki. The main idea of the proposed technique is to divide the attacked rounds into two independent executions such that each execution is affected by a different set of inputs. The outputs of the two executions meet at a matching point where a solution is selected to satisfy both executions. The MitM preimage attack has been applied to MD4 [6,8], MD5 [6], HAS-160 [9], and all functions of the SHA family [5,4,8]. The attack exploits the fact that all the previously mentioned functions are ARX-based and operate in the Davis-Mayer (DM) mode, where the state is initialized by the chaining value and some of the expanded message blocks are used independently each round. Thus, one can determine which message blocks affect each execution for the MitM attack. However, several AES-based hash functions operate in the Miyaguchi-Preneel mode, where the input message is fed to the initial state which undergoes a chain of successive transformations. Consequently, the process of separating independent executions becomes relatively more complicated.

In FSE 2011, Sasaki proposed the first MitM preimage attack on several AES hashing modes [20]. In the same work, a 5-round pseudo preimage attack on the compression function of Whirlpool was presented and used for a second preimage attack on the whole hash function. Afterwards, Wu et al. applied the MitM preimage attack on Grøstl [24] and used a time-memory trade off approach to improve the time complexity of the 5-round attack on the Whirlpool compression function. Lastly, a pseudo preimage attack on the 6-round Whirlpool compression function and a memoryless preimage attack on the reduced hash function were proposed in [21].

In this work, we investigate the security of Stribog and its compression function, assessing their resistance to the MitM preimage attacks. We present a pseudo preimage attack on the compression function reduced to 5 out of 12 rounds by employing the partial matching and initial structure concepts [20]. In particular, we present an execution separation for the compression function that balances the degrees of freedom in both execution directions with their corresponding matching probability [24]. Furthermore, we extend the attack by one round using the guess and determine approach [21], which allows us to guess parts of the state that belongs to one execution. The proposed 6-round chunk separation maximizes the overall complexity of the attack by balancing the adopted degrees of freedom and the guess size. Finally, we show how to generate preimages of the Stribog hash function using the presented pseudo preimage attacks on the compression function. In Table 1, we provide a summary of the current cryptanalytic results on the Stribog hash function.

The rest of the paper is organized as follows. In the next section, the specification of the Stribog hash function along with the notation used throughout the paper are provided. A brief overview of the MitM preimage attack and the

Table 1. Summary of the current cryptanalytic results on Stribog

Target	#Rounds	Time	Memory	Data	Attack	Reference
Internal cipher	5	2^8	2^8	-	Free-start collision	[2]
	8	2^{64}	2^8	-		
Internal permutation	6.5	2^{64}	-	2^{64}	Integral distinguisher	[3]
	7.5	2^{120}	-	2^{120}		
Compression function	7.75	2^{184}	2^8	-	Semi free-start collision	[2]
	4.75	2^8	-	-		
	7.75	2^{72}	2^8	-	Semi free-start near collision	
	8.75	2^{128}	2^8	-		
	9.75	2^{184}	2^8	-		
	5	2^{448}	2^{64}	-	Pseudo preimage	Sec. 4
	6	2^{496}	2^{112}	-		Sec. 5
	6	2^{64}	-	2^{64}	Integral distinguisher	[3]
	7	2^{120}	-	2^{120}		
Hash function	5	2^{481}	2^{256}	-	Preimage	Sec. 6
	6	2^{505}	2^{256}	-		

used approaches are given in Section 3. Afterwards, in Sections 4 and 5, we provide detailed description of the attacks and their corresponding complexity. In Section 6, we show how preimages of the hash function are generated using the attacks presented in Sections 4 and 5. Finally, the paper is concluded and a short discussion is provided in Section 7.

2 Specification of Stribog

Stribog outputs a 512 or 256-bit hash value, where half the last state is truncated when adopting the 256-bit output. The standard specifies two different IVs to be used with the two output lengths. The function can process messages of length up to $2^{512} - 1$. The compression function iterates over 12 rounds of an AES-like cipher with an 8×8 byte internal state and a final round of key mixing. The compression function operates in Miyaguchi-Preneel mode and is plugged in Merkle-Damgård domain extender with a finalization step. The input message M is padded into a multiple of 512 bits by appending one followed by zeros. The message length for MD-strengthening is further included as an extra separate block, followed by a block of a checksum evaluated by the modulo 2^{512} addition of all message blocks as a finalization step. More precisely, let $n = \lfloor \frac{|M|}{512} \rfloor$ and the input message $M = x\|m_n\|..\|m_1\|m_0$, where $|M|$ is length of M, and x is an un-complete or an empty block. The message is padded as follows: let $m_{n+1} = 0^{511-|x|}\|1\|x$, then the padded message $M = m_{n+1}\|m_n\|..\|m_1\|m_0$. Let $\sum = m_{n+1}+..+m_1+m_0$. The compression function g_N is fed with three inputs: the chaining value h_{i-1}, a message block m_{i-1}, and the counter of bits hashed

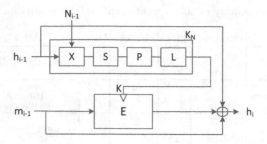

Fig. 1. Stribog's compression function g_N

so far $N_{i-1} = 512 \times i$. (see Figure 1). Let h_i be a 512-bit chaining variable. The first state is loaded with the initial value IV and assigned to h_0. The hash value of M is computed as follows:

$$h_i \leftarrow g_N(h_{i-1}, m_{i-1}, N_{i-1}) \text{ for } i = 1, 2, .., n + 2$$
$$h_{n+3} \leftarrow g_0(h_{n+2}, |M|, 0)$$
$$h(M) \leftarrow g_0(h_{n+3}, \sum, 0),$$

where $h(M)$ is the hash value of M, and g_0 is g_N with $N = 0$. As depicted in Figure 1, the compression function g_N consists of:

- K_N: a nonlinear whitening round of the chaining value. It takes a 512-bit chaining variable h_{i-1} and a counter of the bits hashed so far N_{i-1} and outputs a 512-bit key K.
- E: an AES-based cipher that iterates over the message for 12 rounds in addition to a finalization key mixing round. The cipher E takes a 512-bit key K and a 512-bit message block m as a plaintext. As shown in Figure 2, it consists of two similar parallel flows for the state update and the key scheduling.

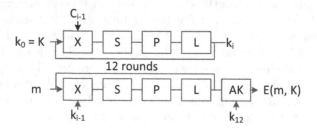

Fig. 2. The internal block cipher (E)

Both K_N and E operate on an 8×8 byte key state K. E updates an additional 8×8 byte message state M. In one round, a given state is updated by the following sequence of transformations:

- AddKey(X): XOR with either a round key, a constant, or the counter of bits hashed so far (N).
- SubBytes (S): A nonlinear byte bijective mapping.
- Transposition (P): Byte permutation.
- Linear Transformation (L): Row multiplication by an MDS matrix in GF(2).

Initially, state K is loaded with the chaining value h_{i-1} and updated by K_N as follows:

$$k_0 = L \circ P \circ S \circ X[N_{i-1}](K).$$

Now K contains the key k_0 to be used by the cipher E. The message state M is initially loaded with the message block m and $E(k_0, m)$ runs the key scheduling function on state K to generate 12 round keys $k_1, k_2, .., k_{12}$ as follows:

$$k_i = L \circ P \circ S \circ X[C_{i-1}](k_{i-1}), \text{ for } i = 1, 2, .., 12,$$

where C_{i-1} is the ith round constant. The state M is updated as follows:

$$M_i = L \circ P \circ S \circ X[k_{i-1}](M_{i-1}), \text{ for } i = 1, 2, ..., 12.$$

The final round output is given by $E(k_0, m) = M_{12} \oplus k_{12}$. The output of g_N in the Miyaguchi Preneel mode is $E(K_N(h_{i-1}, N_{i-1}), m_{i-1}) \oplus m_{i-1} \oplus h_{i-1}$ as shown in Figure 1. For further details, the reader is referred to [1].

2.1 Notation

Let M and K be (8×8)-byte states denoting the message and key state, respectively. The following notation will be used throughout the paper:

- M_i: The message state at the beginning of round i.
- M_i^U: The message state after the U transformation at round i, where $U \in X, S, P, L$.
- $M_i[r, c]$: A byte at row r and column c of state M_i.
- $M_i[\text{row } r]$: Eight bytes located at row r of M_i state.
- $M_i[\text{col } c]$: Eight bytes located at column c of M_i state.

Same notation applies to K.

3   MitM Preimage Attacks on AES-Based Hash Functions

The first preimage attack on AES-based hash functions [20] was proposed for the cryptanalysis of the AES cipher operating in several hashing modes. It is a meet in the middle attack where the attacked rounds are divided at a given round (starting point) into two independent executions called the forward and backward chunks. To maintain the independence constraint, each chunk must be influenced by a different set of inputs. These set of inputs are often called the chunk neutral bytes, e.g., if a change in a given byte affects the forward chunk only, then this byte is known as a forward neutral byte, and consequently, it is a forward degree of freedom as well. Accordingly, the degree of freedom for each execution direction is the number of independent starting values for each execution. Hence, the output of the forward and the backward executions can be independently calculated and stored. Similar to all MitM attacks, the two separated chunks must meet at a common round (matching point) for matching a solution from both the forward and backward directions that satisfies both executions. This is accomplished by adopting the cut and splice technique [6] that employs the mode of operation of the hash functions which chains the input and output states through feedforwarding. More precisely, this technique regards the first and last states as successive rounds. Subsequently, the whole attacked rounds behave in a cyclic manner and one can find a common matching point between the forward and backward executions and one can also select any starting point.

Improvements to this attack aim to stretch the starting and matching points over more than one round state and hence extend the number of the overall attacked rounds. Specifically, the initial structure approach [20] provides the means for the starting point to cover a few successive transformations where bytes in the states belong to both the forward and backward chunks. Although, neutral bytes of both chunks are shared within the initial structure, independence of both executions is achieved in the rounds at the edges of the initial structure. Additionally, the partial matching technique [6] allows only parts of the state to be matched at the matching point. This method is used to extend the matching point further and makes use of the fact that round transformations may update only parts of the state. Thus the remaining unchanged parts can be used for matching. This approach is highly successful in ARX-based hash functions which are characterized by the slow diffusion of their round update functions and so some state variables remain independent in one direction while execution is in the opposite direction. The unaffected parts of the states at each chunk are used for partial matching at the matching point. However, in AES-based hash functions, full diffusion is achieved after two rounds and this approach can be used to extend the matching point of two states for a limited number of transformations. Once a partial match is found, the inputs of both chunks that resulted in the matched values are selected and used to evaluate the remaining undetermined parts of the state at the matching point to check for a full state match.

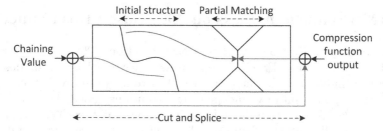

Fig. 3. MitM preimage attack techniques for hash functions operating in MP mode

Figure 3 illustrates the MitM preimage attack approaches when a hash function operates in the Miyaguchi-Preneel mode. The red and blue arrows denote the forward and backward executions on the message state, respectively.

In what follows, we apply the techniques discussed in this section to derive a 5-round pseudo preimage attack on the Stribog compression function.

4   5-Round Pseudo Preimage of the Compression Function

For a compression function CF that operates on a chaining value h and a message block m, a preimage attack is defined as follows: given h and x, where x is the compression function output, find m such that $CF(h, m) = x$. However, in a pseudo preimage attack, only x is given and we must find h and m such that $CF(h, m) = x$. Generally, pseudo preimages of the compression function of some narrow pipe constructions are important because they can be turned to preimages of the hash function with little cost [17]. As for Stribog, the impact of the pseudo preimage attacks on its compression function is demonstrated in Section 6, where we combine these attacks with 2^t multicollision to produce preimages for the hash function. Pseudo preimage attacks are adopted when the compression function operates in Davis-Mayer mode where the first state is initialized by the chaining value. Subsequently, using the cut and splice technique enforces changes in the first state through the feedforward. Additionally, the initial phase of MitM preimage attack usually produces pseudo preimages when the function operates in the Miyaguchi-Preneel mode and the complexity of finding a preimage is higher than the available bits that can be chosen freely in the message. Consequently, the chaining value is utilized as a source of randomization to satisfy the number of multiple restarts required by the attack. As a result, we end up with a pseudo preimage rather than a preimage of the compression function output.

The attack on the compression function starts by chunk separation. Specifically, we divide five rounds of Stribog execution into a forward chunk and a backward chunk around a starting point (initial structure). The adopted chunk separation is shown in Figure 4. The forward chunk starts at M_3 and ends at M_4^P which is the input state to the matching point. The backward chunk starts at M_1^P and ends after the feedforward at M_4^L which is the output state of the matching point. The red bytes are the neutral bytes for the forward chunk and after choosing them in the initial structure, all other red bytes can be independently calculated. White bytes in the forward chunk are the ones whose values depend on the neutral bytes of the backward chunk which are the blue bytes in the initial structure. Accordingly, their values are undetermined, these bytes cannot be evaluated until a partial match is found. Same rationale applies to the backward chunk and the blue bytes. Grey bytes are constants which are either given (compression function output) or chosen (chaining value and constants in the initial structure).

In the initial structure, we try to balance the degrees of freedom in each direction and the number of known bytes at the end of each chunk. The degrees

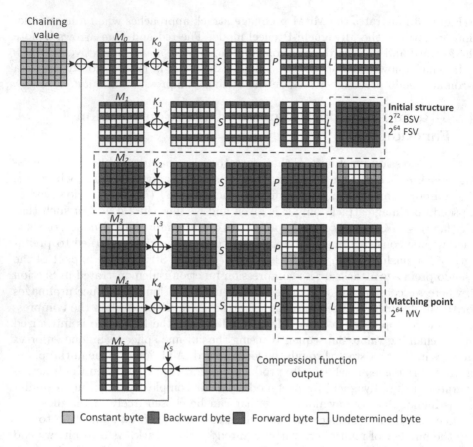

Fig. 4. Chunk separation for a 5-round MitM preimage attack on Stribog compression function. BSV: Backward starting value, FSV: Forward starting value, MV: Matching value.

of freedom in both directions should produce candidate pairs at the matching point to satisfy the matching probability. More precisely, to minimize the complexity, the total degrees of freedom in both chunks must be greater than the matching size. For further clarification, we first explain the idea behind the initial structure. The main point is to choose several bytes as neutral bytes so that the number of output bytes of the L and L^{-1} transformations at the start of each chunk that are constant or relatively constant is maximized. A relatively constant byte is a byte whose value is affected by the degrees of freedom in one execution direction but remains constant from the opposite execution perspective. The initial structure for the 5-round MitM preimage attack on the compression function of Stribog is shown in Figure 5. We start by randomly choosing the five constant bytes in d[row 0] and then determine the values of blue bytes in c[row 0] so that after applying L on c[row 0], we maintain the chosen five constants. Since we need five constant bytes in d[row 0], we only need

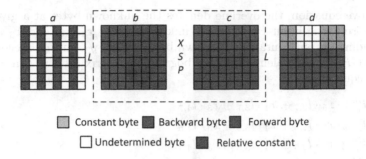

Fig. 5. Initial structure for the 5-round attack on the Stribog compression function

five free variables in $c[\text{row } 0]$ to solve a system of five equations when the other three bytes are fixed. Accordingly, for any of the first three rows in state c, we can randomly choose any three blue bytes and compute the remaining five so that the output of L maintains the previously chosen five constants at $d[\text{row } 0]$. To this end, we have nine free blue bytes (three for each row in state c). Thus the backward degrees of freedom is 2^{72} which means that we can start the backward execution by 2^{72} different starting values and hence 2^{72} different output values at the matching point M_4^L. Similarly, we choose 32 constants in state a and for each row in state b we randomly choose one red byte and compute the other four bytes such that, after the L^{-1} transformation, we get the predetermined constants at each row in a. However, the value of the four shaded blue bytes in each row of state a depends also on the three blue bytes in the rows of state b. We call these bytes relative constants because their final values cannot be determined until the backward execution starts and these values are different for each execution iteration. Specifically, their final values are the predetermined constants XORed with the corresponding blue bytes multiplied by the L^{-1} coefficients. In the sequel, we have eight free bytes (one for each row in b) which means 2^{64} forward degrees of freedom to start the forward execution and hence 2^{64} different input values to the matching point M_4^P.

At the matching point, we match results at M_4^P from the forward chunk with the values at M_4^L from the backward chunk through the L transformation. As depicted in Figure 4 at the matching point, five bytes are known from the forward computation and four bytes are known from the backward computation in each row. As a result, we can form four linear equations using three unknowns and match the resulting forward and backward values through the remaining equation. More precisely, we use the following equation to compute a given output row y through the linear transformation L given an input row x.

$$\begin{bmatrix} x_7 & x_6 & \overline{x_5} & \overline{x_4} & \overline{x_3} & x_2 & x_1 & x_0 \end{bmatrix} \begin{bmatrix} l_{0,7} & l_{0,6} & l_{0,5} & l_{0,4} & l_{0,3} & l_{0,2} & l_{0,1} & l_{0,0} \\ l_{1,7} & l_{1,6} & l_{1,5} & l_{1,4} & l_{1,3} & l_{1,2} & l_{1,1} & l_{1,0} \\ l_{2,7} & l_{2,6} & l_{2,5} & l_{2,4} & l_{2,3} & l_{2,2} & l_{2,1} & l_{2,0} \\ l_{3,7} & l_{3,6} & l_{3,5} & l_{3,4} & l_{3,3} & l_{3,2} & l_{3,1} & l_{3,0} \\ l_{4,7} & l_{4,6} & l_{4,5} & l_{4,4} & l_{4,3} & l_{4,2} & l_{4,1} & l_{4,0} \\ l_{5,7} & l_{5,6} & l_{5,5} & l_{5,4} & l_{5,3} & l_{5,2} & l_{5,1} & l_{5,0} \\ l_{6,7} & l_{6,6} & l_{6,5} & l_{6,4} & l_{6,3} & l_{6,2} & l_{6,1} & l_{6,0} \\ l_{7,7} & l_{7,6} & l_{7,5} & l_{7,4} & l_{7,3} & l_{7,2} & l_{7,1} & l_{7,0} \end{bmatrix} = \begin{bmatrix} y_7 & \overline{y_6} & y_5 & \overline{y_4} & y_3 & \overline{y_2} & y_1 & \overline{y_0} \end{bmatrix}$$

In the above equation, the overline denotes the unknown bytes at a given row. More precisely, the input contains the unknown bytes x_5, x_4, and x_3 and the corresponding output contains the known bytes y_7, y_5, y_3, and y_1. Accordingly, given the $GF(2^8)$ equivalent of the Stribog binary matrix [11], we can form the following equations:

$$y_7 = t_7^{in} \oplus x_5 \cdot l_{2,7} \oplus x_4 \cdot l_{3,7} \oplus x_3 \cdot l_{4,7} \tag{1}$$

$$y_5 = t_5^{in} \oplus x_5 \cdot l_{2,5} \oplus x_4 \cdot l_{3,5} \oplus x_3 \cdot l_{4,5} \tag{2}$$

$$y_3 = t_3^{in} \oplus x_5 \cdot l_{2,3} \oplus x_4 \cdot l_{3,3} \oplus x_3 \cdot l_{4,3} \tag{3}$$

$$y_1 = t_1^{in} \oplus x_5 \cdot l_{2,1} \oplus x_4 \cdot l_{3,1} \oplus x_3 \cdot l_{4,1}, \tag{4}$$

where t_i^{in} is the total of the known input bytes in the ith row multiplied by their corresponding matrix coefficients. To this end, we calculate x_5, x_4, and x_3 from equations 1, 2, and 3 and substitute their values in equation 4. Consequently, the two sides of equation 4 are all known from both input and output directions. Hence, the matching size per row is one byte and hence the matching probability for the whole state is 2^{-64}. The choice of the number forward and backward values directly affects the matching probability as their number determines the number of red and blue bytes at a given row at the matching point. If the number of blue and red bytes are not properly chosen at the initial structure, one might have no value to match at the matching point. In other words, we cannot have a MitM matching value if the total number of red and blue bytes in a given row at the matching point is less than or equal to eight. The attack can be summarized as follows:

1. Randomly choose the chaining value and the constants at the initial structure.
2. For each forward starting value fw_i in the 2^{64} forward starting values at M_2, compute the forward matching value fm_i at M_4^P and store (fw_i, fm_i) in a lookup table T.
3. For each backward starting value bw_j in the 2^{72} backward starting values in M_2^P compute the backward matching value bm_j at M_4^L and check if there exists an $fm_i = bm_j$ in T. If found, then a partial match exists and the full match should be checked using the matched starting points fw_i and bw_i. If a full match exists, then output the chaining value and the message M_0, else go to step 1.

The complexity of the MitM preimage attack is given by $2^n(2^{-r} + 2^{-b} + 2^{-m})$, where n is the state size and r, b, and m are the forward, backward, and matching bit sizes, respectively [24]. The choice of these parameters should minimize the complexity and this can be achieved by keeping r, b and m, as close as possible. In the chunk separation shown in Figure 4, $r = 64$, $b = 72$, and $m = 64$. To further explain the complexity of the attack, we consider the attack procedure. After step 2, we have 2^{64} forward matching values and we need 2^{64} memory to store them. At the end of step 3, we have 2^{72} backward

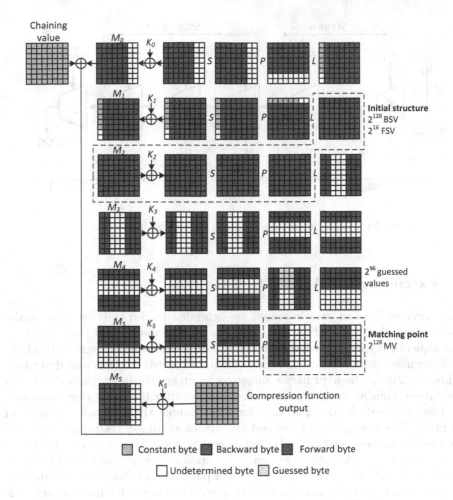

Fig. 6. Chunk separation for a 6-round MitM preimage attack on Stribog compression function. BSV: Backward starting value, FSV: Forward starting value, MV: Matching value.

matching values. Accordingly, we get $2^{64+72} = 2^{136}$ partial matching candidate pairs. Since the probability of a partial match is 2^{-64}, we expect 2^{72} partially matching pairs. The probability that a partial match results in a full match is $2^{64-512} = 2^{-448}$. Consequently, the expected number of fully matching pairs is 2^{-376}. Thus we need to repeat the attack 2^{376} times to get a fully matching pair. The time complexity for one repetition of the attack is 2^{64} for the forward computation, 2^{72} for the backward computation, and 2^{72} to check that partially matching pairs fully match. Consequently, the overall complexity of the attack is $2^{376}(2^{64} + 2^{72} + 2^{72}) \approx 2^{448}$ time and 2^{64} memory

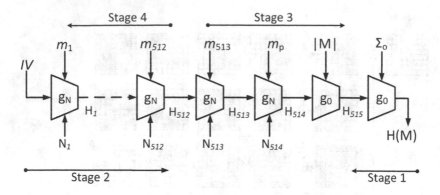

Fig. 7. Preimage attack on the Stribog hash function

5   Extending the Attack to 6-Rounds

The previous 5-round attack cannot be extended to 6-rounds because at the end of each chunk execution the state has undetermined bytes at each row. Consequently, applying the linear transformation L to such state results in a fully undetermined state and no matching can be achieved. A guess and determine approach [21] can be used in one direction to guess the undetermined bytes in some rows. Thus we have some known state rows after the linear transformation L. The proposed chunk separation for the 6-round MitM attack is shown in Figure 6. In order to be able extend the attack by one extra round, we guess the twelve undetermined bytes (yellow bytes) in state M_4^P. As a result, we can reach state M_5^P with four determined columns where matching takes place.

Our choice of the separation and guessed parameters is based on our analysis of the attack complexity and enumerating several values. Our main objective is to maximize the attack probability by carefully selecting the forward, backward, and guessed bit values. We aim to maximize the number of forward bits and keep the backward and the matching number of bits larger than the number of guessed bits and as close as possible. For our attack, the chosen forward, backward, and guessed bit sizes are 16, 128, and 96, respectively. Setting these parameters fixes the matching bit size which is equal to 128. In what follows, we give the attack procedure and complexity based on the above chosen parameters:

1. Randomly choose the chaining value and the constants the initial structure.
2. For each forward starting value fw_i and guessed value g_i in the 2^{16} forward starting values and the 2^{96} guessed values, compute the forward matching value fm_i at M_5^P and store (fw_i, g_i, fm_i) in a lookup table T.
3. For each backward starting value bw_j in the 2^{128} backward starting values, compute the backward matching value bm_j at M_5^L and check if there exists an $fm_i = bm_j$ in T. If found, then a partial match exists and the full match should be checked using the matched forward, guessed, and backwards values

fw_i, g_i, and bw_i. If a full match exists, then output the chaining value and the message M_0, else go to step 1.

After step 2, we have $2^{16+96} = 2^{112}$ forward matching values which need 2^{112} memory for the look up table. At the end of step 3, we have 2^{128} backward matching values. Accordingly, we get $2^{112+128} = 2^{240}$ partial matching candidate pairs. Since the probability of a partial match is 2^{-128} and the probability of a correct guess is 2^{-96}, we expect $2^{240-128-96} = 2^{16}$ correctly guessed partially matching pairs. The probability that a partial match is a full match is 2^{-384}. Consequently, the expected number of fully matching pairs is 2^{-368} and hence we need to repeat the attack 2^{368} times to get a full match. The time complexity for one repetition is 2^{112} for the forward computation, 2^{128} for the backward computation, and 2^{16} to check that partially matching pairs fully match. The overall complexity of the attack is $2^{368}(2^{112} + 2^{128} + 2^{16}) \approx 2^{496}$ time and 2^{112} memory.

6 Preimage of the Stribog Hash Function

In this section, we show how the previously presented pseudo preimage attacks on the Stribog compression function can be utilized to produce preimages for the whole hash function. Stribog has a finalization step which is the last compression function call in the hash function. In this step, the compression function operates on the modular addition of the previously processed message blocks. At first glance, this may seem to limit the ability of turning a pseudo preimage of the compression function to a hash function preimage because inverting the last compression function call returns the sum of the message blocks and thus constraints their values. However, a preimage of the hash function can be found when we consider a large set of long messages that produce different sums and a set pseudo preimage attacks on the last compression function call. Hence, another MitM attack can be performed on both sets to find the message that corresponds to the retrieved sum [15]. As depicted in Figure 7, the attack is divided into four stages:

1. Given the hash function output $H(M)$, we produce 2^p pseudo preimages for the last compression function call. The output of this step is 2^p pairs of the last chaining value and the message sum $(H_{515}, \sum_o)$. We store these results in a table T.
2. In this stage, we construct a large set of equal length messages such that all of them collide at H_{512}. This structure is called a multicollision of length 512 [10]. More precisely, a multicollision of length t is a set of 2^t messages where each message consists of exactly t block and every application of the compression function results in the same chaining value. Consequently, all the 2^t messages lead to the same H_t value. Building a multicollision of length t is done with time complexity of $t \cdot 2^{n/2}$ and memory complexity of $t \cdot 2 \cdot n$ to store t 2-message blocks, where n is the state size. In our case, we build 2^{512} multicollision, i.e., $M_i = m_1^j \| m_2^j \| ... \| m_{512}^j$, where $i \in \{1, .., 2^{512}\}$ and

$j \in \{1,2\}$ such that all the $M_i's$ lead to the same H_{512}. To this end, we have 2^{512} different massages stored in $512 \cdot 2 \cdot 512 = 2^{19}$ memory and hence 2^{512} candidate sums $\sum_{M_i}$.

3. At this point, we try to connect the results of stages 1 and 2 using the freedom of choosing m_{513}. Specifically, since we are using messages of 513 complete blocks, then both the padding block m_p and the length block $|M|$ are known constants. We also have one known value of H_{512} produced from the previous stage. In the sequel, we randomly choose m_{513}^*, compute H_{515}^* and check if it exists in T. As T contains 2^p entries, it is expected to find a match after 2^{512-p} evaluations of the following three compression function calls:

$$H_{513} = g_N(H_{512}, m_{513}^*, N_{513})$$
$$H_{514} = g_N(H_{513}, m_p, N_{514})$$
$$H_{515}^* = g_0(H_{514}, |M|)$$

Once a matching H_{515} value is found in T, the corresponding $\sum_o$ is fixed as well. Hence the desired sum at the output of the multicollision $\sum_{M_i}$ is equal to $\sum_o - m_p - m_{513}$.

4. At the last stage of the attack, we try to find a message M_i out of the 2^{512} messages generated in stage 2 that has a sum equal to the sum $\sum_{M_i}$ acquired at the previous stage. This can be achieved by a meet in the middle attack. More precisely, we first calculate all the 2^{256} sums of the first half of all the 2^{256} messages $\sum_{M_1} = m_1^j + m_2^j + ... + m_{256}^j$ and we store them in a table. Afterwards, for each second half message we compute the sum $\sum_{M_2} = m_{266}^j + m_{267}^j + ... + m_{512}^j$ and check if $\sum_{M_i} - \sum_{M_2}$ is in the table. It is expected to find a match after 2^{256} checks. Once a match is found, the concatenation of the two message halves that correspond to the matching sums and m_{513} is the preimage of the given $H(M)$.

The time complexity of the attack is evaluated as follows: we need $2^P \times$ (complexity of pseudo preimage attack) in stage 1, 512×2^{256} to build the multicollision at stage 2, 2^{512-p} evaluations of three compression function calls at stage 3, and finally 2^{256} for the MitM attack in stage 4. The memory complexity for the four stages is as follows: 2^p 2-states to store the pseudo preimages in stage 1, 512 2-message blocks for the multicollision, and 2^{256} for the MitM table in stage 4. Since the time complexity is highly influenced by p, so we have chosen $p = 32$ for the 5-round attack and $p = 8$ for the 6-round attack to obtain the maximum gain. Accordingly, preimages for 5-round Stribog hash function can be produced with a time complexity of $2^{32+448} + 2^{9+256} + 2^{512-32} \times 3 + 2^{256} \approx 2^{481}$. The time complexity for the 6-round attack is $2^{8+496} + 2^{9+256} + 2^{512-8} \times 3 + 2^{256} \approx 2^{505}$, both attacks have a similar memory complexity of 2^{256} dominated by the MitM attack in stage 4.

7 Conclusion and Discussion

In this paper, we have analyzed Stribog and its compression function with respect to preimage attacks. We have shown that with a carefully balanced chunk separation, pseudo preimages for the 5-round reduced compression function are generated with time complexity of 2^{448} and memory complexity of 2^{64}. Additionally, we have adopted a guess and determine technique to obtain a 6-round chunk separation that maximizes the forward degrees of freedom and balances the backward and the guess bit sizes. As a result, we were able to extend the 5-round attack by one more round with time complexity of 2^{496} and memory complexity of 2^{112}. Finally, using 2^{512} multicollision and another MitM attack, the compression function pseudo preimage attacks are used to produce 5 and 6-round hash function preimages with time complexity of 2^{481} and 2^{505}, respectively. The two preimage attacks have equal memory complexity of 2^{256}.

It should be noted that the Stribog compression function key whitening round K_N enhances its resistance to certain attacks. Specifically, the attacks that require similar diffusion of the executions of both the message and the chaining value. The guess and determine approach is more effective in reducing the complexity when similar chunk separation is performed on the key of the internal cipher to provide additional starting values in both directions [21]. However, key separation cannot be achieved because Stribog has an initial nonlinear whitening round that deviates the chaining value (key) from the message by one round. Hence, even if we were able to start from the middle and separate the chaining value execution, we lose all information when we get to the input chaining value because of the wide trail effect. Similar observation has been noted in [2], where the effect of the additional nonlinear round on finding free-start collision has been discussed. Finally, we know that the presented results do not directly impact the practical security of the Stribog hash function. However, they are forward steps in the public cryptanalysis of this new Russian standard that will likely be included in future suites and protocols.

Acknowledgment. The authors would like to thank the anonymous reviewers for their valuable comments and suggestions that helped improve the quality of the paper. This work is supported by the Natural Sciences and Engineering Research Council of Canada (NSERC).

References

1. The National Hash Standard of the Russian Federation GOST R 34.11-2012. Russian Federal Agency on Technical Regulation and Metrology report (2012), https://www.tc26.ru/en/GOSTR34112012/GOST_R_34_112012_eng.pdf
2. AlTawy, R., Kircanski, A., Youssef, A.M.: Rebound attacks on Stribog. In: ICISC (2013), http://eprint.iacr.org/2013/539.pdf
3. AlTawy, R., Youssef, A.M.: Integral distinguishers for reduced-round stribog. Cryptology ePrint Archive, Report 2013/648 (2013), http://eprint.iacr.org/2013/648.pdf

4. Aoki, K., Guo, J., Matusiewicz, K., Sasaki, Y., Wang, L.: Preimages for step-reduced SHA-2. In: Matsui, M. (ed.) ASIACRYPT 2009. LNCS, vol. 5912, pp. 578–597. Springer, Heidelberg (2009)

5. Aoki, K., Sasaki, Y.: Meet-in-the-middle preimage attacks against reduced SHA-0 and SHA-1. In: Halevi, S. (ed.) CRYPTO 2009. LNCS, vol. 5677, pp. 70–89. Springer, Heidelberg (2009)

6. Aoki, K., Sasaki, Y.: Preimage attacks on one-block MD4, 63-step MD5 and more. In: Avanzi, R.M., Keliher, L., Sica, F. (eds.) SAC 2008. LNCS, vol. 5381, pp. 103–119. Springer, Heidelberg (2009)

7. Gauravaram, P., Knudsen, L.R., Matusiewicz, K., Mendel, F., Rechberger, C., Schläffer, M., Thomsen, S.S.: Grøstl a SHA-3 candidate. NIST Submission (2008)

8. Guo, J., Ling, S., Rechberger, C., Wang, H.: Advanced meet-in-the-middle preimage attacks: First results on full Tiger, and improved results on MD4 and SHA-2. In: Abe, M. (ed.) ASIACRYPT 2010. LNCS, vol. 6477, pp. 56–75. Springer, Heidelberg (2010)

9. Hong, D., Koo, B., Sasaki, Y.: Improved preimage attack for 68-step HAS-160. In: Lee, D., Hong, S. (eds.) ICISC 2009. LNCS, vol. 5984, pp. 332–348. Springer, Heidelberg (2010)

10. Joux, A.: Multicollisions in iterated hash functions. application to cascaded constructions. In: Franklin, M. (ed.) CRYPTO 2004. LNCS, vol. 3152, pp. 306–316. Springer, Heidelberg (2004)

11. Kazymyrov, O., Kazymyrova, V.: Algebraic aspects of the russian hash standard GOST R 34.11-2012. In: CTCrypt, pp. 160–176 (2013), http://eprint.iacr.org/2013/556

12. Lai, X., Massey, J.L.: Hash function based on block ciphers. In: Rueppel, R.A. (ed.) EUROCRYPT 1992. LNCS, vol. 658, pp. 55–70. Springer, Heidelberg (1993)

13. Matyukhin, D., Rudskoy, V., Shishkin, V.: A perspective hashing algorithm. In: RusCrypto (2010) (in Russian)

14. Matyukhin, D., Shishkin, V.: Some methods of hash functions analysis with application to the GOST P 34.11-94 algorithm. Mat. Vopr. Kriptogr 3, 71–89 (2012) (in Russian)

15. Mendel, F., Pramstaller, N., Rechberger, C.: A (second) preimage attack on the GOST hash function. In: Nyberg, K. (ed.) FSE 2008. LNCS, vol. 5086, pp. 224–234. Springer, Heidelberg (2008)

16. Mendel, F., Pramstaller, N., Rechberger, C., Kontak, M., Szmidt, J.: Cryptanalysis of the GOST hash function. In: Wagner, D. (ed.) CRYPTO 2008. LNCS, vol. 5157, pp. 162–178. Springer, Heidelberg (2008)

17. Menezes, A.J., Van Oorschot, P.C., Vanstone, S.A.: Handbook of applied cryptography. CRC press (2010)

18. NIST. Announcing request for candidate algorithm nominations for a new cryptographic hash algorithm (SHA-3) family. In: Federal Register, vol. 72(212) (November 2007), http://csrc.nist.gov/groups/ST/hash/documents/FR_Notice_Nov07.pdf

19. Rijmen, V., Barreto, P.S.L.M.: The Whirlpool hashing function. NISSIE Submission (2000)

20. Sasaki, Y.: Meet-in-the-middle preimage attacks on AES hashing modes and an application to Whirlpool. In: Joux, A. (ed.) FSE 2011. LNCS, vol. 6733, pp. 378–396. Springer, Heidelberg (2011)

21. Sasaki, Y., Wang, L., Wu, S., Wu, W.: Investigating fundamental security requirements on Whirlpool: Improved preimage and collision attacks. In: Wang, X., Sako, K. (eds.) ASIACRYPT 2012. LNCS, vol. 7658, pp. 562–579. Springer, Heidelberg (2012)
22. Wang, X., Yin, Y.L., Yu, H.: Finding collisions in the full SHA-1. In: Shoup, V. (ed.) CRYPTO 2005. LNCS, vol. 3621, pp. 17–36. Springer, Heidelberg (2005)
23. Wang, X., Yu, H.: How to break MD5 and other hash functions. In: Cramer, R. (ed.) EUROCRYPT 2005. LNCS, vol. 3494, pp. 19–35. Springer, Heidelberg (2005)
24. Wu, S., Feng, D., Wu, W., Guo, J., Dong, L., Zou, J.: (Pseudo) preimage attack on round-reduced Grøstl hash function and others. In: Canteaut, A. (ed.) FSE 2012. LNCS, vol. 7549, pp. 127–145. Springer, Heidelberg (2012)

Breaking the IOC Authenticated Encryption Mode

Paul Bottinelli, Reza Reyhanitabar, and Serge Vaudenay

EPFL, Lausanne, Switzerland
{paul.bottinelli,reza.reyhanitabar,serge.vaudenay}@epfl.ch

Abstract. In this paper we cryptanalyse a block cipher mode of operation, called Input Output Chaining (IOC), designed by Recacha and submitted to NIST in 2013 for consideration as a lightweight authenticated encryption mode. We present an existential forgery attack against IOC which makes only one chosen message query, runs in a small constant time, and succeeds with an overwhelming probability $1 - 3 \times 2^{-n}$, where n is the block length of the underlying block cipher. Therefore, this attack fully breaks the integrity of IOC.

Keywords: authenticated encryption, confidentiality, integrity, block cipher, existential forgery.

1   Introduction

An Authenticated Encryption (AE) scheme is a symmetric-key cryptographic scheme whose goal is to guarantee both confidentiality (privacy) and integrity (authenticity) of data. Even though authenticated encryption had been used for many years, it was only in the early 2000s that the security notions for AE, as a distinct cryptographic goal, were formalized in [3, 5, 8]. Having been explored for about two decades, analysis and design of AE schemes yet remains a highly active and interesting area of research, as evidenced by the currently running Competition for Authenticated Encryption: Security, Applicability, and Robustness ("CAESAR") [6].

A popular approach to constructing an AE scheme, adopted by widely-used security protocols such as SSH, SSL/TLS, and IPsec, is to generically combine a confidentiality-only encryption scheme (such as the CBC mode of operation for a block cipher) with a message authentication code (MAC). This approach is neither very efficient as it requires processing the message twice (once to achieve privacy and once to ensure authenticity) nor robust against implementation attacks [7, 18]. A comprehensive analysis of generically composed AE schemes is provided by Bellare and Namprempre [3, 4].

The strive for more efficient AE schemes has been the incentive for considerable effort on constructing several AE algorithms, including block cipher modes of operation for authenticated encryption [1] and dedicated AE designs [2, 10].

Several schemes have tried to use a simple, classical approach known as the "encrypt-with-redundancy" paradigm, where a non-cryptographic checksum is

D. Pointcheval and D. Vergnaud (Eds.): AFRICACRYPT 2014, LNCS 8469, pp. 126–135, 2014.

appended to the message before encrypting it, but almost all such schemes have been fully or partially broken [9, 11–13]. Generic attacks on a large class of such schemes are described by Preneel in [14].

Recently, Mitchell in ACISP 2013 [13] analyzed an AE scheme called Input Output Block Chaining (IOBC), which is based on the encrypt-with-redundancy paradigm. IOBC was proposed in 1996 by Recacha in his PhD thesis and published (in Spanish) in [15]; however, only recently the author provided an English description of IOBC, which was then analyzed by Mitchell [13], who showed a known-plaintext-based forgery attack with a complexity of about $2^{n/3}$ (where n is the block length of the underlying block cipher).

To fix the weakness pointed out by the attack in [13], Recacha revised IOBC and proposed the Input Output Chaining (IOC) to NIST in April 2013 for consideration as a (lightweight) authenticated encryption mode [16]. The designer has recently published the latest revised version of IOC in January 2014 [17], to fix the flaws in its original guideline for fresh IV generation for IOC.

In this paper, we present an attack on IOC which applies both to the original version as submitted to NIST [16] and the most recently revised version available from [17]. Our attack is an existential forgery attack that can forge a ciphertext, having queried just one short chosen message (of length 5 blocks) within a small constant time; therefore, it completely breaks the integrity of IOC as an AE scheme.

Organization of the Paper. In Section 2 we provide a brief description of the IOC scheme. In Section 3 we explain our forgery attack against IOC. The paper is concluded in Section 4.

2   Description of IOC

In this section, we first describe the latest version of IOC [17] and then we will point out the changes made by the designer to get this version from the originally submitted version in [16]. The differences between the two versions do not affect the way that our attack works, hence we will only briefly overview them.

IOC [17] is a mode of operation for a block cipher. Let $E_K(\cdot)$ denote the encryption operation of the block cipher under the key K. Let the block length of the block cipher be n bits. During the lifetime of a key K (called a "security session") each message P to be encrypted gets a unique sequence number denoted by S. When such a security session is initiated, the sequence number S is reset to 1 and then is incremented synchronously for each message exchanged between the sender and the receiver. To encrypt a message $P = P_1 \cdots P_m$ with message number S, the encryption algorithm of IOC uses two random and secret initial values, IV_a and IV_b, which are renewed for each message. It is assumed that the message length is a multiple of the block length n, otherwise a length indicating padding can be used to make it so.

The encryption and decryption processes of IOC are described in Algorithm 1 and Algorithm 2 and depicted in Fig. 1. The value ICV (Integrity Check

Vector) is a secret value computed using the secret initial values IV_a and IV_b, the message number S, and the length of the message in blocks, i.e. m. The tag value MDC (Modification Detection Code) aims to detect any forged ciphertext. $\langle x \rangle_n$ denotes the representation of an integer x as an n-bit string and $+$ denotes the regular arithmetic addition modulo 2^n.

When a security session corresponding to a key K is established, the values IV_a and IV_b are computed as follows:

$$IV_a = E_{K'}(\langle 0 \rangle_n)$$
$$IV_b = E_{K'}(IV_a)$$

where $K' = K + S$.

When subsequent messages are encrypted, the following two possibilities are presented. Either use the last inner vectors, namely $IV_a = O_{m+1}$ and $IV_b = I_{m+1}$, or derive new values with the method presented above, with an updated value of S.

Algorithm 1: $\text{IOC.Encrypt}_K(S, IV_a, IV_b, P)$

1: Let $P = P_1 P_2 \cdots P_m$ where $|P_i| = n$
2: $O_0 \leftarrow IV_a$
3: $I_0 \leftarrow IV_b$
4: $ICV \leftarrow (\langle S \rangle_n \oplus IV_a) + (\langle m \rangle_n \oplus IV_b)$
5: **for** $i \leftarrow 1$ **to** m **do**
6:      $I_i \leftarrow P_i \oplus O_{i-1}$
7:      $O_i \leftarrow E_K(I_i)$
8:      $C_i \leftarrow O_i + I_{i-1}$
9: **end for**
10: $MDC \leftarrow I_m + E_K(ICV \oplus O_m)$
11: **return** $C||MDC = C_1 \ldots C_m||MDC$

2.1   Differences between Two Versions of IOC

The two versions of IOC, namely the one submitted to NIST in [16] (the "old" version) and the latest version in [17] (as described in Algorithm 1 and Algorithm 2) have the same iteration structure for the encryption and decryption parts except the final application of the block cipher for generation and verification of the tag MDC. They also differ in the way that the secret and random initial values IV_a and IV_b are generated. The latter difference, i.e. generation of the IVs, does not make any change as far as our attack is concerned, so we do not detail this change.

As for the first difference, we note that MDC in the old version of IOC [16] is computed as below:

$$MDC = E_{O_m \oplus \langle S \rangle_n}(\langle m \rangle_n \oplus I_m)$$

Algorithm 2: IOC.Decrypt$_K(S, IV_a, IV_b, C'\|MDC')$

1: Let $C' = C_1'C_2' \cdots C_m'$ where $|C_i'| = n$
2: $Q_0 \leftarrow IV_a$
3: $Y_0 \leftarrow IV_b$
4: $ICV \leftarrow (\langle S \rangle_n \oplus IV_a) + (\langle m \rangle_n \oplus IV_b)$
5: **for** $i \leftarrow 1$ **to** m **do**
6:     $Q_i \leftarrow C_i' - Y_{i-1}$
7:     $Y_i \leftarrow D_K(Q_i)$
8:     $R_i \leftarrow Y_i \oplus Q_{i-1}$
9: **end for**
10: $ICV' \leftarrow Q_m \oplus D_K(MDC' - Y_m)$
11: **if** $ICV' = ICV$ **then**
12:     **return** $R = R_1 R_2 \ldots R_m$
13: **else**
14:     **return** INVALID
15: **end if**

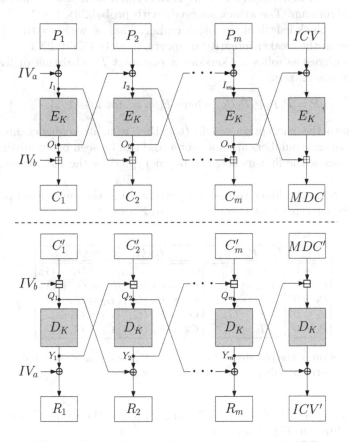

Fig. 1. Encryption and decryption processes in IOC

where $\langle x \rangle_n$ denotes the representation of x as an n-bit string.

As we will show in the following section, our attack is not prevented by this MDC generation; that is, we have a successful forgery attack against both the new and old versions of IOC.

3   Forgery Attack against IOC

Two notions of integrity (authenticity) can be considered for AE schemes: integrity of plaintexts (INT-PTXT) and integrity of ciphertexts (INT-CTXT) [3, 4]. INT-PTXT requires that producing a ciphertext that is decrypted to a *new* message (never encrypted by the sender) must be computationally infeasible. INT-CTXT requires that producing any new ciphertext that can be decrypted (i.e. not rejected) must be computationally infeasible, regardless of whether or not the decrypted plaintext is new. Clearly, INT-CTXT is a stronger property than INT-PTXT and AE schemes aim to provide INT-CTXT.

We show an attack which only needs a single chosen message to be encrypted and then forges a new ciphertext that is decrypted to a *new* plaintext by the decryption algorithm. The attack succeeds with probability $1 - 3 \times 2^{-n}$ (≈ 1), where n is the block length of the block cipher. That is, we show that IOC does not satisfy even the weaker integrity property, namely INT-PTXT.

The attack goes as follows. Consider a plaintext P consisting of five blocks whose values are all zero:

$$P = P_1 P_2 P_3 P_4 P_5, \text{ where } P_i = 0^n \text{ for } i \in \{1, 5\},$$

and send it to the encryption oracle (together with its sequence number S). Note that sequence numbers are not secret and can be seen by an adversary. For example, if this is the first message to be encrypted by the IOC using a key K then $S = 1$.

Figure 2 shows the inner values computed during the encryption process. A picture of the encryption of P can be seen in Figure 3.

P_i	$I_i = P_i \oplus O_{i-1}$	$O_i = E_K(I_i)$	$C_i = O_i + I_{i-1}$
$P_1 = 0^n$	$I_1 = IV_a$	$O_1 = E_K(IV_a)$	$C_1 = O_1 + IV_b$
$P_2 = 0^n$	$I_2 = O_1$	$O_2 = E_K(O_1)$	$C_2 = O_2 + IV_a$
$P_3 = 0^n$	$I_3 = O_2$	$O_3 = E_K(O_2)$	$C_3 = O_3 + O_1$
$P_4 = 0^n$	$I_4 = O_3$	$O_4 = E_K(O_3)$	$C_4 = O_4 + O_2$
$P_5 = 0^n$	$I_5 = O_4$	$O_5 = E_K(O_4)$	$C_5 = O_5 + O_3$

Fig. 2. Encryption of the plaintext P with IOC. The columns show the different inner values and the corresponding ciphertext blocks obtained.

We get the ciphertext $C_1 C_2 C_3 C_4 C_5 \| MDC$, where the values C_i are as in the rightmost column in Fig. 2, and we have

$$MDC = I_5 + E_K(ICV \oplus O_5). \tag{1}$$

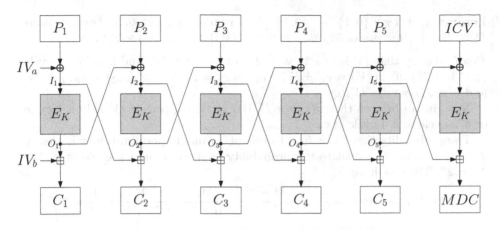

Fig. 3. Encryption process in IOC with a plaintext of 5 blocks

Now, we modify C to forge a ciphertext $C' = C_1'C_2'C_3'C_4'C_5'\|MDC$ as follows

$$C_1' = C_5 - C_3 + C_1$$
$$C_2' = C_4$$
$$C_3' = C_3$$
$$C_4' = C_4$$
$$C_5' = C_5.$$

First, we need to show that the forged ciphertext C' is different from C. This is easy to verify as follows. If $C' = C$ then we must have $C_1' = C_1$ and $C_2' = C_2$, implying that

$$C_1' = C_1 \iff C_5 - C_3 + C_1 = C_1 \iff C_5 = C_3$$
$$\iff O_5 + O_3 = O_3 + O_1 \iff O_5 = O_1$$
$$\iff E_K(E_K(E_K(E_K(E_K(IV_a))))) = E_K(IV_a)$$
$$\iff E_K(E_K(E_K(E_K(IV_a)))) = IV_a$$

and

$$C_2' = C_2 \iff C_4 = C_2 \iff O_4 + O_2 = O_2 + IV_a$$
$$\iff O_4 = IV_a \iff E_K(E_K(E_K(E_K(IV_a)))) = IV_a$$

As the underlying block cipher E is assumed to be a secure block cipher; i.e. $E_K(.)$ (for a secret random key K) is indistinguishable from a random permutation $\pi : \{0,1\}^n \to \{0,1\}^n$, it can be shown (Lemma 1) that the event specified by the condition above will occur with a (negligible) probability 3×2^{-n}; hence, with an overwhelming probability $1 - 3 \times 2^{-n}$ (≈ 1) the forged ciphertext C' is new.

Lemma 1. *Let $\pi : \{0,1\}^n \to \{0,1\}^n$ be a random permutation. For any value $IV \in \{0,1\}^n$, the probability that $\pi(\pi(\pi(\pi(IV)))) = IV$ is 3×2^{-n}.*

Proof. We say that IV is in a 1-cycle of π iff $\pi(IV) = IV$, and it is in a k-cycle $(1 < k \leq 2^n)$ iff: $\pi(IV) = v_1 \neq IV, \pi(v_1) = v_2 \neq IV, \cdots, \pi(v_{k-2}) = v_{k-1} \neq IV$ and $\pi(v_{k-1}) = v_k = IV$.

Now, it can be seen that $\pi(\pi(\pi(\pi(IV)))) = IV$ happens if IV is in a 1-cycle or a 2-cycle or a 4-cycle of π.

The probability that IV is in a 1-cycle of a random permutation π is clearly 2^{-n}. It remains to calculate the probability that IV is in a k-cycle of π. Let $M = 2^n$. Then we have

$$\Pr[IV \text{ is in a k-cyle}] = \frac{M-1}{M}\frac{M-2}{M-1}\cdots\frac{M-k}{M-k+1}\frac{1}{M-k} = \frac{1}{M} = 2^{-n}$$

So, $\Pr\left[\pi(\pi(\pi(\pi(IV)))) = IV\right] = 3 \times 2^{-n}$ □

The attack proceeds as follows. Send $C' = C_1'C_2'C_3'C_4'C_5'||MDC$ to the decryption oracle (receiver) together with *the same sequence number S* (corresponding to the queried plaintext P). In other words, just modify the original ciphertext C to get C', then forward C' *in place of C* to the receiver for decryption.

Note that C' is forwarded in the same security session as the one in which P was queried. As a result, the key K has not changed. Thus, when sending the sequence number S corresponding to P, the values IV_a and IV_b derived in the decryption process for the forged ciphertext C' are the same that the ones used to protect the chosen plaintext P.

Fig. 4 shows the inner values computed during the decryption process of the ciphertext C'. A picture of the decryption of C' can be seen in Fig. 5.

C_i'	$Q_i = C_i' - Y_{i-1}$	$Y_i = D_K(Q_i)$	$R_i = Y_i \oplus Q_{i-1}$
$C_1' = O_5 + IV_b$	$Q_1 = O_5$	$Y_1 = D_K(O_5) = O_4$	$R_1 = O_4 \oplus IV_a$
$C_2' = O_4 + O_2$	$Q_2 = O_2$	$Y_2 = D_K(O_2) = O_1$	$R_2 = O_1 \oplus O_5$
$C_3' = O_3 + O_1$	$Q_3 = O_3$	$Y_3 = D_K(O_3) = O_2$	$R_3 = O_2 \oplus O_2 = 0$
$C_4' = O_4 + O_2$	$Q_4 = O_4$	$Y_4 = D_K(O_4) = O_3$	$R_4 = O_3 \oplus O_3 = 0$
$C_5' = O_5 + O_3$	$Q_5 = O_5$	$Y_5 = D_K(O_5) = O_4$	$R_5 = O_4 \oplus O_4 = 0$

Fig. 4. Decryption of the ciphertext C' with IOC. The columns show the different inner values and the corresponding plaintext blocks obtained.

The computation of the value ICV' as in line 10 of Algorithm 2 goes as follows

$$ICV' = Q_5 \oplus D_K(MDC - Y_5) \tag{2}$$
$$= Q_5 \oplus D_K((I_5 + E_K(ICV \oplus O_5)) - Y_5) \tag{3}$$
$$= Q_5 \oplus D_K(E_K(ICV \oplus O_5)) \tag{4}$$
$$= Q_5 \oplus (ICV \oplus O_5) \tag{5}$$
$$= ICV \tag{6}$$

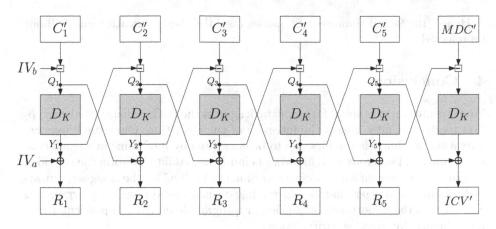

Fig. 5. Decryption process in IOC with a ciphertext of 5 blocks and the MDC

where equalities (2) and (3) come from Equation 1 and equalities (3) and (4) are driven from the fact that $Y_5 = O_4$ and $O_4 = I_5$ as it can be seen in Fig. 4 and Fig. 2.

This shows that the verification step in line 11 of Algorithm 2 will pass, and thus the ciphertext C' will be considered as valid. Finally, we will get the plaintext $R = R_1R_2R_3R_4R_5$ as computed in the rightmost column in Fig. 4 and depicted in Fig. 5.

It is easy to verify that R is a new message, i.e. different from the queried message P. To do so, note that $R_1 \neq P_1 = 0$ with probability $1 - 3 \times 2^{-n}$, because $R_1 = 0$ iff $O_4 = IV_a$ iff $E_K(E_K(E_K(E_K(IV_a)))) = IV_a$ and this happens only with probability 3×2^{-n} when the block cipher is secure (PRP).

Hence, with probability $1 - 3 \times 2^{-n}$, we can forge a new ciphertext which is decrypted to a new plaintext.

Attack on the Old Version of IOC. Interestingly, our attack also applies to the old version of IOC [16] (see subsection 2.1). We remind that the two versions of IOC have the same iteration structure for the encryption and decryption processes except the final application of the block cipher for generation and verification of MDC.

In the old version of IOC [16] the MDC is computed as below:

$$MDC = E_{O_m \oplus \langle S \rangle_n}(\langle m \rangle_n \oplus I_m)$$

Hence, the tag MDC for the queried message $P = P_1 \cdots P_5$, where $P_i = 0^n$ for all i, is computed (see Fig. 2) as

$$MDC = E_{O_5 \oplus \langle S \rangle_n}(\langle 5 \rangle_n \oplus I_5) = E_{O_5 \oplus \langle S \rangle_n}(\langle 5 \rangle_n \oplus O_4).$$

The tag computed during the decryption of $C' = C_1'C_2'C_3'C_4'C_5'||MDC$ is computed (see Fig. 4) as

$$MDC' = E_{Q_5 \oplus \langle S \rangle_n}(\langle 5 \rangle_n \oplus Y_5) = E_{O_5 \oplus \langle S \rangle_n}(\langle 5 \rangle_n \oplus O_4) = MDC.$$

Hence, the forged ciphertext C' passes the MDC tag verification and will not be rejected.

4   Conclusion

We presented an existential forgery attack against the IOC authenticated encryption mode that fully breaks the integrity of the algorithm. Our attack requires only a single, short chosen message to be encrypted by IOC, runs in a small constant time, and succeeds with an overwhelming probability. Despite the provided lengthy but rather informal security arguments for IOC by the designer, our attack shows the danger that lies in relying on non-rigorous security arguments and stresses the importance of backing cryptographic designs, in particular new AE schemes, by sound security proofs.

References

1. Authenticated Encryption Modes. National Institute of Standards and Technology, http://csrc.nist.gov/groups/ST/toolkit/BCM/modes_development.html#01
2. ISO/IEC 19772:2009: Information technology – Security techniques – Authenticated encryption. International Organization for Standardization, Geneva, Switzerland (2009)
3. Bellare, M., Namprempre, C.: Authenticated Encryption: Relations among Notions and Analysis of the Generic Composition Paradigm. In: Okamoto, T. (ed.) ASIACRYPT 2000. LNCS, vol. 1976, pp. 531–545. Springer, Heidelberg (2000)
4. Bellare, M., Namprempre, C.: Authenticated Encryption: Relations among Notions and Analysis of the Generic Composition Paradigm. J. Cryptology 21(4), 469–491 (2008)
5. Bellare, M., Rogaway, P.: Encode-Then-Encipher Encryption: How to Exploit Nonces or Redundancy in Plaintexts for Efficient Cryptography. In: Okamoto, T. (ed.) ASIACRYPT 2000. LNCS, vol. 1976, pp. 317–330. Springer, Heidelberg (2000)
6. Bernstein, D.J.: Cryptographic competitions: CAESAR, http://competitions.cr.yp.to
7. Canvel, B., Hiltgen, A.P., Vaudenay, S., Vuagnoux, M.: Password Interception in a SSL/TLS Channel. In: Boneh, D. (ed.) CRYPTO 2003. LNCS, vol. 2729, pp. 583–599. Springer, Heidelberg (2003)
8. Katz, J., Yung, M.: Unforgeable Encryption and Chosen Ciphertext Secure Modes of Operation. In: Schneier, B. (ed.) FSE 2000. LNCS, vol. 1978, pp. 284–299. Springer, Heidelberg (2001)
9. Kohl, J.T.: The use of Encryption in Kerberos for Network Authentication. In: Brassard, G. (ed.) CRYPTO 1989. LNCS, vol. 435, pp. 35–43. Springer, Heidelberg (1990)
10. Krovetz, T., Rogaway, P.: The software performance of authenticated-encryption modes. In: Joux, A. (ed.) FSE 2011. LNCS, vol. 6733, pp. 306–327. Springer, Heidelberg (2011)
11. Menezes, A., van Oorschot, P.C., Vanstone, S.A.: Handbook of Applied Cryptography. CRC Press (1996)

12. Mitchell, C.J.: Cryptanalysis of Two Variants of PCBC Mode When Used for Message Integrity. In: Boyd, C., González Nieto, J.M. (eds.) ACISP 2005. LNCS, vol. 3574, pp. 560–571. Springer, Heidelberg (2005)
13. Mitchell, C.J.: Analysing the IOBC Authenticated Encryption Mode. In: Boyd, C., Simpson, L. (eds.) ACISP. LNCS, vol. 7959, pp. 1–12. Springer, Heidelberg (2013)
14. Preneel, B.: Cryptographic Primitives for Information Authentication - State of the Art. In: Preneel, B., Rijmen, V. (eds.) State of the Art in Applied Cryptography. LNCS, vol. 1528, pp. 49–104. Springer, Heidelberg (1998)
15. Recacha, F.: IOBC: Un nuevo modo de encadenamiento para cifrado en bloque. In: Proceedings: IV Reunion Espanola de Criptologia, Valladolid, pp. 85–92 (September 1996)
16. Recacha, F.: IOC: The Most Lightweight Authenticated Encryption Mode? National Institute of Standards and Technology, Modes Development, Proposed Modes (April 2013),
    http://csrc.nist.gov/groups/ST/toolkit/BCM/modes_development.html
17. Recacha, F.: Input Output Chaining (IOC) AE Mode Revisited (January 2014),
    http://inputoutputblockchaining.blogspot.ch/
18. Vaudenay, S.: Security Flaws Induced by CBC Padding - Applications to SSL, IPSEC, WTLS... In: Knudsen, L.R. (ed.) EUROCRYPT 2002. LNCS, vol. 2332, pp. 534–546. Springer, Heidelberg (2002)

New Treatment of the BSW Sampling and Its Applications to Stream Ciphers[*]

Lin Ding[1], Chenhui Jin[1], Jie Guan[1], and Chuanda Qi[2]

[1] Information Science and Technology Institute, 450000 Zhengzhou, China
[2] Xinyang Normal University, 464000 Xinyang, China
{dinglin_cipher,guanjie007}@163.com, jinchenhui@126.com,
qichuanda@sina.com

Abstract. By combining the time-memory-data tradeoff (TMDTO) attack independently proposed by Babbage and Golić (BG) with the BSW sampling technique, this paper explores to mount a new TMDTO attack on stream ciphers. The new attack gives a wider variety of trade-offs, compared with original BG-TMDTO attack. It is efficient when multiple data is allowed for the attacker from the same key with different IVs, even though the internal state size is twice the key size. We apply the new attack to MICKEY and Grain stream ciphers, and improves the existing TMDTO attacks on them. Our attacks on Grain v1 and Grain-128 stream ciphers are rather attractive in the respect that the online time, offline time and memory complexities are all better than an exhaustive key search, and the amount of keystream needed are completely valid. Finally, we generalize the new attack to a Guess and Determine-TMDTO attack on stream ciphers, and mount a Guess and Determine-TMDTO attack on SOSEMANUK stream cipher with the online time and offline time complexities both equal to 2^{128}, which achieves the best time complexity level compared with all existing attacks on SOSEMANUK so far.

Keywords: Cryptanalysis, Time-memory-data tradeoff attack, BSW sampling, Guess and Determine attack, Stream cipher, MICKEY, Grain, SOSEMANUK.

1 Introduction

Stream ciphers can be described as keyed generators of pseudo random sequences over a finite field. Usually, the problem of recovering the secret key of stream cipher can be generalized as the problem of inverting a one-way function $y = f(x)$. Exhaustive key search and table lookup attack are two extreme examples of generic attacks to invert one-way functions. Time-Memory tradeoff (TMTO) attack is a method combining the exhaustive key search and the table lookup attack, and offers a generic technique to invert one-way functions, where one

[*] This work is supported in part by the National Natural Science Foundation of China (No. 61202491, 61272041, 61272488) and Foundation of Science and Technology on Information Assurance Laboratory (Grant No. KJ-13-007).

D. Pointcheval and D. Vergnaud (Eds.): AFRICACRYPT 2014, LNCS 8469, pp. 136–146, 2014.

can trade off time and memory costs. A typical TMTO attack consists of two phases, i.e., the offline (or pre-computation) phase and the online phase. The complexities of TMTO attack can be evaluated by looking at three main parameters, i.e., the online time complexity T, the memory cost M, and the offline time complexity P.

The idea of TMTO attack was originally proposed by Hellman [1] for attacking the DES block cipher. The attack has a lower time complexity (in online phase) than the exhaustive key search and a lower memory complexity than the table lookup attack. Its tradeoff curve is obtained as $TM^2 = N^2$ and $P = N$, where N is the number of possible keys. Hence, a reasonable choice of M and T is $T = M = N^{2/3}$, which is lower than the exhaustive key search. However, the offline time complexity of Hellmans attack is always no less than the time complexity of the exhaustive key search.

Babbage [2] and Golić [3] independently proposed a simple time-memory-data tradeoff (TMDTO) attack on stream ciphers. The tradeoff curve of Babbage-Golić (BG-TMDTO) attack can be represented as $TM = N$, $P = M$ and $T = D$, where D is the amount of data available to the attacker. Here, N is the number of possible internal states. In ASIACRYPT 2000, Biryukov and Shamir [4] found that TMTO attacks against stream ciphers can be extended to TMDTO attacks by utilizing multiple data points. The idea of Biryukov-Shamir (BS-TMDTO) attack was similar to the original attack by Hellman. The tradeoff curve of this attack can be represented as $TM^2D^2 = N^2$ and $P = N/D$, while the restriction $1 \leq D^2 \leq T$ has to be satisfied. The attack was applied to one of the most widely deployed stream ciphers, GSM's A5/1. In these attacks, the attacker tries to invert the function mapping the internal state of stream cipher to a segment of the keystream output. These attacks imply that the state size should be at least twice the key size, which is widely considered as an essential design principle for modern stream ciphers.

On a different attack scenario, TMDTO attacks can also be used to invert the function mapping the initial inputs (e.g., Key and IV) of stream cipher to a segment of the keystream output. In the TMDTO attack proposed by Hong and Sarkar [5], they treat both the secret key and IV as unknown in the offline phase. The trade-off curve for Hong-Sarkar (HS-TMDTO) attack is the same as the BS-TMDTO attack with $N = K \times V$. Here, denote K and V the numbers of possible keys and IVs respectively.

In [6], Dunkelman and Keller presented a new approach to TMDTO attacks against stream ciphers. They did not treat the IV as part of the secret key material, after exploiting the fact that the IV is known during an online attack. The Dunkelman-Keller (DK-TMDTO) attack get the same trade-off curve as $TM^2D^2 = N^2$ and $P = N/D$, but with the restriction $1 \leq D^2 \leq T$ replaced by the restrictions $V \geq D$ and $T \geq D$. The attack implies that if the key length is n bits, the cipher offers n-bit security with respect to DK-TMDTO attacks only if the IV length is at least $1.5n$ bits. However, its offline time complexity cannot be faster than exhaustive key search irrespective of the IV length.

The concept of BSW Sampling was introduced by Biryukov, Shamir and Wagner [7] at FSE 2000. It helps the BS-TMDTO attack to get a wider choice of parameters by relaxing its restriction. However, a view exists that the BG-TMDTO attack is not really helped as much by the BSW sampling technique [8]. This paper explores the possibility of combining the BG-TMDTO attack with the BSW sampling technique to mount a new TMDTO attack on stream ciphers. The new attack can be considered as a generalization of BG-TMDTO attack. It is rather efficient when multiple data is allowed for the attacker from the same key with different IVs, even though the internal state size is twice the key size. As applications, we mount new TMDTO attacks on MICKEY and Grain stream ciphers. Finally, a general Guess and Determine-TMDTO attack on stream ciphers is presented by exploiting the fact that the BSW sample technique can be generalized to a simple Guess and Determine attack. As an application, we give a Guess and Determine-TMDTO attack on SOSEMANUK stream cipher, which achieves the best time complexity level compared with all existing attacks on SOSEMANUK so far.

This paper is organized as follows. Our new attack is proposed in Section 2. Applications of the new attack and some discussions on it are given in Section 3. The paper is concluded in Section 4.

2   New Time-Memory-Data Tradeoff Attack

2.1   BSW Sampling Technique

Inverting a one-way function has an important role in the security of most encryption schemes. The problem of recovering the internal state of a stream cipher, given multiple keystream segments, can be generalized as an inversion problem, as shown as follows.

Given a one-way function $f : X \to Y$ and a set $D = \{y_i\}_i \subset Y$, to identify an internal state $x \in X$ such that $f(x) = y_i$ for some i.

Where $N = |X|$ denotes the number of possible internal states, and Y the set of enough keystream segments.

The BSW sampling aims at obtaining wider choices of tradeoff parameters for the tradeoff curve of BS-TMDTO attack. Its main idea is to find an efficient way to generate and enumerate special cipher states, from which the first subsequent keystream output bits of the cipher are a fixed string (such as a run of consecutive 1 or 0 bits). If this can be done for a run of l bits, the sampling resistance of the cipher is defined to be $R = 2^{-l}$. Usually, the BSW sampling tradeoff works when the following assumption is satisfied for a given stream cipher.

Assumption 1. For a given stream cipher with the internal state size $n = \log_2 N$, given the value of $n - l$ particular state bits of it and the first l keystream bits produced from that state, the remaining l internal state bits may be deduced directly.

It is easy to see that the sampling resistance of the given stream cipher is $R = 2^{-l}$ under the assumption. Given the sampling resistance, the attacker can apply the BS tradeoff to the problem of inverting the restricted function $f' : X' \to Y'$, rather than the function $f : X \to Y$. The restricted function is obtained as follows.

---

1. Fix a specific function by choosing an l-bit string S.
2. Given an $(n - l)$-bit input value x, treat S as the first l bits of keystream, compute the remaining l state bits according to the assumption above, and then expand it to n bits.
3. Clock the stream cipher n steps, generating an n-bit keystream segment $S|y$.
4. Output y.

Clearly, the inversion problem of inverting the function f mapping n-bit states to n-bit keystream segments is equivalent to the inversion problem of inverting the restricted function $f' : \{0,1\}^{n-l} \to \{0,1\}^{n-l}$. Thus, the attacker would consider the cost of inverting f' rather than the full stream cipher.

The trade-off curve for BSW sampling is the same as the BS-TMDTO attack, i.e., $TM^2D^2 = N^2$ and $P = N/D$, while a wider choice of parameters by relaxing the restriction $1 \leq D^2 \leq T$ to $1 \leq R^2D^2 \leq T$. The BSW sampling technique has been applied to MICKEY and Grain stream ciphers, see [9,8] for more details. The BSW sampling technique allows mounting the BS-TMDTO attack in a larger variety of settings by relaxing the restriction, though the obtained tradeoff curve keeps unchanged.

2.2 New Treatment of the BSW Sampling Technique

As shown above, the BSW sampling technique helps the BS-TMDTO attack to get a wider choice of parameters by relaxing its restriction. However, claimed by [8], the BG tradeoff is not really helped as much by the BSW sampling, as the amount of keystream needed remains prohibitive. In fact, if the keystream segments available for cryptanalysis are generated by only one single (K, IV), the required amount of keystream is indeed prohibitive. However, if the proposed cryptanalysis does not require a long keystream generated by one single (K, IV) pair but a (large) number of short keystreams generated by the same key with different IVs, the required amount of keystream may be completely valid in this respect. This motivates us to explore the possibility of combining the BG-TMDTO attack with the BSW sampling technique to mount a new TMDTO attack. In this attack, the keystream segments available for cryptanalysis are generated by the same key with different IVs.

Assume that an attacker can collect a set of d keystream sequences generated by the given stream cipher for different IVs, and that the length of each sequence is d'. Accordingly, we assume that the set of samples available for the cryptanalysis consists of approximately $D = d \cdot d'$ $2n$-bit keystream segment samples.

Like the typical TMDTO attack, the new attack consists of two phases, i.e., the offline phase and the online phase. The attack utilizes an integer parameter r satisfying the restriction $1 \leq r \leq R^{-1}$.

The offline phase is to construct some tables consisting of pairs of internal state and corresponding keystream segment. The algorithm for the offline phase is described as follows.

The Offline Algorithm

Choose r strings $S_1, \cdots, S_r$ randomly, and each consists of l bits. For each fixed string S_i, do the followings.

1. Choose N' strings $I_1, \cdots, I_{N'}$ randomly, and each consists of $n - l$ bits.
2. Treat S_i as the first l bits of keystream and I_j as the $n-l$ particular state bits, compute the remaining l bits, clock the stream cipher n steps to generate an n-bit keystream segment, and then memory the (n-bit internal state, n-bit keystream segment) pair in the table T_i.

The online phase is to recover an internal state which has generated a keystream segment in one table. The algorithm for the online phase is described as follows.

The Online Algorithm

For each $2n$-bit keystream segment sample available for the cryptanalysis, check if the first l bits of the sample match one of r strings $S_1, \cdots, S_r$. If a matching is not found, go to check the next $2n$-bit keystream segment sample. If a matching is found, do the followings.

1. For each matching (say S_i) found, check if the first n-bit keystream segment exists in the second column of the corresponding table (i.e., T_i). If it does not exist, move to consideration of the next $2n$-bit keystream segment sample. If it exists, read the corresponding n-bit internal state in the first column of the table, clock the stream cipher $2n$ steps to generate $2n$ keystream bits, and then match them with the sample. If the match passes, go to the output (a). Otherwise, move to consideration of the next $2n$-bit keystream segment sample.
2. If no more keystream segment samples, go to the output (b).

Output: (a) recovered n-bit internal state; (b) a flag that the algorithm has failed.

The complexities of the proposed TMDTO attack are calculated as follows.

In the offline phase, for each fixed string S_i, N' strings should be chosen randomly to execute the Step 2. Thus, the time complexity of the offline phase (denoted as P) is mainly determined by the number of (S_i, I_j), which implies that $P = rN'$. At the same time, the memory complexity of the offline phase (denoted as M) is the same with the offline time complexity, i.e., $M = P = rN'$, since the offline phase has to construct r tables, and that the size of each table is N'.

Clearly, a total of rN' (n-bit internal state, n-bit keystream segment) pairs has been stored in the offline phase, since N' strings are chosen randomly for each of all r strings $S_1, \cdots, S_r$ in this phase. According to the birthday paradox, the expected number of keystream segment samples available for the online phase should be $D = N/rN'$. Since the probability that the first l bits of a given sample match one of r strings $S_1, \cdots, S_r$ is $p = r \cdot 2^{-l} = rR$, the expected number of matchings found among all D keystream segment samples should be $D \cdot p = RN/N'$. It is easy to see that the time complexity of the online phase (denoted as T) is mainly determined by the number of matchings found among all samples, which implies that $T = D \cdot p = RN/N'$. Note that the unit of the online time complexity is one table lookup. Therefore, the trade-off curve of our attack is given as follows.

$$MT = rRN, \ MD = N, \ P = M \text{ and } D = d \cdot d'$$

Where r is integer parameter satisfying the restriction $1 \leq r \leq R^{-1}$.

Clearly, the tradeoff curve of our attack is the same as the BG-TMDTO attack when $r = R^{-1}$ holds. Thus, the BG-TMDTO attack may be considered as a special case of our attack. By introducing the integer parameter, our attack gives a wider variety of trade-offs. Let $D_{\max} = d_{\max} \cdot d'_{\max}$ be the maximum number of keystream bits generated by the key K, where $d'_{\max}$ denotes the maximum number of keystream bits generated by a single (K, IV), and $d_{\max}$ denotes the maximum number of keystream sequences generated by the same key K and different IVs. A lemma is obtained as follows.

Lemma 1. For a given stream cipher with the internal state size of $\log_2 N$ bits, when $N^{1/2} < D_{\max} < (rR)^{-1} N^{1/2}$ is allowed for the attacker, $1 \leq r < R^{-1}$, there certainly exists a TMDTO attack with the online time, offline time and memory complexities all faster than an exhaustive key search on the cipher, even though the internal state size is twice the key size.

Proof. Recall the trade-off curve of our attack as follows.

$$MT = rRN, \ MD = N, \ P = M \text{ and } D = d \cdot d'$$

When $N^{1/2} < D_{\max} < (rR)^{-1} N^{1/2}$ is allowed for the attacker, we may choose the data complexity such that $N^{1/2} < D \leq D_{\max}$, which implies

$$rRN^{1/2} < P = M = \frac{N}{D} < N^{1/2}$$
$$T = \frac{rRN}{M} < \frac{rRN}{rRN^{1/2}} = N^{1/2}$$

Thus, this Lemma follows directly.   ∎

Our attack can be considered as a generalization of the BG-TMDTO attack. It is rather efficient when multiple data is allowed for the attacker from the same key with different IVs, even though the internal state size is twice the key size.

3   Applications and Discussions

3.1   Previous Works on MICKEY and Grain Stream Ciphers

At FSE 2000, particularly efficient attacks on A5/1 were proposed by using the BSW sampling technique [7]. After then, the BS-TMDTO attack with BSW sampling had been applied to MICKEY 1.0 and Grain stream ciphers, see [9,8].

MICKEY 1.0 [10] is a hardware-oriented stream cipher proposed by Babbage and Dodd in 2005. Its strengthened version, named MICKEY 2.0 [11], had been selected as one of the seven finalists of eSTREAM project. In [9], Hong and Kim showed that MICKEY 1.0 stream cipher has a sampling resistance of at most 2^{-27}. Since MICKEY 1.0 has an internal state size of 160 bits, the BS-TMDTO attack with BSW sampling on the cipher has the online and offline time complexities of 2^{67} and 2^{100} respectively, while the time complexity of an exhaustive key search is 2^{80}. The attack is not applicable to MICKEY 2.0, because the internal state size was increased from 160 to 200 bits. MICKEY-128 2.0 [11] is a variant of MICKEY 2.0, which supports a key size of 128 bits, and an IV varying between 0 and 80 bits in length. Its internal state size is of 320 bits.

Grain v1 [12], a hardware-oriented stream cipher proposed by Hell, Johansson and Meier, is also one of the seven eSTREAM finalists. It has a key size of 80 bits, an IV size of 64 bits, and an internal state size of 160 bits. In [8], Bjørstad showed that Grain v1 stream cipher has a sampling resistance of at most 2^{-21}. The BS-TMDTO attack with BSW sampling on the cipher has the online and offline time complexities of 2^{70} and 2^{104} respectively. Grain-128 [12] is a variant of Grain, which supports a key size of 128 bits, and an IV size of 96 bits. Its internal state size is of 256 bits.

3.2   New Attacks on MICKEY and Grain Stream Ciphers

Now, we will apply our attack to MICKEY and Grain stream ciphers. MICKEY 1.0 has a key size of 80 bits, and supports an IV varying between 0 and 80 bits in length. As for MICKEY 1.0 stream cipher, we get that $N = 2^{160}$ and $R = 2^{-27}$. In the specification of MICKEY 1.0, there exists a restriction that the maximum length of keystream sequence that may be generated with a single (K, IV) pair is 2^{40} bits, and it is acceptable to generate 2^{40} such sequences, all from the same key K but with different values of IV. This restriction implies $d \leq 2^{40}$ and $d' \leq 2^{40}$. Here, we choose $d = d' = 2^{40}$, and $r = 1$. Since $N = 2^{160}$ and $R = 2^{-27}$ are known, we can mount a TMDTO attack on MICKEY 1.0 with $M = D = P = N^{1/2} = 2^{80}$ and $T = 2^{53}$.

Table 1. The results and comparisons with the existing attacks

Stream ciphers	R	Attacks	Parameter	T	M	D	P
MICKEY 1.0	2^{-27}	[9]	-	2^{67}	2^{67}	2^{60}	2^{100}
		This paper	$r = 1$	2^{53}	2^{80}	$d = d' = 2^{40}$	2^{80}
MICKEY 2.0	2^{-33}	This paper	$r = 1$	2^{47}	2^{120}	$d = d' = 2^{40}$	2^{120}
MICKEY-128 2.0	2^{-54}	This paper	$r = 1$	2^{74}	2^{192}	$d = d' = 2^{64}$	2^{192}
Grain v1	2^{-21}	[8]	-	2^{70}	2^{59}	2^{56}	2^{104}
		This paper	$r = 1$	$2^{69.5}$	$2^{69.5}$	$d = d' = 2^{45.25}$	$2^{69.5}$
		This paper	$r = 2^{11}$	2^{75}	2^{75}	$d = d' = 2^{42.5}$	2^{75}
Grain-128	2^{-22}	This paper	$r = 1$	2^{117}	2^{117}	$d = d' = 2^{69.5}$	2^{117}
		This paper	$r = 2^{12}$	2^{123}	2^{123}	$d = d' = 2^{66.5}$	2^{123}

Similarly, we can mount TMDTO attacks on MICKEY 2.0, MICKEY-128 2.0, Grain v1 and Grain-128 stream ciphers. The results and comparisons with the existing attacks are summarized in Table 1.

Note that similar to MICKEY 2.0, there also exists a restriction for MICKEY-128 2.0 that the maximum length of keystream sequence that may be generated with a single (K, IV) pair is 2^{64} bits, and it is acceptable to generate 2^{64} such sequences, all from the same key K but with different values of IV. Thus, the restrictions $d \leq 2^{64}$ and $d' \leq 2^{64}$ should hold simultaneously. While, as for Grain v1 and Grain-128 stream ciphers, their specifications do not specify any limits to the amount of keystream that may be generated by the same key K with different IVs, so our attacks on them are completely valid in this respect.

According to Table 1, our TMDTO attacks have one remarkable advantage in the online time complexity, which is always better than an exhaustive key search. Since the offline phase only performs once, the attack with a low online time complexity works, particularly when the attacker wants to recover many internal states generated by different secret keys in the online phase. Furthermore, as for Grain v1 and Grain-128 stream ciphers, our attacks are rather attractive in the respect that the online time, offline time and memory complexities are all better than an exhaustive key search, and the amount of keystream needed are completely valid.

3.3 General Guess and Determine-TMDTO Attack

The Guess and Determine (GD) attack is a common attack on stream ciphers. Its main idea is to guess a portion of the internal state, and then to recover the remaining internal state by using a small amount of known keystream. Clearly, the BSW sample technique can be generalized to a simple Guess and Determine attack. The Assumption 1 can be rewritten as follows.

Assumption 2. For a given stream cipher with internal state size $n = \log_2 N$, the attacker guesses the values of $n-l$ particular internal state bits, the remaining l internal state bits may be determined directly by using the first l keystream bits produced from that state.

Assume that for a given stream cipher, the attacker obtains a simple Guess and Determine attack, described as the Assumption 2. The attack has a time complexity of 2^{n-l}, requiring l keystream bits. It can be transformed into a Guess and Determine-TMDTO attack by fitting it into the model showed in Subsection 2.2.

Take the SOSEMANUK stream cipher [13], one of seven eSTREAM finalists, for example. SOSEMANUK has an internal state size of 384 bits. The internal state of SOSEMANUK at time t can be showed as $s_{t+1}, \cdots, s_{t+10}, R1_{t+1}, R2_{t+1}$, and that each contains 32 bits. Its key length is variable between 128 and 256 bits. It accommodates a 128-bit IV. Any key length is claimed to achieve 128-bit security. The best Guess and Determine attack on the cipher so far has been proposed by Feng et al. [14] in ASIACRYPT 2010, with a time complexity of 2^{176}. In their attack, the attacker guesses a total of 176 internal state bits, and then determine the remaining 208 internal state bits by using eight 32-bit keystream words (i.e., a total of 256 keystream bits). The attack consists of five phases. Their attack can not be transformed into a Guess and Determine-TMDTO attack, since it requires too many keystream bits that it does not fit into our model. Here, we give a new Guess and Determine attack on SOSEMANUK as follows. We follow the notes used in [14]. Let x be a 32-bit word. Denote $x^{(i)}$ the i-th byte component of x, $0 \leq i \leq 3$, i.e., $x = x^{(3)}||x^{(2)}||x^{(1)}||x^{(0)}$, where each $x^{(i)}$ is a byte, and $||$ is the concatenation of two bit strings. For simplicity we write $x^{(1)}||x^{(0)}$ as $x^{(0,1)}$ and $x^{(2)}||x^{(1)}||x^{(0)}$ as $x^{(0,1,2)}$.

In the new attack, we should first guess $s_1, s_2, s_3, s_4^{(0)}, R2_1^{(0,1,2)}$ and $R1_1$ (a total of 160 bits). Note that in their attack, they make an assumption on the least significant bit of $R1_1$. As showed in the Section 6 of [14], it shows that the assumption is not necessary for their attack to work. For convenience, we guess the least significant bit of $R1_1$ directly instead of making one assumption. After then, we determine $s_4, s_5, s_6, R2_1$ and s_{10} by executing the Phase 1-3 of their attack. Finally, the attacker should guess s_7, s_8 and s_9 (a total of 96 bits). Up to now, we have recovered all 384 internal state bits of SOSEMANUK. The attack only uses four 32-bit keystream words, i.e., z_1, z_2, z_3 and z_4.

In the new attack, we should guess $s_1, s_2, s_3, s_4^{(0)}, R2_1^{(0,1,2)}, s_7, s_8, s_9$ and $R1_1$ (a total of 256 bits), and then recover the remaining 128 internal state bits by using four 32-bit keystream words (a total of 128 bits). The new attack fits into our model quite well. Thus, we have $N = 2^{384}$ and $R = 2^{-128}$ for SOSEMANUK stream cipher. Our results and comparison with the existing attacks are summarized in Table 2.

Note that the specification of SOSEMANUK does not specify any limits to the amount of keystream that may be generated under a single (K, IV), so our attacks are completely valid in this respect. According to Table 2, the attacker can mount a Guess and Determine-TMDTO attack with the online time, offline time and memory complexities all equal to 2^{128}, which achieves the best time complexity level compared with all existing attacks on SOSEMANUK so far. Of course, one can also mount a Guess and Determine-TMDTO attack with an

Table 2. The results and comparisons with the existing attacks on SOSEMANUK

Attacks	Parameter	T	M	D	P
GD attack [14]	-	2^{176}	-	2^4	-
Linear Cryptanalysis [15]	-	$2^{147.9}$	$2^{147.1}$	$2^{145.5}$	-
Linear Cryptanalysis [16]	-	$2^{147.4}$	$2^{146.8}$	$2^{135.7}$	-
This paper	$r = 1$	2^{136}	2^{120}	$d = 2^{128}, d' = 2^{136}$	2^{120}
This paper	$r = 1$	2^{128}	2^{128}	$d = d' = 2^{128}$	2^{128}
This paper	$r = 1$	2^{116}	2^{140}	$d = d' = 2^{122}$	2^{140}

online time complexity of 2^{116}, which is significantly better than an exhaustive key search, at the cost of increased offline time and memory complexities.

4   Conclusions

By combining the BG-TMDTO attack with the BSW sampling technique, this paper proposes a new TMDTO attack on stream ciphers. The results show that the new attack gives a wider variety of trade-offs, compared with original BG-TMDTO attack, and is efficient when multiple data is allowed for the attacker from the same key with different IVs, even though the internal state size is twice the key size. As applications, we mount TMDTO attacks on MICKEY and Grain stream ciphers. Particularly, as for Grain v1 and Grain-128 stream ciphers, our TMDTO attacks are rather attractive in the respect that the online time, offline time and memory complexities are all better than an exhaustive key search, and the amount of keystream needed are completely valid. The results are sufficient evidences of validity of our attack. Finally, we generalize our attack to a Guess and Determine-TMDTO attack, and apply it to SOSEMANUK stream cipher, which achieves the best time complexity level compared with all existing attacks on SOSEMANUK so far. We hope that our attack provides some new insights on TMDTO attacks on stream ciphers.

Acknowledgements. The authors would like to thank the anonymous reviewers and Dr. Long Wen for their valuable comments and suggestions.

References

1. Hellman, M.: A cryptanalytic time-memory trade-off. IEEE Transactions on Information Theory 26(4), 401–406 (1980)
2. Babbage, S.: Improved exhaustive search attacks on stream ciphers. In: European Convention on Security and Detection 1995. IEE Conference Publication, pp. 161–166. IEEE Press, New York (1995)
3. Golić, J.D.: Cryptanalysis of alleged A5 stream cipher. In: Fumy, W. (ed.) EUROCRYPT 1997. LNCS, vol. 1233, pp. 239–255. Springer, Heidelberg (1997)

4. Biryukov, A., Shamir, A.: Cryptanalytic time/memory/data tradeoffs for stream ciphers. In: Okamoto, T. (ed.) ASIACRYPT 2000. LNCS, vol. 1976, pp. 1–13. Springer, Heidelberg (2000)
5. Hong, J., Sarkar, P.: New Applications of Time Memory Data Tradeoffs. In: Roy, B. (ed.) ASIACRYPT 2005. LNCS, vol. 3788, pp. 353–372. Springer, Heidelberg (2005)
6. Dunkelman, O., Keller, N.: Treatment of the initial value in Time-Memory-Data Trade-off attacks on stream ciphers. Information Processing Letters 107(5), 133–137 (2008)
7. Biryukov, A., Shamir, A., Wagner, D.: Real time cryptanalysis of A5/1 on a PC. In: Schneier, B. (ed.) FSE 2000. LNCS, vol. 1978, pp. 1–18. Springer, Heidelberg (2001)
8. Bjørstad, T.E.: Cryptanalysis of Grain using Time/Memory/Data Tradeoffs. ECRYPT Stream Cipher Project Report 2008/012 (2008), http://www.ecrypt.eu.org/stream
9. Hong, J., Kim, W.-H.: TMD-Tradeoff and State Entropy Loss Considerations of Streamcipher MICKEY. In: Maitra, S., Veni Madhavan, C.E., Venkatesan, R. (eds.) INDOCRYPT 2005. LNCS, vol. 3797, pp. 169–182. Springer, Heidelberg (2005)
10. Babbage, S., Dodd, M.: The stream cipher MICKEY (version 1). ECRYPT Stream Cipher Project Report 2005/015 (2005), http://www.ecrypt.eu.org/stream
11. Babbage, S., Dodd, M.: The MICKEY Stream Ciphers. In: Robshaw, M., Billet, O. (eds.) New Stream Cipher Designs. LNCS, vol. 4986, pp. 191–209. Springer, Heidelberg (2008)
12. Hell, M., Johansson, T., Maximov, A., Meier, W.: The Grain Family of Stream Ciphers. In: Robshaw, M., Billet, O. (eds.) New Stream Cipher Designs. LNCS, vol. 4986, pp. 179–190. Springer, Heidelberg (2008)
13. Berbain, C., et al.: Sosemanuk, A Fast Software-Oriented Stream Cipher. In: Robshaw, M., Billet, O. (eds.) New Stream Cipher Designs. LNCS, vol. 4986, pp. 98–118. Springer, Heidelberg (2008)
14. Feng, X., Liu, J., Zhou, Z., Wu, C., Feng, D.: A Byte-Based Guess and Determine Attack on SOSEMANUK. In: Abe, M. (ed.) ASIACRYPT 2010. LNCS, vol. 6477, pp. 146–157. Springer, Heidelberg (2010)
15. Lee, J.-K., Lee, D.-H., Park, S.: Cryptanalysis of SOSEMANUK and SNOW 2.0 using linear masks. In: Pieprzyk, J. (ed.) ASIACRYPT 2008. LNCS, vol. 5350, pp. 524–538. Springer, Heidelberg (2008)
16. Cho, J.Y., Hermelin, M.: Improved Linear Cryptanalysis of SOSEMANUK. In: Lee, D., Hong, S. (eds.) ICISC 2009. LNCS, vol. 5984, pp. 101–117. Springer, Heidelberg (2010)

Multidimensional Zero-Correlation Linear Cryptanalysis of E2

Long Wen[1], Meiqin Wang[1,*], and Andrey Bogdanov[2,*]

[1] Key Laboratory of Cryptologic Technology and Information Security,
Ministry of Education, Shandong University, Jinan 250100, China
longwen@mail.sdu.edu.cn, mqwang@sdu.edu.cn
[2] Technical University of Denmark, Denmark
anbog@dtu.dk

Abstract. E2 is a block cipher designed by NTT and was a first-round AES candidate. E2's design principles influenced several more recent block ciphers including Camellia, an ISO/IEC standard cipher. So far the cryptanalytic results for round-reduced E2 have been concentrating around truncated and impossible differentials. At the same time, rather recently at SAC'13, it has been shown how to improve upon the impossible differential cryptanalysis of Camellia with the zero-correlation linear cryptanalysis. Due to some similarities between E2 and Camellia, E2 might also render itself more susceptible to this type of cryptanalysis.

In this paper, we investigate the security of E2 against zero-correlation linear cryptanalysis. We identify zero-correlation linear approximations over 6 rounds of E2. With these linear approximations, we can attack 8-round E2-128 and 9-round E2-256 without IT and FT. The attack on 8-round E2-128 requires $2^{124.1}$ known plaintexts (KPs), $2^{119.3}$ encryptions and 2^{99} bytes memory. The attack on 9-round E2-256 requires $2^{124.6}$ KPs, $2^{225.5}$ encryptions and 2^{99} bytes memory. In contrast, the previous attacks on 8-round E2-128 had an uncertain time complexity and one could only attack 8-round E2-256. Besides, for the first time, we propose a key recovery attack on reduced-round E2 with both IT and FT taken into consideration. More concretely, we can attack 6-round E2-128 with $2^{123.7}$ KPs, $2^{119.1}$ encryptions and 2^{29} bytes and 7-round E2-256 requires $2^{124.7}$ KPs, $2^{252.8}$ encryptions and 2^{91} bytes when both IT and FT are considered.

Keywords: Block cipher, zero-correlation, multidimensional linear cryptanalysis, E2.

1 Introduction

Zero-Correlation Linear Cryptanalysis. The concept of zero-correlation linear cryptanalysis is proposed by Bogdanov and Rijmen in [3]. The foundation of this new kind of cryptanalytic technique is the availability of numerous

---

* Corresponding Authors.

D. Pointcheval and D. Vergnaud (Eds.): AFRICACRYPT 2014, LNCS 8469, pp. 147–164, 2014.
© Springer International Publishing Switzerland 2014

key-independent unbiased linear approximations with correlation zero in many ciphers. (If a linear approximation holds with probability p, then its correlation is defined by $c = 2p - 1$). Despite the novelty of this new cryptanalysis, its application is limited due to the data complexity required to mount the attack, where almost the whole codebook is needed to distinguish right key guess and wrong key guess. Yet, this drawback is overcome at FSE'12 [4] by Bogdanov and Wang. They constructed a more data-efficient distinguisher utilizing the existence of multiple linear approximations with correlation zero in the target ciphers. Improved attacks on TEA and XTEA are presented in [4] using the new distinguisher. However, the distinguisher in [4] is constructed based on the assumption that all obtained zero-correlation linear approximations are independent, which is not met in certain target ciphers. In a follow-up work at ASIACRYPT'12 [5], fundamental links of integral cryptanalysis to zero-correlation cryptanalysis have been revealed. Namely, integrals (similar to saturation or multiset distinguishers) have been demonstrated to be essentially a special case of the zero-correlation property. Moreover, the multidimensional zero-correlation linear distinguisher has been constructed for the zero-correlation property, which removed the unnecessary independency assumptions on the distinguishing side [4]. This new model can be seen as multidimensional linear cryptanalysis [13] with capacity equal to zero and its validity is verified by experiments on small variant of LBlock with 32-bit block at WCC'13 by Soleimany and Nyberg [6]. At SAC'13 [8] the Discrete Fast Fourier Transform technique is applied in the zero-correlation linear cryptanalysis resulting in improved attacks on Camellia-128 and Camellia-192. As zero-correlation linear cryptanalysis is considered as the counter part of impossible differential cryptanalysis in the domain of linear cryptanalysis, a mathematical link between impossible differential distinguisher and zero-correlation linear distinguisher is revealed by Blondeau and Nyberg [7] at EUROCRYPT'13.

E2 Block Cipher and Existing Cryptanalysis. E2 [10] is a 128-bit block cipher proposed by NTT and is one of the fifteen candidates in the first round of AES project. Although E2 was not selected as AES, its design principle has been used in Camellia [1], which is adopted as one of the ISO block ciphers [9]. Classical Feistel structure is adopted in E2. The key size could be 128, 192 or 256-bit, and are denoted as E2-128, E2-192 and E2-256, respectively. All three version of E2 have the same round number, 12. The round function employs Substitution-Permutation-Substitution (SPS) structure with byte table lookups and byte XOR operation. Moreover, the initial transformation IT and the final transformation FT consisting of XOR operation and modular multiplication operation with subkeys and byte permutation are applied before the first round and after the last round, respectively.

The security of E2 has been evaluated with truncated differential cryptanalysis and impossible differential cryptanalysis. At FSE'99 [14], Matsui and Tokita identified a truncated differential characteristic for 7-round E2 and mounted key recovery attack on 8-round E2-128 without IT and FT under data complexity

2^{100} chosen plaintexts and indefinite time complexity[1]. At SAC'99 [15], Moriai et al. found another 7-round truncated differential characteristic with higher probability than that identified by Matsui et al., and a possible key recovery attack on 8-round E2-128 without IT and FT is given under data complexity 2^{94} chosen plaintexts and uncertain time complexity[2]. Then they proposed the distinguishing attack (other than key recovery attack) on 7-round E2-128 with IT and FT.

In the case of impossible differential cryptanalysis of E2, Wei et al. [16] identified 6-round impossible differential characteristics of E2, yet they didn't report key recovery attack on E2. In [17], Wei et al. presented key recovery attack on 7-round E2-128 without IT and FT requiring 2^{120} chosen plaintexts and $2^{115.5}$ encryptions, and the key recovery attack on 8-round E2-256 without IT and FT is reported with data complexity 2^{121} chosen plaintexts and time complexity 2^{214} encryptions.

Our Contributions. In this paper, we present zero-correlation cryptanalysis for E2 block cipher. Our contributions are three-fold and various key recovery attacks on E2-128 and E2-256 are summarized in Table 1.

- Zero-correlation linear approximations over 6-round E2 are derived.
- Key recovery attacks on 8-round E2-128 and 9-round E2-256 without considering IT and FT are mounted using multidimensional zero-correlation linear cryptanalytic technique. Compared with the previous impossible differential attacks, our attacks can work one more round. In comparison with the previous truncated differential attacks, the time complexity of our attack on 8-round E2-128 is explicitly lower than that of exhaustive search.
- We present the first key recovery attacks on 6-round E2-128 and 7-round E2-256 with both IT and FT taken into consideration. The previous impossible differential attack in [15] can work on 8-round E2-128 with only IT or FT taken into consideration. Moreover, whether this attack's time complexity is lower than exhaustive search is still unknown.

Organization of the Paper. The remainder of this paper is organized as follows. Section 2 describes basis zero-correlation linear cryptanalysis, multidimensional zero-correlation linear cryptanalysis and E2 block cipher. Section 3 presents how to identify zero-correlation linear approximations for 6-round E2. Section 4 deals the multidimensional zero-correlation linear cryptanalysis of 8-round E2-128 and 9-round E2-256 without IT and FT. Section 5 reports the

---

[1] They claim : *"The straightforward method for realizing the algorithm above requires complexity more than 2^{128}, but by discarding impossible pairs and introducing a counting method for the second layer subkey with 2^{64} counters, we can reduce the complexity to less than 2^{128}."*

[2] They claim: *"Note that the complexity of the procedure above for deriving the last round keys (128 bits) exceeds the complexity of exhaustive search $\mathcal{O}(2^{128})$. We've not confirmed whether an improved attack with complexity less than $\mathcal{O}(2^{128})$ is possible."*

Table 1. Summary of Key Recovery Attacks on E2-128 and E2-256

Attack	Round	IT/FT	Data	Time	Memory	Ref.
		E2-128				
Impossible Differential	7	none	2^{120}CPs	$2^{115.5}$	–	[17]
Truncated Differential*	8	none	2^{100}CPs	$< 2^{128}$	–	[14]
Multidimensional Z.C.	**8**	**none**	$2^{124.1}$**KPs**	$2^{119.3}$	2^{99}	**Sect. 4**
Truncated Differential[†]	8	one[‡]	2^{94}CPs	–	–	[15]
Multidimensional Z.C.	**6**	**both**	$2^{123.7}$**KPs**	$2^{119.1}$	2^{29}	**Sect. 5**
		E2-256				
Impossible Differential	8	none	2^{121}CPs	2^{214}	–	[17]
Multidimensional Z.C.	**9**	**none**	$2^{124.6}$**KPs**	$2^{225.5}$	2^{99}	**Sect. 4**
Multidimensional Z.C.	**7**	**both**	$2^{124.7}$**KPs**	$2^{252.8}$	2^{91}	**Sect. 5**

[*] No detailed attack procedure is described in [14]. See Footnote 1 on the previous page.

[†] Authors of [15] are not sure whether it is possible to mount the attack with time complexity of less than 2^{128} encryptions. See Footnote 2 on the previous page. Besides, a distinguishing attack on 7-round E2 with IT and FT reported in [15] is not listed in the table.

[‡] "one" means only IT or FT is taken into consideration, but not both.

key recovery attacks on 6-round E2-128 and 7-round E2-256 with IT and FT. We conclude the paper in Section 6.

2   Preliminaries

2.1   Basics of Zero-Correlation Linear Cryptanalysis [3]

Consider an n-bit block cipher f_K with key K. Let P denote a plaintext which is mapped to a ciphertext C under key K, $C = f_K(P)$. If Γ_P and Γ_C are nonzero plaintext and ciphertext linear masks of n-bit each, we denote the linear approximation $\Gamma_P^T \cdot P \oplus \Gamma_C^T \cdot C = 0$ as $\Gamma_P \to \Gamma_C$. Here, $\Gamma_A^T \cdot A$ denotes the multiplication of the transposed bit vector Γ_A (linear mask for A) by a column bit vector A over $\mathbb{F}_2$. The linear approximation $\Gamma_P \to \Gamma_C$ has probability

$$p_{\Gamma_P, \Gamma_C} = \Pr_{P \in \mathbb{F}_2^n} \{ \Gamma_P^T \cdot P \oplus \Gamma_C^T \cdot C = 0 \}.$$

The value $c_{\Gamma_P, \Gamma_C} = 2p_{\Gamma_P, \Gamma_C} - 1$ is called the *correlation* of linear approximation $\Gamma_P \to \Gamma_C$. Note that $p_{\Gamma_P, \Gamma_C} = 1/2$ is equivalent to *zero correlation* $c_{\Gamma_P, \Gamma_C} = 0$.

Given a distinguisher of zero-correlation linear approximation(s) over a part of the cipher, the basic key recovery can be done with a technique similar to that of Matsui's Algorithm 2 [11]. That is the attacker partially encrypts/decrypts the plantext-ciphertext (denoted as PC) pairs to the boundaries of the distinguisher and then verifies the distinguisher property to distinguish between right key guess and wrong key guess.

2.2   Multidimensional Zero-Correlation Linear Cryptanalysis [5]

Suppose that we can obtain ℓ zero-correlation linear approximations over a part of a cipher and these linear approximations are a linear space spanned by m base zero-correlation linear approximations, $\ell = 2^m - 1$. For each of the 2^m values $z \in \mathbb{F}_2^m$, the attacker initializes a counter $V[z]$, $z = 0, 1, 2, \ldots, 2^m - 1$, to zero. The attacker partially encrypts and decrypts each distinct PC pair to the boundaries of zero-correlation linear approximations by guessing some key values and compute the corresponding data value in $\mathbb{F}_2^m$ by evaluating the m basis linear approximations and increments the corresponding counter $V[z]$ by one. Then the attacker computes the statistic T:

$$T = \sum_{i=0}^{2^m-1} \frac{(V[z] - N2^{-m})^2}{N2^{-m}(1 - 2^{-m})}. \tag{1}$$

The statistic T for the right key guess follows a χ^2-distribution with mean $\mu_0 = (\ell - 1)\frac{2^n - N}{2^n - 1}$ and variance $\sigma_0^2 = 2(\ell - 1)\left(\frac{2^n - N}{2^n - 1}\right)^2$, while for the wrong key guess it follows a χ^2-distribution with mean $\mu_1 = \ell - 1$ and variance $\sigma_1^2 = 2(\ell - 1)$.

We denote the type-I error probability as β_0 (the probability to wrongfully discard the right key guess), the type-II error probability as β_1 (the probability that a wrong key guess survives the filteration). If we consider the decision threshold as $\tau = \mu_0 + \sigma_0 z_{1-\beta_0} = \mu_1 - \sigma_1 z_{1-\beta_1}$ where $z_{1-\beta_0}$ and $z_{1-\beta_1}$ are the respective quantiles of the standard normal distribution, then the number of known plaintexts N should satisfy

$$N = \frac{(2^n - 1)(z_{1-\beta_0} + z_{1-\beta_1})}{\sqrt{(\ell - 1)/2} + z_{1-\beta_0}} + 1. \tag{2}$$

General Attack Procedure. Decompose the target cipher E as a cascade $E = E_f \circ E_d \circ E_b$, where E_d is covered by the ℓ zero-correlation linear approximations (can be generated by m base zero-correlation linear approximations) that we obtained and E_f and E_b are the rounds that are added before and appended after E_d.

The attacker partially encrypts and decrypts the N PC pairs through E_f and E_b by guessing some key values to the boundaries of E_d. Then the counters $V[z]$ is constructed from N PC pairs under each possible key value and the statistic T are computed according to Equation (1). By choosing proper β_0 and β_1, the attacker can compute the threshold value τ. If $T \leq \tau$ then guessed key values are right key candidates. All right key candidates are then tested against a few PC pairs and in the end only the right key will survive.

Normally the partial sum technique is used during the partial encryption and decryption phase to help reduce the time complexity of the partial encryption and decryption phase because only a part of state values are involved in this phase. Also, there usually exists some data-time trade-off as the number of PC pairs is related to the β_0 and β_1 and the value of β_1 (the probability that a wrong key guess surviving the filteration) would affect the time complexity of the final exhaustive search for the right key value.

2.3   Description of E2

E2 [10] is a 128-bit block cipher proposed by NTT in 1998 and is selected as one of the fifteen candidates in the first round of AES project. A 12-round Feistel network along with an initial transformation IT and a final transformation FT are adopted in E2, see Figure 1(a). The key size can be 128, 192 or 256 bits. The round function uses SPS structure including the XOR operation with the first round subkey, the first nonlinear transformation consisting of eight parallel 8×8 S-boxes, the linear transformation P, the XOR operation with the second round subkey and the second nonlinear transformation consisting of eight parallel S-boxes (8×8). The details of the round function are illustrated in Figure 1(b).

The linear transformation $P : \mathbb{F}_2^{64} \to \mathbb{F}_2^{64}$ in F can also be expressed with matrix-vector product where $z_i, z_i' \in \mathbb{F}_2^8, 1 \le i \le 8$:

$$
\begin{pmatrix} z_1' \\ z_2' \\ z_3' \\ z_4' \\ z_5' \\ z_6' \\ z_7' \\ z_8' \end{pmatrix} = \begin{pmatrix} 0 1 1 1 1 1 1 0 \\ 1 0 1 1 0 1 1 1 \\ 1 1 0 1 1 0 1 1 \\ 1 1 1 0 1 1 0 1 \\ 1 1 0 1 1 1 0 0 \\ 1 1 1 0 0 1 1 0 \\ 0 1 1 1 0 0 1 1 \\ 1 0 1 1 1 0 0 1 \end{pmatrix} \cdot \begin{pmatrix} z_1 \\ z_2 \\ z_3 \\ z_4 \\ z_5 \\ z_6 \\ z_7 \\ z_8 \end{pmatrix}.
$$

If $X, Y, A, B \in \mathbb{F}_2^{128}$, the initial transformation IT and the final transformation FT can be shown as follows where the byte permutation BP and BP^{-1} are shown in Figure 1(c):

$$
\begin{aligned}
IT &: \mathbb{F}_2^{128} \times \mathbb{F}_2^{128} \times \mathbb{F}_2^{128} \to \mathbb{F}_2^{128}; (X, A, B) \mapsto BP((X \oplus A) \otimes B), \\
FT &: \mathbb{F}_2^{128} \times \mathbb{F}_2^{128} \times \mathbb{F}_2^{128} \to \mathbb{F}_2^{128}; (X, A, B) \mapsto (BP^{-1}(X) \oslash B) \oplus A.
\end{aligned}
$$

As to operation $\otimes$ and $\oslash$, if we represent $X = (x_1, x_2, x_3, x_4)$, $Y = (y_1, y_2, y_3, y_4)$, and $B = (b_1, b_2, b_3, b_4)$, where $x_i, y_i, b_i \in \mathbb{F}_2^{32}, 1 \le i \le 4$, and use $\vee 1$ to denote bitwise logical OR with $1 \in \mathbb{F}_2^{32}$, then we have:

$$
\begin{aligned}
Y = X \otimes B &:= y_i = x_i(b_i \vee 1) \bmod 2^{32} \ (i = 1, 2, 3, 4), \\
Y = X \oslash B &:= x_i = y_i(b_i \vee 1)^{-1} \bmod 2^{32} \ (i = 1, 2, 3, 4).
\end{aligned}
$$

As E2's key schedule is somewhat complex and our attacks do not utilize the key relation, we omit the details of E2's key schedule. For the complete specification of E2, we refer to [10].

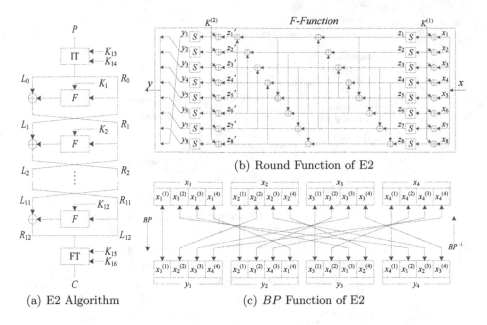

(a) E2 Algorithm   (b) Round Function of E2

(c) BP Function of E2

Fig. 1. E2 Block Cipher

3   Zero-Correlation Linear Approximations over 6-Round E2

Following the properties on the propagation of linear masks over basic block cipher operations proposed in [2,3], we can derive several types of zero-correlation linear approximations for 6-round E2.

Property 1. If the input mask is $(0|0|0|0|0|0|0|0, 0|0|0|0|0|0|0|b)$ and the output mask after 6-round E2 is $(0|0|0|0|0|0|h|0, 0|0|0|0|0|0|0|0)$, where $b, h \in \mathbb{F}_2^8, b \neq 0, h \neq 0$, then the correlation of these linear approximations is zero, see Figure 2.

Proof. As discussed in [2,12], when analyzing mask values in linear cryptanalysis, each XOR operation is replaced by a branch operation and each branch operation is replaced by an XOR operation. This means that for an XOR operation, the values of the two input mask values is equal to the output mask value. On the other hand, for a branch operation, the XOR of the two output mask values should be equal to the input mask value. This duality is the basic difference between linear and differential cryptanalysis.

Therefore, for the linear transformation P of E2, the relation between the input mask and the output mask can be obtained with the dual of linear transformation P, denoted as LP. LP can be obtained by replacing the XOR operation and the branch operation with each other and reversing direction of arrows. Details of P and LP are illustrated in Figure 3 and Figure 4, respectively.

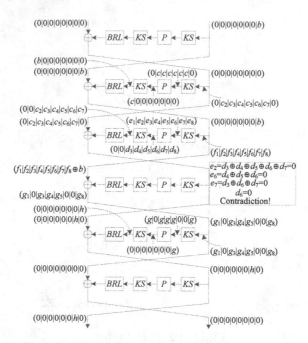

Fig. 2. Zero-Correlation Linear Approximations for 6-Round E2

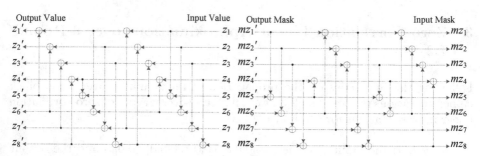

Fig. 3. Linear Transformation P **Fig. 4.** Dual of Linear Transformation LP

At the same time, Figure 3 and Figure 4 could be expressed with matrix-vector products as follows:

$$
\begin{pmatrix} z_1' \\ z_2' \\ z_3' \\ z_4' \\ z_5' \\ z_6' \\ z_7' \\ z_8' \end{pmatrix} = \begin{pmatrix} 0&1&1&1&1&1&1&0 \\ 1&0&1&1&0&1&1&1 \\ 1&1&0&1&1&0&1&1 \\ 1&1&1&0&1&1&0&1 \\ 1&1&0&1&1&1&0&0 \\ 1&1&1&0&0&1&1&0 \\ 0&1&1&1&0&0&1&1 \\ 1&0&1&1&1&0&0&1 \end{pmatrix} \cdot \begin{pmatrix} z_1 \\ z_2 \\ z_3 \\ z_4 \\ z_5 \\ z_6 \\ z_7 \\ z_8 \end{pmatrix}, \quad \begin{pmatrix} mz_1 \\ mz_2 \\ mz_3 \\ mz_4 \\ mz_5 \\ mz_6 \\ mz_7 \\ mz_8 \end{pmatrix} = \begin{pmatrix} 0&1&1&1&1&1&0&1 \\ 1&0&1&1&1&1&1&0 \\ 1&1&0&1&0&1&1&1 \\ 1&1&1&0&1&0&1&1 \\ 1&0&1&1&1&0&0&1 \\ 1&1&0&1&1&1&0&0 \\ 1&1&1&0&0&1&1&0 \\ 0&1&1&1&0&0&1&1 \end{pmatrix} \cdot \begin{pmatrix} mz_1' \\ mz_2' \\ mz_3' \\ mz_4' \\ mz_5' \\ mz_6' \\ mz_7' \\ mz_8' \end{pmatrix}.
$$

As shown in Figure 2, in the forward direction we can get that the mask for L_3 is $(f_1|f_2|f_3|f_4|f_5 \oplus b|f_6|f_7|f_8)$ from the input mask of the first round $(0|0|0|0|0|0|0|0, 0|0|0|0|0|0|0|b)$ where $f_i(i = 1, \ldots, 8)$ are unknown masks. And in the backward direction, the mask for L_3 that we deduce from the output mask of the last round $(0|0|0|0|0|0|h|0, 0|0|0|0|0|0|0|0)$ is $(g_1|0|g_3|g_4|g_5|0|0|g_8)$, where $g_i(i = 1, 3, 4, 5, 8)$ are non-zero masks. Then we have $f_2 = 0, f_6 = 0$ and $f_7 = 0$ and therefore $e_2 = 0, e_6 = 0$ and $e_7 = 0$, which give us the following equations:

$$e_2 = d_3 \oplus d_4 \oplus d_5 \oplus d_6 \oplus d_7 = 0, e_6 = d_4 \oplus d_5 \oplus d_6 = 0, e_7 = d_3 \oplus d_6 \oplus d_7 = 0.$$

From these equations, we can derive $d_6 = 0$. This contradicts with the fact that $d_6 \neq 0$. In this case, we claim that linear approximations shown in Figure 2 have zero correlation.   □

Similarly, other zero-correlation linear approximations over 6-round E2 can be derived. For example, Figure 5 illustrates another type of zero-correlation linear approximations over 6-round E2.

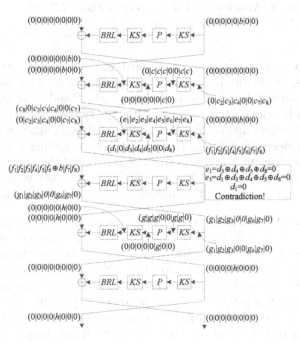

Fig. 5. Another Type of Zero-Correlation Linear Approximations for 6-Round E2

4   Multidimensional Zero-Correlation Linear Cryptanalysis of E2 without *IT* and *FT*

4.1   Key Recovery Attack on 8-Round E2-128 without *IT* and *FT*

With the zero-correlation linear approximations over 6-round E2 presented in Figure 5, we can attack 8-round E2 without *IT* and *FT* by adding one round

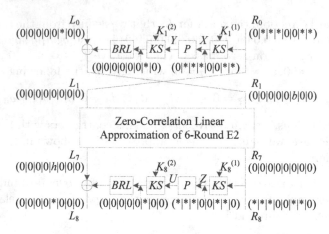

Fig. 6. Attack on 8-Round E2-128 without IT and FT

before and appending one round after the 6-round E2 zero-correlation linear approximations, see Figure 6. In Figure 6, state bytes that are involved in the partial encryption and decryption are denoted with '*' and those bytes that have nothing to do with the partial encryption and decryption are denoted by '0'.

As described in the general procedure earlier, the attack starts with the partial encryption and decryption where the partial sum technique is used. X, Y, Z and U are the intermediate states in the first round and the last round shown in Figure 6. From now on, we use "$[j]$" referring the jth byte of a variable containing multiple bytes and '|' denoting the concatenation of two binary strings.

1. Allocate 32-bit counters $V_1[s_1]$ for all possible values of 96-bit $s_1 = L_0[6]|$ $R_0[2,3,4,7,8]|L_8[5]|R_8[1,2,3,6,7]$ and initialize these counters to zero. Extract s_1 from each of N PC pairs and increment the corresponding counter $V_1[s_1]$ by one. In this step, no more that 2^{128} PC pairs are divided into 2^{96} distinct values of s_1, the expected counter for each s_1 is no more than 2^{32}, so 32-bit counter is sufficient. The time complexity of this step is N memory accesses to process N PC pairs. If we assume that processing one PC pair is equivalent to $1/4$ round encryption, then the time complexity of this step is about $N \cdot 1/4 \cdot 1/8$ encryptions.

2. Allocate 32-bit counters $V_2[s_2]$ for all possible values of 96-bit $s_2 = L_0[6]$ $|R_0[3,4,7,8]|X[2]|L_8[5]|R_8[1,2,3,6,7]$ and initialize them to zero. Guess 8-bit $K_1^{(1)}[2]$ and partially encrypt s_1 to compute s_2 and add the corresponding $V_1[x_1]$ to $V_2[x_2]$, that is $V_2[s_2]+ = V_1[s_1]$. The time complexity of this step is about $2^8 \cdot 2^{96} \cdot 1/4 \cdot 1/8 = 2^{99}$ encryptions.

The following steps in the partial encryption and decryption phase are similar to Step 2, thus to be consistent and to be more clear we use Table 2 to show the details of each partial encryption and decryption step. In Table 2, the second column stands for the subkey bytes that have to be guessed in each step, the third column denotes the time complexity of corresponding step measured in $1/8$ round encryption. In each step, we compute the values of the intermediate state s_i, $3 \leq i \leq 13$, which are shown in column "Computed States". For each possible value of s_i, the counter $V_i[s_i]$ will record how many PC pairs could produce the corresponding intermediate state s_i. The counter along with its size are shown in the last column.

Table 2. Partial Encryption and Decryption of the Attack on 8-Round E2-128

Step	Guess	Complexity	Computed States	Counter-Size
3	$K_1^{(1)}[3]$	$2^{16} \cdot 2^{96}$	$s_3 = L_0[6]\|R_0[4,7,8]\|X[2] \oplus X[3]\|L_8[5]\|R_8[1,2,3,6,7]$	$V_3 - 2^{88}$
4	$K_1^{(1)}[4]$	$2^{24} \cdot 2^{88}$	$s_4 = L_0[6]\|R_0[7,8]\|X[2] \oplus X[3] \oplus X[4]\|L_8[5]\|R_8[1,2,3,6,7]$	$V_4 - 2^{80}$
5	$K_1^{(1)}[7]$	$2^{32} \cdot 2^{80}$	$s_5 = L_0[6]\|R_0[8]\|X[2] \oplus X[3] \oplus X[4] \oplus X[7]\|L_8[5]\|R_8[1,2,3,6,7]$	$V_5 - 2^{72}$
6	$K_1^{(1)}[8]$	$2^{40} \cdot 2^{72}$	$s_6 = L_0[6]\|Y[7]\|L_8[5]\|R_8[1,2,3,6,7]$	$V_6 - 2^{64}$
7	$K_1^{(2)}[7]$	$2^{48} \cdot 2^{64}$	$s_7 = R_1[6]\|L_8[5]\|R_8[1,2,3,6,7]$	$V_7 - 2^{56}$
8	$K_8^{(1)}[1]$	$2^{56} \cdot 2^{56}$	$s_8 = R_1[6]\|L_8[5]\|R_8[2,3,6,7]\|Z[1]$	$V_8 - 2^{56}$
9	$K_8^{(1)}[2]$	$2^{64} \cdot 2^{56}$	$s_9 = R_1[6]\|L_8[5]\|R_8[3,6,7]\|Z[1] \oplus Z[2]$	$V_9 - 2^{48}$
10	$K_8^{(1)}[3]$	$2^{72} \cdot 2^{48}$	$s_{10} = R_1[6]\|L_8[5]\|R_8[6,7]\|Z[1] \oplus Z[2] \oplus Z[3]$	$V_{10} - 2^{40}$
11	$K_8^{(1)}[6]$	$2^{80} \cdot 2^{40}$	$s_{11} = R_1[6]\|L_8[5]\|R_8[7]\|Z[1] \oplus Z[2] \oplus Z[3] \oplus Z[6]$	$V_{11} - 2^{32}$
12	$K_8^{(1)}[7]$	$2^{88} \cdot 2^{32}$	$s_{12} = R_1[6]\|L_8[5]\|U[6]$	$V_{12} - 2^{24}$
13	$K_8^{(2)}[6]$	$2^{96} \cdot 2^{24}$	$s_{13} = R_1[6]\|L_7[5]$	$V_{13} - 2^{16}$

After Step 13, we have reached the boundaries of the zero-correlation linear approximations over 6-round E2. We then proceed the following steps to recover the right key.

14. Allocate 128-bit counters $V[z]$ for 16-bit z and initialize them to zero where z is the concatenation of evaluations of 16 basis zero-correlation masks. Compute z from s_{13} with 16 basis zero-correlation masks, then $V[z]+ = V_{13}[s_{13}]$. Compute the statistic T according to Equation (1). The computation in this step is to construct $V[z]$ from V_{13} and then compute T. If we assume constructing $V[z]$ from V_{13} as one encryption and computing T from $V[a]$ as another encryption, then the time complexity of this step is about $2^{96} \cdot 2 = 2^{97}$ encryptions because after Step 13, we have guessed 96-bit key values.

15. If $T \leq \tau$, then the guessed key value is a right key candidate. All key values compatible to the guessed key value are exhaustively searched.

The dominant time complexity of the partial encryption and decryption phase lies in Step 1 and Step 9 to Step 13. Step 1 is about $N \cdot 2^{-5}$ encryptions and Step 9 to Step 13 requires about $5 \cdot 2^{120} \cdot 1/8 \cdot 1/8 \approx 2^{116.3}$ encryptions. By construct the time complexity of Step 14 is negligible. The time complexity of Step 15, where we exhaustively search the right key value, is related to the value of β_1 that we choose. If we set $\beta_0 = 2^{-2.7}, \beta_1 = 2^{-98}$, then $z_{1-\beta_0} \approx 1.0$, $z_{1-\beta_1} = 11.4$. Since $n = 128$, $\ell = 2^{16}$, then according to Equation (2), the data complexity N is about $2^{124.1}$. In this case, the time complexity of Step 1 is about $2^{119.1}$ encryptions and the time complexity of Step 15 is negligible since only the right key guess is expected to survive the filteration. The overall time complexity is then about $2^{119.1} + 2^{116.3} \approx 2^{119.3}$ encryptions. The memory requirements are dominated by Step 1 and Step 2, where we need $2 \cdot 4 \cdot 2^{96} = 2^{99}$ bytes to store V_1 and V_2. Thus, our key recovery attack on 8-round E2-128 without IT and FT needs $2^{124.1}$ known plaintexts, $2^{119.3}$ encryptions and 2^{99} bytes memory.

4.2   Key Recovery Attack on 9-Round E2-256 without IT and FT

If the zero-correlation linear approximations over 6-round E2 in Figure 5 cover round 2 to round 7, by adding one round before and appending two rounds after the linear approximations, we can attack 9-round E2-256, illustrated in Figure 7.

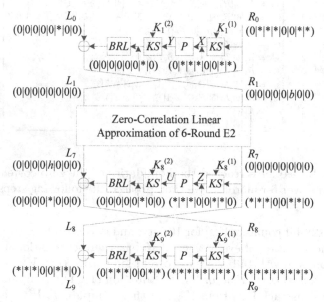

Fig. 7. Attack on 9-Round E2-256 without IT and FT

Comparing Figure 6 and Figure 7, we can find that the attack on 9-round E2-256 is quite similar as the attack on 8-round E2-128 except that we have to partially decrypt one more round in the partial decryption phase. Thus, if we firstly guess all values of the bytes in $K_9^{(1)}$ and $K_9^{(2)}$ that are involved partial decryption over round 9, and the remaining procedure is basically the same as that of the attack on 8-round E2-128. To be more concrete, the first step in the attack on 9-round E2-256 is:

1. Guess all possible values of 13-byte key $K_9^{(2)}[2,3,4,7,8]$ and $K_9^{(1)}$. Under each possible value allocate 32-bit counters $V_1[s_1]$ for all possible values of 96-bit $s_1 = L_0[6]|R_0[2,3,4,7,8]|R_8[5]|L_8[1,2,3,6,7]$ and initialize them to zero. Partially decrypt N PC pairs through round 9 to get value of

$R_8[5]|L_8[1,2,3,6,7]$ and increment the corresponding $V_1[s_1]$ by one where $L_0[6]|R_0[2,3,4,7,8]$ is extracted from plaintext. Note that in the attack on 9-round E2-256, s_1 is slightly different from the one in the attack on 8-round E2-128 due to the swap operation after round 8. The time complexity of this step is about $2^{104} \cdot N \cdot 1/9$ encryptions.

The following procedure of the attack on 9-round E2-256 is basically the same as Step 2 to Step 15 of the attack on 8-round E2-128. The only difference is introduced by the slightly different s_1 in the two attacks. As this difference is only a matter of different notations and does not affect the data, time complexities or memory requirements. Thus, we omit the details of the following steps. Just as the attack on 8-round E2-128, the partial encryption and decryption are done after Step 13, statistic T is computed in Step 14 and the exhaustively searching the right key is still done in Step 15.

As we guessed 13-byte key values in Step 1, the time complexities of Step 2 to Step 13 now should be multiplied by 2^{104} while the dominant time complexity of the partial encryption and decryption still lies in Step 1 and Step 9 to Step 13. The time complexity of Step 9 to Step 13 is now about $2^{104} \cdot 5 \cdot 2^{120} \cdot 1/8 \cdot 1/9 \approx 2^{220.2}$. The time complexity of Step 14 is about $2^{104} \cdot 2^{97} = 2^{221}$ encryptions. Still the time complexity of Step 15 is related to value of β_1 we choose. If we set $\beta_0 = 2^{-2.7}, \beta_1 = 2^{-202}$, then $z_{1-\beta_0} \approx 1.0$, $z_{1-\beta_1} \approx 16.5$. Since $n = 128$, $\ell = 2^{16}$, then according to Equation (2), the data complexity N is about $2^{124.6}$. As we have guessed $104 + 96 = 200$ bits key value after Step 13 and $\beta_1 = 2^{-202}$, only the right key value is expected to survive the filteration making the time complexity of Step 15 negligible. Thus the time complexity of our attack on 9-round E2-256 is about $2^{104} \cdot 2^{124.6} \cdot 1/9 + 2^{220.2} + 2^{221} \approx 2^{225.5}$ encryptions. The memory requirements is still dominated by Step 1 and Step 2 to store V_1 and V_2. In all, our attack on 9-round E2-256 requires $2^{124.6}$ known plaintexts, $2^{225.5}$ encryptions and 2^{99} bytes memory.

5 Multidimensional Zero-Correlation Linear Cryptanalysis of E2 with IT and FT

In this section, we present zero-correlation attacks on 6-round E2-128 and 7-round E2-256 with IT and FT using the zero-correlation linear approximations in Figure 2. We firstly show a property about the modular multiplication over $\mathbb{F}_2^{32}$ that is adopted both in IT and FT.

Property 2. Denote 32-bit input, output and subkey of the modular multiplication over $\mathbb{F}_2^{32}$ in IT function as (x_1, x_2, x_3, x_4), (y_1, y_2, y_3, y_4) and (k_1, k_2, k_3, k_4) respectively, where x_4, y_4 and k_4 are the most significant bytes. Then the output byte $y_i, 1 \leq i \leq 4$ is only related to $x_1, \ldots, x_i$ and $k_1, \ldots, k_i$.

According to the property of modular multiplication operation in $\mathbb{F}_2^{32}$, it is easy to prove Property 2. The meaning of Property 2 is that when we want to compute y_i, we only need to guess $k_1, \ldots, k_i$ other than guess all 32-bit subkey values. Meanwhile, there is no need to obtain all 32-bit input value either, the knowledge of the value $x_1, \ldots, x_i$ is sufficient to compute y_i. Note that in FT function, it is the inverse of subkey that are involved in the the modular multiplication operation. In this case, we will guess the value of the inverse of the subkey other than directly guessing the value of the subkey. Thus when partially encrypt over IT function and partially decrypt over FT function, Property 2 can be utilized to decrease the number of guessed key bytes.

5.1   Key Recovery Attack on 6-Round E2-128 with IT and FT

To attack 6-round E2-128 with IT and FT, the 6-round zero-correlation linear approximations from Figure 2 now start from the input of round 1 and ends at the output of round 6. The IT function is added before and FT function is appended after the linear approximations, refer to Figure 8(a).

The partial encryption and decryption using the partial sum technique are proceeded as follows.

1. Allocate 128-bit counters $V_1[s_1]$ for 2^{24} possible values of $s_1 = P_R[4]|C_R[7,8]$ and initialize them to zero. For each one of the N PC pairs, extract the value of s_1 and increment the corresponding counter $V_1[s_1]$. The time complexity of step is N memory accesses to process N PC pairs. Still we assume that processing each PC pair is equivalent to $1/4$ round encryption, then the time complexity of this step is about $N \cdot 1/4 \cdot 1/6$ encryptions.

2. Allocate 128-bit counters $V_2[s_2]$ for 2^{24} possible values of $s_2 = R_0[8]|C_R[7,8]$ and initialize them to zero. Guess $K_7[12]$ and $K_8[12]$ and partially encrypt s_1 to get the value of s_2, then update the corresponding counter by $V_2[s_2]+ = V_1[s_1]$. The computation in this step is much simpler than one round encryption and is proceeded about $2^{16} \cdot 2^{24} = 2^{40}$ times.

3. Allocate 128-bit counters $V_3[s_3]$ for 2^{16} possible values of $s_3 = R_0[8]|L_6[7]$ and initialize them to zero. Guess $K_9'[15,16]$ ($K_9' = (K_9 \vee 1)^{-1}$) and $K_{10}[15,16]$ and partially decrypt s_2 to obtain the value of s_3, then update the corresponding counter by $V_3[s_3]+ = V_2[s_2]$. The computation in this step is also much simpler than one round encryption and is proceeded about $2^{48} \cdot 2^{24} = 2^{72}$ times.

After Step 3, we have reached the boundaries of the zero-correlation linear approximations over the first six rounds. We then construct $V[z]$ and compute statistic T and filter out the wrong key guess. The right key is then recovered by exhaustively search all right key candidates. The detailed procedure is described Step 4 and Step 5.

4. Allocate 128-bit counter $V[z]$ for 2^{16} values of z and initialize them to zero, where z is the concatenation of evaluations of 16 basis zero-correlation masks. Compute z from s_3 with 16 basis zero-correlation masks, then $V[z]+ = V_3[x_3]$. Compute statistic T according to Equation (1). Again we assume the computation in this step is equivalent to 2 encryptions, then the time complexity of this step is about $2^{48} \cdot 2 = 2^{49}$ encryptions.

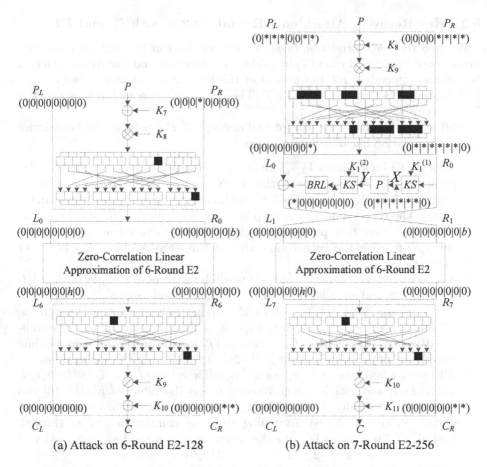

Fig. 8. Attacks on 6-Round E2-128 and 7-Round E2-256 with IT and FT

5. If $T \leq \tau$, then the guessed key value is a possible right key candidate and all key values can be recovered by exhaustively search.

Set $\beta_0 = 2^{-2.7}, \beta_1 = 2^{-50}$, then $z_{1-\beta_0} \approx 1.0$, $z_{1-\beta_1} = 7.96$. Since $n = 128$, $\ell = 2^{16}$, then according to Equation (2), the data complexity N is about $2^{123.7}$ known plaintexts and the time complexity of Step 1 is about $2^{123.7} \cdot 1/4 \cdot 1/6 \approx 2^{119.1}$ encryptions. By contrast, the time complexity of Step 2 and Step 3 is negligible. As only 48-bit key values are guessed in the partial encryption and decryption phase and $\beta_1 = 2^{-50}$, only the right key value is supposed to survive the filteration. Thus the time complexity of the exhaustively search in Step 5 is also negligible. The memory requirements are dominated by Step 1 and Step 2, where we need about $2 \cdot 16 \cdot 2^{24} = 2^{29}$ bytes to store the counters V_1 and V_2. In one word, our attack on 6-round E2-128 with IT and FT requires $2^{123.7}$ known plaintexts, $2^{119.1}$ encryptions and 2^{29} bytes memory.

5.2   Key Recovery Attack on 7-Round E2-256 with IT and FT

This time the zero-correlation linear approximations of 6-round E2 from Figure 2 cover round 2 to round 7, by adding IT function and one round function before and appending FT function after the linear approximations, we can attack 7-round E2-256 with IT and FT. The details of our attack is illustrated in Figure 8(b).

Still we need to partially encrypt and decrypt N PC pairs to the boundaries of zero-correlation linear approximations.

1. Allocate 64-bit counters $V_1[s_1]$ for 2^{88} possible values of $s_1 = $ and initialize them to zero. For each one of the N PC pairs, extract the value of $s_1 = P_L[2,3,4,7,8]\|P_R[5,6,7,8]\|C_R[7,8]$ and increment the corresponding counter $V_1[s_1]$. The time complexity of step is N memory accesses to process N PC pairs. Assuming that processing each PC pair is equivalent to $1/4$ round encryption, then the time complexity of this step is about $N \cdot 1/4 \cdot 1/7$ encryptions.

2. Allocate 64-bit counters $V_2[s_2]$ for 2^{72} possible values of $s_2 = L_0[8]\|R_0[2,3,4,5,6,7]\|C_R[7,8]$ and initialize them to zero. Guess the values of $K_8[2,3,4,7,8,13,14,15,16]$ and $K_9[2,3,4,7,8,13,14,15,16]$, partially encrypt s_1 through the IT function to get s_2 and update $V_2[s_2]+ = V_1[s_1]$. Assuming that the computation of s_2 through the IT function as $1/2$ round encryption, then the time complexity of this step is about $2^{144} \cdot 2^{88} \cdot 1/2 \cdot 1/7 \approx 2^{228.2}$ encryptions.

3. Allocate 64-bit counters $V_3[s_3]$ for 2^{64} possible values of $s_3 = L_0[8]\|R_0[2,3,4,5,6,7]\|L_7[7]$ and initialize them to zero. Guess the values of $K_{10}[15,16]$ and $K_{12}[15,16]$, partially decrypt through the FT function to obtain s_3 and update $V_3[s_3]+ = V_2[s_2]$. Assuming that the computation of s_3 through partial FT function as $1/4$ round encryption. Then the time complexity of this step is about $2^{176} \cdot 2^{72} \cdot 1/4 \cdot 1/7 \approx 2^{243.2}$ encryptions.

The remaining steps in the partial encryption and decryption phase are similar to Step 2 and Step 3. Thus we summarize the remaining steps in Table 3, where the columns have the same meaning as those of Table 2 and the "Complexity" is still measured in $1/8$ round encryption.

11. Allocate 128-bit counter $V[z]$ for 16-bit z and initialize to zero where z is the concatenation of evaluations of 16 basis zero-correlation masks. Compute

Table 3. Partial Encryption and Decryption on 7-Round E2-256

Step	Guess	Complexity	Computed States	Counter-Size
4	$K_1^{(1)}[2]$	$2^{184} \cdot 2^{64}$	$s_4 = L_0[8]\|R_0[3,4,5,6,7]\|X[2]\|L_7[7]$	$V_4 - 2^{64}$
5	$K_1^{(1)}[3]$	$2^{192} \cdot 2^{64}$	$s_5 = L_0[8]\|R_0[4,5,6,7]\|X[2] \oplus X[3]\|L_7[7]$	$V_5 - 2^{56}$
6	$K_1^{(1)}[4]$	$2^{200} \cdot 2^{56}$	$s_6 = L_0[8]\|R_0[5,6,7]\|X[2] \oplus X[3] \oplus X[4]\|L_7[7]$	$V_6 - 2^{48}$
7	$K_1^{(1)}[5]$	$2^{208} \cdot 2^{48}$	$s_7 = L_0[8]\|R_0[6,7]\|X[2] \oplus X[3] \oplus X[4] \oplus X[5]\|L_7[7]$	$V_7 - 2^{40}$
8	$K_1^{(1)}[6]$	$2^{216} \cdot 2^{40}$	$s_8 = L_0[8]\|R_0[7]\|X[2] \oplus X[3] \oplus X[4] \oplus X[5] \oplus X[6]\|L_7[7]$	$V_8 - 2^{32}$
9	$K_1^{(1)}[7]$	$2^{224} \cdot 2^{32}$	$s_9 = L_0[8]\|Y[1]\|L_7[7]$	$V_9 - 2^{24}$
10	$K_1^{(2)}[1]$	$2^{232} \cdot 2^{24}$	$s_{10} = R_1[8]\|L_7[7]$	$V_{10} - 2^{16}$

z from x_{10} with 16 basis zero-correlation masks, then $V[z]+ = V_{10}[s_{10}]$. Compute statistic T according to Equation (1). The time complexity of this step is about $2^{232} \cdot 2 = 2^{233}$ encryptions.

12. If $T \leq \tau$, then the guessed key value is a possible right key candidate. Then exhaustively search all compatible key values to recover the right key value.

As N is no more than 2^{128}, thus the time complexity of Step 1 is negligible. The time complexity of the partial encryption and decryption phase is about $2^{243.2}+2^{248}\cdot1/8\cdot1/7+6\cdot2^{256}\cdot1/8\cdot1/7 \approx 2^{252.8}$ encryptions. By contrast, the time complexity of Step 11 is negligible. Set $\beta_0 = 2^{-2.7}, \beta_1 = 2^{-234}$, then $z_{1-\beta_0} \approx 1.0$, $z_{1-\beta_1} \approx 17.8$. In this case only the right key guess is supposed to survive the filteration, making the time complexity of exhaustive search negligible. Since $n = 128$, $\ell = 2^{16}$, then according to Equation (2), the data complexity N is about $2^{124.7}$. In all, the time complexity of our attack on 7-round E2-256 with IT and FT requires $2^{124.7}$ known plaintexts, $2^{252.8}$ encryptions and $8 \cdot 2^{88} = 2^{91}$ bytes memory.

6 Conclusion

In this paper, we evaluate the security of E2 with respect to the technique of multidimensional zero-correlation linear cryptanalysis. As a result, we can attack 8-round E2-128 and 9-round E2-256 without IT and FT. We further propose the first key recovery attacks on reduced-round E2 with both IT and FT taken into consideration, where 6-round E2-128 and 7-round E2-256 can be attacked. Our attacks on E2 without IT and FT is the best attack with explicit time complexity better than exhaustive search in term of round number.

Acknowledgements. This work has been supported by the National Basic Research 973 Program of China under Grant No. 2013CB834205, the National Natural Science Foundation of China under Grant Nos. 61133013, 61103237, the Program for New Century Excellent Talents in University of China under Grant No. NCET-13-0350, as well as the Interdisciplinary Research Foundation of Shandong University of China under Grant No. 2012JC018.

References

1. Aoki, K., Ichikawa, T., Kanda, M., Matsui, M., Moriai, S., Nakajima, J., Tokita, T.: Camellia: A 128-Bit Block Cipher Suitable for Multiple Platforms-Design and Analysis. In: Stinson, D.R., Tavares, S. (eds.) SAC 2000. LNCS, vol. 2012, pp. 39–56. Springer, Heidelberg (2001)
2. Biham, E.: On Matsui's Linear Cryptanalysis. In: De Santis, A. (ed.) EURO-CRYPT 1994. LNCS, vol. 950, pp. 341–355. Springer, Heidelberg (1995)
3. Bogdanov, A., Rijmen, V.: Linear Hulls with Correlation Zero and Linear Cryptanalysis of Block Ciphers. Accepted to Designs, Codes and Cryptography (2012) (in press)

4. Bogdanov, A., Wang, M.: Zero Correlation Linear Cryptanalysis with Reduced Data Complexity. In: Canteaut, A. (ed.) FSE 2012. LNCS, vol. 7549, pp. 29–48. Springer, Heidelberg (2012)
5. Bogdanov, A., Leander, G., Nyberg, K., Wang, M.: Integral and Multidimensional Linear Distinguishers with Correlation Zero. In: Wang, X., Sako, K. (eds.) ASIACRYPT 2012. LNCS, vol. 7658, pp. 244–261. Springer, Heidelberg (2012)
6. Soleimany, H., Nyberg, K.: Zero-Correlation Linear Cryptanalysis of Reduced-Round LBlock. In: WCC 2013 (2013)
7. Blondeau, C., Nyberg, K.: New Links Between Differential and Linear Cryptanalysis. In: Johansson, T., Nguyen, P.Q. (eds.) EUROCRYPT 2013. LNCS, vol. 7881, pp. 388–404. Springer, Heidelberg (2013)
8. Bogdanov, A., Geng, H., Wang, M., Wen, L., Collard, B.: Zero-Correlation Linear Cryptanalysis with FFT and Improved Attacks on ISO Standards Camellia and CLEFIA. In: SAC 2013. LNCS. Springer (2014)
9. ISO/IEC 18033-3:2005, Information technology – Security techniques – Encryption algrithm – Part 3: Block Ciphers (July 2005)
10. Kanda, M., Moriai, S., Aoki, K., Ueda, H., Takashima, Y., Ohta, K., Matsumoto, T.: E2-a new 128-bit block cipher. IEICE Transactions Fundamentals of Electronics, Communications and Computer Sciences E83-A(1), 48–59 (2000)
11. Matsui, M.: Linear Cryptanalysis Method for DES cipher. In: Helleseth, T. (ed.) EUROCRYPT 1993. LNCS, vol. 765, pp. 386–397. Springer, Heidelberg (1994)
12. Matsui, M.: On Correlation between the Order of S-boxes and the Strength of DES. In: De Santis, A. (ed.) EUROCRYPT 1994. LNCS, vol. 950, pp. 366–375. Springer, Heidelberg (1995)
13. Hermelin, M., Cho, J.Y., Nyberg, K.: Multidimensional Extension of Matsui's Algorithm 2. In: Dunkelman, O. (ed.) FSE 2009. LNCS, vol. 5665, pp. 209–227. Springer, Heidelberg (2009)
14. Matsui, M., Tokita, T.: Cryptanalysis of a Reduced Version of the Block Cipher E2. In: Knudsen, L. (ed.) FSE 1999. LNCS, vol. 1636, pp. 71–80. Springer, Heidelberg (1999)
15. Moriai, S., Sugita, M., Aoki, K., Kanda, M.: Security of E2 against Truncated Differential Cryptanalysis. In: Heys, H.M., Adams, C.M. (eds.) SAC 1999. LNCS, vol. 1758, pp. 106–117. Springer, Heidelberg (2000)
16. Wei, Y., Li, P., Sun, B., Li, C.: Impossible Differential Cryptanalysis on Feistel Ciphers with SP and SPS Round Functions. In: Zhou, J., Yung, M. (eds.) ACNS 2010. LNCS, vol. 6123, pp. 105–122. Springer, Heidelberg (2010)
17. Wei, Y., Yang, X., Li, C., Du, W.: Impossible Differential Cryptanalysis on Tweaked E2. In: Xu, L., Bertino, E., Mu, Y. (eds.) NSS 2012. LNCS, vol. 7645, pp. 392–404. Springer, Heidelberg (2012)

Further Improvement of Factoring RSA Moduli with Implicit Hint

Liqiang Peng[1,2], Lei Hu[1,2], Jun Xu[1,2], Zhangjie Huang[1,2], and Yonghong Xie[1,2]

[1] State Key Laboratory of Information Security, Institute of Information Engineering, Chinese Academy of Sciences, Beijing 100093, China
[2] Data Assurance and Communication Security Research Center, Chinese Academy of Sciences, Beijing 100093, China
{lqpeng,hu,jxu,zjhuang,xyxie}@is.ac.cn

Abstract. We investigate the problem of factoring RSA moduli with implicit hint, which was firstly proposed by May and Ritzenhofen in 2009 where unknown prime factors of several RSA moduli shared some number of least significant bits (LSBs) and was considered by Faugère et al. in 2010 where some most significant bits (MSBs) were shared between the primes. In this paper, we further consider this factorization with implicit hint problem, present a method to deal with the case when the number of shared LSBs or MSBs is not large enough to satisfy the bound proposed by May et al. and Faugère et al. by making use of a result from Herrmann and May for solving linear equations modulo unknown divisors, and finally get a better lower bound on the the number of shared LSBs or MSBs. To the best of our knowledge, our lower bound is better than all known results and we can theoretically deal with the implicit factorization for the case of balanced RSA moduli.

Keywords: RSA modulus, factorization with implicit hint, Copper-smith's technique.

1 Introduction

Factoring large integers efficiently is a problem of most concern in algorithmic number theory and also in practical cryptographic applications since the RSA public key cryptosystem based on the factorization problem has been widely used. However, due to practical reasons, e.g., for achieving high implementation efficiency, specific RSA parameters are often adopted and the security of such an RSA cryptosystem may be threatened by cryptanalysis such as small private exponent attack [4,20], small CRT-exponent (Chinese-remainder-theorem-exponent) attack [12] and so on. Recently, Lenstra et al. [13] and Bernstein et al. [3] utilized the weakness of pseudo random number generators to successfully factor some RSA moduli which are used in the real world. Hence, the problem of factoring RSA moduli with some specific hint is worthy of investigation.

In the PKC'2009 conference, May and Ritzenhofen proposed an efficient method to factor RSA moduli with an implicit hint [16]. More precisely, for two

D. Pointcheval and D. Vergnaud (Eds.): AFRICACRYPT 2014, LNCS 8469, pp. 165–177, 2014.

n-bit RSA moduli $N_1 = p_1q_1$ and $N_2 = p_2q_2$ where p_1 and p_2 share tn least significant bits (LSBs) and q_1 and q_2 are (αn)-bit prime integers, it has been proved in [16] that if $tn \geq 2\alpha n+3$, then (q_1, q_2) is a shortest vector in a two-dimensional lattice and it can be found by a lattice basis reduction algorithm. Thus, the two RSA moduli can be factored. May et al. [16] also gave a heuristic generalization for the factorization of multiple RSA moduli $N_1 = p_1q_1, \cdots, N_k = p_kq_k$, where the number of shared LSBs, tn, is at least $\frac{k}{k-1}\alpha n$. Shortly later, Faugère et al. [7] made an extension analysis to deal with the case that $p_1, \cdots, p_k$ share most significant bits (MSBs) or bits in the middle.

In 2011, Sarkar and Maitra [19] transformed the factorization with implicit hint problem to the approximate integer common divisor problem [10,5], and lower bounds on the number of LSBs or MSBs required to be shared is improved in theory and experimentally [19]. Sarkar and Maitra used Coppersmith's lattice-based technique to find out the desired roots of modular equation, and the lower bound they obtained was improved to

$$
\begin{cases}
t > \max\{\alpha, \frac{\alpha k^2 - (2\alpha+1)k+1+\sqrt{k^2+2\alpha^2 k-\alpha^2 k^2-2k+1}}{k^2-3k+2}\}, \text{for } k > 2, \\
t > 2\alpha - \alpha^2, \text{for } k = 2.
\end{cases}
$$

Based on this result, Lu et al. [15] modified the polynomials in the construction of the lattice and the bound was further improved as $1 - (1 - \alpha)^{\frac{k}{k-1}}$.

In this paper, we firstly reconsider the problem of factoring RSA moduli with primes sharing LSBs, which has been discussed by May et al. [16]. As it has been shown in [16], if there are enough shared LSBs, the desired factorization can be directly obtained from the L^3 lattice basis reduction algorithm. We present a method to deal with the case where the shared LSBs are not enough to ensure that the desired factorization is included in the output of the L^3 algorithm. The idea is that we represent the vector which we desire to find out as an integer linear combination of the reduced basis vectors of the lattice and obtain a modular equation system, then we transform the modular equation system to a modular equation with unknown modulus by applying the Chinese remainder theorem, and finally, we solve this modular equation by a method of Herrmann and May in [8]. Note that, our method does not require the constraint that $t \geq \alpha$ in [16,7,19,15], which means for multiple RSA moduli we can for the first time theoretically deal with the implicit factorization for the case of balanced RSA moduli (i.e., p_i and q_i have the same bitlength). The factorization of RSA moduli with primes sharing MSBs is also revisited in this paper.

Table 1 lists a comparison of our result with the previous results in [16], [7], [19] and [15], where

$$
F(\alpha, k) = \begin{cases}
\frac{\alpha k^2 - (2\alpha+1)k+1+\sqrt{k^2+2\alpha^2 k-\alpha^2 k^2-2k+1}}{k^2-3k+2}, \text{for } k > 2, \\
2\alpha - \alpha^2, \text{for } k = 2,
\end{cases}
$$

$$
G(\alpha, k) = \frac{k}{k-1}(\alpha - 1 + (1 - \alpha)^{\frac{k+1}{k}} + (k + 1)(1 - (1 - \alpha)^{\frac{1}{k}})(1 - \alpha)),
$$

Table 1. Comparison with existing results on t

	[16]	[7]	[19]	[15]	this paper
LSB	$\frac{k}{k-1}\alpha$	-	$F(a,k)$	$1-(1-\alpha)^{\frac{k}{k-1}}$	$G(\alpha,k)$
MSB	-	$\frac{k}{k-1}\alpha + \frac{6}{n}$	$F(a,k)$	$1-(1-\alpha)^{\frac{k}{k-1}}$	$G(\alpha,k)$

and the curves of $G(\alpha,k)$ and $1-(1-\alpha)^{\frac{k}{k-1}}$ as functions on α can be seen in Figures 1 and 2 in Sections 3 and 4 which show $G(\alpha,k) < 1-(1-\alpha)^{\frac{k}{k-1}}$. To the best of our knowledge, our lower bound on the number of the shared bits is theoretically better than all known results and experimental results also show this improvement.

2   Preliminaries

Let $w_1, w_2, \cdots, w_k$ be k linearly independent vectors in $\mathbb{R}^n$. They span a k-dimensional lattice L which is the set of all integer linear combinations, $c_1 w_1 + \cdots + c_k w_k$, of $w_1, \cdots, w_k$, where $c_1, \cdots, c_k \in \mathbb{Z}$. The vectors $w_1, \cdots, w_k$ form a basis of the lattice L. Any lattice of dimension larger than 1 has infinitely many bases [18].

Calculating the shortest vectors in a lattice is known to be an NP-hard problem under randomized reductions [2]. However, some approximations of shortest vectors in a lattice can be found out in polynomial time and the famous L^3 lattice basis reduction algorithm is invented thirty years ago for attending such a goal [14,18], and since then lattice becomes a fundamental tool to analyze the security of public key cryptosystems.

Lemma 1. (L^3, [14,18]) Let L be a lattice of dimension k. Applying the L^3 algorithm to L, the outputted reduced basis vectors $v_1, \cdots, v_k$ satisfy that

$$\|v_1\| \leq \|v_2\| \leq \cdots \leq \|v_i\| \leq 2^{\frac{k(k-i)}{4(k+1-i)}} \det(L)^{\frac{1}{k+1-i}}, \text{for any } 1 \leq i \leq k.$$

Lattices are used to find small roots of univariate modular equations and bivariate equations [6], and this strategy is now usually called Coppersmith's technique. In [11], Jochemsz and May extended the technique and gave a general result to find roots of multivariate polynomials.

Given a polynomial $g(x_1, \cdots, x_k) = \sum_{(i_1, \cdots, i_k)} a_{i_1, \cdots, i_k} x_1^{i_1} \cdots x_k^{i_k}$, define the norm of g by

$$\|g(x_1, \cdots, x_k)\| = \Big(\sum_{(i_1, \cdots, i_k)} a_{i_1, \cdots, i_k}^2 \Big)^{1/2}.$$

The following lemma due to Howgrave-Graham [9] gives a sufficient condition under which roots of a modular equation also satisfy an integer equation.

Lemma 2. *(Howgrave-Graham, [9]) Let* $g(x_1, \cdots, x_k) \in \mathbb{Z}[x_1, \cdots, x_k]$ *be an integer polynomial with at most w monomials. Suppose that*

$$1. g(y_1, \cdots, y_k) \equiv 0 \pmod{p^m} \text{ for } |y_1| \leq X_1, \cdots, |y_k| \leq X_k, \text{ and}$$

$$2. \|g(x_1 X_1, \cdots, x_k X_k)\| < \frac{p^m}{\sqrt{w}}$$

Then $g(y_1, \cdots, y_k) = 0$ *holds over the integers.*

Lattice based approaches of solving small roots of a modular or integer equation are first to construct a lattice from the polynomial of the equation, then by lattice basis reduction algorithm obtain new short lattice vectors which correspond to new polynomials with small norms and with the same roots as the original polynomial. These approaches usually rely on the following heuristic assumption.

Assumption 1. *The common roots of the polynomials yielded by lattice based constructions can be efficiently computed by using numerical method, symbolic method or exploiting the special structure of these polynomials.*

In our analysis, we will use the following theorem proposed by Herrmann and May in [8]. Based on Coppersmith's technique, they gave upper bounds on the size of solutions of a bivariate linear equation modulo an unknown divisor of a known composite integer.

Theorem 1. *(Herrmann and May, [8]) Let* $\epsilon > 0$, N *be a sufficiently large composite integer with an unknown divisor* $p \geq N^\beta$, *and* $f(x_1, x_2) \in \mathbb{Z}[x_1, x_2]$ *be a bivariate linear polynomial. Under Assumption 1, one can find all solutions* (y_1, y_2) *of the equation* $f(x_1, x_2) = 0 \pmod p$ *with* $|y_1| \leq N^\gamma$ *and* $|y_2| \leq N^\delta$ *if*

$$\gamma + \delta \leq 3\beta - 2 + 2(1 - \beta)^{\frac{3}{2}} - \epsilon. \tag{1}$$

The above theorem 1 has been extended to a modular equation with $k \geq 3$ variables [8].

Theorem 2. *(Herrmann and May, [8]) Let* $\epsilon > 0$, N *be a sufficiently large composite integer with an unknown divisor* $p \geq N^\beta$, $f(x_1, \cdots, x_k) \in \mathbb{Z}[x_1, \cdots, x_k]$ *be a monic linear polynomial in k variables. Under Assumption 1, one can find all solutions* $(y_1, \cdots, y_k)$ *of the equation* $f(x_1, \cdots, x_k) = 0 \pmod p$ *with* $|y_1| \leq N^{\gamma_1}, \cdots, |y_k| \leq N^{\gamma_k}$ *if*

$$\sum_{i=1}^{k} \gamma_i \leq 1 - (1 - \beta)^{\frac{k+1}{k}} - (k + 1)(1 - \sqrt[k]{1 - \beta})(1 - \beta) - \epsilon. \tag{2}$$

More details about the theorems can be referred to [8]. Note that, in our experiments, the equations obtained by calculation of the resultant or finding a Gröbner basis are not univariate polynomials, however we can exploit the structure of these polynomials to solve out the desired small roots.

3   Factoring Two RSA Moduli with Implicitly Common LSBs

Recall in the implicit factoring of two RSA moduli in [16], there are two different n-bit RSA moduli $N_1 = p_1 q_1$ and $N_2 = p_2 q_2$, where p_1 and p_2 satisfy that $p_1 \equiv p_2 (\mathrm{mod}\, 2^{tn})$, where $0 < t < \log_{N_i} p_i$ for $i = 1, 2$.

Since $p_1 \equiv p_2 (\mathrm{mod}\, 2^{tn})$, we let $p_1 = p + 2^{tn} \widetilde{p_1}$ and $p_2 = p + 2^{tn} \widetilde{p_2}$. We have

$$(p + 2^{tn} \widetilde{p_1}) q_1 = N_1,$$
$$(p + 2^{tn} \widetilde{p_2}) q_2 = N_2,$$

which means

$$p q_1 = N_1 \ (\mathrm{mod}\, 2^{tn}),$$
$$p q_2 = N_2 \ (\mathrm{mod}\, 2^{tn}).$$

Moreover, we get the following linear equation

$$(N_1^{-1} N_2) q_1 - q_2 \equiv 0 \ (\mathrm{mod}\, 2^{tn}), \tag{3}$$

where N_1^{-1} is the inverse of N_1 modulo 2^{tn}.

In [16], the authors have proved that the vector (q_1, q_2) is the shortest vector of the two-dimensional lattice L_1 generated by the row vectors of the following matrix

$$\begin{pmatrix} 1 & N_1^{-1} N_2 \\ 0 & 2^{tn} \end{pmatrix} \tag{4}$$

when q_1 and q_2 are both (αn)-bit numbers and $tn > 2(\alpha n + 1)$, where $\alpha \approx 1 - \log_{N_i} p_i$ for $i = 1, 2$. Note that $t < \log_{N_i} p_i \approx 1 - \alpha$. Once q_1 and q_2 are obtained by the L^3 algorithm in polynomial time, N_1 and N_2 are factored.

However, when $tn \leq 2(\alpha n + 1)$ the vector (q_1, q_2) is not the shortest vector of L_1, which means (q_1, q_2) is generally not included in the outputted basis (λ_1, λ_2) of the L^3 algorithm. Write the vector (q_1, q_2) as a linear combination of λ_1 and λ_2. Below we present a method to find out the linear combination by solving linear equations modulo unknown RSA factors. Once the linear combination is found, a better bound on t than that in [16] is obtained.

Let $\lambda_1 = (l_{11}, l_{12})$ and $\lambda_2 = (l_{21}, l_{22})$ be the basis vectors of L_1 obtained from the L^3 algorithm. Then we have a rough estimation on the l_{ij}, with overwhelming probability, the minima of a lattice are all asymptotically close to the Gaussian heuristic [1], hence we have $\|\lambda_1\| \approx \|\lambda_2\| \approx \sqrt{\frac{2}{2\pi e}} \det(L)^{\frac{1}{2}}$. Thus, the sizes of $l_{11}, l_{12}, l_{21}, l_{22}$ can be estimated from $\det(L_1)^{\frac{1}{2}} = 2^{\frac{tn}{2}}$.

Let (q_1, q_2) be represented as $(q_1, q_2) = x_1 \lambda_1 + x_2 \lambda_2$ with integral coefficients x_1 and x_2. Then we get two modular equations modulo unknown prime numbers

$$\begin{cases} x_1 l_{11} + x_2 l_{21} = q_1 \equiv 0 \ (\mathrm{mod}\, q_1), \\ x_1 l_{12} + x_2 l_{22} = q_2 \equiv 0 \ (\mathrm{mod}\, q_2). \end{cases} \tag{5}$$

Since $l_{11}, l_{12}, l_{21}, l_{22}$ have roughly the same size, the desired coefficients x_1 and x_2 can be roughly estimated as $\frac{q_j}{2l_{ij}}$ for any i and j.

Using the Chinese remainder theorem, from (5) we get an equation with the form of

$$ax_1 + bx_2 \equiv 0 \,(\mathrm{mod}\ q_1 q_2), \tag{6}$$

where a is an integer satisfying $a \equiv l_{11} \,(\mathrm{mod}\ N_1)$ and $a \equiv l_{12} \,(\mathrm{mod}\ N_2)$, and b is an integer satisfying $b \equiv l_{21} \,(\mathrm{mod}\ N_1)$ and $b \equiv l_{22} \,(\mathrm{mod}\ N_2)$. Clearly, a and b can be calculated from $l_{11}, l_{12}, l_{21}, l_{22}, N_1$ and N_2 by the extended Euclidean algorithm.

Since $q_1 \approx q_2 \approx 2^{\alpha n}$, we have $q_1 q_2 \approx (N_1 N_2)^{\alpha}$. By Theorem 1, we can find all solutions (y_1, y_2) of equation (6) with $|y_1| \leq (N_1 N_2)^{\delta_1} \approx 2^{2\delta_1 n}$ and $|y_2| \leq (N_1 N_2)^{\delta_2} \approx 2^{2\delta_2 n}$ if

$$\delta_1 + \delta_2 \leq 3\alpha - 2 + 2(1-\alpha)^{\frac{3}{2}} - \epsilon.$$

When $\delta_1 \approx \delta_2$, we have

$$2\delta_1 \approx 2\delta_2 \leq 3\alpha - 2 + 2(1-\alpha)^{\frac{3}{2}} - \epsilon. \tag{7}$$

From (5), there is a good possibility that the desired solution of (5) can be estimated with $\frac{q_1}{2l_{11}} \approx 2^{(\alpha - \frac{t}{2})n}$. Hence, when

$$\alpha - \frac{t}{2} \leq 3\alpha - 2 + 2(1-\alpha)^{\frac{3}{2}} - \epsilon,$$

or equivalently,

$$t \geq 4 - 4\alpha - 4(1-\alpha)^{\frac{3}{2}} + \epsilon,$$

the desired solution can be solved out.

Comparing with the works of [16], [19] and [15], we can get the following Figure 1.

Experimental Results:
We have implemented the experiment program in Sage 5.12 computer algebra system on a PC with Intel(R) Core(TM) Duo CPU(2.53GHz, 1.9GB RAM ubuntu 13.10) and carried out the L^2 algorithm [17]. In all experiments, we obtained several integer equations with desired roots (y_1, y_2) over $\mathbb{Z}$ and found that these equations had a common factor with the form of $ax_1 + bx_2$. In these situations, $ay_1 + by_2$ always equals to 0 and $\gcd(y_1, y_2)$ is small. Hence, the solution (y_1, y_2) can be solved out.

The following Table 2 lists some theoretical and experimental results on factoring two 1024-bit RSA moduli with shared LSBs.

4   Extending to Factoring Multiple RSA Moduli with Implicitly Common LSBs

In the case of multiple RSA moduli with implicit common LSBs, let $N_i = p_i q_i$, $i = 1, 2, \cdots, k$, be k different n-bit RSA moduli and p_i share tn least significant

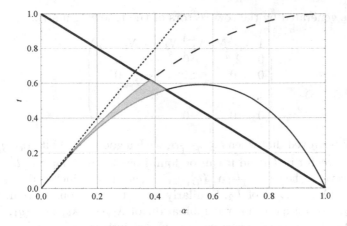

Fig. 1. Comparison with previous ranges on t with respect to α. Since $t \leq 1 - \alpha$, any valid range is under the thick solid diagonal line. Here the dotted line denotes the lower bound on t in [16], the dashed line denotes that in [19] and [15], and the thin solid line denotes that in this paper. The grey shaded area is a new improvement presented in this paper.

Table 2. Theoretical and experimental results of factoring 1024-bit RSA moduli with LSBs. Here dim denotes the dimension of the lattice.

k	bitsize of (p_i, q_i), i.e., $((1-\alpha)\log_2 N_i, \alpha \log_2 N_i)$	no. of shared LSBs in p_i ([19])				no. of shared LSBs in p_i (this paper)			
		theo.	expt.	dim	time (sec)	theo.	expt.	dim	time (sec)
2	$(874, 150)$	278	–	–	–	267	278	190	1880.10
2	$(824, 200)$	361	–	–	–	340	357	190	1899.21
2	$(774, 250)$	439	–	–	–	405	412	190	2814.84
2	$(724, 300)$	513	–	–	–	461	470	190	2964.74

bits. Let q_i be of (αn)-bit. Write the moduli as

$$N_1 = (p + 2^{tn}\widetilde{p_1})q_1,$$

$$\cdots$$

$$N_k = (p + 2^{tn}\widetilde{p_k})q_k.$$

Then,

$$N_1 \equiv pq_1 \pmod{2^{tn}},$$

$$\cdots$$

$$N_k \equiv pq_k \pmod{2^{tn}}.$$

Similarly as in the analysis in the previous section for the case $k = 2$, we have $\frac{N_1}{q_1} \equiv \frac{N_i}{q_i} \pmod{2^{tn}}$, for $i = 2, 3, \cdots, k$. Since the modular equation $N_1^{-1}N_i q_1 - q_i \equiv 0 \pmod{2^{tn}}$ holds, we get a vector $(q_1, q_2, \cdots, q_k)$ in a k-dimensional lattice

L_2 which is generated by the row vectors of the following matrix

$$\begin{pmatrix} 1 & N_1^{-1}N_2 & N_1^{-1}N_3 & \cdots & N_1^{-1}N_k \\ 0 & 2^{tn} & 0 & \cdots & 0 \\ 0 & 0 & 2^{tn} & \cdots & 0 \\ \vdots & \vdots & \vdots & \ddots & \vdots \\ 0 & 0 & 0 & \cdots & 2^{tn} \end{pmatrix}.$$

In [16], it is proved that when $t \geq \frac{k}{k-1}\alpha$, with a good possibility $(q_1, q_2, \cdots, q_k)$ is a shortest vector in L_2 and it can be found out by applying the L^3 algorithm to L_2. However, when $t < \frac{k}{k-1}\alpha$, $(q_1, q_2, \cdots, q_k)$ is not included in the L^3 reduced basis $\{\lambda_1, \cdots, \lambda_k\}$ of L_2. Similarly as in the previous section, we represent $(q_1, q_2, \cdots, q_k)$ as a linear combination of $\lambda_1, \cdots, \lambda_k$, i.e., $(q_1, \cdots, q_k) = x_1\lambda_1 + x_2\lambda_2 + \cdots + x_k\lambda_k$, where $x_1, \cdots, x_k$ are integers.

Hence, we have the following modular equation system

$$\begin{cases} x_1 l_{11} + x_2 l_{21} + \cdots + x_k l_{k1} = q_1 \equiv 0 \pmod{q_1}, \\ \quad\quad\quad \cdots \\ x_1 l_{1k} + x_2 l_{2k} + \cdots + x_k l_{kk} = q_k \equiv 0 \pmod{q_k}, \end{cases} \tag{8}$$

where $\lambda_i = (l_{i1}, l_{i2}, \cdots, l_{ik})$, $i = 1, 2, \cdots, k$.

The lengths of the output vectors of the L^3 algorithm can be estimated based on the Gaussian heuristic and experimental experience of the L^3 algorithm. We roughly estimate the sizes of $|\lambda_1|, \cdots, |\lambda_k|$ and the entries of λ_i as $\det(L_2)^{\frac{1}{k}} = 2^{\frac{nt(k-1)}{k}}$ and the solution of (8) as $|x_i| \approx \frac{q_i}{k l_{ij}} \approx 2^{\alpha n - \frac{nt(k-1)}{k} - \log_2 k} \leq 2^{\alpha n - \frac{nt(k-1)}{k}}$.

Similarly as in the previous section, we can obtain an equation with the form of

$$a_1 x_1 + a_2 x_2 + \cdots + a_k x_k \equiv 0 \pmod{q_1 q_2 \cdots q_k} \tag{9}$$

from equation system (8) by using the Chinese remainder theorem, where a_i is an integer satisfying $a_i \equiv l_{ij} \pmod{N_j}$ for $1 \leq j \leq k$ and it can be calculated from the l_{ij} and N_j.

For this linear polynomial equation in k variables modulo the unknown integer $q_1 q_2 \cdots q_k \approx (N_1 N_2 \cdots N_k)^\alpha$, by Theorem 2 the variables with $|x_i| \leq (N_1 N_2 \cdots N_k)^{\delta_i} \approx 2^{k\delta_i n}$, $i = 1, 2, \cdots, k$, can be solved out if

$$\sum_{i=1}^{k} \delta_i \leq 1 - (1-\alpha)^{\frac{k+1}{k}} - (k+1)(1 - \sqrt[k]{1-\alpha})(1-\alpha) - \epsilon,$$

or equivalently,

$$k\delta_i \leq 1 - (1-\alpha)^{\frac{k+1}{k}} - (k+1)(1 - \sqrt[k]{1-\alpha})(1-\alpha) - \epsilon$$

when $\delta_1 \approx \delta_2 \approx \cdots \approx \delta_k$.

Hence, when

$$\alpha - \frac{t(k-1)}{k} \leq 1 - (1-\alpha)^{\frac{k+1}{k}} - (k+1)(1 - \sqrt[k]{1-\alpha})(1-\alpha) - \epsilon,$$

or namely,

$$t \geq \frac{k}{k-1}(\alpha - 1 + (1-\alpha)^{\frac{k+1}{k}} + (k+1)(1-(1-\alpha)^{\frac{1}{k}})(1-\alpha)) + \epsilon, \qquad (10)$$

the desired solution can be solved out.

To the best of our knowledge, the previous best theoretical bound on t is given in [15]: $t \geq 1 - (1-\alpha)^{\frac{k}{k-1}}$. We make a comparison between our theoretical bound (10) and this bound, see Figure 2 for the cases of $k = 3$ and $k = 4$. We shall note that when $k \geq 3$, there exists t satisfying $t \leq 1 - \alpha$ and the inequality (10), which removes the requirement that $t \geq \alpha$ in [16,7,19,15] and means for multiple RSA moduli we can for the first time theoretically deal with the implicit factorization for the case of balanced RSA moduli (i.e., p_i and q_i have the same bitlength and $\alpha = \frac{1}{2}$).

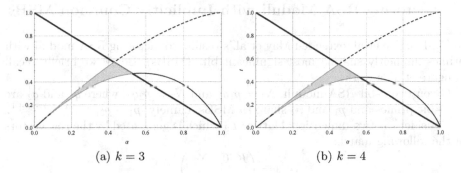

(a) $k = 3$ (b) $k = 4$

Fig. 2. The comparison between the bound (10) and the known best bound in [15]. As in Figure 1, any valid range is under the thick solid diagonal line. Here the dashed line denotes the lower bound on t in [15], the thin solid line denotes that in this paper, the grey shaded area on the figure is a new improvement presented in this paper.

Experimental Results:
We have implemented the program in Sage 5.12 computer algebra system on a PC with Intel(R) Core(TM) Duo CPU (2.53GHz, 1.9GB RAM ubuntu 13.10).

In all experiments for the case $k = 3$ and 1000-bit RSA moduli, we obtained several integer equations with desired roots (y_1, y_2, y_3) over $\mathbb{Z}$. To find out the roots, we used the technique of calculation of resultants and we always obtained a homogeneous equation of the form of $c_1 x_2^4 + c_2 x_2^3 x_3 + c_3 x_2^2 x_3^2 + c_4 x_2 x_3^3 + c_5 x_3^4 = 0$ which has the desired roots. Then we transformed these homogeneous bivariate equations to univariate equations over $\mathbb{Q}$ and obtained the ratio of $\frac{y_2}{y_3}$ by solving univariate equations. Similarly as in the experiments in the previous section, the common divisor of the desired roots is always small, hence we can obtain the desired roots (y_1, y_2, y_3). See Table 3 for the comparison with the previous bounds on t.

Table 3. For 1000-bit RSA moduli, theoretical and experimental bounds on t

k	bitsize of q_i	[16]		[19]		this paper	
		theo.	expt.	theo.	expt.	theo.	expt.
3	250	375	378	352	367	309	350
3	300	450	452	416	431	354	420
3	350	525	527	478	499	392	440
3	400	600	–	539	562	423	480

We notice that, when k is increasing, the lower bound on t will decrease, however, the dimension of the lattice constructed for solving the roots of the polynomials will be also increase. Due to the restriction of our computing ability, it is hard to evaluate the experimental results for larger k.

5   Factoring RSA Moduli with Implicitly Common MSBs

In [7], Faugère et al. extended May et al.'s results to factoring RSA moduli with primes implicitly sharing most significant bits (MSBs). Below we briefly recall Faugère et al.'s work.

Given two n-bit RSA moduli, $N_1 = p_1 q_1$ and $N_2 = p_2 q_2$, where q_1 and q_2 are (αn)-bit primes and p_1 and p_2 share tn MSBs, namely $|p_1 - p_2| \leq 2^{n-\alpha n-tn+1}$.

Consider the two-dimensional lattice L_3 which is generated by the row vectors of the following matrix

$$M_3 = \begin{pmatrix} K & 0 & N_2 \\ 0 & K & -N_1 \end{pmatrix}$$

where $K = \lfloor 2^{n-tn+\frac{1}{2}} \rfloor$. It has been proved in [7] that when $tn \geq 2\alpha n + 3$, or for simplicity $t \geq 2\alpha$ for efficiently large n, the vector $(q_1 K, q_2 K, q_1 q_2 (p_2 - p_1))$ is the shortest vector in L_3. Similarly, when $t \leq 2\alpha$ the vector $(q_1 K, q_2 K, q_1 q_2 (p_2 - p_1))$ that we wanted is not the shortest vector of L_3 and q_1 and q_2 can not be obtained directly from the basis vectors λ_1 and λ_2 of L_3 which are outputted by applying the L^3 algorithm.

In order to enable our result succinct, we make a rough estimation on the sizes of $\lambda_1 = (l_{11}, l_{12}, l_{13})$ and $\lambda_2 = (l_{21}, l_{22}, l_{23})$ and their entries. Since

$$\det(L_3) = \det(M_3 M_3^T) = K\sqrt{N_1^2 + N_2^2 + K^2} \approx 2^{2n-tn+1},$$

the length of $|\lambda_1|$ and $|\lambda_2|$ can be estimated as $\det(L_3)^{\frac{1}{2}} \approx 2^{n-\frac{tn}{2}+\frac{1}{2}}$, hence the entries can be bounded as $|l_{ij}| \approx 2^{n-\frac{tn}{2}}, i = 1, 2, j = 1, 2, 3$.

Since $(q_1 K, q_2 K, q_1 q_2 (p_2 - p_1)) \in L_3$, there exist integers x_1 and x_2 such that $(q_1 K, q_2 K, q_1 q_2 (p_2 - p_1)) = x_1 \lambda_1 + x_2 \lambda_2$. Hence, we obtain a modular equation system

$$\begin{cases} x_1 l_{11} + x_2 l_{21} = q_1 K \equiv 0 \pmod{q_1}, \\ x_1 l_{12} + x_2 l_{22} = q_2 K \equiv 0 \pmod{q_2}. \end{cases} \tag{11}$$

Since $|l_{ij}| \approx 2^{n-\frac{tn}{2}}$, the solutions to (11) can be estimated roughly by $x_i \approx \frac{q_j K}{2l_{ij}} \approx 2^{\alpha n + n - tn - n + \frac{tn}{2}} \approx 2^{\alpha n - \frac{tn}{2}}$.

Using the Chinese remainder theorem, from (11) we get a modular equation with the form of

$$ax_1 + bx_2 \equiv 0 \ (\mathrm{mod} \ q_1 q_2). \qquad (12)$$

On the other hand, since $q_1 q_2 \approx (N_1 N_2)^\alpha$, from Theorem 1 the solution of (12) with $|y_1| < (N_1 N_2)^{\delta_1} \approx 2^{2\delta_1 n}$ and $|y_2| < (N_1 N_2)^{\delta_2} \approx 2^{2\delta_2 n}$ can be found if

$$\delta_1 + \delta_2 \leq 3\alpha - 2 + 2(1-\alpha)^{\frac{3}{2}} - \epsilon.$$

With $\delta_1 \approx \delta_2$, we have

$$2\delta_1 \approx 2\delta_2 \leq 3\alpha - 2 + 2(1-\alpha)^{\frac{3}{2}} - \epsilon.$$

Hence, when

$$\alpha - \frac{t}{2} \leq 2\delta_1 \leq 3\alpha - 2 + 2(1-\alpha)^{\frac{3}{2}} - \epsilon,$$

or equivalently,

$$t \geq 4 - 4\alpha - 4(1-\alpha)^{\frac{3}{2}} + \epsilon,$$

the desired solution can be solved out.

The above method can be extended to factoring multiple RSA moduli with primes implicitly sharing MSBs. In a similar way, we can prove that one can factor k RSA moduli with primes implicitly sharing (tn)-bit MSBs if

$$t \geq \frac{k}{k-1}(\alpha - 1 + (1-\alpha)^{\frac{k+1}{k}} + (k+1)(1 - (1-\alpha)^{\frac{1}{k}})(1-\alpha)) + \epsilon.$$

To illustrate our optimization on the lower bounds on t, we list in Table 4 some numerical values for comparison with the results in [16], [7], [19] and [15]. It can be seen that our improvement with previous results increases as α increases.

Table 4. Comparison with previous results on the theoretical bounds on t

k	α	[16]([7])	[19]	[15]	this paper	α	[16]([7])	[19]	[15]	this paper
5	0.20	0.2500	0.2437	0.2434	0.2182	0.30	0.3750	0.3606	0.3597	0.3012
5	0.40	0.5000	0.4740	0.4719	0.3642	0.45	0.5625	0.5292	0.5264	0.3874
5	0.50	–	–	–	0.4045	–	–	–	–	–
10	0.20	0.2222	0.2197	0.2196	0.1962	0.30	0.3333	0.3276	0.3272	0.2725
10	0.40	0.4444	0.4341	0.4331	0.3320	0.45	0.5000	0.4868	0.4853	0.3546
10	0.50	–	–	–	0.3720	–	–	–	–	–
50	0.20	0.2041	0.2037	0.2036	0.1818	0.30	0.3061	0.3052	0.3051	0.2539
50	0.40	0.4082	0.4064	0.4062	0.3112	0.45	0.4592	0.4570	0.4567	0.3335
50	0.50	–	–	–	0.3512	–	–	–	–	–

Experimental Results:
We implemented our analysis in Sage 5.12 computer algebra system on a PC with Intel(R) Core(TM) Duo CPU(2.53GHz, 1.9GB RAM ubuntu 13.10). We present some numerical values for comparison with [19] in Table 5.

Table 5. For 1024-bit RSA moduli, theoretical and experimental results on factoring RSA moduli with implicitly common MSBs

k	bitsize of (p_i, q_i) $((1-\alpha)\log_2 N_i, \alpha\log_2 N_i)$	no. of shared MSBs in p_i ([19])				no. of shared MSBs in p_i (this paper)			
		theo.	expt.	dim	time (sec)	theo.	expt.	dim	time (sec)
2	(874,150)	278	289	16	1.38	267	278	190	1974.34
2	(824,200)	361	372	16	1.51	340	358	190	2030.92
2	(774,250)	439	453	16	1.78	405	415	190	2940.35
2	(724,300)	513	527	16	2.14	461	474	190	3105.79
3	(874,150)	217	230	56	29.24	203	225	220	5770.99
3	(824,200)	286	304	56	36.28	260	288	220	6719.03
3	(774,250)	352	375	56	51.04	311	343	220	6773.48
3	(724,300)	417	441	56	70.55	356	395	220	7510.86
3	(674,350)	480	505	56	87.18	395	442	220	8403.91
3	(624,400)	540	569	56	117.14	428	483	220	9244.42

6 Conclusion

In this paper, we presented a further method for factoring RSA moduli with implicitly common LSBs or MSBs, and got a more lower bound on the number of the bits shared by the unknown primes of the RSA moduli. Our improvement can deal with some situations where the number of shared LSBs or MSBs does not satisfy the lower bounds proposed by May and Ritzenhofen in [16] and Faugère et al. in [7]. It is nice to see our theoretical bound and experimental results both have an improvement on existing results.

Acknowledgements. The authors would like to thank anonymous reviewers for their helpful comments and suggestions. The work of this paper was supported by the National Key Basic Research Program of China (2013CB834203), the National Natural Science Foundation of China (Grant 61070172), the Strategic Priority Research Program of Chinese Academy of Sciences under Grant XDA06010702, and the State Key Laboratory of Information Security, Chinese Academy of Sciences.

References

1. Ajtai, M.: Generating random lattices according to the invariant distribution. Draft of March (2006)
2. Ajtai, M.: The shortest vector problem in L_2 is NP-hard for randomized reductions. In: Proceedings of the Thirtieth Annual ACM Symposium on Theory of Computing, pp. 10–19. ACM (1998)
3. Bernstein, D.J., Chang, Y.-A., Cheng, C.-M., Chou, L.-P., Heninger, N., Lange, T., van Someren, N.: Factoring RSA keys from certified smart cards: Coppersmith in the wild. In: Sako, K., Sarkar, P. (eds.) ASIACRYPT 2013, Part II. LNCS, vol. 8270, pp. 341–360. Springer, Heidelberg (2013)
4. Boneh, D., Durfee, G.: Cryptanalysis of RSA with private key d less than $N^{0.292}$. IEEE Transactions on Information Theory 46(4), 1339–1349 (2000)

5. Cohn, H., Heninger, N.: Approximate common divisors via lattices. arXiv preprint arXiv:1108.2714 (2011)
6. Coppersmith, D.: Small solutions to polynomial equations, and low exponent RSA vulnerabilities. Journal of Cryptology 10(4), 233–260 (1997)
7. Faugère, J.-C., Marinier, R., Renault, G.: Implicit factoring with shared most significant and middle bits. In: Nguyen, P.Q., Pointcheval, D. (eds.) PKC 2010. LNCS, vol. 6056, pp. 70–87. Springer, Heidelberg (2010)
8. Herrmann, M., May, A.: Solving linear equations modulo divisors: On factoring given any bits. In: Pieprzyk, J. (ed.) ASIACRYPT 2008. LNCS, vol. 5350, pp. 406–424. Springer, Heidelberg (2008)
9. Howgrave-Graham, N.: Finding small roots of univariate modular equations revisited. In: Darnell, M. (ed.) Cryptography and Coding 1997. LNCS, vol. 1355, pp. 131–142. Springer, Heidelberg (1997)
10. Howgrave-Graham, N.: Approximate integer common divisors. In: Silverman, J.H. (ed.) CaLC 2001. LNCS, vol. 2146, pp. 51–66. Springer, Heidelberg (2001)
11. Jochemsz, E., May, A.: A strategy for finding roots of multivariate polynomials with new applications in attacking RSA variants. In: Lai, X., Chen, K. (eds.) ASIACRYPT 2006. LNCS, vol. 4284, pp. 267–282. Springer, Heidelberg (2006)
12. Jochemsz, E., May, A.: A polynomial time attack on RSA with private CRT-exponents smaller than $N^{0.073}$. In: Menezes, A. (ed.) CRYPTO 2007. LNCS, vol. 4622, pp. 395–411. Springer, Heidelberg (2007)
13. Lenstra, A.K., Hughes, J.P., Augier, M., Bos, J.W., Kleinjung, T., Wachter, C.: Public keys. In: Safavi-Naini, R., Canetti, R. (eds.) CRYPTO 2012. LNCS, vol. 7417, pp. 626–642. Springer, Heidelberg (2012)
14. Lenstra, A.K., Lenstra, H.W., Lovász, L.: Factoring polynomials with rational coefficients. Mathematische Annalen 261(4), 515–534 (1982)
15. Lu, Y., Zhang, R., Lin, D.: Improved bounds for the implicit factorization problem. Advances in Mathematics of Communications 7(3), 243–251 (2013)
16. May, A., Ritzenhofen, M.: Implicit factoring: On polynomial time factoring given only an implicit hint. In: Jarecki, S., Tsudik, G. (eds.) PKC 2009. LNCS, vol. 5443, pp. 1–14. Springer, Heidelberg (2009)
17. Nguên, P.Q., Stehlé, D.: Floating-point LLL revisited. In: Cramer, R. (ed.) EUROCRYPT 2005. LNCS, vol. 3494, pp. 215–233. Springer, Heidelberg (2005)
18. Nguyen, P.Q., Valle, B.: The LLL algorithm: Survey and applications. Springer Publishing Company, Incorporated (2009)
19. Sarkar, S., Maitra, S.: Approximate integer common divisor problem relates to implicit factorization. IEEE Transactions on Information Theory 57(6), 4002–4013 (2011)
20. Wiener, M.J.: Cryptanalysis of short RSA secret exponents. IEEE Transactions on Information Theory 36(3), 553–558 (1990)

New Attacks on the RSA Cryptosystem

Abderrahmane Nitaj[1], Muhammad Rezal Kamel Ariffin[2,3],
Dieaa I. Nassr[4], and Hatem M. Bahig[4]

[1] Laboratoire de Mathématiques Nicolas Oresme
Université de Caen Basse Normandie, France
abderrahmane.nitaj@unicaen.fr
[2] Al-Kindi Cryptography Research Laboratory,
Institute for Mathematical Research, Malaysia
[3] Department of Mathematics, Faculty of Science,
Universiti Putra Malaysia (UPM), Selangor, Malaysia
rezal@upm.edu.my
[4] Computer Science Division, Department of Mathematics,
Faculty of Science, Ain Shams University, Cairo, Egypt
hmbahig@sci.asu.edu.eg, diaa_rsa@yahoo.com

Abstract. This paper presents three new attacks on the RSA cryptosystem. The first two attacks work when k RSA public keys (N_i, e_i) are such that there exist k relations of the shape $e_i x - y_i \phi(N_i) = z_i$ or of the shape $e_i x_i - y \phi(N_i) = z_i$ where $N_i = p_i q_i$, $\phi(N_i) = (p_i - 1)(q_i - 1)$ and the parameters x, x_i, y, y_i, z_i are suitably small in terms of the prime factors of the moduli. We show that our attacks enable us to simultaneously factor the k RSA moduli N_i. The third attack works when the prime factors p and q of the modulus $N = pq$ share an amount of their least significant bits (LSBs) in the presence of two decryption exponents d_1 and d_2 sharing an amount of their most significant bits (MSBs). The three attacks improve the bounds of some former attacks that make RSA insecure.

Keywords: RSA, Cryptanalysis, Factorization, LLL algorithm, Simultaneous diophantine approximations, Coppersmith's method.

1 Introduction

The RSA cryptosystem [14] is currently the most widely known and widely used public key cryptosystem. The main parameters in RSA are the RSA modulus N and the public exponent e. The modulus $N = pq$ is the product of two large primes of equal bit-size and e satisfies $\gcd(e, \phi(N)) = 1$ where $\phi(N) = (p-1)(q-1)$ is the Euler totient function. The integer d satisfying $ed \equiv 1 \pmod{\phi(N)}$ is the private exponent. The RSA cryptosystem is deployed in various application systems for encryption, signing and for providing privacy and ensuring authenticity of digital data. Therefore, most research is focused on reducing the encryption/decryption execution time or the signature verification/generation time. For example, to reduce the decryption time or the

D. Pointcheval and D. Vergnaud (Eds.): AFRICACRYPT 2014, LNCS 8469, pp. 178–198, 2014.
© Springer International Publishing Switzerland 2014

signature generation time, one may wish to use a small private exponent d. Unfortunately, based on the convergents of the continued fraction expansion of $\frac{e}{N}$, Wiener [19] showed that the RSA cryptosystem is insecure when $d < N^{1/4}$. Boneh and Durfee [3] proposed an extension of Wiener's attack that allows the RSA cryptosystem to be broken when $d < N^{0.292}$. Their Method is based on lattice basis reduction techniques. Similarly, Blömer and May [2] proposed an extension of Wiener's attack and showed that the RSA cryptosystem is insecure if there exist three integers x, y and z satisfying $ex - y\phi(N) = z$ with $x < \frac{1}{3}N^{1/4}$ and $|z| < exN^{-3/4}$. Their method combines lattice basis reduction techniques and the continued fraction algorithm. In general, the use of short secret exponent encounters serious security problem in various instances of RSA. A typical example is when a single user generates many instances of RSA (N, e_i) with the same modulus and small private exponents [8]. Another example is when a single user generates k instances of RSA (N_i, e_i), each with the same small private exponent d. Using k equations $e_i d - k_i \phi(N_i) = 1$, Hinek [6] showed that it is possible to factor the k modulus N_i if $d < N^\delta$ with $\delta = \frac{k}{2(k+1)} - \varepsilon$ where ε is a small constant depending on the size of $\max N_i$. Similarly, to improve the computational efficiency of server-aided signature generation (see [16]), one may use RSA with a modulus $N = pq$ such that the prime factors p and q share a large number of least significant bits (LSBs). The security of this variant of RSA has been analyzed under the partial key exposure attacks in [16], [17], [20], and [18]. In [18], Sun et al. showed that RSA is more vulnerable in the situation when p and q share a large number of LSBs than the standard scenario when the prime factors p and q differ in the first LSBs. When $e = N^\gamma$, they showed that RSA is vulnerable if $|p - q| = 2^m x$ with $2^m = N^\alpha$ and $d < N^\delta$ whenever $\delta < \frac{7}{6} - \frac{2}{3}\alpha - \frac{1}{3}\sqrt{(1 - 4\alpha)(1 - 4\alpha + 6\gamma)}$. For example, if $\gamma = 1$, and $\alpha = 0.2$, then $\delta < 0.662$, that is, RSA is insecure if the private exponent is such that $d < N^{0.662}$. In [8], Howgrave-Graham and Seifert extended Wiener's attack in the presence of many decryption exponents for a single RSA modulus. They showed that RSA is insecure if one knows two public exponents e_1 and e_2 such that the corresponding private exponents d_1 and d_2 satisfy $d_1, d_2 < N^{0.357}$. In [11], Sarkar and Maitra improved this bound up to $d_1, d_2 < N^{0.416}$.

In this paper, we present three new attacks on RSA. The first attack works for $k \geq 2$ moduli $N_i = p_i q_i$, $i = 1, \ldots, k$, when k instances (N_i, e_i) are such that there exist an integer x, k integers y_i, and k integers z_i satisfying $e_i x - y_i \phi(N_i) = z_i$. We show that the k RSA moduli N_i can be factored in polynomial time if $N = \min_i N_i$ and

$$x < N^\delta, \ y_i < N^\delta, \ |z_i| < \frac{p_i - q_i}{3(p_i + q_i)} y_i N^{1/4} \text{ where } \delta = \frac{k}{2(k+1)}.$$

The second attack works when the k instances (N_i, e_i) of RSA are such that there exist an integer y, and k integers x_i, and k integers z_i satisfying $e_i x_i - y\phi(N_i) = z_i$. Similarly, we show that the k RSA moduli N_i can be factored in polynomial time if $N = \min_i N_i$, $\min_i e_i = N^\alpha$, and

$$x_i < N^\delta, \ y < N^\delta, \ |z_i| < \frac{p_i - q_i}{3(p_i + q_i)} y N^{1/4} \quad \text{where} \quad \delta = \frac{(2\alpha - 1)k}{2(k + 1)}.$$

In both scenarios, we transform the equations into a simultaneous diophantine problem and apply lattice basis reduction techniques to find the parameters (x, y_i) or (y, x_i). This leads to a suitable approximation of $p_i + q_i$ which allows us to apply Coppersmith's method [4] to compute the prime factors p_i and q_i of the moduli N_i.

The third attack enables us to factor an RSA modulus $N = pq$ when the prime factors share their LSBs in the presence of two public exponents e_1 and e_2 such that the corresponding decryption exponents d_1 and d_2 share their MSBs. To be more precise, suppose that $e = N^\gamma$, $|p - q| = 2^m x$ with $2^m = N^\alpha$, and $|d_1 - d_2| < N^\beta$. We show that one can factor the RSA modulus if $d_1, d_2 < N^\delta$ under the condition

$$\delta < \frac{5}{2} - 2\alpha - \beta - \frac{1}{4} \sqrt{6(1 - 4\alpha)(5 + 4\gamma - 4\alpha - 4\beta)}. \tag{1}$$

As an example, observe that, in the situation that $\gamma = 1$, $\alpha = 0.2$, and $\beta = \delta$, that is d_1 and d_2 differ in the first MSBs, then the condition (1) gives $\delta < 0.736$ which improves the bound $\delta < 0.662$ obtained in [18]. On the other hand, in the standard situation $\gamma = 1$, $\alpha = 0$, and $\beta = \delta$, that is when the prime integers p, q do not share any LSBs and d_1, d_2 do not share any MSBs, the condition (1) gives $\delta < 0.422$ which also improves the bound $\delta < 0.416$ found in [11]. Our method is based on Coppersmith's method for solving polynomial equations

The remainder of this paper is organized as follows. In Section 2, we review the tools that we apply in the scenarios, namely Coppersmith's method, lattice basis reduction and simultaneous diophantine approximations. We also present some useful results that will be used through the paper. In Section 3, we present the first attack. In Section 4, we present the second attack and in Section 5, we present the third attack. We conclude in Section 6.

2   Preliminaries

In this section, we give some basics on Coppersmith's method, lattice basis reduction techniques and simultaneous diophantine equations that will be used in this paper.

2.1   Coppersmith's Method

At Eurocrypt'96, Coppersmith [4] proposed an algorithm for finding small roots of bivariate integer polynomial equations in polynomial-time. The algorithm is based on the LLL algorithm [10] for lattice reduction. A clever application of Coppersmith's algorithm is to factor an RSA modulus $N = pq$ when half of the least significant or most significants bits of p are known.

Theorem 1 (Coppersmith). *Let $N = pq$ be the product of two unknown integers such that $q < p < 2q$. Given an approximation of p with additive error term at most $N^{\frac{1}{4}}$, then p and q can be found in polynomial time.*

Coppersmith's method has been heuristically extended to many variables. To find the small roots of a multivariate polynomial $f(x_1, \cdots, x_n)$, we construct a set of coprime polynomials with small coefficients which contain the same roots over the integers. This can be done by applying the LLL algorithm to a lattice that can be built using the strategy of Jochemsz and May [9]. To this end, a practical way is the use the following result of Howgrave-Graham [7].

Theorem 2 (Howgrave-Graham). *Let $h(x_1, \cdots, x_n) \in \mathbb{Z}[x_1, \cdots, x_n]$ be a polynomial with at most ω monomials. Suppose that $h\left(x_1^{(0)}, \cdots, x_n^{(0)}\right) \equiv 0 \pmod{R}$ where $|x_i^{(0)}| < X_i$ for $i = 1, \ldots, n$, and*

$$h(x_1 X_1, \cdots, x_n X_n) < \frac{R}{\sqrt{\omega}}.$$

Then $h\left(x_1^{(0)}, \cdots, x_n^{(0)}\right) = 0$ holds over the integers.

To find the small roots of the first polynomials of the LLL-reduced basis, we can use Gröbner bases or evaluation of resultants.

2.2 Lattice Reductions and Simultaneous Diophantine Approximations

Let $u_1 \ldots, u_d$ be d linearly independent vectors of $\mathbb{R}^n$ with $d \leq n$. The set of all integer linear combinations of the vectors $u_1 \ldots, u_d$ is called a lattice and is in the form

$$\mathcal{L} = \left\{ \sum_{i=1}^{d} x_i u_i \mid x_i \in \mathbb{Z} \right\}.$$

The set $(u_1, \ldots, u_d)$ is called a basis of $\mathcal{L}$ and d is its dimension. The determinant of $\mathcal{L}$ is defined as $\det(\mathcal{L}) = \sqrt{\det(U^T U)}$ where U is the the matrix of the u_i's in the canonical basis of $\mathbb{R}^n$. Define $\|v\|$ to be the Euclidean norm of a vector $v \in \mathcal{L}$. A central problem in lattice reduction is to find a short non-zero vector in $\mathcal{L}$. The LLL algorithm of Lenstra, Lenstra, and Lovász [10] produces a reduced basis and answers positively but partially this problem. The following result fixes the sizes of the reduced basis vectors (see [12]).

Theorem 3. *Let L be a lattice of dimension ω with a basis $\{v_1, \ldots, v_\omega\}$. The LLL algorithm produces a reduced basis $\{b_1, \cdots, b_\omega\}$ satisfying*

$$\|b_1\| \leq \|b_2\| \leq \cdots \leq \|b_i\| \leq 2^{\frac{\omega(\omega-1)}{4(\omega+1-i)}} \det(L)^{\frac{1}{\omega+1-i}},$$

for all $1 \leq i \leq \omega$.

One important application of the LLL algorithm is that it provides a solution to the simultaneous diophantine approximations problem which is defined as follows. Let $\alpha_1, \ldots, \alpha_n$ be n real numbers and ε a real number such that $0 < \varepsilon < 1$. A classical theorem of Dirichlet asserts that there exist integers $p_1, \cdots, p_n$ and a positive integer $q \leq \varepsilon^{-n}$ such that

$$|q\alpha_i - p_i| < \varepsilon \quad \text{for} \quad 1 \leq i \leq n.$$

In 1982, Lenstra, Lenstra and Lovász[10] described a method to find simultaneous diophantine approximations to rational numbers. In their work, they considered a lattice with real entries. We state below a similar result for a lattice with integer entries.

Theorem 4 (Simultaneous Diophantine Approximations). *There is a polynomial time algorithm, for given rational numbers $\alpha_1, \ldots, \alpha_n$ and $0 < \varepsilon < 1$, to compute integers $p_1, \cdots, p_n$ and a positive integer q such that*

$$\max_i |q\alpha_i - p_i| < \varepsilon \quad \text{and} \quad q \leq 2^{n(n-3)/4} \cdot 3^n \cdot \varepsilon^{-n}.$$

Proof. See Appendix A. □

2.3 Primes Sharing LSBs

The following lemma is reformulation of a result of [15]. It concerns an RSA modulus $N = pq$ when the prime factors p and q share an amount of their LSBs.

Lemma 1. *Let $N = pq$ be an RSA modulus with $q < p < 2q$. Suppose that $p - q = 2^m u$ for a known value m. Then $p = 2^m p_1 + u_0$ and $q = 2^m q_1 + u_0$ where u_0 is a solution of the equation $x^2 \equiv N \pmod{2^m}$ and $p + q = 2^{2m} v + v_0$ with*

$$v_0 \equiv 2u_0 + \left(N - u_0^2\right) u_0^{-1} \pmod{2^{2m}}.$$

Proof. See Appendix B. □

2.4 Approximations of the Primes in RSA

Let $N = pq$ be an RSA modulus with $q < p < 2q$. Then $p + q$ satisfies the following inequalities (see [13])

$$2\sqrt{N} < p + q < \frac{3\sqrt{2}\sqrt{N}}{2}. \tag{2}$$

The following result shows that any approximation of $p + q$ will lead to an approximation of p.

Lemma 2. *Let $N = pq$ be an RSA modulus with $q < p < 2q$. Suppose we know an approximation S of $p + q$ such that $S > 2N^{\frac{1}{2}}$ and*

$$|p + q - S| < \frac{p - q}{3(p + q)} N^{\frac{1}{4}}.$$

Then $\tilde{P} = \frac{1}{2}\left(S + \sqrt{S^2 - 4N}\right)$ is an approximation of p satisfying $|p - \tilde{P}| < N^{\frac{1}{4}}$.

Proof. See Appendix C.   □

Remark 1. Notice that in Section 4.1.2 of the ANSI X9.31:1998 standard for public key cryptography [1], there are a number of recommendations for the generation of the primes in $N = pq$. One criteria is that the primes p, q shall satisfy $p - q > 2^{-100}\sqrt{N}$. Combining with (2) when $q < p < 2q$ and $N > 2^{1024}$, this implies that the term $\frac{p-q}{3(p+q)}N^{\frac{1}{4}}$ satisfies

$$\frac{p-q}{3(p+q)}N^{\frac{1}{4}} > \frac{2^{-100}\sqrt{N}}{9\frac{\sqrt{2}}{2}\sqrt{N}} \cdot 2^{256} = \frac{2^{157}}{9\sqrt{2}}.$$

This shows that, when $N = pq > 2^{1024}$ and the prime factors p and q are chosen following the ANSI X9.31:1998 standard, the approximation extra term $\frac{(p-q)}{3(p+q)}N^{\frac{1}{4}}$ of $p + q$ is not too small.

3   The First Attack on k RSA Moduli

In this section, we are given $k \geq 2$ moduli $N_i = p_iq_i$ with the same size N. We suppose in this scenario that the RSA moduli satisfy k equations $e_ix - y_i\phi(N_i) = z_i$. Notice that the parameters $\phi(N_i) = (p_i - 1)(q_i - 1)$ are also unknown. We show that it is possible to factor the RSA moduli N_i if the unknown parameters x, y_i and z_i are suitably small.

Theorem 5. *For $k \geq 2$, let $N_i = p_iq_i$, $1 \leq i \leq k$, be k RSA moduli. Let $N = \min_i N_i$. Let e_i, $i = 1, \ldots, k$, be k public exponents. Define $\delta = \frac{k}{2(k+1)}$. If there exist an integer $x < N^{\delta}$ and k integers $y_i < N^{\delta}$ and $|z_i| < \frac{p_i - q_i}{3(p_i + q_i)}y_iN^{1/4}$ such that $e_ix - y_i\phi(N_i) = z_i$ for $i = 1, \ldots, k$, then one can factor the k RSA moduli $N_1, \cdots N_k$ in polynomial time.*

Proof. For $k \geq 2$ and $i = 1, \ldots, k$, the equation $e_ix - y_i\phi(N_i) = z_i$ can be rewritten as $e_ix - y_i(N_i + 1) = z_i - y_i(p_i + q_i)$. Hence

$$\left|\frac{e_i}{N_i + 1}x - y_i\right| = \frac{|z_i - y_i(p_i + q_i)|}{N_i + 1}. \tag{3}$$

Let $N = \min_i N_i$ and suppose that $y_i < N^{\delta}$ and $|z_i| < \frac{p_i - q_i}{3(p_i + q_i)}y_iN^{1/4}$. Then $|z_i| < y_iN^{1/4} < N^{\delta + \frac{1}{4}}$. Since by (2) we have $p_i + q_i < \frac{3\sqrt{2}}{2}\sqrt{N}$, we will get

$$\frac{|z_i - y_i(p_i + q_i)|}{N_i + 1} \leq \frac{|z_i| + y_i(p_i + q_i)}{N}$$

$$< \frac{N^{\delta + 1/4} + \frac{3\sqrt{2}}{2}N^{\delta + 1/2}}{N}$$

$$< \frac{\sqrt{5}N^{\delta + 1/2}}{N}$$

$$= \sqrt{5}N^{\delta - 1/2}.$$

Plugging in (3), we get

$$\left| \frac{e_i}{N_i + 1} x - y_i \right| < \sqrt{5} N^{\delta - 1/2}.$$

We now proceed to prove the existence of the integer x. Let $\varepsilon = \sqrt{5} N^{\delta - 1/2}$, $\delta = \frac{k}{2(k+1)}$. We have

$$N^{\delta} = N^{k/2 - k\delta} < 2^{k(k-3)/4} \cdot 3^k \cdot \left(\sqrt{5} N^{\delta - 1/2} \right)^{-k} = 2^{k(k-3)/4} \cdot 3^k \cdot \varepsilon^{-k}.$$

It follows that if $x < N^{\delta}$, then $x < 2^{k(k-3)/4} \cdot 3^k \varepsilon^{-k}$. Summarizing, for $i = 1, \ldots, k$, we have

$$\left| \frac{e_i}{N_i + 1} x - y_i \right| < \varepsilon, \quad x < 2^{k(k-3)/4} \cdot 3^k \cdot \varepsilon^{-k}.$$

It follows that the conditions of Theorem 4 are fulfilled which will find x and y_i for $i = 1, \ldots, k$. Next, using the equation $e_i x - y_i \phi(N_i) = z_i$, we get

$$p_i + q_i = N_i + 1 - \frac{e_i x}{y_i} + \frac{z_i}{y_i}.$$

Since $|z_i| < \frac{p_i - q_i}{3(p_i + q_i)} y_i N^{1/4}$, then $\frac{z_i}{y_i} < \frac{p_i - q_i}{3(p_i + q_i)} N^{1/4}$ and $S_i = N_i + 1 - \frac{e_i x}{y_i}$ is an approximation of $p_i + q_i$ with an error of at most $\frac{p_i - q_i}{3(p_i + q_i)} N^{1/4}$. Hence, using Lemma 2, we can find an approximation $\tilde{P}_i = \frac{1}{2} \left(S_i + \sqrt{S_i^2 - 4N_i} \right)$ of p_i such that $|p_i - \tilde{P}_i| < N^{1/4}$. Then, for each $i = 1, \ldots, k$, we find p_i using Theorem 1. This leads to the factorization of the k RSA moduli $N_1, \ldots, N_k$. □

Remark 2. It is conjectured in [3] that an RSA instance with a modulus $N = pq$ and a public exponent e is insecure if $ed - y\phi(N) = 1$ with $d < N^{1/2}$. This conjecture can be related to Theorem 5 as follows. Suppose that k RSA moduli $N_1, \cdots, N_k$ and k public exponents $e_1, \ldots, e_k$ satisfy $e_1 d - y_1 \phi(N_1) = 1$ and $e_i d - y_i \phi(N_i) = z_i$, $i = 2, \ldots, k$, where $d < N^{\delta}$, $y_i < N_1^{\delta}$ and $|z_i| < \frac{p_i - q_i}{3(p_i + q_i)} y_i N^{1/4}$ with $\delta = \frac{k}{2(k+1)}$. Then, by Theorem 5, one can factor the RSA moduli $N_1, \cdots, N_k$. Observe that, for sufficiently large k, we have $\delta \approx \frac{1}{2}$, which answer positively the conjecture in this case.

Example 1. Consider the following 3 RSA moduli and public exponents

$N_1 = 13393545150918238591518012241, \ e_1 = 10501852846143160024884092263,$

$N_2 = 5761318740014279657192789531, \ e_2 = 14921528533564369531595992262$

$N_3 = 12579369006828790258496914469, \ e_3 = 1039188969087059416671255887.$

Then $N = \max(N_1, N_2, N_3) = 5761318740014279657192789531$. Since $k = 3$, we get $\delta = \frac{k}{2(k+1)} = 0.375$ and $\varepsilon = \sqrt{5} N^{\delta - 1/2} \approx 0.000757$. Using (11) with $n = k = 3$, we find

$$C = \left[3^{n+1} \cdot 2^{\frac{(n+1)(n-4)}{4}} \cdot \varepsilon^{-n-1} \right] = 123330787675873.$$

Consider the lattice $\mathcal{L}$ spanned by the matrix

$$
M = \begin{bmatrix} 1 - \lceil Ce_1/(N_1+1) \rceil & -\lceil Ce_2/(N_2+1) \rceil & -\lceil Ce_3/(N_3+1) \rceil \\ 0 & C & 0 & 0 \\ 0 & 0 & C & 0 \\ 0 & 0 & 0 & C \end{bmatrix}.
$$

Then, applying the LLL algorithm to $\mathcal{L}$, we get a reduced basis with the matrix

$$
K = \begin{bmatrix} -3779027519, & -18311525449, & -3194797920, & -5032583842 \\ 7689269805, & -1894087712, & 24623557005, & -10208017761 \\ 33347077827, & -5532195789, & -23880055457, & -2777199762 \\ 1955330759, & -28195205997, & 36977018712, & 75348896931 \end{bmatrix}.
$$

Next, computing $K \cdot M^{-1}$, we observe that the first row is

$$[-3779027519, -2963128168, -978749302, -312187655],$$

from which we deduce $x = 3779027519$, $y_1 = 2963128168$, $y_2 = 978749302$ and $y_3 = 312187655$. Using x and y_i for $i = 1, 2, 3$, define $S_i = \left[N_i + 1 - \frac{e_i x}{y_i} \right]$. We get

$$S_1 = 73202632183869, \quad S_2 - 152156156125070, \quad S_3 = 102878795201660.$$

For $i = 1, 2, 3$, let $D_i = \left[\sqrt{S_i^2 - 4N_i} \right]$. We get

$$D_1 = 1098771258961, \quad D_2 = 10306351764921, \quad D_3 = 74513749733949.$$

By Lemma 2, for $i = 1, 2, 3$, $\tilde{P}_i = \frac{1}{2}(S_i + D_i)$ is a candidate for an approximation of p_i. Applying Coppersmith's method 1 with $\tilde{P}_i$ for $i = 1, 2, 3$, we get

$$p_1 = 37150702190747, \quad p_2 = 81231254125183, \quad p_3 = 88696272470797.$$

This leads us to the factorization of the 3 RSA moduli N_1, N_2 and N_3. Observe that $x > N^{0.344}$ is much larger than Blömer-May's bound $x < \frac{1}{3}N^{1/4}$. This shows that Blömer-May's attack will not give the factorization of the RSA moduli in this example.

4   The Second Attack on k RSA Moduli

In this section, we consider the second scenario when the k RSA moduli satisfy k equations of the shape $e_i x_i - y\phi(N_i) = z_i$ where the parameters x_i, y and z_i are suitably small unknown parameters.

Theorem 6. *For $k \geq 3$, let $N_i = p_i q_i$, $1 \leq i \leq k$, be k RSA moduli with the same size N. Let e_i, $i = 1, \ldots, k$, be k public exponents with $\min_i e_i = N^\alpha$. Let $\delta = \frac{(2\alpha-1)k}{2(k+1)}$. If there exist an integer $y < N^\delta$ and k integers $x_i < N^\delta$ and $|z_i| < \frac{p_i - q_i}{3(p_i + q_i)} y N^{1/4}$ such that $e_i x_i - y\phi(N_i) = z_i$ for $i = 1, \ldots, k$, then one can factor the k RSA moduli $N_1, \cdots N_k$ in polynomial time.*

Proof. For $i = 1, \ldots, k$, the equation $e_i x_i - y\phi(N_i) = z_i$ can be transformed into $e_i x_i - y(N_i + 1) = z_i - y(p_i + q_i)$. Hence

$$\left| \frac{N_i + 1}{e_i} y - x_i \right| = \frac{|z_i - y(p_i + q_i)|}{e_i}. \tag{4}$$

Let $N = \max_i N_i$. Suppose that $y < N^\delta$ and $|z_i| < \frac{p_i - q_i}{3(p_i + q_i)} y N^{1/4}$. Also, suppose that $\min_i e_i = N^\alpha$. Since by (2) we have $p_i + q_i < \frac{\sqrt{3}}{2}\sqrt{N_i}$, then we get

$$\frac{|z_i - y(p_i + q_i)|}{e_i} \leq \frac{|z_i| + y(p_i + q_i)}{N^\alpha}$$

$$< \frac{N^{\delta + \frac{1}{4}} + \frac{3\sqrt{2}}{2} N_i^{\delta + \frac{1}{2}}}{N^\alpha}$$

$$< \frac{\sqrt{5} N^{\delta + \frac{1}{2}}}{N^\alpha}$$

$$= \sqrt{5} N^{\delta + \frac{1}{2} - \alpha}.$$

Using this in (4), we get

$$\left| \frac{N_i + 1}{e_i} y - x_i \right| < \sqrt{5} N^{\delta + \frac{1}{2} - \alpha}.$$

We now proceed to prove the existence of y and the integers x_i. Let $\varepsilon = \sqrt{5} N^{\delta + \frac{1}{2} - \alpha}$, $\delta = \frac{(2\alpha - 1)k}{2(k+1)}$. We have

$$N^\delta \varepsilon^k = 5^{\frac{k}{2}} N^{\delta + k\delta + \frac{k}{2} - k\alpha} = 5^{\frac{k}{2}}.$$

Then, since $5^{\frac{k}{2}} < 2^{\frac{k(k-3)}{4}} \cdot 3^k$ for $k \geq 2$, we get $N^\delta \varepsilon^k < 2^{k(k-3)/4} \cdot 3^k$. It follows that if $y < N^\delta$, then $y < 2^{k(k-3)/4} \cdot 3^k \varepsilon^{-k}$. Summarizing, we have

$$\left| \frac{N_i + 1}{e_i} y - x_i \right| < \varepsilon, \quad y < 2^{k(k-3)/4} \cdot 3^k \cdot \varepsilon^{-k}, \quad \text{for} \quad i = 1, \ldots, k,$$

It follows that the conditions of Theorem 4 are fulfilled and we will obtain y and x_i for $i = 1, \ldots, k$. Next, by utilizing the equation $e_i x_i - y\phi(N_i) = z_i$, we get

$$p_i + q_i = N_i + 1 - \frac{e_i x_i}{y} + \frac{z_i}{y}.$$

Since $|z_i| < \frac{p_i - q_i}{3(p_i + q_i)} y N^{1/4}$, then $\frac{|z_i|}{y} < \frac{p_i - q_i}{3(p_i + q_i)} N^{1/4}$ and $S_i = N_i + 1 - \frac{e_i x_i}{y}$ is an approximation of $p_i + q_i$ with an error of at most $\frac{p_i - q_i}{3(p_i + q_i)} N^{1/4}$. Hence, using Lemma 2, we can find an approximation $\tilde{P}_i = \frac{1}{2}\left(S_i + \sqrt{S_i^2 - 4N_i} \right)$ of p_i such that $|p_i - \tilde{P}_i| < N^{1/4}$. Then, using Theorem 1, we find p_i for $i = 1, \ldots, k$. This leads to the factorization of the k RSA moduli $N_1, \ldots, N_k$. □

Example 2. Consider the following three RSA moduli and three public exponents

$$N_1 = 701404527220444023808491592451,$$
$$e_1 = 598872437015970469816654047240,$$
$$N_2 = 287595248854210987719090191831,$$
$$e_2 = 166801923182837419445821944696,$$
$$N_3 = 431174708848373283683684641751,$$
$$e_3 = 373743791338260494286817160907.$$

Then $N = \max(N_1, N_2, N_3) = 701404527220444023808491592451$. We also get $\min(e_1, e_2, e_3) = N^\alpha$ with $\alpha \approx 0.9791$. Since $k = 3$, we get $\delta = \frac{k(2\alpha-1)}{2(k+1)} = 0.359325$ and $\varepsilon = \sqrt{5}N^{\delta+1/2-\alpha} \approx 0.000595$. Using (11) with $n = k = 3$, let

$$C = \left[3^{n+1} \cdot 2^{\frac{(n+1)(n-4)}{4}} \cdot \varepsilon^{-n-1}\right] = 323072188568099.$$

Consider the lattice $\mathcal{L}$ spanned by the the rows of the matrix

$$M = \begin{bmatrix} 1 & -\lceil C(N_1+1)/e_1\rceil & -\lceil C(N_2+1)/e_2\rceil & -\lceil C(N_3+1)/e_3\rceil \\ 0 & C & 0 & 0 \\ 0 & 0 & C & 0 \\ 0 & 0 & 0 & C \end{bmatrix}.$$

Then, applying the LLL algorithm to $\mathcal{L}$, we get a reduced basis with the matrix

$$K = \begin{bmatrix} 9963214223 & -13283752558 & -23775330798 & -12098528625 \\ -23587427317 & -20479775765 & -11829398252 & 7542788188 \\ -80616201478 & 123609103667 & -102601176821 & -4090837289 \\ -641512285490 & -64738610576 & -108985237738 & -147068239663 \end{bmatrix}.$$

Next, we get

$$K \cdot M^{-1} = \begin{bmatrix} 9963214223 & 11669001827 & 17178297583 & 11494200282 \\ -23587427317 & -27625796886 & -40668787863 & -27211962691 \\ -80616201478 & -94418385601 & -138996218289 & -93003999013 \\ -41512285490 & -48619544294 & -71574331088 & -47891223947 \end{bmatrix}.$$

From the first row, we deduce $y = 9963214223$, $x_1 = 11669001827$, $x_2 = 17178297583$, and $x_3 = 11494200282$. Using y and x_i for $i = 1, 2, 3$, define $S_i = \left[N_i + 1 - \frac{e_i x_i}{y}\right]$. We get

$$S_1 = 1677562597323852, \quad S_2 = 1169977613299368, \quad S_3 = 1377024442150848.$$

For $i = 1, 2, 3$, let $D_i = \left[\sqrt{S_i^2 - 4N_i}\right]$. We get

$$D_1 = 92726258730590, \quad D_2 = 467404129426390, \quad D_3 = 414122540907110.$$

By Lemma 2, for $i = 1, 2, 3$, $\tilde{P}_i = \frac{1}{2}(S_i + D_i)$ is a candidate for an approximation of p_i. Applying Coppersmith's method 1 with $\tilde{P}_i$ for $i = 1, 2, 3$, we get

$$p_1 = 885144428027221, \quad p_2 = 818690871362879, \quad p_3 = 895573491528979.$$

This leads to the factorization of the three RSA moduli N_1, N_2 and N_3. Observe that $\min(x_1, x_2, x_3) > N^{0.337}$ is much larger than Blömer-May's bound $x < \frac{1}{3} N^{1/4}$. This shows that Blömer-May's attack does not work in this case.

5 The Third Attack on RSA With Primes and Decryption Exponents Sharing Bits

In this section, we present the attack which applies when the prime factors of an RSA modulus share an amount of their LSBs in the presence of two decryption exponents d_1 and d_2 sharing an amount of their MSBs.

5.1 The Attack

Theorem 7. *Let $N = pq$ be an RSA modulus such that $p - q = 2^m u$ where $2^m \approx N^\alpha$. Let e_1 and e_2 be two public exponents satisfying $e_1, e_2 \approx N^\gamma$, $e_1 d_1 - k_1 \phi(N) = 1$, and $e_2 d_2 - k_2 \phi(N) = 1$. Suppose that $d_1, d_2 < N^\delta$ and $|d_1 - d_2| < N^\beta$. Then one can factor N in polynomial time if*

$$\delta < \frac{5}{2} - 2\alpha - \beta - \frac{1}{4}\sqrt{6(1 - 4\alpha)(5 + 4\gamma - 4\alpha - 4\beta)}.$$

Proof. Suppose that e_1 and e_2 are two public exponents satisfying $e_1 d_1 - k_1 \phi(N) = 1$, $e_2 d_2 - k_2 \phi(N) = 1$. Multiplying the first equation by e_2 and the second by e_1 and subtracting, we get

$$e_1 e_2(d_1 - d_2) - e_2 k_1 \phi(N) + e_1 k_2 \phi(N) = e_2 - e_1. \tag{5}$$

Suppose that $p - q = 2^m u$. Then, Lemma 1 shows that $p + q$ is in the form $p + q = v_0 + 2^{2m} v$ where $v_0 \equiv 2u_0 + \left((N - u_0^2) u_0^{-1} \pmod{2^{2m}}\right)$ and u_0 is a solution of the modular equation $x^2 \equiv N \pmod{2^m}$. Hence

$$\phi(N) = N + 1 - (p + q) = N + 1 - v_0 - 2^{2m} v.$$

Plugging this in (5), we get

$$e_1 e_2(d_1 - d_2) - e_2 k_1 \left(N + 1 - v_0 - 2^{2m} v\right) + e_1 k_2 \left(N + 1 - v_0 - 2^{2m} v\right)$$
$$= e_2 - e_1.$$

which can be rewritten as

$$e_1 e_2(d_1 - d_2) - e_2(N + 1 - v_0)k_1 + 2^{2m} e_2 k_1 v + e_1(N + 1 - v_0)k_2$$
$$- 2^{2m} e_1 k_2 v + (e_1 - e_2) = 0. \tag{6}$$

Fix the known and the unknown parameters as follows

$$\begin{cases} a_1 = e_1 e_2, \\ a_2 = -e_2(N+1-v_0), \\ a_3 = 2^{2m}e_2, \\ a_4 = e_1(N+1-v_0), \\ a_5 = -2^{2m}e_1, \\ a_6 = e_1 - e_2, \end{cases} \quad \text{and} \quad \begin{cases} x_1 = d_1 - d_2, \\ x_2 = k_1, \\ x_3 = k_2, \\ x_4 = v. \end{cases}$$

Hence, the equation (6) becomes $a_1 x_1 + a_2 x_2 + a_3 x_2 x_4 + a_4 x_3 + a_5 x_3 x_4 + a_6 = 0$. Consider the polynomial

$$f(x_1, x_2, x_3, x_4) = a_1 x_1 + a_2 x_2 + a_3 x_2 x_4 + a_4 x_3 + a_5 x_3 x_4 + a_6.$$

Then $(d_1 - d_2, k_1, k_2, v)$ is a root of $f(x_1, x_2, x_3, x_4)$ which can be small enough to be found by Coppersmith's technique. To find the small roots of $f(x_1, x_2, x_3, x_4)$ using this method, we use the extended strategy of Jochemsz and May [9]. We will need the following bounds.

- $\max(e_1, e_2) = N^{\gamma}$,
- $\max(d_1, d_2) < N^{\delta}$,
- $|d_1 - d_2| < X_1 = N^{\beta}$,
- $k_1 = \frac{e_1 d_1 - 1}{\phi(N)} < X_2 = N^{\gamma+\delta-1}$,
- $k_2 = \frac{e_2 d_2 - 1}{\phi(N)} < X_3 = N^{\gamma+\delta-1}$,
- $p - q = 2^m u$ with $2^m = N^{\alpha}$ and $\alpha < \frac{1}{4}$.
- By (2) and Lemma 1, $p + q = 2^{2m} v + v_0$ with $v < X_4 = 3N^{1/2-2\alpha}$.

Observe that $\alpha < \frac{1}{4}$, otherwise p and q can be found using Coppersmith's metho [4]. Let us fix the bounds of the unknown parameters

$$X_1 = N^{\beta}, \quad X_2 = N^{\gamma+\delta-1}, \quad X_3 = N^{\gamma+\delta-1}, \quad X_4 = 3N^{1/2-2\alpha}. \tag{7}$$

Let m and t be two positive integers. Define the set

$$S = \bigcup_{0 \le j \le t} \{x_1^{i_1} x_2^{i_2} x_3^{i_3} x_3^{i_3+j} \mid x_1^{i_1} x_2^{i_2} x_3^{i_3} x_4^{i_4} \text{ monomial of } f^{m-1}\}.$$

and the set

$$M = \{\text{monomials of } x_1^{i_1} x_2^{i_2} x_3^{i_3} x_4^{i_4} f \mid x_1^{i_1} x_2^{i_2} x_3^{i_3} x_4^{i_4} \in S\}.$$

Neglecting the coefficients, it is easy to find that $f^{m-1}(x_1, x_2, x_3, x_4)$ satisfies

$$f^{m-1}(x_1, x_2, x_3, x_4) = \sum_{i_1=0}^{m-1} \sum_{i_2=0}^{m-1-i_1} \sum_{i_3=0}^{m-1-i_1-i_2} \sum_{i_4=0}^{i_2+i_3} x_1^{i_1} x_2^{i_2} x_3^{i_3} x_4^{i_4}.$$

This leads to the characterization of the monomials $x_1^{i_1} x_2^{i_2} x_3^{i_3} x_4^{i_4}$ of S:

$$x_1^{i_1} x_2^{i_2} x_3^{i_3} x_4^{i_4} \in S \quad \text{if} \quad \begin{cases} i_1 = 0, \dots, m-1, \\ i_2 = 0, \dots, m-1-i_1, \\ i_3 = 0, \dots, m-1-i_1-i_2, \\ i_4 = 0, \dots, i_2 + i_3 + t. \end{cases}$$

We also easily find

$$x_1^{i_1} x_2^{i_2} x_3^{i_3} x_4^{i_4} \in M \quad \text{if} \quad \begin{cases} i_1 = 0, \dots, m, \\ i_2 = 0, \dots, m-i_1, \\ i_3 = 0, \dots, m-i_1-i_2, \\ i_4 = 0, \dots, i_2 + i_3 + t. \end{cases}$$

Define

$$W = \| f(x_1 X_1, x_2 X_2, x_3 X_3, x_4 X_4) \|_\infty$$
$$= \max(|a_1| X_1, |a_2| X_2, |a_3| X_2 X_3, |a_4| X_4, |a_5| X_4 X_3, |a_6|).$$

Then W satisfies

$$W \geq |a_2| X_2 = e_2 (N + 1 - v_0) N^{\gamma + \delta - 1} \approx N^{2\gamma + \delta}. \tag{8}$$

Next, define

$$R = W X_1^{m-1} X_2^{m-1} X_3^{m-1} X_4^{m-1+t}.$$

Without loss of generality, suppose that $a_6 = e_1 - e_2$ is coprime with R. We define $f'(x_1, x_2, x_3, x_4) = a_6^{-1} f(x_1, x_2, x_3, x_4) \pmod{R}$ so that $f'(0,0,0,0) = 1$. Next, define the polynomials

$$g_{i_1,i_2,i_3,i_4} = x_1^{i_1} x_2^{i_2} x_3^{i_3} x_4^{i_4} f' X_1^{m-1-i_1} X_2^{m-1-i_2} X_3^{m-1-i_3} X_4^{m-1+t-i_4},$$
$$\text{with} \quad x_1^{i_1} x_2^{i_2} x_3^{i_3} x_4^{i_4} \in S,$$
$$h_{i_1,i_2,i_3,i_4} = x_1^{i_1} x_2^{i_2} x_3^{i_3} x_4^{i_4} R,$$
$$\text{with} \quad x_1^{i_1} x_2^{i_2} x_3^{i_3} x_4^{i_4} \in M \backslash S.$$

The monomials of $M \backslash S$ reduce to $x_1^{i_1} x_2^{i_2} x_3^{i_3} x_4^{i_4}$ with $(x_1, x_2, x_3, x_4) \in S_i$ for $i = 1, 2, 3$ where

$$S_1 = \{ x_1^{i_1} x_2^{i_2} x_3^{i_3} x_4^{i_4} \} \text{ for } \begin{cases} i_1 = m, \\ i_2 = 0, \dots, m-i_1, \\ i_3 = 0, \dots, m-i_1-i_2, \\ i_4 = 0, \dots, i_2 + i_3 + t. \end{cases}$$

$$S_2 = \{ x_1^{i_1} x_2^{i_2} x_3^{i_3} x_4^{i_4} \} \text{ for } \begin{cases} i_1 = 0, \dots, m-1, \\ i_2 = m - i_1, \\ i_3 = 0, \dots, m-i_1-i_2, \\ i_4 = 0, \dots, i_2 + i_3 + t. \end{cases}$$

$$S_3 = \{ x_1^{i_1} x_2^{i_2} x_3^{i_3} x_4^{i_4} \} \text{ for } \begin{cases} i_1 = 0, \dots, m-1, \\ i_2 = 0, \dots, m-1-i_1, \\ i_3 = m - i_1 - i_2, \\ i_4 = 0, \dots, i_2 + i_3 + t. \end{cases}$$

As shown in [9], we use the coefficients of $g_{i_1,i_2,i_3,i_4}(x_1X_1, x_2X_2, x_3X_3, x_4X_4)$ and $h_{i_1,i_2,i_3,i_4}(x_1X_1, x_2X_2, x_3X_3, x_4X_4)$ to build a basis of a lattice L with dimension

$$\omega = \sum_{x_1^{i_1} x_2^{i_2} x_3^{i_3} x_4^{i_4} \in M} 1 = \frac{1}{12}(m+1)(m+2)(m+3)(m+2t+2).$$

The following ordering of the monomials is performed to construct an upper triangular matrix: if $\sum i_j < \sum i'_j$ then $x_1^{i_1} x_2^{i_2} x_3^{i_3} x_4^{i_4} < x_1^{i'_1} x_2^{i'_2} x_3^{i'_3} x_4^{i'_4}$ and if $\sum i_j = \sum i'_j$ then the monomials are lexicographically ordered. The diagonal entries of the matrix are of the form

$$\begin{cases} (X_1X_2X_3)^{m-1}X_4^{m-1+t} & \text{for the polynomials } g. \\ WX_1^{m-1+i_1} X_2^{m-1+i_2} X_3^{m-1+i_3} X_4^{m-1+t+i_4} & \text{for the polynomials } h. \end{cases}$$

Define

$$s_j = \sum_{x_1^{i_1} x_2^{i_2} x_3^{i_3} x_4^{i_4} \in M \backslash S} i_j, \text{ for } j = 1, \ldots, 4. \tag{9}$$

The determinant of L is then

$$\det(L) = W^{|M \backslash S|} X_4^{(m-1+t)|S|+(m-1+t)|M \backslash S|+s_4} \prod_{j=1}^{3} X_j^{(m-1)|S|+(m-1)|M \backslash S|+s_j}$$

$$= W^{|M \backslash S|} x_4^{(m-1+t)\omega+s_4} \prod_{j=1}^{3} X_j^{(m-1)\omega+s_j}.$$

All the polynomials $g(x_1, x_2, x_3, x_4)$ and $h(x_1, x_2, x_3, x_4)$ and their combinations share the root $(d_1 - d_2, k_1, k_2, v)$ modulo R. Applying the LLL algorithm to the lattice L with the basis spanned by the polynomials $g(x_1X_1, x_2X_2, x_3X_3, x_4X_4)$ and $h(x_1X_1, x_2X_2, x_3X_3, x_4X_4)$, we get a new basis with short vectors. Let $f_i(x_1X_1, x_2X_2, x_3X_3, x_4X_4)$, $i = 1, 2, 3$ be three short vectors of the reduced basis. Each f_i is a combination of g and h, and then share the root (d_1-d_2, k_1, k_2, v). Then, by Theorem 3, we have for $i = 1, 2, 3$

$$\|f_i(x_1X_1, x_2X_2, x_3X_3, x_4X_4)\| < 2^{\frac{\omega(\omega-1)}{4(\omega-2)}} \det(L)^{\frac{1}{\omega-2}}.$$

For $i = 1, 2, 3$, we force the polynomials f_i to satisfy Howgrave-Graham's bound $\|f_i(x_1X_1, x_2X_2, x_3X_3, x_4X_4)\| < \frac{R}{\sqrt{\omega}}$. A sufficient condition is

$$2^{\frac{\omega(\omega-1)}{4(\omega-2)}} \det(L)^{\frac{1}{\omega-2}} < \frac{R}{\sqrt{\omega}},$$

which can be transformed into $\det(L) < R^\omega$, that is

$$W^{|M \backslash S|} x_4^{(m-1+t)\omega+s_4} \prod_{j=1}^{3} X_j^{(m-1)\omega+s_j} < \left(WX_1^{m-1}X_2^{m-1}X_3^{m-1}X_4^{m-1+t} \right)^\omega.$$

Using $\omega = |M|$ and $|M| - |M\backslash S| = |S|$, we get

$$\prod_{j=1}^{4} X_j^{s_j} < W^{|S|}. \tag{10}$$

Using (9), we easily get

$$s_1 = \frac{1}{12}m(m+1)(m+2)(m+2t+1),$$

$$s_2 = \frac{1}{24}m(m+1)(m+2)(3m+4t+5),$$

$$s_3 = \frac{1}{24}m(m+1)(m+2)(3m+4t+5),$$

$$s_4 = \frac{1}{24}(m+1)(m+2)(3m^2+5m+8tm+6t+6t^2).$$

Similarly, we get

$$|S| = \sum_{x_1^{i_1} x_2^{i_2} x_3^{i_3} x_4^{i_4} \in S} 1 = \frac{1}{12}m(m+1)(m+2)(m+2t+1).$$

Set $t = \tau m$, then,

$$s_1 = \frac{1}{12}(2\tau+1)m^4 + o(m^4),$$

$$s_2 = \frac{1}{24}(4\tau+3)m^4 + o(m^4),$$

$$s_3 = \frac{1}{24}(4\tau+3)m^4 + o(m^4),$$

$$s_4 = \frac{1}{24}(6\tau^2+8\tau+3)m^4 + o(m^4),$$

$$|S| = \frac{1}{12}(2\tau+1)m^4 + o(m^4).$$

Using this, and after simplifying by m^4, the inequation (10) transforms into

$$X_1^{\frac{1}{12}(2\tau+1)} X_2^{\frac{1}{24}(4\tau+3)} X_3^{\frac{1}{24}(4\tau+3)} X_4^{\frac{1}{24}(6\tau^2+8\tau+3)} < W^{\frac{1}{12}(2\tau+1)}.$$

Substituting the values of X_1, X_2, X_3, X_4 from (7) and W from (8), we get

$$\frac{1}{12}(2\tau+1)\beta + \frac{1}{24}(4\tau+3)(\gamma+\delta-1) + \frac{1}{24}(4\tau+3)(\gamma+\delta-1)$$

$$+ \frac{1}{24}(6\tau^2+8\tau+3)\left(\frac{1}{2}-2\alpha\right) < \frac{1}{12}(2\tau+1)(2\gamma+\delta),$$

or equivalently,

$$(6-24\alpha)\tau^2 + (8\beta+8\delta-8-32\alpha)\tau + 4\gamma+4\beta+8\delta-9-12\alpha < 0.$$

For the optimal value $\tau = \frac{2(1+4\alpha-\beta-\delta)}{3(1-4\alpha)}$, this reduces to

$$-8\delta^2 + (40 - 32\alpha - 16\beta)\delta + 16\alpha^2 - 48\gamma\alpha + 16\beta\alpha + 8\alpha + 28\beta - 35 + 12\gamma - 8\beta^2 < 0,$$

which is valid if

$$\delta < \frac{5}{2} - 2\alpha - \beta - \frac{1}{4}\sqrt{6(1 - 4\alpha)(5 + 4\gamma - 4\alpha - 4\beta)}.$$

Under this condition, we find four polynomials, namely f, f_1, f_2 and f_3 with the root $(d_1 - d_2, k_1, k_2, v)$. Using the resultant technique, we find the solution $(d_1 - d_2, k_1, k_2, v)$. Using v, we compute $p - q = 2^m v$. Since $N = pq$, we get $p^2 - 2^m v p - N = 0$ which leads to the factorization of the RSA modulus $N = pq$. This terminates the proof.   □

5.2   Comparison with Former Attacks

We compare the bound on δ of Theorem 7 with two former bounds, namely the bound obtained by Sarkar and Maitra in [11] and the bound obtained by Sun et al. in [18].

5.2.1   Comparison with the Bound of Sarkar and Maitra

In [11], Sarkar and Maitra showed that for $d_1, d_2 < N^\delta$, and $|d_1 - d_2| < N^\beta$, RSA is insecure if $\delta < \frac{5}{8} - \frac{1}{2}\beta$. To compare this with the bound of Theorem 7, we consider $\gamma = 1$ and $\alpha = 0$ in the next result. This corresponds to the situation when $e_1 \approx e_2 \approx N$ and p and d differ in their first LSBs.

Corollary 1. *Let $N = pq$ be an RSA modulus. Let e_1 and e_2 be two public exponents satisfying $e_1 d_1 - k_1 \phi(N) = 1$, $e_2 d_2 - k_2 \phi(N) = 1$. Suppose that $d_1, d_2 \leq N^\delta$ and $|d_1 - d_2| < N^\beta$. Then one can factor N in polynomial time if*

$$\delta < \frac{5}{2} - \beta - \frac{1}{4}\sqrt{6(9 - 4\beta)}.$$

Proof. This is a direct application of Theorem 7 with $\gamma = 1$ and $\alpha = 0$.   □

In Table 1, we compare the bound $\delta < \frac{5}{8} - \frac{1}{2}\beta$. of Sarkar and Maitra and the bound of Corollary 1 for various values of $\beta = \log_N(|d_1 - d_2|)$.

Table 1. Comparison of the new method with the method of [11]

$\beta = \log_N(|d_1 - d_2|)$	$\beta = 0.6$	$\beta = 0.5$	$\beta = 0.4$	$\beta = 0.3$	$\beta = 0.25$
Bound for δ in [11]	0.325	0.375	0.425	0.475	0.5
Bound for δ in Corollary 1	0.326	0.379	0.434	0.489	0.517

One may note that when d_1 and d_2 differ in their first MSBs, then $\beta = \delta$ and the bound of Sarkar and Maitra is valid if $\delta < \frac{5}{12} \approx 0.416$, while the bound of Corollary 1 gives $\delta < 0.422$.

5.2.2   Comparison with the Bound in Sun et al.

In [18], Sun et al. showed that RSA is insecure when $e = N^\gamma$, $p - q = 2^m v$ with $2^m = N^\alpha$, and $d < N^\delta$, if $\delta < \frac{7}{6} - \frac{2}{3}\alpha - \frac{1}{3}\sqrt{(1 - 4\alpha)(1 - 4\alpha + 6\gamma)}$. To compare our method with the method of Sun et al., we consider Theorem 7 with $\beta = \delta$, that is when d_1 and d_2 do not share any amount of their MSBs. We get the following corollary.

Corollary 2. *Let $N = pq$ be an RSA modulus such that $p - q = 2^m u$ where $2^m \approx N^\alpha$. Let e_1 and e_2 be two public exponents satisfying $e_1, e_2 \approx N^\gamma$, and $e_1 d_1 - k_1 \phi(N) = 1$, $e_2 d_2 - k_2 \phi(N) = 1$. Suppose that $d_1, d_2 \leq N^\delta$. Then one can factor N in polynomial time if*

$$\delta < \frac{17}{16} - \frac{1}{4}\alpha - \frac{1}{16}\sqrt{3(1 - 4\alpha)(3 + 32\gamma - 12\alpha)}.$$

Proof. In the bound of δ in Theorem 7, if we plug $\beta = \delta$ and solve the inequation for δ, we get the desired bound on δ.   □

In Table 2, we compare the largest values of δ of Corollary 2 and the the largest values obtained in [18] for various values of $\gamma = \log_N(e)$ and $\alpha = \log_N(2^m)$.

Table 2. Comparisons of the new method with the method of [18] for $\alpha = \log_N(2^m)$

$\gamma = \log_N(e)$	$\gamma = 1$	$\gamma = 0.9$	$\gamma = 0.8$	$\gamma = 0.7$	$\gamma = 0.6$
Bound for δ in [18] with $\alpha = 0$	0.284	0.323	0.363	0.406	0.451
New bound for δ with $\alpha = 0$	0.422	0.452	0.483	0.516	0.552
Bound for δ in [18] with $\alpha = 0.1$	0.436	0.467	0.500	0.534	0.570
New bound for δ with $\alpha = 0.1$	0.550	0.573	0.598	0.625	0.653
Bound for δ in [18] with $\alpha = 0.2$	0.662	0.680	0.699	0.720	0.742
New bound for δ with $\alpha = 0.2$	0.736	0.750	0.764	0.780	0.797
Bound for δ in [18] with $\alpha = 0.25$	1	1	1	1	1
New bound for δ with $\alpha = 0.25$	1	1	1	1	1

6   Conclusion

For $k \geq 2$ and $i = 1, \ldots, k$, let (N_i, e_i) be k RSA instances with k moduli $N_i = p_i q_i$ and k public exponents e_i. In this paper, we proposed a new method to factor all the RSA moduli $N_1, \ldots, N_k$ in the scenario that the RSA instances satisfy k equations of the shape $e_i x - y_i \phi(N_i) = z_i$ or of the shape $e_i x_i - y\phi(N_i) = z_i$ with suitably small parameters x_i, y_i, z_i, x, y where $\phi(N_i) = (p_i - 1)(q_i - 1)$. We also proposed an attack on RSA when the prime factors p and q of the RSA modulus $N = pq$ are of the same bit-size. The attack factors N when p and q share a number of their least significant bits (LSBs) in the presence of two public exponents e_1 and e_2 with decryption exponents d_1 and d_2 sharing an amount of their most significant bits (MSBs).

References

1. ANSI Standard X9.31-1998, Digital Signatures Using Reversible Public Key Cryptography for the Financial Services Industry (rDSA)
2. Blömer, J., May, A.: A generalized Wiener attack on RSA. In: Bao, F., Deng, R., Zhou, J. (eds.) PKC 2004. LNCS, vol. 2947, pp. 1–13. Springer, Heidelberg (2004)
3. Boneh, D., Durfee, G.: Cryptanalysis of RSA with private key d less than $N^{0.292}$. In: Stern, J. (ed.) EUROCRYPT 1999. LNCS, vol. 1592, pp. 1–11. Springer, Heidelberg (1999)
4. Coppersmith, D.: Small solutions to polynomial equations, and low exponent RSA vulnerabilities. Journal of Cryptology 10(4), 233–260 (1997)
5. Håstad, J.: On using RSA with low exponent in a public key network. In: Williams, H.C. (ed.) CRYPTO 1985. LNCS, vol. 218, pp. 403–408. Springer, Heidelberg (1986)
6. Hinek, J.: On the Security of Some Variants of RSA, Phd. Thesis, Waterloo, Ontario, Canada (2007)
7. Howgrave-Graham, N.: Finding small roots of univariate modular equations revisited. In: Darnell, M.J. (ed.) Cryptography and Coding 1997. LNCS, vol. 1355, pp. 131–142. Springer, Heidelberg (1997)
8. Howgrave-Graham, N., Seifert, J.-P.: Extending Wiener's attack in the presence of many decrypting exponents. In: Baumgart, R. (ed.) CQRE 1999. LNCS, vol. 1740, pp. 153–166. Springer, Heidelberg (1999)
9. Jochemsz, E., May, A.: A strategy for finding roots of multivariate polynomials with new applications in attacking RSA variants. In: Lai, X., Chen, K. (eds.) ASIACRYPT 2006. LNCS, vol. 4284, pp. 267–282. Springer, Heidelberg (2006)
10. Lenstra, A.K., Lenstra, H.W., Lovász, L.: Factoring polynomials with rational coefficients. Mathematische Annalen 261, 513–534 (1982)
11. Sarkar, S., Maitra, S.: Cryptanalysis of RSA with two decryption exponents. Information Processing Letters 110, 178–181 (2010)
12. May, A.: New RSA Vulnerabilities Using Lattice Reduction Methods. Ph.D. thesis, University of Paderborn (2003)
13. Nitaj, A.: Another generalization of Wiener's attack on RSA. In: Vaudenay, S. (ed.) AFRICACRYPT 2008. LNCS, vol. 5023, pp. 174–190. Springer, Heidelberg (2008)
14. Rivest, R., Shamir, A., Adleman, L.: A Method for Obtaining digital signatures and public-key cryptosystems. Communications of the ACM 21(2), 120–126 (1978)
15. Steinfeld, R., Zheng, Y.: On the Security of RSA with Primes Sharing Least-Significant Bits. Appl. Algebra Eng. Commun. Comput. 15(3-4), 179–200 (2004)
16. Steinfeld, R., Zheng, Y.: An advantage of Low-Exponent RSA with Modulus Primes Sharing Least Significant Bits. In: Naccache, D. (ed.) CT-RSA 2001. LNCS, vol. 2020, pp. 52–62. Springer, Heidelberg (2001)
17. Steinfeld, R., Zheng, Y.: On the Security of RSA with Primes Sharing Least-Significant Bits. Appl. Algebra Eng. Commun. Comput. 15(3-4), 179–200 (2004)
18. Sun, H.-M., Wu, M.-E., Steinfeld, R., Guo, J., Wang, H.: Cryptanalysis of Short Exponent RSA with Primes Sharing Least Significant Bits. In: Franklin, M.K., Hui, L.C.K., Wong, D.S. (eds.) CANS 2008. LNCS, vol. 5339, pp. 49–63. Springer, Heidelberg (2008)
19. Wiener, M.: Cryptanalysis of short RSA secret exponents. IEEE Transactions on Information Theory 36, 553–558 (1990)
20. Zhao, Y.-D., Qi, W.-F.: Small private-exponent attack on RSA with primes sharing bits. In: Garay, J.A., Lenstra, A.K., Mambo, M., Peralta, R. (eds.) ISC 2007. LNCS, vol. 4779, pp. 221–229. Springer, Heidelberg (2007)

A   Proof of Theorem 4

Proof. Let $\varepsilon \in (0, 1)$. Set

$$C = \left\lceil 3^{n+1} \cdot 2^{\frac{(n+1)(n-4)}{4}} \cdot \varepsilon^{-n-1} \right\rceil, \tag{11}$$

where $\lceil x \rceil$ is the integer greater than or equal to x. Consider the lattice $\mathcal{L}$ spanned by the rows of the matrix

$$M = \begin{bmatrix} 1 & -[C\alpha_1] & -[C\alpha_2] & \cdots & -[C\alpha_n] \\ 0 & C & 0 & \cdots & 0 \\ 0 & 0 & C & \cdots & 0 \\ \vdots & \vdots & \vdots & \ddots & \vdots \\ 0 & 0 & 0 & \cdots & C \end{bmatrix},$$

where $[x]$ is the nearest integer to x. The determinant of $\mathcal{L}$ is $\det(\mathcal{L}) = C^n$ and the dimension is $n + 1$. Applying the LLL algorithm, we find a reduced basis $(b_1, \cdots, b_{n+1})$ with

$$\|b_1\| \leq 2^{n/4} \det(\mathcal{L})^{1/(n+1)} = 2^{n/4} C^{n/(n+1)}.$$

Since $b_1 \in \mathcal{L}$, we can write $b_1 = \pm[q, p_1, p_2, \ldots, p_n]M$, that is

$$b_1 = \pm \left[q, Cp_1 - q[C\alpha_1], Cp_2 - q[C\alpha_2], \cdots, Cp_n - q[C\alpha_n] \right], \tag{12}$$

where $q > 0$. Hence, the norm of b_1 satisfies

$$\|b_1\| = \left(q^2 + \sum_{i=1}^{n} |Cp_i - q[C\alpha_i]|^2 \right)^{1/2} \leq 2^{n/4} C^{n/(n+1)},$$

which leads to

$$q \leq \left\lfloor 2^{n/4} C^{n/(n+1)} \right\rfloor \quad \text{and} \quad \max_i |Cp_i - q[C\alpha_i]| \leq 2^{n/4} C^{n/(n+1)}. \tag{13}$$

Let us consider the entries $q\alpha_i - p_i$. We have

$$|q\alpha_i - p_i| = \frac{1}{C} |Cq\alpha_i - Cp_i|$$

$$\leq \frac{1}{C} \left(|Cq\alpha_i - q[C\alpha_i]| + |q[C\alpha_i] - Cp_i| \right)$$

$$= \frac{1}{C} \left(q|C\alpha_i - [C\alpha_i]| + |q[C\alpha_i] - Cp_i| \right)$$

$$\leq \frac{1}{C} \left(\frac{1}{2} q + |q[C\alpha_i] - Cp_i| \right).$$

Using the two inequalities in (13), we get

$$|q\alpha_i - p_i| \leq \frac{1}{C} \left(\frac{1}{2} \cdot 2^{n/4} C^{n/(n+1)} + 2^{n/4} C^{n/(n+1)} \right) = \frac{3 \cdot 2^{(n-4)/4}}{C^{1/(n+1)}}$$

Observe that (11) gives

$$3^{n+1} \cdot 2^{\frac{(n+1)(n-4)}{4}} \cdot \varepsilon^{-n-1} \leq C \leq\leq 3^{n+1} \cdot 2^{\frac{(n+1)(n-3)}{4}} \varepsilon^{-n-1}, \tag{14}$$

which leads to $\varepsilon \geq \frac{3 \cdot 2^{(n-4)/4}}{C^{1/(n+1)}}$. As a consequence, we get $|q\alpha_i - p_i| \leq \varepsilon$. On the other hand, using (13) and (14) , we get

$$q \leq \left\lfloor 2^{n/4} C^{n/(n+1)} \right\rfloor \leq 2^{n/4} C^{n/(n+1)} \leq 2^{n(n-3)/4} \cdot 3^n \cdot \varepsilon^{-n}.$$

To compute the vector $[q, p_1, p_2, \ldots, p_n]$, we use (12)

$$[q, p_1, p_2, \ldots, p_n] = \pm [q, Cp_1 - q\,[C\alpha_1], Cp_2 - q\,[C\alpha_2], \cdots, Cp_n - q\,[C\alpha_n]]\, M^{-1}.$$

This terminates the proof.   □

B   Proof of Lemma 1

Proof. Suppose that $p - q = 2^m u$. Then $p = q + 2^m u$ and $N = q^2 + 2^m uq$. Hence $q^2 \equiv N \pmod{2^m}$. Let u_0 be a solution of the congruence $x^2 \equiv N \pmod{2^m}$. For $m \leq 2$, this equation has only one solution and for $m \geq 3$, there are four solutions that can be found in polynomial time using Hensel's Lemma. Then $q \equiv u_0 \pmod{2^m}$ for one of the solutions u_0 which implies that $q = 2^m q_1 + u_0$ for a positive integer q_1. Now, we have

$$p = q + 2^m u = 2^m q_1 + u_0 + 2^m u = 2^m (q_1 + u) + u_0 = 2^m p_1 + u_0,$$

where $p_1 = q_1 + u$. Using $N = pq$, we get

$$N = (2^m p_1 + u_0)(2^m q_1 + u_0) = 2^{2m} p_1 q_1 + 2^m u_0(p_1 + q_1) + u_0^2.$$

From this, we deduce $2^m u_0(p_1 + q_1) + u_0^2 \equiv N \pmod{2^{2m}}$. Since u_0 is odd, we obtain

$$2^m (p_1 + q_1) \equiv (N - u_0^2) u_0^{-1} \pmod{2^{2m}},$$

which can be rewritten as $2^m (p_1 + q_1) = 2^{2m} v + t_0$ with

$$t_0 \equiv (N - u_0^2) u_0^{-1} \pmod{2^{2m}}.$$

Finally, we get

$$\begin{aligned}
p + q &= 2^m p_1 + u_0 + 2^m q_1 + u_0 \\
&= 2^m (p_1 + q_1) + 2u_0 \\
&= 2^{2m} v + t_0 + 2u_0 \\
&= 2^{2m} v + v_0,
\end{aligned}$$

where $v_0 = t_0 + 2u_0$. This terminates the proof.   □

C   Proof of Lemma 2

Proof. Suppose that $S > 2N^{\frac{1}{2}}$ and let $D = \sqrt{S^2 - 4N}$. We have

$$\left|(p-q)^2 - D^2\right| = \left|(p-q)^2 - S^2 + 4N\right| = \left|(p+q)^2 - S^2\right|.$$

Dividing by $p - q + D$, we get

$$|p - q - D| = \frac{(p+q+S)|p+q-S|}{p-q+D}$$

Next, suppose $|p + q - S| < \frac{p-q}{3(p+q)}N^{\frac{1}{4}}$. Since $\frac{p-q}{3(p+q)}N^{\frac{1}{4}} < N^{\frac{1}{4}}$, then

$$p + q + S < 2(p+q) + N^{\frac{1}{4}} < 3(p+q).$$

Combining with $p - q + D > p - q$, we deduce

$$|p - q - D| < \frac{3(p+q)|p+q-S|}{p-q} < \frac{3(p+q)}{p-q} \cdot \frac{p-q}{3(p+q)}N^{\frac{1}{4}} = N^{\frac{1}{4}}.$$

Now, set $\tilde{P} = \frac{1}{2}(S + D)$. We have

$$\begin{aligned}
\left|p - \tilde{P}\right| &= \left|p - \frac{1}{2}(S+D)\right| \\
&= \frac{1}{2}|p+q-S+p-q-D| \\
&\leq \frac{1}{2} \cdot |p+q-S| + \frac{1}{2}|p-q-D| \\
&< \frac{1}{2} \cdot \frac{p-q}{3(p+q)}N^{\frac{1}{4}} + \frac{1}{2}N^{\frac{1}{4}} \\
&< N^{\frac{1}{4}},
\end{aligned}$$

where we used $\frac{1}{2} \cdot \frac{p-q}{3(p+q)} < \frac{1}{2}$. This terminates the proof.   □

Formulae for Computation of Tate Pairing on Hyperelliptic Curve Using Hyperelliptic Nets

Christophe Tran

IRMAR, UMR CNRS 6625, Université de Rennes 1, Campus de Beaulieu F-35042,
France

Abstract. Stange has showed how to compute the Tate pairing on an elliptic curve using elliptic nets. After that, Uchida and Uchiyama gave a generalization of elliptic nets to hyperelliptic curves. They also gave an algorithm to compute the Tate pairing on a hyperelliptic curve of genus 2. In this paper, we extend their algorithm for curves of all genus. In a computational point of view, we also study the optimality of these algorithms.

Keywords: Tate pairing, hyperelliptic curve, hyperelliptic net.

1 Introduction

In the years 2000, the bilinear property of pairings permitted the construction of new cryptographic protocols. The most famous examples are certainly the identity based encryption schemes.

The Tate pairing is usually computed by the Miller's algorithm, but Stange ([11]) proposed a new tool to compute the Tate pairing on an elliptic curve : the elliptic nets. In 2012, Uchida and Uchiyama ([13]) gave a generalization of the notion of elliptic nets to hyperelliptic curves, and explained how to use these nets to compute pairings on a curve of genus 2.

The contributions of this paper are the following:

- a simplification of the formula to compute pairings using hyperelliptic nets (theorem 9),
- a generalization of the algorithms based on hyperelliptic nets given in [13] for genus 2 to all genus (sections 5 and 6),
- the proof of the optimality of these algorithms, in a sense defined in section 5.1.

In section 2, we recall some notions of arithmetic on hyperelliptic curves; in section 3, we recall some properties of the sigma function used in the construction of the elliptic and hyperelliptic nets; in section 4 we recall the theory of elliptic nets of Stange, and the theory of hyperelliptic nets of Uchida and Uchiyama; in sections 5 and 6 we give our version of the formulae to compute pairings, and the derived algorithms; finally, we illustrate them in a genus 3 example in section 7.

D. Pointcheval and D. Vergnaud (Eds.): AFRICACRYPT 2014, LNCS 8469, pp. 199–214, 2014.

2   Background on Hyperelliptic Curves

Here we fix some notations and terminology for the rest of the paper (more informations could be found in [1], [2], or [3] for example).

Let $\mathcal{C}$ be the hyperelliptic curve of genus g defined over a field $\mathbb{K}$ by the equation

$$\mathcal{C} : y^2 + H(x)y = F(x),$$

with H and $F \in \mathbb{K}[x]$, F monic of degree $2g + 1$, and $deg(H) \leq g + 1$. So $\mathcal{C}$ has one point at infinity, denoted $\mathcal{O}$.

Its Jacobian $\mathcal{J}$ is the abelian variety defined over $\mathbb{K}$ whose points are classes of divisors of degree 0 modulo the principal divisors.

Let $\lambda : \mathcal{C} \to \mathcal{J}$ be an embedding such that $\lambda(\mathcal{O})$ is the neutral element of $\mathcal{J}$. The *theta divisor* Θ on $\mathcal{J}$ is defined by $\Theta = \lambda(\mathcal{C}) + \cdots + \lambda(\mathcal{C})$ ($g - 1$ times).

Definition 1. *For a divisor D and a positive integer s, the scalar multiplication of D by s is denoted $[s]D$. The* Miller function *$f_{s,D}$ is the uniquely defined up to scalar multiplication by elements of $\mathbb{K}^*$ rational function with divisor*

$$(f_{s,D}) = sD - ([s]D).$$

Remark 1. If $D \in \mathcal{J}(\mathbb{K})[r]$, then $f_{r,D}$ is the unique (up to scalar multiplication) function of divisor $[r]D$.

We now consider the case where $\mathbb{K}$ is a finite field $\mathbb{F}_q$. Let r prime such that $r | \sharp J(\mathbb{F}_q)$. The entire group μ_r of r-th roots of unity is contained in $\mathbb{F}_{q^k}$, where $k \in \mathbb{N}$ is the minimal with $r | q^k - 1$ (k is called the *embedding degree*).

Theorem 1. *The application*

$$\tau : \mathcal{J}(\mathbb{F}_{q^k})[r] \times \mathcal{J}(\mathbb{F}_{q^k})[r] \to \mu_n \subset \mathbb{F}_{q^k}^*$$

$$(D_1, D_2) \mapsto (f_{r,D_1}(D_2))^{\frac{q^k - 1}{r}}.$$

is a well-defined pairing called the modified Tate-Lichtenbaum pairing.

Let π denote the q-power of the Frobenius endomorphism. In practice, we will take $D_1 \in \mathbb{G}_1$ and $D_2 \in \mathbb{G}_2$, with $\mathbb{G}_1 = \mathcal{J}(\mathbb{F}_q)[r]$ and $\mathbb{G}_2 = \mathcal{J}[r] \cap ker(\pi - [q])$. This choice ensures that $\mathbb{G}_1$ is 1-dimensional over $\mathbb{Z}/r\mathbb{Z}$, $\mathbb{G}_2 \subset \mathcal{J}(\mathbb{F}_{q^k})$ and $\mathbb{G}_1 \neq \mathbb{G}_2$.

3   Sigma Function

The *sigma function* gives the equations defining $\mathcal{J}$ as an algebraic variety over a projective space, and the group law in these coordinates. We begin by recalling the situation in genus 1 (see for example [10]).

3.1   The Weierstrass Sigma Funtion

Let $\mathcal{E}$ be an elliptic curve defined over $\mathbb{C}$ by the lattice Λ (i.e. $\mathcal{E} \simeq \mathbb{C}/\Lambda$).

Definition 2. *The infinite product*

$$z \prod_{\substack{\omega \in \Lambda \\ \omega \neq 0}} \left(1 - \frac{z}{\omega}\right) e^{(z/\omega) + \frac{1}{2}(z/\omega)^2}$$

defines a holomorphic function on all of $\mathbb{C}$ called the Weierstrass σ function (relative to $\mathcal{E}$). It has simple zeros at each point z of the lattice Λ and no other zeros.

The group law on $\mathcal{E}(\mathbb{C})$ can be describe by:

Proposition 1. *For all u and v of $\mathbb{C}$,*

$$\frac{\sigma(u+v)\sigma(u-v)}{\sigma(u)^2\sigma(v)^2} = \wp(v) - \wp(u)$$

(for a recall of the definition of the Weierstrass $\wp$ function, see [10] for example).

Using this relation, we can prove the following property of the σ function:

Proposition 2. *For all u, v, w and $z \in \mathbb{C}$, we have*

$$\begin{aligned}
\sigma(u+v)\sigma(u-v)\sigma(w+z)\sigma(w-z) & \\
+ \sigma(u+w)\sigma(u-w)\sigma(z+v)\sigma(z-v) & \\
+ \sigma(u+z)\sigma(u-z)\sigma(v+w)\sigma(v-w) &= 0.
\end{aligned}$$

We will see in section 4.1 this is with this property Stange defined her elliptic nets. We now examine the situation for curves of higher genus.

3.2   The Kleinian Sigma Funtion

The equivalent for the hyperelliptic curves is the *Kleinian sigma function*. It is a function on $\mathbb{C}^g$. As its formula is a little complicated (and not very useful in this work), we do not recall its exact definition, which can be found in [4] and [6] for example. Imitating the genus 1 case, we begin by studying the zero locus of σ:

Proposition 3. *Let $\mathcal{C}$ be a hyperelliptic curve of genus g defined over $\mathbb{C}$. Its Jacobian $\mathcal{J}$ is isomorphic to a torus $\mathbb{C}^g/\Lambda$.*
The zero locus of the σ function relative to $\mathcal{C}$ is well-defined over $\mathbb{C}^g/\Lambda$: σ has simple zeros on the points of Θ and nowhere else.

The following is a property of σ we will use in the rest of this work:

Proposition 4. *σ is an odd function when $g \equiv 1$ or 2 (mod 4) and even in the other cases.*

As for $g = 1$, the group law on $\mathcal{J}$ can be described by the σ function:

Proposition 5. *For $i, j, k \in \{1, 2, \dots, g\}$ and $u \in \mathbb{C}^g$, we define the hyperelliptic $\wp$–functions:*

$$\wp_{ij}(u) = -\frac{\partial^2}{\partial u_i \partial u_j} log\sigma(u), \quad \wp_{ijk}(u) = -\frac{\partial^3}{\partial u_i \partial u_j \partial u_k} log\sigma(u),$$

which are well defined over $\mathcal{J} \simeq \mathbb{C}^g/\Lambda$.

   Let $D_1, D_2 \in \mathcal{J}$. There are well-defined polynomials $\mathcal{F}_g(D_1, D_2)$ depending only on the $(\wp_{ij}(D_1))_{i,j}$, $(\wp_{ijk}(D_1))_{i,j,k}$, the $(\wp_{ij}(D_2))_{i,j,k}$ and the $(\wp_{ij}(D_2))_{i,j,k}$ such that

$$\frac{\sigma(D_1 + D_2)\sigma(D_1 - D_2)}{\sigma(D_1)^2 \sigma(D_2)^2} = \mathcal{F}_g(D_1, D_2).$$

We now give the equivalent of proposition 2 for all genus:

Proposition 6. *Let $m > 2^g$ be an integer and $u^{(1)}, \dots u^{(m)} \in \mathbb{C}^g$. The matrix*

$$A = \left(\sigma(u^{(i)} + u^{(j)})\sigma(u^{(i)} - u^{(j)}) \right)_{1 \leq i, j \leq m}$$

has $det(A) = 0$.

This property is the starting point of the hyperelliptic nets theory.

Remark 2. The traditional way to make computations on $\mathcal{J}$ is to use the *Mumford representation* of the elements of $\mathcal{J}$ (see for example [1]). The Mumford representation (u, v) of $D \in \mathcal{J} \backslash \Theta$ can be expressed by:

$$u = x^g - \sum_{l=1}^{g} \wp_{lg}(D)x^{l-1}, \quad v = \sum_{l=1}^{g} \frac{\wp_{lgg}(D)}{2} x^{l-1}.$$

4   Background on Hyperelliptic Nets

We begin by recalling the genus 1 case, i.e. the Stange's elliptic net theory ([11]). We then will see that the work of Uchida and Uchiyama in [13] follows the same program as Stange : definition of the nets on $\mathbb{C}$, transport of the nets and their properties to general fields, and development of a formula linking the Tate pairing and the nets. They also gave an algorithm to compute the Tate pairing on hyperelliptic curves of genus 2 using hyperelliptic nets.

   Once we have recalled all these results, we will simplify the Uchida and Uchiyama's formula (theorem 9) and extend their algorithm to all genus (section 5 and 6).

4.1   The Genus One Case

We begin by recalling the definition of an elliptic net:

Definition 3. *Let A be a finitely generated free abelian group, and R be an integral domain. An elliptic net is a map $W : A \to R$ such that for all a, b, c and d in A:*

$$W(a+b)W(a-b)W(c+d)W(c-d)$$
$$+ W(a+c)W(a-c)W(d+b)W(d-b)$$
$$+ W(a+d)W(a-d)W(b+c)W(b-c) = 0,$$

In the aim of computing pairing, A will be $\mathbb{Z}^2$ and R will be the field we are working on : $\mathbb{C}$ in a first time, $\mathbb{F}_{q^k}$ for cryptographic use.

Let $\mathcal{E}$ be an elliptic curve defined over $\mathbb{C}$. With the sigma function attached to $\mathcal{E}$, Stange built a first elliptic net:

Proposition 7. *For all $\mathbf{v} = (v_1, v_2) \in \mathbb{Z}^2$,*

$$\Psi_{\mathbf{v}}(z_1, z_2) = \frac{\sigma(v_1 z_1 + v_2 z_2)}{\sigma(z_1)^{v_1^2 - v_1 v_2} \sigma(z_1 + z_2)^{v_1 v_2} \sigma(z_2)^{v_2^2 - v_1 v_2}}$$

is a well-defined function on $\mathcal{E} \times \mathcal{E}$.

Remark 3. In her original work, Stange defined $\Psi_{\mathbf{v}}$ for $\mathbf{v}$ in general $\mathbb{Z}^n$. We specialize $n = 2$ because it is the only interesting case to compute Tate pairings.

Proposition 8. *The function $W : \mathbb{Z}^2 \to \mathbb{C}$ defined by $W(v_1, v_2) = \Psi_{\mathbf{v}}(z_1, z_2)$ is an elliptic net. For computational purpose, it has the interesting property $W(1,0) = W(0,1) = W(1,1) = 1$.*

As we know the divisor of σ, we can compute the divisor of $\Psi_{\mathbf{v}}$.

Theorem 2 ([11]). *Let $s : \mathcal{E}^2 \to \mathcal{E}$ and, for $i = 1$ or 2, $p_i : \mathcal{E}^2 \to \mathcal{E}$ denote respectively the sum of all components and the projection onto the i-th component, and s^* and p_i^* their pullbacks. The divisor of $\Psi_{\mathbf{v}}$ is:*

$$D_{\mathcal{E},\mathbf{v}} = (([v_1] \times [v_2])^* s^* \mathcal{O}) - v_1 v_2 ((p_1^* \times p_2^*) s^* \mathcal{O}) - (v_1^2 - v_1 v_2) p_1^* \mathcal{O} - (v_2^2 - v_1 v_2) p_2^* \mathcal{O}.$$

This theorem has two important consequences. Firstly, we can transport the elliptic net theory over arbitrary fields:

Theorem 3. *Let $\mathcal{E}$ be an elliptic curve defined over a field $\mathbb{K}$. Let P_1 and P_2 two points of $\mathcal{E}$. There is an elliptic net $W_{\mathcal{E},(P_1,P_2)} : \mathbb{Z}^2 \to \overline{\mathbb{K}}$ such that*

- $W_{\mathcal{E},(P_1,P_2)}(0,1) = W_{\mathcal{E},(P_1,P_2)}(1,0) = W_{\mathcal{E},(P_1,P_2)}(1,1) = 1;$
- $W_{\mathcal{E},(P_1,P_2)}(v_1, v_2) = 0 \Leftrightarrow [v_1]P_1 + [v_2]P_2 = \mathcal{O}$ *on the curve $\mathcal{E}$.*

Such an elliptic net is said associated to *the curve $\mathcal{E}$ and the points P_1 and P_2.*

Secondly, we can link these nets to the Tate pairing:

Theorem 4. *If $\mathcal{E}$ is defined over the finite field $\mathbb{F}_q$ of characteristic p, let r be prime with p, and k be the embedding degree of $\mathcal{E}\,(\mathbb{F}_q)$. Let $P_1 \in \mathbb{G}_1$ and $P_2 \in \mathbb{G}_2$ (the two groups defined in section 2). Let W be an elliptic net associated to $\mathcal{E}$, P_1 and P_2. The Tate pairing of P_1 and P_2 is*

$$\tau(P_1, P_2) = \left(\frac{W(r+1,1)W(1,0)}{W(r+1,0)W(1,1)} \right)^{\frac{q^k-1}{r}} = (W(r+1,1))^{\frac{q^k-1}{r}}$$

Remark 4. The first equality is due to Stange, the second was established in [9], using the final exponentiation and the fact that for all integers a, $W(a,0) \in \mathbb{F}_q$.

4.2   The General Case

The definition of an hyperelliptic net is postponed for a while. The first step of Uchida and Uchiyama is the equivalent of proposition 7 : the construction of $\Psi_{\mathbf{v}}$.

Proposition 9. *Let $\mathcal{C}/\mathbb{C}$ be an hyperelliptic curve of genus g, $\mathcal{J}$ be its Jacobian and σ be its sigma function.*
For $\mathbf{v} = (v_1, v_2) \in \mathbb{Z}^2$,

$$\Psi_{\mathbf{v}}(z_1, z_2) = \frac{\sigma(v_1 z_1 + v_2 z_2)}{\sigma(z_1)^{v_1^2 - v_1 v_2} \sigma(z_1 + z_2)^{v_1 v_2} \sigma(z_2)^{v_2^2 - v_1 v_2}}$$

is a well-defined function on $\mathcal{J} \times \mathcal{J}$.

As in the previous section, we only study the case $\mathbf{v} \in \mathbb{Z}^2$, because our aim is the computation of pairings.

Theorem 5. *Let $s : \mathcal{J}^2 \to \mathcal{J}$ and, for $i = 1$ or 2, $p_i : \mathcal{J}^2 \to \mathcal{J}$ denote respectively the sum of all components and the projection onto the i-th component, and s^* and p_i^* their pullbacks. For $\mathbf{v} \neq 0$, the divisor of $\Phi_{\mathbf{v}}$ is:*

$$D_{\mathcal{C}, \mathbf{v}} = (([v_1] \times [v_2])^* s^* \Theta) - v_1 v_2 ((p_1^* \times p_2^*) s^* \Theta) - (v_1^2 - v_1 v_2) p_1^* \Theta - (v_2^2 - v_1 v_2) p_2^* \Theta.$$

The next theorem is the recurrent property which extends the equation defining the elliptic net:

Theorem 6. *Let $m > 2^g$ be an integer and for $1 \leq i \leq m$, $v_i \in (1/2\mathbb{Z})^2$ such that for all $1 \leq i, j \leq m$, $v_i + v_j$ and $v_i - v_j \in \mathbb{Z}^2$.*
The matrix

$$A = (\Phi_{v_i + v_j} \Phi_{v_i - v_j})_{1 \leq i, j \leq m}$$

has $\det(A) = 0$.

Remark 5. When $g \equiv 1$ or $2 \pmod 4$, this theorem can only be used with even values of m. Indeed, as σ is an odd function in these cases, the matrix A is antisymmetric, and the determinant of an antisymmetric matrix of odd size is identically 0. If $m = 2n$, its determinant is the square of a degree n polynomial called the *Pfaffian* of A and denoted $Pf(A)$:

$$Pf(A) = \frac{1}{2^n n!} \sum_{\sigma \in S_{2n}} sgn(\sigma) \prod_{i=1}^{n} a_{\sigma(2i-1),\sigma_{2i}}$$

where S_{2n} is the symmetric group and $sgn(\sigma)$ is the signature of σ.

For example, for $g = 1$, theorem 6 with $m = 4$ (the minimal possible value) gives the elliptic nets equation. So this equation can be used as a definition of an hyperelliptic net. As in the genus 1 case, using theorem 5, Uchida and Uchiyama could define hyperelliptic nets over all fields:

Theorem 7. *Let $C/\mathbb{K}$ be an hyperelliptic curve defined over an arbitrary field $\mathbb{K}$, and D_1 and D_2 two divisors of C. There is a map $W_{C,(D_1,D_2)} : \mathbb{Z}^2 \to \overline{\mathbb{K}}$ such that*

- $W_{C,(D_1,D_2)} = 0 \Leftrightarrow [v_1]D_1 + [v_2]D_2 \in \Theta$.
- *W satisfies the relation of theorem 6, i.e. for all integer $m > 2^g$ and $v_1,\ldots v_m, w_1,\ldots w_m$ in $1/2\mathbb{Z}$ such that for all i and j, $v_i \pm v_j$ and $w_i \pm w_j \in \mathbb{Z}$, the matrix*

$$A = (W(v_i + v_j, w_i + w_j)W(v_i - v_j, w_i - w_j))_{1 \leq i,j \leq m}$$

has $det(A) = 0$.

$W_{C,(D_1,D_2)}$ *is called an* hyperelliptic net *associated to C and (D_1, D_2).*

Using theorem 5, Uchida and Uchiyama established the following relation between the Tate pairing and the hyperelliptic net:

Theorem 8. *Let $D_1 \in \mathbb{G}_1$ and $D_2 \in \mathbb{G}_2$. Let W be an hyperelliptic net associated to D_1 and D_2. The Tate pairing of D_1 and D_2 is*

$$\tau(D_1, D_2) = \left(\frac{W(r+1,1)W(1,0)}{W(r+1,0)W(1,1)}\right)^{\frac{q^k-1}{r}}$$

As Ogura et al. did in [9] for $g = 1$, we can simplify this formula:

Theorem 9. *With D_1, D_2 and W as in the previous theorem, the Tate pairing of D_1 and D_2 is*

$$\tau(D_1, D_2) = \left(\frac{W(r+1,1)}{W(1,1)}\right)^{\frac{q^k-1}{r}}.$$

Proof. The key ingredient of this proof is the following : as $D_1 \in \mathcal{J}(\mathbb{F}_q)$, for all $a \in \mathbb{Z}$, $W(a,0) \in \mathbb{F}_q$.

Indeed, by proposition 6, we have for all a and $b \in \mathbb{Z}$,

$$\frac{W(a+b,0)W(a-b,0)}{W(a,0)^2 W(b,0)^2} = \mathcal{F}_g([a]D_1,[b]D_1),$$

where the right-hand side of this equality is a polynomial with integer coefficients in the Mumford coordinates of $[a]D_1$ and $[b]D_1$. In particular, it is an element of $\mathbb{F}_q$. As W is defined by the divisor given in theorem 5, we can fix $W(1,0) \in \mathbb{F}_q$ (in practice, it is convenient to choose $W(1,0) = 1$, as in the elliptic case). So by induction, for all integers a, $W(a,0) \in \mathbb{F}_q$.

We now can use the final exponentiation to simplify the expression of theorem 8 : as r is prime, $r|q^k - 1$ and $r \nmid q - 1$, we have $q - 1|\frac{q^k-1}{r}$ and

$$\left(\frac{W(1,0)}{W(r+1,0)}\right)^{\frac{q^k-1}{r}} = 1.$$

$\square$

We now explain how we can use theorem 6 to compute this term $W(r+1,1)$ needed in theorem 9. We have to make the distinction between the case $g \equiv 1,2$ (mod 4) and the case $g \equiv 0,3$ (mod 4) : in the first case, we have to compute pfaffians, which are polynomials of degree $m/2$ in the coefficients of the matrix, whereas in the second case we only have the determinant tool, which is a polynomial of degree m. An other important difference between the two situations is that if $g \equiv 1,2$ (mod 4), then $\sigma(0) = 0$.

5   The Hyperelliptic Net Algorithm in the Case $g \equiv 1,2$ (mod 4)

5.1   Definitions of the Blocks and the Settings

Let $D_1 \in \mathbb{G}_1$ and $D_2 \in \mathbb{G}_2$. Let W be an hyperelliptic net associated to D_1 and D_2. The aim is to compute $W(r+1,1)$ using theorem 6. Before we give the formularies for this, let's explain how we can manipulate this theorem.

As $m > 2^g$ has to be even, we take $m = 2^g + 2$. The matrix A of theorem 6 becomes

$$A = (W(v_i + v_j, w_i + w_j)W(v_i - v_j, w_i - w_j))_{1 \le i,j \le m},$$

with $v_1, \ldots v_m, w_1, \ldots w_m$ of our choice in $1/2\mathbb{Z}$ such that for all i and j, $v_i \pm v_j$ and $w_i \pm w_j \in \mathbb{Z}$.

As $W(0,0) = 0$, we can already remark that the diagonal terms of A are all 0. Moreover, we have to choose for all $i \neq j$, $(v_i, w_i) \neq (v_j, w_j)$, or $det(A)$ will be the null polynomial, and so theorem 6 will become useless.

So our goal is to find values for the $(v_i)_{1 \le i \le m}$ and the $(w_i)_{1 \le i \le m}$ such that

- for all $i \neq j$, $(v_i, w_i) \neq (v_j, w_j)$;
- there are indexes $i_0 \neq j_0$ such that $(v_{i_0} + v_{j_0}, w_{i_0} + w_{j_0}) = (r+1,1)$ and $W(v_{i_0} - v_{j_0}, w_{i_0} - w_{j_0})$ is known;

- for all indexes $i \neq i_0$ and $j \neq j_0$, the $W(v_i + v_j, w_i + w_j)W(v_i - v_j, w_i - w_j)$
  are known.

For example, let $M = \lfloor r + 1 \rfloor + 1$, we can choose $(v_1, w_1) = (M, 1)$ and
$(v_2, w_2) = (r + 1 - M, 0) = (M - 1, 0)$ or $(M - 2, 0)$ depending on the parity of r. If we choose the other v_i and w_i to be as small as possible, then we
obtain a polynomial relationship involving the $W(r + 1, 1)$ we want, some terms
$W(M + a, b)$ with small a and b, and some terms $W(a', b')$ with small a' and b'.
We then see that before to obtain $W(r + 1, 1)$, we have to compute several terms
$W(M + a, b)$.

So, we will build a double-and-add algorithm. Following Stange's idea, we define
the notion of block centered in an integer k. Then, we have to write down the
formulae explaining how, if we know the block centered in k, we can obtain the
block centered in $2k$ or in $2k + 1$. So finally, we will have $W(r + 1, 1)$.

Theorem 10. *We define the initial terms as*

- *the $2m - 7$ terms $\{W(i, 0) \mid 1 \leq i \leq 2m - 7\}$ (or the two terms $W(1, 0)$ and*
  $W(2, 0)$ if $g = 1$ and $m = 4$),
- *the $2m - 2$ terms $\{W(i, 1) \mid 1 - m \leq i \leq m - 2\}$.*

Let $k \in \mathbb{N}$. We define a block centered on k to be

- *the $4m - 8$ values $\{W(k + i, 0) \mid -2m + 4 \leq i \leq 2m - 5\}$,*
- *the $2m - 5$ values $\{W(k + i, 1) \mid 3 - m \leq i \leq m - 3\}$.*

*If we have the initial terms and the block centered on k, then we can compute
the block centered on $2k$ and the block centered on $2k + 1$.*

Remark 6. A block is composed of two levels : the terms $W(k + i, 0)$, which only
involve D_1 and are in $\mathbb{F}_q$, and the terms $W(k + i, 1)$, which involve both D_1 and
D_2, and are in $\mathbb{F}_{q^k}$.

Proof. We start with the terms independent of D_2.
In the matrix $A = (W(v_i + v_j, w_i + w_j)W(v_i - v_j, w_i - w_j))_{1 \leq i, j \leq m} = (a_{ij})_{1 \leq i, j \leq m}$,
we take the following:

- for all $1 \leq i \leq m$, $w_i = 0$,
- $v_1 = l + l' + 1$,
- $v_2 = l - 1$,
- for $3 \leq i \leq m$, $v_i = m - i$.

These settings will now be called **settings** $(\mathbf{l}, \mathbf{l'}, \mathbf{0})$ for

$$k - (m - 2) \leq l \leq k + (m - 2), \quad l' = 0 \text{ or } -1.$$

By these settings, we compute the $W(2k + c, 0)$ for $4 - 2m \leq c \leq 2m - 4$ (with
$l' = 0$ if c is even and $l' = -1$ if c is odd), using the terms of the block centered
in k and the initial terms independent of D_2.

For the terms involving both D_1 and D_2, we use the settings **settings (k, c, 1)** for $-(m-3) \le c \le m-2$:

- $w_1 = 1$ and the other w_i are 0,
- $v_1 = k$,
- $v_2 = k + c$,
- for $3 \le i \le m$, $v_i = m - i$.

□

For $g = 1$ and $g = 2$, we obtain the same formulae than Stange ([11]) and Uchida and Uchiyama ([13]) respectively. We now see in the next proposition that these formulae are "optimal" in the following sense:

Proposition 10. *If someone want to compute $W(r+1,1)$ by a double-and-add algorithm using theorem 8, then our choice of definition of block is of minimal size.*

Proof. We explain this statement for the settings $(l, l', 0)$, the same ideas being used for the settings $(k, c, 1)$.

We want to compute $W(2k+a, 0)$, for some value of a, by choosing in A the right v_i and w_i. Obviously, we prefer to do all computations in the base fields $\mathbb{F}_q$, so we choose $w_i = 0$ for all i. As we have seen in the discussion at the beginning of this section, we cannot have $v_i = v_j$ for some $i \ne j$. So we have to chose

- $v_1 + v_2 = 2k + a$ and $v_1 - v_2$ small : we write $v_1 = k + a_1$ and $v_2 = k + a_2$ with $a_1 + a_2 = a$ and for example $a_1 > a_2$;
- $\{v_3, \ldots, v_m\} = \{0, 1, \ldots, m-3\}$.

With this choice of setting, we obtain $W(2k+a, 0)$ as a function of the $\{W(k+i, 0) \mid a_2 - (m+3) \le i \le a_1 + m - 3\}$. To minimize the number of the inputs, we have to take $a_1 - a_2 > 0$ as small as possible, i.e. $a_1 - a_2 = 1$ if $a = a_1 + a_2$ is odd, and $a_1 - a_2 = 2$ if a is even. This is exactly the setting $(l, l', 0)$ we described in the proof of theorem 10.

□

5.2　Formulae

We now exhibit the formulae given by the different settings $(l, 0, 0)$, $(l, -1, 0)$ and $(k, c, 1)$ we have defined.

For $1 \le i, j, k, l \le m$, let A_{ij} denote the matrix A with the i-th and j-th rows and columns removed, and A_{ijkl} be the matrix A with the i-th, j-th, k-th and l-th rows and columns removed. Then the pfaffian of A can be developed as follows:

$$0 = Pf(A) = a_{12}Pf(A_{12}) + \sum_{i=3}^{m}(-1)^i a_{1i}Pf(A_{1i})$$

$$= a_{12}Pf(A_{12}) + \sum_{i=3}^{m}\sum_{\substack{j \ne i \\ 3 \le j \le m}}(-1)^{i+j}a_{1i}a_{2j}Pf(A_{12ij}).$$

The $Pf(A_{12})$ and $Pf(A_{12ij})$ are the same for all the settings, and so are precomputed. So finally we obtain the formulae:

- for $-m + 2 \leq c \leq m - 2$,

$$W(2k + 2c, 0) = \sum_{\substack{j \neq i \\ 3 \leq i, j \leq m}} \frac{(-1)^{i+j+1} Pf(A_{12ij})}{Pf(A_{12}) W(2,0)} a_{1i} a_{2j}$$

with $a_{1i} = W(k + c + m + 1 - i, 0)W(k + c + 1 + i - m, 0)$ and $a_{2j} = W(k + c + m - 1 - j, 0)W(k + c - 1 + j - m, 0)$;
- for $-m + 3 \leq c \leq m - 2$,

$$W(2k + 2c + 1, 0) = \sum_{\substack{j \neq i \\ 3 \leq i, j \leq m}} \frac{(-1)^{i+j+1} Pf(A_{12ij})}{Pf(A_{12}) W(1,0)} a_{1i} a_{2j}$$

with $a_{1i} = W(k + c + m + 1 - i, 0)W(k + c + 1 + i - m, 0)$ and $a_{2j} = W(k + c + m - j, 0)W(k + c + j - m, 0)$;
- for $-m + 3 \leq c \leq m - 2$,

$$W(2k + c, 1) = \sum_{\substack{j \neq i \\ 3 \leq i, j \leq m}} \frac{(-1)^{i+j+1} Pf(A_{12ij})}{Pf(A_{12}) W(-c, 1)} a_{1i} a_{2j}$$

with $a_{1i} = W(k + m - i, 1)W(k + i - m, 1)$ and $a_{2j} = W(k + c + m - j, 0)W(k + c + j - m, 0)$;

Remark 7. In genus 1, as $m = 4$, all the $Pf(A_{12ij})$ are equal to 1, as are the $W(1,0)$, $W(0,1)$ and $W(1,1)$. With these simplifications, we have the formulae given by Stange in [11].

So at each step of the double-and-add algorithm, the work is twofold : compute the coefficients a_{1i} and a_{2j}, $3 \leq i, j \leq m$, for all the settings $(l, 0, 0)$, $(l, -1, 0)$ and $(k, c, 1)$, and then put them in these formulae to compute all the terms of the block.

But, we have to notice that some coefficients are redundant:

- the a_{1j} of the setting $(l + 1, -1, 0)$ are the same than those of the setting $(l, 0, 0)$,
- the a_{2j} of the setting $(l, -1, 0)$ are the same than those of the setting $(l, 0, 0)$,
- the a_{2j} of the setting $(l + 2, -1, 0)$ are the a_{1j} of the setting $(l, 0, 0)$,
- the a_{1j} of the setting $(k, c, 1)$ for all c are the same,
- the a_{2j} of the setting $(k, c, 1)$ are the same than those of the setting $(k + c + 1, 0, 0)$.

So, the coefficients we still have to compute are:

- the a_{1j}, $3 \leq j \leq m$, for all the $2m - 4$ settings $(l, 0, 0)$;
- the a_{2j}, $3 \leq j \leq m$, for the settings $(k - m + 2, 0, 0)$ and $(k - m + 3, 0, 0)$ or $(k - m + 3, 0, 0)$ and $(k - m + 4, 0, 0)$;
- the a_{1j}, $3 \leq j \leq m$, for the setting $(k, -m + 3, 1)$.

6 The Hyperelliptic Net Algorithm in the Case $g \equiv 0, 3 \pmod 4$

6.1 Definitions of the Blocks and the Settings

The strategy is the same as in the case $g \equiv 1, 2 \pmod 4$: given a block centered in k, we build the block centered in $2k$ or $2k + 1$ by using theorem 6, and so by a double-and-add algorithm we finally obtain $W(r+1, 1)$. The difference with the previous section is that now W is even, and we don't have $W(0, 0) = 0$ anymore, so:

- the matrix $A = (W(v_i + v_j, w_i + w_j)W(v_i - v_j, w_i - w_j))_{1 \le i, j \le m}$ is now symmetric,
- we don't have to take m even, so we choose $m = 2^g + 1$,
- unfortunately, we don't have the tool pfaffian anymore but only the determinant, so the computations will be longer.

Theorem 11. *We define the initial terms as*

- *the $2m - 3$ terms $\{W(i, 0) \mid 0 \le i \le 2m - 4\}$,*
- *the $2m - 2$ terms $\{W(i, 1) \mid 0 \le i \le 2m - 3\}$.*

Let $k \in \mathbb{N}$. We define a block centered on k to be

- *the $4m - 6$ values $\{W(k + i, 0) \mid -2m + 4 \le i \le 2m - 3\}$,*
- *the $2m - 2$ values $\{W(k + i, 1) \mid 0 \le i \le 2m - 3\}$.*

If we have the initial terms and the block centered on k, then we can compute the block centered on $2k$ and the block centered on $2k + 1$.

Proof. In the matrix $A = (W(v_i + v_j, w_i + w_j)W(v_i - v_j, w_i - w_j))_{1 \le i, j \le m}$, we take the following:

- for all $1 \le i \le m$, $w_i = \epsilon/2$,
- $v_1 = k + l$,
- for $2 \le i \le m$, $v_i = m - i$.

These settings will be called **settings** $(1, 0, \epsilon)$, for $\epsilon = 0, 1, -m + 2 \le l \le m - 1$ if $\epsilon = 0$ and $0 \le l \le m - 1$ if $\epsilon = 1$. With them, we compute the $W(2k + 2l, \epsilon)$.

   Then, by the **settings** $(1, 1, \epsilon)$, $\epsilon = 0, 1, -m + 2 \le l \le m - 2$ if $\epsilon = 0$ and $0 \le l \le m - 2$ if $\epsilon = 1$:

- for all $1 \le i \le m$, $w_i = \epsilon/2$,
- $v_1 = k + l + 1/2$,
- for $2 \le i \le m$, $v_i = m - i + 1/2$,

we compute the $W(2k + 2l + 1, \epsilon)$.   □

As in the case $g \equiv 1, 2 \pmod 4$, these settings are "optimal".

6.2   Formulae

As in the genus $1, 2$ (mod 4) case, we now exhibit the formulae given by the settings we have defined.

For $1 \leq i, j, k, l \leq m$, let $\hat{A}_{i,j}$ denote the matrix A with the i-th row and the j-th column removed, and $\hat{A}_{ij,kl}$ be the matrix A with the i-th and j-th rows, and the k-th and l-th columns removed. Then the determinant of A can be developed as follows:

$$0 = a_{11} \det\left(\hat{A}_{1,1}\right) + \sum_{1 \leq i,j \leq m} (-1)^{i+j} a_{1i} a_{1j} \det\left(\hat{A}_{1i,1j}\right)$$

with $\det\left(\hat{A}_{1,1}\right)$ and the $\det\left(\hat{A}_{1i,1j}\right)$ precomputed. Unlike the $g \equiv 1, 2$ (mod 4) case, these determinants are different for each setting. So, we will note them with the letters B, C, D and E respectively for the settings $(l,0,0)$, $(l,1,0)$, $(l,0,1)$ and $(l,1,1)$:

- for $-m + 2 \leq l \leq m - 1$,

$$W(2k + 2l, 0) = \sum_{2 \leq i,j \leq m} \frac{(-1)^{i+j+1} \det\left(\hat{B}_{1j,1i}\right)}{\det\left(\hat{B}_{1,1}\right) W(0,0)} a_{1i} a_{1j}$$

with $a_{1i} = W(k + l + m - i, 0) W(k + l + i - m, 0)$;
- for $-m + 2 \leq l \leq m - 2$,

$$W(2k + 2l + 1, 0) = \sum_{2 \leq i,j \leq m} \frac{(-1)^{i+j+1} \det\left(\hat{C}_{1j,1i}\right)}{\det\left(\hat{C}_{1,1}\right) W(0,0)} a_{1i} a_{1j}$$

with $a_{1i} = W(k + l + m + 1 - i, 0) W(k + l + i - m, 0)$;
- for $0 \leq l \leq m - 1$,

$$W(2k + 2l, 1) = \sum_{2 \leq i,j \leq m} \frac{(-1)^{i+j+1} \det\left(\hat{D}_{1j,1i}\right)}{\det\left(\hat{D}_{1,1}\right) W(0,0)} a_{1i} a_{1j}$$

with $a_{1i} = W(k + l + m - i, 1) W(k + l + i - m, 0)$;
- for $0 \leq l \leq m - 2$,

$$W(2k + 2l + 1, 1) = \sum_{2 \leq i,j \leq m} \frac{(-1)^{i+j+1} \det\left(\hat{E}_{1j,1i}\right)}{\det\left(\hat{E}_{1,1}\right) W(0,0)} a_{1i} a_{1j}$$

with $a_{1i} = W(k + l + m + 1 - i, 1) W(k + l + i - m, 0)$.

There is an other unpleasant difference with the $g \equiv 1, 2$ (mod 4) case : we do not have any redundant coefficients among the a_{1i} and a_{2j}, and so we have to compute all of them.

7 Genus 3 Case

We now present the algorithm in genus 3 case. We suppose we have a genus 3 hyperelliptic curve

$$\mathcal{C} \; : \; y^2 = x^7 + \lambda_6 x^6 + \cdots + \lambda_0$$

defined on a field $\mathbb{F}_q$ of characteristic p. Let $\mathcal{J}$ be its jacobian, we are given two divisors D and D' by their Mumford representation:

$$D = [U, V] \text{ with } U = x^3 + U_2 x^2 + U_1 + U_0, \; V = V_2 x^2 + V_1 X + V_0,$$

$$D' = [U', V'] \text{ with } U' = x^3 + U'_2 x^2 + U'_1 + U'_0, \; V' = V'_2 x^2 + V'_1 X + V'_0.$$

The first thing to do is to generate the hyperelliptic net W associated to $\mathcal{C}$, D and D'. Then we use formulae of section 6.2 in an example.

7.1 Initialization

For $g = 3$, we have $m = 9$, so the initial values we need are the $\{W(i, 0) \mid i = 0, \ldots, 14\}$ and the $\{W(i, 1) \mid i = 0, \ldots, 15\}$.

We are able to set $W(1, 0) = W(0, 1) = 1$, and for the other values we use proposition 5:

$$\forall a, b, i \in \mathbb{Z}, \; \frac{W(a+b, i) W(a-b, i)}{W(a, i)^2 W(b, 0)^2} = \mathcal{F}_3([a] D_1 + [i] D_2, [b] D_1), \; \frac{W(2a, 0)}{W(a, 0)^4} = \mathcal{G}_3([a] D_1),$$

with the polynomials $\mathcal{F}_3$ and $\mathcal{G}_3$ given by (see [5] and [12]):

$$\mathcal{F}_3(u, v) = (\wp_{31}(v) - \wp_{31}(u))(\wp_{22}(v) - \wp_{22}(u)) - (\wp_{31}(v) - \wp_{31}(u))^2$$
$$+ (\wp_{32}(v) - \wp_{32}(u))(\wp_{21}(v) - \wp_{21}(u)) + (\wp_{33}(v) - \wp_{33}(u))(\wp_{11}(v) - \wp_{11}(u)),$$
$$\mathcal{G}_3(u) = \wp_{113}(u) \wp_{223}(u) + \wp_{133}(u) \wp_{122}(u) - 2 \wp_{133}(u) \wp_{113}(u)$$
$$- \wp_{123}(u)^2 - \wp_{233}(u) \wp_{112}(u) + \wp_{133}(u) \wp_{113}(u) + \wp_{333}(u) \wp_{111}(u).$$

We recall that the $\wp_{i3}$ and $\wp_{i33}$, $1 \le i \le 3$, of a divisor D are directly given by its Mumford coordinates (see remark 2). For the others $\wp$-functions, we use the formulae given in the appendix C of [8]. In details, we use the three first formulae to compute successively $\wp_{22}$, $\wp_{12}$ and $\wp_{11}$. Then we use the formulae of weight (-18) to compute the seven products $\wp_{ijk} \wp_{lmn}$ we need in the expression of $\mathcal{G}_3$.

7.2 An Example

Let $\mathcal{C}$ be the curve

$$y^2 = x^7 + 3x^6 + 2x^5 + 10x^4 + 9x^3 + 3x^2 + 11$$

defined over $\mathbb{F}_{29}$ and $\mathcal{J}$ its jacobian. With $r = 41$, the embedding degree is 40, so the exponent in the computation of the pairing is

$$e = \frac{29^{40} - 1}{41} = 7640751216319753516158033810725590182075467567587589888800.$$

Let $a \in \overline{\mathbb{F}}_{29}$ a root of

$$X^{40} + X^5 + 4 \in \mathbb{F}_{29}[X]$$

then $\mathbb{F}_{29^{40}} \simeq \mathbb{F}_{29}(a)$.

Let $D_1 \in \mathcal{J}(\mathbb{F}_{29})[41]$ be given by its Mumford representation $[x^3 + 2x^2 + 9x + 24, 23x^2 + 24x + 4]$. Let $D_2 \in \mathcal{J}(\mathbb{F}_{29^{40}})$ be:

$$U = x^3 + (27a^5 + 27a^2 + 28a + 16)x^2 + (4a^7 + 2a^6 + 21a^5 + 2a^3 + 21a^2 + 5a + 10)x + 25a^8 + 26a^7 + 24a^6 + 18a^5 + 24a^3 + 18a^2 + a + 8$$

$$\begin{aligned}
V = &(26a^{39} + 16a^{38} + 14a^{37} + 7a^{36} + 27a^{35} + 19a^{34} + 7a^{33} + 19a^{32} + 15a^{31} + a^{30} + \\
&21a^{29} + 2a^{28} + 5a^{27} + 22a^{26} + 27a^{25} + 21a^{24} + 4a^{23} + 5a^{22} + 6a^{21} + 27a^{20} + 6a^{19} + \\
&27a^{18} + 13a^{17} + 12a^{16} + 15a^{15} + 10a^{14} + 23a^{13} + 23a^{12} + 25a^{11} + 2a^{10} + 4a^9 + 14a^8 + \\
&a^{26} + 21a^{25} + 26a^{24} + 28a^{23} + 10a^{22} + 14a^{21} + 3a^{20} + 23a^{19} + 14a^{18} + 26a^{17} + 7a^{16} + \\
&23a^7 + 3a^6 + 28a^5 + 2a^4 + 26a^3 + 12a^2 + 27a + 4)x^2 + (a^{39} + 15a^{38} + 6a^{37} + 13a^{36} + \\
&20a^{35} + 5a^{34} + 23a^{33} + 28a^{32} + 8a^{31} + 20a^{30} + 16a^{29} + a^{28} + 12a^{27} + 13a^{15} + 28a^{14} + \\
&15a^{13} + 25a^{12} + a^{11} + 19a^{10} + 11a^9 + 17a^8 + 21a^7 + 11a^6 + 12a^5 + 16a^4 + 8a^2 + 21a + \\
&22)x + 15a^{39} + 14a^{38} + 23a^{37} + 25a^{36} + 27a^{35} + 14a^{34} + 13a^{33} + 10a^{31} + 3a^{30} + \\
&14a^{29} + 13a^{28} + 27a^{27} + 14a^{26} + 21a^{25} + 13a^{24} + 8a^{23} + 25a^{22} + 27a^{21} + 23a^{19} + \\
&24a^{18} + 11a^{17} + 6a^{16} + a^{15} + 3a^{14} + 18a^{13} + a^{12} + 21a^{11} + 2a^{10} + 26a^9 + \\
&26a^8 + 11a^7 + 4a^6 + 18a^5 + 9a^4 + 28a^3 + 5a^2 + 4a + 6).
\end{aligned}$$

We denote $W_{D1,D2}$ the hyperelliptic net build as in section 7.1. Then we compute:

$$(W_{D1,D2}(42, 1))^e = 4a^{39} + 14a^{38} + 21a^{37} + \cdots + 5a^2 + 4a + 16.$$

We have to verify we obtain have a well defined bilinear map

$$\tau : \mathcal{J}(\mathbb{F}_q)[r] \times \mathcal{J}(\mathbb{F}_{q^k})/r\mathcal{J}(\mathbb{F}_{q^k}) \to \qquad \mu_r$$
$$(D_1, D_2) \qquad\qquad \mapsto (W_{D_1,D_2}(r + 1, 1))^e$$

- we choose a random $D_3 \in \mathcal{J}(\mathbb{F}_{q^k})$ and verify that

$$(W_{D_1,D_2+rD_3}(r + 1, 1))^e = (W_{D_1,D_2}(r + 1, 1))^e;$$

- we choose random integers m and n and verify that

$$(W_{mD_1,nD_2}(r + 1, 1))^e = (W_{D_1,D_2}(r + 1, 1))^{emn}.$$

8   Conclusion

The point of this article is to extend the algorithm to compute pairings using hyperelliptic nets to all genus. This algorithm can be an interesting alternative to the traditional Miller method. In particular, the hyperelliptic nets can take advantage of all the improvements made on Miller algorithm in the last decade : computation of pairing with shorter length of loop as the ate pairing, exploitation of efficiently computable automorphisms, or use of degenerate divisors... The effects of all these technics on hyperelliptic nets have to be study in future work.

References

1. Avanzi, R., Cohen, H., Doche, C., Frey, G., Lange, T., Nguyen, K., Vercauteren, F.: Handbook of elliptic and hyperelliptic curve cryptography. Discrete Mathematics and Its Applications, vol. 34, pp. 115–123. CRC Press (2006)
2. Balakrishnan, J., Belding, J., Chisholm, S., Eeisenträger, K., Stange, K.E., Teske, E.: Pairings on hyperelliptic curves. ArXiv e-prints, 09083731 (2009)
3. Blake, I., Seroussi, G., Smart, N.: Advances in elliptic curve cryptography. London Mathematical Society Lecture Note Series, vol. 317, pp. 183–212. Cambridge University press (2005)
4. Buchstaber, V., Enolskii, V.: Explicit algebraic description of hyperelliptic jacobians on the basis of the Klein σ-functions. Functional Analysis and Its Applications 30(1), 44–47 (1996)
5. Buchstaber, V., Enolskii, V., Leykin, D.: A recursive family of differential polynomials generated by the Sylvester identity and additions theorems for hyperelliptic Kleinian functions. Functional Analysis and Its Applications 34(4), 240–251 (1997)
6. Buchstaber, V., Enolskii, V., Leykin, D.: Hyperelliptic Kleinian functions and applications. Solitons Geometry and Topology: On the Crossroad, Advances in Math. Sciences, Am. Math. Soc. Transl, Series 2 179, 1–34 (1997)
7. Cantor, D.: Computing in the Jacobian of a hyperelliptic curve. Mathematics of Computation 48(177), 95–101 (1987)
8. Eilbeck, J., England, M., Ônishi, Y.: Abelian functions associated with genus three algebraic curves. LMS J. Comput. Math. 14, 291–326 (2011)
9. Ogura, N., Kanayama, N., Uchiyama, S., Okamoto, E.: Cryptographic pairings based on elliptic nets. In: Iwata, T., Nishigaki, M. (eds.) IWSEC 2011. LNCS, vol. 7038, pp. 65–78. Springer, Heidelberg (2011)
10. Silverman, J.: The arithmetic of elliptic curves, pp. 157–178. Springer, New York (1985)
11. Stange, K.E.: The Tate pairing via elliptic nets. In: Takagi, T., Okamoto, T., Okamoto, E., Okamoto, T. (eds.) Pairing 2007. LNCS, vol. 4575, pp. 329–348. Springer, Heidelberg (2007)
12. Uchida, Y.: Division polynomials and canonical local heights on hyperelliptic Jacobians. Manuscrypta Mathematica 134(3-4), 273–308 (2011)
13. Uchida, Y., Uchiyama, S.: The Tate-Lichtenbaum pairing on a hyperelliptic curve via hyperelliptic nets. In: Abdalla, M., Lange, T. (eds.) Pairing 2012. LNCS, vol. 7708, pp. 218–233. Springer, Heidelberg (2013)

New Speed Records for Montgomery Modular Multiplication on 8-Bit AVR Microcontrollers

Zhe Liu and Johann Großschädl

University of Luxembourg,
Laboratory of Algorithmics, Cryptology and Security (LACS),
6, rue Richard Coudenhove-Kalergi, L–1359 Luxembourg
{zhe.liu,johann.groszschaedl}@uni.lu

Abstract. Modular multiplication of large integers is a performance-critical arithmetic operation of many public-key cryptosystems such as RSA, DSA, Diffie-Hellman (DH) and their elliptic curve-based variants ECDSA and ECDH. The computational cost of modular multiplication and related operations (e.g. exponentiation) poses a practical challenge to the widespread deployment of public-key cryptography, especially on embedded devices equipped with 8-bit processors (smart cards, wireless sensor nodes, etc.). In this paper, we describe basic software techniques to improve the performance of Montgomery modular multiplication on 8-bit AVR-based microcontrollers. First, we present a new variant of the widely-used hybrid method for multiple-precision multiplication that is 10.6% faster than the original hybrid technique of Gura et al. Then, we discuss different hybrid Montgomery multiplication algorithms, including Hybrid Finely Integrated Product Scanning (HFIPS), and introduce a novel approach for Montgomery multiplication, which we call Hybrid Separated Product Scanning (HSPS). Finally, we show how to perform the modular subtraction of Montgomery reduction in a regular fashion without execution of conditional statements so as to counteract Simple Power Analysis (SPA) attacks. Our AVR implementation of the HFIPS and HSPS method outperforms the Montgomery multiplication of the MIRACL Crypto SDK by up to 21.58% and 14.24%, respectively, and is twice as fast as the modular multiplication of the TinyECC library.

Keywords: AVR architecture, multi-precision arithmetic, hybrid multiplication, modular reduction, SPA countermeasure.

1   Introduction

Long integer modular arithmetic, in particular modular multiplication, is at the heart of many practical public-key cryptosystems, including "traditional" ones that operate in a large ring or group (e.g. RSA [23], DSA [22], Diffie-Hellman [7]), as well as elliptic curve schemes (e.g. ECDSA [22], ECDH [14]) if they use a prime field $\mathbb{F}_p$ as underlying algebraic structure. The major operation of the former class of cryptosystems is exponentiation in either $\mathbb{Z}_n$ or $\mathbb{Z}_p^*$, which can be carried out through modular multiplications and modular squarings [9]. On the

D. Pointcheval and D. Vergnaud (Eds.): AFRICACRYPT 2014, LNCS 8469, pp. 215–234, 2014.

other hand, elliptic curve schemes perform scalar multiplication in an additive group, an operation that in turn is composed of additions, multiplications, and inversions in the underlying field [14]. However, most software implementations use projective coordinates to represent points on the curve, thereby trading inversions for multiplications in $\mathbb{F}_p$ to reduce the overall execution time. In this case, the performance of a scalar multiplication is primarily determined by the efficiency of the multiplication in the prime field $\mathbb{F}_p$. Modular multiplication is also a performance-critical arithmetic operation of pairing-based cryptosystems (e.g. identity-based encryption, short signature schemes) [3].

It is common practice in Elliptic Curve Cryptography (ECC) to use primes of a "special" form so as to facilitate the modular reduction [14]. A well-known example are pseudo-Mersenne primes, i.e. primes that are slightly smaller than a power of two and can be written as $p = 2^n - c$ where c is typically chosen to fit into a single register of the target processor. The computational complexity of reduction modulo such primes grows linearly with their length, whereas the reduction operation for general primes has quadratic complexity [14]. A second example of primes that allow one to perform a reduction in linear time are the so-called generalized-Mersenne primes, which are standardized by the National Institute of Standards and Technology (NIST) [22]. Software implementations of ECC often follow a dual approach and support both fast modular reduction techniques for a small set of special primes (e.g. the NIST primes) and a generic reduction routine for "arbitrary" primes. Many cryptographic libraries, such as TinyECC [18] and OpenSSL, take this approach to combine high performance with high flexibility. Therefore, generic modular multiplication techniques, like those introduced by Barrett [4] and Montgomery [21] roughly 30 years ago, are not only important for RSA but also for ECC.

Formally, a modular multiplication $A \cdot B \bmod M$ involves multiplying two n-bit operands A and B, yielding a $2n$-bit product $P = A \cdot B$, followed by the reduction of P modulo M to get a final result in the range of $[0, M - 1]$. The latter operation, i.e. the reduction of P with respect to a given modulus M, has a major impact on the execution time of a modular multiplication. A straightforward way to obtain the residue $P \bmod M$ is to divide P by M and find the remainder of this division. However, performing integer division in software is extremely expensive for large operands, which makes this approach unpractical for cryptographic applications. In 1985, Peter Montgomery [21] introduced an efficient (and nowadays widely-used) technique to accomplish a modular reduction without trial division. The basic idea is to replace the modular reduction $P \bmod M$ by a computation of the form $P \cdot 2^{-n} \bmod M$ (where n denotes the bitlength of M), which is much cheaper than computing the actual residue via division. In general, when implemented in software, the Montgomery reduction of a $2n$-bit product P with respect to an n-bit modulus M is just slightly more costly than the multiplication of two n-bit operands [10].

The efficient implementation of multiplication, reduction and other computation-intensive arithmetic operations is particularly challenging for embedded processors with limited resources. The root of the problem is the length of the operands (e.g. 160 bits for an elliptic curve cryptosystem, 1024 bits in the case of RSA), which exceeds the word-size of a small 8 or 16-bit processor by up to two orders of magnitude. Recent research in the area of long-integer arithmetic for such processors focused on the 8-bit AVR architecture [1] (e.g. ATmega128 [2]) as target platform. In 2004, Gura et al published a landmark paper [13] on optimizing modular arithmetic for AVR processors in which they introduce the idea of *hybrid multiplication*. By exploiting the large register file to store (parts of) the operands, the hybrid method allows for a considerable reduction of the number of load instructions compared to a conventional (i.e. column-wise) implementation of multiple-precision multiplication [6,13]. Gura et al reported an execution time of 3106 clock cycles for a (160×160)-bit multiplication on the ATmega128, a result that was subsequently further improved by Uhsadel et al (2881 cycles [27]), Liu et al (2865 cycles [19]), Zhang et al (2845 cycles [32]), as well as Scott et al (2651 cycles with "unrolled" loops [24]).

In this paper, we continue the line of research described above and advance the state-of-the-art in efficient modular arithmetic for 8-bit AVR processors in three directions. First, we introduce a new variant of the hybrid multiplication technique that is roughly 10% faster than Gura et al's original hybrid method [13]. Our hybrid technique is similar to the one of Zhang et al [32], but benefits from better register allocation and reduced loop overhead (i.e. improved initialization of pointers and more efficient testing of branch conditions). Thanks to our sophisticated register allocation, only 30 (out of 32) AVR working registers are actually occupied during execution of a hybrid multiplication, which allows for easy integration of Montgomery reduction[1]. The second contribution of this paper is a comprehensive performance analysis and comparison of six methods for software implementation of Montgomery multiplication; five are described in [17] and the sixth variant is from [19]. Our results shed some new light on the relative performance of the different Montgomery multiplication methods since they contradict the findings of the current literature, e.g. [17]. Finally, as third contribution, we describe how to perform the final subtraction of M (which is required when a Montgomery product is not fully reduced) in a regular fashion so as to thwart side-channel attacks [20]. Our approach tolerates incompletely-reduced operands and ensures that always the same sequence of instructions is executed, regardless of the actual value of the Montgomery product.

2 Montgomery Modular Multiplication

Montgomery multiplication (named after Peter Montgomery) was originally introduced in 1985 [21] and has since then become one of the most-widely used techniques for high-speed implementation of modular multiplication [8]. In the

---

[1] The integration of Montgomery reduction into hybrid multiplication (using e.g. the so-called FIOS or FIPS method [17]) can significantly increase the register pressure since two registers are necessary to accommodate the 16-bit pointer to the modulus M. We designed our hybrid multiplication to take this into account by leaving two registers for M, which helps to prevent register spills in the FIPS inner loop.

Algorithm 1. Calculation of the Montgomery product

Input: An odd n-bit modulus M, Montgomery radix $R = 2^n$, two operands A, B in
the range $[0, M - 1]$, and pre-computed constant $M' = -M^{-1} \bmod R$
Output: Montgomery product $Z = \mathrm{MonPro}(A, B) = A \cdot B \cdot R^{-1} \bmod M$

1: $T \leftarrow A \cdot B$
2: $Q \leftarrow T \cdot M' \bmod R$
3: $Z \leftarrow (T + Q \cdot M)/R$
4: **if** $Z \geq M$ **then** $Z \leftarrow Z - M$ **end if**
5: **return** Z

---

following, we use M to denote an odd modulus consisting of n bits and A, B to
denote two residues modulo M, i.e. $0 \leq A, B < M$. Rather than computing the
residue of $A \cdot B \bmod M$ directly, Montgomery's algorithm returns the so-called
Montgomery product of A and B as result, which is defined as follows.

$$\mathrm{MonPro}(A, B) = A \cdot B \cdot R^{-1} \bmod M \tag{1}$$

The factor R in Equation (1) is often referred to as *Montgomery radix* and can
be any integer that is bigger than M and relatively prime to it, i.e. R needs to
satisfy $\gcd(N, R) = 1$. However, for reasons of implementation efficiency, R is
in general a power of two, e.g. $R = 2^n$. The central idea of Montgomery multiplication is to replace the reduction modulo M (which would normally require
a costly division by M) by a division by R and a reduction mod R, which are
cheap operations when R is a power of two. More precisely, a division by 2^n is
merely an n-bit right-shift operation, while a reduction modulo 2^n requires the
truncation of all high-order bits above the n-th position. Algorithm 1 specifies
the computation of the Montgomery product in detail. In addition to the three
operands A, B, and M, the algorithm needs M' as input, which is the inverse
of $-M$ (or, more precisely, the inverse of $R - M$) modulo R. However, M' can
be pre-computed (using e.g. the Euclidean algorithm as described in [17]) since
it depends only on M and R, i.e. M' is fixed for a given M.

Based on Algorithm 1, the Montgomery product $A \cdot B \cdot R^{-1} \bmod M$ can be
obtained as follows. First, the n-bit operand A is multiplied by n-bit operand
B, giving a $2n$-bit product T. Then, in line 2, the quotient $Q = -\frac{T}{M} \bmod R$ is
calculated, which is simply a multiplication of the low-order n bits of T by the
pre-computed constant $M' = -M^{-1} \bmod R$ [8]. Note that we actually need to
calculate only the lower half (i.e. the n least significant bits) of $T \cdot M'$ because
our Montgomery radix R is 2^n. In line 3, a multiplication and a division by R is
performed; the latter is just an n-bit right-shift since $R = 2^n$. Thus, we have to
calculate only the upper half of the product $Q \cdot M$. The n least significant bits
of $T + Q \cdot M$ are 0, which means the division by R (i.e. the n-bit right-shift) in
line 3 does not destroy any information. The result Z obtained so far may be
not fully reduced (i.e. Z may not be the least non-negative residue modulo M)
so that a "final subtraction" of M becomes necessary (line 4). In summary, the
computational cost of Algorithm 1 amounts to one conventional multiplication
of n-bit operands (line 1) and two "half" multiplications where only either the

lower part (line 2) or the upper part (line 3) of the product is really needed. As a consequence, computing the Montgomery product is just slightly more costly than two conventional multiplications.

Software implementations of Algorithm 1 generally store the large integers A, B, and M in arrays of single-precision words (i.e. arrays of unsigned int in C and similar programming languages). Assuming a processor with a word-size of w bits, an n-bit integer X consists of $s = \lceil n/w \rceil$ single-precision (i.e. w-bit) words. Throughout this paper, we will use uppercase letters to represent large integers, whereas lowercase letters, usually with a numerical index, will denote individual w-bit words. The most and least significant word of an integer X are x_{s-1} and x_0, respectively, i.e. we have $X = (x_{s-1}, \ldots, x_1, x_0)$. There exist several implementation options and optimization techniques to efficiently perform a Montgomery multiplication in software; they can be categorized according to the order in which the words of the operands (resp. product) are accessed and whether multiplication and modular reduction are carried out *separately* or in an *integrated* fashion (see e.g. [17] for details). In brief, when using the so-called *operand scanning* method, the words of the operands are loaded sequentially, in ascending order, starting with the least significant word. On the other hand, the main characteristic of the *product scanning* technique is that each word of the result is stored (i.e. written to memory) only once, which happens in ascending order [6]. Both methods can be used to implement Montgomery multiplication in either a separated way (i.e. the modular reduction is accomplished after the multiplication) or an integrated way by alternating multiplication and reduction steps. In the latter case, we can further distinguish between a coarse and a fine integration of multiplication and modular reduction. Combinations of all these techniques allow for a multitude of algorithms for calculating the Montgomery product, six of which we briefly describe in the following subsections.

2.1 Separated Operand Scanning (SOS)

In Koç et al's original description of the SOS method, both the multiplication and the reduction are carried out according to the operand-scanning technique [17]. The inner loop of the multiplication (and also that of the reduction) performs operations of the form $(u, v) \leftarrow a \cdot b + c + d$, whereby a, b, c, and d are single-precision integers (i.e. w-bit words) and (u, v) denotes a double-precision (i.e. $2w$-bit) quantity. Each execution of this inner loop on a general-purpose RISC processor, e.g. the ATmega128, involves a mul and four add (resp. adc) instructions[2]. Assuming s-word operands, the operand-scanning multiplication of the SOS method executes s^2 mul, $4s^2$ add (or adc), $2s^2 + s$ load, as well as $s^2 + s$ store instructions (see Algorithm 1 in [10] for a detailed analysis). The original operand-scanning approach for Montgomery reduction as described in

---

[2] Note that we count the number of add *instructions* (in the same way as [10]), while Koç et al [17] assess the number of add *operations*. Adding a single-precision word to a double-precision quantity (u, v) counts for one add operation, but requires two add instructions, one of which is actually an adc (add-with-carry).

Algorithm 2. Montgomery reduction (operand scanning form)

Input: An s-word modulus $M = (m_{s-1}, \ldots, m_1, m_0)$, operand $P = (p_{2s-1}, \ldots, p_1, p_0)$
  with $P < 2M - 1$, and pre-computed constant $m_0' = -m_0^{-1} \bmod 2^w$
Output: Montgomery residue $Z = P \cdot 2^{-n} \bmod M$

1: $t \leftarrow 0$
2: **for** i from 0 by 1 to $s - 1$ **do**
3:    $u \leftarrow 0$
4:    $q \leftarrow p_i \cdot m_0' \bmod 2^w$
5:    **for** j from 0 by 1 to $s - 1$ **do**
6:       $(u, v) \leftarrow m_j \cdot q + p_{i+j} + u$
7:       $p_{i+j} \leftarrow v$
8:    **end for**
9:    $(u, v) \leftarrow p_{i+s} + u + t$
10:   $p_{i+s} \leftarrow v$
11:   $t \leftarrow u$
12: **end for**
13: **for** j from 0 by 1 to $s - 1$ **do**
14:   $z_j \leftarrow p_{j+s}$
15: **end for**
16: $z_s \leftarrow t$
17: **if** $Z \geq M$ **then** $Z \leftarrow Z - M$ **end if**

Section 4 of [17] employs a special ADD function to propagate a carry bit up to the most significant word. Our implementation simply holds the carry bit in an extra register t and adds it in the next iteration of the outer loop as shown in Algorithm 2. In this way, the operand-scanning form of Montgomery reduction consists of $s^2 + s$ mul, $4s^2 + 2s$ add or adc, $2s^2 + 2s + 1$ load, and $s^2 + 2s + 1$ store instructions, which means the SOS method (excluding final subtraction) needs to execute $2s^2 + s$ mul, $8s^2 + 2s$ add (resp. adc), $4s^2 + 3s + 1$ load, and $2s^2 + 3s + 1$ store instructions altogether.

2.2   Finely Integrated Product Scanning (FIPS)

The FIPS method (Algorithm 1 in [11]), originally introduced in [8], performs multiplication and reduction steps in a "finely" interleaved fashion in the same inner loop. From an algorithmic viewpoint, the FIPS technique consists of two nested loops; both inner loops compute (parts of) the product $A \cdot B$ and then add (parts of) the product $Q \cdot M$ to it. After the first inner loop, a word of the quotient Q is calculated with help of the least-significant word of M' (i.e. the pre-computed constant $m_0' = -m_0^{-1} \bmod 2^w$ [17]) and temporarily stored in the array of the final result. The least-significant word of the intermediate sum obtained at the end of the second inner loop is always zero, which means it can be right-shifted by w bits without "destroying" any information. In each iteration of the second outer loop, a word of the result (i.e. the Montgomery product) is obtained and written to memory. Note that this result consists of $s + 1$ words (whereby the MSW is either 0 or 1) since it may be incompletely reduced.

In each iteration of one of the inner loops, two multiply-accumulate (MAC) operations of the form $(t, u, v) \leftarrow (t, u, v) + a \cdot b$ are carried out, i.e. two words are multiplied and the double-precision product is added to a cumulative sum held in the three registers v, u and t. Note that Koç et al [17] employ a special ADD function to process carries (similar to the SOS method), but we avoid this by using three registers to hold the cumulative sum. The inner-loop operation of our FIPS method is identical to that of the product-scanning multiplication [14] and needs one mul and three add instructions. In total, the FIPS method requires $2s^2 + s$ mul, $6s^2$ add/adc, $4s^2 - s$ load, and $2s + 1$ store instructions altogether (excluding final subtraction) [10].

2.3 Coarsely Integrated Operand Scanning (CIOS)

Instead of computing the complete multiplication first and doing the reduction afterwards (like in Section 2.1), the CIOS method performs multiplication and reduction in an interleaved fashion, similar to Section 2.2. Algorithm 4 in [10] describes the CIOS method in detail; it consists of an outer loop that contains two inner loops. The first inner loop calculates parts of the product $A \cdot B$ and stores the intermediate result in an array in RAM. After the first inner loop, a word of the quotient Q is determined, which is subsequently used in the second inner loop to get a multiple of M to be added to the intermediate result. This addition zeroes out the least significant word of the intermediate result and so contributes to the modular reduction. A w-bit right-shift operation is implicitly performed in the second inner loop through indexing, i.e. by writing a word with index i to the $(i-1)$-th position in the target array. The two inner loops execute exactly the same operation as the SOS method, namely a computation of the form $(u, v) \leftarrow a \cdot b + c + d$. We eventually obtain a result that consists of $s + 1$ words (with the most-significant word being either 0 or 1), which means a final subtraction of M may be necessary to get a fully reduced result [17]. In total, the CIOS method requires $2s^2 + s$ mul, $8s^2 + 4s$ add, $4s^2 + 5s$ load, and $2s^2 + 3s$ store instructions (see [10] for further details[3]).

2.4 Coarsely Integrated Hybrid Scanning (CIHS)

This method, introduced in [17, Section 8], is related to both the SOS and the CIOS approach sketched before. It is called "hybrid scanning" method because it mixes operand scanning and product scanning for multiplication, while the reduction operation is accomplished solely in operand-scanning form. The CIHS method consists of two outer loops and three inner loops. The first outer loop computes a part of the product $A \cdot B$, while the second outer loop contributes to the reduction operation and the rest of the multiplication. Furthermore, the second outer loop shifts the intermediate result one word (i.e. by w bits) to the right in each iteration. The "splitting" of the multiplication is possible since, in

---

[3] Note that the number of add (resp. adc) instructions for the CIOS method specified in Table 4 of [10] is wrong; the correct number is $8s^2 + 4s$ for s-word operands.

the course of Montgomery modular reduction, the variable m computed at the beginning of the second outer loop only depends on t_0. The operation executed by the first two inner loops is exactly the same as that of the SOS and CIOS method, respectively. However, the third inner loop is slightly simpler because it performs an operation of the form $(u, v) \leftarrow a \cdot b + c$, each execution of which costs one mul and two add (resp. adc) instructions. Putting it all together, the CIHS method requires $2s^2 + s$ mul, $9s^2 + 5s$ add/adc, $11s^2/2 + 7s/2$ load, as well as $3s^2 + 2s$ store instructions (excluding the final subtraction).

2.5 Finely Integrated Operand Scanning (FIOS)

The last operand-scanning variant of Montgomery multiplication we discuss in this paper is the Finely Integrated Operand Scanning (FIOS) method, given in [12, Algorithm 1]. Compared to the four methods outlined before, the structure of this algorithm is very simple as it comprises just an outer loop with a single inner loop. The inner loop of the FIOS variant described in [12] executes two operations of the form $(u, v) \leftarrow a \cdot b + c + d$, one contributes to the calculation of the product of A and B, and the other to the Montgomery reduction of this product. Similar to the CIOS method, the quality of the implementation of the inner-loop operation has a major impact on the algorithm's overall execution time. In summary, the FIOS method of Montgomery multiplication requires to perform $2s^2 + s$ mul, $8s^2$ add, $3s^2 + 4s$ load, and $s^2 + s$ store instructions.

2.6 Separated Product Scanning (SPS)

The Montgomery multiplication methods sketched in the previous five subsections were first described and analyzed by Koç et al [17]. In this subsection, we present a sixth method, which we call *Separated Product Scanning (SPS)*. The SPS method separates multiplication steps and reduction steps (similar to the SOS method), i.e. the Montgomery reduction is carried out as a self-contained operation *after* the multiplication. As its name suggests, the SPS technique is based on the product scanning approach for multiplication (see Algorithm 2 in [10]) and then uses the product-scanning form of Montgomery reduction shown in Algorithm 3. More details on this product-scanning based Montgomery reduction can be found in [10,19]. The SPS method was originally introduced in [19] as a product-scanning variant of the SOS technique, but we feel that the name "Separated Product Scanning" better denotes the characteristics of this method. As per [10], a product-scanning multiplication of two s-word operands consists of s^2 mul, $3s^2$ add, $2s^2$ load, and $2s$ store instructions. Algorithm 3 requires $s^2 + s$ mul, $3s^2 + 6s$ add, $2s^2 + 2s$ load, and $2s + 1$ store instructions [10], which amounts to $2s^2 + s$ mul, $6s^2 + 6s$ add (or adc), $4s^2 + 2s$ load, and $4s + 1$ store instructions for the complete SPS method.

2.7 Analysis and Comparison

Table 1 summarizes and compares the base instruction counts of all six Montgomery multiplication techniques considered in this section. The two variants

---

Algorithm 3. Montgomery reduction (product scanning form) [10, Algorithm 5]

Input: An s-word modulus $M = (m_{s-1}, \ldots, m_1, m_0)$, a product P in the range of $[0, 2M - 2]$, pre-computed constant $m_0' = -m_0^{-1} \bmod 2^w$

Output: Montgomery residue $Z = P \cdot 2^{-n} \bmod M$

1: $(t, u, v) \leftarrow 0$
2: **for** i from 0 by 1 to $s - 1$ **do**
3:     **for** j from 0 by 1 to $i - 1$ **do**
4:        $(t, u, v) \leftarrow (t, u, v) + z_j \cdot m_{i-j}$
5:     **end for**
6:     $(t, u, v) \leftarrow (t, u, v) + p_i$
7:     $z_i \leftarrow v \cdot m_0' \bmod 2^w$
8:     $(t, u, v) \leftarrow (t, u, v) + z_i \cdot m_0$
9:     $v \leftarrow u, \; u \leftarrow t, \; t \leftarrow 0$
10: **end for**
11: **for** i from s by 1 to $2s - 2$ **do**
12:     **for** j from $i - s + 1$ by 1 to $s - 1$ **do**
13:        $(t, u, v) \leftarrow (t, u, v) + z_j \cdot m_{i-j}$
14:     **end for**
15:     $(t, u, v) \leftarrow (t, u, v) + p_i$
16:     $z_{i-s} \leftarrow v$
17:     $v \leftarrow u, \; u \leftarrow t, \; t \leftarrow 0$
18: **end for**
19: $(t, u, v) \leftarrow (t, u, v) + p_{2s-1}$
20: $z_{s-1} \leftarrow v, \; z_s \leftarrow u$
21: **if** $Z \geq M$ **then** $Z \leftarrow Z - M$ **end if**

Table 1. Comparison of base instructions for Multiplication modular multiplications (excluding final subtraction)

Algorithm	# mul	# add	# load	# store
FIPS	$2s^2 + s$	$6s^2$	$4s^2 - s$	$2s + 1$
SPS	$2s^2 + s$	$6s^2 + 6s$	$4s^2 + 2s$	$4s + 1$
CIOS	$2s^2 + s$	$8s^2 + 4s$	$4s^2 + 5s$	$2s^2 + 3s$
SOS	$2s^2 + s$	$8s^2 + 2s$	$4s^2 + 3s + 1$	$2s^2 + 3s + 1$
CIHS	$2s^2 + s$	$9s^2 + 5s$	$11s^2/2 + 7s/2$	$3s^2 + 2s$
FIOS	$2s^2 + s$	$8s^2$	$3s^2 + 4s$	$s^2 + s$

based on the product-scanning method (i.e. FIPS and SPS) execute multiply-accumulate operations of the form $(t, u, v) \leftarrow (t, u, v) + a \cdot b$ in the inner loops [10], whereby each operation involves three add or adc instructions to add the product $a \cdot b$ to a cumulative sum. Consequently, the FIPS and SPS technique execute three add (resp. adc) per one mul instruction. On the other hand, the operand-scanning variants feature a common inner-loop operation of the form $(u, v) \leftarrow a \cdot b + c + d$, which costs four add/adc per mul instruction. A second major difference between the product-scanning variants and their counterparts based on the operand-scanning technique is the number of store instructions as shown in the last column of Table 1. The former execute store instructions solely in the outer loops, whereas the latter perform stores in the inner loop(s) [10]. Therefore, the number of store instructions carried out by FIPS and SPS increases linearly with the number of words. The operand-scanning variants, on the other hand, exhibit a quadratic growth of the number of stores.

Our analysis of the base instructions indicates a clear advantage of the two product-scanning methods, which will be confirmed by implementation results in Section 4. However, our analysis is not in agreement with that of Koç et al [17], who clearly identified the CIOS method as the most efficient one on basis of both their theoretical cost model and measured results. As stated in Section 2.1, this deviation can be explained by differences in the underlying cost model since Koç et al consider the number of basic operations, whereas we count the number of basic instructions as this is more accurate. Furthermore, Koç et al use a special ADD function to propagate carries in their SOS, FIOS, and FIPS method, which we do not need since we hold all carries in registers.

3   Our Implementation

In this section, we first introduce a novel variant of the hybrid multiplication method, which saves 10.6% execution time compared to the original one from [13]. Then, we combine our hybrid multiplication with Montgomery's algorithm to obtain different variants of a hybrid Montgomery multiplication. Finally, we describe an efficient implementation of the conditional subtraction of M.

3.1   Optimized Hybrid Multiplication

A straightforward implementation of the product-scanning method processes a single word of operand A and operand B at a time; therefore, in each iteration of the inner loop, a word of each A and B is loaded from RAM, multiplied, and added to a cumulative sum [6]. Gura et al [13] observed that the performance of the product-scanning method can be significantly improved if several words of the operands are processed in each iteration. This approach is, in essence, a special form of loop unrolling and particularly efficient on processors featuring a large number of registers. Taking the 8-bit AVR platform [1] as example, we can easily process $d = 4$ (or even $d = 5$) bytes of the operands at a time, and so reduce the number of loop iterations by a factor of d. In each iteration of the inner loop, four bytes (i.e. 32 bits) of A and B are loaded from memory and multiplied together to yield a 8-byte (i.e. 64-bit) result, which is then added to a cumulative sum held in nine registers. Gura et al used the operand-scanning approach for the 4-byte-by-4-byte (i.e. $(32 \times 32\text{-bit})$-bit) multiplications in the inner loop as illustrated on the left of Figure 1. This multiplication technique is referred to as "hybrid multiplication" because it combines product scanning in the outer loop with operand scanning in the inner loop(s). The main advantage of hybrid multiplication is a reduced number of load instructions compared to the straightforward product-scanning method (see [13] for details).

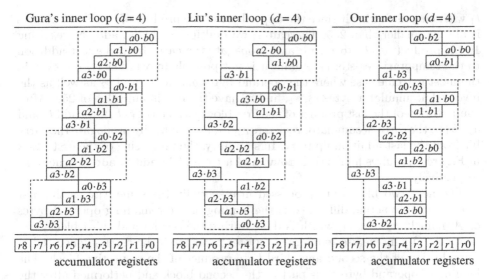

Fig. 1. Comparison of inner-loop operation for hybrid multiplication

In recent years, there have been several attempts to improve the inner-loop operation of the hybrid method, taking the properties of the AVR architecture into account[4]. For example, Liu et al re-arranged in [19] the order of the multiplications in the inner loop (depicted in the middle of Figure 1), which allowed them to decrease the number of mov (resp. movw) instructions compared to the original hybrid method. Scott et al [24] used so-called "carry catcher" registers to limit the propagation of carries and totally unrolled the loops to achieve an extra speed-up. Our implementation of the inner loop, shown on the right side of Figure 1, is inspired by both Liu et al and Scott et al. Just like Liu et al, we schedule the mul instructions in a special order with the goal of reducing the computational cost of the inner loop. If we assume $d = 4$, the 16 byte products are calculated as shown in Figure 1, whereby the execution time elapses from top to bottom, i.e. $a_0 \cdot b_2$ is the first byte product we generate and $a_3 \cdot b_2$ the last. Our variant of the inner-loop operation borrows the idea of catching carry bits from [24], but we do not use separate registers for that purpose.

To simplify the explanation of our inner loop, we split the 16 byte-products into four blocks, indicated by dashed boxes in Figure 1. At the beginning, four bytes of operand B (labeled b_0, b_1, b_2 and b_3 in Figure 1) along with two bytes of A (namely a_0 and a_1) are loaded from RAM. We first multiply a_0 by b_2 and copy the 16-bit product to two temporary registers, t_0 and t_1, with help of the movw instruction. The register t_1 holds the "upper" (i.e. more significant) byte of the product and t_0 the "lower" byte. Next, we form the product $a_0 \cdot b_0$ and add it along with the content of t_0 to the three accumulator registers r_0, r_1 and r_2. A potential carry from this addition can be safely added into the temporary register

---

[4] A special "feature" of AVR is that the mul instruction modifies the carry flag, which complicates the implementation of multi-precision multiplication.

t_1 without overflowing it since the upper byte of the product of two 8-bit integers is always smaller than 255. Thereafter, we multiply a_0 by b_1, add the resulting 16-bit product $a_0 \cdot b_1$ to r_1, r_2, and propagate the carry from the last addition to the temporary register t_1. Again, it is not possible to overflow t_1, not even in the most extreme case where the operand bytes a_0, b_0, b_1, and b_2 as well as the involved accumulator bytes r_0, r_1, and r_2 have the maximum value of 255. After computation of the last product of the first block (which is $a_1 \cdot b_3$), we add t_1 and $a_1 \cdot b_3$ to the three accumulator registers r_3, r_4, r_5, and finally propagate the carry bit from the last addition up to r_8. In summary, the processing of the first block in Figure 1 requires four `mul`, a `movw`, and a total of 13 `add` or `adc` instructions, respectively.

The next two blocks are processed in essentially the same way as the first block; the only actual difference is the loading of the remaining operand bytes of A, namely a_2 and a_3, which is done during the second and third block, respectively. Again, we use temporary register t_1 to catch the carries generated in the addition of the second and third byte-product of the respective block. The loading of operand byte a_2 is part of second block and performed after the multiplication of a_0 by b_3. Note that the byte a_0 is not needed anymore once $a_0 \cdot b_3$ has been produced, which means we can load a_2 into the register holding a_0. The operand byte a_3 gets loaded after the multiplication of a_1 by b_2 in the third block. At that time, the byte a_1 is not needed anymore, and hence we can load a_3 into the same register, thereby overwriting a_1. In summary, the second and third block execute 12 and 11 `add` (or `adc`) instructions, respectively. The number of `mul` and `movw` instructions are the same as for the first block.

The fourth block, in which the remaining four byte-products are generated and added to the accumulator registers, differs a bit from the former three. We first multiply a_3 by b_1 and move the resulting 16-bit product to the temporary register pair t_1, t_0. Then, we compute the product $a_2 \cdot b_1$, add its lower byte to the accumulator register r_3 and the upper byte to the two temporary registers holding $a_3 \cdot b_1$. The last addition does not produce a "carry out," which means this addition can not overflow the temporary register pair. Next in schedule is the third product $a_3 \cdot b_0$; it is processed in the same way as before and can also not overflow the registers t_1, t_0. After finally multiplying a_3 by b_2, the temporary register t_0 is added to r_4, and a possible carry bit is added with t_1 to the product $a_2 \cdot b_3$. The obtained sum is then added to the accumulator registers r_5, r_6 and the carry from the last addition is propagated to r_8. All in all, the fourth block requires to execute 13 `add` (resp. `adc`) instructions, very similar to the first block. The complete inner-loop operation for $d = 4$ consists of a total of 46 `add` (or `adc`), 16 `mul`, eight `ld` (i.e. load), and four `movw` instructions. On an ATmega128 processor [2], these instruction counts translate to an execution time of 101 clock cycles per iteration of the inner loop (including update of the loop-control variable and branch instruction). Another property of our loop is its economic register usage; it occupies only 30 out of the 32 available registers [1], which simplifies the implementation of Montgomery multiplication.

Table 2. Comparison of instruction counts for 160-bit multi-precision multiplication on the ATmega128 (without function call overhead)

Instruction type	add	mul	ld	st	mov	Other	Total
CPI	1	2	2	2	1	cycles	cycles
Classic Comba	1200	400	800	40	81	44	3805
Gura et al [13]	1360	400	167	40	355	197	3106
Uhsadel et al [27]	986	400	238	40	355	184	2881
Liu et al [19]	1194	400	200	40	212	179	2865
Zhang et al [32]	1092	400	200	20	202	271	2845
Our work (parameterised)	1213	400	200	40	100	185	**2778**
Hutter et al [15] (looped)	1252	400	92	66	41	276	2685
Scott et al [24] (unrolled)	1263	400	200	40	70	38	2651
Hutter et al [15] (unrolled)	1240	400	80	60	2	68	2395
Seo et al [25] (unrolled)	1240	400	70	60	n/a	56	2356
Seo et al [26] (unrolled)	1230	400	70	60	n/a	56	2346

3.2   Evaluation of Our Optimized Hybrid Multiplication

Table 2 shows the instruction counts and total execution time (in clock cycles) of our improved hybrid method for a (160×160)-bit multiplication on an ATmega128 processor [2]. We use (160×160)-bit multiplication as benchmark to allow for a direct comparison with past work that targeted ECC. Note that the instruction numbers in the columns labeled with add, ld, and mov also include adc, ldd, and movw, respectively (i.e. we do not differentiate between add and adc as they both require a single cycle on AVR processors). Our variant of the hybrid method executes a (160×160)-bit multiplication in just 2778 cycles on the ATmega128, which is approximately 10.6% faster than the original hybrid method of Gura et al [13]. This saving in execution time is mainly due to the fact that we have to carry out only 100 mov (resp. movw) instructions, whereas Gura et al need 355 mov or movw instructions. Furthermore, our special scheduling of the multiplications in the inner loops reduces the number of add (and adc) instructions, similar to the implementations described in [19] or [32]. The hybrid multiplication technique of Uhsadel et al [27] requires 2881 cycles, even though their implementation (as well as the one of Gura et al [13]) is based on $d = 5$ for 160-bit operands instead of $d = 4$ as in our work.

In general, when analyzing different software libraries for multiple-precision arithmetic, one has to distinguish three implementation options with respect to the processing of loops: unrolled, looped, and parameterized. Loop unrolling is well known to improve performance as it eliminates the loop overhead (such as the updating of a loop counter or execution of a branch instruction) and allows for some extra optimizations. For example, the first and last iteration of a loop often differs from the middle iterations and can, therefore, be specifically tuned when the loop is unrolled. The drawbacks of loop unrolling are large code size (i.e. increased program memory) and poor flexibility (resp. scalability) since an

Table 3. Comparison of code size (in bytes) of "conventional" multiplication (without reduction) for operand lengths ranging from 160 to 1024 bits

Implementation	160	192	224	256	512	1024
Hutter et al [15] (looped)	1562	1866	1538	1766	1544	1572
Hutter et al [15] (unrolled)	3778	5436	7340	9558	37884	151044
Our work (parameterised)	514	514	514	514	514	514

unrolled implementation supports just a single operand length. At the opposite end of the design space are parameterized implementations, which allow one to pass the operand length as a parameter to a function call. Such parameterized implementations are very flexible since one and the same function can process operands of any size, but this flexibility comes at the expense of decreased performance due to the fact that (full) loop unrolling and other optimizations are not possible anymore. Somewhere in the middle between these two approaches are looped implementations, which have "rolled" loops but still support only a single operand length. Looped implementations outperform their parameterized counterparts since they provide more avenues for optimization. Having a fixed operand helps to improve the performance as the number of loop iterations is constant and can therefore be "hard-coded." Thus, it is not necessary to waste a register for storing the operand length, which leaves more registers available for the actual computation.

Even though our implementation of the hybrid method is parameterized, it compares very well with looped and unrolled implementations. For example, the looped version of Hutter et al's operand caching technique [15] is just 93 cycles faster than our work (2685 vs. 2778 cycles, see Table 2), even though their code is optimized for 160-bit operands, while our implementation supports operands of any length. However, this slight performance gain comes at the cost of three times larger codes size, which can be seen from Table 3. Furthermore, one has to consider that Hutter et al achieved their execution time of 2685 clock cycles by using all 32 available registers[5] of the ATmega128. The unrolled implementations from [15,24,25,26], while being fast, suffer from a prohibitively large code size, especially for operands exceeding 256 bits in size (see Table 3). Full loop unrolling may be a viable optimization for ECC, but not for RSA.

3.3   Hybrid Montgomery Multiplication

Similar to the "ordinary" multiplication (without modular reduction), also the six Montgomery multiplication techniques described in this paper can be made

---

[5] Note that the fastest implementation of a conventional multiplication (i.e. a multiplication without reduction) does not necessarily lead to the fastest implementation of Montgomery multiplication. Generic algorithms for modular multiplication have three input operands (namely A, B, and M), which increases the register pressure compared to an ordinary multiplication. Our variant of the hybrid method occupies only 30 registers and, thus, allows for easy integration of Montgomery reduction.

significantly faster by applying the hybrid method in order to take advantage of the large register file of the AVR platform [1]. Processing several bytes of the operands in each inner-loop iteration yields a performance gain by reducing the number of loads/stores and loop overhead. By combining the hybrid technique with the six Montgomery variants, we get six hybrid Montgomery multiplication methods, which we call hybrid SOS (HSOS), hybrid FIPS (HFIPS), hybrid CIOS (HCIOS), hybrid CIHS (HCIHS), hybrid FIOS (HFIOS), and hybrid SPS (HSPS). Our implementations of these six algorithms have in common that, in each iteration of the inner loop, four bytes of the operands are loaded into the register file and the total number of loop iterations is accordingly reduced by a factor of four compared to the corresponding straightforward (i.e. non-hybrid) Montgomery multiplication technique.

The hybrid product-scanning techniques, namely HFIPS and HSPS, execute operations of the form $(t, u, v) \leftarrow (t, u, v) + a \cdot b$ in the inner loops, whereby the two operand words a and b consist of four bytes each. A total of nine registers is necessary to hold the cumulative sum (t, u, v). Therefore, we can employ the highly-optimized hybrid implementation of the inner-loop operation shown on the right of Figure 1 and explained in detail in Section 3.1. Unlike HSPS, the HFIPS method has to keep four pointers (namely the pointers to the arrays in which the two operands A, B, the result Z, and the modulus M are stored) in registers during the execution of the inner loop to reach top performance. The inner-loop implementation from Subsection 3.1 is ideally suited for the HFIPS method since it needs only 30 registers so that the remaining two registers can be used to hold the pointer to M. The four hybrid Montgomery multiplication methods based on operand-scanning (i.e. HSOS, HCIOS, HCIHS, and HFIOS) have a slightly different inner loop due to the fact that they execute operations of the form $(u, v) \leftarrow a \cdot b + c + d$ and $(u, v) \leftarrow a \cdot b + c$. We implemented these operations to process four bytes at once (i.e. per loop iteration) and optimized them following exactly the same strategies as discussed in Section 3.1.

3.4 Regular Execution of Final Subtraction

As shown in Algorithm 1, the calculation of the Montgomery product may require a final subtraction of the modulus M to get a fully reduced result in the range of $[0, M - 1]$. However, this final subtraction is not carried out when the intermediate result after step 3 of Algorithm 1 is already smaller than M. It is well known that such a conditional execution of a subtraction typically entails observable differences in the power consumption profile, which can be exploited to mount an SPA attack as described in [30] for RSA and in [29] for an elliptic curve cryptosystem. Walter proposed in [28] a smart approach to eliminate the final subtraction by using a larger Montgomery radix of e.g. $R = 2^{n+2}$ instead of $R = 2^n$ and adapting the Montgomery algorithm accordingly. However, this approach requires to calculate the Montgomery product with longer operands (since, as in our case, the operand length must be a multiple of 32), which can severely degrade performance. To overcome this problem, we implemented the final subtraction in an unconditional way by "zeroing out" the words m_i of the

---

Algorithm 4. Final subtraction without conditional statements

---

Input: $(s+1)$-word Montgomery product $Z = (z_s, z_{s-1}, \ldots, z_1, z_0)$ with $z_s \in \{0, 1\}$
  and s-word modulus $M = (m_{s-1}, \ldots, m_1, m_0)$
Output: $Z = Z - M$ if $z_s = 1$, otherwise, $Z = Z - 0$
 1: $mask \leftarrow -z_s \bmod 2^w$ {w is the bitlength of a word}
 2: $(\varepsilon, z_0) \leftarrow z_0 - (m_i \,\&\, mask)$
 3: **for** i from 1 by 1 to $s - 1$ **do**
 4:     $(\varepsilon, z_i) \leftarrow z_i - (m_i \,\&\, mask) - \varepsilon$
 5: **end for**
 6: **return** $Z = (z_{s-1}, \ldots, z_1, z_0)$

---

modulus M, if necessary, as shown in Algorithm 4. The notation in Algorithm 4 follows that of [14], i.e. the word-subtractions are carried out with help of an "subtract with borrow" instruction whereby ε represents the borrow bit.

Based on the concept of incomplete modular arithmetic [31], we do not perform an exact comparison between Z and M, but rather use the value of the most significant word z_s of Z to determine whether Z is too big or not. More precisely, we use z_s to derive a mask that is either a zero word (if $z_s = 0$) or an "all 1" word (if $z_s = 1$). As shown in line 1 of Algorithm 4, such a mask can be simply generated by forming the two's complement of z_s. The mask is applied to the bytes of M (i.e. each m_i is logically ANDed with the mask) before they are subtracted from the words z_i using subtract-with-borrow instructions. In this way, we either subtract the modulus M from product Z (if $z_s = 1$) or we subtract 0 (if $z_s = 0$) so that Z remains the same. The final result may not be the least non-negative residue, but is always in the range $[0, 2^n - 1]$ and hence fits into s words. This incomplete reduction does not introduce any problems in practice since the n-bit result, even if not fully reduced, can still be used as operand in a subsequent Montgomery multiplication (see [31] for details).

4   Performance Evaluation and Comparison

We implemented the six hybrid Montgomery multiplication algorithms in AVR assembly language and evaluated their performance for operands ranging from 160 to 1024 bits. Table 4 shows the simulated execution times we obtained on an ATmega128 processor [2]; these figures include time for the unconditional final subtraction introduced in Section 3.4. Our fastest method, HFIPS, only needs 6080 clock cycles to perform a full 160-bit Montgomery multiplication, which is approximately 1.4 times faster than the slowest algorithm, namely HFIOS. All obtained execution times are visualized on the left of Figure 2.

Besides the computational complexity of algorithms themselves, there are a few other factors affecting the actual performance of the various multiplication methods. For example, the overhead for controlling the loop or the cost to find the correct start address of arrays also impact the execution time. Our results indicate that the interleaved versions of hybrid Montgomery multiplication are sightly faster than the separated versions, e.g. HFIPS outperforms HSPS, and

Table 4. Execution time (in clock cycles) of six hybrid Montgomery multiplication techniques for different operand lengths

Algorithm	160	192	224	256	512	768	1024
HFIPS	6080	8539	11420	14723	56339	124964	220596
HSPS	6648	9171	12110	15465	57281	125722	221044
HCIOS	7140	9983	13310	17121	65033	143922	253787
HSOS	7921	10956	14500	18553	69301	152626	268788
HCIHS	8127	11385	15197	19563	74435	164764	290549
HFIOS	8216	11660	15716	20384	79760	178315	316018

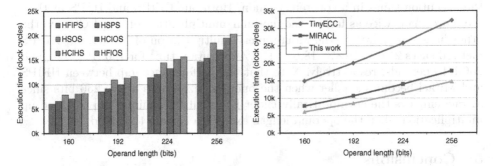

Fig. 2. Performance comparison of our six Montgomery algorithms (left) and comparison or our HFIPS method with Miracl and TinyECC (right)

HCIOS is faster than HSOS. This is mainly because the interleaved versions, in general, incur less overhead than the separated versions (i.e. reduced overhead for controlling loops, handling pointers, and calculating start addresses).

The HCIHS and HFIOS method are the slowest of the six hybrid Montgomery multiplication techniques shown in Table 4. The poor performance of the HCIHS approach is primarily due to the overhead caused by frequent loadings of operands into registers. On the other hand, HFIOS uses a lot of time for the pointer arithmetic required to obtain the correct start address of the operands at the beginning of a loop. Another disadvantage of this method is that it has to handle six variables, namely a_j, b_i, m_j, q, t, and z_j, in the inner loop. Since the hybrid multiplication of $a_j \cdot b_i$ occupies almost all of the 32 working registers, a number of expensive **push** and **pop** operations are required to save pointers on the stack. The cost of the stack operations in HFIOS is higher than cost of the frequent operand loadings in HCIHS; thus, HFIOS is slower than HCIHS.

Table 5 compares our hybrid product-scanning methods, namely HSPS and HFIPS, with the two popular cryptographic libraries TinyECC [18] and Miracl [5] for operands ranging from 160 to 1024 bits in size. The right side of Figure 2 visualizes the execution times of TinyECC, Miracl, and HFIPS, which is the fastest of our six implementations of Montgomery multiplication. To ensure a fair comparison, we downloaded the source code of TinyECC and Miracl from

Table 5. Montgomery Multiplication timings (in clock cycles) of TinyECC, Miracl, and our implementation of the HSPS and HFIPS method

Implementation	160	192	224	256	512	1024
TinyECC [18]	14929	20060	25765	n/a	n/a	n/a
Miracl [5]	7753	10653	14033	17761	58806	221329
This work (HSPS)	6648	9171	12110	15465	57281	221044
This work (HFIPS)	6080	8539	11420	14723	56339	220596

the corresponding home pages, compiled them with AVR studio, and simulated the execution times in a coherent fashion. Both our HFIPS and HSPS method are more than twice as fast as the modular multiplication of TinyECC. On the other hand, compared to the Montgomery multiplication of Miracl, our HFIPS method saves 21.6%, 19.8%, 18.6%, 17.1% execution time for 160, 192, 224, and 256-bit operands, respectively. Note that the performance gap between HFIPS and Miracl becomes smaller when the operand size grows above 256 bits since Miracl employs the asymptotically faster Karatsuba technique [16] to speed up multiplication when the operand length exceeds a certain threshold.

5   Conclusions

The contribution of this work is threefold. First, we presented a new approach to implement hybrid multiplication, saving 10.6% execution time compared to the original method of Gura et al (CHES 2004). This performance gain is achieved by re-ordering the sequence of multiplications in the inner loop along with an efficient way of catching carries, thereby reducing the total number of add and mov (resp. movw) instructions. Another advantage of our hybrid technique is its suitability to implement interleaved variants Montgomery multiplication since it occupies only 30 registers of an AVR processor. Our second contribution is a through analysis and comparison of six hybrid variants of Montgomery modular multiplication. Based on a more precise cost model along with some small optimizations (e.g. elimination of the ADD function for carry propagation), we conclude that the FIPS and SPS method reach the best performance, which is contradicting previous results of Koç et al, who found the CIOS method to be superior. A detailed benchmarking on an 8-bit ATmega128 processor confirms our theoretical evaluation and shows that the hybrid FIPS technique requires merely 6080 clock cycles to execute a 160-bit Montgomery multiplication. This result sets a new speed record for modular multiplication on an 8-bit platform and outperforms the Miracl library by more than 20%. Our implementation is parameterized and very compact in terms of code size. The third contribution of this paper is a simple yet efficient approach to perform the conditional final subtraction in an unconditional way by "zeroing out" the words of the modulus if the intermediate result is already smaller than 2^n. This ensures that always exactly the same sequence of instructions is executed, regardless of the actual value of the operands, which helps to thwart certain side-channel attacks.

References

1. Atmel Corporation. 8-bit ARV$^{\circledR}$ Instruction Set. User Guide (July 2008), http://www.atmel.com/dyn/resources/prod_documents/doc0856.pdf
2. Atmel Corporation. 8-bit ARV$^{\circledR}$ Microcontroller with 128K Bytes In-System Programmable Flash: ATmega128, ATmega128L. Datasheet (June 2008), http://www.atmel.com/dyn/resources/prod_documents/doc2467.pdf
3. Barreto, P.S., Kim, H.Y., Lynn, B., Scott, M.: Efficient algorithms for pairing-based cryptosystems. In: Yung, M. (ed.) CRYPTO 2002. LNCS, vol. 2442, pp. 354–368. Springer, Heidelberg (2002)
4. Barrett, P.: Implementing the rivest shamir and adleman public key encryption algorithm on a standard digital signal processor. In: Odlyzko, A.M. (ed.) CRYPTO 1986. LNCS, vol. 263, pp. 311–323. Springer, Heidelberg (1987)
5. CertiVox Corporation. CertiVox MIRACL SDK. Source code (June 2012), http://www.certivox.com
6. Comba, P.G.: Exponentiation cryptosystems on the IBM PC. IBM Systems Journal 29(4), 526–538 (1990)
7. Diffie, W., Hellman, M.E.: New directions in cryptography. IEEE Transactions on Information Theory 22(6), 644–654 (1976)
8. Dussé, S.R., Kaliski Jr., B.S.: A cryptographic library for the Motorola DSP 56000. In: Damgård, I.B. (ed.) EUROCRYPT 1990. LNCS, vol. 473, pp. 230–244. Springer, Heidelberg (1991)
9. Gordon, D.M.: A survey of fast exponentiation methods. Journal of Algorithms 27(1), 129–146 (1998)
10. Großschädl, J., Avanzi, R.M., Savaş, E., Tillich, S.: Energy-efficient software implementation of long integer modular arithmetic. In: Rao, J.R., Sunar, B. (eds.) CHES 2005. LNCS, vol. 3659, pp. 75–90. Springer, Heidelberg (2005)
11. Großschädl, J., Kamendje, G.-A.: Architectural enhancements for montgomery multiplication on embedded RISC processors. In: Zhou, J., Yung, M., Han, Y. (eds.) ACNS 2003. LNCS, vol. 2846, pp. 418–434. Springer, Heidelberg (2003)
12. Großschädl, J., Kamendje, G.-A.: Optimized RISC architecture for multiple-precision modular arithmetic. In: Hutter, D., Müller, G., Stephan, W., Ullmann, M. (eds.) Security in Pervasive Computing 2003. LNCS, vol. 2802, pp. 253–270. Springer, Heidelberg (2004)
13. Gura, N., Patel, A., Wander, A., Eberle, H., Shantz, S.C.: Comparing elliptic curve cryptography and RSA on 8-bit cPUs. In: Joye, M., Quisquater, J.-J. (eds.) CHES 2004. LNCS, vol. 3156, pp. 119–132. Springer, Heidelberg (2004)
14. Hankerson, D.R., Menezes, A.J., Vanstone, S.A.: Guide to Elliptic Curve Cryptography. Springer (2004)
15. Hutter, M., Wenger, E.: Fast multi-precision multiplication for public-key cryptography on embedded microprocessors. In: Preneel, B., Takagi, T. (eds.) CHES 2011. LNCS, vol. 6917, pp. 459–474. Springer, Heidelberg (2011)
16. Karatsuba, A.A., Ofman, Y.P.: Multiplication of multidigit numbers on automata. Soviet Physics - Doklady 7(7), 595–596 (1963)
17. Koç, Ç.K., Acar, T., Kaliski, B.S.: Analyzing and comparing Montgomery multiplication algorithms. IEEE Micro 16(3), 26–33 (1996)
18. Liu, A., Ning, P.: TinyECC: A configurable library for elliptic curve cryptography in wireless sensor networks. In: Proceedings of the 7th International Conference on Information Processing in Sensor Networks (IPSN 2008), pp. 245–256. IEEE Computer Society Press (2008)

19. Liu, Z., Großschädl, J., Kizhvatov, I.: Efficient and side-channel resistant RSA implementation for 8-bit AVR microcontrollers. In: Proceedings of the 1st International Workshop on the Security of the Internet of Things, SECIOT 2010 (2010)
20. Mangard, S., Oswald, E., Popp, T.: Power Analysis Attacks: Revealing the Secrets of Smart Cards. Springer (2007)
21. Montgomery, P.L.: Modular multiplication without trial division. Mathematics of Computation 44(170), 519–521 (1985)
22. National Institute of Standards and Technology (NIST). Digital Signature Standard (DSS). FIPS Publication 186-4 (July 2013), http://nvlpubs.nist.gov/nistpubs/FIPS/NIST.FIPS.186-4.pdf
23. Rivest, R.L., Shamir, A., Adleman, L.M.: A method for obtaining digital signatures and public key cryptosystems. Communications of the ACM 21(2), 120–126 (1978)
24. Scott, M., Szczechowiak, P.: Optimizing multiprecision multiplication for public key cryptography. Cryptology ePrint Archive, Report 2007/299 (2007), http://eprint.iacr.org
25. Seo, H., Kim, H.: Multi-precision multiplication for public-key cryptography on embedded microprocessors. In: Lee, D.H., Yung, M. (eds.) WISA 2012. LNCS, vol. 7690, pp. 55–67. Springer, Heidelberg (2012)
26. Seo, H., Kim, H.: Optimized multi-precision multiplication for public-key cryptography on embedded microprocessors. International Journal of Computer and Communication Engineering 2(3), 255–259 (2013)
27. Uhsadel, L., Poschmann, A., Paar, C.: Enabling full-size public-key algorithms on 8-bit sensor nodes. In: Stajano, F., Meadows, C., Capkun, S., Moore, T. (eds.) ESAS 2007. LNCS, vol. 4572, pp. 73–86. Springer, Heidelberg (2007)
28. Walter, C.D.: Montgomery exponentiation needs no final subtractions. Electronics Letters 38(21), 1831–1832 (1999)
29. Walter, C.D.: Simple power analysis of unified code for ECC double and add. In: Joye, M., Quisquater, J.-J. (eds.) CHES 2004. LNCS, vol. 3156, pp. 191–204. Springer, Heidelberg (2004)
30. Walter, C.D., Thompson, S.: Distinguishing exponent digits by observing modular subtractions. In: Naccache, D. (ed.) CT-RSA 2001. LNCS, vol. 2020, pp. 192–207. Springer, Heidelberg (2001)
31. Yanık, T., Savaş, E., K. Koç, Ç.: Incomplete reduction in modular arithmetic. IEE Proceedings – Computers and Digital Techniques 149(2), 46–52 (2002)
32. Zhang, Y., Großschädl, J.: Efficient prime-field arithmetic for elliptic curve cryptography on wireless sensor nodes. In: Proceedings of the 1st International Conference on Computer Science and Network Technology (ICCSNT 2011), vol. 1, pp. 459–466. IEEE (2011)

Minimizing S-Boxes in Hardware by Utilizing Linear Transformations

Sebastian Kutzner, Phuong Ha Nguyen,
Axel Poschmann, and Marc Stöttinger

Physical Analysis and Cryptographic Engineering (PACE),
Temasek Laboratories at Nanyang Technological University, Singapore
{skutzner,aposchmann,mstottinger}@ntu.edu.sg

Abstract. Countermeasures against side-channel analysis attacks are increasingly considered already during the design/implementation step of cryptographic algorithms for embedded devices. An important challenge is to reduce the overhead (area, time) introduced by the countermeasures, and, consequently, in the past years a lot of progress has been achieved in this direction. In this contribution we propose a further optimization of decomposing 4-bit S-boxes by exploiting affine transformations and a single shared quadratic permutation. Thereby many various S-boxes can be merged into one component and thus reduce the resource overhead. We applied our proposed scheme on a Threshold Implementation masked PRESENT S-box and its inverse in order to construct a merged masked S-box, which can be used for both encryption and decryption. This design saves up to 24% resources on a Virtex-5 FPGA platform and up to 28% for an ASIC implementation compared to previously published designs. It is noteworthy to stress that our technique is not restricted to the TI countermeasure, but also allows to reduce the resource requirements of the non-linear layer of cryptographic algorithms with a set of different S-boxes, such as SERPENT or DES, amongst others.

1 Introduction

Since the introduction of side-channel analysis attacks by Kocher in [9], these attacks gained more and more attention in the area of circuit design for security-critical applications. Commonly, these non-invasive implementation attacks are used to reveal secret parameters or values from devices running cryptographic algorithms by exploiting their physical behavior during runtime. More precisely, the physical characteristic of the implementation, such as the power consumption [9], the electromagnetic emission [2], or the execution duration [8] of an implemented algorithm, is used to gather additional information about the not directly accessible, but still exploitable intermediate values. Among these attacks differential power analysis attacks have become the most popular and commonly conducted side-channel analysis attacks. They exploit the data-dependent dynamic power consumption of CMOS-technology based circuits by using sophisticated statistical methods in order to verify hypotheses about the used secret key. For further information about power analysis attacks we refer to [13].

D. Pointcheval and D. Vergnaud (Eds.): AFRICACRYPT 2014, LNCS 8469, pp. 235–250, 2014.
© Springer International Publishing Switzerland 2014

Of course countermeasures have been proposed meanwhile to cope with such kind of attacks. For instance, the masking countermeasure randomizes the intermediate values to gain resistance against power analysis attacks on software and hardware implementations. But even so the implementation of the masking scheme itself has to be carried out carefully in order to avoid data-dependent glitches. It was shown in [7,14] that these data-dependent glitches especially in hardware implementations can be exploited and by that compromise the masking countermeasure. One promising efficient masking scheme, which is provable secure against glitches and resistant against first-order power analysis attacks, is the *Threshold Implementation* (*TI*) countermeasure, cf. [16,17].

In this paper we further investigate optimizations of the TI masking scheme. We focus on a general optimization for all 4-bit S-boxes,which can be decomposed into quadratic S-boxes. These S-boxes are grouped together under the term A_{16}. The term and classification of A_{16} was introduced by Bilgin et al in [3]. Our optimization aims at minimizing the resource overhead of a 4-bit S-box with 3-shares, which is generated by a newly proposed decomposition step based on the original 4-bit S-box. In particular, the proposed decomposition procedure enables us to combine any of the S-boxes in A_{16} into one component using the same quadratic permutation core. Hence we are able to merge several 4-bit S-boxes of different S-box classes[1] together into one shared and TI masked component. As an illustrative example how the S-box lookup and its inverse can be merged efficiently, we applied the proposed decomposition scheme on the S-box and the inverse of S-box of the PRESENT [4], a lightweight block cipher standardized in ISO/IEC-29192 [1]. We are merging the PRESENT S-box $S_{\text{PRESENT}}(\cdot)$ and the inverse PRESENT S-box $S^{-1}_{\text{PRESENT}}(\cdot)$ into one component, sharing the same non-linear function for encryption and decryption. By doing so the resource overhead of the over all PRESENT design can be reduced even further.

The remainder of this paper is organized as follows: In Sect. 2 we discuss in more in detail how various 4-bit S-boxes can be decomposed for the TI countermeasure by using the same quadratic permutation. In Sect. 3 we show how to apply the proposed decomposition on the PRESENT S-box for performing encryption and decryption. Before we conclude this paper in Sect. 6, we discuss an FPGA implementation of the merged S-box in Sect. 4, as well as its side-channel resistance in Sect. 5.

2 Decomposing S-boxes for TI Scheme

The core idea of the TI masking scheme is to share the computation of a non-linear Boolean function of dth order into d+1 shares. Hence, the computation of the intermediate values is masked and not directly exploitable by a straightforward power analysis attack. Initially, a 4-bit S-box as for example in PRESENT[2] has to be shared by four shares. Due to a sophisticated decomposition of the

---

[1] Members of A_{16}.

[2] The non-linear permutation of that S-box is cubic.

S-box into quadratic and linear Boolean terms, only three shares are required, as it has been demonstrated in [20].

In this section we investigate the decomposability of so-called optimal 4-bit S-boxes in a more general way. First, we recall the definition of optimal S-boxes and some background about decomposing non-linear functions for usage in the TI-scheme. After that we demonstrate how to decompose a subset of all optimal 4-bit S-boxes into quadratic Boolean functions by using the same non-linear core permutation G and affine transformations.

2.1   Optimal 4-Bit S-boxes

A 4-bit S-box is considered as cryptographically optimal, if it fulfills the natural requirement to be resistant against linear and differential cryptanalysis as best as possible. According to the definitions in [12] an optimal 4-bit S-box has the following properties:

Definition 1. *Let S: $\mathbb{F}_2^4 \to \mathbb{F}_2^4$ be an S-box. If $S(\cdot)$ fulfills the following conditions we call $S(\cdot)$ an optimal S-box:*

1. *$S(\cdot)$ is a bijection,*
2. *$\mathrm{Lin}(S(\cdot)) = 8$,*
3. *$\mathrm{Diff}(S(\cdot)) = 4$.*

The values $\mathrm{Lin}(S(\cdot))$ and $\mathrm{Diff}(S(\cdot))$ are measures for the resilience of an S-box against linear attacks and differential attacks, respectively. The chosen values of $\mathrm{Lin}(S)$ and $\mathrm{Diff}(S)$ are discussed in [12] based on observation by Nyberg in [18]. Furthermore, the authors of [12] categorize all possible optimal 4-bit S-boxes into 16 classes of non-equivalent optimal S-boxes. Based on the observation and discussion in [6] and [19] resistance against linear attacks $\mathrm{Lin}(S(\cdot))$ of an S-box and the resistance against differential attacks $\mathrm{Diff}(S(\cdot))$ of an S-box does not change if an affine transformation is applied on the S-box. This linear equivalence between two optimal S-boxes (based on Definition 1) can be expressed formally with Definition 2:

Definition 2. *Two S-boxes $S(\cdot), S'(\cdot)$ are linearly equivalent iff there exist two 4×4-bit invertible matrices A, B and two 4-bit vector c, d such that*

$$S(x) = A(S'(Bx \oplus c) \oplus d), \forall x \in \{0, \ldots, 15\}.$$

Based on Definition 2 two linear equivalent S-boxes $S(\cdot), S'(\cdot)$ with the same properties can be transformed into each other and thus belong to the same categorization of S-box class. For sake of convenience and readability with the reference, we follow the notations in [12], so the affine 4×4-bit transformation matrices will be noted in a hexadecimal form from now on:

$$A = \begin{pmatrix} 1 & 0 & 1 & 0 \\ 0 & 1 & 0 & 0 \\ 1 & 0 & 0 & 0 \\ 1 & 0 & 1 & 1 \end{pmatrix} = \begin{pmatrix} a \\ 4 \\ 8 \\ b \end{pmatrix} = (0xb84a), \tag{1}$$

2.2   Decomposition of Optimal 4-bit S-boxes for the TI Scheme

The TI masking scheme is based on secret sharing and distributed computation. In particular the non-linear element of the cipher, namely the S-box of degree d, is distributed with the d+1 shares. The final result can be obtained by an XORing the d+1 shares, thereby the *correctness* property of the masked permutation is fulfilled. In each part of the computation only d shares are involved, the computation is fully independent from the original secret and assure the *non-completeness* property of the scheme. The last property which has been fulfilled for the shared computation is the so-called *uniformity*. It assures that the output distribution of the shared computed permutation is identical to the unshared one.

The reason to decompose 4-bit S-boxes into a combination of quadratic and linear Boolean functions is to reduce the number of shares required for a TI-protected implementation, which in turn reduces the resources required. A decomposition of an optimal 4-bit S-box is given by the following definition:

Definition 3. *If a vectorial Boolean function such as a permutation $S(\cdot)$ can be written as a composition of several lower degree vectorial Boolean functions $f_1(\cdot), f_2(\cdot), \ldots, f_n(\cdot)$, i.e $S(\cdot) = f_n(\ldots f_2(f_1(\cdot)))$, then $f_1(\cdot), f_2(\cdot), \ldots, f_n(\cdot)$ is called the* decomposition *of $S(\cdot)$.*

After a given 4-bit S-box has been decomposed into several 4-bit quadratic permutations we can apply the TI countermeasure by using each of the 4-bit quadratic permutations to construct a 12-bit permutation, cf. [17]. In order to assure a proper masking of the intermediate calculations, this 12-bit quadratic permutation has to be so-called *shareable* by definition 4.

Definition 4. *A 4-bit linear or quadratic permutation is called* shareable *if it can be converted to a 12-bit permutation which fulfils the following properties of TI: correctness, non-completeness and uniformity, cf. [17].*

Note 1. The composition of a quadratic permutation and a linear permutation is again quadratic. Hence, a quadratic permutation is able to be described as a composition of linear and quadratic permutations. Recalling the equivalence observation in Definition 2, one can easily see that if one S-box of an optimal 4-bit S-box class is decomposable and sharable, then all S-boxes of this class are shareable, see [3].

Hence, all 4-bit S-boxes of the S-box classes 0, 1, 2, 4, 5, 7, 8, 13 out of the 16 classes defined in [12] are sharable and forming a cluster of S-box classes refereed to as group A_{16}, cf. [10].

2.3   Decomposing all S-Boxes of A_{16} with One Quadratic Permutation

The same quadratic permutation can be used for sharing the 4-bit permutation to construct a 12-bit permutation for each S-box of the same S-box class,

Table 1. Core permutation G

x	0	1	2	3	4	5	6	7	8	9	10	11	12	13	14	15
$G(x)$	0	4	1	5	2	15	11	6	8	12	9	13	14	3	7	10

if definition 2 and definition 4 are exploited. Furthermore, due to careful selection of the quadratic permutation, it can be reused for each share in an iterative manner to reduce resource overhead, see [10]. As reported in [10], we can also construct an arbitrary S-box by exploiting a consecutive execution of an appropriate quadratic permutation G' so that $S'(\cdot) = G'(G'(\cdot))$. We investigated if it is possible to find a generic procedure to generate all shareable S-boxes of A_{16} in a similar manner. As result we came up with a special decomposition structure in order to further generalize the idea of sharing a basic quadratic 4-bit permutation as a non-linear mapping core for shareable S-boxes of A_{16}:

$$S(\cdot) = f_n(\ldots f_2(f_1(\cdot))) = M_n \cdot G(M_{n-1}G(\ldots M_1(G(\cdot))), \tag{2}$$

where $M_n, \ldots, M_0$ are invertible matrices and G is a quadratic, sharable permutation of A_{16}. This new S-box transformation utilizes definition 2, definition 3, and the idea behind lemma 3 and lemma 6 in [3], that each S-box of the group A_{16} can be composed by a sequence of other S-boxes. This means by using this appropriate core permutation G, listed in Tab. 1, we can construct each S-box belonging to A_{16} with the core permutation G and affine transformation. So each S-box belonging to one of the classes within the group A_{16} can be constructed by using the decomposition description of Eq. (2) with an appropriate transposition matrix M^{cls} and G as given in Tab. 2.

Because G is a permutation within the group A_{16}, it is a sharable quadratic permutation and can be shared by a 3-share TI-scheme. Hence, based on Definition 2 and Note 1 we are able to construct all possible optimal S-boxes of A_{16}. By using

Table 2. S-box representation of A_{16} using core G

S-box class	Transformation matrix	Transformation to the S-box S(x)
0	$M^{cls0} = 0x1249$	$S(\cdot) = G(M^{cls0}G(\cdot))$
1	$M^{cls1} = 0x1248$	$S(\cdot) = G(M^{cls1}G(\cdot))$
2	$M^{cls2} = 0x1259$	$S(\cdot) = G(M^{cls2}G(\cdot))$
4	$M^{cls4} = 0x12e6$	$S(\cdot) = G(M^{cls4}G(G(\cdot)))$
5	$M^{cls5} = 0x14a7$	$S(\cdot) = G(M^{cls5}G(G(G(\cdot))))$
7	$M^{cls7} = 0x1843$	$S(\cdot) = G(M^{cls7}G(G(\cdot)))$
8	$M^{cls8} = 0x1295$	$S(\cdot) = G(M^{cls8}G(\cdot))$
13	$M^{cls13} = 0x134b$	$S(\cdot) = G(M^{cls13}G(G(\cdot)))$

the decomposition construction in Tab. 2 and appropriate affine transformation due to Definition 2 we can implement any TI-masked 4-bit optimal S-box which is equivalent to one of the S-boxes in A_{16}. Utilizing these two tricks we can reduce the time complexity to find a decomposition for a given S-box to 2^{19}.

Furthermore, we can use this trick to share the resources of an S-box and its inversion in order to use only one component to perform the substitution for the encryption and decryption of $S(\cdot)$ and $S^{-1}(\cdot) \in A_{16}$. For example, the S-box $S(\cdot)$ used in PRESENT and its inverse $S^{-1}(\cdot)$ of can be merged together and used in a 3-share masked TI countermeasure implementation.

3 Application on the PRESENT Block Cipher

In this section we apply the proposed decomposition method for sharing G on the PRESENT S-box $S_{\text{PRESENT}}(\cdot)$ and its inverse $S^{-1}_{\text{PRESENT}}(\cdot)$. First, we present the decomposed design for the S-box and the inverse S-box of PRESENT separately by means of determining the correct parameter A, B, c, and d of the affine transformations in definition 2. Second, we show how to construct a merged S-box by combining $S_{\text{PRESENT}}(\cdot)$ and $S^{-1}_{\text{PRESENT}}(\cdot)$, while using the same core function G. At the end of this section we describe how to share the commonly used quadratic core function G for its application in the TI masking scheme.

3.1 S-Box Decomposition

Encryption S-box. The decomposition of $S_{\text{PRESENT}}(\cdot)$ in Eq. (2) uses $M^{cls1} = 0x1248$, because $S_{\text{PRESENT}}(\cdot)$ belongs to class 1, see Tab. 2. Please note that in this case $M^{cls1} = \mathbf{I}$ and thus, the matrix multiplication of M^{cls1} can be spared in Eq. 3 without changing the description of the S-box. This S-box description is similar to the S-box decomposition in [10].

$$S_{\text{PRESENT}}(x) = A(S'(Bx \oplus c) \oplus d) = A(G(M^{cls1}G(Bx \oplus c)) \oplus d),$$
$$= A(G(G(Bx \oplus c)) \oplus d), \quad \forall x \in \{0, \ldots, 15\}, \tag{3}$$

The parameter for affine transformation from $S'(\cdot)$ to $S(\cdot)$ is given in the appendix at Subsect. A.

Decryption S-box. The decomposition of $S^{-1}_{\text{PRESENT}}(\cdot)$ has the same basic structure as the decomposition of $S_{\text{PRESENT}}(\cdot)$, because the inverse of the PRESENT S-box also belongs to class 1 of A_{16}. Hence, the $S^{-1}_{\text{PRESENT}}(\cdot)$ can also be expressed by a linear combination of the linearly equivalent by. Eq. (4).

$$S^{-1}_{\text{PRESENT}}(x) = \widetilde{A}(S'^{-1}(\widetilde{B}x \oplus \widetilde{c}) \oplus \widetilde{d}) = \widetilde{A}(G(M^{cls1}G(\widetilde{B}x \oplus \widetilde{c})) \oplus \widetilde{d}),$$
$$= \widetilde{A}(G(G(\widetilde{B}x \oplus \widetilde{c})) \oplus \widetilde{d}), \quad \forall x \in \{0, \ldots, 15\} \tag{4}$$

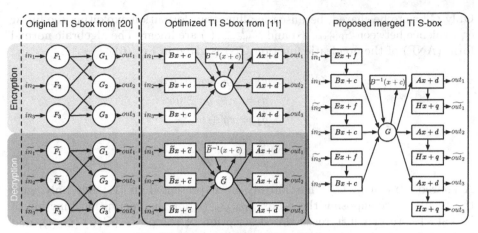

Fig. 1. S-box decomposition schemes

Actually, the same quadratic mapping function $G(\cdot)$, which was used to decompose $S_{\text{PRESENT}}(\cdot)$, can be used again to construct $S^{-1}_{\text{PRESENT}}(\cdot)$ as described in Sect. 2. We derived the appropriate parameter values of $\widetilde{A}, \widetilde{B}, \widetilde{c}$ and $\widetilde{d}$ by hand in order to satisfy Eq. (4), see Subsect. A.

Merged S-box. Furthermore, by employing the proposed decomposition structure in Sect. 2, the S-box properties of $S_{\text{PRESENT}}(\cdot)$ and $S^{-1}_{\text{PRESENT}}(\cdot)$ can be exploited to find a suitable linear transformation between them. In that manner the $S(\cdot)$ can be reused in Eq. (4) with suitable parameters for both the affine transformation, inside and outside of the non-linear Boolean mapping function $G(\cdot)$, cf. Eq. (5). Suitable parameters for fulfilling the Eq. (5) are provided in the appendix, given in Eq. (10) and in Eq. (12).

$$
\begin{aligned}
S^{-1}_{\text{PRESENT}}(x) &= \widetilde{A}(S'^{-1}(\widetilde{B}x \oplus \widetilde{c}) \oplus \widetilde{d}) = H(S(Ex \oplus f) \oplus q) \\
&= H(A(G(G(B(Ex \oplus f) \oplus c)) \oplus d) \oplus q), \\
&\forall x \in \{0, \ldots, 15\}
\end{aligned}
\tag{5}
$$

Applying the decomposition methods of $S_{\text{PRESENT}}(\cdot)$ and $S^{-1}_{\text{PRESENT}}(\cdot)$ in Eq. (5) the two S-box modules can efficiently be merged and thus, the resource overhead is reduced, cf. Fig. 1.

3.2 Decomposed S-box Structure for Threshold Implementation

The TI masking scheme can be applied to the merged design of the PRESENT S-box of $S(\cdot)$ and $S^{-1}(\cdot)$ straightforward, if the core permutation G is sharable.

Only G has to be shared, because the affine transformation exploiting the linear equivalence between $S_{\text{PRESENT}}(\cdot)$ and $S_{\text{PRESENT}}^{-1}(\cdot)$ are linear. The algebraic normal form (ANF) of the quadratic function $G(\cdot)$ is given in Eq. (6).

$$
\begin{aligned}
G(x, y, z, w) &= (g_3, g_2, g_1, g_0) \\
g_3 &= x \oplus yz \oplus yw \\
g_2 &= xy \oplus w \\
g_1 &= y \\
g_0 &= z \oplus yw.
\end{aligned}
\tag{6}
$$

A shared version of $G(\cdot)$ has already been presented in [11] but only with the focus of decomposing the S-box $S_{\text{PRESENT}}(\cdot)$ for the encryption function of PRESENT. We can just reuse this description of shared G ($G_1(\cdot)$, $G_2(\cdot)$, and $G_3(\cdot)$) for our design, because we use the same core function G for our decomposition in Eq. (5):

$$
\begin{aligned}
G_1(x_2, y_2, z_2, w_2, x_3, y_3, z_3, w_3) &= (g_{13}, g_{12}, g_{11}, g_{10}) \\
g_{13} &= x_2 \oplus y_2 z_2 \oplus y_2 z_3 \oplus y_3 z_2 \oplus y_2 w_2 \oplus y_2 w_3 \oplus y_3 w_2 \\
g_{12} &= w_2 \oplus x_2 y_2 \oplus x_2 y_3 \oplus x_3 y_2 \\
g_{11} &= y_2 \\
g_{10} &= z_2 \oplus y_2 w_2 \oplus y_2 w_3 \oplus y_3 w_2 \\
G_2(x_1, y_1, z_1, w_1, x_3, y_3, z_3, w_3) &= (g_{23}, g_{22}, g_{21}, g_{20}) \\
g_{23} &= x_1 \oplus y_1 z_1 \oplus y_1 z_3 \oplus y_3 z_1 \oplus y_1 w_1 \oplus y_1 w_3 \oplus y_3 w_1 \\
g_{22} &= w_1 \oplus x_1 y_1 \oplus x_1 y_3 \oplus x_3 y_1 \\
g_{21} &= y_1 \\
g_{20} &= z_1 \oplus y_1 w_1 \oplus y_1 w_3 \oplus y_3 w_1 \\
G_3(x_1, y_1, z_1, w_1, x_2, y_2, z_2, w_2) &= (g_{33}, g_{32}, g_{31}, g_{30}) \\
g_{33} &= x_1 \oplus y_1 z_1 \oplus y_1 z_2 \oplus y_2 z_1 \oplus y_1 w_1 \oplus y_1 w_2 \oplus y_2 w_1 \\
g_{32} &= w_1 \oplus x_1 y_1 \oplus x_1 y_2 \oplus x_2 y_1 \\
g_{31} &= y_1 \\
g_{30} &= z_1 \oplus y_1 w_1 \oplus y_1 w_2 \oplus y_2 w_1
\end{aligned}
$$

The equation labels (7), (8), (9) appear to the right of the corresponding groups.

The index numbers $_1$, $_2$, $_3$ mark each of the share input parameters of the shared functions of G_1, G_2, and G_3, while the letters x, y, z, and w denote the bit position of each of the three shared input nibbles.

4   Implementation of PRESENT with Merged S-boxes

We applied our previous observations on the 3-share PRESENT S-box for a TI countermeasure scheme by implementing the proposed merged S-box design, denoted as $S_{GG,mer}$ from now on, on a Virtex-5 FPGA platform. The right hand side of Fig. 1 depicts the architecture of the proposed merged TI S-box design $S_{GG,mer}$. For comparison reasons we also implemented a straightforward TI 3-share-PRESENT S-box $(S_{G'F,enc})$ based on the design in [20]. Six different quadratic permutations are needed for the S-box and its inverse as well, following the design in [20], as it is clearly illustrated on the left hand side of Fig. 1. We also implemented the PRESENT S-box $S(\cdot)$ $(S_{GG,enc})$ and the inverse PRESENT S-box $S^{-1}(\cdot)$ $(S_{GG,dec})$ based on the proposed optimized shared S-box scheme of Kutzner et al. in [11]. The middle section of Fig. 1 shows the general structure of the implementation of $S_{GG,enc}$ and $S_{GG,dec}$.

Furthermore, we also implemented several versions of a round based PRESENT block cipher for performing encryption and decryption on the Virtex-5 platform. For those designs we used the optimized three shares and the proposed merged three shares TI PRESENT S-box design of Sect. 3. All of the 4 presented S-box designs use a pipeline register to assure that only two out of three shares are processed at the same time, as it is recommended by the concept description of the TI masking scheme in [17].

4.1 Resource Consumption of Merged Designs

The implementation of the merged 3-share S-box design $S_{GG,mer}$ follows the previously stated design idea. The originally twelve non-linear shared function, which are needed for calculating $S(\cdot)$ and $S^{-1}(\cdot)$, are decomposed into only one shared non-linear function and five different linear affine transformations. The implementation of the sharable version of the S-box uses the shared version of $G(\cdot)$ in Eq. (7), and not a shared implementation of $G(G(\cdot))$. The implemented function $G(G(\cdot))$ is cubic and hence does not fulfill the requirements of a sharable permutation function using only three shares. Therefore, an additional affine transformation is required in a feedback path of the S-box design to perform $G(G(\cdot))$ iteratively, as depicted in the upper part of Fig. 2.

The additionally required iterative step is hidden by the pipelined structure of the S-box and thus, will not cause any additional clock cycle influencing the latency of the S-box. The efficient implementation of the additional functionality of $S_{GG,mer}$ to calculate the inverse of the S-box $S^{-1}(\cdot)$ is established by reusing three of the five affine transformations for performing the calculation of $S^{-1}(\cdot)$. Hence, only 4 additional slices are needed for the additional affine transformations of $S_{GG,mer}$ compared to the design of $S_{GG,enc}$. One additional slices is need for the additional multiplexer input of the $S_{GG,mer}$ design. The different slice occupation between $S_{GG,enc}$ and $S_{GG,dec}$ is caused by the different affine transformations, which are necessary to map the sharable quadratic permutation $G(\cdot)$ to $S(\cdot)$ or $S^{-1}(\cdot)$, respectively. A straightforward combined implementation of $S_{GG,enc}$ and $S_{GG,dec}$ can be estimated with a resource occupation of approximately 21 slices without considering further optimization due to the synthesis step, cf. Tab. 3. Hence, this amount of resource utilization is by far larger (approx. 24%) than the resource utilization of the proposed $S_{GG,mer}$ design.

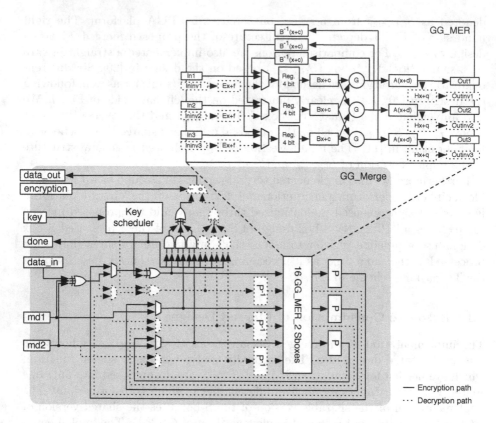

Fig. 2. Design of the merge PRESENT TI implementation

We integrated the three different shared $G(\cdot)$ function based S-box designs in four different PRESENT block cipher designs to evaluate the impact of the the GG_MER S-box design on the resource utilization. At first, we implement a round-oriented PRESENT ($P_{GG,enc}$) based on the optimized design in [11]. Second, we exchanged the affine representation according to Eq. (11) and extended the key scheduler to perform only decryptions with the second design ($P_{GG,dec}$). The third design $P_{GG,com}$ is a joined implementation of $P_{GG,enc}$ and $P_{GG,dec}$, which can perform both encryption and decryption by using the same secret key as input. One encryption run with 63 clock cycles has to be performed in advance to perform a decryption with the correct round keys, which are derived from the initial secret key. After one encryption run is conducted several decryptions can be performed sequently, which also takes 63 clock cycles by reusing the stored least round key.

The fourth implemented design $P_{GG,mer}$ is identical to $P_{GG,com}$ except for the used S-box design, instead of using the two S-boxes $S_{GG,enc}$ and $S_{GG,enc}$, the proposed merged S-box GG_MER is used. For better comparison the implementation properties of the various four designs are summarized in Tab. 4. The analysis results clearly indicate that the design $P_{GG,mer}$ is more efficient than

Table 3. Resource Utilization of PRESENT TI-S-box designs on Virtex-5

	Design			Performance properties		
Name	Based on	S-box mode $S(\cdot)$	$S^{-1}(\cdot)$	Area [Slices]	Latency [clk. cyc.]	Max. Freq. [MHz]
$S_{G'F,enc}$	[20]	x		14	1	410
$S_{GG,enc}$	[11]	x		11	1	374
$S_{GG,dec}$	Sect. 3.1		x	10	1	449
$S_{GG,mer}$	Sect. 3.1	x	x	16	1	346

$P_{GG,com}$. The resource utilization is reduced by 23%, while the throughput is increased by 4 Mbit/s, if the S-box GG_MER is used instead of a combined version of $P_{GG,enc}$ and $P_{GG,dec}$.

4.2 Merged Present Design for Encryption and Decryption

To provide a holistic view, we also provide results for ASIC implementations. We used *Synopsys Design Compiler* version *E-2010.12-SP2* to synthesize our designs to the standard cell library *UMCL18G212T3*, which is based on a 180nm process. Tab. 5 shows that the design $P_{GG,mer}$ requires 28% less resources than the $P_{GG,com}$ design. We used both resource optimization option *simple* and *ultra* of the synthesis tool for a more general comparison between the designs.

5 Side-Channel Evaluation

In this section we present practical side-channel evaluations of our new design to prove that it matches the same security level as previous designs, i.e., resistance against first-order attacks even in the presence of glitches. We mount two attacks, i.e., Correlation-Power-Analysis (CPA) [5] targeting the register update, and a correlation-enhanced collision attack [15] targeting glitches in the S-box calculation.

Table 4. Resource Utilization of TI-masked PRESENT designs on Virtex-5

	Design			Performance properties		
Name	Mode ENC	DEC	Area [Slices]	Freq. [Mhz]	Duration [clk cyc.]	T'put [Mbit/s]
$P_{GG,enc}$	x		278	265	63	269
$P_{GG,dec}$		x	251	266	63(+63)	270
$P_{GG,com}$	x	x	463	252	63(+63)	256
$P_{GG,mer}$	x	x	355	256	63(+63)	260
Difference			-23.0%	+1.6%	–	+1.6%

Table 5. Resource Utilization of TI-masked PRESENT designs on UMC180

| Name | Design | | Duration | Synth. opt. for area | |
| | Mode | | [clk cyc.] | Simple | Ultra |
	ENC	DEC		[GE]	[GE]
$P_{GG,com}$	x	x	63(+63)	14,592	12,259
$P_{GG,mer}$	x	x	63(+63)	10,550	8,830
Difference			–	-27.7%	-28.0%

We implemented our design on a SASEBO-GII side-channel evaluation board (hosting a Xilinx Virtex-5 FPGA) running at 2 MHz. For the analysis we recorded 5,000,000 traces with a sampling rate of 1GS/s using a LeCroy WaveRunner 610Zi oscilloscope. Every trace consists of 2,000 samples and covers the first three clock cycles of an encryption, see Fig. 3.

Since our design differs from previous designs we had to modify the attack models. As mentioned before we target the register update of the three 4-bit registers within the S-box. Note that the registers save the S-box input, unlike many other implementations which save the S-box output. In the first clock cycle of an encryption the values data_in $\oplus$ md1 $\oplus$ md2 $\oplus$ key, md1, and md2 are saved in the registers and are overwritten in the second clock cycle by the intermediate (shared) S-box state $B^{-1}(G(B(x) \oplus c) \oplus c)$, cf. Fig. 2. Hence, we chose our attack model to be the Hamming distance between these two register values of two consecutive clock cycles, i.e., the transition count of the register update. To verify our setup as well as our attack models we first mounted the two aforementioned attacks with known masks on 50,000 traces, targeting the highest nibble (CPA) and the highest and second highest nibble, respectively (collision). Figure 4(a) and 4(b) show the two successful attacks yielding clear correlations in the second clock cycle (time frame from samples 1300 to 1400 cf.

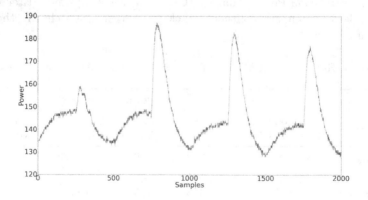

Fig. 3. Power Trace of the first three clock cycles of a PRESENT encryption

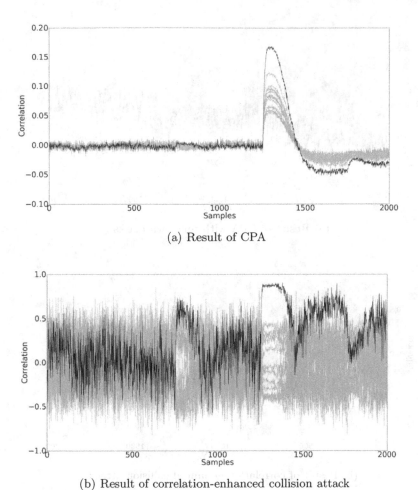

(a) Result of CPA

(b) Result of correlation-enhanced collision attack

Fig. 4. Attacks on the proposed TI PRESENT design with known masks

Fig. 3), as expected. Approximately 800 traces are needed to recover the correct key nibble.

Finally, we mounted the two attacks on the 5,000,000 traces. As we can see in Fig. 5(a) and 5(b), the correct key hypothesis does not yield the highest correlation, neither in the CPA nor in the correlation-enhanced collision attack. Hence, we have practically shown that our design matches the same security level as previous designs, while being significantly smaller, cf. [11, 20] The results of the side-channel analysis resistance against first order attacks of the encryption can be directly transferred to the decryption of GG_Merge, because the same secured non-linear 3-share function $G(\cdot)$ is used.

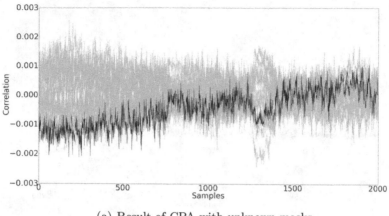

(a) Result of CPA with unknown masks

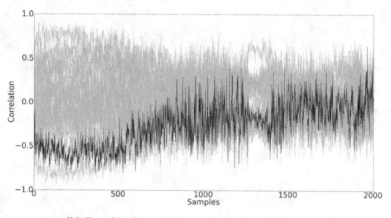

(b) Result of correlation-enhanced collision attack

Fig. 5. Attacks on the proposed TI PRESENT design with unknown masks

6 Conclusion

In this paper we presented an efficient way to merge at least two different S-box descriptions into one design by using a single quadratic function as the non-linear core together with some affine transformations. We applied our theoretic investigation on the PRESENT S-box $S(\cdot)$ and its inverse $S^{-1}(\cdot)$ as a demonstrative example. Hence, we optimized the 4-bit substitution element of PRESENT by using only 5 linear affine transformations and a single non-linear mapping core to perform both $S(\cdot)$ and $S^{-1}(\cdot)$. Due to this decomposed structure the Threshold Implementation countermeasure [17] can be applied straightforward on the proposed merged S-box design. Additionally, we implemented several round based TI masked versions of PRESENT capable to perform encryption and decryption. For FPGA implementations, the overall resource utilization of the entire PRESENT implementation using the proposed merged S-box design is at least 23% smaller than using an S-box design based on the optimal S-box design proposed in [11] while at the same time achieving a higher throughput. For ASIC implementation our approach saves 28% of the area. Our evaluation shows that the proposed optimization does not cause any new exploitable design flaws and thus it is still resistant against first-order attacks using up to 5,000,000 traces. For future work we will investigate the usage of the proposed decomposition for merging S-boxes of different classes.

References

1. ISO/IEC 29192-2: Information Technology – Security Techniques – Lightweight Cryptography – Part 2: Block Ciphers
2. Agrawal, D., Archambeault, B., Rao, J.R., Rohatgi, P.: The EM side-channel(s). In: Kaliski Jr., B.S., Koç, Ç.K., Paar, C. (eds.) CHES 2002. LNCS, vol. 2523, pp. 29–45. Springer, Heidelberg (2003)
3. Bilgin, B., Nikova, S., Nikov, V., Rijmen, V., Stütz, G.: Threshold Implementations of All 3 3 and 4 4 S-Boxes. In: Prouff, E., Schaumont, P. (eds.) CHES 2012. LNCS, vol 7428, pp. 76–91. Springer, Heidelberg (2012)
4. Bogdanov, A.A., Knudsen, L.R., Leander, G., Paar, C., Poschmann, A., Robshaw, M., Seurin, Y., Vikkelsoe, C.: PRESENT - An Ultra-Lightweight Block Cipher. In: Paillier, P., Verbauwhede, I. (eds.) CHES 2007. LNCS, vol. 4727, pp. 450–466. Springer, Heidelberg (2007)
5. Brier, E., Clavier, C., Olivier, F.: Correlation Power Analysis With a Leakage Model. In: Joye, M., Quisquater, J.-J. (eds.) CHES 2004. LNCS, vol. 3156, pp. 16–29. Springer, Heidelberg (2004)
6. Carlet, C., Charpin, P., Zinoviev, V.: Codes, bent functions and permutations suitable for des-like cryptosystems. Des. Codes Cryptography 15(2), 125–156 (1998)
7. Fischer, W., Gammel, B.M.: Masking at gate level in the presence of glitches. In: Rao, J.R., Sunar, B. (eds.) CHES 2005. LNCS, vol. 3659, pp. 187–200. Springer, Heidelberg (2005)
8. Kocher, P.C.: Timing attacks on implementations of Diffie-Hellman, RSA, DSS, and other systems. In: Koblitz, N. (ed.) CRYPTO 1996. LNCS, vol. 1109, pp. 104–113. Springer, Heidelberg (1996)
9. Kocher, P.C., Jaffe, J., Jun, B.: Differential Power Analysis. In: Wiener, M. (ed.) CRYPTO 1999. LNCS, vol. 1666, pp. 388–397. Springer, Heidelberg (1999)
10. Kutzner, S., Nguyen, P.H., Poschmann, A.: Enabling 3-share Threshold Implementations for any 4-bit S-box. IACR Cryptology ePrint Archive 2012, 510 (2012)
11. Kutzner, S., Nguyen, P.H., Poschmann, A., Wang, H.: On 3-share Threshold Implementations for 4-bit S-boxes. In: Prouff, E. (ed.) COSADE 2013. LNCS, vol. 7864, pp. 99–113. Springer, Heidelberg (2013)
12. Leander, G., Poschmann, A.: On the Classification of 4 Bit S-Boxes. In: Carlet, C., Sunar, B. (eds.) WAIFI 2007. LNCS, vol. 4547, pp. 159–176. Springer, Heidelberg (2007), http://dx.doi.org/10.1007/978-3-540-73074-3_13
13. Mangard, S., Popp, T., Oswald, M.E.: Power Analysis Attacks - Revealing the Secrets of Smart Cards. Springer (2007)

14. Mangard, S., Pramstaller, N., Oswald, E.: Successfully attacking masked aes hardware implementations. In: Rao, J.R., Sunar, B. (eds.) CHES 2005. LNCS, vol. 3659, pp. 157–171. Springer, Heidelberg (2005)
15. Moradi, A., Mischke, O., Eisenbarth, T.: Correlation-Enhanced Power Analysis Collision Attack. In: Mangard, S., Standaert, F.-X. (eds.) CHES 2010. LNCS, vol. 6225, pp. 125–139. Springer, Heidelberg (2010)
16. Nikova, S., Rijmen, V., Schläffer, M.: Secure Hardware Implementation of Nonlinear Functions in the Presence of Glitches. In: Lee, P.J., Cheon, J.H. (eds.) ICISC 2008. LNCS, vol. 5461, pp. 218–234. Springer, Heidelberg (2009)
17. Nikova, S., Rijmen, V., Schläffer, M.: Secure Hardware Implementation of Nonlinear Functions in the Presence of Glitches. J. Cryptology 24(2), 292–321 (2011)
18. Nyberg, K.: Perfect nonlinear s-boxes. In: Davies, D.W. (ed.) EUROCRYPT 1991. LNCS, vol. 547, pp. 378–386. Springer, Heidelberg (1991)
19. Nyberg, K.: Differentially uniform mappings for cryptography. In: Helleseth, T. (ed.) EUROCRYPT 1993. LNCS, vol. 765, pp. 55–64. Springer, Heidelberg (1994)
20. Poschmann, A., Moradi, A., Khoo, K., Lim, C., Wang, H., Ling, S.: Side-Channel Resistant Crypto for less than 2,300 GE. Journal of Cryptology, 1–24 (2010)

A   Affine Transformation Parameters

Affine transformation parameters for the PRESENT S-box used for encryption:

$$A = \begin{pmatrix} 1 & 0 & 1 & 0 \\ 0 & 1 & 0 & 0 \\ 1 & 0 & 0 & 0 \\ 1 & 0 & 1 & 1 \end{pmatrix}_2, \quad B = \begin{pmatrix} 1 & 1 & 0 & 0 \\ 0 & 1 & 1 & 0 \\ 0 & 0 & 1 & 0 \\ 1 & 0 & 1 & 0 \end{pmatrix}_2,$$

$$c = (0001)_2, \text{ and } d = (0101)_2. \tag{10}$$

Affine transformation parameters for the PRESENT S-box used for decryption:

$$\widetilde{A} = \begin{pmatrix} 1 & 0 & 0 & 0 \\ 0 & 1 & 1 & 1 \\ 0 & 1 & 0 & 1 \\ 1 & 0 & 0 & 1 \end{pmatrix}_2, \quad \widetilde{B} = \begin{pmatrix} 0 & 1 & 0 & 1 \\ 1 & 0 & 1 & 0 \\ 1 & 1 & 0 & 1 \\ 1 & 1 & 0 & 0 \end{pmatrix}_2,$$

$$\widetilde{c} = (1101)_2, \text{ and } \widetilde{d} = (0010)_2. \tag{11}$$

Affine transformation parameters for the decryption functionality of the merged PRESENT S-box design:

$$H = \begin{pmatrix} 0 & 0 & 1 & 0 \\ 1 & 1 & 0 & 1 \\ 0 & 1 & 1 & 1 \\ 1 & 0 & 1 & 1 \end{pmatrix}_2, \quad E = \begin{pmatrix} 1 & 0 & 0 & 0 \\ 1 & 1 & 0 & 1 \\ 0 & 1 & 1 & 1 \\ 1 & 0 & 1 & 1 \end{pmatrix}_2,$$

$$f = (0011)_2, \text{ and } q = (0011)_2. \tag{12}$$

Efficient Masked S-Boxes Processing
– A Step Forward –

Vincent Grosso[1], Emmanuel Prouff[2], and François-Xavier Standaert[1]

[1] ICTEAM/ELEN/Crypto Group, Université catholique de Louvain, Belgium
[2] ANSSI, 51 Bd de la Tour-Maubourg, 75700 Paris 07 SP, France

Abstract. To defeat side-channel attacks, the implementation of block cipher algorithms in embedded devices must include dedicated countermeasures. To this end, security designers usually apply secret sharing techniques and build masking schemes to securely operate an shared data. The popularity of this approach can be explained by the fact that it enables formal security proofs. The construction of masking schemes thwarting *higher-order* side-channel attacks, which correspond to a powerful adversary able to exploit the leakage of the different shares, has been a hot topic during the last decade. Several solutions have been proposed, usually at the cost of significant performance overheads. As a result, the quest for efficient masked S-box implementations is still ongoing. In this paper, we focus on the scheme proposed by Carlet *et al* at FSE 2012, and latter improved by Roy and Vivek at CHES 2013. This scheme is today the most efficient one to secure a generic S-box at any order. By exploiting an idea introduced by Coron *et al* at FSE 2013, we show that Carlet *et al*'s scheme can still be improved for S-boxes with input dimension larger than four. We obtain this result thanks to a new definition for the addition-chain exponentiation used during the masked S-box processing. For the AES and DES S-boxes, we show that our improvement leads to significant efficiency gains.

1 Introduction

Side-channel attacks (SCA) are a class of attacks, where the attacker has access to some *leakages* about the internal state during the computation [16]. In practice, such scenario makes it possible to attack implementations that are believed secure against classical (black-box) cryptanalyses. To defeat SCA, implementations of cryptographic algorithms must embed appropriate countermeasures. This is actually mandatory for implementations dedicated to smart card payments, pay-TV applications or citizen authentication with e-Passports.

Securing block cipher implementations has been a long-standing issue for the embedded systems industry. A sound approach is to use *secret sharing* [2,24], often called *masking* in the context of side-channel attacks [6]. The principle is to split every *sensitive* variable[1] x occurring during the computation into $d + 1$

---

[1] A variable is said to be *sensitive* in an SCA context if it functionally depends on both a public variable and a secret whose size is small enough to enable exhaustive search.

D. Pointcheval and D. Vergnaud (Eds.): AFRICACRYPT 2014, LNCS 8469, pp. 251–266, 2014.

shares $x_0, \ldots, x_d$ in such a way that the following relation is satisfied for a group operation $\perp$:

$$x_0 \perp x_1 \perp \cdots \perp x_d = x \ . \tag{1}$$

In the rest of the paper, we shall consider that $\perp$ is the addition over some field of characteristic 2 (*i.e.* $\perp$ will be the bitwise addition $\oplus$). Usually, the d shares $x_1, \ldots, x_d$ (called *the masks*) are randomly picked up and the last one x_0 (called *the masked variable*) is processed such that it satisfies (1). The full tuple $(x_i)_i$ is further called a *dth-order encoding of x*. When d random masks are involved per sensitive variable, the masking is said to be *of order d*. It has been shown that the complexity of mounting a successful side-channel attack against a masked implementation increases exponentially with the masking order [6,10,19]. Starting from this observation, the design of efficient masking schemes for different ciphers has become a foreground issue.

Higher-Order Masking Schemes. A *higher-order secure (masking) scheme* must ensure that the final shares correspond to the expected ciphertext on the one hand, and it must ensure the *dth-order security property* for the chosen order d on the other hand. The latter property states that every tuple of d or less intermediate variables is independent of any sensitive variable. When satisfied, it guarantees that no attack exploiting information on less than d intermediate results can succeed. As argued in several previous works (see *e.g.* [21] or [23]), the main difficulty in designing higher-order secure schemes for block ciphers lies in masking the *S-box(es)*, which are the only internal primitives that perform non-linear operations.

Masking and S-Boxes. Whereas many solutions have been proposed to deal with the case of first-order masking (see *e.g.* [3,18]), only few solutions exist for the higher-order case. A scheme has been proposed by Schramm and Paar in [23] which generalizes the (first-order) table re-computation method described in [18]. Although the authors apply their method in the particular case of an AES implementation, it is generic and can be applied to protect any S-box. Unfortunately, this scheme has been shown to be vulnerable to a 3rd-order attack whatever the chosen masking order [8]. In other words, it only provides 2nd-order security. Further schemes were proposed by Rivain, Dottax and Prouff in [20] with formal security proofs but still limited to 2nd-order security.

To the best of our knowledge, four approaches currently exist which enable the design of dth-order secure masking schemes for any arbitrary chosen d. One is due to Genelle *et al* and consists in mixing additive and multiplicative sharings (namely to use alternatively (1) for $\perp = \oplus$ and $\perp = \times$). This scheme is primarily dedicated to the AES algorithm and seems difficult to generalize efficiently to other block ciphers where the S-box is not affinely equivalent to a power function. The second one is due to Prouff and Roche and it relies on solutions developed in secure *multi-party computation* [1]. It is much less efficient than the other schemes (see *e.g.* [12]) but, contrary to them, remains secure even in presence of hardware glitches [17]. The third approach has been recently proposed by Coron in [7]. The core idea is to represent the S-box by several look-up tables which are regenerated from fresh random masks and the S-box truth table, each time a new S-box processing must be done. It extends the *table re-computation technique* introduced in the original paper by Kocher *et al* [16]. The security of Coron's scheme against higher-order SCA is formally proved under the assumption that the variable shares leak independently. Its asymptotic timing complexity is quadratic in the number of shares and can be applied to any S-box. However, the RAM memory consumption to secure (at order d) an S-box with input (resp. output) dimension n (resp. m) is $m(d+1)2^n$ bits, which can quickly exceed the memory capacity of the hosted device.

The three methods recalled in previous paragraph have important limitations which strongly impact their practicability: the first method is hardly generalizable to any S-box and the two other ones have a large extra cost (in terms of either processing or memory complexity). Actually, when the S-box to secure is not a power function and has input/output dimensions close to 8, the fourth approach is the most practical one when d is greater than or equal to 3. This approach, proposed in [5], generalizes the study conducted in [21] for power functions. The core idea is to split the S-box processing into a sequence of field multiplications and $\mathbb{F}_2$-linear operations, and then to secure both operations independently. The complexity of the masking schemes for the multiplication and a $\mathbb{F}_2$-linear operation[2] is $\mathcal{O}(d^2)$ and $\mathcal{O}(d)$ respectively. Moreover, the constant terms in these complexities are (usually) significantly greater for the multiplication than for the $\mathbb{F}_2$-linear operations. Based on this observation, the authors of [5] propose to look for S-box representations that minimize the number of field multiplications which are not $\mathbb{F}_2$-linear[3] (this kind of multiplication shall be called *non-linear* in this paper). This led them to introduce the notion of *S-box masking complexity*, which corresponds to the minimal number of non-linear multiplications needed to evaluate the S-box. This complexity is evaluated for any power function defined in $\mathbb{F}_{2^n}$ with $n \leq 10$ (in particular, the complexity of $x \in \mathbb{F}_{2^8} \mapsto x^{254}$, which is the non-linear part of the AES S-box, is shown to be equal to 4). Tight upper bounds on the masking complexity are also given for any random S-box. The work of Carlet *et al* has been further improved by Roy and Vivek in [22], where it is in particular shown that the masking complexity of the DES S-boxes is lower-bounded by 3. The authors of [22] also present a method that requires 7 non-linear multiplications. Another improvement of [5] has been proposed in [9], where it is shown that it is possible to improve the processing of the non-linear multiplications with the particular form $x \times g(x)$

---

[2] A function f is $\mathbb{F}_2$-linear if it satisfies $f(x \oplus y) = f(x) \oplus f(y)$ for any pair (x, y) of elements in its domain. This property must not be confused with linearity of a function which is defined such that $f(ax \oplus by) = af(x) \oplus bf(y)$. A linear function is $\mathbb{F}_2$-linear but the converse is false in general (the homogeneity of degree 1 must indeed be also satisfied for the converse to be true).

[3] A multiplication over a field of characteristic 2 is $\mathbb{F}_2$-linear if it corresponds to a squaring.

with g being $\mathbb{F}_2$-linear. This type of multiplication is called *bilinear* in the rest of the paper[4].

Our Contribution. In this paper we refine the notion of S-box masking complexity introduced in [5] and further studied in [22]. We still link it to the minimum number of non-linear multiplications needed to evaluate the S-box, but we don't include bilinear multiplications in this counting. We justify this choice thanks to the analysis in [9] which shows that the complexity of the latter multiplications is between that of general non-linear multiplications (costly) and that of $\mathbb{F}_2$-linear multiplications (cheap). For all exponentiations in $\mathbb{F}_{2^n}$, with $n \in \{4, 6, 8\}$, we give the new masking complexities and we afterwards illustrate, for the AES and DES S-boxes, the effective gain obtained by using the corresponding new *addition-chain exponentiation* [11]. This works raises the need for new polynomial evaluation algorithms minimizing the number of multiplications which are neither linear nor bilinear. It could also be of interest to study whether specialized (efficient) schemes cannot be dedicated to the secure processing of other types of non-linear multiplications (which are not linear or bilinear but have some helpful properties).

2   Existing Schemes for Elementary Operations

In this section, it is assumed that the S-box to protect manipulates data of bit-length n (typically $n \in \{4, 8, 16\}$). Depending on the kind of operation to process, these data can be viewed as elements of the vector space $\mathbb{F}_2^n$ defined over the field $(\mathbb{F}_2, \oplus, \&)$, where $\oplus$ is the XOR operation and $\&$ the AND operator. Or, they can be defined as elements of the field $\mathbb{F}_{2^n} \cong (\mathbb{F}_2[X]/p(X), \oplus, \times)$, where $p(X)$ is an irreducible polynomial of degree n and $\times$ denotes the polynomial multiplication modulo $p(X)$.

As recalled in the previous section, the most efficient solution which today exists to secure an S-box against higher-order SCA is to rewrite it as a polynomial function over $\mathbb{F}_{2^n}$ and to split its evaluation as a sequence of $\mathbb{F}_2$-linear operations and multiplications. Indeed, whatever d, d^{th}-order secure schemes exist for these two types of operations. We recall them hereafter. Note that, some mask refreshing must sometimes be done between different calls to these algorithms in order to guaranty the security of the whole process. Since, mask refreshing has a minor impact on the efficiency improvement proposed in this paper we do not recall here the mask refreshing algorithm and we exclude it from the description of the S-box secure evaluation procedures in Section 3 (for more details about this point we suggest the reading of [9,21]).

---

[4] We chose this term because the multiplication $y \times g(x)$, viewed as a function over $\mathbb{F}_{2^n} \times \mathbb{F}_{2^n}$, is indeed $\mathbb{F}_2$-bilinear when g is $\mathbb{F}_2$-linear. For such a function g, it may be checked that the *algebraic degree* [4, Chapter 9] of $x \mapsto x \times g(x)$, viewed as a vectorial function, is quadratic.

$\mathbb{F}_2$-*linear operation.* To securely process a $\mathbb{F}_2$-linear function g on a data x encoded by the tuple $(x_i)_i$, we just need to evaluate the function on each share x_i separately. The sharing $(g(x_i))_i$ is indeed an encoding of $g(x)$.

Algorithm 1. Secure evaluation of a $\mathbb{F}_2$-linear function g

Require: Shares $(x_i)_i$ satisfying $\oplus_i x_i = x$.
Ensure: Shares $(y_i)_i$ satisfying $\oplus_i y_i = g(x)$.
 1: **for** i from 0 to d **do**
 2:     $y_i \leftarrow g(x_i)$
 3: **end for**

In particular, Algorithm 1 can be applied to secure the Frobenius endomorphism over the field $\mathbb{F}_{2^n}$ (i.e. the squaring in characteristic 2) as this operation is $\mathbb{F}_2$-linear.

Multiplication. To securely process the multiplication between two sensitive variables x and y encoded by $(x_i)_i$ and $(y_i)_i$ respectively, the following algorithm has been proposed in [13] (and generalized in [21]).

Algorithm 2. Multiplication of two masked secrets x and y

Require: Shares $(x_i)_i$ and $(y_i)_i$ satisfying $\oplus_i x_i = x$ and $\oplus_i y_i = y$
Ensure: Shares $(w_i)_i$ satisfying $\oplus_i w_i = x \times y$
 1: **for** i from 0 to d **do**
 2:     **for** j from $i + 1$ to d **do**
 3:         $r_{i,j} \in_R \mathbb{F}_{2^n}$
 4:         $r_{i,j} \leftarrow (r_{i,j} \oplus x_i \times y_j) \oplus x_j \times y_i$
 5:     **end for**
 6: **end for**
 7: **for** i from 0 to d **do**
 8:     $w_i \leftarrow x_i \times y_i$
 9:     **for** j from 0 to d, $j \neq i$ **do**
10:         $w_i \leftarrow w_i \oplus r_{i,j}$
11:     **end for**
12: **end for**

Remark 1. The order of the XORs operations in Step 4 must be respected for the security guarantee to hold.

Starting from Lagrange's interpolation formula, [5] and [22] introduce S-box evaluation techniques which are only based on Algorithms 1 and 2 (and a third algorithm used to refresh the sharings when the input sharings correspond to dependent variables). Because the complexity of the dth-order secure multiplication is quadratic, whereas that of an $\mathbb{F}_2$-linear function is linear, the polynomial evaluation strategies try to minimize the number of calls to Algorithm 2. However, Coron *et al* have recently shown that multiplications of the form $x \times g(x)$, with g being $\mathbb{F}_2$-linear, can be securely evaluated more efficiently than standard multiplications [9]. This observation naturally raises the following new question: can we improve the complexities of the S-box evaluation strategies in [5,22] by replacing, as much as possible, standard multiplications by multiplications in the form $x \times g(x)$. Before dealing with this question, let us first recall the particular multiplication proposed in [9].

Multiplications of the form $x \times g(x)$, with g $\mathbb{F}_2$-linear. To securely process this type of multiplication, the following algorithm is proposed in [9].

Algorithm 3. Secure evaluation of a product of $h(x) = x \times g(x)$

Require: shares $(x_i)_i$ satisfying $\oplus_i x_i = x$.
Ensure: shares $(y_i)_i$ satisfying $\oplus_i y_i = h(x)$.
1: **for** i from 0 to d **do**
2:     **for** j from $i + 1$ to d **do**
3:         $r_{i,j} \in_R \mathbb{F}_{2^n}$
4:         $r'_{i,j} \in_R \mathbb{F}_{2^n}$
5:         $t \leftarrow r_{i,j}$
6:         $t \leftarrow t \oplus h(x_i \oplus r'_{i,j})$
7:         $t \leftarrow t \oplus h(x_j \oplus r'_{i,j})$
8:         $t \leftarrow t \oplus h((x_i \oplus r'_{i,j}) \oplus x_j)$
9:         $t \leftarrow t \oplus h(r'_{i,j})$
10:        $r_{j,i} \leftarrow t$
11:     **end for**
12: **end for**
13: **for** i from 0 to d **do**
14:    $y_i \leftarrow h(x_i)$
15:    **for** j from 0 to d, $j \neq i$ **do**
16:       $y_i \leftarrow y_i \oplus r_{i,j}$
17:    **end for**
18: **end for**

Notation. In the particular case where g is an exponentiation by a power of 2, say $g(x) = x^{2^s}$, the function h is denoted by h_{s+1}.

The complexity of Algorithm 3 is still quadratic but, for many typical application contexts, the constant terms are much smaller than in Algorithm 2. Indeed, the processing of h can be tabulated on standard embedded processors as long as $n \leqslant 10$, whereas the field multiplications $\times$ occurring in Algorithm 2 cannot if $n \geqslant 5$. In the following, functions/operations which can be evaluated thanks to Algorithm 1 or Algorithm 2 will be said to be of Type-I or Type-III respectively. Functions of the form $x \times g(x)$ with g $\mathbb{F}_2$-linear will be said to be of Type-II. Table 1 summarizes the cost of the three algorithms in term of XORs, field multiplications and look-up table accesses (referred to as LUT access).

Table 1. Cost of different algorithms

	XOR	Multiplication	LUT access
Algorithm 1	0	0	$d + 1$
Algorithm 2	$2d^2 + 2d$	$d^2 + 2d + 1$	0
Algorithm 3	$5d^2 + 5d$	0	$2d^2 + 3d + 1$

In most of classical architectures, a memory access (or a XOR) can be done in 1 or 2 CPU clock cycles, whereas the processing of a field multiplication with the CPU instructions set only requires between 20 and 40 cycles (we recall some classical field multiplication algorithms in Appendix A). This explains why the replacing of Type-III operations by Type-II ones leads to a significant efficiency improvement when $n \in [5; 10]$. Based on this observation, we propose in the next section new sequences of operations that lead to practically more efficient processing of power functions than the state of the art solutions [5,22].

3 New Proposal for Power Functions Evaluation

Considering the fact that the processing of power functions in the form x^{1+2^s} (which corresponds to the Type-II operation $x \times x^{2^s}$) is more efficient than that of other power functions, we followed an approach close to [5] in order to exhibit the most efficient processing for any power function defined in $\mathbb{F}_{2^n}$ for $n \leq 8$. Namely, for every power function $x \mapsto x^\alpha$, we exhibit by exhaustive search a sequence of operations of types I, II and III, which minimizes first the number of Type III operations, and then the number of Type II operations. This amounts to find, for each exponent α, the shortest *addition chain*[5] [15] with the supplementary constraint that multiplications by 2^t, for any integer t, or additions in the form $v + 2^t v$ are for free. We recall that an addition chain for $\alpha \in \mathbb{N}$ is an increasing sequence of integers $v_0, ..., v_s$ such that $v_0 = 1$, $v_s = \alpha$ and for any $j \neq 0$ there exist two indices $i < j$ and $k < j$ (not necessary different) s.t. $v_j = v_i + v_k$. The length of such a sequence is defined as the total number of additions (including multiplications by 2) needed to get $v_s = \alpha$ from $v_0 = 1$, with only operations between elements of the sequence. The definition of length used in [5,22] excludes multiplications by 2. For the reasons discussed previously, we extend the classical definition of the addition chain by adding the operation $v \mapsto (1 + 2^t)v$ for any integer t. We moreover assume that this operation is also excluded from the sequence length definition (it indeed corresponds to the function $h_{t+1} : x^v \mapsto x^{(1+2^t)v}$). The corresponding new length definition is referred to as *extended length* in the following. Our purpose is to minimize it. This point is the main (and important) difference with the (shortest) sequences investigated in [5]. Our results are given[6] in Table 2 for $n = 8$, where the exponents are grouped into classes. Each class, say C_j, corresponds to the set of exponents which can be obtained by multiplying j by a power of 2 (modulo $2^n - 1$).

---

[5] In the context of exponentiation processing, these chains are sometimes also referred to as *addition-chain exponentiation* (see for instance [11]).

[6] Tables for the cases $n = 4, 6$ are given in Appendix B.

Table 2. Smallest cost to process x^α with operations of types II and III

# Type-II	# Type-III	Exponent α
0	0	$C_0 = \{0\}, C_1 = \{1, 2, 4, 8, 16, 32, 64, 128\}$
1	0	$C_3 \quad = \quad \{3, 6, 12, 24, 48, 96, 192, 129\}$, $C_5 \quad = \quad \{5, 10, 20, 40, 80, 160, 65, 130\}$, $C_9 = \{9, 18, 36, 72, 144, 33, 66, 132\}$, $C_{17} = \{17, 34, 68, 136\}$
2	0	$C_{15} \quad = \quad \{15, 30, 60, 120, 240, 225, 195, 135\}$, $C_{21} \quad = \quad \{21, 42, 84, 168, 81, 162, 69, 138\}$, $C_{25} \quad = \quad \{25, 50, 100, 200, 145, 35, 70, 140\}$, $C_{27} \quad = \quad \{27, 54, 108, 216, 177, 99, 198, 141\}$, $C_{45} \quad = \quad \{45, 90, 180, 105, 210, 165, 75, 150\}$, $C_{51} = \{51, 102, 204, 153\}$, $C_{85} = \{85, 170\}$
3	0	$C_{63} \quad = \quad \{63, 126, 252, 249, 243, 231, 207, 159\}$, $C_{95} \quad = \quad \{95, 190, 125, 250, 245, 235, 215, 175\}$, $C_{111} = \{111, 222, 189, 123, 246, 237, 219, 183\}$
4	0	$C_{39} \quad = \quad \{39, 78, 156, 57, 114, 228, 201, 147\}$, $C_{55} \quad = \quad \{55, 110, 220, 185, 115, 230, 205, 155\}$, $C_{87} = \{87, 174, 93, 186, 117, 234, 213, 171\}$
1	1	$C_7 \quad = \quad \{7, 14, 28, 56, 112, 224, 193, 131\}$, $C_{11} \quad = \quad \{11, 22, 44, 88, 176, 97, 194, 133\}$, $C_{13} \quad = \quad \{13, 26, 52, 104, 208, 161, 67, 134\}$, $C_{19} \quad = \quad \{19, 38, 76, 152, 49, 98, 196, 137\}$, $C_{37} = \{37, 74, 148, 41, 82, 164, 73, 146\}$
2	1	$C_{23} \quad = \quad \{23, 46, 92, 184, 113, 226, 197, 139\}$, $C_{29} \quad = \quad \{29, 58, 116, 232, 209, 163, 71, 142\}$, $C_{31} \quad = \quad \{31, 62, 124, 248, 241, 227, 199, 143\}$, $C_{43} \quad = \quad \{43, 86, 172, 89, 178, 101, 202, 149\}$, $C_{47} \quad = \quad \{47, 94, 188, 121, 242, 229, 203, 151\}$, $C_{53} \quad = \quad \{53, 106, 212, 169, 83, 166, 77, 154\}$, $C_{59} \quad = \quad \{59, 118, 236, 217, 179, 103, 206, 157\}$, $C_{61} \quad = \quad \{61, 122, 244, 233, 211, 167, 79, 158\}$, $C_{91} \quad = \quad \{91, 182, 109, 218, 181, 107, 214, 173\}$, $C_{119} = \{119, 238, 221, 187\}$
3	1	$C_{127} = \{127, 254, 253, 251, 247, 239, 223, 191\}$

Remark 2. As the cost of Type-I operations is negligible compared to the cost of operations of types II and III, we chose to not give them in Table 2.

Remark 3. The costs given in Table 2 have been obtained by first minimizing the global number of Type-II and Type-III operations, and then by minimizing the number of Type-III multiplications. It can be noticed that other minimization strategies could be applied. For instance, if the goal is to minimize the number of Type-III multiplications, then it can be checked that x^{254} can be evaluated without such operation: first process x^{63}, then $(x+x^{63})^3 = x^{189} + x^{127} + x^{65} + x^3$, end eventually process x^{189}, x^{65} and x^3, and subtract them to $(x + x^{63})^3$ to get

$x^{254} = (x^{127})^2$ (which gives a processing without Type-III operations and 9 Type-II operations).

For the exponentiation $x \mapsto x^{254}$ (the non-linear part of the AES S-box), we found the extended addition chain $(1, 2, 5, 25, 125, 127, 254)$ whose extended length is 1. This sequence indeed requires only 1 operation of Type-III($+$) (to get 127), 2 operations of Type-I ($\times$) (to get and 2 and 254) and 3 operations of Type-II ($\times(1 + 2^2)$) (to get 5, 25 and 125)). It may moreover be observed that the sequence involves the same operation $v \mapsto (1 + 2^t)v$ (for $t = 2$) each time, which reduces the memory required to implement the solution.

The extended addition chain used for the AES S-box is represented in Figure 1.

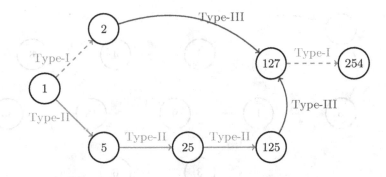

Fig. 1. AES S-box extended addition chain

Algorithm 4 shows how to use the extended addition chain to calculate invert in the field $\mathbb{F}_{2^8}$.

Algorithm 4. Exponentiation to the 254

Require: x.
Ensure: $y = x^{254}$.

$y \leftarrow x^2$	Type-I
$x \leftarrow x \times x^4$	Type-II
$x \leftarrow x \times x^4$	Type-II
$x \leftarrow x \times x^4$	Type-II
$y \leftarrow y \times x$	Type-III
$y \leftarrow y^2$	Type-I

For DES, we take advantage of the S-box representation proposed in [22]. In that paper it has been shown that all DES S-boxes can be calculated with 7 non-linear multiplications. They indeed can be represented by a polynomial of the form:

$$P_{\text{DES}}(x) = (x^{36} + p_1(x)) \times (((x^{18} + p_2(x)) \times p_3(x)) + (x^9 + p_4(x)))$$
$$+ ((x^{18} + p_5(x)) \times p_6(x) + (x^9 + p_7(x))),$$

where the polynomials $p_i(x)$ are of degree at most 9, and can be obtained by successive Euclidean polynomial divisions. Hence, only monomials of degree lower than 9 plus x^{18} and x^{36} are required to calculate any DES S-box. To compute these powers, we found an extended addition chain of extended length 1. It is represented in Figure 2 allows to calculate monomials x, x^2, x^3, x^4, x^5, x^6, x^7, x^8, x^9, x^{18} and x^{36}, where it can be checked that only 3 Type-II operations and 1 Type-III operation are needed.

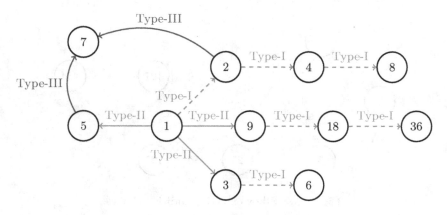

Fig. 2. DES monomials extended addition chain

Eventually, once monomials are calculated, we can evaluate the different polynomials $p_i(x)$. Then, 3 more operations of Type-III are required to calculate P_{DES}. These 3 multiplications are the 3 polynomial multiplications in P_{DES}. As a result, any of the DES S-boxes can be computed using 4 Type-III operations and 3 of Type-II operations. Full description of computation of DES S-boxes is given in Algorithm 5.

Remark 4. Remark that having an optimal representation of a polynomial that has exponents in different classes (as for the DES S-boxes) is quite challenging, which explains the poorer efficiency compared to the AES S-box. Quite naturally, we expect that further research should allow improving this (since smaller S-boxes should generally be easier to mask).

4   Efficiency Comparisons and Simulations

In this section, we compare, for different orders $d = 1, 2, 3$, the efficiency of our new extended addition chain with that of previous techniques to securely process

---

Algorithm 5. DES S-boxes

Require: x_1, $p_{i,j}$ coefficients of polynomial p_i.
Ensure: $p_2 = P_{\text{DES}}(x)$.

$x_2 \leftarrow x_1^2$	Type-I
$x_4 \leftarrow x_2^2$	Type-I
$x_8 \leftarrow x_4^2$	Type-I
$x_3 \leftarrow x_1 \times x_1^2$	Type-II
$x_6 \leftarrow x_3^2$	Type-I
$x_9 \leftarrow x_1 \times x_1^8$	Type-II
$x_{18} \leftarrow x_9^2$	Type-I
$x_{36} \leftarrow x_{18}^2$	Type-I
$x_5 \leftarrow x_1 \times x_1^4$	Type-II
$x_7 \leftarrow x_2 \times x_5$	Type-III
$p_1 \leftarrow \sum p_{1,j} x_j$	
$p_2 \leftarrow \sum p_{2,j} x_j$	
$p_3 \leftarrow \sum p_{3,j} x_j$	
$p_4 \leftarrow \sum p_{4,j} x_j$	
$p_5 \leftarrow \sum p_{5,j} x_j$	
$p_6 \leftarrow \sum p_{6,j} x_j$	
$p_7 \leftarrow \sum p_{7,j} x_j$	
$p_2 \leftarrow p_2 + x_{18}$	
$p_2 \leftarrow p_2 \times p_3$	Type-III
$p_2 \leftarrow p_2 + x_9$	
$p_2 \leftarrow p_2 + p_4$	
$p_1 \leftarrow p_1 + x_{36}$	
$p_2 \leftarrow p_2 \times p_1$	Type-III
$p_5 \leftarrow p_5 + x_{18}$	
$p_5 \leftarrow p_5 \times p_6$	Type-III
$p_5 \leftarrow p_5 + x_9$	
$p_5 \leftarrow p_5 + p_7$	
$p_2 \leftarrow p_2 + p_5$	

---

the AES S-box and the first DES S-box (similar results can be obtained for the other ones).

For the AES S-box, we implemented the schemes proposed in [21] and [9]. We also implemented the scheme in [14], which follows a Tower Fields approach to improve the timing complexities. Essentially, it consists in using the isomorphism between $\mathbb{F}_{2^8}$ and $(\mathbb{F}_{2^4})^2$ to have processing only in $\mathbb{F}_{2^4}$ where the multiplication can be tabulated. Since our approach, described in the previous section, is advantageous when the field multiplication cannot be tabulated, we did not consider to combine it with Tower Fields approach.

The implementations are done in C and compile for ATMEGA644p microcontroller thanks to the compiler avr_gcc with optimisation flag -o2. We also did some implementations directly in assembler for the same micro-controller.

For the AES the results are given in Table 3.

Table 3. Secure AES S-box for ATMEGA644p

Solution	[C] $d = 1$	[C] $d = 2$	[C] $d = 3$	[Assembly] $d = 1$
Addition chain [21]	753	1999	3702	623
Addition chain + Tower Fields [14]	897	1805	3077	565
Addition chain + Type-II op.[9]	540	1376	2554	431
Extended addition chain	488	1227	2319	338

For the DES the results are given in Table 4.

Table 4. Secure DES S-box for ATMEGA644p

Solution	C $d = 1$	C $d = 2$	C $d = 3$
Addition chain [22]	2001	4646	8182
Extended addition chain	1623	3574	7413

In Table 5, we compare, on ATMEGA644p, the practical costs of the Type-II and Type-III multiplications. For $d = 1$, it can be seen that Type-II multiplications are around 2.5 (when implemented in Assembly) or 2.4 (when implemented in C) faster than Type-III multiplications. This means that, on ATMEGA644p, replacing N Type-III multiplications by N' Type-II multiplications leads to a global efficiency gain as long as $N'/N \leq 2.4$. This ratio becomes 2.8 for $d = 2$ and 2.3 for $d = 3$.

Table 5. Costs comparison (in cycles) for Type-II and Type-III operations over $\mathbb{F}_{2^8}$

Operation	C $d = 1$	C $d = 2$	C $d = 3$	[Assembly] $d = 1$
Type-III	146	430	802	136
Type-II	61	152	344	54

As already pointed out, the interest of exchanging Type-III multiplications by Type-II ones is only advantageous when the field (Type-III) multiplications cannot been tabulated (i.e. when $n \geq 5$). Hence, for the 4-bit PRESENT S-box, our approach does not lead to practical efficiency improvement.

Conclusion

By exploiting an idea introduced by Coron *et al* at FSE 2013, we have shown in this paper that Carlet *et al*'s masking scheme can be improved when the S-box dimensions are too large to allow the tabulation of field multiplications. For this purpose, we introduced a new type of addition-chain exponentiation which combine three operations (multiplications by 2^s, multiplications by $1 + 2^s$ and additions) instead of two. For the AES and DES S-boxes, our improvement leads to an efficiency gain between 35% and 55%. Our work also opens avenues for further research of polynomial evaluation techniques minimizing the number of multiplications which are neither $\mathbb{F}_2$-linear nor $\mathbb{F}_2$-bilinear.

Acknowledgements. Work funded in parts by the European Commission through the ERC project 280141 (acronym CRASH) and the European ISEC action grant HOME/2010/ISEC/AG/INT-011 B-CCENTRE project. F.-X. Standaert is an associate researcher of the Belgian Fund for Scientific Research (FNRS-F.R.S.). We also thank Jean-Sébastien Coron for pointing us out the minimization strategy discussed in Remark 3.

References

1. Bellare, M., Goldwasser, S., Micciancio, D.: "Pseudo-random" number generation within cryptographic algorithms: The DSS case. In: Kaliski Jr., B.S. (ed.) CRYPTO 1997. LNCS, vol. 1294, pp. 277–291. Springer, Heidelberg (1997)
2. Blakely, G.: Safeguarding cryptographic keys. In: National Comp. Conf. vol. 48, pp. 313–317. AFIPS Press, New York (1979)
3. Blömer, J., Guajardo, J., Krummel, V.: Provably secure masking of AES. In: Handschuh, H., Hasan, M.A. (eds.) SAC 2004. LNCS, vol. 3357, pp. 69–83. Springer, Heidelberg (2004)
4. Carlet, C.: Boolean functions for cryptography and error correcting codes. Boolean Methods and Models, pp. 257 (2010)
5. Carlet, C., Goubin, L., Prouff, E., Quisquater, M., Rivain, M.: Higher-order masking schemes for S-boxes. In: Canteaut, A. (ed.) FSE 2012. LNCS, vol. 7549, pp. 366–384. Springer, Heidelberg (2012)
6. Chari, S., Jutla, C., Rao, J., Rohatgi, P.: Towards sound approaches to counteract power-analysis attacks. In: Wiener (ed.) [25], pp. 398–412
7. Coron, J.-S.: Higher Order Masking of Look-up Tables. In: Nguyen, P.Q., Oswald, E. (eds.) Advances in Cryptology – EUROCRYPT 2014. LNCS. Springer (to appear, 2014)
8. Coron, J.-S., Prouff, E., Rivain, M.: Side channel cryptanalysis of a higher order masking scheme. In: Paillier, P., Verbauwhede, I. (eds.) CHES 2007. LNCS, vol. 4727, pp. 28–44. Springer, Heidelberg (2007)
9. Coron, J.-S., Prouff, E., Rivain, M., Roche, T.: Higher-order side channel security and mask refreshing. In: Moriai, S. (ed.) Fast Software Encryption – FSE 2013. LNCS. Springer (2013)(to appear)
10. Duc, A., Dziembowski, S., Faust, S.: Unifying Leakage Models: from Probing Attacks to Noisy Leakage. In: Nguyen, P.Q., Oswald, E. (eds.) Advances in Cryptology – EUROCRYPT 2014. LNCS, Springer (to appear, 2014)
11. Gordon, D.M.: A survey of fast exponentiation methods. J. Algorithms 27(1), 129–146 (1998)
12. Grosso, V., Standaert, F.-X., Faust, S.: Masking vs. multiparty computation: How large is the gap for AES? In: Bertoni, G., Coron, J.-S. (eds.) CHES 2013. LNCS, vol. 8086, pp. 400–416. Springer, Heidelberg (2013)
13. Ishai, Y., Sahai, A., Wagner, D.: Private Circuits: Securing Hardware against Probing Attacks. In: Boneh, D. (ed.) CRYPTO 2003. LNCS, vol. 2729, pp. 463–481. Springer, Heidelberg (2003)
14. Kim, H., Hong, S., Lim, J.: A fast and provably secure higher-order masking of AES S-box. In: Preneel, B., Takagi, T. (eds.) CHES 2011. LNCS, vol. 6917, pp. 95–107. Springer, Heidelberg (2011)
15. Knuth, D.: The Art of Computer Programming, 3rd edn. vol. 2. Addison-Wesley (1988)

16. Kocher, P., Jaffe, J., Jun, B.: Differential Power Analysis. In: Wiener (ed.) [25], pp. 388–397.
17. Mangard, S., Popp, T., Gammel, B.M.: Side-Channel Leakage of Masked CMOS Gates. In: Menezes, A. (ed.) CT-RSA 2005. LNCS, vol. 3376, pp. 351–365. Springer, Heidelberg (2005)
18. Messerges, T.S.: Securing the AES finalists against power analysis attacks. In: Schneier, B. (ed.) FSE 2000. LNCS, vol. 1978, pp. 150–164. Springer, Heidelberg (2001)
19. Prouff, E., Rivain, M.: Masking against side-channel attacks: A formal security proof. In: Johansson, T., Nguyen, P.Q. (eds.) EUROCRYPT 2013. LNCS, vol. 7881, pp. 142–159. Springer, Heidelberg (2013)
20. Rivain, M., Dottax, E., Prouff, E.: Block ciphers implementations provably secure against second order side channel analysis. In: Nyberg, K. (ed.) FSE 2008. LNCS, vol. 5086, pp. 127–143. Springer, Heidelberg (2008)
21. Rivain, M., Prouff, E.: Provably secure higher-order masking of AES. In: Mangard, S., Standaert, F.-X. (eds.) CHES 2010. LNCS, vol. 6225, pp. 413–427. Springer, Heidelberg (2010)
22. Roy, A., Vivek, S.: Analysis and improvement of the generic higher-order masking scheme of FSE 2012. In: Bertoni, G., Coron, J.-S. (eds.) CHES 2013. LNCS, vol. 8086, pp. 417–434. Springer, Heidelberg (2013)
23. Schramm, K., Paar, C.: Higher order masking of the AES. In: Pointcheval, D. (ed.) CT-RSA 2006. LNCS, vol. 3860, pp. 208–225. Springer, Heidelberg (2006)
24. Shamir, A.: How to share a secret. Commun. ACM 22(11), 612–613 (1979)
25. Wiener, M. (ed.): CRYPTO 1999. LNCS, vol. 1666. Springer, Heidelberg (1999)

A   Field Multiplication Algorithms

In the algorithm hereafter, we recall how a multiplication over an extension of $\mathbb{F}_2$ can be done. Since we consider extensions of the form $\mathbb{F}_{2^n} \cong \mathbb{F}_2[X]/p(X)$ where the coefficients of $p(X)$ are in $\mathbb{F}_2$, we denote by $\boldsymbol{p}$ the binary vector whose coordinates are the coefficients of $p(X)$ (from MSB to LSB). The operation $\ll t$ stands for the shift of t bits and the ith bit of a binary vector b is denoted by $b(i)$.

For fields of small dimension (e.g. $n \leq 4$), the multiplication can be tabulated. Then, only one access to a double entry table is required to perform the multiplication in an efficient manner. If the field is composed of 2^n elements, the table will have 2^{2n} elements of size n. For larger fields (e.g. $n > 4$) the size of such a table becomes larger than the memory available in embedded system. Hence, other evaluation methods are applied, such that the so-called *log/alog tables* method. It is based on the fact that the non-zero elements of $\mathbb{F}_{2^n}$ can all be represented as a power of a primitive element which is a root of $p(X)$. The *log* table is used to get this power for each $x \in \mathbb{F}_2[X]/p(X)$, whereas the *alog* table is used to get the element of $\mathbb{F}_2[X]/p(X)$ that corresponds to a given power. Under this representation, multiplying two non-zero elements x and y, simply consists in processing $alog(log(x) + log(y) \bmod 2^n - 1)$.

Algorithm 6. Field multiplication naive way

Require: Field elements a, b in $\mathbb{F}_{2^n} \cong \mathbb{F}_2[X]/p(X)$, the binary representation $\boldsymbol{p}$ of $p(X)$.
Ensure: The field element c such that $c = a \times b$
1: $tmp \leftarrow a$
2: $c \leftarrow 0$
3: **for** i from 0 to $degree(p(X))$ **do**
4:     **if** $b(i) = 1$ **then**
5:         $c \leftarrow c \oplus tmp$
6:     **end if**
7:     $tmp \leftarrow tmp \ll 1$
8:     **if** $tmp(degree(p(X))) = 1$ **then**
9:         $tmp \leftarrow tmp \oplus \boldsymbol{p}$
10:     **end if**
11: **end for**

Algorithm 7. Field multiplication with log/alog tables

Require: Field elements a, b.
Ensure: c such that $c = a \times b$
1: $d \leftarrow log[a]$
2: $e \leftarrow log[b]$
3: $c \leftarrow d + e \bmod 2^n - 1$
4: $c \leftarrow alog[c]$

B Masking Complexity of Power Functions

For exponentiation in $\mathbb{F}_{2^4}$, we report on the cost of our extended addition chain in Table 6.

Table 6. Smallest cost to process x^α with operations of types II and III in $\mathbb{F}_{2^4}$

# Type-II	# Type-III	Exponent α
0	0	$C_0 = \{0\}, C_1 = \{1, 2, 4, 8\}$
1	0	$C_3 = \{3, 6, 12, 9\}, C_5 = \{5, 10\}$
1	1	$C_7 = \{7, 14, 13, 11\}$

For the case of operations in $\mathbb{F}_{2^6}$, like for the DES S-boxes. We report on the cost of our extended addition chain in Table 7.

Table 7. Smallest cost to process x^α with operations of types II and III in $\mathbb{F}_{2^6}$

# Type-II	# Type-III	Exponent α
0	0	$C_0 = \{0\}, C_1 = \{1, 2, 4, 8, 16, 32\}$
1	0	$C_3 = \{3, 6, 12, 24, 48, 33\}$, $C_5 = \{5, 10, 20, 40, 17, 34\}$, $C_9 = \{9, 18, 36\}$
2	0	$C_{11} = \{11, 22, 44, 25, 50, 37\}, C_{15} = \{15, 30, 60, 57, 51, 39\}$ $C_{27} = \{27, 54, 45\}$
1	1	$C_7 = \{7, 14, 28, 56, 49, 35\}$ $C_{13} = \{13, 26, 52, 41, 19, 38\}, C_{21} = \{21, 42\}$ $C_{31} = \{31, 62, 61, 59, 55, 47, \}$
2	1	$C_{23} = \{23, 46, 29, 58, 53, 43\}$

A More Efficient AES Threshold Implementation

Begül Bilgin[1,2], Benedikt Gierlichs[1], Svetla Nikova[1],
Ventzislav Nikov[3], and Vincent Rijmen[1]

[1] KU Leuven, ESAT-COSIC and iMinds, Belgium
{name.surname}@esat.kuleuven.be
[2] University of Twente, EEMCS-DIES, The Netherlands
[3] NXP Semiconductors, Belgium
venci.nikov@gmail.com

Abstract. Threshold Implementations provide provable security against
first-order power analysis attacks for hardware and software implementations. Like masking, the approach relies on secret sharing but it differs
in the implementation of logic functions. At EUROCRYPT 2011 Moradi
et al. published the to date most compact Threshold Implementation
of AES-128 encryption. Their work shows that the number of required
random bits may be an additional evaluation criterion, next to area and
speed. We present a new Threshold Implementation of AES-128 encryption that is 18% smaller, 7.5% faster and that requires 8% less random
bits than the implementation from EUROCRYPT 2011. In addition, we
provide results of a practical security evaluation based on real power
traces in adversary-friendly conditions. They confirm the first-order attack resistance of our implementation and show good resistance against
higher-order attacks.

Keywords: Threshold Implementation, First-order DPA, Glitches, Sharing, AES, S-box.

1 Introduction

Embedded devices seem to be easily protected by modern ciphers in a black-box
scenario. However, in the late 90s [10] the security of such devices has been shown
to depend on the algorithm implementation. During the computation of an algorithm the device leaks information. Side channel attacks (SCA) are among
the most relevant threats for the security of implementations of cryptographic
algorithms. Certain countermeasures aim at introducing noise in the side channel, e.g. random delays, random order execution, dummy operations, etc., while
masking conceals all sensitive intermediate values of a computation with random
data and allows one to formally argue the security such a protection provides.
Different masking schemes, like additive [8,9] and multiplicative [14], have been
proposed in order to provide security against differential power analysis (DPA)
attacks. However, it was shown [11,12,17] that masked hardware implementations can still be vulnerable to first-order DPA due to the presence of glitches.
One can try to eliminate the security relevant glitches by carefully balancing

D. Pointcheval and D. Vergnaud (Eds.): AFRICACRYPT 2014, LNCS 8469, pp. 267–284, 2014.

signal propagation delays, but this requires expertise, time, iterations of design and testing, and hence is expensive. As an alternative, new masking schemes have been developed that provide provable security even if glitches occur.

In 2006 Nikova et al. proposed such a scheme called Threshold Implementation (TI) [19]. It is based on secret-sharing and provably secure against first-order DPA [20]. In 2012 Prouff and Roche proposed a dth-order masking scheme [24], based on Shamir's secret sharing, for which they claim security even against higher-order attacks. It is a general method that replaces every field multiplication by $4d^3$ field multiplications and $4d^3$ additions, using $2d^2$ bytes of randomness. In some cases this may prove too costly or inefficient. And [16] has shown that the multivariate leakages can be exploitable in univariate attacks.

Related Work. The Threshold Implementation technique is based on a specific type of multi-party computation and applies Boolean masking. Interesting properties of the technique are that it provides provable security against first-order side-channel attacks, that it requires few assumptions on the hardware leakage behavior, and that it allows to construct realistic-size circuits without intervention and design iterations. However, threshold implementations can still be broken by univariate mutual information analysis (MIA) [2,20] or univariate higher-order attacks [15].

It has been shown that all 3×3 and 4×4 have a TI sharing with 3, 4 or 5 shares [5]. The TI approach has been applied to only few entire algorithms: PRESENT [21], AES [18], KECCAK [3] and Fides [4]. In AES, the S-box is the by far most challenging part to share. Moradi et al. [18] have proposed a TI of this S-box that constantly uses 3 shares based on the tower field approach.

Contribution. We propose a more compact and faster Threshold Implementation of AES-128 encryption that requires less random bits compared to the one by Moradi et al. from EUROCRYPT 2011. For the S-box we use the tower field approach over $GF(2^4)$ and for each block in the S-box computation we adapt the number of shares. This reduces the area by 13% and the clock cycles by 40%. However, our main focus is to optimize not only the S-box but the whole cipher. Our implementation of AES is 18% smaller, 7.5% faster and requires 8% less random bits than the implementation from EUROCRYPT 2011. We investigate the uniformity problem and the need for re-masking in more detail. We prove that under certain circumstances, it is enough to re-mask only a fraction of the shares. We provide results of a practical security evaluation based on real power traces in adversary-friendly conditions. They confirm the theoretically guaranteed first-order attack resistance and show good security against higher-order attacks.

2   Threshold Implementation

TIs use sharings with the following properties: correctness, incompleteness and uniformity. The last property is often the most difficult to achieve. We propose implementations where not every function satisfies the property of uniformity and use fresh randomness instead to do a re-masking. In this section, we recall the TI properties defined in [19] and describe how circuit complexity can be traded off for fresh random bits.

2.1   Notation and Definitions

We denote by upper-case characters stochastic variables, and by lower-case characters the values they can take, i.e. elements of a finite field. Let X, taking values in $\mathcal{F}^m$, denote the input of the (unshared) function f. A *masking* takes as inputs a value x and some auxiliary values (*random masks*), and outputs a vector $(x_1, \ldots, x_{s_x})$ such that the XOR-sum of the s_x shares equals x. For all values x with $\Pr(X = x) > 0$, let $\mathrm{Sh}(x)$ denote the set of valid share vectors $(x_1, \ldots, x_{s_x})$ for x:

$$\mathrm{Sh}(x) = \{(x_1, \ldots, x_{s_x}) \in \mathcal{F}^{m s_x} \mid x_1 + \cdots + x_{s_x} = x\}.$$

$\Pr((X_1, \ldots, X_{s_x}) = (x_1, \ldots, x_{s_x}) | X = x)$ denotes the probability that $(X_1, \ldots, X_{s_x}) = (x_1, \ldots, x_{s_x})$ when the input of the masking equals x, taken over all auxiliary inputs of the masking. Similarly, we denote the output of the unshared function by Y, taking values in $\mathcal{F}^n$, $(y_1, \ldots, y_{s_y})$ and $\mathrm{Sh}(y)$. Let F denote the vector function with input $(X_1, \ldots, X_{s_x})$ and output $(Y_1, \ldots, Y_{s_y})$; we will call it a *sharing*. TIs, like most other masking schemes, require that the masking is *uniform*, in the sense of the following definition.

Definition 1 (Uniform masking). *A masking is* uniform *if and only if for all x we have:*

$$\Pr((X_1, \ldots, X_{s_x}) = (x_1, \ldots, x_{s_x}) | X = x) = |\mathcal{F}|^{-m(s_x - 1)}$$

if $(x_1, \ldots, x_{s_x}) \in \mathrm{Sh}(x)$, else it is 0.

In words, we call a masking uniform if for each value x of the variable X, the corresponding vectors with masked values occur with the same probability.

Threshold implementations use sharings that satisfy the following properties. Firstly, the sharing F of f needs to be *correct*:

$$\forall y \in \mathcal{F}^n, \forall (x_1, \ldots, x_{s_x}) \in \mathrm{Sh}(x), \forall (y_1, \ldots, y_{s_y}) \in \mathrm{Sh}(y):$$
$$F(x_1, \ldots, x_{s_x}) = (y_1, \ldots, y_{s_y}) \Leftrightarrow f(x) = y.$$

Secondly, the sharing needs to be *incomplete*: every component function of F that outputs Y_i should be independent of at least one share X_i. The third property is *uniformity of the sharing* [19]. Although the main point of this section is that also sharings which do not satisfy the third property can be used in threshold implementations, we provide the definition already now.

Definition 2 (Uniform sharing). *The sharing F of f is* uniform *if and only if*

$$\forall y \in \mathcal{F}^n, \forall (y_1, \ldots, y_{s_y}) \in \mathrm{Sh}(y), \forall x \in \mathcal{F}^m \text{ with } f(x) = y:$$
$$\left| \{(x_1, \ldots, x_{s_x}) \in \mathrm{Sh}(x) | F(x_1, \ldots, x_{s_x}) = (y_1, \ldots, y_{s_y})\} \right| = \frac{|\mathcal{F}|^{m(s_x - 1)}}{|\mathcal{F}|^{n(s_y - 1)}}.$$

If $s_x = s_y$ and $m = n$, this simplifies to:

$$\forall y \in \mathcal{F}^n, \ \forall (y_1, y_2, \ldots, y_{s_y}) \in \mathrm{Sh}(y) \forall x \in \mathcal{F}^n \text{ with } f(x) = y :$$
$$\left| \left\{ (x_1, x_2, \ldots, x_{s_x}) \in \mathrm{Sh}(x) \middle| F(x_1, x_2, \ldots, x_{s_x}) = (y_1, y_2, \ldots, y_{s_y}) \right\} \right| = 1 \ .$$

It follows that in this case a uniform sharing F is invertible if and only if f is invertible.

2.2 Security from Correctness and Incompleteness

The security of threshold implementations against first-order side-channel attacks follows from two intuitively easy steps. If the masking is uniform and the sharing F is incomplete, then

1. any single component function of F does not get the information to determine the value of X (*it does not know* x), hence cannot leak any information on X, and
2. the expected value (average) of any leakage signal of an implementation of the sharing F, be it instantaneous or summed over an arbitrary period of time, is constant.

Note that the only assumption on the physical behavior of the hardware or software implementation of F that is needed for this reasoning, is that it should be possible to implement the component functions in such a way that they are each independent of one share X_i. In other words, the cross-talk between implementations of different components should be negligible.

2.3 Uniformity for the Cascaded and Parallel Functions

If the threshold implementation technique is used to protect cascaded functions, then extra measures need to be taken, such that the input for the next non-linear operation is again a uniform masking. A similar situation occurs when the threshold implementation technique is used to protect several functional blocks acting in parallel on (partially) dependent inputs. This occurs for example in implementations of the AES S-box using the tower field approach. If no special care is taken, then "local uniformity" of the distributions of the inputs of the individual blocks will not lead to "global uniformity", i.e. for the joint distributions of the inputs of all blocks. For example, let g and h be two functions acting on the same input X. Then, even if G and H are uniform sharings, producing uniform $Y = G(X)$ and $Y' = H(X)$, this does not imply that (Y, Y') is uniform. If each of the parallel blocks satisfies the properties of correctness and incompleteness, there will be no leakage of signals within the parallel blocks. However, the lack of uniformity in the joint distribution of the masking of the outputs can lead to information leakage if the outputs are combined as inputs to a next function.

We can take different types of actions to remedy this problem. The *first approach* is to require uniformity of the sharing F (Definition 2). We can show that if the sharing is uniform and the masking of its input is uniform, then also the masking of its output is uniform.

Theorem 1. *If the masking of X is uniform and the sharing F is uniform, then the masking of $Y = f(X)$, defined by $(y_1, \ldots, y_{s_y}) = F(x_1, \ldots, x_{s_x})$, is uniform.*

The proof is omitted here to save space. Practice shows that adding the uniformity requirement to a sharing tends to blow up the mathematical complexity of the sharing, as well as the cost of its implementation. In some applications, it might be better to consider a *second approach*: re-masking as for example done by Moradi et al. [18]. Indeed, by adding new random masks to the shares, we can make the distribution uniform.

2.4 Reducing the Randomness Used in a Re-masking Step

The following theorem allows to reduce the amount of random bits used by re-masking steps of threshold implementations: under certain circumstances, only a fraction of the shares needs to be re-masked.

Theorem 2. *Let $(X_1, \ldots, X_s)$ be a sharing of a variable $X \in \mathcal{F}^m$, where $\Pr(X_1 = x_1, \ldots, X_t = x_t) = |\mathcal{F}|^{-tm}, \forall (x_1, \ldots, x_t)$ for some t with $1 \le t \le s$. Then the sharing $(Y_1, \ldots, Y_s)$, defined by $Y_i = X_i$ for $1 \le i \le t$ and $Y_i = X_i + R_i$ for $t < i \le s$, is a* uniform *sharing for X, i.e.: $\Pr(Y_1 = y_1, \ldots, Y_s = y_s | X = y_1 + \cdots y_s) = |\mathcal{F}|^{(1-s)m}$, provided that the R_i, $i = t+1, \ldots, s-1$ are independently and uniformly distributed random variables and that $R_s = -(R_{t+1} + \cdots + R_{s-1})$.*

Proof. We give here a sketch of the proof. We have:

$$
\begin{aligned}
&\Pr(Y_1 = y_1, \ldots, Y_s = y_s | X = y_1 + \cdots y_s) \\
&= \Pr(Y_1 = y_1, \ldots, Y_t = y_t | X = y_1 + \cdots y_s) \\
&\quad \cdot \Pr(Y_{t+1} = y_{t+1}, \ldots, Y_s = y_s | X = y_1 + \cdots y_s, Y_1 = y_1, \ldots, Y_t = y_t).
\end{aligned} \tag{1}
$$

Since $Y_i = X_i$ for $1 \le i \le t$, the first factor equals $|\mathcal{F}|^{-tm}$. For the second factor we recall the definition of Y_{t+1} to obtain that:

$$
\Pr(Y_{t+1} = y_{t+1}) = \sum_{x_{t+1}} \Pr(X_{t+1} = x_{t+1}) \underbrace{\Pr(R_{t+1} = y_{t+1} - x_{t+1})}_{|\mathcal{F}|^{-m}}.
$$

The same holds for $Y_{t+2}, \ldots, Y_{s-1}$ and since the R_i have independent distributions, we can equate the second factor of (1) to:

$$
|\mathcal{F}|^{(1-s-t)m} \sum_{x_{t+1}, \ldots, x_{s-1}} \Pr(X_{t+1} = x_{t+1}, \ldots, X_{s-1} = x_{s-1}, Y_s = y_s |
$$

$$
X = y_1 + \cdots + y_s, X_1 = x_1, \ldots, X_t = x_t).
$$

Recalling the definition of Y_s completes the proof. □

Note that generating the extra randomness required by the re-masking approach may become a bigger challenge in some cases than the blow-up in gate count caused by the uniform sharing approach.

Conclusion. Assume that we have an input that is uniformly masked. Section 2.2 explains that single circuits are secure against first-order side-channel attacks, if they satisfy the incompleteness property. Section 2.3 explains that for cascaded circuits we need to ensure that the inputs of all circuits are uniformly masked. This can be done either by using uniform sharings (Def. 2) or by re-masking. The point that we want to stress here, however is that we do not need to do both: an implementation that uses re-masking, does not need uniform sharings in order to resist first-order attacks.

By relinquishing the uniformity requirement, it is often possible to reduce the number of shares and the size of the implementation. This will be used in the next section in order to reduce the number of shares in the subblocks of the AES S-box and improve on the implementation of [18].

3 Implementation

In this section, we will discuss the new TI of AES in detail. We will first describe the general data flow of our implementation. Then we will introduce a new approach to apply the TI to the S-box of AES which is the only non-linear layer of the block cipher. We used ModelSim to verify the functionality of the proposed design and Synopsys Design Vision D-201-.03-SP4 with Faraday Standard Cell Library FSA0A_C_Generic_Core, which is based on UMC 0.18μm GenericII Logic Process with 1.8V voltage, for synthesis. We will conclude this section by providing the performance of our design together with the comparison with the previous work in [18]. We should note that the work in [18] uses a similar standard cell library based on UMC 0.18μm logic process with 1.8V voltage.

3.1 General Data Flow

Our main goal for this implementation is to minimize the area and randomness overhead caused by the sharing. To achieve this, we use a serial implementation as proposed in [18] which requires only one S-box instance and loads the plaintext and key byte-wise in row-wise order. Moreover, we adapt the number of shares used in each operation in the block cipher. That is, we use two shares which is the minimum number of shares possible for the affine operations such as MixColumns or Key XOR and increase or decrease the number of shares when required for the non-linear layer. This can also be seen in Fig. 1, as the key and the state registers are 256 bits implying the two shares. With this approach we already decrease a significant part of the register cost since one bit register costs 5.33 GE in our library.

The TI of the S-box, for which the details will be given in the following section, requires four input shares and 20 bits of randomness and outputs three shares. Therefore our initial sharing for the plaintext is also with four shares. However, it is enough to initialize the sharing of the key with two shares. More details about the key scheduling will be given later in this section. The two shares of the key are XORed with two of the plaintext shares before the S-box operation. After three

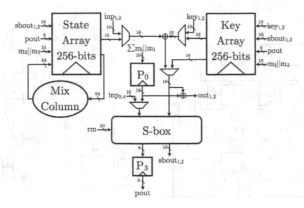

Fig. 1. Architecture of the serialized TI of AES-128.

clock cycles the first output share of the S-box operation is written to the register P_3 whereas the remaining two shares are written to the state register S_{33}. The data in P_3 is XORed with the second share of the S-box output, in the state register S_{33}, after one clock cycle to be able to continue with two shares for the linear operations. In the following AES rounds, we increase the number of shares from two to four by using 24 bits of randomness one clock cycle before the S-box operation. We store the additional two shares in P_0 to achieve the non-completeness property in the following combinational logic. The registers P_0 and P_3 are used both for the round transformations and the key scheduling.

State Array (Fig. 2a). The state array consists of sixteen 16-bit registers each corresponding to the two shares of a byte in the state. From the first to the sixteenth clock cycle, the four input shares (first round) or the shares in the registers S_{00} and P_0 (later rounds) are sent to the S-box module. The corresponding three output shares are written to the registers S_{33} and P_3. The signal sig_2 is active from the fourth to the nineteenth clock cycle to reduce the number of shares from three to two in the state such that one of the shares in S_{33} is XORed with P_3 and the other share stays the same. The state is shifted to the left horizontally from the third to the eighteenth clock cycle. The Shift Rows operation is also completed in the nineteenth clock cycle with an irregular horizontal shift. In the next four clock-cycles, the data in the registers S_{00}, S_{10}, S_{20} and S_{30} are sent to MixColumns operation, the rest of the registers are shifted to the left horizontally and the output of the MixColumns operation is written to the registers S_{03}, S_{13}, S_{23} and S_{33}. The MixColumns operation is implemented column-wise as in [18] and with two shares working in parallel. The registers except S_{10}, S_{11} and S_{12} are implemented as scan flip-flops (SFF) that are D-flip-flops (DFF) combined with 2-to-1 MUXes and can operate with two inputs to reduce the area since a single 2-to-1 MUX costs 3.33 GE in our library whereas one bit SFF costs 6.33 GE. One round of AES takes 23 clock cycles. The signal sig_1 is active for sixteen clock cycles, starting from the last clock-cycle of each round, to increase the number of shares from two to four.

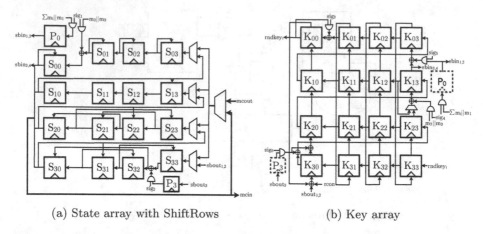

(a) State array with ShiftRows   (b) Key array

Fig. 2. Architecture of the registers where S_i, K_i and P_0 hold two shares and P_3 holds one share. The registers P_0 and P_3 are shared by the state and the key array. The XOR of the value in P_3 and S_{33} (resp. K_{30}) is on one share of the value in register S_{33} (resp. K_{30}) whereas all the other combinational operations are on two shares.

Key Array (Fig. 2b). Similar to the state array, the key array also consists of sixteen 16-bit registers implemented as SFFs each corresponding to the two shares of a byte in the key schedule. The round key is inserted from the register K_{33} in the first sixteen clock cycles of each round. For the next three clock cycles, the registers except K_{03}, K_{13}, K_{23} and K_{33} are not clocked. The registers K_{03}, K_{23} and K_{33} are also not clocked in the seventeenth clock cycle. In that clock cycle, we increase the number of shares in the register K_{13}. In the following three clock cycles this re-sharing is done during the vertical shift from the register K_{23} to K_{13}. Hence the re-sharing signal sig_4 is active from the seventeenth to the twentieth clock cycle. Signal sig_5 is active from the eighteenth to the twenty-first clock cycle to reduce the number of shares back to two. The registers K_{03}, K_{13}, K_{23} and K_{33} are not clocked in the remaining two clock cycles of each round. We choose this way of irregular clocking to avoid using extra MUXes in our design. Two shares of the S-box output are XORed to the data in K_{00} in the last four clock cycles of each round. In the twentieth clock cycle the round counter $rcon$ is additionally XORed to one of these shares. The number of shares is reduced back to two by XORing the share in P_3 to one of the shares in K_{30}. Signal sig_3 is active in the first sixteen clock cycles except the fourth, eighth, twelfth and sixteenth clock cycles. The roundkey is taken from the register K_{00} to be XORed with the corresponding plaintext before going to the S-box operation.

3.2 TI of the AES S-box

The S-box (Fig. 3) is instantiated only once to be used by both the key schedule and the state update. In the first sixteen clock cycles, it gets its inputs from

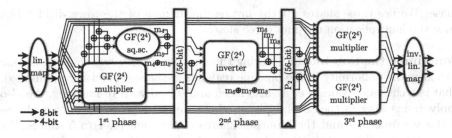

Fig. 3. The Sbox of our implementation.

the state. The input is taken from the key array in clock cycles eighteen to twenty-one.

The S-box implementation in [18] uses the tower field approach up to $GF(2^2)$ for a smaller implementation. Therefore, the only non-linear operation is $GF(2^2)$ multiplication which must be followed by registers to avoid first order leakages.

We also chose to use the tower field approach, however, we decided to go to $GF(2^4)$ instead of $GF(2^2)$. With this approach, the $GF(2^4)$ inverter can be seen as a four bit permutation and the $GF(2^4)$ multiplier as a four bit multiplication both of which are well studied in [6]. Therefore, we can find uniform TIs for these non-linear blocks individually which implies using less fresh random bits during the combination of these uniformly implemented pieces. Moreover, with this approach the S-box calculation takes three clock cycles instead of five.

The algebraic normal form of the multiplier in $GF(2^4)$ is given in Appendix A.1. This multiplication can be shared uniformly as in Appendix A.3 with four input and three output shares. The required area is 625 GE without any optimization.

The $GF(2^4)$ inverter, on the other hand, can be represented with the formula in Appendix A.2. To have a uniform sharing for this function, which belongs to class C_{282}^4 [5], we consider two options. Either using four shares which is the minimum number of shares necessary for a uniform implementation in that class and decomposing the function into three uniform sub-functions as $Inv(x) = F(G(H(x)))$, or using five shares without any decomposition. Our experiments show that both versions have similar area requirements but a different number of clock cycles. To reduce the number of cycles, we chose the version with five shares, with the formula in Appendix A.4, which requires 618 GE. The sharing for this module is found by using the method described in [20] which is slightly different from the direct sharing [5]. We chose this formula since it can be implemented with less logic gates in hardware compared to the direct sharing.

Even though it is enough to use only two shares for linear operations, we sometimes chose to work on more than two shares to avoid the need for extra random bits. The linear map of the tower-field S-box operates on four shares since the multiplication needs four input shares. The inverter requires five input shares and the multiplication outputs only three shares, therefore we use two shares for the square scalar to have five shares in the beginning of the second phase. We use three shares for the inverse linear map of the tower-field S-box since the multiplication outputs three shares.

Combining the Sub-blocks. During this process we face two challenges. One is to keep the uniformity in the pipeline registers as the sub-blocks are combined. That is a challenge Moradi et al. also faced and solved with re-masking. We also apply re-masking in the second phase where we combine the 2 output shares of the square scaler and the 3 output shares of the multiplier to 5 shares. We must note that this combination also acts as the XOR of the output of the square scaler and multiplier in the unshared case. By theorem 2, it is enough to re-mask only the output shares from one function to achieve uniformity. We choose to re-mask the output of the square scaler since it operates on less shares hence requires less random bits. The correction mask, i.e. the XOR of the masks, is XORed to one of the output shares of the multiplier to achieve correctness and non-completeness.

The second challenge is to keep the uniformity as we increase or decrease the number of shares. This is achieved by introducing new masks before the S-box operation to increase from two to four shares and at the end of the second phase to decrease from five to four shares. The output of the third phase is not uniform when the three shares are considered together. However, we verified by simulation that each share individually is uniform which implies that there is no first-order leakage in the following register. We combine the first two shares with an XOR and keep the third share as it is to go back to two shares. We also verified that, when we decrease the number of shares to two, the output shares are uniform.

We always keep the XOR of the masks in the pipeline registers and complete the re-masking in the next clock cycle as in [18]. Overall, we need 44 fresh random bits per S-box operation which is less than what was required in [18].

3.3   Performance

Like other countermeasures TIs require extra area and randomness. In this work we minimize these needs for a more efficient implementation. In Table 1, we show the area, randomness and timing requirements of our implementation and compare them with [18]. The area cost for the state and the key arrays include the ANDs and XORs that are in Fig. 2. An expected observation is that the cost of the state and key array together with the MixColumns is reduced by one third compared to [18] since we use two shares instead of three. The area cost of the S-box is the sum of the combinational logic in three phases and the registers required. For the three phases, we use four linear maps (each 42 GE), two square scalers (each 9 GE), three multipliers (each 625 GE), one inverter (618 GE), three inverse linear maps (each 33 GE) and some additional XORs for re-masking. The registers P_0 and P_3 are also counted in the cost of the S-box together with the pipelining registers P_1 and P_2.

Table 1. Synthesis results for different versions of AES TI

	State Array	Key Array	S-box	MixCol Col	Contr.[1]	Key XOR	MUX	Other	Total	cycles	rand bits[2]
[18]	2529	2526	4244	1120	166	64	376	89	11114/11031[3]	266	48
This paper	1698	1890	3708	770	221	48	746	21	9102	246	44
This paper[3]	1698	1890	3003	544	221	48	746	21	8171	246	44

[1] including round constant   [2] per S-box   [3] *compile_ultra*

In this implementation, the S-box occupies 40% of the total area. When compared to the previous implementation by Moradi et al., the S-box is 13% smaller and the overall area is 18% smaller. Moreover it is faster and requires less randomness. The numbers provided in Table 1 are taken from the Synopsys tool with *compile* command. We use these numbers for a fair quantitative comparison. On the other hand, it is also possible to compile each function that is provided in Appendix A.3 and A.4 *individually* with the *compile_ultra* command to let the tool optimize these functions and use the generated optimized descriptions of these functions. This reduces the cost of TI of AES to 8171 GE. However, the results for *compile_ultra* mainly reflect how good the tools are at optimizing and a comparison may not be fair.

4 Power Analysis

To evaluate the security of our design in practice we implement it on a SASEBO-G board [1] using Xilinx ISE version 10.1. We use the "keep hierarchy" constraint to prevent the tools from optimizing over module boundaries (see the last paragraph of Sect. 2.2). The board features two Xilinx Virtex-II Pro FPGA devices: we implement the TI AES and a PRNG on the crypto FPGA (xc2vp7) while the control FPGA (xc2vp30) handles I/O with the measurement PC and other equipment. The PRNG that generates all random bits is implemented as AES-128 in CTR mode.

We measure the power consumption of the crypto FPGA during the first 1.5 rounds of TI AES as the voltage drop over a 1Ω resistor in the FPGA core GND line. The output of the passive probe is sampled with a Tektronix DPO 7254C digital oscilloscope at 1GS/s sampling rate.

Methodology. We define two main goals for our practical evaluation. First, we want to verify our implementation's resistance against first-order attacks. But in practice adversaries are of course not restricted to applying such attacks. Therefore, our second goal is to assess the level of security our implementation provides against other, e.g. higher-order, power analysis attacks.

Since there is no single, all-embracing test to evaluate the security of an implementation, we follow the approach of [18] and test its resistance against state-of-the-art attacks. We narrow the evaluation to univariate attacks because our implementation processes all shares of a value in parallel. Estimating the information-theoretic metric by Standaert et al. [25] is out of reach. It would require estimation of up to 2^{56} Gaussian templates.

We make several choices that are in favor of an adversary and make attacks easier. First, to minimize algorithmic noise the PRNG and the TI AES do not operate in parallel, i.e. the PRNG generates and stores a sufficient number of random bits before each TI AES operation. In practice, running them in parallel will increase the level of noise and thus the number of measurements needed for an attack to succeed. Second, we provide the crypto FPGA with a stable 3MHz clock frequency to ensure that the traces are well aligned and the power peaks of adjacent clock cycles do not overlap (this would also help to assign a possibly identified leak to a specific clock cycle). In practice, clocking the device at a faster or unstable clock will make attacks harder. Note that the "combining effect" of the measurement setup or a faster clock described in [16] does not apply to our situation. In our implementation all shares are processed and leak at the same time, in contrast to the implementation analyzed in [16] where all shares are processed and leak separated in time. Hence we expect the effect to not ease an attack. Third, we let the adversary know the implementation. Specifically, if the PRNG was switched off the adversary would be able to correctly compute bit values and bit flips under the correct key hypothesis. In practice, obscurity is often used as an additional layer of security. Fourth, we use synchronous sampling [13] to avoid clock drift and achieve the best possible alignment. In practice, secure devices use an internal (and unstable) clock source which prevents synchronous sampling and increases the number of measurements needed for an attack to succeed.

PRNG Switched Off. To confirm that our setup works correctly and to get some reference values we first attack the implementation with the PRNG switched off. We expect that the implementation can be broken with many first-order attacks. As example, Fig. 4 shows the result of a correlation DPA attack [7] that uses the Hamming distance of two consecutive S-box outputs as power model. The attacks require $2 \cdot 2^8$ key hypotheses. To reduce the computational complexity we let the adversary know one key byte and aim to recover the second one.

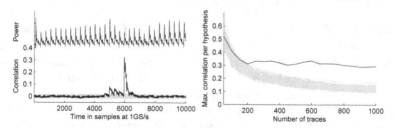

Fig. 4. Results of DPA attacks using HD model over 3/2/1 registers with PRNG off; left: correlation traces for all key hypotheses computed using 50 000 power traces, correct hypothesis in black, and a scaled power trace; right: max. correlation coefficient per key hypothesis (from the overall time span) over number of traces used.

Since the adversary knows the implementation, he can choose to compute the Hamming distance over three 8-bit registers (S_{33} and P3; output of the S-box in three shares), two 8-bit registers (S_{32}; one cycle later; two shares) or ignore the details and compute the distance over a single 8-bit register as if it was a plain implementation. The results for all three options are identical. This is a property of our implementation that vanishes when the PRNG is switched on. Only a few hundred traces are required to recover the key with one of these attacks. It is worth noticing that the highest correlation peak does not occur at the S-box output registers, but three resp. two clock cycles later when the bit-flips occur in register S_{30}. This register drives the MixColumns logic and therefore has a much greater fanout.

Fig. 5 shows the result of a correlation collision attack [17] that targets combinational logic. The attack computes two sets of mean traces for the values of two processed plaintext bytes and shifts the mean traces in the time domain to align them. It aims to recover the linear difference between the two key bytes involved. To do so, it permutes one set of mean traces according to a hypothesis on the linear difference and then correlates both sets of mean traces. The result shows that this attack is successful with a few thousand measurements.

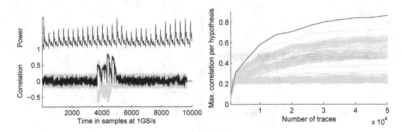

Fig. 5. Result of a correlation collision attack with PRNG off; left: correlation traces for all hypotheses on the linear difference computed using 50 000 power traces, correct hypothesis in black, and a scaled power trace; right: max. correlation coefficient per hypothesis on the linear difference (from the overall time span) over number of traces used.

PRNG Switched On. Next we repeat the evaluation with the PRNG switched on, i.e. the TI AES uses unknown and unpredictable random bits. However, for the DPA attacks using the Hamming distance over two or three registers as power model we again suppose these bits were zero. Fig. 6 shows the results of the first-order attacks against the protected implementation using 10 million measurements. The results show that the attacks fail.

We proceed with higher-order attacks to assess the level of security our implementation provides. For our second-order DPA attacks we use the same power models as before but center and then square the traces (for each time sample) before correlating [8,23,26]. Second-order correlation collision attacks work as above with mean traces replaced by variance traces [15].

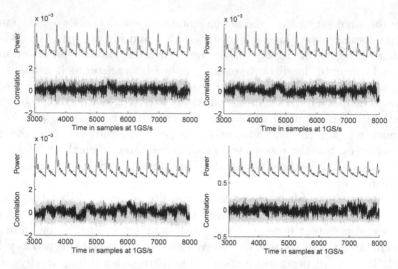

Fig. 6. Results of first-order DPA and correlation collision attacks with PRNG on computed using 10 million traces; top, left: HD over 1 register; top, right: HD over 2 registers; bottom, left: HD over 3 registers; bottom, right: correlation collision.

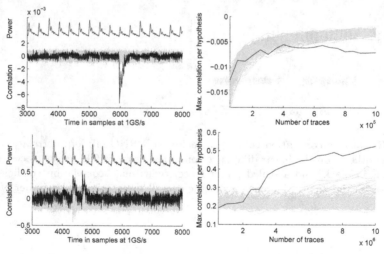

Fig. 7. Results of second-order DPA (top) and correlation collision (bottom) attacks with PRNG on computed using 10 million traces; right: min./max. correlation coefficient per hypothesis (from the overall time span) over number of traces used.

Fig. 7 (top) shows the results of the second-order DPA attack that uses the Hamming distance in a single register as power model (as if it was a plain implementation). The attack requires about 600 000 traces to succeed. We note that the highest correlation peak occurs again when the bitflips happen in register S_{30}, cf. Fig. 4. Second-order DPA attacks using the Hamming distance over two resp. three registers as power model failed to recover the key.

Fig. 7 (bottom) shows the results of the second-order correlation collision attack. The attack requires about 3.5 million traces to succeed. A third-order correlation collision attack works as above with mean traces replaced by skewness traces [15]. This attack fails using 10 million measurements.

Discussion. The first goal of our evaluation is to verify our implementation's resistance against first-order attacks. But this goal is always limited by the number of measurements at hand. It is simply not possible to demonstrate resistance against attacks with an infinite number of traces. We have shown that our implementation resists state-of-the-art first-order attacks with 10 million traces in conditions that are strongly in favor of the adversary (no algorithmic noise from the PRNG, knowledge of the implementation, slow and stable clock, best possible alignment). Given the theoretical foundations of TI and the correctness of our implementation, we are convinced that our implementation resists first-order attacks with any number of measurements, but we have no way to demonstrate that.

The second goal of our evaluation is to assess the level of security our implementation provides against other attacks. In the same adversary-friendly conditions, the most trace-efficient second-order attack in our evaluation requires about 600 000 traces. Recall that our evaluation focuses on univariate attacks, so that the computational overhead is limited to estimating second-order moments and does not involve the notoriously more costly search over pairs of points in time. However, regarding second-order attacks it is well known that the number of traces required for an attack to succeed scales quadratically in the noise standard deviation [8,22]. Therefore, second-order attacks against our implementation in less favorable and more realistic, i.e. much more noisy, conditions (algorithmic noise from the PRNG, no knowledge of the implementation, faster and unstable clock, worse alignment) will require many more traces.

It is tempting to compare the results of our evaluation to the results of the evaluation in [18]. However, not only the implementations but also the measurement platforms and the conditions differ, so that any difference must not be credited to an implementation alone. Already the numbers of traces required for attacks against the implementations with PRNG switched off differ by roughly two orders of magnitude. In addition, the analysis in [18] is limited to four clock cycles during the S-box computation.

Acknowledgements. This work has been supported in part by the Research Council of KU Leuven (OT/13/071), by the FWO (G.0550.12), by the Hercules foundation and by GOA (tense). B. Bilgin was partially supported by the FWO project G0B4213N, V. Nikov was supported by the European Commission (FP7) within the Tamper Resistant Sensor Node (TAMPRES) project with contract number 258754 and Benedikt Gierlichs is a Postdoctoral Fellow of the Research Foundation - Flanders (FWO).

References

1. AIST. Side-channel Attack Standard Evaluation BOard,
   http://staff.aist.go.jp/akashi.satoh/SASEBO/en/
2. Batina, L., Gierlichs, B., Prouff, E., Rivain, M., Standaert, F.-X., Veyrat-Charvillon,
   N.: Mutual Information Analysis: A Comprehensive Study. J. Cryptol. 24(2),
   269–291 (2011)
3. Bertoni, G., Daemen, J., Peeters, M., Van Assche, G.: Building power analysis resistant implementations of Keccak. In: Second SHA-3 Candidate Conference (August
   2010)
4. Bilgin, B., Bogdanov, A., Knežević, M., Mendel, F., Wang, Q.: FIDES: Lightweight
   authenticated cipher with side-channel resistance for constrained hardware. In:
   Bertoni, G., Coron, J.-S. (eds.) CHES 2013. LNCS, vol. 8086, pp. 142–158.
   Springer, Heidelberg (2013)
5. Bilgin, B., Nikova, S., Nikov, V., Rijmen, V., Stütz, G.: Threshold implementations
   of all 3 × 3 and 4 × 4 S-boxes. In: Prouff, E., Schaumont, P. (eds.) CHES 2012.
   LNCS, vol. 7428, pp. 76–91. Springer, Heidelberg (2012)
6. Bilgin, B., Nikova, S., Nikov, V., Rijmen, V., Stütz, G.: Threshold implementations
   of all 3 × 3 and 4 × 4 S-boxes. Cryptology ePrint Archive, Report 2012/300 (2012),
   http://eprint.iacr.org/
7. Brier, E., Clavier, C., Olivier, F.: Correlation power analysis with a leakage model.
   In: Joye, M., Quisquater, J.-J. (eds.) CHES 2004. LNCS, vol. 3156, pp. 16–29.
   Springer, Heidelberg (2004)
8. Chari, S., Jutla, C.S., Rao, J.R., Rohatgi, P.: Towards sound approaches to counteract power-analysis attacks. In: Wiener, M. (ed.) CRYPTO 1999. LNCS, vol. 1666,
   pp. 398–412. Springer, Heidelberg (1999)
9. Goubin, L., Patarin, J.: DES and differential power analysis the "duplication"
   method. In: Koç, Ç.K., Paar, C. (eds.) CHES 1999. LNCS, vol. 1717, pp. 158–172.
   Springer, Heidelberg (1999)
10. Kocher, P.C., Jaffe, J., Jun, B.: Differential power analysis. In: Wiener, M. (ed.)
    CRYPTO 1999. LNCS, vol. 1666, pp. 388–397. Springer, Heidelberg (1999)
11. Mangard, S., Popp, T., Gammel, B.M.: Side-channel leakage of masked CMOS
    gates. In: Menezes, A. (ed.) CT-RSA 2005. LNCS, vol. 3376, pp. 351–365. Springer,
    Heidelberg (2005)
12. Mangard, S., Pramstaller, N., Oswald, E.: Successfully attacking masked AES hard-
    ware implementations. In: Rao, J.R., Sunar, B. (eds.) CHES 2005. LNCS, vol. 3659,
    pp. 157–171. Springer, Heidelberg (2005)
13. Messerges, T.S.: Power analysis attacks and countermeasures on cryptographic
    algorithms. PhD thesis, University of Illinois at Chicago (2000)
14. Messerges, T.S.: Securing the AES finalists against power analysis attacks. In:
    Schneier, B. (ed.) FSE 2000. LNCS, vol. 1978, pp. 150–164. Springer, Heidelberg
    (2001)
15. Moradi, A.: Statistical tools flavor side-channel collision attacks. In: Pointcheval,
    D., Johansson, T. (eds.) EUROCRYPT 2012. LNCS, vol. 7237, pp. 428–445.
    Springer, Heidelberg (2012)
16. Moradi, A., Mischke, O.: On the simplicity of converting leakages from multivariate
    to univariate. In: Bertoni, G., Coron, J.-S. (eds.) CHES 2013. LNCS, vol. 8086, pp.
    1–20. Springer, Heidelberg (2013)
17. Moradi, A., Mischke, O., Eisenbarth, T.: Correlation-enhanced power analysis col-
    lision attack. In: Mangard, S., Standaert, F.-X. (eds.) CHES 2010. LNCS, vol. 6225,
    pp. 125–139. Springer, Heidelberg (2010)

18. Moradi, A., Poschmann, A., Ling, S., Paar, C., Wang, H.: Pushing the limits: A very compact and a threshold implementation of AES. In: Paterson, K.G. (ed.) EUROCRYPT 2011. LNCS, vol. 6632, pp. 69–88. Springer, Heidelberg (2011)
19. Nikova, S., Rechberger, C., Rijmen, V.: Threshold implementations against side-channel attacks and glitches. In: Ning, P., Qing, S., Li, N. (eds.) ICICS 2006. LNCS, vol. 4307, pp. 529–545. Springer, Heidelberg (2006)
20. Nikova, S., Rijmen, V., Schläffer, M.: Secure hardware implementation of nonlinear functions in the presence of glitches. J. Cryptology 24(2), 292–321 (2011)
21. Poschmann, A., Moradi, A., Khoo, K., Lim, C.-W., Wang, H., Ling, S.: Side-channel resistant crypto for less than 2300 GE. J. Cryptology 24(2), 322–345 (2011)
22. Prouff, E., Rivain, M.: Masking against side-channel attacks: A formal security proof. In: Johansson, T., Nguyen, P.Q. (eds.) EUROCRYPT 2013. LNCS, vol. 7881, pp. 142–159. Springer, Heidelberg (2013)
23. Prouff, E., Rivain, M., Bevan, R.: Statistical analysis of second order differential power analysis. IEEE Trans. Computers 58(6), 799–811 (2009)
24. Prouff, E., Roche, T.: Higher-order glitches free implementation of the AES using secure multi-party computation protocols. In: Preneel, B., Takagi, T. (eds.) CHES 2011. LNCS, vol. 6917, pp. 63–78. Springer, Heidelberg (2011)
25. Standaert, F.-X., Malkin, T.G., Yung, M.: A unified framework for the analysis of side-channel key recovery attacks. In: Joux, A. (ed.) EUROCRYPT 2009. LNCS, vol. 5479, pp. 443–461. Springer, Heidelberg (2009)
26. Waddle, J., Wagner, D.: Towards efficient second-order power analysis. In: Joye, M., Quisquater, J.-J. (eds.) CHES 2004. LNCS, vol. 3156, pp. 1–15. Springer, Heidelberg (2004)

A   Equations

A.1   Multiplier in $GF(2^4)$

$$(f_1, f_2, f_3, f_4) = (x_1, x_2, x_3, x_4) \times (x_5, x_6, x_7, x_8)$$

$$f_1 = x_1x_5 \oplus x_3x_5 \oplus x_4x_5 \oplus x_2x_6 \oplus x_3x_6 \oplus x_1x_7 \oplus x_2x_7 \oplus x_3x_7 \oplus x_4x_7 \oplus x_1x_8 \oplus x_3x_8$$

$$f_2 = x_2x_5 \oplus x_3x_5 \oplus x_1x_6 \oplus x_2x_6 \oplus x_4x_6 \oplus x_1x_7 \oplus x_3x_7 \oplus x_2x_8 \oplus x_4x_8$$

$$f_3 = x_1x_5 \oplus x_2x_5 \oplus x_3x_5 \oplus x_4x_5 \oplus x_1x_6 \oplus x_3x_6 \oplus x_1x_7 \oplus x_2x_7 \oplus x_3x_7 \oplus x_1x_8 \oplus x_4x_8$$

$$f_4 = x_1x_5 \oplus x_3x_5 \oplus x_2x_6 \oplus x_4x_6 \oplus x_1x_7 \oplus x_4x_7 \oplus x_2x_8 \oplus x_3x_8 \oplus x_4x_8$$

A.2   Inverter in $GF(2^4)$

$$(f_1, f_2, f_3, f_4) = Inv(x_1, x_2, x_3, x_4)$$

$$f_1 = x_3 \oplus x_4 \oplus x_1x_3 \oplus x_2x_3 \oplus x_2x_3x_4$$

$$f_2 = x_4 \oplus x_1x_3 \oplus x_2x_3 \oplus x_2x_4 \oplus x_1x_3x_4$$

$$f_3 = x_1 \oplus x_2 \oplus x_1x_3 \oplus x_1x_4 \oplus x_2x_2x_4$$

$$f_4 = x_2 \oplus x_1x_3 \oplus x_1x_4 \oplus x_2x_4 \oplus x_1x_2x_3$$

A.3 Sharing Multiplier in $GF(2^4)$ with 4 Input 3 Output Shares

$$f = xy, \quad \text{where}$$
$$f = f_1 \oplus f_2 \oplus f_3$$
$$x = x_1 \oplus x_2 \oplus x_3 \oplus x_4$$
$$y = y_1 \oplus y_2 \oplus y_3 \oplus y_4$$

$$f_1 = (x_2 \oplus x_3 \oplus x_4)(y_2 \oplus y_3) \oplus y_4$$
$$f_2 = ((x_1 \oplus x_3)(y_1 \oplus y_4)) \oplus x_1 y_3 \oplus x_4$$
$$f_3 = ((x_2 \oplus x_4)(y_1 \oplus y_4)) \oplus x_1 y_2 \oplus x_4 \oplus y_4$$

A.4 Sharing Inverter in $GF(2^4)$ with 5 Input 5 Output Shares

$$f = xyz \oplus xy \oplus z, \quad \text{where}$$
$$f = f_1 \oplus f_2 \oplus f_3 \oplus f_4$$
$$x = x_1 \oplus x_2 \oplus x_3 \oplus x_4 \oplus x_5$$
$$y = y_1 \oplus y_2 \oplus y_3 \oplus y_4 \oplus y_5$$
$$z = z_1 \oplus z_2 \oplus z_3 \oplus z_4 \oplus z_5$$

$f_1 = ((x_2 \oplus x_3 \oplus x_4 \oplus x_5)(y_2 \oplus y_3 \oplus y_4 \oplus y_5)(z_2 \oplus z_3 \oplus z_4 \oplus z_5))$
$\quad \oplus ((x_2 \oplus x_3 \oplus x_4 \oplus x_5)(y_2 \oplus y_3 \oplus y_4 \oplus y_5)) \oplus z_2$

$f_2 = (x_1(y_3 \oplus y_4 \oplus y_5)(z_3 \oplus z_4 \oplus z_5) \oplus y_1(x_3 \oplus x_4 \oplus x_5)(z_3 \oplus z_4 \oplus z_5)$
$\quad \oplus z_1(x_3 \oplus x_4 \oplus x_5)(y_3 \oplus y_4 \oplus y_5) \oplus x_1 y_1(z_3 \oplus z_4 \oplus z_5) \oplus x_1 z_1(y_3 \oplus y_4 \oplus y_5)$
$\quad \oplus y_1 z_1(x_3 \oplus x_4 \oplus x_5) \oplus x_1 y_1 z_1) \oplus (x_1(y_3 \oplus y_4 \oplus y_5) \oplus y_1(x_3 \oplus x_4 \oplus x_5) \oplus x_1 y_1) \oplus z_3$

$f_3 = (x_1 y_1 z_2 \oplus x_1 y_2 z_1 \oplus x_2 y_1 x_1 \oplus x_1 y_2 z_2 \oplus x_2 y_1 z_2 \oplus x_2 y_2 z_1 \oplus x_1 y_2 z_4 \oplus x_2 y_1 z_4 \oplus x_1 y_4 z_2$
$\quad \oplus x_2 y_4 z_1 \oplus x_4 y_1 z_2 \oplus x_4 y_2 z_1 \oplus x_1 y_2 z_5 \oplus x_2 y_1 z_5 \oplus x_1 y_5 z_2 \oplus x_2 y_5 z_1 \oplus x_5 y_1 z_2 \oplus x_5 y_2 z_1)$
$\quad \oplus (x_1 y_2 \oplus y_1 x_2) \oplus z_4$

$f_4 = (x_1 y_2 z_3 \oplus x_1 y_3 z_2 \oplus x_2 y_1 z_3 \oplus x_2 y_3 z_1 \oplus x_3 y_1 z_2 \oplus x_3 y_2 z_1) \oplus 0 \oplus z_5$

$f_5 = 0 \oplus 0 \oplus z_1$

Constant Rounds Almost Linear Complexity Multi-party Computation for Prefix Sum

Kazuma Ohara[1], Kazuo Ohta[1], Koutarou Suzuki[2], and Kazuki Yoneyama[2]

[1] The University of Electro-Communications
1-5-1 Chofugaoka Chofu Tokyo 182-8585, Japan
[2] NTT Secure Platform Laboratories
3-9-11 Midori-cho Musashino-shi Tokyo 180-8585, Japan
kazma.ohara@gmail.com, kazuo.ohta@uec.ac.jp,
{suzuki.koutarou,yoneyama.kazuki}@lab.ntt.co.jp

Abstract. One of research goals on multi-party computation (MPC) is to achieve both perfectly secure and efficient protocols for basic functions or operations (e.g., equality, comparison, bit decomposition, and modular exponentiation). Recently, for many basic operations, MPC protocols with constant rounds and linear communication cost (in the input size) are proposed. In this paper, we propose the first MPC protocol for prefix sum in general semigroups with constant $2d + 2dc$ rounds and almost linear $O(l \log^{*(c)} l)$ communication complexity, where c is a constant, d is the round complexity of subroutine protocol used in the MPC protocol, l is the input size, and $\log^{*(c)}$ is the iterated logarithm function. The prefix sum protocol can be seen as a generalization of the postfix comparison protocol proposed by Toft. Moreover, as an application of the prefix sum protocol, we construct the first bit addition protocol with constant rounds and almost linear communication complexity.

Keywords: multi-party computation, constant round, prefix sum, bit addition, bit decomposition.

1 Introduction

Multi-party computation (MPC) protocols for general circuits have been studied in many researches [13,14,7,2,5]. However, such MPC protocol for general circuits is often less efficient than MPC protocol for specific function, which is optimized for the function. Therefore, there are many studies for dedicated MPC protocols for specific functions. In this paper, we focus on the MPC protocol [2,5] using a threshold secret sharing scheme, that is information theoretic secure. The MPC protocol can efficiently compute arithmetic operations in finite field, including multiplication and addition. However, several basic operations, including equality check, comparison, and modulo arithmetic, are hard to implement efficiently by using the arithmetic operations. The *bit decomposition protocol*, that converts a sharing of a finite field element to a sharing of bit representation of the element, makes it possible to compute these basic operations efficiently.

D. Pointcheval and D. Vergnaud (Eds.): AFRICACRYPT 2014, LNCS 8469, pp. 285–299, 2014.

The efficiency of the MPC is evaluated by two perspectives, round complexity and communication complexity. Especially, the round complexity of a protocol is significant, since it strongly affect the execution time for the protocol. So, the constant round MPC protocols have been studied intensively as follows.

Damgård et al. [6] proposed constant round MPC protocols for several important operations, including comparison, equality, exponentiation, and bit decomposition. The bit decomposition protocol that requires constant round $O(l \log l)$ communication complexity, where l is the length of the input.

Nishide and Ohta [10] proposed MPC protocols for comparison and equality check, and Ning and Xu [8,9] proposed MPC protocols for exponentiation and modulo arithmetic. These protocols achieve constant round and linear communication complexity by avoiding the use of bit decomposition.

Toft [12] proposed bit decomposition protocol with constant round and *almost linear* communication complexity, that is based on the subprotocol called postfix comparison. The postfix comparison protocol takes two bit strings of l bit, and outputs the results of comparison of most significant k bits for $k = l, l - 1, ..., 1$. The postfix comparison protocol and the bit decomposition protocol using it require constant rounds and almost linear $O(l \log^{*(c)} l)$ communication complexity, where c is a constant, l is the input size, and $\log^{*(c)}$ is the iterated logarithm function.

Only the *bit addition* protocol in [6] is left with constant rounds and superlinear $O(l \log l)$ communication complexity. We propose a MPC protocol for *prefix sum* and a bit addition protocol as its application with constant round and almost linear $O(l \log^{*(c)} l)$ communication complexity.

Our Contribution: In this paper, we propose a MPC protocol for *prefix sum in general semigroups* (an algebraic structure consisting of a set together with an *associative* binary operation ∘), i.e., it takes l elements $(a_0, ..., a_{l-1})$ and outputs the sums $(e_i = \circ_{k=0}^{i} a_k)_{i=0,...,l-1}$ of prefix i elements of the inputs. The prefix sum is an important basic operation, that is widely used in various algorithms such as carry propagation, counting sort, list ranking, and recurrence relation [3]. Especially, in the MPC setting, carry propagation is necessary to achieve bit addition protocol.

Our prefix sum protocol requires constant $2d + 2dc$ round and almost linear $O(l \log^{*(c)} l)$ communication complexity, where d is a round complexity of the subroutine protocol (BlockSum in Sect. 3), c is an any integer, l is the input size, and $\log^{*(c)}$ is the iterated logarithm function. The main tool of the proposed protocol is the notion of $(G(x), l)$-tree (see Definition 1), that is a generalization of binary tree $(= (G(x) = x/2, l)$-tree). In the proposed protocol, prefix sum is computed along the path in $(G(x) = \log^{*(c-1)} x, l)$-tree with $\log^{*(c)} l$-depth, and we can realize $O(l \log^{*(c)} l)$ communication complexity. The proposed protocol can be seen as a generalization of the postfix comparison by Toft [12].

By applying the prefix sum to the carry calculation of bit addition [6], we propose the first bit addition protocol that achieves constant $2d + 2dc$ round and almost linear $O(l \log^{*(c)} l)$ communication complexity as in Table 2. As described above, in previous works, only the bit addition protocol is left with

Table 1. Known protocols

Protocol	References	Rounds	#Mul
$[ab \mod p]_p \leftarrow \mathsf{Mul}([a]_p, [b]_p)$	[2]	1	1
$a \leftarrow \mathsf{Reveal}([a]_p)$	[2]	1	$1/n$
$[a^{-1} \mod p]_p \leftarrow \mathsf{Inv}([a]_p)$	[1,6,10]	2	2
$[r]_p \leftarrow \mathsf{Rand}()$	[6]	1	1
$[a_b]_p \leftarrow \mathsf{Cond}([b]_p, [a_1]_p, [a_0]_p)$	[11]	1	1
$[a \overset{?}{<} b]_p \leftarrow \mathsf{BitLessThan}([a]_B, [b]_B)$	[12,8]	6	$14l$
$([r]_B, [r]_p) \leftarrow \mathsf{SolvedBits}()$	[12]	7	$56l$

$O(l \log l)$ communication complexity. Since it seems hard to realize $O(l)$ communication complexity, the almost linear communication complexity is reasonable for bit addition. Since the bit addition protocol is versatile as building block for constructing many MPC protocols, our result is useful for making these MPC protocols more efficient.

In addition, by applying our bit addition protcol to the construction of Damgård et al. [6], we have almost linear bit decomposition protocol. The detailed estimation and comparison is shown in Table 2.

Organization: The rest of this paper is structured as follows. In section 2, we described known MPC protocols, in particular, Toft's postfix comparison protocol. In section 3, we propose a MPC protocol for prefix sum in general semigroups with constant rounds and almost linear communication complexity. In section 4, we construct a bit addition protocol with constant rounds and almost linear communication complexity, and we compare with the existing protocols.

2 Preliminaries

2.1 Known MPC Techniques

There exist n parties $\mathcal{P}_1, ..., \mathcal{P}_n$. Let p be a prime of bit length l. We use a linear secret sharing scheme on $\mathbb{Z}_p$ and secure MPC for addition and multiplication in $\mathbb{Z}_p$ based on the linear secret sharing scheme (LSSS). We denote by $[a]_p$ that $a \in \mathbb{Z}_p$ is shared among n parties $\mathcal{P}_1, ..., \mathcal{P}_n$ by the LSSS. We also denote by $[a]_B = ([a_1]_p, ..., [a_l]_p)$ that bit representation $a_i \in \{0, 1\} \subset \mathbb{Z}_p$ of $a = \sum_{i=1}^l a_i 2^{i-1} \in \mathbb{Z}_p$ is shared among n parties $\mathcal{P}_1, ..., \mathcal{P}_n$ by the LSSS. The addition protocol Add and the (public) multiplication protocol PubMul are fundamental tools. Add takes shared values $[a]_p, [b]_p$ ($a, b \in \mathbb{Z}_p$) as input and outputs shared addition $[a + b \mod p]_p$, i.e.,

$$[a + b \mod p]_p \leftarrow \mathsf{Add}([a]_p, [b]_p) \text{ (denoted as } [a + b]_p \text{ for simplicity)}.$$

PubMul takes public value $k \in \mathbb{Z}_p$ and shared value $[a]_p$ ($a \in \mathbb{Z}_p$) as input and outputs shared multiplication $[ka \mod p]_p$, i.e.,

$$[ka \mod p]_p \leftarrow \mathsf{PubMul}(k, [a]_p) \text{ (denoted as } [ka]_p \text{ for simplicity)}.$$

We assume both Add and PubMul are computable without communication due to the underlying LSSS. Also, we use an important protocol to compute multiplication of two shared values, denoted as Mul. Mul takes shared values $[a]_p, [b]_p$ $(a, b \in \mathbb{Z}_p)$ as input and outputs shared multiplication $[ab \mod p]_p$, i.e.,

$$[ab \mod p]_p \leftarrow \mathsf{Mul}([a]_p, [b]_p) \text{ (denoted as } [a \cdot b]_p \text{ for simplicity)}.$$

We measure the round complexity by the number of rounds of parallel invocations of Mul, that is described below, and the communication complexity by the number of invocations of Mul, i.e., Mul requires 1 round and 1 multiplication. The reveal protocol Reveal is also useful. Reveal takes a shared value $[a]_p$ $(a \in \mathbb{Z}_p)$ as input and outputs the value $a \in \mathbb{Z}_p$, i.e.,

$$a \leftarrow \mathsf{Reveal}([a]_p).$$

It is known that Reveal requires 1 round and $1/n$ multiplication. Our protocol uses following known protocols [6,11,10,12,8,9]:

- The **Unbounded Fan-In Multiplication Protocol** Mul* takes a shared values $[a_0]_p, \ldots, [a_{l-1}]_p$ $(a_i \in \mathbb{Z}_p$ for $i = 0, \ldots, l-1)$ as input and outputs shared products $[\Pi_{i=k}^{l-1} a_i]_p$ for $k = 0, \ldots, l-1$, and requires 5 rounds and 6 multiplication [6], i.e.,

$$([a_{l-1}]_p, [a_{l-1} a_{l-2}]_p, \ldots, [a_{l-1} a_{l-2} \cdots a_0]_p) \leftarrow \mathsf{Mul}^*([a_{l-1}]_p, [a_{l-2}]_p, \ldots, [a_0]_p).$$

- The **secure inversion protocol** Inv takes a shared value $[a]_p$ $(a \in \mathbb{Z}_p)$ as input and outputs shared inversion $[a^{-1} \mod p]_p$, and requires 2 round and 2 multiplication [1,6,10], i.e.,

$$[a^{-1} \mod p]_p \leftarrow \mathsf{Inv}([a]_p).$$

- The **shared random value generation protocol** Rand takes no input and outputs shared random value $[r]_p$ $(r \in \mathbb{Z}_p)$, requires 1 round and 1 multiplication [6], i.e.,

$$[r]_p \leftarrow \mathsf{Rand}().$$

- The **conditional selection protocol** Cond takes a shared bit $[b]_p$ and shared values $[a_1]_p, [a_0]_p$ $(a_1, a_0 \in \mathbb{Z}_p)$ as input and outputs shared value $[a_b]_p$, and requires 1 round and 1 multiplication [11], i.e.,

$$[a_b]_p \leftarrow \mathsf{Cond}([b]_p, [a_1]_p, [a_0]_p).$$

- The **bitwise less-than protocol** BitLessThan takes bitwise shared values $[a_1]_B, [a_0]_B$, $(a \in \mathbb{Z}_p)$ as input and outputs shared bit $[a \overset{?}{<} b]_p$, requires 6 round and $14l$ multiplication [12,8], i.e.,

$$[a \overset{?}{<} b]_p \leftarrow \mathsf{BitLessThan}([a]_B, [b]_B).$$

- The **bitwise shared random value generation protocol** SolvedBits takes no input and outputs bitwise shared random value $[r]_B = ([r_1]_p, ..., [r_l]_p)$ $(r_i \in_U \{0,1\} \subset \mathbb{Z}_p)$ and shared random value $[r]_p$ $(r = \sum_{i=1}^{l} r_i 2^{i-1} \in \mathbb{Z}_p)$, requires 7 round and $56l$ multiplication [12], i.e.,

$$([r]_B, [r]_p) \leftarrow \mathsf{SolvedBits}().$$

We summarize known protocols as in Table.1.

2.2   Toft's Almost Linear Bit Decomposition Protocol

Toft [11,12] proposed a MPC protocol BitDecomp for bit decomposition with constant $O(c)$ round and almost linear $O(l \log^{*(c)} l)$ communication complexity. To construct the bit decomposition protocol BitDecomp, Toft proposed a MPC protocol PostComp for postfix comparison with constant $O(c)$ round and $O(l \log^{*(c)} l)$ communication complexity. The postfix comparison protocol PostComp takes two shared values $[a]_B = ([a_{l-1}]_p, ..., [a_0]_p)$ and $[b]_B = ([b_{l-1}]_p, ..., [b_0]_p)$ where $a = \sum_{i=0}^{l-1} 2^i a_i$ and $b = \sum_{i=0}^{l-1} 2^i b_i$ as input, and outputs $[c_i]_p$ $(i = 0, ..., l-1)$ where $c_i = 1$ if $(a \mod 2^i) > (b \mod 2^i)$ and otherwise $c_i = 0$, i.e.,

$$([c_{l-1}]_p, ..., [c_0]_p) \leftarrow \mathsf{PostComp}([a]_B, [b]_B).$$

Toft [12] introduced parallelized computation technique using $l \log^{*(c)} l$-depth tree to construct the postfix comparison protocol PostComp with constant $O(c)$ round and $O(l \log^{*(c)} l)$ communication complexity. In this paper, we generalize the Toft's technique to prefix sum in general semigroup in section 3. So, the Toft's construction of postfix comparison protocol PostComp can be described using our prefix sum protocol in section 3 as follows.

We define semigroup $\mathcal{X} = \{o_>, o_<, o_=\}$ with associative product

$$\forall x \in \mathcal{X}, o_> \circ x = o_>,$$
$$\forall x \in \mathcal{X}, o_< \circ x = o_<,$$
$$\forall x \in \mathcal{X}, o_= \circ x = x.$$

Intuitively, $o_>$, $o_<$, and $o_=$ means $a_k > b_k$, $a_k < b_k$, and $a_k = b_k$.

For input $(a_{l-1}, ..., a_0)$ and $(b_{l-1}, ..., b_0)$ of PostComp, we define $d_k = o_>$ if $a_k > b_k$, $d_k = o_<$ if $a_k < b_k$, and $d_k = o_=$ if $a_k = b_k$. Then, output c_i of PostComp can be computed from prefix sum

$$e_i = d_{i-1} \circ \cdots \circ d_1 \circ d_0,$$

for $i = l-1, ..., 0$. Since output $([c_{l-1}]_p, ..., [c_0]_p)$ of PostComp can be computed from prefix sum in semigroup $\mathcal{X}$, we can apply prefix sum protocol in section 3 and obtain PostComp protocol with constant $O(c)$ round and almost linear $O(l \log^{*(c)} l)$ communication complexity.

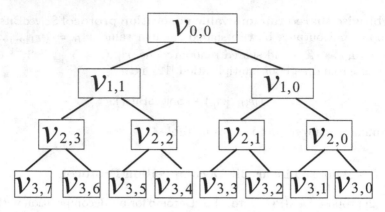

Fig. 1. $(x/2, l)$-tree, i.e., binary tree for PrefixSum in case of $l=8$

3   Proposed MPC Protocol for Generic Prefix Sum

In this section, we propose an unconditionally secure MPC protocol to compute prefix sum in general semigroup, that is an generalization of Toft's technique [12]. The proposed MPC protocol runs with constant $2d+2dc$ rounds and almost linear $O(l \log^{*(c)} l)$ communication complexity.

Prefix sum can be computed trivially by sequential algorithm, but especially in the model of MPC, sequential computation increases the round complexity, that is running time of the MPC protocol. In this section, we describe the MPC protocol for prefix sum operation by using parallel computing.

Let $\mathcal{X}$ be a finite semigroup and let $\circ : \mathcal{X} \times \mathcal{X} \to \mathcal{X}$ denote associative product of $\mathcal{X}$ (i.e., $\forall x_1, x_2, x_3 \in \mathcal{X}$, $(x_1 \circ x_2) \circ x_3 = x_1 \circ (x_2 \circ x_3)$). Notice that we do not require the existence of inverse element and the commutativity for $\mathcal{X}$. We define the prefix sum of $a_{l-1}, ..., a_0 \in \mathcal{X}$ with respect to product $\circ$ as

$$(e_i = a_i \circ \cdots \circ a_0 = \circ_{k=0}^{i} a_k)_{i=0,...,l-1}$$
$$= (a_0, a_1 \circ a_0, a_2 \circ a_1 \circ a_0, ..., a_{l-1} \circ \cdots \circ a_0).$$

We define iterated logarithm $\log^{*(c)} l$ as follows.

$$\log^{*(0)} l = \log l = \log_2 l$$
$$\log^{*(c)} l = \begin{cases} 1 + \log^{*(c)}(\log^{*(c-1)} l) \ (l > 1) \\ 0 \qquad\qquad\qquad\qquad (l \leq 1) \end{cases}$$

We denote $\log^* l = \log^{*(1)} l$. The iterated logarithm $\log^{*(c)} l$ is the number of times function $\log^{*(c-1)}$ must be iteratively applied to l before the result is less than or equal to 1. For instance, we have $\log^* l = \log^{*(1)} l = 5$ for $l = 2^{2^{2^{2^2}}} = 2^{65536}$. Notice that $\log^{*(c)}(\cdot)$ is equivalent to the inverse of Ackermann function $A(c+3, \cdot)$ up to constant.

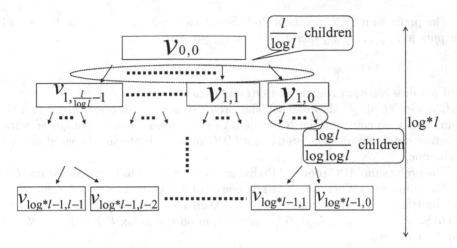

Fig. 2. $(\log x, l)$-tree for PrefixSum*

Next, we define the notion of $(G(x), l)$-tree that is main tool of our proposed protocol.

Definition 1 $((G(x), l)$-**tree**)**.** *For function G such that $G(x) < x$ and integer $l > 1$, we define that tree T is $(G(x), l)$-tree if each node at depth i has $G^i(l)/G^{i+1}(l)$ child nodes for $i = 0, 1, ..., d$ where $d = \max\{i : G^{i+1}(l) > 1\}$.*

In $(G(x), l)$-tree T, at depth i, there are

- $l/G^i(l)$ nodes,
- $l/G^{i-1}(l)$ groups of brother nodes with the same parent node,
- $G^{i-1}(l)/G^i(l)$ brother nodes in a group.

Note that $(x/2, l)$-tree is binary tree and has depth $\log l$, $(\log x, l)$-tree has depth $\log^* l$, and $(\log^{*(c-1)} x, l)$-tree has depth $\log^{*(c)} l$.

In the following, we recall MPC protocol PrefixSum [6] for prefix sum with $O(l \log l)$ communication complexity that is based on $(x/2, l)$-tree. Then, we propose MPC protocol PrefixSum* for prefix sum with $O(l \log^* l)$ communication complexity that is based on $(\log x, l)$-tree, and MPC protocol PrefixSum*$^{(c)}$ for prefix sum with $O(l \log^{*(c)} l)$ communication complexity that is based on $(\log^{*(c-1)} x, l)$-tree.

We assume that there exists MPC protocol $[e_i]_p \leftarrow \mathsf{BlockSum}([a_i]_p, ...[a_0]_p)$ to compute $[e_i]_p = \circ_{k=0}^i [a_k]_p$ with constant d rounds and linear $O(i)$ communication complexity. BlockSum is used in the following MPC protocols PrefixSum, PrefixSum*, and PrefixSum*$^{(c)}$.

3.1 MPC Protocol for Prefix Sum with $O(l \log l)$ Complexity

In this section, we recall MPC protocol PrefixSum [6], which is based on $(x/2, l)$-tree, to compute prefix sum with constant rounds and $O(l \log l)$ communication complexity.

The prefix sum MPC protocol PrefixSum takes $[a_{l-1}]_p, \ldots, [a_0]_p$ as input and outputs $[e_{l-1}]_p, \ldots, [e_0]_p$ such that $e_i = \circ_{k=0}^{i} a_k$ for $i = 0, \ldots, l-1$, i.e.,

$$([e_{l-1}]_p, \ldots, [e_0]_p) \leftarrow \mathsf{PrefixSum}([a_{l-1}]_p, \ldots, [a_0]_p),$$

and requires constant rounds and $O(l \log l)$ multiplications.

Chandra et al. [4] showed that the prefix sum w.r.t an operation $\circ$ can be computed with constant depth circuit if the operation $\circ$ can be computed with constant depth circuit. The prefix sum MPC protocol PrefixSum is based on this technique.

The prefix sum MPC protocol PrefixSum is provided as follows. We assume $l = 2^k$ for simplicity. (When $l = N$ is not power of 2, let $l' = \min\{2^k \mid k \in \mathbb{N}, 2^k > N\}$ and $[a_{l'}]_p = \cdots = [a_{N+1}]_p = [0]_p$, then compute $([e_{l'-1}]_p, \ldots, [e_0]_p) \leftarrow \mathsf{PrefixSum}([a_{l'-1}]_p, \ldots, [a_0]_p)$. The calculation on the index $l', l' - 1, \ldots, N + 1$ can be ignored.)

1. Consider $(x/2, l)$-tree, i.e., binary tree and denote by $v_{i,j}$ the j-th node from right at depth i for $i = 0, \ldots, \log l$, $j = 0, \ldots, 2^i - 1$ as in Fig. 1. For each depth $i = 0, \ldots, \log l$, assign block $[a_{i,j}]_p = ([a_{j \cdot (l/2^i) + (l/2^i) - 1}]_p, \ldots, [a_{j \cdot (l/2^i)}]_p)$ to node $v_{i,j}$ for $j = 0, \ldots, 2^i - 1$. For each node $v_{i,j}$, parallely compute value

$$[e_{i,j}]_p = \circ_{k=j \cdot (l/2^i)}^{j \cdot (l/2^i) + (l/2^i) - 1}[a_k]_p$$

using MPC protocol BlockSum.

2. For $j = 0, \ldots, l-1$, parallely perform the following procedure to compute $[e_j]_p$.
   - Set node $v_{\log l, j}$ as current node and append $[e_{\log l, j}]_p$ to list L.
   - If the right brother node $v_{i,j}$ of the current node exists, append value $[e_{i,j}]_p$ to list L. Set the parent node of the current node as current node, and repeat this procedure until reaching the root node.
   - If the current node is the root node, compute product of all elements in list L using MPC protocol BlockSum to obtain $[e_j]_p$.

Example: In the example of Fig. 1, in step 1, we compute

$$[e_{0,0}]_p = [a_7 \circ a_6 \circ a_5 \circ a_4 \circ a_3 \circ a_2 \circ a_1 \circ a_0]_p,$$
$$[e_{1,1}]_p = [a_7 \circ a_6 \circ a_5 \circ a_4]_p, [e_{1,0}]_p = [a_3 \circ a_2 \circ a_1 \circ a_0]_p,$$
$$[e_{2,3}]_p = [a_7 \circ a_6]_p, \ldots, [e_{2,0}]_p = [a_1 \circ a_0]_p,$$
$$[e_{3,7}]_p = [a_7]_p, \ldots, [e_{3,0}]_p = [a_0]_p.$$

In step 2, we compute, for instance, $[e_7]_p = [e_{3,7}]_p \circ [e_{3,6}]_p \circ [e_{2,2}]_p \circ [e_{1,0}]_p$ and $[e_4]_p = [e_{3,4}]_p \circ [e_{1,0}]_p$.

Complexity: Recall that BlockSum computes $[e_i]_p = \circ_{k=0}^{i}[a_k]_p$ with constant d rounds and linear $O(i)$ communication complexity.

In step 1, $[e_{i,j}]_p$ are computed with constant d rounds and $O(l) + 2 \cdot O(l/2) + 4 \cdot O(l/4) + \cdots + l \cdot O(1) = \log l \cdot O(l) = O(l \log l)$ communication complexity.

In step 2, $[e_i]_p$ are computed with constant d rounds $l \cdot O(\log l) = O(l \log l)$ communication complexity, since the number of elements in list L is $O(\log l)$.

Thus, PrefixSum runs with constant $2d$ rounds and $O(l \log l)$ communication complexity.

3.2 Proposed Protocol for Prefix Sum with $O(l \log^* l)$ Complexity

In this section, we propose MPC protocol PrefixSum*, which is based on $(\log x, l)$-tree, to compute prefix sum with constant rounds and $O(l \log^* l)$ communication complexity, that is an generalization of Toft's technique [12].

The proposed prefix sum MPC protocol PrefixSum* takes $[a_{l-1}]_p, \ldots, [a_0]_p$ as input and outputs $[e_{l-1}]_p, \ldots, [e_0]_p$, i.e.,

$$([e_{l-1}]_p, \ldots, [e_0]_p) \leftarrow \text{PrefixSum}^*([a_{l-1}]_p, \ldots, [a_0]_p),$$

and requires constant rounds and $O(l \log^* l)$ communication complexity.

The prefix sum MPC protocol PrefixSum* is provided as follows. We assume $l = 2^{\cdot^{\cdot^2}}$ for simplicity.

1. Consider $(\log x, l)$-tree and denote by $v_{i,j}$ the j-th node from right at depth i for $i = 0, \ldots, \log^* l - 1$, $j = 0, \ldots, l/G^i(l) - 1$ as in Fig. 1. For each depth $i = 0, \ldots, \log^* l$, assign block $[a_{i,j}]_p = ([a_{j \cdot G^i(l) + G^i(l) - 1}]_p, \ldots, [a_{j \cdot G^i(l)}]_p)$ to node $v_{i,j}$ for $j = 0, \ldots, l/G^i(l) - 1$. For each node $v_{i,j}$, parallely compute value

$$[e_{i,j}]_p = \circ_{k=j \cdot G^i(l)}^{j \cdot G^i(l) + G^i(l) - 1} [a_k]_p$$

   using MPC protocol BlockSum.

2. For each depth $i = 0, \ldots, \log^* l - 2$, for each group of brother nodes $(v_{i,n+j})$ $j = 0, \ldots, \frac{G^{i-1}(l)}{G^i(l) - 1}$ with the same parent node, parallely execute prefix sum MPC protocol

$$([e'_{i,n+j}]_p)_{j=0, \ldots, G^{i-1}(l)/G^i(l)-1}$$
$$\leftarrow \text{PrefixSum}(([e_{i,n+j}]_p)_{j=0, \ldots, G^{i-1}(l)/G^i(l)-1}),$$

   where $[e'_{i,n+j}]_p = \circ_{k=0}^{j}[e_{i,n+k}]_p$, and assign value $e'_{i,n+j}$ to node $v_{i,n+j}$ for $j = 0, \ldots, G^{i-1}(l)/G^i(l) - 1$.

3. For $j = 0, \ldots, l - 1$, parallely perform the following procedure to compute $[e_j]_p$.
   - Set node $v_{\log^* l, j}$ as current node and append $[e'_{\log l, j}]_p$ to list L.
   - If the right *next* brother node $v_{i,j}$ of the current node exists, append value $[e'_{i,j}]_p$ to list L. Set the parent node of the current node as current node, and repeat this procedure until reaching the root node.
   - If the current node is the root node, compute product of all elements in list L using MPC protocol BlockSum to obtain $[e_j]_p$.

Complexity: Recall that BlockSum computes $[e_i]_p = \circ_{k=0}^{i}[a_k]_p$ with constant d rounds and linear $O(i)$ communication complexity.

In step 1, $[e_{i,j}]_p$ are computed with constant d rounds and

$$O(l) + l/\log l \cdot O(\log l) + l/\log\log l \cdot O(\log\log l) + \cdots$$
$$= \log^* l \cdot O(l) = O(l \log^* l)$$

communication complexity.

In step 2, each depth $i = 0, ..., \log^* l - 2$, total communication complexity of executions of PrefixSum is $O(l)$, since $O((l/\log l) \cdot \log(l/\log l)) \leq O((l/\log l) \cdot \log(l))$ in level 1, $(l/\log l) \cdot O((\log l/\log\log l) \cdot \log(\log l/\log\log l)) \leq (l/\log l) \cdot O((\log l/\log\log l) \cdot \log(\log l))$ in level 2,... So, $[e'_{i,j}]_p$ are computed with constant $2d$ rounds and

$$O\left(\frac{l}{\log l}\log\frac{l}{\log l}\right) + \frac{l}{\log l}\cdot O\left(\frac{\log l}{\log\log l}\log\frac{\log l}{\log\log l}\right) + \cdots$$
$$\leq \log^* l \cdot O(l) = O(l \log^* l)$$

communication complexity.

In step 3, $[e_i]_p$ are computed with constant d rounds $l \cdot O(\log^* l) = O(l \log^* l)$ communication complexity, since the number of elements in list L is $O(\log^* l)$.

Thus, PrefixSum* runs with constant $2d + 2d$ rounds and $O(l \log^* l)$ communication complexity.

3.3   Proposed Protocol for Prefix Sum with $O(l \log^{*(c)} l)$ Complexity

In this section, we propose MPC protocol PrefixSum$^{*(c)}$, which is based on $(\log^{*(c-1)} x, l)$-tree, to compute prefix sum with constant $2d + 2dc$ rounds and $O(l \log^{*(c)} l)$ communication complexity, that is also an generalization of Toft's technique [12].

The proposed prefix sum MPC protocol PrefixSum$^{*(c)}$ takes $[a_{l-1}]_p, \ldots, [a_0]_p$ as input and outputs $[e_{l-1}]_p, \ldots, [e_0]_p$, i.e.,

$$([e_{l-1}]_p, \ldots, [e_0]_p) \leftarrow \text{PrefixSum}^{*(c)}([a_{l-1}]_p, \ldots, [a_0]_p),$$

and requires constant rounds and $O(l \log^{*(c)} l)$ multiplications.

The prefix sum MPC protocol PrefixSum$^{*(c)}$ is same as PrefixSum* except the following points.

- We use $(\log^{*(c-1)} x, l)$-tree of depth $\log^{*(c)} l$.
- We use PrefixSum$^{*(c-1)}$ recursively in step 2. (We denote PrefixSum$^{*(1)}$ = PrefixSum*.)

Complexity: Recall that BlockSum computes $[e_i]_p = \circ_{k=0}^{i}[a_k]_p$ with constant d rounds and linear $O(i)$ communication complexity. We can assume PrefixSum$^{*(c-1)}$ computes prefix sum with constant $2d + 2d(c-1)$ rounds and linear $O(l \log^{*(c-1)} l)$ communication complexity. We denote $G(x) = \log^{*(c-1)} x$.

Protocol: BitAdd($[a]_B, [b]_B$).
Inputs: $[a]_B, [b]_B$
Outputs: $([d]_B) = ([d_l]_p, \ldots, [d_0]_p)$.

1. $([c_l]_p, \ldots, [c_1]_p) \leftarrow \mathsf{SetCarry}([a]_B, [b]_B)$.
2. $[d_0]_p \leftarrow [a_0]_p + [b_0]_p - 2[c_1]_p$,
   $[d_l]_p \leftarrow [c_l]_p$
3. For $i = 1, \ldots, l - 1$ in parallel : $[d_i]_p \leftarrow [a_i]_p + [b_i]_p + [c_i]_p - 2[c_{i+1}]_p$.
4. Output $([d_l]_p, \ldots, [d_0]_p)$.

Fig. 3. Description of BitAdd protocol [6]

In step 1, $[e_{i,j}]_p$ are computed with constant d rounds and

$$O(l) + l/G(l) \cdot O(G(l)) + l/G^2(l) \cdot O(G^2(l)) + \cdots$$
$$= \log^{*(c)} l \cdot O(l) = O(l \log^{*(c)} l)$$

communication complexity.

In step 2, each depth $i = 0, \ldots, \log^{*(c)} l - 2$, total communication complexity of executions of $\mathsf{PrefixSum}^{*(c-1)}$ is $O(l)$, since $O((l/G(l)) \cdot G(l/G(l))) \leq O((l/G(l)) \cdot G(l))$ in level 1, $(l/G(l)) \cdot O((G(l)/G^2(l)) \cdot G(G(l)/G^2(l))) \leq (l/G(l)) \cdot O((G(l)/G^2(l)) \cdot G(G(l)))$ in level 2,... So, $[e'_{i,j}]_p$ are computed with constant $2d + 2d(c - 1)$ rounds and

$$O\left(\frac{l}{G(l)} G\left(\frac{l}{G(l)}\right)\right) + \frac{l}{G(l)} \cdot O\left(\frac{G(l)}{G^2(l)} G\left(\frac{G(l)}{G^2(l)}\right)\right) + \cdots$$
$$\leq \log^{*(c)} l \cdot O(l) = O(l \log^{*(c)} l)$$

communication complexity.

In step 3, $[e_i]_p$ are computed with constant d rounds $l \cdot O(\log^{*(c)} l) = O(l \log^{*(c)} l)$ communication complexity, since the number of elements in list L is $O(\log^{*(c)} l)$.

Thus, $\mathsf{PrefixSum}^{*(c)}$ runs with constant $2d + 2dc$ rounds and $O(l \log^{*(c)} l)$ communication complexity.

4 Proposed MPC Protocol for Bit Addition

In this section, we propose an unconditionally secure MPC protocol BitAdd with constant $2d + 2dc + 1$ rounds and almost linear $O(l \log^{*(c)} l)$ communication complexity, that securely compute bit addition $[a+b]_B = [d]_B = ([d_l]_p, \ldots, [d_0]_p)$ of shared values $[a]_B = ([a_{l-1}]_p, \ldots, [a_0]_p)$ and $[b]_B = ([b_{l-1}]_p, \ldots, [b_0])$, i.e.,

$$[d]_B \leftarrow \mathsf{BitAdd}([a]_B, [b]_B).$$

BitAdd protocol is shown in Fig. 3, and uses carry calculation protocol SetCarry. BitAdd protocol runs with R round and C multiplications, where SetCarry protocol runs with R round and C multiplications.

Protocol: SetSPK($[a]_B, [b]_B$).
Inputs: $[a]_B, [b]_B$
Outputs: $([e_{l-1}]_p, \ldots, [e_0]_p)$
  $= (([s_{l-1}]_p, [p_{l-1}]_p, [k_{l-1}]_p), \ldots, ([s_0]_p, [p_0]_p, [k_0]_p))$.
   1. For $i = 0, \ldots, l-1$ in parallel : $[s_i]_p \leftarrow \mathsf{Mul}([a_i]_p, [b_i]_p)$.
   2. For $i = 0, \ldots, l-1$ in parallel : $[p_i]_p \leftarrow [a_i]_p + [b_i]_p - 2[s_i]_p$.
   3. For $i = 0, \ldots, l-1$ in parallel : $[k_i]_p \leftarrow 1 - [s_i]_p - [p_i]_p$.
   4. For $i = 0, \ldots, l-1$: $[e_i]_p := ([s_i]_p, [p_i]_p, [k_i]_p)$.
   5. Output $([e_{l-1}]_p, \ldots, [e_0]_p)$.

Fig. 4. SetSPK Protocol [6]

4.1 Computation of Carry Bits

In this section, we recall carry calculation protocol SetCarry shown in Fig. 5 and introduced in [6].

To compute carry bits, we define the operation $\circ : \{S, P, K\} \times \{S, P, K\} \to \{S, P, K\}$ as

$$\forall x, S \circ x = S, \quad \forall x, K \circ x = K, \quad \forall x, P \circ x = x.$$

The operation $\circ$ is associative, and we have semigroup $\{S, P, K\}$ with product $\circ$.

The operation $\circ$ handle propagation of carry bits as follows. For example, if $e_i = P$ and $e_{i-1} = x$, $e_i \circ e_{i-1} = P \circ x = x$. This means that i-th carry bit depends on $i-1$-th carry bit, i.e., P propagates carry bit. If $e_i = S$ (or $e_i = K$), i-th carry bit is set (or killed) regardless of the $i-1$-th carry bit.

Let denote $e'_i = \circ^i_{j=0} e_j = e_i \circ \cdots \circ e_0 \circ e_{-1}$ where $e_{-1} = K$. Note that e'_i is equal to S or K, and $e'_i = S$ means i-th carry bit is 1 and $e'_i = K$ means i-th carry bit is 0. Thus, the computation of carry bits can be reduced to the computation of prefix sum in semigroup $\{S, P, K\}$ with product $\circ$.

SetSPK protocol, shown in Fig. 4, takes $[a]_B$ and $[b]_B$ as input and outputs $([e_{l-1}]_p, \ldots, [e_0]_p)$, where $[e_i]_p = ([s_i]_p, [p_i]_p, [k_i]_p)$, i.e.,

$$([e_{l-1}]_p, \ldots, [e_0]_p) \leftarrow \mathsf{SetSPK}([a]_B, [b]_B).$$

SetSPK sets $s_i = 1$ (i.e. $e_i = S$) iff $a_i + b_i = 2$, $p_i = 1$ (i.e. $e_i = P$) iff $a_i + b_i = 1$, and $k_i = 1$ (i.e. $e_i = K$) iff $a_i + b_i = 0$. SetSPK protocol runs with 1 round and l multiplications.

SetCarry protocol, shown in Fig. 5, takes $[a]_B$ and $[b]_B$ as input and outputs carry bits $([c_l]_p, \ldots, [c_0]_p)$, i.e.,

$$([c_l]_p, \ldots, [c_1]_p) \leftarrow \mathsf{SetCarry}([a]_B, [b]_B).$$

Note that if $[e_{-1}]_p$ is not given as input of PrefixCarry protocol and $e_k = e_{k-1} = \cdots = e_0 = P$ for some k, $e'_k, e'_{k-1}, \ldots, e'_0$ is not equal to K but P. However, it is no matter because $c_{i+1} = s_i$ for all i and $s_i = 0$ when $e'_i = K$ or P.

Protocol: SetCarry($[a]_B, [b]_B$).
Inputs: $[a]_B, [b]_B$
Outputs: $([c_l]_p, \ldots, [c_1]_p)$.

1. $([e_{l-1}]_p, \ldots, [e_0]_p) \leftarrow \mathsf{SetSPK}([a_i]_p, [b_i]_p)$.
2. $([e'_{l-1}]_p, \ldots, [e'_0]_p) \leftarrow \mathsf{PrefixCarry}([e_{l-1}]_p, \ldots, [e_0]_p)$.
3. For $i = 0, \ldots, l-1$:
   $([s_i]_p, [p_i]_p, [k_i]_p) := [e'_i]_p, \quad [c_{i+1}] \leftarrow [s_i]$.
4. Output $([c]_B = ([c_l]_p, \ldots, [c_1]_p)$.

Fig. 5. SetCarry Protocol [6]

Protocol: BlockCarry$(([e_m]_p, \ldots, [e_1]_p))$.
Inputs: $([e_m]_p, \ldots, [e_1]_p) = (([s_m]_p, [p_m]_p, [k_m]_p), \ldots, ([s_1]_p, [p_1]_p, [k_1]_p))$.
Outputs: $[e']_p = ([s']_p, [p']_p, [k']_p)$.

1. $([f_m]_p, \ldots, [f_1]_p) \leftarrow \mathsf{Mul}^*([p_{[m]}]_p, \ldots, [p_{[1]}]_p)$.
2. For $i = 1, \ldots, m$ in parallel : $[g_i]_p \leftarrow \mathsf{Mul}([f_i]_p, [s_{i-1}]_p)$.
3. $[s']_p \leftarrow \vee_{i=1}^m [g_i] \vee [s_{m-1}]$.
4. $[p']_p \leftarrow [g_m]_p$.
5. $[k']_p \leftarrow 1 - [s']_p - [p']_p$.
6. $[e']_p := ([s']_p, [p']_p, [k']_p)$
7. Output $[e']_p$.

Fig. 6. BlockCarry Protocol [6]

Complexity: SetCarry protocol uses SetSPK protocol and PrefixCarry protocol. SetCarry protocol runs with $R + 1$ round and $C + l$ multiplications, where PrefixCarry protocol runs with R round and C multiplications.

4.2 Proposed Prefix Carry Protocol

In this section, we propose carry calculation protocol PrefixCarry with constant $2d + 2dc + 1$ rounds and almost linear $O(l \log^{*(c)} l)$ communication complexity. PrefixCarry protocol takes $([e_{l-1}]_p, \ldots, [e_0]_p)$ as input and outputs prefix sum $[e'_{l-1}]_p, \ldots, [e'_0]_p$, where $e'_i = \circ_{j=0}^i e_j = e_i \circ \cdots \circ e_0$, i.e.,

$$([e'_{l-1}]_p, \ldots, [e'_0]_p) \leftarrow \mathsf{PrefixCarry}([e_{l-1}]_p, \ldots, [e_0]_p).$$

To construct PrefixCarry protocol, we prepare BlockCarry protocol [6] shown in Fig. 6, that takes $([e_m]_p, \ldots, [e_1]_p)$ as input and outputs block sum $[e']_p = \circ_{i=1}^m [e_i]_p$, i.e.,

$$[e']_p \leftarrow \mathsf{BlockCarry}([e_m]_p, \ldots, [e_1]_p).$$

BlockCarry protocol runs with 11 rounds and $13m$ multiplications.

To compute carry bits, PrefixCarry protocol computes prefix sum $[e'_{l-1}]_p, \ldots, [e'_0]_p$, where $e'_i = \circ_{j=0}^i e_j = e_i \circ \cdots \circ e_0$, in semigroup $\{S, P, K\}$ with product $\circ$.

Table 2. Comparison with the existing bit addition and bit decomposition protocols. In the row of BitAdd, bit addition protocol of [6] and the proposed bit addition protocols are compared. In the row of BitDecomp, bit decomposition protocols of [6] and of [10], bit decomposition protocols of [12], and bit decomposition protocols based on the combination of [10] and the proposed bit addition protocols are compared.

Protocol	Scheme	Rounds	The number of Mul	Reference
BitAdd	Damgård et al. [6]	37	$55l \log l$	[6]
	proposed 1: $(\log x, l)$-tree	45	$52l \log^* l + l$	Sec. 3.2
	proposed 2: $(\log^{*(c-1)}, l)$-tree	$23 + 22c$	$52l \log^{*(c)} l + l$	Sec. 3.3
BitDecomp	Damgård et al. [6]	114	$110l \log l + 118l$	[6]
	Nishide and Ohta [10]	25	$47l \log l + 63l + 30\sqrt{l}$	[10]
	Toft [12]	33	$57l \log^* l + 71l + 14\sqrt{l} \log^* l + 30\sqrt{l}$	[12]
	Toft [12]	$23 + 10c$	$57l \log^{*(c)} l + 71l + 14\sqrt{l} \log^{*(c)} l + 30\sqrt{l}$	[12]
	[10] + proposed 1	52	$52l \log^* l + 64l + 30\sqrt{l}$	This paper
	[10] + proposed 2	$30 + 22c$	$52l \log^{*(c)} l + 64l + 30\sqrt{l}$	This paper

Thus, we apply the proposed MPC protocol for prefix sum in general semigroup to specific semigroup $\{S, P, K\}$. The proposed PrefixCarry protocol can be implemented as PrefixSum* protocol in section 3.2 or PrefixSum$^{*(c)}$ protocol in section 3.3 using BlockCarry protocol as BlockSum protocol.

Complexity: As described in section 3.2, PrefixSum* protocol uses BlockCarry protocol in twice. The first BlockCarry protocol takes l-size input, and is computed for each level of $(\log x, l)$-tree in parallel. The second BlockCarry protocol takes $\log^* l$-size input, and is computed for each carry bit in parallel. Therefore, the complexity of PrefixSum* is 44 rounds and $52l \log^* l$ multiplications. Since the BitAdd protocol consist of PrefixSum* protocol and SetSPK protocol, the resulting BitAdd protocol using PrefixSum* runs with 45 rounds and $52l \log^* l + l$ multiplications.

The PrefixSum$^{*(c)}$ protocol uses Prefix$^{*(c-1)}$ recursively, and for all Prefix$^{*(i)}$, two BlockCarry protocols are required. Therefore, the complexity of PrefixSum* is $22 + 22c$ rounds and $52l \log^{*(c)} l$ multiplications and the resulting BitAdd protocol using PrefixSum$^{*(c)}$ runs with $23 + 22c$ rounds and $52l \log^{*(c)} l + l$ multiplications.

So, we can improve communication complexity of BitAdd from $O(l \log l)$ multiplications [6] to $O(l \log^{*(c)} l)$ multiplications.

4.3   Comparison

We compare the proposed protocols with the existing constant rounds bit addition and bit decomposition protocols in Table 2.

The proposed bit addition protocol BitAdd is the first protocol that realizes both of constant rounds and almost linear $O(l \log^{*(c)} l)$ communication complexity. However, on the round complexity, BitAdd protocol of [6] is better than the proposed.

The proposed bit decomposition protocol BitDecomp based on the combination of bit decomposition protocol of [10] and the proposed bit addition protocol

BitAdd realizes both of constant rounds and almost linear $O(l \log^{*(c)} l)$ communication complexity. However, on the round complexity, BitDecomp protocol of [12] is better than the proposed.

References

1. Bar-Ilan, J., Beaver, D.: Non-Cryptographic Fault-Tolerant Computing in Constant Number of Rounds of Interaction. In: PODC 1989, pp. 201–209 (1989)
2. Ben-Or, M., Goldwasser, S., Wigderson, A.: Completeness Theorems for Non-Cryptographic Fault-Tolerant Distributed Computation. In: STOC 1988, pp. 1–10 (1988)
3. Blelloch, G.E.: Prefix sums and their applications. In: Reif, J.H. (ed.) Synthesis of Parallel Algorithms. Morgan Kaufmann (1991)
4. Chandra, A.K., Fortune, S., Lipton, R.J.: Unbounded fan-in circuits and associative functions. In: STOC 1983, pp. 52–60 (1983)
5. Chaum, D., Crépeau, C., Damgård, I.: Multiparty Unconditionally Secure Protocols (Extended Abstract). In: STOC 1988, pp. 11–19 (1988)
6. Damgård, I.B., Fitzi, M., Kiltz, E., Nielsen, J.B., Toft, T.: Unconditionally Secure Constant-Rounds Multi-party Computation for Equality, Comparison, Bits and Exponentiation. In: Halevi, S., Rabin, T. (eds.) TCC 2006. LNCS, vol. 3876, pp. 285–304. Springer, Heidelberg (2006)
7. Goldreich, O., Micali, S., Wigderson, A.: How to Play any Mental Game or A Completeness Theorem for Protocols with Honest Majority. In: STOC 1987, pp. 218–229 (1987)
8. Ning, C., Xu, Q.: Multiparty Computation for Modulo Reduction without Bit-Decomposition and a Generalization to Bit-Decomposition. In: Abe, M. (ed.) ASIACRYPT 2010. LNCS, vol. 6477, pp. 483–500. Springer, Heidelberg (2010)
9. Ning, C., Xu, Q.: Constant-Rounds, Linear Multi-party Computation for Exponentiation and Modulo Reduction with Perfect Security. In: Lee, D.H., Wang, X. (eds.) ASIACRYPT 2011. LNCS, vol. 7073, pp. 572–589. Springer, Heidelberg (2011)
10. Nishide, T., Ohta, K.: Multiparty Computation for Interval, Equality, and Comparison Without Bit-Decomposition Protocol. In: Okamoto, T., Wang, X. (eds.) PKC 2007. LNCS, vol. 4450, pp. 343–360. Springer, Heidelberg (2007)
11. Toft, T.: Primitives and Applications for Multi-party Computation. PhD thesis, University of Aarhus (2007)
12. Toft, T.: Constant-Rounds, Almost-Linear Bit-Decomposition of Secret Shared Values. In: Fischlin, M. (ed.) CT-RSA 2009. LNCS, vol. 5473, pp. 357–371. Springer, Heidelberg (2009)
13. Yao, A.C.-C.: Protocols for Secure Computations (Extended Abstract). In: FOCS 1982, pp. 160–164 (1982)
14. Yao, A.C.-C.: How to Generate and Exchange Secrets (Extended Abstract). In: FOCS 1986, pp. 162–167 (1986)

Position-Based Cryptography
from Noisy Channels*

Stefan Dziembowski[1] and Maciej Zdanowicz[2]

[1] Institute of Computer Science, University of Warsaw, Poland
[2] Institute of Mathematics, University of Warsaw, Poland

Abstract. We study the problem of constructing secure positioning protocols (Sastry et. al, 2003). Informally, the goal of such protocols is to enable a party P to convince a set of verifiers about P's location in space, using information about the time it takes P to respond to queries sent from different points. It has been shown by Chandran et al (2009) that in general such task is impossible to achieve if the adversary can position his stations in multiple points in space. Chandran et al proposed to overcome this impossibility result by moving to Maurer's bounded-storage model. Namely, they construct schemes that are secure under the assumption that the memory of the adversary is bounded. Later Buhrman et al (2010) considered secure positioning protocols schemes in quantum settings.

In this paper we show how to construct secure positioning schemes in the so-called noisy channel scenario, i.e.: in the setting where the parties participating in a protocol have access to a source of random bits sent to them via independent noisy channels. We argue that for some practical applications such assumptions may be more realistic than those used before.

Keywords: Position-based cryptography, information theoretic security.

1 Introduction

The problem of *secure positioning* [11,12,3] can informally be described as follows. Suppose a party P wants to convince a verifier V that it is situated in a certain geographic location. We are interested in the settings where verifier does not trust P, and hence it is not enough that P simply determines its position $\hat{P}$ (using a GPS device, say), and sends it to the verifier. Therefore, our goal is to construct protocols that allow P to *prove* to V that it really is in the position $\hat{P}$. Moreover, in many cases it would be useful to have a key-agreement protocol on top of such a proof, i.e. to make the parties conclude the protocol with a secret key k that can be used for future secure communication.

---

* This work was supported by the WELCOME/2010-4/2 grant founded within the framework of the EU Innovative Economy (National Cohesion Strategy) Operational Programme.

D. Pointcheval and D. Vergnaud (Eds.): AFRICACRYPT 2014, LNCS 8469, pp. 300–317, 2014.

There are several potential applications of such protocols. For example, one could use them to grant free access to a wifi network to all users that are inside of some building, or to provide an additional layer of security in the communication between military bases: to communicate with a personnel of such a base one would verify not only the knowledge of a secret key, but also the fact that the party is located physically within the base. For a descryption of other potential applications the reader may consult e.g. [4].

A standard, non-cryptographic, way to verify someone's position $\hat{P}$ is to go to $\hat{P}$ and physically check that P is indeed there. Obviously for several applications, including those mentioned above, this solution is infeasible and hence one needs a protocol that is purely based on the communication between P and V. In this case the only known way to construct such protocols is to base them on the fact that the speed of the electromagnetic signals is constant, and equal to the speed of light c. For example, if V sends a message to P and P replies within time t then this implies that P is in a distance at most $ct/2$ from V (assume for a moment that we are interested only in verifying the position of P, not in the key-agreement). Unfortunately, this method, known under the name *distance bounding protocols* [1], gives only a very rough estimate of the position of P, as a cheating P can be in fact closer to V than $ct/2$ and delay his answer to convince V that he is further away.

A natural solution to this problem is to use a standard geometric technique called *triangulation* (see e.g. [3]). More precisely, instead of considering just one verifier, use 4 verifiers $V^1, \ldots, V^4$, and let each V^i independently check (via the distance bounding technique) that P is within the distance $\|V^i\hat{P}\|$ from him (where $\hat{P}$ is the position claimed by P and $\|V^i\hat{P}\|$ denotes the geometric distance of V_i from $\hat{P}$). It is easy to see that such information uniquely determines the position of the prover within the tetrahedron determined by $V^1, \ldots, V^4$. Hence, intuitively, P can succeed in convincing the verifiers only if he really is in position $\hat{P}$. This argument is correct (see, e.g. [3]) as long as the cheater, who wants to falsely claim that he is in position $\hat{P}$, is only a single entity, located in one geographic position. Unfortunately, the security of such protocols breaks completely if there is a larger number of cheaters that may collude, or if one cheater can appear in several copies (spread over different locations). To see why it is the case, simply imagine a situation when an adversary A^i is placed next to each verifier V^i — in this case A^i can reply to the messages coming from V^i in the right moment in time, and hence $A^1, \ldots, A^4$, can jointly convince the verifiers that there is a prover in position $\hat{P}$.

This, of course, shows only that the particular protocol considered above (a simple combination of triangulation and distance bounding) does not work, and one could hope to find other, more sophisticated solutions to this problem. Unfortunately it turns out that in general in this model no secure positioning protocol exists, unless one makes some additional assumptions about the power of the cheaters. This impossibility result was shown by Chandran et al. [4], who pioneered the theoretical study of positioning protocols, and coined the term *position-based cryptography*.

In their paper Chandran et al. pose a question if there exist natural assumptions about the power of the adversary that one could introduce into the model in order to bypass their impossibility result. They answer it affirmatively, by showing constructions of the position-based authentication and key agreement schemes secure in the bounded-storage model of Maurer [9]. In this model one assumes that the parties can send huge random strings that are too large to fit into adversary's memory. While from the theoretical point of view it is a beautiful result, it is not clear how realistic this assumption is in practice, especially given the fact that storage becomes increasingly cheaper nowadays. Hence the random strings need to be really large, and it may be hard to generate them and to perform computations on them. What makes things additionally difficult is that [4] assume that all the random bits are broadcast at once (they call the process of generating it an "explosion"), which would require the machines to operate on very high rates (both to generate and to compute on these bits). Moreover the bounded memory assumption can be violated by using "mirrors" - in order to "store" some string R at some point B in space, simply make the route of the signal longer, by placing a mirror in some point B' in such a way that the mirrored signal arrives to B with a certain delay (hence [4] need to assume that such mirrors do not exist). Finally, another problem with the model of [4] is that they assume that the adversary that simultaneously observes two random strings R_1 and R_2 cannot compute an arbitrary function of (R_1, R_2), but is restricted to functions with low communication complexity (this assumption may be just an artifact of their proof, but it is completely unclear how to remove it).

All of these issues are a very good motivation to look for other models where the position-based cryptography is possible, and the authors of [4] leave it as an open research direction. One natural idea is to move to the quantum settings. Unfortunately, recently, Buhrman et al [2] extended the impossibility result of [4] also to this case ([2] contains also some positive results, for more restricted quantum models).

2   Our Contribution

We continue this line of research. Our main result is positive, namely we propose an information-theoretically-secure position-based authentication protocol and a computationally-secure position-based key agreement in Maurer's noisy channels model [10]. Our protocols work even if the adversary has a much better antenna than the honest parties. Unlike the protocol of [4], our scheme works only if the adversary does not enter what we call a "prohibited region" (which, very roughly speaking is the line segment connecting the prover and the source of randomness, plus some margin around it). In Section 2.4 we explain why this restriction makes sense for several practical applications. On the other hand, our protocol enjoys several advantages over the one of [4], in particular it is much more efficient, it should be much easier to implement in practice, and its security proof does not put any artificial restrictions on the power of the adversary. We discuss this further in Section 2.3.

Recall that in Maurer's noisy channels model, one assumes the existence of a publicly available source of random bits that is subject of distortions, i.e., some bits sent by this source are randomly altered. This broadcast channel might be, for instance, realized as a satellite transmitting bits from space without application of any error-correction mechanism. Alternatively, the bits can come from observations of natural phenomenons happening in deep space.

Maurer [10] showed (under additional, mild assumptions, i.e., existence of a noise-less public channel) that in this model two honest parties can determine a secret key k based on the satellite signal, i.e., any adversary eavesdropping the communication and receiving satellite signal has essentially no (information-theoretic) knowledge about k. This holds even if the adversary has a much stronger antenna than the honest parties, i.e. when the transmission error is much higher for the users than for the adversary. There has been lots of follow-up works building on Maurer's original idea, including some very interesting implementations proposals coming from the systems community (e.g. [8]).

In this work, we apply the above noisy source of randomness scenario to the problem of position-based authentication and key agreement. In order to do it we extend Maurer's model with the necessary geometric and timing information. Let us first informally describe our security model (the formal description appears in Section 4). In a typical deployment scenario the source of noisy randomness (Maurer's public satellite broadcaster), transmitting messages at the speed of light c, would be located high in the space, while the verifiers would be placed close to the ground level.

To keep this informal introduction simple assume for a moment that the randomness source is positioned exactly above the prover, and the prover lies somewhere within the triangle determined by the verifiers. Our protocol uses only three verifiers, denoted $\mathcal{V}^1, \mathcal{V}^2$, and $\mathcal{V}^3$. Let $\hat{\mathcal{V}}^1, \hat{\mathcal{V}}^2$ and $\hat{\mathcal{V}}^3$ be their respective positions. The verifiers can receive the noisy signal and securely communicate with each other, but not with the signal source. Similarly to [4] we assume that the antennas are not directional, and we use an assumption (that is standard in Maurer's model) that the noise is independent for each receiver. We would also like to stress that our results do not rely on the fact that the noise can be larger if the signal travels on longer distances.

The protocol is attacked by a set of adversaries, each receiving the chunk of a noisy signal from the randomness source. As already mentioned, there are some restrictions about the positions in which the adversaries can be placed in order for the protocol to be secure. We will discuss them in a moment.

2.1 Position-Based Authentication

Let us first informally describe our position-based authentication protocol (the formal description appears in Section 5). Following the previous work in this area we assume that the computation takes no time. When implementing this protocol in real life one would of course need to take into account the processing time of the prover (which would result in a scheme that proves the location within some limited precision). Our protocol is fairly simple. Let $\hat{\mathcal{S}}$ and $\hat{\mathcal{P}}$ denote the respective positions of the randomness source and the prover. Denote the bits broadcasted by the source by $S = (S_1, \ldots, S_n)$, each S_j being sent in some time t_j (with $t_1 < \cdots < t_n$) specified in advance. Hence t_j arrives to P in time $t_j + \|\hat{S}\hat{P}\|/c$ (where $\|\hat{S}\hat{P}\|$ denotes the length of a segment $\overline{\hat{S}\hat{P}}$, and c is the speed of light), and to each $\mathcal{V}^i$ in time $t_j + \|\hat{S}\hat{V}^i\|/c$. We also assume that the difference between each consecutive times t_{i+1} and t_i is large compared to the time the light needs to travel between the satellite and the verifiers. The consequence is that execution can be divided into n steps, each step corresponding to one bit being sent by a randomness source, and the adversary's behavior in step i cannot depend on the "future" bits $S_{i+1}, \ldots, S_n$. We have this assumption for the following reasons: (1) it makes the proofs in Section 6 simpler, and (2) in the practical implementations this condition can be satisfied easily, without any significant loss in efficiency. Actually it would probably take an extra effort to violate this assumption, as one would need the source S to produce the random bits at a very high rate.

Let S_j^P denote the noisy version of S_j received by P. To keep the exposition simple we assume that only one verifier, $\mathcal{V}^1$, say, listens to the satellite. Let S_j^V denote the version of S_j received by $\mathcal{V}^1$. We note that slightly better parameters could be achieved by making more verifiers listen to the randomness source and computing the bits S_j^V using the majority voting. Denote $S^P := (S_1^P, \ldots, S_n^P)$ and $S^V := (S_1^V, \ldots, S_n^V)$. Note also that there is no communication from the verifiers to the prover.

The party P, claiming to be in position $\hat{P}$ simply sends to every verifier $\mathcal{V}^i$ (via a noiseless channel) each noisy bit S_j received from the source. This is done without any delay and therefore this bit should arrive to each $\mathcal{V}^i$ in time $t_j + \|\hat{S}\hat{P}\|/c + \|\hat{V}^i\hat{P}\|/c$. If it does not arrive there precisely in this moment, then the verifier rejects the proof. Let S_j^i be the bit received by the verifier $\mathcal{V}^i$ from P as the bit S_j. Of course if P is honest then $S_j^1 = S_j^2 = S_j^3$. The verifiers check jointly (by communicating via their private channels) if this is indeed the case (this is called the "consistency check"). The verifier $\mathcal{V}^1$ also checks if the received string of bits $(S_1^1, \ldots, S_n^1)$ is "correlated" with $(S_1^V, \ldots, S_n^V)$, i.e., if the fraction of positions on which these two vectors are equal is substantially greater than $1/2$ (this is called the "correlation check"). These two checks can be done offline, and hence the time needed for them is irrelevant.

The basic idea behind this protocol is the observation that any honest user P claiming to be in position $\hat{P}$ sends his message based on a *single* version of satellite signal, and therefore every verifier receives the same message from him. On the other hand, a group of adversaries not present in $\hat{P}$ receive different versions of the noisy message. Later in the security proof we show that in this case it is unlikely that the adversaries send consistent messages to all the verifiers, the reason being that it is hard for them to pass both the consistency and the correlation check. Clearly passing each of these test independently is easy: in particular to pass the correlation check it is enough to position an adversary $\mathcal{A}^i$ close to each $\mathcal{V}^i$, and instruct him to forward to $\mathcal{V}^i$ each bit that the receives, adding some delay. More concretely: assume $\mathcal{A}^i$ is positioned exactly in $\hat{V}^i$, then $\mathcal{A}^i$ can send each S_j to $\mathcal{V}^i$

in time $t_j + \|\hat{\mathcal{S}}\hat{\mathcal{V}}^i\|/c$ by delaying it by time $\|\hat{\mathcal{P}}\hat{\mathcal{V}}^i\|/c + \|\hat{\mathcal{S}}\hat{\mathcal{P}}\|/c - \|\hat{\mathcal{S}}\hat{\mathcal{V}}^i\|/c$, which, by the triangle inequality is always non-negative.

It is also easy to construct a set of adversaries that make the verifiers accept the consistency check with probability 1: again position an adversary $\mathcal{A}^i$ close to each $\mathcal{V}^i$ and let him send as every S_i some fixed constant (0, say). In this case every $\mathcal{V}^i$ receives the same value, although, obviously, there is no correlation between the string received by the verifiers from the adversary and from the satellites.

Intuitively, what we would like to say now is that for an adversary it is hard to obtain both correlation and consistency, as long as he is not physically in position $\hat{\mathcal{P}}$.

Unfortunately, it is not true if we allow the adversaries to be put in arbitrary locations. Firstly, it is easy to see that our protocol can be broken if there is an adversary very close to the satellite (say: he is exactly in point $\hat{\mathcal{S}}$): such an adversary can simply receive the noisy satellite signal and forward it via a noiseless channel to every verifier. This has to be done after an appropriate delay, but it is always possible since, by the triangle inequality the value of $\|\hat{\mathcal{S}}\hat{\mathcal{P}}\| + \|\hat{\mathcal{P}}\hat{\mathcal{V}}^i\|$ (the total length of the route $\hat{\mathcal{S}} \to \hat{\mathcal{P}} \to \hat{\mathcal{V}}^i$) cannot be smaller than $\|\hat{\mathcal{S}}\hat{\mathcal{V}}^i\|$ (the length of the route $\hat{\mathcal{S}} \to \hat{\mathcal{V}}^i$). Therefore both the correlation and the consistency conditions will be satisfied, and hence the verifiers will accept this proof.

More generally, it is easy to see that it is enough to position such a "forwarding adversary" A at any point $\hat{\mathcal{A}}$ on a line connecting $\hat{\mathcal{P}}$ and $\hat{\mathcal{S}}$. To see why it works, observe that the only thing that needs to be checked is if A has enough time to send each bit S_i to every verifier $\mathcal{V}^j$. This is done by the following simple calculation. First observe that the length of the route $\hat{\mathcal{S}} \to \hat{\mathcal{A}} \to \hat{\mathcal{V}}^i$ is equal to $(*) = \|\hat{\mathcal{S}}\hat{\mathcal{A}}\| + \|\hat{\mathcal{A}}\hat{\mathcal{V}}^i\|$. On the other hand the length of $\hat{\mathcal{S}} \to \hat{\mathcal{P}} \to \hat{\mathcal{V}}^i$ is equal to the length of $\hat{\mathcal{S}} \to \hat{\mathcal{A}} \to \hat{\mathcal{P}} \to \hat{\mathcal{V}}^i$ (since $\hat{\mathcal{A}}$ is on a line form $\hat{\mathcal{S}}$ to $\hat{\mathcal{P}}$), and hence it is equal to $\|\hat{\mathcal{S}}\hat{\mathcal{A}}\| + \|\hat{\mathcal{A}}\hat{\mathcal{P}}\| + \|\hat{\mathcal{P}}\hat{\mathcal{V}}^i\|$, which is clearly larger than $(*)$ (from the triangle inequality).

It is also easy to see that the attack above can be performed by any adversary $\hat{\mathcal{A}}$ that is sufficiently close to the line connecting $\hat{\mathcal{S}}$ and $\hat{\mathcal{P}}$ (as long as $\hat{\mathcal{S}} \to \hat{\mathcal{A}} \to \hat{\mathcal{V}}^i$ is not greater than $\hat{\mathcal{S}} \to \hat{\mathcal{P}} \to \hat{\mathcal{V}}^i$, for every $\mathcal{V}^i$). In Lemma 4 we fully characterize the area where the adversary has to be in order to make the verifiers accept. We call it a "prohibited region" $\mathcal{Q}$. Very informally speaking $\mathcal{Q}$ is equal to the segment $\overline{\hat{\mathcal{S}}\hat{\mathcal{P}}}$ plus some "margin" around it. Just to get a general impression about how large $\mathcal{Q}$ is, denote by $\mathcal{Q}_H$ (for some parameter h) the set of points in $\mathcal{Q}$ that are at height h above the ground, and let d_H denote the diameter of $\mathcal{Q}_H$. The first good news is that $d_0 = 0$, which corresponds to the fact that if the adversaries are on the ground level then the only point from which the adversary can convince the verifiers is exactly in point $\hat{\mathcal{P}}$ (and hence the protocol is completely secure in this case). Since the shape of $\mathcal{Q}_H$ becomes quite complicated for $H > 0$ we only performed some numerical experiments to estimate d_H, that show that d_H is linear in H, for small H's and linear in $\sqrt{H}$ for larger H's. The details of this analysis will be provided in a full version of this paper.

2.2 Position-Based Key Agreement

In this section we discuss how to construct the key-agreement protocol in our model. As remarked in the introduction, for practical purposes the key-agreement is much more important than the authentication. The main difference is that we want the prover and the verifiers to conclude the protocol with a secret key k known only to them. The difficulty comes from the requirement that the adversary should not be able to learn any information about k at any point after the protocol has concluded. Hence, e.g., using the bits S_i directly to produce the secret key (even after the so-called "privacy amplification") will not work, as the adversary can at some later moment learn those bits, no matter in which physical location he is.

Fortunately, [4] show a generic method for converting any position-based authentication protocol into a key-agreement protocol. The main idea is as follows. The verifiers first generate a public key - secret pair (pk, sk) for some CCA2 secure public key encryption scheme, and send pk to the prover[1]. The parties then execute a standard non-authenticated key agreement protocol. Let k be the agreed key, and let T denote the transcript of the communication. Then they execute the authentication protocol, with the following modification: instead of sending a bit S_j to a verifier $\mathcal{V}^i$, the provers send the following ciphertext: $E(pk, (T, S_i, i))$. The security of this method is based on the non-malleability [5] of the encryption scheme, that follows from its CCA2 security (for more details see [4]). We also note that in the original [4] approach all the bits were sent at once, i.e., the prover sent one message $E(pk, (T, S_1, \ldots, S_n))$ to each verifier. The problem with this is that the prover needs to compute very quickly the ciphertexts in the CCA2-secure encryption scheme. Our approach of sending the bits separately has the advantage of being easier to implement from this point of view, as the prover can precompute $E(pk, (T, b, j))$ for all $b \in \{0, 1\}, j \in \{1, \ldots, n\}$, and then simply choose, after learning each S_j whether to send $E(pk, (T, 0, j))$ to $E(pk, (T, 1, j))$ to the verifiers. For the lack of space we skip the details of the key-agreement protocol. It will be presented in the full version of this paper. Hence, from now on we concentrate only on the authentication protocol.

2.3 Comparison with the Previous Work

Our protocol is very simple to implement: the prover needs only to broadcast the messages he observes from the satellite, and the verifiers need to compare equality of the strings they received from P (the "consistency check"), and compute the Hamming distance between S^V and S^P. Hence it is probably simpler to implement than the protocol of [4] that involves computing a chain of locally-computable randomness extractors. Recall that this computation has to take very short time (much shorter than the time needed for light to travel between the parties), and therefore implementing it may be challenging, especially,

---

[1] The assumption that the prover knows the public key of the verifiers can be actually removed (see [4] for more on this), although, in most of the practical applications it is reasonable to simply assume it.

since the inputs are huge, in order to satisfy the assumption that they do not fit into adversary's memory. Moreover the protocol of [4] requires the verifiers to send huge random strings, while in our case the verifiers can be completely passive (except of some small communication in the key agreement case).

An obvious drawback of our protocol, compared to the one of [4] is that it allows the adversary to cheat the verifiers by placing himself within the prohibited region $\mathcal{Q}$. In the next section we argue why is some applications it may be ok, and propose some security improvements.

2.4 Implementation Ideas

We believe that the paradigm introduce in our paper can potentially be implemented in practice, possibly in combination with other techniques (as an additional layer of security). We argue that for some scenarios the restrictions that we put on the position of the adversaries may be realistic. In particular, they make sense if the honest users can control the airspace above the protected area (plus some margins around it), which can be the case for the military applications. Also, in some cases, like granting free wifi to users within some building, the effort needed to position the adversary above the building may not be worth the potential gains.

Also, the users of the protocol can use more than one source of randomness, e.g, one can fix a large set of astronomical objects $S_1, \ldots, S_\ell$ to observe and agree on a different key k_i using each S_i and then use a hash of all keys for secure communication. This would force the adversary to put several antennas above the building. We leave the geometric analysis of this idea as an open research direction.

Another, perhaps more intriguing approach is to use randomness coming not from above the ground, but from the underground (like the electromagnetic radiation of Earth's core). In this case, in order to break the system by entering a prohibited region, the adversary would need to go deep underground, which in many situations would be too expensive to do.

If, instead of a satellite, we choose another source of randomness in space, say: coming from some natural phenomena, then the authenticity of the bits has to be verified in some other way, e.g., by using a directional antenna pointed on a specific astronomical object. Observe also that the verifiers could use one trusted server (available remotely) that listens to this object, and, say, publishes the results of these observations online.

3 Notation and Assumptions

By $\mathbb{R}^3$ we denote 3-dimensional space representing the Universe and by x_1, x_2, x_3 we mean usual Euclidean coordinates. The set $\mathcal{E} = \{x_3 = 0\}$ represents Earth's, assumed planar, surface and $\mathcal{E}_H$ is a set $\{x_3 = H\}$ parallel to $\mathcal{E}$. Moreover, by a letter with a subscript i, e.g. A_i, we mean the ith coordinate of a point $A \in \mathbb{R}^3$ (We use the same convention for referring to vector's coordinates). We say that a vector V is hooked in a point P if it leads from P to $P + V$. Sometimes we identify a point with a vector hooked in the centre of the coordinate system.

We will also use the Chernoff bound in the following form (see, e.g., [6], Theorem 1.1):

Lemma 1 (Chernoff bound). *Let $X := \sum_i^n X_i$ where X_i's are independently distributed in $[0, 1]$. Then for all $t > 0$ we have that $\mathbb{P}(X > \mathbb{E}(X) + t) \le e^{-2t^2/n}$.*

4   Security Definition

In this section we describe in details the model that was already informally discussed in Section 2. Formally, a *secure position-based authentication protocol* is a set $\Pi(\hat{\mathcal{P}})$ (where $\hat{\mathcal{P}}$ is a point in space) consisting of the following types of machines positioned in a three-dimensional space:

1. the *verifiers* $\mathcal{V}^1, \mathcal{V}^2$, and $\mathcal{V}^3$,
2. the *prover* $\mathcal{P}$ (positioned in $\hat{\mathcal{P}}$), and
3. the *randomness source* $\mathcal{S}$.

The protocol will be attacked be a set of adversaries $\{\mathcal{A}^1, \ldots, \mathcal{A}^t\}$, each of them positioned somewhere in the space. We assume that all the machines are equipped with perfect clocks and that their computation takes no time. Each machine is aware of its own position in space (more formally: it gets it as an auxiliary input). The position of each verifier $\mathcal{V}^i$ is denoted by $\hat{\mathcal{V}}^i$ and the position of the randomness source is denoted with $\hat{\mathcal{S}}$. Additionally, the verifiers get as input a position $\hat{\mathcal{P}}$ where the prover "claims to be". Their goal is to check if he indeed is in this position. The decision (yes/no) of the verifiers is communicated at the end of the protocol by one of then ($\mathcal{V}^1$, say).

The only messages that are sent are of a broadcast type (i.e. there are no directional antennas). A message sent by a machine or a natural source positioned in point U arrives to a machine in point U' in time $\|UU'\|/c$. We assume that the messages sent by the randomness source are noisy. If S is a bit sent by $\mathcal{S}$, then $\mathcal{V}_1$ receives[2] a bit S^V equal to S with probability $1 - \epsilon_V/2$ (for both $S = 0, 1$), and $\mathcal{P}$ receives a bit S^P equal to S with probability $1 - \epsilon_P/2$ (for both $S = 0, 1$), where $\epsilon_P, \epsilon_V \in [0, 1]$. These events are independent for every value of S.

It is a little bit trickier to define what it means that the bits received by the adversaries are noisy. One method of doing it would be to define an error of an antenna of each individual adversary. The problem with this approach is that, of course, the adversaries can communicate with each other and jointly "correct" the errors, by using, for example, the majority voting. Hence, a much more natural approach is to assume that the adversaries *jointly* cannot guess the bit S without some error, no matter what strategy they use. To make it precise, assume that each adversary $\mathcal{A}^i$ receives a bit S^i. The bits received by the adversaries are defined by a conditional distribution $p_{(\mathcal{A}^1, \ldots, \mathcal{A}^t)|S}$, also called a

---

[2] Recall that, as described in the introduction, in our protocols only one verifier, namely $\mathcal{V}^1$, listens to the satellite signal.

channel end denoted $S \to (A^1, \ldots, A^t)$. We assume that this channel is ϵ_A-noisy (for $\epsilon_A \in [0,1]$), by which we mean the following:

1. for both $s \in \{0,1\}$ the events $\{A^i = s\}_{i=1}^t$ are independent conditioned on $S = s$ and
2. for any $f : \{0,1\}^t \to \{0,1\}$ we have that

$$\left| \mathbb{P}\left(f(A_0^1, \ldots, A_0^t) = 0 | S = 0\right) - \mathbb{P}\left(f(A_1^1, \ldots, A_1^t) = 0 | S = 1\right) \right| \leq 1 - \epsilon_A. \quad (1)$$

Any function f of a type $\{0,1\}^t \to \{0,1\}$ will be called a *guessing strategy*. Note that we do not give any concrete bounds for transmission errors for the individual antennas. The only thing that we assume is that the adversaries *jointly* cannot guess S with a high probability: it is actually easy to see that Point 2 is equivalent to requiring that for S distributed uniformly over $\{0,1\}$ and for any guessing strategy $f : \{0,1\}^t \to \{0,1\}$ we have $\mathbb{P}\left(f(A_0^1, \ldots, A_0^t) = S\right) \leq 1 - \epsilon_A/2$.

Of course this means that the error rates of individual antennas need to be much larger than $\epsilon_A/2$, especially if t is large. While at the first sight it may look unrealistic, we would like to note that implicitly this assumption appears in every paper that constructs protocols in the noisy channels model: obviously the adversary can always get a "better antenna" by simply investing in a large number of weaker antennas, in order to correct the errors.

The communication links between the verifiers are secure (secret and authenticated) and every participant of the protocol can verify the authenticity of the messages sent by a satellite (the case when, instead of an artificial satellite, we use some natural object was already discussed in Section 2.4). Obviously, this can be achieved by standard cryptographic techniques. Observe that there is no formal reason to assume that the messages sent by the prover to the verifiers are secret (as, if there exists an honest prover in $\hat{\mathcal{P}}$, then the outcome of the protocol should anyway be positive).

We also assume that the adversary cannot block or delay the messages sent between the honest participants. It is clear that such an assumption is unavoidable, as, by blocking all the messages, the adversary can always prevent any protocol from succeeding.

As described in the introduction, our protocols work only when the prover is placed within some subset $\mathcal{G}$ of a three-dimensional space (called the set of *admissible positions*), and when there is no adversary positioned in a subset $\mathcal{Q}$ (without loss of generality assume that the position of P is in $\mathcal{Q}$). Moreover, we accept that with some small probability ξ an honest prover fails to convince the verifiers, and with a small probability ρ the adversaries manage to make the verifiers accept, even if no adversary is placed within $\mathcal{Q}$. More formally, we say that $\Pi(\hat{\mathcal{P}})$ (with $\hat{\mathcal{P}} \in \mathcal{G}$) is an $(\sigma, \rho, \mathcal{Q})$-*secure position-based authentication protocol* if the following two conditions hold:

σ-**correctness** If the prover P is placed in the claimed position $\hat{\mathcal{P}} \in \mathcal{G}$ then the verifiers output "yes" with probability at least $1 - \sigma$,

ρ-**security** If the prover is not in position $\hat{\mathcal{P}}$ and there is no adversary in set $\mathcal{Q}$ then the verifiers output "yes" with probability at most ρ.

If σ and ρ are negligible in n then we will also simply say that π is an $\mathcal{Q}$-secure. We will also assume that the difference between each consecutive times t_{j+1} and t_j is greater than $\max_i \|\hat{\mathcal{S}}\hat{\mathcal{V}}^i\|$ and hence the execution can be divided into n rounds, and the adversary's behavior in step j cannot depend on the bits $S_{j+1}, \ldots, S_n$.

5   Protocol

In this section we describe formally our main position-based authentication protocol PosAuth that has already been discussed informally in Section 2.1. Let $n \in \mathbb{N}$ be a security parameter, let $\kappa \in (0, 1/2)$ be some parameter whose value will be determined later, and let $\hat{\mathcal{P}}$ be the position where the prover claims to be. The protocol $\mathsf{PosAuth}_n^\kappa(\hat{\mathcal{P}})$ consists of the following steps:

1. For $j = 1, \ldots, n$ do:
   (a) In time t_j the randomness source $\mathcal{S}$ broadcasts a random bit S_j.
   (b) Let S_j^P be the version of S_j that the prover receives (this happens in time $t_j + \|\hat{\mathcal{S}}\hat{\mathcal{P}}\|/c$).
   (c) Immediately after receiving S_j^P the prover P broadcasts (S_j^P, j) to all the verifiers.
   (d) Each verifier $\mathcal{V}^i$ checks if in time $t_j + \|\hat{\mathcal{S}}\hat{\mathcal{P}}\|/c + \|\hat{\mathcal{P}}\hat{\mathcal{V}}^i\|/c$ he received a pair (S_j^P, j) from the prover. If not, then he rejects the proof and halts. Let S_j^i be equal to the bit that the verifier $\mathcal{V}^i$ received as S_j^P. The verifiers perform the consistency check by verifying if for every j they received the same value. If not then the verifier rejects the proof and halts.
   (e) In time $t_j + \|\hat{\mathcal{S}}\hat{\mathcal{V}}^1\|/c$ the verifier $\mathcal{V}^1$ receives his noisy version S_j^V of S_j (note that this usually happens chronologically before $\mathcal{V}^1$ executes Step (1d) above).
2. Denote $\overrightarrow{S}^1 = (S_1^1, \ldots, S_n^1)$ and $\overrightarrow{S}^V = (S_1^V, \ldots, S_n^V)$. The verifier $\mathcal{V}^1$ performs the correlation check, by computing the Hamming distance between $\overrightarrow{S}^1$ and $\overrightarrow{S}^V$. He outputs "yes" if this value is smaller than $\kappa \cdot n$. Otherwise he outputs "no".

For every verifier $\mathcal{V}^i$ let $\mathcal{X}^i$ denote the set of all positions in space that have the following property: if $\mathcal{A}$ is positioned in $\mathcal{X}$ then $\mathcal{A}$ can send (his version of) a bit S_j to $\mathcal{V}^i$ in such a way that it reaches $\mathcal{V}^i$ exactly at the same time as the bit S_j^P reaches $\mathcal{V}^i$. It is easy to see that this protocol can be broken if an adversary can send to all the verifiers an identical signal S^A that is correlated with $\mathcal{S}$. Obviously, it is always possible if the adversary can position himself in the intersection $\mathcal{X}^1 \cap \mathcal{X}^2 \cap \mathcal{X}^3$. Therefore, in order to hope for any security, we need to assume that there is no adversary in $\mathcal{X}^1 \cap \mathcal{X}^2 \cap \mathcal{X}^3$. In the next section we show that this assumption is sufficient. We postpone the geometric analysis of the shape of $\mathcal{X}^1 \cap \mathcal{X}^2 \cap \mathcal{X}^3$ until Section 7.

6   Security without the Geometric Analysis

In this section we show the security of the protocol from Section 5 abstracting from the geometric information. As already mentioned, the only thing that we will assume is that there is no adversary in the set $\mathcal{X}^1 \cap \mathcal{X}^2 \cap \mathcal{X}^3$ (where the $\mathcal{X}^j$'s were defined above). The main lemma that we prove is as follows.

Lemma 2. *Let ϵ_A, ϵ_P, and ϵ_V be as in Section 4. Let $\kappa \geq (\epsilon_V + \epsilon_P - \epsilon_V \epsilon_P)/2$, let α be such that $\sqrt{\alpha} \leq \epsilon_A/12$, and let $\mathcal{Q} = \mathcal{X}^1 \cap \mathcal{X}^2 \cap \mathcal{X}^3$. Then the protocol $\mathsf{PosAuth}_n^\kappa(\hat{\mathcal{P}})$ from Section 5 is $(\sigma, \rho, \mathcal{Q})$-secure with*

- $\rho = e^{-n(2\kappa - \epsilon_V - \epsilon_P + \epsilon_V \epsilon_P)^2/2}$, *and*
- $\sigma = e^{-n(1/2 - \kappa - 5\sqrt{\alpha})^2/2} + (1 - \alpha)^{(1/2 - \kappa - 5\sqrt{\alpha})n/2}$

Note that the value $1/2 - \kappa - 5\sqrt{\alpha}$ is the gap between $1 - \kappa$, i.e., the desired prover's accuracy and $1/2 + 5\sqrt{\alpha}$.

As an example of an application of Lemma 2 for concrete parameters assume that the error of the adversary is small, e.g.: $\epsilon_A := 0.1$, and the error of the honest participants is large, $\epsilon_P = \epsilon_V = 0.5$, say. If we then set $\kappa = 0.4$ and $\alpha = 10^{-5}$ then we obtain $\rho = e^{-0.00125n}$ and $\sigma \leq 0.9999996^n$. For $n = 2 \cdot 10^8$ (i.e.: around 20MB) we get $\rho \leq 10^{-83332}$ and $\sigma \leq 10^{-34}$. It is very likely that these parameters can be improved, as we did not try to optimize them.

6.1   Single-Bit Case

As the first step towards proving Lemma 2 we consider the single-bit case, i.e., we analyse the possible strategies of the adversary for an individual bit S sent by the randomness source. Recall that a guessing strategy is an arbitrary function of a type $\{0,1\}^t \to \{0,1\}$.

Let $\mathcal{Z} = \{\mathcal{A}^{i_1}, \ldots, \mathcal{A}^{i_{t'}}\}$ (with $i_1 < \cdots < i_t$) be some subset of antennas. We say that f is a *guessing strategy based $\mathcal{Z}$* if it depends only on the inputs corresponding to antennas in $\mathcal{Z}$. More precisely: for any two vectors $\overrightarrow{a} = (a^1, \ldots, a^t)$ and $\overrightarrow{b} = (b^1, \ldots, b^t)$ such that $(a^{i_1}, \ldots, a^{i_t}) = (b^{i_1}, \ldots, b^{i_t})$ we have $f(\overrightarrow{a}) = f(\overrightarrow{b})$.

The following lemma shows that if the guessing strategies: f^1, f^2 and f^3 are based on sets that have no antenna in common, then the only way to keep them consistent with each other (i.e.: make their outputs equal) is to make them (almost) constant. This fact is useful, since, obviously, no function that is close to constant can guess S with probability significantly greater than $1/2$.

Lemma 3. *For any ϵ_A-noisy channel $S \to (A^1, \ldots, A^t)$ consider a set of guessing strategies: f^1, f^2, and f^3, each f^i based on subset of antennas $\mathcal{Y}^i$. Suppose that:*

- *no antenna belongs to every set in the family $\{\mathcal{Y}^i\}_{i=1}^3$, i.e.,*

$$\cap_{i=1}^3 \mathcal{Y}^i = \emptyset, \tag{2}$$

and

– *except with probability α, for some parameter α such that*

$$\sqrt{\alpha} \leq \epsilon_A/12, \tag{3}$$

the strategies agree with each other, i.e., for every bit s we have

$$\mathbb{P}\left(f^1(A^1,\ldots,A^t) = f^2(A^1,\ldots,A^t) = f^3(A^1,\ldots,A^t)|S = s\right) \geq 1 - \alpha. \tag{4}$$

Then the strategies have to be "almost constant", i.e., there exists a bit $c \in \{0,1\}$ such that for every bit s we have

$$\mathbb{P}\left(f^1(A^1,\ldots,A^t) = c|S = s\right) \geq 1 - 9 \cdot \sqrt{\alpha}, \tag{5}$$

Proof. For the lack of space the proof appears in the full version of paper [7]. □

6.2   Proof of Lemma 2

We now present the proof of the main lemma of this section. Let us first address the ρ-correctness. Let H denote the expected Hamming distance between $\vec{S}^1$ and $\vec{S}^V$. Clearly for each j we have

$$\begin{aligned}
\mathbb{P}\left(S_j^1 \neq S_j^V\right) &= (\epsilon_V/2)(1 - \epsilon_P/2) + (1 - \epsilon_V/2)(\epsilon_P/2) \\
&= \epsilon_V/2 + \epsilon_P/2 - \epsilon_V\epsilon_P/2
\end{aligned}$$

Therefore $\mathbb{E}(H) = n \cdot (\epsilon_V + \epsilon_P - \epsilon_V\epsilon_P)/2$, and hence from the Chernoff bound (Lemma 1) we get that the probability that the verifiers reject the honest prover is at most:

$$\begin{aligned}
\mathbb{P}(H \geq n\kappa) &\leq e^{-2(n(2\kappa-\epsilon_V-\epsilon_P+\epsilon_V\epsilon_P)/2)^2/n} \\
&= e^{-n(2\kappa-\epsilon_V-\epsilon_P+\epsilon_V\epsilon_P)^2/2} \\
&= \rho.
\end{aligned}$$

Hence, the ρ-correctness is proven. Let us now consider the σ-security. Recall that S_j^1 denotes the bit received by the verifiers from the prover as S_j. Assume that if the bits received as S_j are not identical for every verifier, then $S_j^1 = \bot$. Without loss of generality, assume that in every $j+1$st step the adversaries learn the bit S_j (so, they know if their guesses in the previous rounds were correct). The goal of the adversary is to minimize the Hamming distance between S^1 and S^V, without being disqualified. In other words: we can assume that his goal is to earn a certain number of point in the following game. At the beginning he has 0 points. For every $j = 1,\ldots,n$ if $S_j^1 = S_j^V$ then he earns 1 point, and otherwise he earns nothing. If $S_j^1 = \bot$ then he gets disqualified and the game is halted. Let *out* denote the total number of points earned by the adversary. He wins the game if $out \geq (1 - \kappa)n$. We now show that for any strategies of the adversary we have

$$\mathbb{P}\left(out \geq (1 - \kappa)n \text{ and the adversary did not get disqualified}\right) \leq \sigma \tag{6}$$

From the assumption that the prohibited region Q is equal to $\mathcal{X}^1 \cap \mathcal{X}^2 \cap \mathcal{X}^3$ we know that there is no antenna in the intersection $\mathcal{X}^1 \cap \mathcal{X}^2 \cap \mathcal{X}^3$. Therefore we can use Lemma 3, form which it follows that in each jth step the adversary can choose one of the following strategies. The first one, that we call a "green strategy" has the following properties: the probability that $S_j^1 \neq \perp$ is large, more precisely

$$\mathbb{P}\left(S_j^1 \neq \perp\right) \geq 1 - \alpha,$$

but on the other hand, the probability that he guesses S_j is small, i.e.

$$\mathbb{P}\left(S_j^1 = S_j\right) \leq \frac{1}{2} + \frac{9}{2} \cdot \sqrt{\alpha}$$
$$\leq \frac{1}{2} + 5\sqrt{\alpha}$$

Alternatively, he can take a "red strategy", where probability that he guesses S_j is large, but the probability that $S_j^1 = \perp$ is large, more precisely:

$$\mathbb{P}\left(S_j^1 = \perp\right) \geq \alpha,$$

and

$$\mathbb{P}\left(S_j^1 = S_j\right) \geq \frac{1}{2} + \frac{9}{2} \cdot \sqrt{\alpha}.$$

Suppose for a moment that the adversary uses the green strategy in each step. Without loss of generality assume that S_j^1 is never equal to $\perp$ and that $\mathbb{P}\left(S_j^1 = S_j\right)$ is actually equal to $1/2 + 5\sqrt{\alpha}$. Let $g = (1 - \kappa) - (1/2 + 5\sqrt{\alpha}) = 1/2 - \kappa - 5\sqrt{\alpha}$ be the gap between the required prover's accuracy $1 - \kappa$ and his average accuracy $1/2 + 5\sqrt{\alpha}$ using green strategy, which was mentioned in the statement of Lemma 2. From the Chernoff bound (Lemma 1) we get

$$Pr[out > (1 - \kappa - g/2)n] = Pr[out > (1/2 + 5\sqrt{\alpha} + g/2)n] \leq e^{-2(ng/2)^2/n} = e^{-ng^2/2}.$$

Now consider an adversary that behaves exactly like the one above, except that he uses the red strategy m times. Moreover, we allow the adversary to first play the green strategy in each step, and *then* choose m steps in which he "gets another chance" and plays the red strategy. Denote the outcome of this game by out'. Since obviously in this way the adversary can earn at most m extra points, hence by the previous inequality, it is easy to see that if $m < gn/2$ then

$$Pr[out' > (1 - \kappa)n] \leq Pr[out > (1 - \kappa - g/2)n]$$
$$\leq e^{-ng^2/2}$$

On the other hand, each time he plays the red strategy, his probability of getting disqualified is at least α and therefore

$$\mathbb{P}\,(\text{the adversary did not get disqualified}) \leq (1 - \alpha)^m,$$

which, if $m \geq gn/2$ is at most $(1 - \alpha)^{gn/2}$. Hence (6) is proven.   □

7   Geometric Analysis

What remains now is to perform the analysis of the protocol $\mathsf{PosAuth}_n^\kappa(\hat{\mathcal{P}})$ from Section 6 to find the geometric assumptions that are sufficient to satisfy the requirements on $\mathcal{Q}$ and $\mathcal{G}$ that are needed in Lemma 2. Obviously, the region $\mathcal{Q}$ where the adversary is not allowed to put his antennas depends on the region $\mathcal{G}$ where $\hat{\mathcal{P}}$ can be, and we would like to have $\mathcal{G}$ as large as possible, and $\mathcal{Q}$ as small as possible. Unfortunately, for large $\mathcal{G}$'s the description of $\mathcal{Q}$ becomes very complicated. Therefore we make some simplifying assumptions. First of all we will be only considering sets $\mathcal{G}$ that lie on the plane $\mathcal{E}$ on which the verifiers are. Let α be the angle between this plane and the satellite. The definition of $\mathcal{G}$ depends on α in the following way: we define $\mathcal{G}_{\hat{\mathcal{S}},\alpha}^{\hat{\mathcal{V}}^1\hat{\mathcal{V}}^2\hat{\mathcal{V}}^3}$ to be the set of points $\hat{\mathcal{P}}$ within the triangle $\triangle\hat{\mathcal{V}}^1\hat{\mathcal{V}}^2\hat{\mathcal{V}}^3$ such that each of the angles $\angle\hat{\mathcal{V}}^1\hat{\mathcal{P}}\hat{\mathcal{V}}^2$, $\angle\hat{\mathcal{V}}^2\hat{\mathcal{P}}\hat{\mathcal{V}}^3$, and $\angle\hat{\mathcal{V}}^1\hat{\mathcal{P}}\hat{\mathcal{V}}^3$ is less than 2α. Hence, e.g., if S is directly above P, then $\alpha = 90°$ and therefore $\mathcal{G}$ is simply equal to the entire interior of $\triangle\hat{\mathcal{V}}^1\hat{\mathcal{V}}^2\hat{\mathcal{V}}^3$, but if $\alpha < 90°$ the area of $\mathcal{G}$ will get smaller, as it will not contain some margins around the edges of the triangle.

Figure 1 illustrates the margins excluded from the triangle.

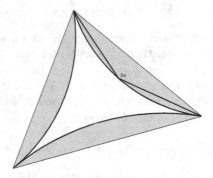

Fig. 1. The gray areas indicate the margins excluded from the triangle (in the admissible region) for $\alpha = 70°$

Defining $\mathcal{Q}$ is a bit more tricky. As discussed in the introduction, it is obvious that if there is an adversary $\mathcal{A}^i$ located on a line segment that connects $\hat{\mathcal{S}}$ and $\hat{\mathcal{P}}$ then any such scheme can be broken, as $\mathcal{A}^i$ can simply listen to S and broadcast his own noise-less version of the randomness source's signal. Hence $\mathcal{Q}$ needs to contain at least the line segment $\overline{\hat{\mathcal{S}}\hat{\mathcal{P}}}$. Our protocol provides security when $\mathcal{Q}$ contains a little bit more than this. Namely, $\mathcal{Q}_{\hat{\mathcal{S}},\hat{\mathcal{P}}}^{\hat{\mathcal{V}}^1\hat{\mathcal{V}}^2\hat{\mathcal{V}}^3}$ will be defined as an intersection of interiors of ellipsoids with foci in $\hat{\mathcal{S}}$ and $\hat{\mathcal{P}}$ and an appropriately chosen major radius (not much larger than $\|\hat{\mathcal{S}}\hat{\mathcal{P}}\|$). More precisely for any points $\hat{\mathcal{S}}, \hat{\mathcal{P}}$ and $\hat{\mathcal{V}}$ in 3-dimensional space define a set $\mathsf{Ellipse}_{\hat{\mathcal{S}},\hat{\mathcal{P}}}(\hat{\mathcal{V}})$ of points as:

$$\text{Ellipse}_{\hat{S},\hat{P}}(\hat{V}) := \{\hat{A} \in \mathbb{R}^3 : \|\hat{S}\hat{A}\| + \|\hat{A}\hat{V}\| \leq \|\hat{S}\hat{P}\| + \|\hat{P}\hat{V}\|\}.$$

It is clear that the set defined this way is the interior of the ellipsoid with foci in $\hat{S}$ and $\hat{P}$ and the major radius equal to $\|\hat{S}\hat{P}\| + \|\hat{P}\hat{V}\|$.

Intuitively the set $\text{Ellipse}_{\hat{S},\hat{P}}(\hat{V})$ consists of all points Q such that the signal broadcasted from S can be transmitted from Q to V before an analogous transmission from P. We now define the set $\mathcal{Q}_{\hat{S},\hat{P}}^{\hat{V}^1\hat{V}^2\hat{V}^3}$ as follows:

$$\mathcal{Q}_{\hat{S},\hat{P}}^{\hat{V}^1\hat{V}^2\hat{V}^3} := \bigcap_{i=1}^{3} \text{Ellipse}_{\hat{S},\hat{P}}(\hat{V}^i).$$

Lemma 4. *Consider the protocol* $\mathsf{PosAuth}_n^\kappa(\hat{P})$, *and let* $\mathcal{X}^1, \mathcal{X}^2$ *and* $\mathcal{X}^3$ *be as in Section 5. Let* $\mathcal{G}_{\hat{S},\alpha}^{\hat{V}^1\hat{V}^2\hat{V}^3}$ *and* $\mathcal{Q}_{\hat{S},\hat{P}}^{\hat{V}^1\hat{V}^2\hat{V}^3}$ *be as above. Then the protocol* $\mathsf{PosAuth}_n^\kappa(\hat{P})$ *is* $\left(\sigma, \rho, \mathcal{Q}_{\hat{S},\hat{P}}^{\hat{V}^1\hat{V}^2\hat{V}^3}\right)$-*secure for* σ *and* δ *as in Lemma 2.*

Proof. Suppose $\hat{P} \in \mathcal{G}_{\hat{S},\alpha}^{\hat{V}^1\hat{V}^2\hat{V}^3}$ and there is no adversary in set $\mathcal{Q}_{\hat{S},\hat{P}}^{\hat{V}^1\hat{V}^2\hat{V}^3}$. Recall that each $\mathcal{X}^i$ was defined as a set of all positions $\hat{A}$ such that if $\mathcal{A}$ is positioned in $\hat{A}$ then he can send his version of a bit S_j to $\mathcal{V}^i$ in such a way that it reaches $\mathcal{V}^i$ exactly at the same time as the bit S_j^P reaches $\mathcal{V}^i$. Translating it into distances we get that $\mathcal{X}^i = \text{Ellipse}_{\hat{S},\hat{P}}(\hat{V}^i)$, and therefore $\mathcal{Q}_{\hat{S},\hat{P}}^{\hat{V}^1\hat{V}^2\hat{V}^3} = \mathcal{X}^1 \cap \mathcal{X}^2 \cap \mathcal{X}^3$, which is exactly what we need to apply Lemma 2. □

7.1 Geometric Properties

We will now analyse the geometric properties of the regions $\mathcal{G}_{\hat{S},\alpha}^{\hat{V}^1\hat{V}^2\hat{V}^3}$ and $\mathcal{Q}_{\hat{S},\hat{P}}^{\hat{V}^1\hat{V}^2\hat{V}^3}$ defined above. In particular, we show that $\hat{P}$ is the only prohibited point on the plane defined by $\hat{V}^1$, $\hat{V}^2$, and $\hat{V}^3$. This implies, e.g., that if the verifiers and the adversary are on the ground level, then there are essentially no prohibited points (as the adversary positioned in $\hat{P}$ can anyway always win). For $H > 0$ we only performed some numerical experiments to estimate d_H. These experiments show that d_H is linear in H, for small H's and linear in $\sqrt{H}$ for larger H's. The details of this analysis will be provided in a full version of this paper.

Lemma 5. *Let* $\mathcal{E}$ *be the plane determined by* $\hat{V}^1$, $\hat{V}^2$ *and* $\hat{V}^3$ *(call it a "ground level"). Let* α *be the angle between* $\hat{S}\hat{P}$ *and* $\mathcal{E}$. *For* $\mathcal{G}_{\hat{S},\alpha}^{\hat{V}^1\hat{V}^2\hat{V}^3}$ *and* $\mathcal{Q}_{\hat{S},\hat{P}}^{\hat{V}^1\hat{V}^2\hat{V}^3}$ *as above we have*

$$\mathcal{Q}_{\hat{S},\hat{P}}^{\hat{V}^1\hat{V}^2\hat{V}^3} \cap \mathcal{E} = \{\hat{P}\}.$$

The proof together with the rest of geometrical analysis appears in Appendix A of full version of the paper [7]. We also note that by increasing the number of verifiers we can cover more general areas than the "triangle without the margins". In particular, imagine that the verifiers $V^1, \ldots, V^\ell$ are placed regularly on a circle. Then, a prover $\mathcal{P}$ can prove that he is in $\hat{P}$ if he finds 3 verifiers

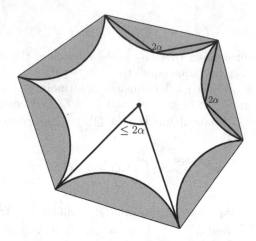

Fig. 2. The gray areas indicate the margins excluded from the polygon (in the admissible region)

V^i, V^j and V^k such that $\mathcal{P} \in \mathcal{G}_{\hat{\mathcal{S}}, \alpha}^{\hat{\mathcal{V}}^i \hat{\mathcal{V}}^j \hat{\mathcal{V}}^k}$. Hence, the admissible set $\mathcal{G}$ becomes equal to the polygon with vertices in $\hat{\mathcal{V}}^1, \ldots, \hat{\mathcal{V}}^\ell$, except of some margins around the edges. This is illustrated on Figure 2.

8   Conclusions

As a conclusion, we briefly sketch the contribution of the paper in the perspective of an explicit implementation of the protocol.

In the paper, we proposed a position-based authentication scheme applying assumption of existence of a noisy *passive* source of randomness. It is important to analyse the possible physical candidates for noisy source of randomness which is realistic and convenient to use in our framework. Moreover, we also briefly described some details of implementation (see, e.g., Section 2.4). They also require additional empirical and technical justification, which might lead to a practical deployment of the scheme.

References

1. Brands, S., Chaum, D.: Distance bounding protocols. In: Helleseth, T. (ed.) EUROCRYPT 1993. LNCS, vol. 765, pp. 344–359. Springer, Heidelberg (1994)
2. Buhrman, H., Chandran, N., Fehr, S., Gelles, R., Goyal, V., Ostrovsky, R., Schaffner, C.: Position-based quantum cryptography: Impossibility and constructions. In: Rogaway, P. (ed.) CRYPTO 2011. LNCS, vol. 6841, pp. 429–446. Springer, Heidelberg (2011)
3. Capkun, S., Hubaux, J.P.: Secure positioning of wireless devices with application to sensor networks. In: INFOCOM 2005, Proceedings of the IEEE 24th Annual Joint Conference of the IEEE Computer and Communications Societies, vol. 3, pp. 1917–1928. IEEE (2005)

4. Chandran, N., Goyal, V., Moriarty, R., Ostrovsky, R.: Position based cryptography. In: Halevi, S. (ed.) CRYPTO 2009. LNCS, vol. 5677, pp. 391–407. Springer, Heidelberg (2009)
5. Dolev, D., Dwork, C., Naor, M.: Nonmalleable cryptography. SIAM Review 45(4), 727–784 (2003)
6. Dubhashi, D.P., Panconesi, A.: Concentration of Measure for the Analysis of Randomized Algorithms. Cambridge University Press (2009)
7. Dziembowski, S., Zdanowicz, M.: Position-Based Cryptography from Noisy Channels. In: ePrint archive (2014)
8. Mathur, S., Trappe, W., Mandayam, N., Ye, C., Reznik, A.: Secret key extraction from level crossings over unauthenticated wireless channels. In: Liu, R., Trappe, W. (eds.) Securing Wireless Communications at the Physical Layer, pp. 201–230. Springer, US (2010)
9. Maurer, U.: Conditionally-perfect secrecy and a provably-secure randomized cipher. Journal of Cryptology 5(1) (1992)
10. Maurer, U.: Secret key agreement by public discussion from common information. IEEE Transcations on Information Theory 39 (May 1993)
11. Sastry, N., Shankar, U., Wagner, D.: Secure verification of location claims. In: Proceedings of the 2nd ACM workshop on Wireless security, pp. 1–10. ACM (2003)
12. Singelee, D., Preneel, B.: Location verification using secure distance bounding protocols. In: IEEE International Conference on Mobile Adhoc and Sensor Systems Conference, p. 7. IEEE (2005)

A Comparison of the Homomorphic Encryption Schemes FV and YASHE

Tancrède Lepoint[1,*] and Michael Naehrig[2]

[1] CryptoExperts, École Normale Supérieure and University of Luxembourg
tancrede.lepoint@cryptoexperts.com
[2] Microsoft Research
mnaehrig@microsoft.com

Abstract. We conduct a theoretical and practical comparison of two Ring-LWE-based, scale-invariant, leveled homomorphic encryption schemes – Fan and Vercauteren's adaptation of BGV and the YASHE scheme proposed by Bos, Lauter, Loftus and Naehrig. In particular, we explain how to choose parameters to ensure correctness and security against lattice attacks. Our parameter selection improves the approach of van de Pol and Smart to choose parameters for schemes based on the Ring-LWE problem by using the BKZ-2.0 simulation algorithm.

We implemented both encryption schemes in C++, using the arithmetic library FLINT, and compared them in practice to assess their respective strengths and weaknesses. In particular, we performed a homomorphic evaluation of the lightweight block cipher SIMON. Combining block ciphers with homomorphic encryption allows to solve the gargantuan ciphertext expansion in cloud applications.

1 Introduction

In 2009, Gentry proposed the first fully homomorphic encryption scheme [16]. A fully homomorphic encryption (FHE) scheme is an encryption scheme that allows, from ciphertexts $E(a)$ and $E(b)$ encrypting bits a, b, to obtain encryptions of $\neg a$, $a \wedge b$ and $a \vee b$ without using the secret key. Clearly, this allows to publicly evaluate any Boolean circuit given encryptions of the input bits. This powerful primitive has become an active research subject in the last four years. Numerous schemes based on different hardness assumptions have been proposed [16,12,5,4,30,20] and have improved upon previous approaches.

In all of the aforementioned schemes, a ciphertext contains a noise that grows with each homomorphic operation. The noise is minimal when the ciphertext is a fresh encryption of a plaintext bit and has not yet been operated on. Homomorphic operations as those above can be (and are often) expressed as homomorphic addition and multiplication operations, *i.e.* addition and multiplication in the binary field $\mathbb{F}_2$. Both increase the noise in ciphertexts, which means that the noise

---

* This work was started when the first author was an intern in the Cryptography Research group at Microsoft Research.

D. Pointcheval and D. Vergnaud (Eds.): AFRICACRYPT 2014, LNCS 8469, pp. 318–335, 2014.
© Springer International Publishing Switzerland 2014

in a resulting encryption is larger than the noise in the respective input encryptions. In particular, homomorphic multiplication increases the noise term significantly.

After a certain amount of such homomorphic computations have been carried out, the noise reaches a certain maximal size after which no more homomorphic operations can be done without losing correctness of the encryption scheme. At this point, the ciphertext needs to be publicly refreshed to allow subsequent homomorphic operations. This refreshing procedure is called bootstrapping and is very costly. As a consequence, only few of the FHE schemes have been *fully* implemented [17,11,9] and the resulting performances are rather unsatisfactory.

However, real-world applications do not necessarily need to handle any input circuit. One might avoid using the bootstrapping procedure if the multiplicative depth of the circuit to be evaluated is known in advance and small enough (*cf.* [33,21,24,3] and even [19]). Unfortunately, for the schemes of [17,11,9] the noise grows exponentially with the depth of the circuit being evaluated, severely limiting the circuits that can be evaluated with reasonable parameters. To mitigate this noise growth, Brakerski, Gentry and Vaikuntanathan introduced the notion of *leveled* homomorphic encryption schemes [5]. In such a scheme, the noise grows only linearly with the multiplicative depth of the circuit being evaluated. Therefore for a given circuit of reasonable depth, one can select the parameters of the scheme to homomorphically evaluate the circuit in a reasonable time. They describe a leveled homomorphic encryption scheme called BGV using a modulus switching technique. Furthermore, this scheme and other ring-based homomorphic encryption schemes allow the use of larger plaintext spaces, where bits are replaced by polynomials with coefficients modulo a plaintext modulus possibly different from 2. Such plaintext spaces allow the encryption of more information in a single ciphertext, for example via batching of plaintext bits. Unfortunately, to homomorphically evaluate a circuit of multiplicative depth d using the modulus switching technique, the public key needs to contain d distinct versions of a so-called evaluation key.

At Crypto 2012, Brakerski proposed the new notion of *scale-invariance* [4] for leveled homomorphic encryption schemes. In contrast to a scheme that uses modulus switching, the ciphertexts for a scale-invariant scheme keep the same modulus during the whole homomorphic evaluation and only one copy of the scale-invariant evaluation key has to be stored. This technique has been adapted to the BGV scheme [5] by Fan and Vercauteren [14], and to López-Alt, Tromer and Vaikuntanathan's scheme [30] by Bos, Lauter, Loftus and Naehrig [3].[1] The resulting schemes are called FV and YASHE, respectively. No implementation of the FV scheme is known (except for a proof-of-concept implementation in a computer algebra system that is used in [21]). The YASHE scheme [3] was the first (and only) scale-invariant leveled homomorphic encryption scheme implemented so far. Very satisfactory timings are claimed for a small modulus (then able to

---

[1] This technique was also adapted to the homomorphic encryption scheme over the integers [12] by Coron, Lepoint and Tibouchi [10].

handle only circuits of multiplicative depth at most 2) on a personal computer. Unfortunately the implementation is not openly available for the community.

Sending Data to the Cloud. In typical real-world scenarios for using FHE with cloud applications, one or more clients communicate with a cloud service. They upload data encrypted with an FHE scheme under the public key of a specific user. The cloud can process this data homomorphically and return an encrypted result. Unfortunately, ciphertext expansion (*i.e.* the ciphertext size divided by the plaintext size) of current FHE schemes is prohibitive (thousands to millions). For example using techniques in [11] (for 72 bits of claimed security), sending 4MB of data on which the cloud is allowed to operate, would require to send more than 73TB of encrypted data over the network.

To solve this issue, it was proposed in [33] to instead send the data encrypted *with a block cipher* (in particular AES). The cloud service then encrypts the ciphertexts with the FHE scheme and the user's public key and *homomorphically decrypts* them before they are processed. Therefore, network communication is lowered to the data size (which is optimal) plus a costly *one-time setup* that consists of sending the FHE public key and an FHE encryption of the block cipher secret key.

The AES circuit was chosen as a standard circuit to evaluate because it is nontrivial (but still reasonably small) and has an algebraic structure that works well with the plaintext space of certain homomorphic encryption schemes [19]. However, there might be other ciphers that are more suitable for being evaluated under homomorphic encryption. In June 2013, the U.S. National Security Agency unveiled a family of lightweight block ciphers called SIMON [2]. These block ciphers were engineered to be extremely small, easy to implement and efficient in hardware. SIMON has a classical Feistel structure and each round only contains one AND. This particularly simple structure is a likely candidate for homomorphic cryptography.

Our Contributions. In this work, we provide a concrete comparison of the supposedly most practical leveled homomorphic encryption schemes FV and YASHE. (To our knowledge, this is the first comparison of leveled homomorphic encryption schemes.) In particular, we revisit and provide precise upper bounds for the norm of the noises in the FV scheme, as done for the YASHE scheme in [3]. It appears from our work that the FV scheme has a theoretical smaller noise growth than YASHE.

We revisit van de Pol and Smart's approach [35] to derive secure parameters for these schemes. They use the BKZ-2.0 simulation algorithm [7,8] (the most up-to-date lattice basis reduction algorithm) to determine an upper bound on the modulus to ensure a given level of security. We show that their methodology has some small limitations and we describe how to resolve them. The resulting method yields a more conservative but meaningful approach to select parameters for lattice-based cryptosystems.

Finally, we propose proof-of-concept implementations of both FV and YASHE in C++ using the arithmetic library FLINT [23]. This allows us to *practically* compare the noise growth and the performances of the FV and YASHE schemes. The implementations provide insights into the behavior of these schemes for circuits of multiplicative depth larger than 2 (contrary to the implementation described in [3]). For this purpose, we implemented SIMON-32/64 using FV, YASHE and the batch integer-based scheme from [10]. Our implementations are publicly available for the community to reproduce our experiments [26]. Due to the similarity in the design of the FV and YASHE schemes and the common basis of our implementations, we believe that our comparison gives meaningful insights into which scheme to use according to the desired application, and on the achievable performance of leveled homomorphic encryption.

2   Preliminaries

In this section, we provide a succinct background on lattices, the (Ring) Learning With Errors problem and recall the FV [14] and YASHE [3] leveled homomorphic encryption schemes.

2.1   Lattices

A (full-rank) *lattice* of dimension m is a discrete additive subgroup of $\mathbb{R}^m$. For any such lattice $L \neq \{0\}$, there exist linearly independent vectors $\mathbf{b}_1, \ldots, \mathbf{b}_m \in \mathbb{R}^m$ such that $L = \mathbf{b}_1 \mathbb{Z} \oplus \cdots \oplus \mathbf{b}_m \mathbb{Z}$. This set of vectors is called a *basis* of the lattice. Thus a lattice can be represented by its basis matrix $\mathbf{B} \in \mathbb{R}^{m \times m}$, *i.e.* the matrix consisting of the rows $\mathbf{b}_i$ in the canonical basis of $\mathbb{R}^m$. In particular, we have $L = \{\mathbf{z} \cdot \mathbf{B} : \mathbf{z} \in \mathbb{Z}^m\}$. The determinant (or volume) of a lattice is defined as $\det(L) = (\det(\mathbf{B}\mathbf{B}^t))^{1/2} = |\det(\mathbf{B})|$, where $\mathbf{B}$ is any basis of L. This quantity is well-defined since it is independent of the choice of basis.

Among all the bases of a lattice L, some are 'better' than others. The goal of lattice basis reduction is to shorten the basis vectors and thus, since the determinant is invariant, to make them more orthogonal. In particular, any basis $\mathbf{B} = (\mathbf{b}_1, \ldots, \mathbf{b}_m)$ can be uniquely written as $\mathbf{B} = \mu \cdot \mathbf{D} \cdot \mathbf{Q}$ where $\mu = (\mu_{ij})$ is lower triangular with unit diagonal, $\mathbf{D}$ is diagonal with positive coefficients and $\mathbf{Q}$ has orthogonal row vectors. We call $\mathbf{B}^* = \mathbf{D} \cdot \mathbf{Q}$ the Gram-Schmidt orthogonalization of $\mathbf{B}$, and $\mathbf{D} = \mathrm{diag}(\|\mathbf{b}_1^*\|, \ldots, \|\mathbf{b}_m^*\|)$ is the diagonal matrix formed by the ℓ_2-norms $\|\mathbf{b}_i^*\|$ of the Gram-Schmidt vectors.

Following the approach popularized by Gama and Nguyen [15], we say that a specific basis $\mathbf{B}$ has *root Hermite factor* γ if its element of smallest norm $\mathbf{b}_1$ (*i.e.* we assume that basis vectors are ordered by their norm) satisfies

$$\|\mathbf{b}_1\| = \gamma^m \cdot |\det(\mathbf{B})|^{1/m} .$$

By using lattice basis reduction algorithms, one aims to determine an output lattice basis with guaranteed norm and orthogonality properties. A classical lattice basis reduction algorithm is LLL (due to Lenstra, Lenstra and Lovász [25]), which ensures that for all $i < m$, $\delta_{\mathsf{LLL}}\|\mathbf{b}_i^*\|^2 \leqslant \|\mathbf{b}_{i+1}^* + \mu_{i+1 i}\mathbf{b}_i^*\|^2$ for a given parameter $\delta_{\mathsf{LLL}} \in (1/4, 1]$. The LLL algorithm runs in polynomial-time and provides bases of quite decent quality. For many cryptanalytic applications, Schnorr and Euchner's blockwise algorithm BKZ [36] is the most practical algorithm for lattice basis reduction in high dimensions. It provides bases of higher quality but its running time increases significantly with the blocksize. Now if A denotes a lattice basis reduction algorithm, applying it to $\mathbf{B}$ yields a reduced basis $\mathbf{B}' = \mathsf{A}(\mathbf{B})$. Thus we can define $\gamma_{\mathsf{A}(\mathbf{B})}$ as the value such that

$$\|\mathbf{b'}_1\| = \gamma_{\mathsf{A}(\mathbf{B})}^m \cdot |\det(\mathbf{B}')|^{1/m} = \gamma_{\mathsf{A}(\mathbf{B})}^m \cdot |\det(\mathbf{B})|^{1/m} .$$

It is conjectured [15,7] that the value $\gamma_{\mathsf{A}(\mathbf{B})}$ depends mostly on the lattice basis reduction algorithm, and not on the input basis $\mathbf{B}$ (unless it has a special structure and cannot be considered random). Thus, in this paper, we refer to this value as γ_{A}. For example for LLL and BKZ-20 (i.e. BKZ with a blocksize $\beta = 20$), in the literature one can find the well-known values $\gamma_{\mathsf{LLL}} \approx 1.021$ and $\gamma_{\mathsf{BKZ\text{-}20}} \approx 1.013$.

2.2   Ring-LWE

In this section, we briefly introduce notation for stating the Ring-LWE-based homomorphic encryption schemes FV and YASHE, and formulate the Ring Learning With Errors (RLWE) Problem relating to the security of the two schemes. For further details, we refer to [31], [14], and [3].

Let d be a positive integer and let $\Phi_d(x) \in \mathbb{Z}[x]$ be the d-th cyclotomic polynomial. Let $R = \mathbb{Z}[x]/(\Phi_d(x))$, i.e. the ring R is isomorphic to the ring of integers of the d-th cyclotomic number field. The elements of R are polynomials with integer coefficients of degree less than $n = \varphi(d)$. For any polynomial $a = \sum_{i=0}^{n} a_i x^i \in \mathbb{Z}[x]$, let $\|a\|_\infty = \max\{|a_i| : 0 \leqslant i \leqslant n\}$ be the infinity norm of a. When multiplying elements of R, the norm of the product grows at most with a factor $\delta = \sup\{\|ab\|_\infty/\|a\|_\infty\|b\|_\infty : a, b \in R\}$, the so-called expansion factor. For an integer modulus $q > 0$, define $R_q = R/qR$. If t is another positive integer, let $r_t(q)$ be the reduction of q modulo t into the interval $[0, t)$, and let $\Delta = \lfloor q/t \rfloor$, then $q = \Delta t + r_t(q)$. Denote by $[\cdot]_q$ reduction modulo q into the interval $(-q/2, q/2]$ of an integer or integer polynomial (coefficient wise). Fix an integer base w and let $\ell_{w,q} = \lfloor \log_w(q) \rfloor + 1$. Then a polynomial $a \in R_q$ can be written in base w as $\sum_{i=0}^{\ell_{w,q}-1} a_i w^i$, where $a_i \in R$ with coefficients in $(-w/2, w/2]$. Define $\mathsf{WordDecomp}_{w,q}(a) = ([a_i]_w)_{i=0}^{\ell_{w,q}-1} \in R^{\ell_{w,q}}$ and $\mathsf{PowersOf}_{w,q}(a) = ([aw^i]_q)_{i=0}^{\ell_{w,q}-1} \in R^{\ell_{w,q}}$. Note that

$$\langle \mathsf{WordDecomp}_{w,q}(a), \mathsf{PowersOf}_{w,q}(b) \rangle = ab \pmod{q} .$$

Let χ_{key} and χ_{err} be two discrete, bounded probability distributions on R. In practical instantiations, the distribution χ_{err} is typically a truncated discrete Gaussian distribution that is statistically close to a discrete Gaussian. The distribution χ_{key} is chosen to be a very narrow distribution, sometimes even such that the coefficients of the sampled elements are in the set $\{-1, 0, 1\}$. We denote the bounds corresponding to these distributions by B_{key} and B_{err}, respectively.

This means that $\|e\|_\infty < B_{err}$ for $e \leftarrow \chi_{err}$ and $\|f\|_\infty < B_{key}$ for $f \leftarrow \chi_{key}$. With the help of χ_{key} and χ_{err}, we define the Ring-LWE distribution on $R_q \times R_q$ as follows: sample $a \leftarrow R_q$ uniformly at random, $s \leftarrow \chi_{key}$ and $e \leftarrow \chi_{err}$, and output $(a, [as + e]_q)$.

Next, we formulate a version of the Ring-LWE problem that applies to the schemes FV and YASHE considered in this paper.

Definition 1 (Ring-LWE problem). *With notation as above, the* Ring-Learning With Errors Problem *is the problem to distinguish with non-negligible probability between independent samples* $(a_i, [a_i s + e_i]_q)$ *from the Ring-LWE distribution and the same number of independent samples* (a_i, b_i) *from the uniform distribution on* $R_q \times R_q$.

In order for FV and YASHE to be secure, the RLWE problem as stated above needs to be infeasible. We refer to [14] and [3] for additional assumptions and detailed discussions of the properties of χ_{key} and χ_{err}.

2.3 The Fully Homomorphic Encryption Scheme FV

Fan and Vercauteren [14] port Brakerski's scale-invariant FHE scheme introduced in [4] to the RLWE setting. Using the message encoding as demonstrated in an RLWE encryption scheme presented in an extended version of [31] makes it possible to avoid the modulus switching technique for obtaining a leveled homomorphic scheme. We briefly summarize (a slightly generalized version of) the FV scheme in this subsection.

- FV.ParamsGen(λ): Given the security parameter λ, fix a positive integer d that determines R, moduli q and t with $1 < t < q$, distributions χ_{key}, χ_{err} on R, and an integer base $w > 1$. Output $(d, q, t, \chi_{key}, \chi_{err}, w)$.
- FV.KeyGen($d, q, t, \chi_{key}, \chi_{err}, w$): Sample $s \leftarrow \chi_{key}$, $a \leftarrow R_q$ uniformly at random, and $e \leftarrow \chi_{err}$ and compute $b = [-(as + e)]_q$. Sample $\mathbf{a} \leftarrow R_q^{\ell_{w,q}}$ uniformly at random, $\mathbf{e} \leftarrow \chi_{err}^{\ell_{w,q}}$, compute $\boldsymbol{\gamma} = ([\mathsf{PowersOf}_{w,q}(s^2) - (\mathbf{e} + \mathbf{a} \cdot s)]_q, \mathbf{a}) \in R^{\ell_{w,q}}$, and output $(\mathsf{pk}, \mathsf{sk}, \mathsf{evk}) = ((b, a), s, \boldsymbol{\gamma})$.
- FV.Encrypt($(b, a), m$): The message space is R/tR. For a message $m + tR$, sample $u \leftarrow \chi_{key}$, $e_1, e_2 \leftarrow \chi_{err}$, and output the ciphertext $\mathbf{c} = ([\Delta[m]_t + bu + e_1]_q, [au + e_2]_q) \in R^2$.
- FV.Decrypt($s, \mathbf{c} = (c_0, c_2)$): Output $m = [\lfloor t/q \cdot [c_0 + c_1 s]_q \rceil]_t \in R_t$.
- FV.Add($\mathbf{c_1}, \mathbf{c_2}$): Given ciphertexts $\mathbf{c_1} = (c_{1,0}, c_{1,1})$ and $\mathbf{c_2} = (c_{2,0}, c_{2,1})$, output $\mathbf{c_{add}} = ([c_{1,0} + c_{2,0}]_q, [c_{1,1} + c_{2,1}]_q)$.
- FV.ReLin($\tilde{\mathbf{c}}_{mult}, \mathsf{evk}$): Let $(\mathbf{b}, \mathbf{a}) = \mathsf{evk}$ and let $\tilde{\mathbf{c}}_{mult} = (c_0, c_1, c_2)$. Output the ciphertext

$$([c_0 + \langle \mathsf{WordDecomp}_{w,q}(c_2), \mathbf{b} \rangle]_q, [c_1 + \langle \mathsf{WordDecomp}_{w,q}(c_2), \mathbf{a} \rangle]_q).$$

- FV.Mult($\mathbf{c_1}, \mathbf{c_2}, \mathsf{evk}$): Output the ciphertext $\mathbf{c}_{mult} = \text{FV.ReLin}(\tilde{\mathbf{c}}_{mult}, \mathsf{evk})$, where

$$\tilde{\mathbf{c}}_{mult} = (c_0, c_1, c_2) = \left(\left[\left\lfloor \frac{t}{q} c_{1,0} c_{2,0} \right\rceil \right]_q, \left[\left\lfloor \frac{t}{q} (c_{1,0} c_{2,1} + c_{1,1} c_{2,0}) \right\rceil \right]_q, \left[\left\lfloor \frac{t}{q} c_{1,1} c_{2,1} \right\rceil \right]_q \right).$$

2.4 The Fully Homomorphic Encryption Scheme YASHE

In [3], a fully homomorphic encryption scheme is introduced that is based on the modified version of NTRU by Stehlé and Steinfeld [38] and the multi-key fully homomorphic encryption scheme presented in [30]. In this subsection, we state the more practical variant of the leveled homomorphic scheme from [3].

- YASHE.ParamsGen(λ): Given the security parameter λ, fix a positive integer d that determines R, moduli q and t with $1 < t < q$, distributions χ_{key}, χ_{err} on R, and an integer base $w > 1$. Output $(d, q, t, \chi_{key}, \chi_{err}, w)$.
- YASHE.KeyGen($d, q, t, \chi_{key}, \chi_{err}, w$): Sample $f', g \leftarrow \chi_{key}$ and let $f = [tf' + 1]_q$. If f is not invertible modulo q, choose a new f'. Compute the inverse $f^{-1} \in R$ of f modulo q and set $h = [tgf^{-1}]_q$. Sample $\mathbf{e}, \mathbf{s} \leftarrow \chi_{err}^{\ell_{w,q}}$, compute $\boldsymbol{\gamma} = [\mathsf{PowersOf}_{w,q}(f) + \mathbf{e} + h \cdot \mathbf{s}]_q \in R^{\ell_{w,q}}$ and output $(\mathsf{pk}, \mathsf{sk}, \mathsf{evk}) = (h, f, \boldsymbol{\gamma})$.
- YASHE.Encrypt(h, m): The message space is R/tR. For a message $m + tR$, sample $s, e \leftarrow \chi_{err}$, and output the ciphertext $c = [\Delta[m]_t + e + hs]_q \in R$.
- YASHE.Decrypt(f, c): Decrypt a ciphertext c by $m = [\lfloor t/q \cdot [fc]_q \rceil]_t \in R$.
- YASHE.Add(c_1, c_2): Output $c_{\mathrm{add}} = [c_1 + c_2]_q$.
- YASHE.KeySwitch($\tilde{c}_{\mathrm{mult}}, \mathsf{evk}$): Output $[\langle \mathsf{WordDecomp}_{w,q}(\tilde{c}_{\mathrm{mult}}), \mathsf{evk}\rangle]_q$.
- YASHE.Mult(c_1, c_2, evk): Output the ciphertext

$$c_{\mathrm{mult}} = \mathsf{YASHE.KeySwitch}(\tilde{c}_{\mathrm{mult}}, \mathsf{evk}), \text{ where } \tilde{c}_{\mathrm{mult}} = [\lfloor t/q \cdot c_1 c_2 \rceil]_q.$$

3   Parameter Derivation

In this section, we explain how to derive parameters for the fully homomorphic encryption schemes FV [14] and YASHE [3]. For security, we follow van de Pol and Smart's approach to derive the maximal size of the modulus achievable in a given dimension [35] and consider the distinguishing attack against RLWE. In particular, we use Chen and Nguyen's simulation algorithm for the state-of-the-art lattice basis reduction algorithm BKZ-2.0 [7,8]. For correctness, we provide a lower bound on the modulus in a given dimension and for a targeted number of levels (depending on the application), for both schemes FV and YASHE. Therefore for a given application, it suffices to combine these upper and lower bounds to select a suitable modulus.

3.1   Revisiting van de Pol and Smart's Approach

We assume the reader to be familiar with Schnorr and Euchner's blockwise algorithm BKZ [36], and Chen and Nguyen's improved version BKZ-2.0 [7,8]. We provide more details in the full version [27] of this paper. In the following BKZ-2.0$_{N,\beta}$ means that BKZ-2.0 is run with blocksize β and for a maximal number of N rounds. In [35], van de Pol and Smart use the formula of [7,8],

$$\mathsf{cost}(\mathsf{BKZ\text{-}2.0}_{N,\beta}) \leqslant N \times (m - \beta) \times \mathsf{cost}(\text{Enumeration in dimension } \beta) + \mathcal{O}(1) \quad (1)$$

to estimate the cost of $\mathsf{BKZ}\text{-}2.0_{N,\beta}$ (in terms of the number of nodes visited) on an m-dimensional basis, and to *generate* secure parameters.[2] Instead of using BKZ-2.0 to verify heuristically selected parameters, they rather propose a rational method to tackle the parameter selection, which we describe below.

For a given security parameter λ and a dimension m, van de Pol and Smart propose to derive the smallest root Hermite factor $\gamma(m)$ on an m-dimensional lattice achievable using BKZ-2.0 by an adversary limited to a computational cost of at most $\mathsf{cost}(\mathsf{BKZ}\text{-}2.0) \leqslant 2^\lambda$. By Equation (1), this means that for all β and N, we need to have

$$N \times (m - \beta) \times \mathsf{cost}(\text{Enumeration in dimension } \beta) \leqslant 2^\lambda .$$

Thus, for each β and using the enumeration costs in [7] (or [8]), one obtains an upper bound N_{max} on the number of BKZ-2.0 rounds with blocksize β that an adversary bounded as above can afford to run, *i.e.* such that this latter inequality is still verified. Next, the quality of the resulting basis is estimated by running the $\mathsf{BKZ}\text{-}2.0_{N_{\mathsf{max}},\beta}$ simulation algorithm on a random lattice with blocksize β and N_{max} rounds. This yields a root Hermite factor $\gamma(m, \beta)$ for this specific blocksize β. By taking the minimum value over all blocksizes, one obtains the minimum root Hermite factor $\gamma(m)$ achievable in dimension m for the security parameter λ using BKZ-2.0.

Van de Pol and Smart show that, for the homomorphic evaluation of the AES circuit of [19], by using their new approach for a given security level, it is possible to work with significantly smaller lattice dimensions than what previous methods recommended, which affects the performance of the underlying lattice-based homomorphic encryption scheme.

Limitations of [35]. However, the approach presented in [35] has some limitations. First of all, van de Pol and Smart only consider dimensions that are a power of two. They use linear interpolation for the missing values and therefore obtain a simplified model which does not reflect the real behavior of the minimal root Hermite factor. Also, the enumeration costs used in [35] are based on the proceedings version [7] of the BKZ-2.0 paper. Recently a full version [8] with smaller enumeration costs has been published, which forces one to revisit van de Pol and Smart's results. Last but not least, they only consider blocksizes that are a multiple of 10 (due to the tables in [7]). This leads to a phenomenon of *plateaus* (*cf.* Fig 1) and might lead to a choice of parameters ensuring less than λ bits of security.

Overcoming the Limitations of [35]. To overcome these issues, we performed the same experiments as van de Pol and Smart but for *all* dimensions from 1000

---

[2] The term $\mathcal{O}(1)$ occurs due to the fact that in high dimension, the enumeration time is usually dominant compared to the time spent on computing the Gram-Schmidt orthogonalization and LLL reduction [7,8]. Note again that Chen and Nguyen provide an *ideal* simulation algorithm – experimental applications of BKZ-2.0 might yield a basis with a larger root Hermite factor. Therefore, using Equation (1) to estimate parameters is conservative.

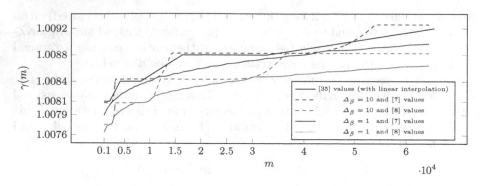

Fig. 1. Minimal root Hermite factor $\gamma(m)$ achievable with a complexity less than 2^{80}, in function of the dimension m

up to 65000. We also considered both the enumeration costs given in Chen and Nguyen's proceedings paper [7] as well as those in the full version [8]. We plotted the results in Fig. 1. As expected, the linear interpolation of [35] does not fully reflect the behavior of the experiments for the other dimensions.

However, when performing the experiments for all dimensions, but only avoiding linear interpolation, we still observe the plateau phenomenon. This can be explained by the fact that the enumeration costs from [7] are only used for blocksizes that are a multiple of $\Delta_\beta = 10$ (which are the only values given in [7]), and only those are considered in [35]. Each plateau consists of the minimal root Hermite factor achievable for a specific blocksize β. Now for the whole plateau, BKZ-2.0$_{N_{max},\beta}$ terminates in less than N_{max} rounds, *i.e.* a fix-point is attained at some round $i < N_{max}$. The next plateau corresponds to a block-size $\beta - \Delta_\beta = \beta - 10$. Between plateaus, the number N_{max} is the limiting factor in BKZ-2.0 (*i.e.* BKZ-2.0$_{N_{max},\beta}$ terminates at round N_{max}) and determines the achievable root Hermite factor (therefore this latter value increases until a blocksize of size $\beta - 10$ instead of β is more useful).

To resolve this issue, we used the least squares method to interpolate the enumeration costs for blocksizes β that are not a multiple of 10 (for more details see the full version [27] of this paper). These new costs allowed us to perform the experiments with *all* blocksizes $\beta \in \{100, \ldots, 250\}$ (*i.e.* with steps $\Delta_\beta = 1$) and we obtained the plain lines in Fig. 1.[3] As one can see there, parameters selected from plateaus might yield attacks of complexity smaller than 2^λ if the attacker chooses a blocksize that actually allows to achieve a smaller root Hermite factor.

Therefore, to be more conservative than [35] in our parameter selection, in the rest of the paper, we use the values of $\gamma(m)$ for $\Delta_\beta = 1$ using the enumeration

---

[3] Note that, without loss of generality, we only considered blocksizes larger than 100. Indeed, for $\beta = 100$ the cost of the enumeration of [8] is 2^{39} and BKZ-2.0 usually reaches a fix point in less than 100 rounds (*cf.* [8, Fig.7]). Therefore for a target security level of 80 bits and dimensions up to $\approx 2^{32}$, one will not be able to obtain a better reduction with a $\beta < 100$.

costs of [8]. Note that there is a significant difference in the achievable values compared to [35].

3.2 Security Requirements for RLWE: The Distinguishing Attack

In this section, we restate and extend the security analysis of [35]. Namely, we consider the distinguishing attack against RLWE (see [32,28]). In the following, we denote by $0 < \epsilon < 1$ the advantage with which we allow the adversary to distinguish an RLWE instance $(a, b = a \cdot s + e) \in R_q^2$ from a uniform random pair $(a, u) \in R_q^2$ (i.e. the advantage of the adversary for solving the Decisional-RLWE problem). For any $a \in R_q$, we denote by $\Lambda_q(a)$ the lattice

$$\Lambda_q(a) = \{y \in R_q : \exists\, z \in R, y = a \cdot z \bmod q\}.$$

Recall that, for an n-dimensional lattice Λ, we denote by $\Lambda^\times$ its dual, i.e. the lattice defined by $\Lambda^\times = \{v \in \mathbb{R}^n : \forall\, b \in \Lambda, \langle v, b \rangle \in \mathbb{Z}\}$. The distinguishing attack consists in finding a small vector $v \in q \cdot \Lambda_q(a)^\times$. Then, for all $y \in \Lambda_q(a)$, $\langle v, y \rangle = 0 \bmod q$. To distinguish whether a given pair (a, u) was sampled according to the RLWE distribution or the uniform distribution, one tests whether the inner product $\langle v, u \rangle$ is 'close' to 0 modulo q (i.e. whether $|\langle v, u \rangle| < q/4$) or not.

Indeed, when u is uniformly distributed in R_q and $n \geqslant 2\lambda + 1$, $\langle v, u \rangle$ is statistically close to the uniform distribution by the leftover hash lemma and the test accepts with probability $1/2 - \mathsf{negl}(\lambda)$. However, when (a, u) is an RLWE sample, i.e. there exists $s \in R_q$ and $e \leftarrow \chi_{\mathrm{err}}$ such that $u = a \cdot s + e$, we have $\langle v, u \rangle = \langle v, e \rangle \bmod q$, which is essentially a sample from a Gaussian (reduced modulo q) with standard deviation $\|v\| \cdot \sigma_{\mathrm{err}}$. Now when this parameter is not much larger than q, $\langle v, e \rangle$ can be distinguished from uniform with advantage $\exp(-\pi\tau^2)$ with $\tau = \|v\| \cdot \sigma_{\mathrm{err}}/q$, for details see [32,28].

The distinguishing attack against LWE is more efficient when working with a $m \times n$ matrix with $m > n$ [32,28,35]. Moreover, it is unknown how to exploit the ring structure of RLWE to improve lattice reduction [15,7]. Therefore, we will embed our RLWE instance into an LWE lattice. Next we apply the distinguishing attack against LWE and the result can be used to distinguish the RLWE instance from uniform. Define an LWE matrix $A \in \mathbb{Z}_q^{m \times n}$ associated to a as the matrix whose first n lines are the coefficient vectors of $x^i \cdot a$ for $i = 0, \ldots, n-1$ and the $m - n$ last lines are small linear combinations of the first n lines. Denote the LWE lattice

$$\Lambda_q(A) = \{y \in \mathbb{Z}^m : \exists\, z \in \mathbb{Z}^n, y = Az \bmod q\}.$$

Now, we use lattice basis reduction in order to find such a short vector $v \in q \cdot \Lambda_q(A)^\times$. An optimal use of BKZ-2.0 would allow us to recover a vector v such that $\|v\| = \gamma(m)^m \cdot q^{n/m}$ (because $\det(q\Lambda_q(A)^\times) = q^n$, cf. [35]). Therefore, to keep the advantage of the BKZ-2.0-adversary small enough, we need to have $\exp(-\pi\tau^2) \leqslant \epsilon$, i.e.

$$\gamma(m)^m \cdot q^{(n/m)-1} \cdot \sigma_{\mathrm{err}} \geqslant \sqrt{-\log(\epsilon)/\pi}\,.$$

Table 1. Maximal values of $\log_2(q)$ to ensure $\lambda = 80$ bits of security, with distinguishing advantage $\epsilon = 2^{-80}$ and standard deviation $\sigma_{\mathrm{err}} = 8$

n	1024	2048	4096	8192	16384
Maximal $\log_2(q)$ (method of [28])	40.6	79.4	157.0	312.2	622.7
Maximal $\log_2(q)$ (our method)	47.5	95.4	192.0	392.1	799.6

Define $\alpha = \sqrt{-\log(\epsilon)/\pi}$. To ensure security for all $m > n$, we obtain the condition

$$\log_2(q) \leqslant \min_{m>n} \frac{m^2 \cdot \log_2(\gamma(m)) + m \cdot \log_2(\sigma/\alpha)}{m - n}. \tag{2}$$

Let us fix the security parameter λ. Following the experiment described in Section 3.1, one can recover the minimal root Hermite factor $\gamma(m)$ for all $m > n$. Therefore, given a target distinguishing advantage ϵ, a dimension n and an error distribution χ_{err}, one can derive the maximal possible value for $\log_2(q)$ using Equation (2). Some interesting values are presented in Table 1. As in [35], it seems that the parameters obtained by using Lindner and Peikert's method [28] are more conservative than those obtained with the BKZ-2.0 simulation.[4]

3.3   Correctness and Noise Growth of YASHE

Any YASHE ciphertext c carries an inherent noise term, which is an element $v \in R$ of minimal norm $\|v\|_\infty$ such that $fc = \Delta[m]_t + v \pmod{q}$. If $\|v\|_\infty$ is small enough, decryption works correctly, which means that it returns the message m modulo t. More precisely, [3, Lemma 1] shows that this is the case if $\|v\|_\infty < (\Delta - r_t(q))/2$. A freshly encrypted ciphertext output by YASHE.Encrypt has an inherent noise term v that can be bounded by $\|v\|_\infty < V = \delta t B_{\mathrm{key}}(2B_{\mathrm{err}} + r_t(q)/2)$, see [3, Lemma 2].

During a homomorphic addition, the inherent noise terms roughly add up such that the resulting noise term is bounded by $\|v_{\mathrm{add}}\|_\infty \leqslant \|v_1\|_\infty + \|v_2\|_\infty + r_t(q)$, where v_1 and v_2 are the respective noise terms in c_1 and c_2. For a multiplication operation, noise growth is much larger. It is shown in [3, Theorem 4 and Lemma 4] that, when $\|v_1\|_\infty, \|v_2\|_\infty < V$ the noise term after multiplication can be bounded by

$$\|v_{\mathrm{mult}}\|_\infty < \delta t(4 + \delta t B_{\mathrm{key}})V + \delta^2 t^2 B_{\mathrm{key}}(B_{\mathrm{key}} + t) + \delta^2 t \ell_{w,q} w B_{\mathrm{err}} B_{\mathrm{key}}.$$

For a homomorphic computation with L levels of multiplications (and considering only the noise growth from multiplications), [3, Corollary 1 and Lemma 9]

---

[4] In [29], Lin and Nguyen obtained significant improvements upon Lindner-Peikert's decoding attack [28] using only pruned enumeration. However, there is no detail on how to compute the success probability, nor on how to estimate the number of nodes to enumerate, nor on how long an enumeration takes. It remains an interesting open problem to adapt van de Pol and Smart's approach to Lin and Nguyen's attack for parameter selection, as it is currently unclear how to compute the above values.

give an upper bound on the inherent noise in the resulting ciphertext as $\|v\|_\infty < C_1^L V + LC_1^{L-1}C_2$, where

$$C_1 = (1 + 4(\delta t B_{\text{key}})^{-1})\delta^2 t^2 B_{\text{key}}, C_2 = \delta^2 t B_{\text{key}} (t(B_{\text{key}} + t) + \ell_{w,q} w B_{\text{err}}).$$

In order to choose parameters for YASHE so that the scheme can correctly evaluate such a computation with L multiplicative levels, the parameters need to satisfy $C_1^L V + LC_1^{L-1}C_2 < (\Delta - r_t(q))/2$. In Table 2(a), we provide some values for power-of-two dimensions n and levels $L = 0, 1, 10, 50$.

3.4 Correctness and Noise Growth of FV

We can treat FV and YASHE ciphertexts similarly by simply interchanging $c_0 + c_1 s$ and fc. Thus, for an FV ciphertext (c_0, c_1) the inherent noise term is an element $v \in R$ of minimal norm such that $c_0 + c_1 s = \Delta[m]_t + v$ (mod q). Since decryption is the same once $[c_0 + c_1 s]_q$ or $[fc]_q$ are computed, respectively, this also means that correctness of decryption is given under the same condition $\|v\|_\infty < (\Delta - r_t(q))/2$ in both schemes. In an FV ciphertext, the value $v = e_1 + e_2 s - eu$ satisfies $c_0 + c_1 s = \Delta[m]_t + v$ (mod q) and therefore, we can bound the noise term in a freshly encrypted FV ciphertext by $\|v\|_\infty < V = B_{\text{err}}(1 + 2\delta B_{\text{key}})$.

The same reasoning shows that noise growth during homomorphic addition can be bounded in the same way by $\|v_{\text{add}}\|_\infty \leqslant \|v_1\|_\infty + \|v_2\|_\infty + r_t(q)$. Following the exact same proofs as for YASHE as in [3] (see the proofs for the more practical variant YASHE', which we use here), one can show that the noise growth during a homomorphic multiplication is bounded by

$$\|v_{\text{mult}}\|_\infty < \delta t(4 + \delta B_{\text{key}})V + \delta^2 B_{\text{key}}(B_{\text{key}} + t^2) + \delta \ell_{w,q} w B_{\text{err}},$$

where as before, it is assumed that $\|v_1\|_\infty, \|v_2\|_\infty < V$. Note that the bound on the multiplication noise growth is smaller than the respective bound for YASHE by roughly a factor t. This means that FV is more robust against an increase of the parameter t. Similarly as above, when doing a computation in L levels of multiplications (carried out in a binary tree without taking into account the noise growth for homomorphic additions), the noise growth can be bounded by $\|v\|_\infty < C_1^L V + LC_1^{L-1}C_2$, where

$$C_1 = (1 + \epsilon_2)\delta^2 t B_{\text{key}}, \quad C_2 = \delta^2 B_{\text{key}}(B_{\text{key}} + t^2) + \delta \ell_{w,q} w B_{\text{err}}, \quad \epsilon_2 = 4(\delta B_{\text{key}})^{-1},$$

and the correctness condition for choosing FV parameters for an L-leveled multiplication is $C_1^L V + LC_1^{L-1}C_2 < (\Delta - r_t(q))/2$ as above. In Table 2(b), we provide some values for power-of-two dimensions n and levels $L = 0, 1, 10, 50$; these values illustrate the smaller theoretical noise growth for FV in comparison to YASHE.

4 Practical Implementations

In order to assess the relative practical efficiency of FV and YASHE, we implemented these leveled homomorphic encryption schemes in C++ using the arithmetic library FLINT [23]; our implementations are publicly available at [26].

Table 2. Minimal value of $\log_2(q)$ to ensure correctness of YASHE and FV, with overwhelming probability, using standard deviation $\sigma_{err} = 8$, plaintext modulus $t = 2$, integer base $w = 2^{32}$, and $B_{key} = 1$

(a) YASHE

n	1024	2048	4096	8192	16384
$L = 0$	20	21	22	23	24
$L = 1$	62	64	66	68	70
$L = 10$	265	286	306	326	346
$L = 50$	1150	1250	1350	1450	1550

(b) FV

n	1024	2048	4096	8192	16384
$L = 0$	19	20	21	22	23
$L = 1$	40	43	46	49	52
$L = 10$	229	250	271	292	313
$L = 50$	1069	1170	1271	1372	1473

Table 3. Timings of YASHE and FV using the same parameters as in [3]: $R = \mathbb{Z}[x]/(x^{4096} + 1)$, $q = 2^{127} - 1$, $w = 2^{32}$, $t = 2^{10}$ on an Intel Core i7-2600 at 3.4 GHz with hyper-threading turned off and over-clocking ('turbo boost') disabled

Scheme	KeyGen	Encrypt	Add	Mult	KeySwitch or ReLin	Decrypt
YASHE	3.4s	16ms	0.7ms	18ms	31ms	15ms
FV	0.2s	34ms	1.4ms	59ms	89ms	16ms
YASHE [3] (estimation)	–	23ms	0.020ms		27ms	4.3ms

Timings. In Table 3, we provide timings using the same parameters as in [3]. As expected from the structure of the ciphertexts, it takes twice more time to Encrypt or Add using FV compared to YASHE and three times longer to multiply two ciphertexts. These parameters also allow us to provide *estimated* timings for the implementation of [3] on the same architecture as an illustration of a possible overhead in performances due to the arithmetic libraries (namely, FLINT) and the C++ wrappers.[5] This corroborates the significant performance gains obtained in recent works in lattice-based cryptography [22,13] using home-made implementations, instead of relying on arithmetic libraries [19,1].

Practical Noise Growth. In Sections 3.3 and 3.4, we provide strict theoretical upper bounds on the noise growth during homomorphic operations in FV and YASHE to ensure correctness with *overwhelming probability*. In practice however, one expects a smaller noise growth on average and one could choose smaller bounds ensuring correctness with high probability only. This yields a huge gain in performance (allowing to reduce q, and thus n) while still ensuring correctness most of the time. In Figure 2, we depict an average noise growth for levels 0 to 10 for FV and YASHE. For example, this figure shows that the real noise growth allows one to reduce the bit size of q by nearly 33% to handle more than 10 levels. Therefore, for optimal performances in practice, one should select a modulus q as small as possible while still ensuring correctness with high probability.

---

[5] Both Intel processors have hyper-threading turned off and over-clocking ('turbo boost') disabled; thus timings were estimated proportionally to the processor speeds of the computers (3.4 GHz versus 2.9 GHz).

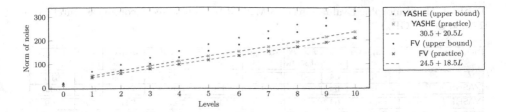

Fig. 2. Evolution of the norm of the noise using a standard deviation $\sigma = 8$, a plaintext modulus $t = 2$, a word $w = 2^{32}$, and $B_{key} = 1$, $R = \mathbb{Z}[x]/(x^{8192} + 1)$, and q a 392-bit prime

4.1   Homomorphic Evaluation of SIMON

The SIMON Feistel Cipher. In June 2013, the U.S. National Security Agency (NSA) unveiled SIMON, a family of lightweight block ciphers. These block ciphers were designed to provide an optimal hardware performance. SIMON has a classical Feistel structure with the round block size of $2n$ bits. For performance reasons, in what follows we focus on SIMON-32/64 having a block size of $2n = 32$ bits, a 64-bit secret key and $N_r = 32$ rounds. At round i, SIMON operates on the left n-bit half $\mathbf{x}_i$ of the block $(\mathbf{x}_i, \mathbf{y}_i)$ and applies a non-linear, non-bijective function $F \colon \mathbb{F}_2^n \to \mathbb{F}_2^n$ to it. The output of F is XORed with the right half along with a round key $\mathbf{k}_i$ and the two halves are swapped. The function F is defined as $F(\mathbf{x}) = ((\mathbf{x} \lll 8) \otimes (\mathbf{x} \lll 1)) \oplus (\mathbf{x} \lll 2)$ where $(\mathbf{x} \lll j)$ denotes left rotation of $\mathbf{x}$ by j positions and $\otimes$ is binary AND. The round keys $\mathbf{k}_i$ are very easily derived from a master key k with shifts and XORs. Details on how these subkeys are generated can be found in [2].

Homomorphic Representation of the State. As in [9,10] for AES, we encrypt the SIMON state state-wise. More precisely, the left half $\mathbf{x} = (x_1, \ldots, x_n) \in \mathbb{F}_2^n$ and the right half $\mathbf{y} = (y_1, \ldots, y_n) \in \mathbb{F}_2^n$ of the SIMON state are encrypted as a set of $2n$ ciphertexts $c_1, \ldots, c_n, c_{n+1}, \ldots, c_{2n}$. For each $1 \leqslant j \leqslant n$, c_j encrypts $x_j \in \mathbb{F}_2$ and c_{n+j} encrypts y_j. In other words, the $2n$ bits of the SIMON state are represented in $2n$ different ciphertexts. Note that the use of batching[6] with ℓ slots allows one to perform ℓ SIMON evaluations in parallel by encoding the corresponding bit of the state of the i-th SIMON plaintext into the i-th slot.

Homomorphic Operations. This state-wise encrypted representation of the steps allows to do the SIMON evaluation easily. Swapping the halves consists in modifying the index of the encrypted state $c_j \leftrightarrow c_{n+j}$. Define encryptions e_{ij} of the bits

---

[6] To evaluate a Boolean circuit, one can select $t = 2$ and encode each plaintext bit as the constant coefficient of a plaintext polynomial. However, if one uses batching with ℓ slots, where each ciphertext can represent a number of ℓ independent plaintexts, one obtains a significant gain in the use of both space and computational resources. Batching was adapted to the BGV scheme in [18], and can be made compatible with both FV and YASHE.

Table 4. Homomorphic Evaluations of SIMON-32/64 on a 4-core computer (Intel Core i7-2600 at 3.4 GHz)

Scheme	Parameter set	λ	$\ell = \#$ of slots	Keygen	Encrypt State	SIMON Evaluation	Relative time	Norm of Noise Final	Maximal
FV	Ib	64	2	4 s	7 s	526 s	263 s	509	516
YASHE	Ia	64	1	64 s	4 s	**200 s**	200 s	561	569
YASHE (1 core)	Ia	64	1	64 s	14 s	**747 s**	747 s	557	569
FV	II	> 80	1800	24 s	209 s	3062 s	1.70 s	918	1024
YASHE	II	> 80	1800	1300 s	104 s	1029 s	**0.57 s**	949	1024
SIBDGHV [10]	–		64	199	1032 s	1 s	628 s	3.15 s	650	704

	λ	d	$n = \phi(d)$	# of slots	$\log_2(q)$	$\log_2(w)$	σ	B_{key}
Set-Ia	64	10501	10500	1	570	70	8	1
Set-Ib	64	9551	9550	2	517	65	8	1
Set-II	> 80	32767	27000	1800	1025	257	8	1

k_{ij} of the round keys $\mathbf{k}_i$, for all i, j. (When using batching, one encrypts the vector $(k_{ij}, \ldots, k_{ij}) \in \{0, 1\}^\ell$.) This simple representation allows to XOR the right half of the state with the key via n homomorphic additions $c_{n+j} \leftarrow c_{n+j} + e_{ij}$. A shift of a positions as used in the function F is obtained by some index swapping $c_{(i+a) \bmod n}$. Finally, the only AND operation in the function F is obtained by n homomorphic multiplications. Therefore to obtain an encrypted state $c'_1, \ldots, c'_{2n}$ from an encrypted state $c_1, \ldots, c_{2n}$, one can perform:

$$c'_{n+j} \leftarrow c_j; \quad c'_j = \left(c_{(j+8) \bmod n} \cdot c_{(j+1) \bmod n} \right) + c_{(j+2) \bmod n} + e_{ij} .$$

Practical Results. We homomorphically evaluated SIMON-32/64 using our C++ implementations of FV and YASHE (and also the implementation of [10] for the leveled homomorphic encryption scheme over the integers).[7]

Results are provided in Table 4. Note that we selected parameters ensuring as many bits of security for the homomorphic encryption schemes as the number of bits of the SIMON key.[8]

4.2 Some Thoughts about Homomorphic Evaluations

Let us define the two notions latency and throughput associated to a homomorphic evaluation. We say that the *latency* of a homomorphic evaluation is the time required to perform the entire homomorphic evaluation. Its *throughput* is the number of blocks processed per unit of time.

The results presented in Table 4 emphasize an important point: different parameter sets can be selected, either to minimize the latency (Set-Ia and Set-Ib), or to maximize the throughput (Set-II). In [10] and [19,9], the parameters were

---

[7] Since each round of SIMON consists of one homomorphic multiplication, the leveled homomorphic encryption schemes need to handle at least as many levels as the number of rounds.

[8] Parameter Set-II ensures more than 80 bits of security (more likely around 120 bits) but the smaller the modulus q, the faster is the computation.

selected to maximize the throughput using batching, and therefore claim a small *relative time per block* – the latency however is several dozens of hours. However, 'real world' homomorphic evaluations (likely to be used in the cloud) should be implemented in a transparent and user-friendly way. It is therefore questionable whether the batching technique (to achieve larger throughput in treating blocks) is suitable for further processing of data. In particular, it might only be suitable when this processing is identical over each block (which is likely *not* to be the case in real world scenarios). Overall, one should rather select parameters to have the latency as small as possible. The throughput can be increased by running the homomorphic evaluations in a cluster.

5   Conclusion

In this work, we revisited van de Pol and Smart's approach to tackle parameter selection for lattice-based cryptosystems. We also conducted both a theoretical and practical comparison of FV and YASHE. We obtained that the noise growth is smaller in FV than in YASHE (both theoretically and practically). Conversely, we obtained that YASHE is, as expected, faster than FV. As a side result, for high performances, it seems interesting to implement all building blocks of the schemes rather than to rely on external arithmetic libraries.

Next, we homomorphically evaluated the lightweight block cipher SIMON, and discussed the notions of throughput and latency. We obtain that SIMON-32/64 can be evaluated *completely* in about 12 minutes on a single core and in about 3 minutes on 4 cores using OpenMP (when optimizing latency). If several blocks are processed in parallel, SIMON-32/64 can be evaluated in about 500ms per block (and less than 20 minutes total); and these timings can be lowered by using additional cores.

Finally, note that our results can certainly be improved further by other optimizations. One could incorporate dynamic scaling during the computation as discussed in [14] such that it is ensured that ciphertexts maintain their minimal size. Another possible variant is to use the Chinese Remainder Theorem to pack each half of the SIMON state into one single ciphertext instead of spreading it out over n ciphertexts. Operations that need to move data between different plaintext slots can be realized by Galois automorphisms as explained in [18]. This can possibly be further combined with batching of several SIMON states into one ciphertext. To explore the application of these to both schemes and possibly further optimizations for realizing a fully home-made and fully optimized implementation of a homomorphic SIMON evaluation is left as future work.

Acknowledgments. We thank the Africacrypt 2013 referees for their interesting reviews, and Frederik Vercauteren for insightful comments on batching.

References

1. Bansarkhani, R.E., Buchmann, J.: Improvement and efficient implementation of a lattice-based signature scheme. Cryptology ePrint Archive, Report 2013/297 (2013), http://eprint.iacr.org/

2. Beaulieu, R., Shors, D., Smith, J., Treatman-Clark, S., Weeks, B., Wingers, L.: The SIMON and SPECK families of lightweight block ciphers. Cryptology ePrint Archive, Report 2013/404 (2013), http://eprint.iacr.org/

3. Bos, J.W., Lauter, K., Loftus, J., Naehrig, M.: Improved security for a ring-based fully homomorphic encryption scheme. In: Stam (ed.) [37], pp. 45–64

4. Brakerski, Z.: Fully homomorphic encryption without modulus switching from classical gapSVP. In: Safavi-Naini, R., Canetti, R. (eds.) CRYPTO 2012. LNCS, vol. 7417, pp. 868–886. Springer, Heidelberg (2012)

5. Brakerski, Z., Gentry, C., Vaikuntanathan, V.: (Leveled) fully homomorphic encryption without bootstrapping. In: Goldwasser, S. (ed.) ITCS, pp. 309–325. ACM (2012)

6. Canetti, R., Garay, J.A. (eds.): CRYPTO 2013, Part I. LNCS, vol. 8042. Springer, Heidelberg (2013)

7. Chen, Y., Nguyen, P.Q.: BKZ 2.0: Better lattice security estimates. In: Lee, D.H., Wang, X. (eds.) ASIACRYPT 2011. LNCS, vol. 7073, pp. 1–20. Springer, Heidelberg (2011)

8. Chen, Y., Nguyen, P.Q.: BKZ 2.0: Better lattice security estimates (2013), http://www.di.ens.fr/~ychen/research/Full_BKZ.pdf

9. Cheon, J.H., Coron, J.-S., Kim, J., Lee, M.S., Lepoint, T., Tibouchi, M., Yun, A.: Batch fully homomorphic encryption over the integers. In: Johansson, T., Nguyen, P.Q. (eds.) EUROCRYPT 2013. LNCS, vol. 7881, pp. 315–335. Springer, Heidelberg (2013)

10. Coron, J.-S., Lepoint, T., Tibouchi, M.: Scale-invariant fully homomorphic encryption over the integers. In: Krawczyk, H. (ed.) PKC 2014. LNCS, vol. 8383, pp. 311–328. Springer, Heidelberg (2014)

11. Coron, J.S., Naccache, D., Tibouchi, M.: Public key compression and modulus switching for fully homomorphic encryption over the integers. In: Pointcheval, Johansson (eds.) [34], pp. 446–464

12. van Dijk, M., Gentry, C., Halevi, S., Vaikuntanathan, V.: Fully homomorphic encryption over the integers. In: Gilbert, H. (ed.) EUROCRYPT 2010. LNCS, vol. 6110, pp. 24–43. Springer, Heidelberg (2010)

13. Ducas, L., Durmus, A., Lepoint, T., Lyubashevsky, V.: Lattice signatures and bimodal gaussians. In: Canetti, Garay (eds.) [6], pp. 40–56

14. Fan, J., Vercauteren, F.: Somewhat practical fully homomorphic encryption. IACR Cryptology ePrint Archive 2012, 144 (2012)

15. Gama, N., Nguyen, P.Q.: Predicting lattice reduction. In: Smart, N.P. (ed.) EUROCRYPT 2008. LNCS, vol. 4965, pp. 31–51. Springer, Heidelberg (2008)

16. Gentry, C.: Fully homomorphic encryption using ideal lattices. In: Mitzenmacher, M. (ed.) STOC, pp. 169–178. ACM (2009)

17. Gentry, C., Halevi, S.: Implementing gentry's fully-homomorphic encryption scheme. In: Paterson, K.G. (ed.) EUROCRYPT 2011. LNCS, vol. 6632, pp. 129–148. Springer, Heidelberg (2011)

18. Gentry, C., Halevi, S., Smart, N.P.: Fully homomorphic encryption with polylog overhead. In: Pointcheval, Johansson (eds.) [34] , pp. 465–482

19. Gentry, C., Halevi, S., Smart, N.P.: Homomorphic evaluation of the AES circuit. In: Safavi-Naini, R., Canetti, R. (eds.) CRYPTO 2012. LNCS, vol. 7417, pp. 850–867. Springer, Heidelberg (2012)
20. Gentry, C., Sahai, A., Waters, B.: Homomorphic encryption from learning with errors: Conceptually-simpler, asymptotically-faster, attribute-based. In: Canetti, Garay (eds.) [6], pp. 75–92
21. Graepel, T., Lauter, K., Naehrig, M.: ML confidential: Machine learning on encrypted data. In: Kwon, T., Lee, M.-K., Kwon, D. (eds.) ICISC 2012. LNCS, vol. 7839, pp. 1–21. Springer, Heidelberg (2013)
22. Güneysu, T., Oder, T., Pöppelmann, T., Schwabe, P.: Software speed records for lattice-based signatures. In: Gaborit, P. (ed.) PQCrypto 2013. LNCS, vol. 7932, pp. 67–82. Springer, Heidelberg (2013)
23. Hart, W.: et al.: Fast Library for Number Theory, Version 2.4 (2013), http://www.flintlib.org
24. Lauter, K.E.: Practical applications of homomorphic encryption. In: Yu, T., Capkun, S., Kamara, S. (eds.) CCSW, pp. 57–58. ACM (2012)
25. Lenstra, A.K., Jr Lenstra, H.W., Lovász, L.: Factoring polynomials with rational coefficients. Math. Ann. 261(4), 515–534 (1982)
26. Lepoint, T.: A proof-of-concept implementation of the homomorphic evaluation of SIMON using FV and YASHE leveled homomorphic cryptosystems (2014), https://github.com/tlepoint/homomorphic-simon
27. Lepoint, T., Naehrig, M.: A comparison of the homomorphic encryption schemes FV and YASHE (full version). Cryptology ePrint Archive, Report 2014/062 (2014), http://eprint.iacr.org/
28. Lindner, R., Peikert, C.: Better key sizes (and attacks) for LWE-based encryption. In: Kiayias, A. (ed.) CT-RSA 2011. LNCS, vol. 6558, pp. 319–339. Springer, Heidelberg (2011)
29. Liu, M., Nguyen, P.Q.: Solving BDD by enumeration: An update. In: Dawson, E. (ed.) CT-RSA 2013. LNCS, vol. 7779, pp. 293–309. Springer, Heidelberg (2013)
30. López-Alt, A., Tromer, E., Vaikuntanathan, V.: On-the-fly multiparty computation on the cloud via multikey fully homomorphic encryption. In: STOC, pp. 1219–1234 (2012)
31. Lyubashevsky, V., Peikert, C., Regev, O.: On ideal lattices and learning with errors over rings. In: Gilbert, H. (ed.) EUROCRYPT 2010. LNCS, vol. 6110, pp. 1–23. Springer, Heidelberg (2010)
32. Micciancio, D., Regev, O.: Lattice-based cryptography. In: Bernstein, D.J., Buchmann, J., Dahmen, E. (eds.) Post-Quantum Cryptography, pp. 147–191. Springer, Heidelberg (2009)
33. Naehrig, M., Lauter, K., Vaikuntanathan, V.: Can homomorphic encryption be practical? In: Cachin, C., Ristenpart, T. (eds.) CCSW, pp. 113–124. ACM (2011)
34. Pointcheval, D., Johansson, T. (eds.): EUROCRYPT 2012. LNCS, vol. 7237. Springer, Heidelberg (2012)
35. van de Pol, J., Smart, N.P.: Estimating key sizes for high dimensional lattice-based systems. In: Stam (ed.) [37], pp. 290–303
36. Schnorr, C.P., Euchner, M.: Lattice basis reduction: Improved practical algorithms and solving subset sum problems. Math. Program. 66, 181–199 (1994)
37. Stam, M. (ed.): IMACC 2013. LNCS, vol. 8308. Springer, Heidelberg (2013)
38. Stehlé, D., Steinfeld, R.: Making NTRU as secure as worst-case problems over ideal lattices. In: Paterson, K.G. (ed.) EUROCRYPT 2011. LNCS, vol. 6632, pp. 27–47. Springer, Heidelberg (2011)

Towards Lattice Based Aggregate Signatures

Rachid El Bansarkhani and Johannes Buchmann

Technische Universität Darmstadt
Fachbereich Informatik
Kryptographie und Computeralgebra,
Hochschulstraße 10, 64289 Darmstadt, Germany
{elbansarkhani,buchmann}@cdc.informatik.tu-darmstadt.de

Abstract. We propose the first lattice-based sequential aggregate signature (SAS) scheme that is provably secure in the random oracle model. As opposed to factoring and number theory based systems, the security of our construction relies on worst-case lattice problems. Generally speaking, SAS schemes enable any group of signers ordered in a chain to sequentially combine their signatures such that the size of the aggregate signature is much smaller than the total size of all individual signatures. This paper shows how to instantiate our construction with trapdoor function families and how to generate aggregate signatures resulting in one single signature. In particular, we instantiate our construction with the provably secure NTRUSign signature scheme presented by Stehlé and Steinfeld at Eurocrypt 2011. This setting allows to generate aggregate signatures being asymptotically as large as individual ones and thus provide optimal compression rates as known from RSA based SAS schemes.

Keywords: Lattice-Based Cryptography, Sequential Aggregate Signatures, Aggregate Signatures.

1 Introduction

There is an inherent threat imposed by the potential production of powerful quantum computers. In the seminal work [Sho97] due to Shor, an algorithm has been presented that enables an attacker featured with a powerful quantum computer to efficiently break most of the classical schemes that base the security on the hardness of number theoretic assumptions. This theoretical result subsequently induced the search for alternatives replacing those affected classical schemes. Lattice-based cryptography constitutes a promising candidate solving most of the problems inherent to classical schemes. As opposed to the discrete log problem and factoring, lattice problems are conjectured to withstand quantum attacks, which can classically be broken in $2^{\tilde{O}(n^{1/3})}$ time. However, the best-known attacks on lattice problems still take exponential time complexity. Beside of the well-studied lattice problems and the underlying rich combinatorial structure, the worst-case to average-case reductions [Ajt96] due to Ajtai furtherly encouraged the design of lattice-based cryptographic primitives leading to conceptually new constructions such as digital signature schemes [DDLL13], [GLP12],

D. Pointcheval and D. Vergnaud (Eds.): AFRICACRYPT 2014, LNCS 8469, pp. 336–355, 2014.

[Lyu12], [GPV08],[MP12], preimage sampleable trapdoor functions [GPV08], [MP12],[AP09],[Pei10],[SS11], encryption schemes [LP11], [HPS98], [SS11], fully homomorphic encryption [BV11],[Gen09],[GSW13], oblivious transfer [PVW08] and multilinear maps [GGH+13b], [GGH13a], just to name a few examples. This breakthrough is particularly of great importance from a practical point of view as it allows to efficiently instantiate lattice-based primitives while enjoying worst-case hardness. Moreover, lattice-based schemes benefit from the simplicity of operations and handling of small integers as compared to classical cryptography, which is characterized by complex algorithms and the generation of huge primes.

In the last couples of years, much efforts have been spent on the construction of lattice-based signature schemes since digital signatures constitute one of the main building blocks in many security applications. As a result, several efficient lattice-based signature schemes arose [DDLL13], [GLP12], [Lyu12], [GPV08],[MP12], which outperform even classical schemes in terms of performance [DDLL13]. This is particularly due to the ring variant, which represents a further step towards practical lattice-based cryptography as it admits very nice constructions with small key sizes and fast processing capabilities such as Fast Fourier Transformations. In general, one differentiates the most common digital signature schemes into Schnorr-like signature schemes and the hash-and-sign approach. The first approach takes advantage of the simplicity of Schnorr-signatures and hence resulted in a sequence of improving works [DDLL13], [GLP12], [Lyu12]. The best known representative for the hash-and-sign approach, however, is given by the GPV-signature scheme in conjunction with a suitable trapdoor. The efficiency of the scheme mainly depends on the quality of the trapdoor and thus inspired the construction of new and improving trapdoors [Pei10],[MP12], [GPV08], [AP09]. Other application areas for trapdoor constructions include CCA-secure and identity based encryption [GPV08], [MP12].

Aggregate Signatures. An aggregate signature (AS) scheme enables a group of signers to combine their signatures on messages of choice such that the combined signature is essentially as large as an individual signature. Aggregate signatures have many application areas such as secure routing protocols [Lyn99] providing path authentication in networks. Moreover, ASs are important mechanisms used in constrained devices, e.g. wireless sensor networks, in order to decrease the amount of transmitted data, which in turn reduces the battery power consumption. Particularly in cluster-based sensor networks, where each cluster consists of a small number of sensor nodes, it is reasonable to apply data aggregation techniques including AS.

The first aggregate signature scheme is due to [BGLS03], which is based on the hardness of the co-Diffie-Hellman problem in the random oracle model. Following this proposal, the aggregation mechanism can be accomplished by any third party since it relies solely on publicly accessible data and the individual signature shares. Conceptually, this scheme is based on bilinear maps. In [LMRS04] Lysyanskaya et al. proposed a new variant of AS, known as sequential aggregate signatures (SAS), which differs from the conventional AS schemes by imposing an order-specific generation of aggregate signatures. In particular, each signer is ordered in a chain and receives the aggregate from its predecessor before the own signature share is added to the aggregate. A characteristical feature of this scheme is to include all previously signed messages and the corresponding public keys in the computation of the aggregate. In practice, one finds, for instance, SAS schemes applied in the S-BGP routing protocol or in certificate chains, where higher level CAs attest the public keys of lower level CAs. The generic SAS construction provided in [LMRS04] is based on trapdoor permutations with proof of security in the random oracle model. However, the SAS scheme suffers from the requirement of certified trapdoor permutations [BY96] for the security proof to go through. But the authors explain how to circumvent the need for certification in the special case of an RSA based instantiation using large exponents e as public keys. This obviously lacks efficiency due to costly exponentiations. Subsequent works provide similar solutions or improve upon existing ones, e.g. the work [BNN07] removes restrictions on the choice of messages imposed by [BGLS03]. In particular, prior to this improvement the messages were forced to be distinct.

First steps towards eliminating random oracles in the security proof were taken by Lu et al. [LOS+06], who proposed a new SAS variant that is based on bilinear pairings while providing provably security in the standard model. In a following work, the authors of [Nev08] address the requirement of certified trapdoor permutations in [LMRS04] and present a new SAS construction removing the need for certification. As a main drawback of both schemes, a potential signer has to verify the actual aggregate signature prior to any modification. This issue is investigated in the works [EFG+10] and [BGR12]. The first proposal successfully solved this problem and hence allows to omit verification beforehand when modifying the aggregate. But this approach works as the BGLS-scheme [BGLS03] in a different setting. The SAS construction with lazy verification [BGR12] has the advantage that each signer does not care about the validity of the intermediate aggregate signatures. Therefore, the messages and the corresponding public keys of precedent signers need not to be requested when generating aggregate signatures. Verification of the aggregates is delayed and can be accomplished at any time afterwards. Interestingly, Hohenberger et al. present in [HSW13] the first unrestricted aggregate signature scheme that is based on leveled multilinear maps. The underlying hardness assumptions are, nevertheless, not directly connected to worst-case lattice problems. In this work we address the question whether it is possible to build (S)AS schemes that can be based on worst-case lattice problems.

1.1   Our Results and Contribution

Sequential Aggregation of Signatures from PSFs. We present the first lattice-based sequential aggregate signature scheme that is secure in the random oracle model [BR93]. It can be instantiated by any collection of preimage sampleable trapdoor functions [GPV08], [AP09], [Pei10], [MP12], [SS11] including identity-based variants such as in [GPV08]. In fact, one can even use different types of trapdoor function families simultaneously. The security model, we adopt, is mainly influenced by [LMRS04],[BNN07], [Nev08] as it requires the scheme to withstand potential attacks even when the forger controls all but at least one secret key. Inspired by the work [Nev08] we prove by means of a sequential forger that breaking the scheme is as hard as solving hard instances of lattice problems. Specifically, we show that solving the SIS problem can be reduced to the task of successfully forging an aggregate signature. We even prove that our scheme is strongly unforgeable under chosen message attacks [Rüc10]. Interestingly, all results immediately transfer to the ring setting, since the proof is based on a more abstract notion of collision-resistant trapdoor functions [GPV08] subsuming both the matrix and ring variant. In terms of performance, the signing costs of each signer are limited to one run of the GPV signature scheme and $(i-1)$ function evaluations, where i denotes the index of signer S_i in the chain. By applying the framework of [SMP08],[BPW12] one can additionally turn any sequential aggregate signature scheme into a proxy signature scheme, where the security is based on the hardness of forging signatures in the underlying SAS scheme.

Instantiation of SAS. As mentioned before, one can principally use any PSF family that is suitable for the GPV signature scheme in order to instantiate the SAS scheme, especially the trapdoor construction presented in [MP12]. Due to the compressing property of lattice-based PSFs the range is in many cases much smaller than the domain $\log_2(B_n) > \log_2(R_n)$, where $\log_2(C) := \max_{c \in C} \lceil \log_2 c \rceil$ denotes the maximum bit size of a set C. Consequently, the size of the aggregate signature is larger than an individual one when instantiating the chain of signers with the same security parameters n and q. However, it is always possible to choose the parameters of the signers in such a way that the resulting aggregate signature has essentially the size of an individual signature. To achieve this, $\log_2(R_{n_{i+1}}) \geq \log_2(B_{n_i})$ must hold for all $1 \leq i \leq k$. Hence, the aggregate signature of any signer completely flows in the computation of the aggregate signature of the next signer in the chain. A way of measuring the quality or suitability of any PSF for use in the proposed SAS scheme is the ratio $\log_2(B_n)/\log_2(R_n)$, where a value close to 1 indicates that the sizes of the domains and ranges are of the same order. Based on this selection criterion, we instantiate our construction with the provably secure NTRUSign signature scheme [SS11]. It furtherly allows to achieve asymptotically optimal compression rates that are known from number-theory based SAS schemes such as the schemes provided in [Nev08, LMRS04]. Moreover, we discuss the potential advantages of our construction over RSA based SAS schemes. In particular, we point out that our proposal is characterized by its flexibility and simplicity of instantiation as compared to the schemes provided in [Nev08, LMRS04]. However, the compressing property of the PSFs leads inherently to some small SAS overheads, if n and q are chosen to be equal for all signers. A solution to this problem consists in selecting the parameters n_i and q_i in such a way that the signature of any signer completely flows in the signature computation of its successor.

1.2 Organization

This paper is structured as follows. In Section 3 we introduce preimage sampleable trapdoor functions and give a sketch of the GPV signature scheme. In Section 4 we provide a detailed description of our SAS construction including an analysis and a proof of security. Afterwards, in Section 5 we focus particularly on the provably secure NTRUSign trapdoor construction presented in [SS11] and give an analysis of the proposed SAS scheme instantiated with NTRUSign. We furtherly compare our setting with classical SAS schemes.

2 Preliminaries

Notation. We denote vectors by lower-case bold letters e.g. $\mathbf{x}$, whereas for matrices we use upper-case bold letters e.g. $\mathbf{A}$. Integers modulo q are denoted by $\mathbb{Z}_q$ and reals by $\mathbb{R}$. By $\vec{v}_i$ we denote a sequence of elements $v_1, \ldots, v_i$ such as vectors or bit strings.

Signature Sizes and Compression Rates. By $\mathsf{size}(\sigma_i)$ we define the size of the individual signature σ_i corresponding to signer S_i for $1 \leq i \leq k$ and $\mathsf{size}(\sigma_{SAS})$ denotes the size of the aggregate signature σ_{SAS}. We express the compression rate $\mathsf{rate}(\cdot)$ of an aggregate signature scheme by

$$\mathsf{rate}(k) = 1 - \tau(k), \quad \tau(k) = \mathsf{size}(\sigma_{SAS}) / \sum_{i=0}^{k} \mathsf{size}(\sigma_i).$$

The size ratio $\tau(\cdot)$ represents the size of the aggregate or compressed signature measured as a percentage of the total size of all individual signatures. Hence, $\mathsf{rate}(k)$ returns the storage savings due to compression or aggregation. The scheme instantiation is said to be optimal in case $\mathsf{rate}(k) = 1 - 1/k$ or $\tau(k) = 1/k$ holds, meaning that in average each signature completely flows in the computation of the next signature in the chain such that the aggregate signature has the size of an individual signature. We use these notions of optimality also for general aggregate signatures rather than for specific ones in order to make the schemes more comparable.

Continuous and Discrete Gaussians. By $\rho : \mathbb{R}^n \to (0, 1]$ we define the n-dimensional Gaussian function $\rho(\mathbf{x}) = e^{-\pi \cdot \|\mathbf{x}\|_2^2}$. It follows $E[\mathbf{x} \cdot \mathbf{x}^\top] = \frac{\mathbf{I}}{2\pi}$. Applying a linear function $\mathbf{B}$ on $\mathbf{x}$ with $\mathbf{y} = \mathbf{B}\mathbf{x}$ leads to the following Gaussian function, where $\mathbf{B}$ is a $n \times n$-matrix with linearly independent columns

$$\rho_{\mathbf{B}}(\mathbf{y}) = \rho(\mathbf{B}^{-1}\mathbf{y}) = e^{-\pi \cdot <\mathbf{B}^{-1}\mathbf{y}, \mathbf{B}^{-1}\mathbf{y}>} = e^{-\pi \cdot \mathbf{y}^\top \Sigma^{-1} \mathbf{y}}, \Sigma = \mathbf{B}^\top \mathbf{B}.$$

One derives the probability density function $f_{\sqrt{\Sigma}}(\mathbf{x}) = \frac{\rho_{\sqrt{\Sigma}}(\mathbf{x})}{\sqrt{\det \Sigma}}$ of the continuous Gaussian distribution $D_{\sqrt{\Sigma}}$ by scaling $\rho_{\sqrt{\Sigma}}$ by its total measure $\int_{-\infty}^{\infty} \rho_{\sqrt{\Sigma}} d\mathbf{x} =$

$\sqrt{\det \mathbf{\Sigma}}$. If $\mathbf{\Sigma} = s^2 \cdot \mathbf{I}$, we simply write $f_s(\mathbf{x}) = \rho_s(\mathbf{x})/s^n$. The conditional probability density function is defined by $f_{\sqrt{\mathbf{\Sigma}}}(\mathbf{x} \mid \mathbf{x} \in C) = \frac{\rho_{\sqrt{\mathbf{\Sigma}}}(\mathbf{x})/\sqrt{\det \mathbf{\Sigma}}}{P[\,C\,]/\sqrt{\det \mathbf{\Sigma}}} = \frac{\rho_{\sqrt{\mathbf{\Sigma}}}(\mathbf{x})}{P[\,C\,]}$ for $\mathbf{x} \in C \subset \mathbb{R}^n$ and $P[\,C\,] = \int_C \rho_{\sqrt{\mathbf{\Sigma}}} d\mathbf{x}$. The discrete Gaussian distribution $D_{\Lambda+c,\sqrt{\mathbf{\Sigma}}}$ is defined to have support $\Lambda + c$, where $c \in \mathbb{R}$ and $\Lambda \subset \mathbb{R}^n$ is a lattice. For $\mathbf{x} \in \Lambda + c$, it basically assigns the probability

$$D_{\Lambda+c,\sqrt{\mathbf{\Sigma}}}(\mathbf{x}) = \frac{\rho_{\sqrt{\mathbf{\Sigma}}}(\mathbf{x})}{\rho_{\sqrt{\mathbf{\Sigma}}}(\Lambda + c)}.$$

Lattice Problems. For the SIS problem we consider the full-rank m-dimensional integer lattices $\Lambda_q^{\perp}(\mathbf{A}) = \{\mathbf{x} \in \mathbb{Z}^m \mid \mathbf{A}\mathbf{x} \equiv 0 \mod q\}$ consisting of all vectors that belong to the kernel of the matrix $\mathbf{A}$. In particular, $SIS_{q,n,\beta}$ is an average-case problem of the approximate shortest vector problem on $\Lambda_q^{\perp}(\mathbf{A})$ for $\beta > 0$. Given is a uniform random matrix $\mathbf{A} \in \mathbb{Z}^{n \times m}$ with $m = poly(n)$, the problem is to find a non-zero vector $\mathbf{x} \in \Lambda_q^{\perp}(\mathbf{A})$ such that $\|\mathbf{x}\| < \beta$. For $q \geq \beta \sqrt{n} \omega(\sqrt{\log n})$ finding a solution to this problem is at least as hard as probabilistically $\tilde{O}(\beta\sqrt{n})$-approximating the Shortest Independent Vector Problem on n-dimensional lattices in the worst-case [GPV08, Ajt96]. Micciancio and Regev introduced the smoothing parameter in [MR04]:

Definition 1. *For any n-dimensional lattice Λ and positive real $\epsilon > 0$, the smoothing parameter $\eta_\epsilon(\Lambda)$ is the smallest real $s > 0$ such that $\rho_{1/s}(\Lambda^* \backslash \{0\}) \leq \epsilon$.*

3   Trapdoor Functions and the Full Domain Hash Scheme

In the following, we recall some basic definitions and properties of trapdoor functions [GPV08], that are required in our security proof. Later, we will particularly focus on collision-resistant preimage sampleable trapdoor functions (PSF) that allow any signer knowing the trapdoor to create signatures in full domain hash schemes such as the GPV signature scheme. According to [GPV08], [AP09, Pei10, MP12] there exists a polynomial-time algorithm TrapGen that on input the security parameter 1^n outputs a public key $\mathbf{A}$ and the corresponding trapdoor $\mathbf{T}$ such that the trapdoor function $f_{\mathbf{A}} : B_n \longrightarrow R_n$ can efficiently be evaluated and satisfies the following properties:

1. The output distribution of $f_{\mathbf{A}}(x)$ is uniform at random over R_n given x is sampled from the domain B_n according to SampleDom(1^n), e.g. $D_{\mathbb{Z}^n,s}$ with $s = \omega(\sqrt{\log n})$ by [GPV08, Lemma 5.2].
2. For a given syndrome $y \in R_n$, anyone knowing the trapdoor can efficiently sample preimages $x \longleftarrow$ SamplePre($\mathbf{T}, y$) such that $f_{\mathbf{A}}(x) = y$, where x is distributed as SampleDom(1^n). By the one-way property the probability to find a preimage $x \in f_{\mathbf{A}}^{-1}(y) \subseteq B_n$ of a uniform syndrome $y \in R_n$ without the knowledge of the trapdoor is negligible.

3. The conditional min-entropy property of $\mathsf{SampleDom}(1^n)$ for a given syndrome $y \in R_n$ implies that two preimages x', x distributed as $\mathsf{SampleDom}(1^n)$ differ with overwhelming property. This is due to the large conditional min-entropy of at least $\omega(\log n)$.
4. The preimage sampleable trapdoor functions are collision resistant, meaning that it is infeasible to find a collision $f_\mathbf{A}(x_1) = f_\mathbf{A}(x_2)$ such that $x_1, x_2 \in B_n$ and $x_1 \neq x_2$.

Due to the results of [Ajt99, GPV08] a lot of research has been made on the construction of preimage sampleable trapdoor functions in recent years resulting in a sequence of improving works [GPV08, AP09, Pei10, MP12]. These constructions often satisfy these properties only statistically, meaning that the statistical distance between the claimed distributions and the provided ones are negligible. As a result, the security proofs of cryptographic schemes involving concrete constructions hold only statistically, which is quite enough for practice. One cryptographic scheme is the GPV-signature scheme, which is secure in the random oracle model and exploits the properties of collision-resistant trapdoor functions. Furthermore, it is stateful, meaning that it does not generate new signatures for messages already signed. This can be attributed to the fact that a potential attacker could otherwise use two signatures of the same message in order to construct an element of the kernel, which solves $SIS_{q,n,\beta}$ and hence allows the attacker to provide a second preimage of any message. One can remove the need for storing message and signature pairs by employing the probabilistic approach [GPV08], where the signer samples an extra random seed, which is appended to the message when calling $H(\cdot)$.

Specifically, the GPV signature scheme consists mainly of sampling a preimage from a hash function endowed with a trapdoor:

KeyGenGPV(1^n) On input 1^n the algorithm $\mathsf{TrapGen}(1^n)$ outputs a key pair $(\mathbf{A}, \mathbf{T})$, where $\mathbf{T}$ is the secret key or trapdoor and $\mathbf{A}$ is the public key describing the preimage sampleable trapdoor function $f_\mathbf{A}$.

SignGPV$(\mathbf{T}, m)$ The signing algorithm computes the hash value $H(m)$ of the message m and looks up $H(m)$ in its table, where $H(\cdot)$ is modeled as a random oracle. If it finds an entry, it outputs σ_m. Otherwise, it samples a preimage $\sigma_m \leftarrow \mathsf{SamplePre}(\mathbf{T}, H(m))$ of $H(m)$ and outputs σ_m as the signature.

VerifyGPV(σ, m) The verification algorithm checks the satisfaction of $H(m) = f_\mathbf{A}(\sigma)$ and $\sigma \in B_m$. If both conditions are valid, it outputs 1, otherwise 0.

Probabilistic Full-Domain Hash Scheme

The probabilistic approach additionally requires the signer to generate a random seed r (e.g. $r \in \{0,1\}^n$) which is appended to the message m. Doing this, we can sign the same message m several times, since the r-part always differs except with negligible probability. Thus, we can consider $m\|r$ as the extended message to be signed.

4 Sequential Aggregate Signatures from Lattices

4.1 Our Basic Signature Scheme

We aim at constructing a SAS scheme from trapdoor functions instead of trap-door permutations. Inspired by the works of [Nev08, LMRS04] we transfer the core ideas from trapdoor permutations to the lattice-based setting. The main obstacle that needs to be handled is the fact that lattice-based trapdoor functions operate in differing domains and ranges. The input bit size is usually larger than the output one. This is due to the need of collisions. Therefore, we use the encoder-technique enc proposed by [Nev08] that takes these properties into account. In particular, it breaks the signature down into two parts, where the first part is injectively mapped to an element of the image space and flows then in the computation of the modified aggregate signature. The second part is simply handed over to the next signer in the chain. The encoder-technique is originally designed to allow for hiding of additional data like messages in RSA like systems in order to decrease not only the signature sizes but also the total amount of data to be send. The following algorithms provide the main steps of the SAS scheme.

Algorithm 1. $\mathsf{AggSign}(\mathbf{T_i}, m_i, \Sigma_{i-1})$: Signing algorithm of the i-th signer	**Algorithm 2.** $\mathsf{AggVerify}(\Sigma_k)$: Verification algorithm

Data: Trapdoor $\mathbf{T_1}$, message m_i, Σ_{i-1}
 1 **if** $i = 1$ **then**
 2    $\Sigma_0 \leftarrow (\epsilon, \epsilon, \epsilon, 0^n)$;
 3 **end**
 4 Parse Σ_{i-1} as
    $(\overrightarrow{\mathsf{f_{A_{i-1}}}}, \overrightarrow{m}_{i-1}, \overrightarrow{\alpha}_{i-1}, \sigma_{i-1}, h_{i-1})$
 5 **if** $\mathsf{AggVerify}(\Sigma_{i-1}) == (\bot, \bot)$ **then**
 6   **return** $\bot$
 7 **end**
 8 $(\alpha_i, \beta_i) \leftarrow \mathsf{enc}_{\mathsf{f_{A_i}}}(\sigma_{i-1})$
 9 $\overrightarrow{\alpha}_i \leftarrow \overrightarrow{\alpha}_{i-1} \mid \alpha_i$
 10 $h_i \leftarrow h_{i-1} \oplus \mathbf{H}(\overrightarrow{\mathsf{f_{A_i}}}, \overrightarrow{m}_i, \overrightarrow{\alpha}_{i-1}, \sigma_{i-1})$
 11 $g_i \leftarrow \mathbf{G}_{\mathsf{f_{A_i}}}(h_i)$
 12 $\sigma_i \leftarrow \mathsf{SamplePre}(\mathbf{T_i}, g_i + \beta_i)$
 13
 14 **Return** $\Sigma_i \leftarrow (\overrightarrow{\mathsf{f_{A_i}}}, \overrightarrow{m}_i, \overrightarrow{\alpha}_i, \sigma_i, h_i)$
	Data: Σ_k
	 1 Parse Σ_k as $(\overrightarrow{\mathsf{f_{A_k}}}, \overrightarrow{m}_k, \overrightarrow{\alpha}_k, \sigma_k, h_k)$
	 2 **for** $i = k \to 1$ **do**
	 3 **if** $\log_2(R_{n_i}) \leq l$ *or* $\sigma_i \notin B_{n_i}$
	    then
	 4 **return** $(\bot, \bot)$
	 5 **end**
	 6 $g_i \leftarrow \mathbf{G}_{\mathsf{f_{A_i}}}(h_i)$
	 7 $\beta_i \leftarrow \mathsf{f}_i(\sigma_i) - g_i$
	 8 $\sigma_{i-1} \leftarrow \mathsf{dec}_{\mathsf{f_{A_i}}}(\alpha_i, \beta_i)$
	 9 $h_{i-1} =$
	      $h_i \oplus \mathbf{H}(\overrightarrow{\mathsf{f_{A_i}}}, \overrightarrow{m}_i, \overrightarrow{\alpha}_{i-1}, \sigma_{i-1})$
	 10 **end**
	 11 **if** $\Sigma_0 = (\epsilon, \epsilon, \epsilon, 0^n)$ **then**
	 12 **return** $(\overrightarrow{\mathsf{f_{A_k}}}, \overrightarrow{m}_k)$
	 13 **end**
	 14 **Else return** $(\bot, \bot)$

Fig. 1. Signing AggSign and Verification algorithm AggVerify of the SAS scheme

Definition 2. *In a sequential aggregate signature (SAS) scheme k distinct sign-ers, that are ordered in a chain, sequentially put their signature on messages of choice to the aggregate signature such that the resulting aggregate signature has the size of an individual signature.*

Definition 3. *We say x-SAS, if the aggregate signature is essentially an individual signature extended by c bits of overhead with $c = (1 - x) \cdot \sum_i size(\sigma_i)$ bits and $x \in [0, 1]$.*

For $x = 1$, we immediately obtain the aforementioned definition of the classical SAS scheme.

4.2   Informal Description

We give an informal description of the sequential aggregate signature scheme. We focus on the signing and verification steps in Figure 1. Each signer S_i in the chain with $1 \leq i \leq k$ follows the same protocol steps. Let $l > 0$ be a public parameter such that $\log_2(R_{n_i}) > l$ for $1 \leq i \leq k$, where R_{n_i} denotes the image space of the trapdoor function $f_{\mathbf{A}_i} : B_{n_i} \to R_{n_i}$ and $\log_2(R_{n_i})$ defines the maximum number of bits needed to represent elements of R_{n_i}.

The input to the signing algorithm $\mathsf{AggSign}(\cdot)$ of the i-th signer S_i is its secret key $\mathbf{T}_i$, the message to be signed and a list of data Σ_{i-1} received from signer S_{i-1}. If S_i corresponds to the first signer, the list of data is empty. Otherwise, Σ_{i-1} parses as a list consisting of a sequence of trapdoor functions $f_{\mathbf{A}_1}, \ldots, f_{\mathbf{A}_{i-1}}$ identified by the public keys $\mathbf{A}_1, \ldots, \mathbf{A}_{i-1}$, a sequence of messages $m_1, \ldots, m_{i-1}$, parts of the encoded signatures $\vec{\alpha}_{i-1}$ from signers S_1 to S_{i-2}, an aggregate signature σ_{i-1} of the predecessor S_{i-1} and a hash value $h_{i-1} \in \{0,1\}^l$. Before adding its own signature to Σ_i, the signer checks the validity of the received aggregate by running the verification algorithm on Σ_{i-1}. If the verification succeeds, S_i continues by invoking the encoder on σ_{i-1} resulting in a breakdown (α_i, β_i). The encoder $enc : \{0,1\}^* \to \{0,1\}^* \times R_{n_i}$ is an injective map that splits up the signature into two parts such that β_i can completely be embedded in the computation of σ_i and can always be recovered. The second part α_i is simply appended to the list $\vec{\alpha}_{i-1}$ and plays an important role when recovering the intermediate aggregate signatures. We give a particular instantiation of the proposed splitting algorithm in Section 5. The next two steps involve two hash functions $H : \{0,1\}^* \to \{0,1\}^l$ and $G_{f_{\mathbf{A}_i}} : \{0,1\}^l \to R_{n_i}$ which are modeled as random oracles. Similar to [BR96], $H(\cdot)$ is considered the compressor that hashes the message down to l bits, whereas $G_{f_{\mathbf{A}_i}}(\cdot)$ is called the generator and outputs random elements from the image space of $f_{\mathbf{A}_i}$. Regarding the proof of security, such a construction involving $H(\cdot)$ and $G(\cdot)$ avoids the need for certified trapdoor functions satisfying the properties specified in Appendix 3. By this means, one gets rid of costly checks, because a potential adversary could generate keys leaving out one of these properties. Finally, the algorithm outputs Σ_i containing the modified aggregate signature. The verification algorithm $\mathsf{AggVerify}(\cdot)$ proceeds in the reverse order and takes Σ_n as input. In each iteration it checks the validity of σ_i and recovers σ_{i-1} with the aid of the decoder $\mathsf{dec}(\cdot)$.

In the following section we present the security model of our scheme including the associated security proof. Subsequently, we show how to instantiate the scheme with PSFs. To this end, we focus on the provably secure NTRUSign preimage sampleable trapdoor function and provide a comparison with RSA-based SAS schemes as proposed in [LMRS04, Nev08]. Finally, we indicate how to build a proxy signature scheme from any SAS scheme.

4.3   Security Model of SAS

We adopt the security model proposed by Neven [Nev08] for sequential aggregate signatures. Moreover, we examine our lattice-based construction in a slightly different setting that is build upon a stronger security assumption $\mathsf{Exp}^{SSAS-SU-CMA}$ and subsumes the former ones [Nev08, LMRS04]. Usually, a sequential aggregate signature scheme is considered to be secure, if it is infeasible to provide existential forgeries of a sequential aggregate signature. The core idea behind these security models is to let the forger F control the private keys and sequential aggregate signatures of all but at least one honest signer. Thus, the forger is allowed to select the public keys of the fake signers. Neven introduces the notion of a sequential forger S that can be built from a forger F with about the same success probability and running time [Nev08, Lemma 5.3]. Therefore, it is more convenient to consider a sequential forger in our proof of security. The way the sequential forger is constructed out of F can directly be transferred to our setting with some minor changes. In the full version of this paper we will provide a detailed description of this transformation. The properties of a sequential forger are as follows:

1. Any input to the random oracles $H(\cdot)$ and $G_{f.}(\cdot)$ is queried once, where $f.$ denotes any preimage sampleable trapdoor function. The signing oracle OAggSign is also queried once for the same input.
2. Each input Q_n to $H(\cdot)$ parses as $Q_n = (\overrightarrow{f_{\mathbf{A}_k}}, \overrightarrow{m_k}, \overrightarrow{\alpha}_{k-1}, \sigma_{k-1})$ such that $\log_2(R_{n_i}) > l$ holds for $1 \leq i \leq k$ and $k \leq k_{max}$.
3. Before any query $Q_k = (\overrightarrow{f_{\mathbf{A}_k}}, \overrightarrow{m_k}, \overrightarrow{\alpha}_{k-1}, \sigma_{k-1})$ to $H(\cdot)$ for $n > 1$, the sequential forger must have made queries Q_i to $H(\cdot)$ for $1 \leq i < k \leq k_{max}$ such that $\mathsf{dec}_{f_{\mathbf{A}_i}}(\alpha_i, f_{\mathbf{A}_i}(\sigma_i) - G_{f_{\mathbf{A}_i}}(h_i)) = \sigma_{i-1}$ for $h_i = h_{i-1} \oplus H(Q_i)$.
4. Preceding any signing query $\mathsf{OAggSign}(\mathbf{T}^*, m_k, \Sigma_{k-1})$ the sequential forger must have made the necessary $H(\cdot)$ and $G_{f.}(\cdot)$ queries in advance with due regard to Property 3. Furthermore, the input query Σ_{k-1} must be valid such that verification algorithm $\mathsf{AggVerify}(\Sigma_{k-1})$ does not fail.
5. Forgeries output by S must be valid and include the challenge public key at some index i such that $f_{\mathbf{A}_i} = f_{\mathbf{A}^*}$ for $1 \leq i \leq k \leq k_{max}$. We explicitly allow S to output forgeries on data Σ_{i-1} that has been signed by the signing oracle. The only required restriction is that the signing oracle responses and the forgery must differ on the same input.

According to an adaptive chosen-message attack we permit S to make arbitrary many sequential aggregate signature queries to the honest signer on messages of its choice. The advantage $\mathsf{AdvAggSign}^*_S$ of S is the success probability in the following experiments.

Setup

The key generation algorithm is invoked in order to produce the challenge key pair $(\mathbf{T}^*, \mathbf{A}^*)$. The challenge key $f_{\mathbf{A}^*}$ is then handed over to the sequential aggregate forger $\mathcal{S}$.

Queries

The adversary $\mathcal{S}$ has access to the signing oracle $\mathsf{OAggSign}(\mathbf{T}^*, *, *)$. $\mathcal{S}$ acts adaptively and provides to the signing oracle a message m_i to be signed, a sequential aggregate signature σ_{i-1} on a sequence of messages $m_1, \ldots, m_{i-1}$ and data $\alpha_1, \ldots, \alpha_{i-1}$ under public keys $f_{\mathbf{A}_1}, \ldots, f_{\mathbf{A}_{i-1}}$. The oracle returns an aggregate signature under the challenge public key $f_{\mathbf{A}^*}$. Furthermore, we allow $\mathcal{S}$ to have random oracle access to some random functions as required in the random oracle model.

Response

$\mathcal{S}$ eventually outputs a sequential aggregate signature σ_k on k distinct public keys $f_{\mathbf{A}_1}, \ldots, f_{\mathbf{A}_k}$, where one of them corresponds to the challenge key. Moreover, $\mathcal{S}$ outputs k messages $m_1, \ldots, m_k$, each corresponding to one public key.

The forger wins the game $\mathsf{Exp}_{\mathcal{A}, SAS}^{SSAS-EU-CMA}(n)$, if he succeeds in outputting a non-trivial valid sequential signature on a sequence of k messages $m_1, \ldots, m_k$ under k distinct public keys $f_{\mathbf{A}_1}, \ldots, f_{\mathbf{A}_k}$ containing the challenge public key $f_{\mathbf{A}_i} = f_{\mathbf{A}^*}$ at some index i. A valid signature is said to be non-trivial, when $\mathcal{S}$ has never made a query to the signing oracle on messages $m_1, \ldots, m_i$ and public keys $f_{\mathbf{A}_1}, \ldots, f_{\mathbf{A}_i}$ before, or he is able to output a forgery that differs from the received signing oracle responses. In the latter case we even allow the forger to use already signed messages to output a forgery as opposed to the security models from [Nev08, LMRS04] focussing on trapdoor permutations. This security notion reflects the strong sequential aggregate signature unforgeability (SAS-SU-CMA), which can be formalized as follows.

Experiment $\mathsf{Exp}_{\mathcal{A}, SAS}^{SSAS-SU-CMA}(n)$
  $(\mathbf{T}^*, \mathbf{A}^*) \longleftarrow \mathsf{KeyGen}(1^n)$
  $\Sigma = (\overrightarrow{f_{\mathbf{A}_i}}, \overrightarrow{m_i}, \overrightarrow{\alpha_i}, \sigma_i, h_i) \longleftarrow \mathcal{A}^{\mathsf{OAggSign}(\mathbf{T}^*, *, *)}(\overrightarrow{f_{\mathbf{A}^*}})$
  Let $f_{\mathbf{A}_i} = f_{\mathbf{A}^*}$ be the challenge public key in $\overrightarrow{f_{\mathbf{A}_i}} = (f_{\mathbf{A}_1}, \ldots, f_{\mathbf{A}_i})$ and
  $\overrightarrow{m_i} = (m_1, \ldots, m_i)$
  Let $((f_{\mathbf{A}_l}, m_l, \Sigma_{l-1}), \Sigma_l)_{l=1}^{Q_{AS}}$ be query-response tuples of $\mathsf{OAggSign}(\mathbf{T}^*, *, *)$
  Return 1 if $\mathsf{AggVerify}(\Sigma) = (\overrightarrow{f_{\mathbf{A}_i}}, \overrightarrow{m_i})$
      and $\Sigma \notin \{\Sigma_l\}_{l=1}^{Q_{AS}}$

The adversary is said to be successful in this experiment if he efficiently provides a valid sequential aggregate signature with non-negligible advantage.

4.4 Security of Our Construction

A collision-finding algorithm $\mathcal{A}$ is said to (t', ϵ')-break a collision-resistant preimage sampleable trapdoor function family (PSF) if it has running time t' and outputs a collision with probability

$$Pr[f_{\mathbf{B}}(x_1) = f_{\mathbf{B}}(x_2) \mid (\mathbf{B}, \mathbf{T}) \leftarrow \mathsf{TrapGen}(1^n), (x_1, x_2) \leftarrow \mathcal{A}(f_{\mathbf{B}})]$$

of at least ϵ'.

Proposition 1. *If there exists a sequential forger S that $(t, q_S, q_H, q_G, k_{max}, \epsilon)$-breaks SAS, then there exists a collision-finding algorithm $\mathcal{A}$ that (t', ϵ')-breaks the collision-resistant PSF for*

$$\epsilon' \geq \epsilon \cdot (1 - \frac{q_H(q_H + q_G)}{2^l}) - \frac{q_H}{2^{\omega(\log n)}}$$

$$t' \leq t + (q_H + k_{max}) \cdot t_{f.} + q_H \cdot t_{\mathsf{SampleDom}}.$$

Proof: By contradiction, we assume that there exists a successful sequential forger S that breaks the SAS with non-negligible probability ϵ. Using S, we construct a poly-time algorithm $\mathcal{A}$ that finds a collision in the collision-resistant trapdoor function $f_{\mathbf{A}_i} : B_{n_i} \longrightarrow R_{n_i}$ with probability negligibly close to ϵ. Given the challenge public key $\mathbf{A}^*$ of the trapdoor function $f_{\mathbf{A}^*}$, $\mathcal{A}$ runs S on public key $\mathbf{A}^*$ with $f_{\mathbf{A}^*} : B_n \longrightarrow R_n$ and simulates the environment as follows:

Setup : At the beginning of this game algorithm $\mathcal{A}$ sets up the empty lists $HT[*]$ and $GT[*, *]$.

H-Random oracle query $H(Q_i)$: After parsing the input Q_i as $(\overrightarrow{f_{\mathbf{A}_i}}, \overrightarrow{m}_i, \overrightarrow{\alpha}_{i-1}, \sigma_{i-1})$, $\mathcal{A}$ checks the index i. If $i = 1$, $\mathcal{A}$ sets $h_0 \leftarrow 0^n$. In case $i > 1$, following Property 3 of a sequential forger there exists a unique sequence of random oracle queries $Q_1, \ldots, Q_{i-1}$ with table entry $H(Q_{i-1}) = (\sigma_{i-1}, h_{i-1})$. If the public key $f_{\mathbf{A}_i}$ does not correspond to the challenge public key $f_{\mathbf{A}^*}$, then $\mathcal{A}$ continues as follows:
  $- h \xleftarrow{\$} \{0,1\}^l$, $h_i = h \oplus h_{i-1}$ and sets $HT[Q_i] \leftarrow (\perp, h_i)$.

Otherwise, if $f_{\mathbf{A}_i} = f_{\mathbf{A}^*}$, then $\mathcal{A}$ performs the following tasks:
  $- h \xleftarrow{\$} \{0,1\}^l$, $h_i = h \oplus h_{i-1}$
  $- (\alpha_i, \beta_i) \leftarrow \mathsf{enc}_{f_{\mathbf{A}_i}}(\sigma_{i-1})$
  $- \sigma_i \leftarrow \mathsf{SampleDom}(1^n)$ and compute $g \leftarrow f_{\mathbf{A}^*}(\sigma_i) - \beta_i \in R_{n_i}$, since R_{n_i} is additive.
  (By Property 1, $f_{\mathbf{A}^*}(\sigma_i) \sim \mathcal{U}(R_{n_i})$)
  If $G_{f_{\mathbf{A}^*}}(h_i)$ has not been defined, he sets $G[f_{\mathbf{A}^*}, h_i] \leftarrow g$, $HT[Q_i] \leftarrow (\sigma_i, h_i)$ and outputs h to S, otherwise BAD_1 occured and $\mathcal{A}$ aborts.

G-Random oracle query 1 $G_{f_{\mathbf{A}_i}}(h)$: On input $f_{\mathbf{A}_i}$ and h algorithm $\mathcal{A}$ checks the entry $GT[f_{\mathbf{A}_i}, h]$. If it is not defined, it selects $g \xleftarrow{\$} R_{n_i}$ uniformly at random, sets $GT[f_{\mathbf{A}_i}, h] = g$ and returns g to S. By Property 1 of a sequential forger it does not make the same query again.

Sequential signing query $\mathsf{OAggSign}(\mathbf{T}^*, m_i, \Sigma_{i-1})$: $\mathcal{A}$ extracts the values $\overrightarrow{f_{\mathbf{A}_{i-1}}}$, $\overrightarrow{m}_{i-1}$, $\overrightarrow{\alpha}_{i-1}$, σ_{i-1}, h_{i-1} from Σ_{i-1}. As per Property 4 he finds a non-empty entry $HT[Q_{i-1}] = (\sigma_i, h_i)$ with $Q_{i-1} = (\overrightarrow{f_{\mathbf{A}_{i-1}}} \mid f_{\mathbf{A}^*}, \overrightarrow{m}_{i-1} \mid m_i, \overrightarrow{\alpha}_{i-1}, \sigma_{i-1})$. Then $\mathcal{A}$ returns $\Sigma_i = (\overrightarrow{f_{\mathbf{A}_i}}, \overrightarrow{m}_i, \overrightarrow{\alpha}_i, \sigma_i, h_i)$ with $(\alpha_i, \beta_i) \leftarrow \mathsf{enc}_{f_{\mathbf{A}_i}}(\sigma_{i-1})$

Finally, the forger $\mathcal{S}$ outputs a valid forgery $\Sigma'_k = (\overrightarrow{f_{\mathbf{A}_k}}, \overrightarrow{m}_k, \overrightarrow{\alpha}_k, \sigma_k)$ with probability ϵ as per property 5. Since Σ'_k is valid, we have $\mathsf{AggVerify}(\Sigma'_k) = (\overrightarrow{f_{\mathbf{A}_k}}, \overrightarrow{m}_k)$ and $\overrightarrow{f_{\mathbf{A}_k}}$ includes the challenge public key $f_{\mathbf{A}_i} = f_{\mathbf{A}^*}$ at index $1 \le i \le k$. During the execution of $\mathsf{AggVerify}(\Sigma'_k)$ we get m_i and Σ'_i containing σ'_i. $\mathcal{A}$ proceeds as follows in order to obtain a collision. We now have to differ two cases:

1. If $\mathcal{S}$ already made a signature query on (m_i, Σ_{i-1}), it received back Σ_i containing the signature σ^*. Since Σ'_k is a forgery, we have $\sigma'_i \ne \sigma^*$ such that $f_{\mathbf{A}^*}(\sigma'_i) = f_{\mathbf{A}^*}(\sigma^*)$

2. In the case, $\mathcal{S}$ did not request a signature on (m_i, Σ_{i-1}) from the signing oracle, by Property 4 there exists an entry $HT[Q_{i-1}] = (\sigma^*, h_i)$ with $\sigma^* \longleftarrow \mathsf{SampleDom}(1^n)$ and $GT[f_{\mathbf{A}^*}, h_i] = g_i$ such that $f_{\mathbf{A}^*}(\sigma^*) = g_i + \beta_i = f_{\mathbf{A}_i}(\sigma'_i)$ and $h = h_i \oplus h_{i-1}$ is returned to $\mathcal{S}$. If $\sigma^* = \sigma'_i$, then BAD_2 occured and $\mathcal{A}$ aborts.

In both cases $\mathcal{A}$ found a collision in $f_{\mathbf{A}^*}$ (which is infeasible according to Property 4 of trapdoor functions).

Analysis and Security: We define by $\neg BAD_i$ the event that BAD_i does not occur. $\mathcal{S}$'s environment is perfectly simulated as in the real system, when the events BAD_1 and BAD_2 do not occur. Thus, we have

$$P[\mathcal{S} \text{ ouputs forgery} \mid \neg BAD_1 \wedge \neg BAD2] = \epsilon.$$

$\mathcal{A}$ wins the game when $\mathcal{S}$ succeeds in providing a valid forgery and the events BAD_1 and BAD_2 do not happen. Therefore, we need to estimate an upper bound for the probability of a successful forger:

$$P[\mathcal{A} \text{ wins}] = \epsilon \cdot P[\neg BAD_1] - P[BAD_2].$$

$$
\begin{aligned}
P[\mathcal{A} \text{ wins}] &= P[\mathcal{S} \text{ outp. forgery} \wedge \neg BAD_1 \wedge \neg BAD_2] \\
&= P[\mathcal{S} \text{outp. forgery} \mid \neg BAD_1 \wedge \neg BAD_2] \cdot P[\neg BAD_1 \wedge \neg BAD_2] \\
&= P[\mathcal{S} \text{ outp. forgery} \mid \neg BAD_1 \wedge \neg BAD_2] \cdot (1 - P[BAD_1 \vee BAD_2]) \\
&\ge P[\mathcal{S} \text{ outp. forgery} \mid \neg BAD_1 \wedge \neg BAD_2] \cdot (1 - \sum_i P[BAD_i]) \\
&= P[\mathcal{S} \text{ outp. forgery} \mid \neg BAD_1 \wedge \neg BAD_2] \cdot (P[\neg BAD_1] - P[BAD_2]) \\
&\ge P[\mathcal{S} \text{ outp. forgery} \mid \neg BAD_1 \wedge \neg BAD_2] \cdot P[\neg BAD_1] - P[BAD_2] \\
&= \epsilon \cdot P[\neg BAD_1] - P[BAD_2]
\end{aligned}
$$

The event BAD_1 occurs when algorithm $\mathcal{A}$ chooses a fresh random value $h \xleftarrow{\$} \{0,1\}^l$ in the H-Random oracle query step and attempts to set a table entry $GT[*, h_k]$ that is already defined, where $h_k = h \oplus h_{k-1}$. The probability of this event is

$$P[BAD_1] = \frac{|GT|}{2^l} \leq \frac{q_H(q_H + q_G)}{2^l}$$

where the last term follows by summation over all H-queries to the simulation. The event BAD_2 occurs when the forger $\mathcal{S}$ outputs a valid forgery σ_i' that is equal to the corresponding table entry $HT(Q_{i-1}) = (\sigma^*, *)$. Based on the conditional min-entropy property of σ^* given $f_{\mathbf{A}^*}(\sigma^*)$ the probability of BAD_2 to happen is

$$P[BAD_2] \leq \frac{q_H}{2^{\omega(\log n)}},$$

which is negligible. Therefore, we obtain

$$\epsilon' \geq \epsilon \cdot (1 - \frac{q_H(q_H + q_G)}{2^l}) - \frac{q_H}{2^{\omega(\log n)}} \, .$$

We derive an upper bound for the running time of $\mathcal{S}$ taking into account only function evaluations and invocations of SampleDom. Each verification requires at most k_{max} function evaluations. Invoking $H(\cdot)$ implies at most one execution of SampleDom and two function evaluations, thereof one evaluation to identify the sequence $Q_{k-1}, \dots, Q_1$. Therefore, the running time is upper bounded by:

$$t' \leq t + (2q_H + k_{max}) \cdot t_f + q_H \cdot t_{\mathsf{SampleDom}} \, .$$

$\square$

Proposition 2. *The proposed sequential aggregate signature scheme is strongly existentially unforgeable under chosen-message attack.*

Proof. By Proposition 1 finding collisions for preimage sampleable trapdoor functions can be reduced to the hardness of forging sequential aggregate signatures in the SAS described above. The authors of [GPV08] give the corresponding algorithms of how to instantiate preimage sampleable trapdoor functions by means of lattices satisfying the required properties and show by [GPV08, Theorem 5.9] that the task of finding collisions is as hard as solving $SIS_{q,n,2s\sqrt{m}}$. $\square$

The security proof of the unstateful probabilistic FDH scheme is almost identical to the stateful one. One notices, that the extended message $m\|r$ to be signed always differs for repeated request queries on the same message m due to the random salt r. As in Proposition 2 one reduces collision-resistance to the unforgeability of sequential aggregate signatures.

5   Instantiation

In general, one can use any collision-resistant trapdoor function that is suitable for the GPV signature scheme. In particular, one can instantiate the SAS scheme with the trapdoor constructions from [GPV08, AP09, Pei10, MP12]. In this section we analyze the proposed sequential aggregate signature scheme in conjunction with NTRUSign.

Therefore, let $f_{A_i} : B_{n_i} \longrightarrow R_{n_i}, 1 \leq i \leq k$ be a family of preimage-sampleable trapdoor functions, each corresponding to the public key A_i of signer S_i. B_{n_i} denotes the domain of the trapdoor function f_{A_i} and can be represented by vectors of bit size $\log_2(B_{n_i}) := \max_{b \in B_i} \lceil \log_2 b \rceil$. Analogously, one defines the maximum bit size $\log_2(R_{n_i})$ of the image space. For instance, if we choose $B_{n_i} = \{z \in \mathbb{Z}^{m_i} \mid \|z\| \leq s_i \sqrt{m_i}\}$ and $R_{n_i} = \mathbb{Z}_{q_i}^{n_i}$, we have $\log_2(R_{n_i}) = n_i \cdot \lceil \log_2(q_i) \rceil$ and respectively $\log_2(B_{n_i}) = m_i \cdot (\lceil \log_2(4.7 \cdot s_i) + 1 \rceil)$ with overwhelming probability. The encoding function $\mathsf{enc}(\cdot)$ can, therefore, be built as follows. The range R_{n_i} is converted into a large bit string that is subsequently split into blocks of size $\lceil \log_2(4.7 \cdot s_i) + 1 \rceil$ bits. Each block is then filled with an entry from the signature σ_i. There are many possibilities to handle the last block as it may contain less bits. Finally, the bit string is converted back to the vector presentation β_{i+1}. The remaining signature bits are stored in the vector α_{i+1}, which is appended to the aggregate signature.

Security and Performance. The bit security of this scheme mainly depends on the bit security of each signers key and the system parameter l. Hence, the security of our construction is upper bounded by $\min_{1 \leq i \leq k}(c_i, l)$, where c_i denotes the bit security of the i-th signer. To determine the performance we ignore all operations beside function evaluations and preimage samplings. The signing costs of the i-th signer amount to one call of $\mathsf{SamplePre}(\cdot)$ and $(i - 1)$ function evaluations f_A. Verification requires k function evaluations.

In what follows, we will focus particularly on the trapdoor construction provided in [SS11] since it has some nice properties which can be utilized in the proposed SAS construction. A crucial factor for our choice is a low ratio $\log_2(B_{n_i}) / \log_2(R_{n_i})$ as compared to other lattice-based PSFs. This ratio implicitly affects the compression rate, since a ratio equal to or smaller than 1 implies optimal compression rates for equal parameters n_i, meaning that signatures completely fit into the image space without wrapping around.

Efficient Instantiation with NTRUSign [SS11]. The provably secure NTRU-Sign signature scheme proposed by Stehlé et al. [SS11] is a full domain hash scheme satisfying the properties of collision-resistant PSFs from Section 3.

$\mathsf{KeyGenGPV}(q, n, 1^n)$ It returns public key $A = g/f \in R_q^\times$ and trapdoor $T = \begin{bmatrix} f & g \\ F & G \end{bmatrix}$ for $f_A(\sigma^{(1)}, \sigma^{(2)}) = A\sigma^{(1)} - \sigma^{(2)}$, where $f_A : B_n \to R_n = R_q$ with $B_n = \{(\sigma^{(1)}, \sigma^{(2)}) \in R^2 : \|(\sigma^{(1)}, \sigma^{(2)})\| \leq s \cdot \sqrt{2n}\}$.

$\mathsf{SignGPV}(T, m)$ The signing algorithm computes the hash value $H(m\|r)$ of the extended message $m\|r$ with a random seed $r \xleftarrow{\$} U(\{0,1\}^d)$. Then it samples $\sigma = (\sigma^{(1)}, \sigma^{(2)}) \leftarrow \mathsf{SamplePre}(T, H(m\|r))$ and outputs $(r, \sigma^{(1)})$ as the signature.

VerifyGPV(σ, m) The verification algorithm computes $t = H(m\|r)$ and determines $\sigma^{(2)} = \mathbf{A}\sigma^{(1)} - t$. If the conditions $\sigma \in B_n$ and $r \in \{0,1\}^d$ are valid, it outputs 1, otherwise 0.

When instantiating the SAS scheme with this trapdoor construction, we obtain compression factors of about 60 % for practical parameters. For the sake of simplicity, assume we have public keys $\mathbf{A}_i \in R_q$ with identical parameters q, $R_q = \mathbb{Z}_q[X]/(X^n + 1)$ for n a power of two, which is obviously different from RSA where the moduli $N = p \cdot q$ have to be distinct since otherwise they would share the same secret. An NTRUSign signature is a vector $(r, \sigma^{(1)}, \sigma^{(2)})$ such that the bit size of $\sigma^{(j)}$ is bounded by $n \cdot (\lceil \log_2(4.7 \cdot s) \rceil + 1) < \log_2(R_n)$ with overwhelming probability and $r \in \{0,1\}^n$. Any vector of the image space occupies at most $\log_2(R_n) = n \cdot \lceil \log_2(q) \rceil$ bits of memory. In general, one can use Algorithm 1 and 2 in order to instantiate the NTRUSign SAS scheme. Since we consider the probabilistic FDH approach using a random seed r, one simply replaces messages m_i by the extended messages $m_i\|r_i$.

5.1 Comparison with RSA Based Sequential Aggregate Signatures

RSA based sequential signatures due to [LMRS04, Nev08] are less flexible compared to the proposed construction. In particular, the public keys $N_i = p_i \cdot q_i$ of RSA based instances have to be distinct and satisfy more restrictive conditions in order to make the scheme work. For example in [LMRS04], the hash space of $H(\cdot)$ requires to be specified before starting aggregation. This can be attributed to the differing domains $\mathbb{Z}_{N_i}^\times$ as a result of different moduli N_i. For instance, the hash space is chosen to be a proper subset of $\mathbb{Z}_{N_1}^\times$. However, this is not the case in our construction, since we can use equal domains and ranges without incurring security. Thus, one allows the corresponding hash functions $\mathbf{G}_{f_{\mathbf{A}_i}}(\cdot)$ to be equal. In order to achieve high compression without blowing up the aggregate signatures too much, the bit sizes of public keys have to be identical or are ordered to be increasing in RSA based SAS schemes. This is due to the fact that the signatures are uniform random elements in $\mathbb{Z}_{N_i}^\times$ and can only be fully embedded in $\mathbb{Z}_{N_{i+1}}^\times$, if $b_i \leq b_{i+1}$ or $N_i < N_{i+1}$ is satisfied for $b_i = \lceil \log(N_i) \rceil$ and $1 \leq i \leq k$. Indeed, this also holds for lattice-based constructions. Specifically, one has to increase the parameters n_{i+1} or q_{i+1} such that $\log_2(D_{n_i}) \leq \log_2(R_{n_{i+1}})$. By this, we have aggregate signatures being as large as individual ones.

5.2 Analysis

We want to derive a measure for the quality of the SAS scheme. Therefore, we consider the compression rate measuring the storage savings due to the SAS scheme. We simply relates the bit size of the aggregate signature to the total size of all individual signatures, which corresponds to the case one does not employ SAS schemes. By [SS11, Theorem 4.2] an NTRUSign signature is distributed as a discrete Gaussian vector with parameter $s = \omega(n^2 \cdot \sqrt{\ln n} \cdot \ln(8nq) \cdot q^{1/2+\epsilon})$ and $\epsilon \in (0, \frac{\ln n}{\ln q})$. In principal, it is possible to choose the parameters q_i and n_i of the signers in such a way that the aggregate signature has the size of an individual signature. Since there is a wide choice of selecting the chain of signers, which result in different compression rates, we restrict to the case, where q_i and n_i are equal for all signers. The aggregate signature is of the form $(\sigma_i, \vec{\alpha}_i, h_i)$ consisting of σ_i of size $2n\lceil\log_2(4.7\cdot s)\rceil$ bits, $\vec{\alpha}_i$ of size $(i-1)\cdot n(2\lceil\log_2(4.7\cdot s)\rceil - \lceil\log_2(q)\rceil)$ bits and h_i occupying l bits of memory. Each signer in the chain produces $n(2\lceil\log_2(4.7\cdot s)\rceil - \lceil\log_2(q)\rceil)$ bits of overhead.

Since the length of the signature strongly depends on q, we consider two cases for the choice of q. First, we let $q = n^{\omega(1)}$ to be slightly superpolynomial in n such that $\log_2(n) = o(\log_2(q))$. Similar to [BPR12, AKPW13] solving γ-Ideal-SVP with slightly superpolynomial factors γ appears to be exponentially hard given present best attack-algorithms. For the compression rate we then have:

$$\mathsf{rate}(i) = 1 - \frac{\lceil 2n\log_2(4.7\cdot s)\rceil + (i-1)\cdot n(2\lceil\log_2(4.7\cdot s)\rceil - \lceil\log_2(q)\rceil) + l}{i\cdot 2n\lceil\log_2(4.7\cdot s)\rceil} \tag{1}$$

$$\geq 1 - \left(\frac{1}{i} + \frac{2n(\lceil\log_2(4.7\cdot s) - \log_2(q^{1/2})\rceil + 1) + l/i}{2n\lceil\log_2(4.7\cdot s)\rceil}\right) \tag{2}$$

$$\geq 1 - \left(\frac{1}{i} + \frac{\lceil\log_2(4.7\cdot n^2\sqrt{\ln(n)\ln(8nq)})\rceil + l/(2\cdot n\cdot i) + 1}{\lceil\log_2(4.7\cdot s)\rceil}\right) \tag{3}$$

$$= 1 - \left(\frac{1}{i} + \frac{o(\log_2(q))}{o(\log_2(q)) + \log_2(q^{1/2})}\right). \tag{4}$$

Thus the compression rate converges towards $1 - 1/i$ which is asymptotically optimal, meaning that in average aggregate signatures and individual signatures are of equal size.

Secondly, we let $q = Poly(n)$. By a trivial computation using $q = n^{2c}$ we have $\mathsf{rate}(\sigma_i) \approx 1 - 1/i - 1/c$, meaning that each signer produces $1/c \cdot \mathsf{size}(\sigma_i)$ of overhead per signature. As a result, we obtain an $(1 - 1/c)$-SAS scheme. So, choosing c large enough returns an almost optimal SAS scheme. Table 1 contains the compression factor for different parameter sets.

5.3   Proxy Signatures

From the aforementioned sequential aggregate signature scheme one can immediately build a proxy signature scheme using the generic construction from [SMP08]. The core idea of a proxy signature scheme is to allow a potential signer, called delegator, to delegate its signing rights to a subentity, called proxy, which is enabled to sign documents on behalf of the delegator. Any verifier can figure out whether a signature is indeed produced by a proxy signer and if he received the signing rights from the delegator. The security of the proxy signature scheme is related to the security of the SAS scheme as stated in Theorem 1.

Table 1. SAS instantiated with secure NTRUSign for different parameter sets

n	q (in bits)	Parameter s (in bits)	Number of signers	Compression rate (in %)
256	70	55.1	20	57
256	100	70.4	20	65
256	160	100.7	20	74
512	70	57.2	20	55
512	100	72.5	20	63
512	160	102.8	20	72
1024	170	109.9	20	71
1024	200	125	20	74
1024	260	155.2	20	78

Theorem 1. *([SMP08, Theorem 2]) Let $\mathcal{AS}$ be a (t, q_s, ϵ)-unforgeable sequential aggregate signature scheme. Then, the above construction provides a $(t_0, q'_s, q'_d, \epsilon')$-unforgeable proxy signature scheme where $\epsilon = \epsilon'/2qd$, $t = t'$ and $qs = q'_s + q'_d$.*

References

[Ajt96]   Ajtai, M.: Generating hard instances of lattice problems (extended abstract). In: 28th Annual ACM Symposium on Theory of Computing, pp. 99–108. ACM Press (May 1996)

[Ajt99]   Ajtai, M.: Generating hard instances of the short basis problem. In: Wiedermann, J., Van Emde Boas, P., Nielsen, M. (eds.) ICALP 1999. LNCS, vol. 1644, pp. 1–9. Springer, Heidelberg (1999)

[AKPW13]   Alwen, J., Krenn, S., Pietrzak, K., Wichs, D.: Learning with rounding, revisited. In: Canetti, R., Garay, J.A. (eds.) CRYPTO 2013, Part I. LNCS, vol. 8042, pp. 57–74. Springer, Heidelberg (2013)

[AP09]   Alwen, J., Peikert, C.: Generating shorter bases for hard random lattices. In: *STACS*. LIPIcs, vol. 3, pp. 75–86. Schloss Dagstuhl - Leibniz-Zentrum fuer Informatik, Germany (2009)

[BGLS03]   Boneh, D., Gentry, C., Lynn, B., Shacham, H.: Aggregate and verifiably encrypted signatures from bilinear maps. In: Biham, E. (ed.) EUROCRYPT 2003. LNCS, vol. 2656, pp. 416–432. Springer, Heidelberg (2003)

[BGR12]   Brogle, K., Goldberg, S., Reyzin, L.: Sequential aggregate signatures with lazy verification from trapdoor permutations. In: Wang, X., Sako, K. (eds.) ASIACRYPT 2012. LNCS, vol. 7658, pp. 644–662. Springer, Heidelberg (2012)

[BNN07]   Bellare, M., Namprempre, C., Neven, G.: Unrestricted aggregate signatures. In: Arge, L., Cachin, C., Jurdziński, T., Tarlecki, A. (eds.) ICALP 2007. LNCS, vol. 4596, pp. 411–422. Springer, Heidelberg (2007)

[BPR12]   Banerjee, A., Peikert, C., Rosen, A.: Pseudorandom functions and lattices. In: Pointcheval, D., Johansson, T. (eds.) EUROCRYPT 2012. LNCS, vol. 7237, pp. 719–737. Springer, Heidelberg (2012)

[BPW12]   Boldyreva, A., Palacio, A., Warinschi, B.: Secure proxy signature schemes for delegation of signing rights. Journal of Cryptology 25(1), 57–115 (2012)

[BR93]   Bellare, M., Rogaway, P.: Random oracles are practical: A paradigm for designing efficient protocols. In: Ashby, V. (ed.) ACM CCS 1993 1st Conference on Computer and Communications Security, pp. 62–73. ACM Press (November 1993)

[BR96]   Bellare, M., Rogaway, P.: The exact security of digital signatures - how to sign with RSA and rabin. In: Maurer, U.M. (ed.) Advances in Cryptology - EUROCRYPT 1996. LNCS, vol. 1070, pp. 399–416. Springer, Heidelberg (1996)

[BV11]   Brakerski, Z., Vaikuntanathan, V.: Efficient fully homomorphic encryption from (standard) LWE. In: Ostrovsky, R. (ed.) 52nd FOCS Annual Symposium on Foundations of Computer Science, pp. 97–106. IEEE Computer Society Press (October 2011)

[BY96]   Bellare, M., Yung, M.: Certifying permutations: Noninteractive zero-knowledge based on any trapdoor permutation. Journal of Cryptology 9(3), 149–166 (1996)

[DDLL13] Ducas, L., Durmus, A., Lepoint, T., Lyubashevsky, V.: Lattice signatures and bimodal gaussians. In: Canetti, R., Garay, J.A. (eds.) CRYPTO 2013, Part I. LNCS, vol. 8042, pp. 40–56. Springer, Heidelberg (2013)

[EFG+10] Eikemeier, O., Fischlin, M., Götzmann, J.-F., Lehmann, A., Schröder, D., Schröder, P., Wagner, D.: History-free aggregate message authentication codes. In: Garay, J.A., De Prisco, R. (eds.) SCN 2010. LNCS, vol. 6280, pp. 309–328. Springer, Heidelberg (2010)

[Gen09]   Gentry, C.: Fully homomorphic encryption using ideal lattices. In: Mitzenmacher, M. (ed.) 41st Annual ACM Symposium on Theory of Computing, pp. 169–178. ACM Press (May/June 2009)

[GGH13a] Garg, S., Gentry, C., Halevi, S.: Candidate multilinear maps from ideal lattices. In: Johansson, T., Nguyen, P.Q. (eds.) EUROCRYPT 2013. LNCS, vol. 7881, pp. 1–17. Springer, Heidelberg (2013)

[GGH+13b] Garg, S., Gentry, C., Halevi, S., Sahai, A., Waters, B.: Attribute-based encryption for circuits from multilinear maps. In: Canetti, R., Garay, J.A. (eds.) CRYPTO 2013, Part II. LNCS, vol. 8043, pp. 479–499. Springer, Heidelberg (2013)

[GLP12]   Güneysu, T., Lyubashevsky, V., Pöppelmann, T.: Practical lattice-based cryptography: A signature scheme for embedded systems. In: Prouff, E., Schaumont, P. (eds.) CHES 2012. LNCS, vol. 7428, pp. 530–547. Springer, Heidelberg (2012)

[GPV08]   Gentry, C., Peikert, C., Vaikuntanathan, V.: Trapdoors for hard lattices and new cryptographic constructions. In: Ladner, R.E., Dwork, C. (eds.) 40th Annual ACM Symposium on Theory of Computing, pp. 197–206. ACM Press (May 2008)

[GSW13]   Gentry, C., Sahai, A., Waters, B.: Homomorphic encryption from learning with errors: Conceptually-simpler, asymptotically-faster, attribute-based. In: Canetti, R., Garay, J.A. (eds.) CRYPTO 2013, Part I. LNCS, vol. 8042, pp. 75–92. Springer, Heidelberg (2013)

[HPS98]   Hoffstein, J., Pipher, J., Silverman, J.H.: Ntru: A ring-based public key cryptosystem. In: Buhler, J.P. (ed.) ANTS-III 1998. LNCS, vol. 1423, pp. 267–288. Springer, Heidelberg (1998)

[HSW13]   Hohenberger, S., Sahai, A., Waters, B.: Full domain hash from (leveled) multilinear maps and identity-based aggregate signatures. In: Canetti, R., Garay, J.A. (eds.) CRYPTO 2013, Part I. LNCS, vol. 8042, pp. 494–512. Springer, Heidelberg (2013)

[LMRS04]   Lysyanskaya, A., Micali, S., Reyzin, L., Shacham, H.: Sequential aggregate signatures from trapdoor permutations. In: Cachin, C., Camenisch, J.L. (eds.) EUROCRYPT 2004. LNCS, vol. 3027, pp. 74–90. Springer, Heidelberg (2004)

[LOS+06]   Lu, S., Ostrovsky, R., Sahai, A., Shacham, H., Waters, B.: Sequential aggregate signatures and multisignatures without random oracles. In: Vaudenay, S. (ed.) EUROCRYPT 2006. LNCS, vol. 4004, pp. 465–485. Springer, Heidelberg (2006)

[LP11]   Lindner, R., Peikert, C.: Better key sizes (and attacks) for LWE-based encryption. In: Kiayias, A. (ed.) CT-RSA 2011. LNCS, vol. 6558, pp. 319–339. Springer, Heidelberg (2011)

[Lyn99]   Lynn, C.: Secure border gateway protocol (s-bgp). In: ISOC Network and Distributed System Security Symposium – NDSS 1999. The Internet Society (February 1999)

[Lyu12]   Lyubashevsky, V.: Lattice signatures without trapdoors. In: Pointcheval, D., Johansson, T. (eds.) EUROCRYPT 2012. LNCS, vol. 7237, pp. 738–755. Springer, Heidelberg (2012)

[MP12]   Micciancio, D., Peikert, C.: Trapdoors for lattices: Simpler, tighter, faster, smaller. In: Pointcheval, D., Johansson, T. (eds.) EUROCRYPT 2012. LNCS, vol. 7237, pp. 700–718. Springer, Heidelberg (2012)

[MR04]   Micciancio, D., Regev, O.: Worst-case to average-case reductions based on Gaussian measures. In: 45th Annual Symposium on Foundations of Computer Science, pp. 372–381. IEEE Computer Society Press (October 2004)

[Nev08]   Neven, G.: Efficient sequential aggregate signed data. In: Smart, N.P. (ed.) EUROCRYPT 2008. LNCS, vol. 4965, pp. 52–69. Springer, Heidelberg (2008)

[Pei10]   Peikert, C.: An efficient and parallel gaussian sampler for lattices. In: Rabin, T. (ed.) CRYPTO 2010. LNCS, vol. 6223, pp. 80–97. Springer, Heidelberg (2010)

[PVW08]   Peikert, C., Vaikuntanathan, V., Waters, B.: A framework for efficient and composable oblivious transfer. In: Wagner, D. (ed.) CRYPTO 2008. LNCS, vol. 5157, pp. 554–571. Springer, Heidelberg (2008)

[Rüc10]   Rückert, M.: Strongly unforgeable signatures and hierarchical identity-based signatures from lattices without random oracles. In: Sendrier, N. (ed.) PQCrypto 2010. LNCS, vol. 6061, pp. 182–200. Springer, Heidelberg (2010)

[Sho97]   Shor, P.W.: Polynomial-time algorithms for prime factorization and discrete logarithms on a quantum computer. SIAM Journal on Computing 26(5), 1484–1509 (1997)

[SMP08]   Schuldt, J.C.N., Matsuura, K., Paterson, K.G.: Proxy signatures secure against proxy key exposure. In: Cramer, R. (ed.) PKC 2008. LNCS, vol. 4939, pp. 141–161. Springer, Heidelberg (2008)

[SS11]   Stehlé, D., Steinfeld, R.: Making ntru as secure as worst-case problems over ideal lattices. In: Paterson, K.G. (ed.) EUROCRYPT 2011. LNCS, vol. 6632, pp. 27–47. Springer, Heidelberg (2011)

A Second Look at Fischlin's Transformation

Özgür Dagdelen[1] and Daniele Venturi[2]

[1] Technical University of Darmstadt, Germany
oezguer.dagdelen@cased.de
[2] Sapienza University of Rome, Italy
venturi@di.uniroma1.it

Abstract. Fischlin's transformation is an alternative to the standard Fiat-Shamir transform to turn a certain class of public key identification schemes into digital signatures (in the random oracle model).

We show that signatures obtained via Fischlin's transformation are existentially unforgeable even if the adversary can get arbitrary (yet bounded) information on the full state of the signer (including the signing key and the random coins used to generate signatures). A similar fact was already known for the Fiat-Shamir transform, however, Fischlin's transformation allows for a significantly higher leakage parameter than Fiat-Shamir.

Moreover, in contrast to signatures obtained via Fiat-Shamir, signatures obtained via Fischlin enjoy a tight reduction to the underlying hard problem. We use this observation to show (via simulations) that Fischlin's transformation, usually considered less efficient, outperforms the Fiat-Shamir transform in verification time for a reasonable choice of parameters. In terms of signing Fiat-Shamir is faster for equal signature sizes. Nonetheless, our experiments show that the signing time of Fischlin's transformation becomes, e.g., 22% of the one via Fiat-Shamir if one allows the signature size to be doubled.

Keywords: Fischlin's transformation, leakage, tightness, random oracle.

1 Introduction

Digital signatures are among the most fundamental primitives in cryptography. The security of a signature scheme (as introduced by Goldwasser, Micali, and Rivest [29]) is usually defined via a game featuring a computationally bounded adversary $\mathcal{A}$, where the game models how the system can be attacked in the real world. More specifically, $\mathcal{A}$ can see valid message/signature pairs for messages of her choice, and must forge a signature on a "fresh" message, i.e., a message $\mathcal{A}$ may choose, but for which she has not seen a valid signature already. Schemes resistant against such attacks are called existentially unforgeable under adaptive chosen message attacks (in short: ufcma).

To prove that a given signature scheme is unforgeable, one typically builds a "reduction" showing that if an *efficient* $\mathcal{A}$ can win the above security game, then an adversary $\mathcal{B}$ can run $\mathcal{A}$ internally to solve a computational problem believed

D. Pointcheval and D. Vergnaud (Eds.): AFRICACRYPT 2014, LNCS 8469, pp. 356–376, 2014.
© Springer International Publishing Switzerland 2014

to be hard. Such a game-based approach is sound if: (i) the security game is a good model of reality; (ii) the constructed reduction is as "tight" as possible. We discuss these issues in detail below.

Abstraction of Reality. One assumption (implicit in the modeling above), is that $\mathcal{A}$ is assumed to interact with the signing oracle in a black-box fashion; this means that all secrets stored "inside the box" are fully hidden to the adversary. Unfortunately this assumption might be too strong and often easy to bypass. In the real world, by exploiting several characteristics of an actual implementation (e.g., timing [34], power consumption [35] and electromagnetic radiation [41]), an attacker can learn *some* information about the secret key, and this information is often sufficient to break otherwise "provably" secure schemes.

Modern-day cryptographic models (starting with [31,37,17,13]) try to formalize side-channel attacks abstractly, with the goal of showing that a scheme has some form of *leakage resilience*.

The standard way of defining leakage-resilient signatures, enhances the unforgeability game by giving $\mathcal{A}$ access to a leakage oracle which outputs bounded (but arbitrary) information about the secret key sk. A signature scheme is existentially unforgeable against λ-leakage attacks if forging signatures on fresh messages is still hard given λ bits of sk-data. Throughout the paper we, in fact, consider a more general setting, where $\mathcal{A}$ leaks information not just about sk, but also about the full randomness used to generate signatures; a scheme secure in this sense is called *fully* leakage-resilient. For a more detailed discussion on leakage models for digital signatures, we refer the reder to Section 1.2.

The value λ is called the leakage parameter of the system. Note that the secret-key size s must be strictly greater than the leakage parameter λ. The quantity λ/s can be thought as the *relative leakage* of the system, with the obvious goal to make it as close to 1 as possible.

Tightness. Suppose that $\mathcal{A}$ takes time t to break the security of a primitive (e.g., a signature scheme) with probability ε. If $\mathcal{B}$ in the reduction has runtime $t' \approx t$ and solves the hard problem with probability $\varepsilon' \approx \varepsilon$, the reduction is *tight*; else it is *loose*. The ratio $(t'\varepsilon)/(t\varepsilon')$ is called the tightness gap of the reduction.

As discussed, e.g., in [33,9], tight reductions are appealing both for theoretical purposes and because they ensure that the primitive is at least as hard to break as the underlying hard problem. A loose reduction, by contrast, only guarantees that a scheme is "plausibly" secure (see [28]). A loose reduction also results in much larger parameters, and thus much slower performances (depending on the tightness gap). In general, many researchers concerned with practice call into question the practical value of non-tight reductionist security proofs.

1.1 Our Contributions

Σ-protocols are a well-studied class of interactive protocols, run between a prover $\mathcal{P}$ and a verifier $\mathcal{V}$, such that $\mathcal{V}$ accepts $\mathcal{P}$ as legitimate if it is convinced that $\mathcal{P}$ knows a witness w to a shared input x. Each protocol run yields a transcript of the form (com, ch, resp), where com is sent by the prover. The Fiat-Shamir transform [23] is a common way of constructing efficient signature schemes (in the random oracle model [4]) from a Σ-protocol for some "hard relation".

Recently, Katz and Vaikuntanathan [32] (building on Alwen *et al.* [3]) showed that the Fiat-Shamir transform yields fully leakage-resilient signatures, provided that the underlying Σ-protocol satisfies two additional properties: (i) each theorem has exponentially many witnesses; (ii) the uncertainty of the witness conditioned on the theorem is high.

The obtained scheme has relative leakage asymptotically approaching $1/2$ and a loose reduction (with a tightness gap of about $1/\varepsilon$), due to the fact that the reduction needs to rewind the adversary in order to extract a valid witness and solve the underlying hard problem.

Fischlin's Transformation & Leakage. Fischlin's transformation [24] is an alternative to get secure signatures schemes from arbitrary Σ-protocols. Roughly, Fischlin's transformation consists of a tuple (of dimension r) of Fiat-Shamir signatures, i.e., $(\mathsf{com}_i, \mathsf{ch}_i, \mathsf{resp}_i)_{1 \le i \le r}$. However, instead of computing the challenge via $\mathsf{ch}_i = H(\mathsf{com}_i, m)$, the prover tries all values in the domain of challenges such that $H(m, x, \mathbf{com}, i, \mathsf{ch}_i, \mathsf{resp}_i) = 0^b$ for all $i \in [r]$ where $\mathbf{com} = (\mathsf{com}_1, \ldots, \mathsf{com}_r)$. If no such challenge can be found the challenge ch_i is chosen with the smallest output in value. Verifying includes now the check the validity of the r Fiat-Shamir signatures and whether the sum of all hash values are below a certain threshold S.

One important feature of signatures obtained via Fischlin, is that the resulting non-interactive protocol has a *straight-line* extractor. Roughly this means that there exists a probabilistic polynomial time algorithm (a.k.a. the extractor) that, upon input the theorem x, a message m, a valid signature σ on m, and all hash queries and answers made to generate σ, outputs a valid witness for x with overwhelming probability (and without further querying the signer).

Our first result is that Fischlin's transformation yields a fully leakage-resilient signature if the underlying Σ-protocol satisfies properties (i) and (ii) above.

Comparing Fischlin and Fiat-Shamir. Even though the above fact is perhaps not very surprising, Fischlin's transformation comes with two important advantages over leakage-resilient signatures obtained via Fiat-Shamir. The first advantage is that for concrete schemes (e.g., the ones based on Okamoto [39] and Guillou-Quisquater [30]), the relative leakage of the resulting signatures asymptotically approaches 1. The second advantage is that the reduction to the security of the underlying hard problem is tight.[1]

As a consequence of the above observations, one might expect that for a pre-fixed level of security, signatures obtained via Fischlin can be instantiated using much smaller parameters, possibly leading to better performances than signatures obtained via Fiat-Shamir. This is surprising, as usually Fischlin's transformation is considered to be less efficient than Fiat-Shamir.

---

[1] We stress that the problem of finding a tight reduction for leakage-resilient signatures obtained via Fiat-Shamir remains open.

Our second contribution is an accurate comparison (supported by simulations) of the performances obtained via Fischlin and Fiat-Shamir, in terms of parameters generation, signing and verification time. The comparison is carried out for the Okamoto scheme [39], whose security relies on the hardness of computing discrete logarithms over a finite field. The main findings of our analysis are sketched below:

- Key generation is always faster in Fischlin's transformation, due to the fact that a tight reduction allows to choose smaller parameters.
- In terms of verification time, signatures obtained via Fischlin are much faster than the ones obtained via Fiat-Shamir. This feature makes Fischlin's transformation particularly interesting for scenarios where one demands fast signature verification (e.g., in car2car and car2X communication [44] one might need to verify 4000-5000 signatures per second).
- In terms of signature generation time, for 80-bit security, signatures obtained via Fischlin are two times slower than the ones obtained via Fiat-Shamir if one insists for the resulting signatures having the same size. However, in case one allows signatures obtained via Fischlin to have twice the size of of the ones obtained via Fiat-Shamir, then signature generation becomes 4.5 times faster and verification reduces to 90% of that from Fiat-Shamir.
- When enforcing a certain amount of leakage resilience, Fischlin's transformation outperforms the Fiat-Shamir transform in terms of security, performance, and signature size.

We remark that, even though in some cases signatures obtained via Fiat-Shamir result in better signing time, Fischlin's transformation might still be preferable in certain scenarios. For instance, an interesting feature of signatures obtained via Fischlin when used in some cryptographic protocol (e.g., for key exchange), is that parties can start verifying the signature (except checking the hash values) before the entire signature is sent. Consequently, the effort to generate and verify a signature consists essentially of the signing time plus $1/r$-th of the verification time. In contrast, signatures obtained via Fiat-Shamir have to be received in full before the verification can start. Taking this feature into account, signing and verifying for Fischlin are indeed faster than for Fiat-Shamir (for 80-bit security).[2] This property and the small key sizes let us find favor with Fischlin's transformation if the resulting signature scheme is deployed on a smartcard.

We conclude that Fischlin's transformation may be a reasonable alternative to the Fiat-Shamir transform, depending on the application scenario.

1.2 Related Work

Tightness. Tight reductions are also discussed in e.g. [27] (for signature schemes based on the family of Diffie-Hellman problems in the random oracle model), in [40,26,45] (for the Schnorr and other related signature schemes in the random oracle model), and in [42] (for signature schemes in the plain model).

---

[2] The same conclusion does not hold for 128-bit security, though.

Leakage. Several leakage models exist so far in the literature. In our work we consider the so-called *bounded leakage* model, where the total amount of leakage is a-priory bounded to some fraction of the secret key length. A more general model is the so-called *continuous leakage model* [14,8], where the leakage is not a priori bounded and there is some efficient procedure to "refresh" the secret key (leaving the corresponding public key unchanged).

Apart from [3,32], other papers on leakage-resilient signatures can be found in [19,15,20,7,38]. These results are either complicated or inefficient, or do not permit optimal relative leakage, and have loose reductions.

2 Preliminaries

2.1 Notation

For $n \in \mathbb{N}$, let $[n] := \{1, 2, \ldots, n\}$. We write log for base-2 logarithms and ln for natural logarithms. We denote vectors by bold lower case letters.

For an algorithm $\mathcal{A}$, $y \leftarrow \mathcal{A}(x)$ denotes that y is output by $\mathcal{A}$ on input x; sometimes we also write $y = \mathcal{A}(x; \omega)$ to make explicit the random coins that $\mathcal{A}$ may use. Also, $\mathcal{A}^{\mathcal{O}}$ denotes that $\mathcal{A}$ has access to oracle $\mathcal{O}$. Algorithm $\mathcal{A}$ is probabilistic polynomial time (PPT) if $\mathcal{A}$ is randomized and for any input $x \in \{0, 1\}^*$ the computation of $\mathcal{A}(x)$ terminates in at most $poly(|x|)$ steps.

The min-entropy of a random variable X is $\mathbb{H}_\infty(X) = -\log \max_x \mathbb{P}[X = x]$, and measures how well X can be predicted by the best (unbounded) predictor. The conditional average min-entropy [16] of X given a random variable Z (over some set $\mathcal{Z}$) possibly dependent on X, is defined as $\widetilde{\mathbb{H}}_\infty(X|Z) := -\log \mathbb{E}_{z \leftarrow Z}[2^{-\mathbb{H}_\infty(X|Z=z)}]$. Following [3], we sometimes rephrase the notion of conditional min-entropy in terms of predictors $\mathcal{A}$ that are given some information Z, so $\widetilde{\mathbb{H}}_\infty(X|Z) = -\log(\max_{\mathcal{A}} \mathbb{P}[\mathcal{A}(Z) = X])$. We recall the following useful lemma, proven in [16], bounding the conditional average min-entropy of a random variable X given λ bits of arbitrary information on X itself.

Lemma 1 ([16]). *For all random variables X, Z and Λ over sets $\mathcal{X}$, $\mathcal{Z}$ and $\{0,1\}^\lambda$ such that $\widetilde{\mathbb{H}}_\infty(X|Z) \geq \beta$, we have*

$$\widetilde{\mathbb{H}}_\infty(X|Z, \Lambda) \geq \widetilde{\mathbb{H}}_\infty(X|Z) - \lambda \geq \beta - \lambda.$$

2.2 Signature Schemes

We recall here the general syntax for digital signatures.

Definition 1 (Signature scheme). *A signature scheme is a triple of algorithms $\mathcal{SS} = (\mathsf{KGen}, \mathsf{Sign}, \mathsf{Vrfy})$ defined as follows.*

Key Generation. *Algorithm KGen is a probabilistic algorithm which, on input a security parameter 1^k, outputs a pair (pk, sk) where pk is the public key, and sk is the secret key.*

Signature. *Algorithm* Sign *is a probabilistic algorithm which, on input a secret key sk together with message m, outputs a signature σ on m under sk.*

Verification. *Algorithm* Vrfy *is a deterministic algorithm which, on input a message m and a signature σ together with the public key pk, outputs either 1 (= valid) or 0 (= invalid).*

We say that $\mathcal{SS}$ has completeness error $\varepsilon_{\text{comp}}$ if $\mathbb{P}[\text{Vrfy}(pk, (m, \text{Sign}(sk, m)) = 0] \leq \varepsilon_{\text{comp}}$, where $(pk, sk) \leftarrow \text{KGen}(1^k)$ and the probability is over the coin tosses of Sign.

Leakage-Resilient Signatures. Consider an oracle $\mathcal{O}_\lambda(x, \cdot)$ taking as input functions $f : \{0,1\}^* \rightarrow \{0,1\}^*$ and returning $f(x)$ for a total of at most λ bits. Roughly, a signature scheme $\mathcal{SS} = (\text{KGen}, \text{Sign}, \text{Vrfy})$ is (fully) leakage-resilient if it is hard to forge a signature even given access to oracle $\mathcal{O}_\lambda(x, \cdot)$, where x contains the secret key, plus the entire history of all random coins tossed by the signing algorithm. More formally, consider the following experiment:

Experiment $\mathbf{Exp}_{\mathcal{SS},\mathcal{A}}^{\text{lkg}-\text{ufcma}}(k, \lambda)$
$(pk, sk) \leftarrow \text{KGen}(1^k)$
$(m^\star, \sigma^\star) \leftarrow \mathcal{A}^{\text{Sign}(sk, \cdot), \mathcal{O}_\lambda(\text{state}, \cdot)}(pk)$
Output 1 iff
    (a) $\text{Vrfy}(pk, m^\star, \sigma^\star) = 1$
    (b) $m^\star \notin \mathcal{Q}$

Set state $= \{sk\}$, and $\mathcal{Q} = \emptyset$
If $\mathcal{A}$ queries $\text{Sign}(sk, m)$:
  - let $\mathcal{Q} := \mathcal{Q} \cup \{m\}$
  - return $\sigma \leftarrow \text{Sign}(sk, m; \omega)$
  - let state := state $\cup \{\omega\}$

Definition 2 (Fully leakage-resilient signature). *We say that* $\mathcal{SS} = (\text{KGen}, \text{Sign}, \text{Vrfy})$ *is* (t, q_s, ε)-*unforgeable against chosen-message attacks (in short: -ufcma) and against λ-leakage attacks if, for every algorithm $\mathcal{A}$ running in time t and asking q_s signing queries, we have:*

$$\mathbb{P}\left[\mathbf{Exp}_{\mathcal{SS},\mathcal{A}}^{\text{lkg}-\text{ufcma}}(k, \lambda) = 1\right] \leq \varepsilon.$$

We say a signature scheme $\mathcal{SS}$ is (t, q_s, ε)-ufcma if the algorithm $\mathcal{A}$ has no access to oracle $\mathcal{O}_\lambda$ in the experiment above.

Note that in case the signature scheme $\mathcal{SS}$ requires the use of a public function H modeled as a random oracle, both the adversary $\mathcal{A}$ and the leakage functions have access to this random oracle.

2.3 Fischlin's Transformation

Let $\mathcal{L} \subseteq \mathbf{NP}$ be a language with a (polynomially computable) relation $\mathcal{R} \subset \{0,1\}^* \times \{0,1\}^*$, i.e., $x \in \mathcal{L}$ if and only if $\exists w$ such that $(x, w) \in \mathcal{R}$ and $|w| = poly(|x|)$. The value w is called a witness for $x \in \mathcal{L}$ (x is sometimes called a "theorem" or statement). Informally, $\mathcal{R}$ is called *hard* if, given $x \in \mathcal{L}$, it is hard to extract a valid witness for x.

Definition 3 (Hard relation). *A relation $\mathcal{R}$ for a language $\mathcal{L}$ is* (t, ε)-*hard if the following holds:*

(i) *There exists an efficient algorithm* Gen *that on input a security parameter k outputs* $(x, w) \leftarrow$ Gen(1^k) *such that* $(x, w) \in \mathcal{R}$ *and* $|w| = poly(|x|)$.

(ii) *For any algorithm* $\mathcal{A}$ *running in time t we have:*

$$\mathbb{P}\left[(x, w') \in \mathcal{R} : \ w' \leftarrow \mathcal{A}(1^k, x); (x, w) \leftarrow \text{Gen}(1^k) \right] \leq \varepsilon.$$

Σ-Protocols. This important class of protocols (run between a prover $\mathcal{P}$ and a verifier $\mathcal{V}$), allows $\mathcal{P}$ to convince $\mathcal{V}$ that it knows a witness w for a shared element $x \in \mathcal{L}$, without giving $\mathcal{V}$ further information. We briefly review Σ-protocols below. Informally, a Σ-protocol consists of three messages (com, ch, resp) (with com sent by $\mathcal{P}$) and satisfies the following properties:

- *Completeness.* Upon interacting with an honest prover holding (x, w), the verifier accepts with overwhelming probability.
- *Special Soundness.* Given accepted proofs (com, ch, resp) and (com, ch', resp') for $x \in \mathcal{L}$ (with ch' $\neq$ ch), there exists a PPT algorithm which outputs a valid witness w for x.
- *Perfect Honest-Verifier Zero Knowledge (HVZK).* There exists a PPT algorithm $\mathcal{Z}$ (the simulator) which, on input $x \in \mathcal{L}$ and a random ch, outputs an accepting conversation of the form (com, ch, resp), with the same probability distribution as conversations between the honest $\mathcal{P}$, $\mathcal{V}$ on input x.[3]

In the following we also assume that com has super-logarithmic min-entropy (in the security parameter k), and that resp is quasi-unique, i.e., it is hard to find $(x, \text{com}, \text{ch}, \text{resp}, \text{resp}')$ such that both (com, ch, resp) and (com, ch, resp') are accepting, with resp $\neq$ resp'.

The Transformation. Let $(\mathcal{P}, \mathcal{V})$ be a Σ-protocol for an **NP**-language $\mathcal{L}$ with hard relation $\mathcal{R}$, and consider a hash function $H : \{0, 1\}^* \rightarrow \{0, 1\}^b$, modeled as a random oracle. As proved in [24], Fischlin's transformation (represented in Fig. 1) describes a non-interactive zero-knowledge proof with a straight-line extractor, and yields an existentially unforgeable signature scheme.

Theorem 1 (Fischlin's transformation). *Consider the scheme in Fig. 1, where the challenge space of $(\mathcal{P}, \mathcal{V})$ has length $l = O(\log k)$. Let b, r, S, μ be functions of k such that $b \cdot r = \omega(\log k)$, $2^{\mu - b} = \omega(\log k)$, $b, r, \mu = O(\log k)$, $S = O(r)$ and $b \leq \mu \leq l$. The following holds:*

(i) *The transformation describes a non-interactive zero-knowledge proof system $(\overline{\mathcal{P}}^H, \overline{\mathcal{V}}^H)$ for language $\mathcal{L}^{msg} = \{((x, m), w) : (x, w) \in \mathcal{R}\}$, where $\overline{\mathcal{P}}$ (resp. $\overline{\mathcal{V}}$) is the signer (resp. verifier) of the signature scheme. In particular, there exists a PPT simulator $\overline{\mathcal{Z}}$ which, on input (x, m) outputs a proof $\sigma = (\text{com}_i, \text{ch}_i^*, \text{resp}_i)_{i=1,\dots,r}$ with the same distribution as a real proof generated via $\overline{\mathcal{P}}^H$ (using x, m, w).*

---

[3] This is also called *special* HVZK, but as argued in [24] can be assumed in general.

Let $\mathcal{R}$ be a hard relation for language $\mathcal{L}$ and $(\mathcal{P}, \mathcal{V})$ be a Σ-protocol for $\mathcal{R}$. For a hash function $H : \{0,1\}^* \rightarrow \{0,1\}^b$ (modeled as a random oracle), let r be the number of repetitions, μ the challenge size, and S the bound on the sum.

Key Generation. Upon input security parameter 1^k compute $(x, w) \leftarrow \mathsf{Gen}(1^k)$.
Output $pk := x$ as public key and $sk := w$ as secret key.

Signature. Upon input a secret key $sk = w$ and a message $m \in \{0,1\}^*$, perform the following steps:

1. For all $i \in [r]$, obtain $\mathsf{com}_i \leftarrow \mathcal{P}(x)$.
2. For all $i \in [r]$ and $\mathsf{ch}_i \in [2^\mu - 1]$ compute $\mathsf{resp}_i := \mathsf{resp}(\mathsf{ch}_i) \leftarrow \mathcal{P}(\mathsf{com}_i, x, w, \mathsf{ch}_i)$. Denote ch_i which satisfies $H(m, x, \mathbf{com}, i, \mathsf{ch}_i, \mathsf{resp}_i) = 0^b$ by ch_i^*, where $\mathbf{com} = (\mathsf{com}_1, \ldots, \mathsf{com}_r)$. If no such ch_i exists, take the one with minimal hash output value.
3. Output $\sigma = (\mathsf{com}_i, \mathsf{ch}_i^*, \mathsf{resp}_i)_{i=1,\ldots,r}$.

Verification. Upon input the public key $pk = x$ and a signature $\sigma = (\mathsf{com}_i, \mathsf{ch}_i^*, \mathsf{resp}_i)_{i=1,\ldots,r}$ for message m, run the verifier of the underlying Σ-protocol to check if $\mathcal{V}(x, (\mathsf{com}_i, \mathsf{ch}_i^*, \mathsf{resp}_i)) = 1$ for all $i \in [r]$. If not, output 0. Furthermore, if $\sum_{i=1}^{r} H(m, x, \mathbf{com}, i, \mathsf{ch}_i^*, \mathsf{resp}_i) \leq S$ output 1; else output 0.

Fig. 1. Fischlin's transformation applied to a Σ-protocol $(\mathcal{P}, \mathcal{V})$ for relation $\mathcal{R}$

(ii) *There exists a PPT straight-line extractor $\mathcal{K}$ and some value ε_{ext}, such that, for any PPT $\mathcal{A}$, if $(x, \sigma) \leftarrow \mathcal{A}^H(1^k)$, then*

$$\mathbb{P}\left[(x, w) \notin \mathcal{R} \wedge \overline{\mathcal{V}}^H(x, \sigma) = 1 \right] \leq \varepsilon_{ext}$$

for $w \leftarrow \mathcal{K}(x, \sigma, \mathcal{Q}_H(\mathcal{A}))$; here $\mathcal{Q}_H(\mathcal{A})$ denotes $\mathcal{A}$'s queries to (resp. answers from) the random oracle H.

(iii) *If the relation $\mathcal{R}$ is (t, ε)-hard, the resulting signature scheme is (t', ε')-ufcma where $t' \approx t$ and $\varepsilon' = \varepsilon + \varepsilon_{ext}$.*

Note that the bound on the challenge space is without loss of generality, as for any l we can easily turn a Σ-protocol with l'-bit challenges into a Σ-protocol with l-bit challenges [11, Lemma 2]. The following corollary follows by inspection of the proof of [24, Theorem 2].

Corollary 1 (Concrete parameters of Fischlin's transformation). *The following holds for the transformation of Fig. 1:*

- *The completeness error is upper-bounded by*

$$\varepsilon_{\mathsf{comp}} \leq e^{r \ln(e \cdot (2S+1)) - (S+1)2^{\mu - b}}.$$

- *The failure probability of the extractor is upper-bounded by*

$$\varepsilon_{\mathsf{ext}} \leq (q_h + 1)(S + 1)2^{(\log(e \cdot (S+r)/(r-1)) - b) \cdot r},$$

where $q_h = |\mathcal{Q}_H(\mathcal{A})|$.
- *The total number of hash function evaluations is upper-bounded by $r \cdot (2^\mu - 1)$ (in worst case).*

3   Leakage Resilience of Fischlin's Transformation

Let $(\mathcal{P}, \mathcal{V})$ be a Σ-protocol for $\mathcal{L} \subset \mathbf{NP}$, with relation $\mathcal{R}$. The main result of this section is that Fischlin's transformation applied to each such protocol yields a fully leakage-resilient signature (in the random oracle model) provided that: (i) each theorem $x \in \mathcal{L}$ has exponentially many witnesses w (and given a valid (x, w) pair is hard to find a valid, distinct (x, w') pair); (ii) the conditional min-entropy of the witness W conditioned on the public theorem X is high.

A similar result is already known for the Fiat-Shamir heuristic [3,32]. The main difference here is that we get a fully tight reduction to the underlying hard problem and relative leakage asymptotically approaching 1—which is optimal. In comparison the best known analysis for Fiat-Shamir has a tigthness gap of roughly $1/\varepsilon$ (where ε is the hardness of the underlying relation), and relative leakage asymptotically approaching $1/2$.

We start by formalizing condition (i) above, by introducing the representation problem for a relation $\mathcal{R} \subset \{0,1\}^* \times \{0,1\}^*$.

Definition 4 (Representation problem). *We say that the representation problem is (t, ε)-hard for a relation $\mathcal{R}$ if for all PPT adversaries $\mathcal{A}$ running in time t, we have:*

$$\mathbb{P}\left[w \neq w' \wedge (x, w), (x, w') \in \mathcal{R} : \ (x, w, w') \leftarrow \mathcal{A}(1^k) \right] \leq \varepsilon.$$

In many cases, the hardness of the representation problem for $\mathcal{R}$ is equivalent to the hardness of the underlying relation $\mathcal{R}$. We comment on two such instantiations, based respectively on the discrete-log and on factoring assumptions, in the concrete instantiations paragraph at the end of this section.

Theorem 2 (Fischlin's transformation is leakage-resilient). *Let $k \in \mathbb{N}$ be a security parameter and let $\mathcal{R} \subset \{0,1\}^* \times \{0,1\}^*$ be an **NP**-relation such that the representation problem is (t, ε)-hard for $\mathcal{R}$. Assume that conditioned on the distribution of the public input $x \in \mathcal{X}$, the witness $w \in \mathcal{W}$ has high average min-entropy β, i.e., $\widetilde{\mathbb{H}}_\infty(W|X) \geq \beta$. Then, the signature scheme of Fig. 1 is $(t', q_s = poly(k), \varepsilon')$-ufcma against λ-leakage attacks, as long as*

$$t' \approx t \qquad \lambda \leq \beta - r\log(3q_h) - k \qquad \varepsilon' \leq \varepsilon + \varepsilon_{\text{ext}} + 2^{-k},$$

where $q_h = poly(k)$ denotes the number of queries to the random oracle.

The proof borrows ideas from [32, Theorem 4]. The original proof requires to rewind $\mathcal{A}$, yielding a loose reduction. By relying on the straight-line extractor of Fischlin's transformation, we avoid rewinding and thus get a tight reduction.

Proof. By contradiction assume there exists a PPT adversary $\mathcal{A}$ running in time t' and having advantage $\varepsilon' > \varepsilon + \varepsilon_{\text{ext}} + 2^{-k}$ in the experiment $\mathbf{Exp}_{\mathcal{SS},\mathcal{A}}^{\text{lkg}-\text{ufcma}}(k, \lambda)$ (for leakage parameter λ as in the theorem statement). Consider all possible states during the execution of $\mathcal{A}$ in the unforgeability experiment, and for any such state i let h_i denote the hash query made at that state. If an execution of $\mathcal{A}$ terminates with a valid forgery $(m^\star, (\mathsf{com}_i, \mathsf{ch}_i^*, \mathsf{resp}_i)_{i \in [r]})$, we say that the forgery is associated with a set of states $\{h_i\}$ where $h_i = H(m^\star, x, \mathsf{com}, i, \mathsf{ch}_i^*, \mathsf{resp}_i)$ for all $i \in [r]$. Note that the size of this set is $\binom{q_h}{r} \leq (\frac{q_h \cdot e}{r})^r < (3q_h)^r$.

We build a PPT adversary $\mathcal{B}$ (running in time $t \approx t'$) breaking the hardness of the representation problem for $\mathcal{R}$ with advantage larger than ε (a contradiction). Without loss of generality, we assume that whenever $\mathcal{A}$ outputs a forgery $(m^\star, \sigma^\star)$: (i) $\mathcal{A}$ queried the random oracle at some point on input $(m^\star, x, \mathsf{com}, i, \mathsf{ch}_i^*, \mathsf{resp}_i)$, for $i \in [r]$; (ii) $\mathcal{A}$ never queried the signing oracle on $m^\star$. For simplicity, we further assume that every leakage query makes the same number of H evaluations. (This can always be achieved adding dummy queries.)

Adversary $\mathcal{B}$ starts by generating $(x, w) \leftarrow \mathsf{Gen}(1^k)$, where $(x, w) \in \mathcal{R}$. Hence, $\mathcal{B}$ gives the public key $pk = x$ to $\mathcal{A}$ and implicitly defines $sk = w$. Note that since $\mathcal{B}$ knows a valid witness w corresponding to x, the reduction can perfectly simulate the experiment $\mathbf{Exp}_{SS,\mathcal{A}}^{\mathrm{lkg-ufcma}}(k, \lambda)$; this includes the answers to $\mathcal{A}$'s queries to both oracles $\mathsf{Sign}(sk, \cdot)$ and $\mathcal{O}_\lambda(\mathsf{state}, \cdot)$, as well as the queries to the random oracle H. Adversary $\mathcal{B}$ keeps also track of all the queries $\mathcal{Q}_H(\mathcal{A})$ of $\mathcal{A}$ to H, and the corresponding answers. Eventually, $\mathcal{A}$ outputs a forgery $(m^\star, \sigma^\star)$. At this point $\mathcal{B}$ runs the straight-line extractor $\mathcal{K}$ on input $(m^\star, \sigma^\star, \mathcal{Q}_H(\mathcal{A}))$ and obtains a value $w' \leftarrow \mathcal{K}(m^\star, \sigma^\star, \mathcal{Q}_H(\mathcal{A}))$. Finally $\mathcal{B}$ outputs (x, w, w') as a solution to the representation problem for $\mathcal{R}$.

By definition, $\mathcal{B}$ solves the representation problem for $\mathcal{R}$ whenever $\mathcal{A}$ succeeds and: (i) the extractor $\mathcal{K}$ does not fail; (ii) the extracted witness w' is different from w. Denote by FAIL the event that the extractor does not return a valid witness and with EQUAL the event that the returned w' is equal to w. Since the event that $\mathcal{A}$ wins and the latter two events are all independent, we can write:

$$\begin{aligned}
\mathbb{P}[\mathcal{B} \text{ wins}] &= \mathbb{P}\left[w' \neq w; (x, w), (x, w') \in \mathcal{R}: (x, w, w') \leftarrow \mathcal{B}(1^k)\right] \\
&= \mathbb{P}[\mathcal{A} \text{ wins} \wedge \neg\mathrm{FAIL} \wedge \neg\mathrm{EQUAL}] \\
&= \mathbb{P}[\mathcal{A} \text{ wins}] \cdot \mathbb{P}[\neg\mathrm{FAIL}] \cdot \mathbb{P}[\neg\mathrm{EQUAL}] \\
&\geq \mathbb{P}[\mathcal{A} \text{ wins}] - \mathbb{P}[\mathrm{FAIL}] - \mathbb{P}[\mathrm{EQUAL}] \\
&\geq \varepsilon' - \varepsilon_{\mathrm{ext}} - \mathbb{P}[\mathrm{EQUAL}],
\end{aligned} \tag{1}$$

where (1) follows by our assumption on $\mathcal{A}$ and by the fact that the probability that the extractor fails is bounded by $\varepsilon_{\mathrm{ext}}$.

Claim. $\mathbb{P}[\mathrm{EQUAL}] \leq 2^{-k}$.

Proof (of claim). We show that the statement holds even in case $\mathcal{A}$ is unbounded. We will argue that $\widetilde{\mathbb{H}}_\infty(W|V) \geq k$, where W is the random variable corresponding to the witness, and V is the random variable corresponding to the view of $\mathcal{A}$ in a run of $\mathbf{Exp}_{SS,\mathcal{A}}^{\mathrm{lkg-ufcma}}(k, \lambda)$. Clearly, this is sufficient as by definition of average min-entropy $\mathbb{P}[\mathrm{EQUAL}] \leq 2^{-\widetilde{\mathbb{H}}_\infty(W|V)}$. The view of $\mathcal{A}$ is of the type $V := (\mathbf{\Sigma}, \mathbf{R}, \Lambda, X)$, where the random variable $\mathbf{\Sigma} = (\Sigma_1, \ldots, \Sigma_{q_s})$ lists the signing queries of $\mathcal{A}$, the random variable $\mathbf{R} = (R_1, \ldots, R_{q_h})$ corresponds to the responses to $\mathcal{A}$'s random oracle queries, Λ corresponds to the leakage queries, and X corresponds to the public key. Now,

$$\widetilde{\mathbb{H}}_\infty(W|\mathbf{\Sigma}, \mathbf{R}, \Lambda, X) \geq \widetilde{\mathbb{H}}_\infty(W|\mathbf{\Sigma}, X) - \lambda - r\log(3q_h) \tag{2}$$

$$\geq \widetilde{\mathbb{H}}_\infty(W|X) - \lambda - r\log(3q_h) \tag{3}$$

$$\geq k. \tag{4}$$

(2) follows by Lemma 1 and the fact that: (i) $\Lambda \in \{0,1\}^\lambda$ and (ii) a forgery reveals at most $r\log(3q_h)$ bits of information on the witness, corresponding to the set of random oracle queries associated with the forgery itself;[4] (3) follows by (perfect) honest-verifier zero-knowledge, as we can compute each Σ_i using only the public key X and the zero-knowledge simulator $\overline{\mathcal{Z}}$ (cf. Theorem 1). Finally, (4) follows by our assumption that $\widetilde{\mathbb{H}}_\infty(W|X) \geq \beta$ and the bound on $\lambda \leq \beta - r\log(3q_h) - k$.

The above claim together with our assumption that $\varepsilon' > \varepsilon + \varepsilon_{\text{ext}} + 2^{-k}$, clearly imply that $\mathbb{P}[\mathcal{B} \text{ wins}] > \varepsilon$, which contradicts the (t,ε)-hardness of the representation problem for $\mathcal{R}$. This finishes the proof.

Concrete Instantiations. Below, we discuss two concrete instantiations of Theorem 2, the first one based on the discrete-log assumption and the second one based on the RSA assumption (and on factoring).

Generalized Okamoto [39]. For a cyclic group $\mathbb{G}$ of prime order p, let $\mathcal{L}_{\text{DL}} := \{(g_1, \ldots, g_\ell, h) : \exists(w_1, \ldots, w_\ell) \text{ s.t. } h = \prod_{i=1}^\ell g_i^{w_i}\}$, where $(g_1, \ldots, g_\ell)$ are generators of $\mathbb{G}$. The tuple $w = (w_1, \ldots, w_\ell)$ is called a representation of h; the ℓ-representation problem asks to compute two distinct representations w, w' for some $x = (g_1, \ldots, g_\ell, h) \in \mathcal{L}$. As argued in [3, Lemma 4.1], the ℓ-representation problem is hard for $\mathcal{R}_{\text{DL}}$ if and only if the discrete-log problem is hard in $\mathbb{G}$.[5]

The standard Σ-protocol $(\mathcal{P}, \mathcal{V})$ to prove knowledge of a representation of an element h goes as follows: (i) $\mathcal{P}$ chooses randomly $a_1, \ldots, a_\ell$ and sets $\text{com} := \prod_{i=1}^\ell g_i^{a_i}$; (ii) $\mathcal{V}$ selects a random $\text{ch} \in \mathbb{Z}_p$; (iii) $\mathcal{P}$ returns $\text{resp} = (\text{ch} \cdot w_1 + a_1, \ldots, \text{ch} \cdot w_\ell + a_\ell)$. Given a proof $\sigma = (\text{com}, \text{ch}, \text{resp})$, the verifier outputs 1 if and only if $\prod_{i=1}^\ell g_i^{\text{resp}_i} = h^{\text{ch}} \cdot \text{com}$.

We obtain the following result:

Corollary 2. *Let $\mathbb{G}$ be a cyclic group of prime order p, such that the ℓ-representation problem is hard for $\mathcal{R}_{\text{DL}}$. Then, the signature scheme obtained by applying Fischlin's transformation to the generalized Okamato Σ-protocol is fully leakage-resilient for leakage parameter $\lambda \leq (1 - o(1)) \cdot n$, where $n = \ell \log p$ is the length of the secret key.*

---

[4] Recall that leakage queries can depend on the random oracle; this could make the set of states associated with a forgery a function of the witness.

[5] Recall that the discrete-log problem requires to compute w such that $g^w = h$, given $(g, h, \mathbb{G}, p)$.

Proof. By Theorem 2, we get that for any desired $\delta > 0$ the leakage bound is $\lambda \leq (1 - 1/\ell - \delta) \cdot n$. Now, choosing $\ell > 1/\delta$ gives $\lambda \leq (1 - \delta) \cdot n$ as desired.

Generalized Guillou-Quisquater [30]. For $N = p \cdot q$, where p and q are primes, let (e, d) be such that $e \cdot d = 1 \mod \phi(N)$ and e is a prime. Consider the language $\mathcal{L}_{\text{RSA}} := \{(g_1, \ldots, g_\ell, h) : \exists(\rho, (w_1, \ldots, w_\ell)) \in \mathbb{Z}_N^* \times \mathbb{Z}_e^\ell \text{ s.t. } h = \prod_{i=1}^{\ell} g_i^{w_i} \cdot \rho^e \mod N\}$, where $(g_1, \ldots, g_\ell)$ are generators of $\mathbb{Z}_N^*$. The tuple $w = (\rho, (w_1, \ldots, w_\ell))$ is called a representation of h; the ℓ-representation problem asks to compute two distinct representations w, w' for some $x = (g_1, \ldots, g_\ell, h) \in \mathcal{L}$. As shown in [39], the ℓ-representation problem is hard for $\mathcal{R}_{\text{RSA}}$ if and only if the RSA problem is hard in $\mathbb{Z}_N^*$.[6]

The standard Σ-protocol $(\mathcal{P}, \mathcal{V})$ to prove knowledge of a representation of an element h goes as follows: (i) $\mathcal{P}$ chooses randomly $a_1, \ldots, a_\ell \leftarrow \mathbb{Z}_e$ and $s \leftarrow \mathbb{Z}_N^*$ and sets $\text{com} := \prod_{i=1}^{\ell} g_i^{a_i} \cdot s^e \mod N$; (ii) $\mathcal{V}$ selects a random $\text{ch} \in \mathbb{Z}_e$; (iii) $\mathcal{P}$ computes $\mathbf{z} = (\text{ch} \cdot w_1 + a_1, \ldots, \text{ch} \cdot w_\ell + a_\ell)$, $u = (s \cdot \rho^{\text{ch}}) \mod N$ and returns $\text{resp} = (\mathbf{z}, u)$. Given a proof $\sigma = (\text{com}, \text{ch}, \text{resp})$, the verifier outputs 1 if and only if $u^e \cdot \prod_{i=1}^{\ell} g_i^{z_i} = h^{\text{ch}} \cdot \text{com} \mod N$.

We obtain the following result:

Corollary 3. *Let (N, e) be such that the ℓ-representation problem is hard for $\mathcal{R}_{\text{RSA}}$. Then, the signature scheme obtained by applying Fischlin's transformation to the generalized Guillou-Quisquater Σ-protocol is fully leakage-resilient for leakage parameter $\lambda \leq (1 - o(1)) \cdot n$, where $n = \ell \log(e) + \log \phi(N)$ is the length of the secret key.*

A similar statement can be obtained based on factoring, following Fischlin and Fischlin [25].

4   Comparison

In this section we investigate the efficiency of signature schemes obtained via the Fiat-Shamir transform and Fischlin's transformation. We do so by comparing the performance of the two paradigms for an implementation of the signature scheme resulting from the Generalized Okamoto scheme [39] (see Section 3). Our implementation was carried out in the Charm cryptographic framework [2] in Python. The experiments were performed on a single core of a 3 GHz Intel Core i7. We instantiated the random oracle by SHA-2.

In our implementation of Fischlin's transformation we do not impose an upper bound on the challenge size (i.e., $\mu = \infty$). As a consequence we do not require a threshold S because given $b \cdot r = \omega(\log k)$ the probability of finding appropriate challenges ch_i mapping $H(m, x, \text{com}, i, \text{ch}_i, \text{resp}_i)$ to 0^b is negligibly close to 1 in the security parameter k. Having no threshold in the loop-clause yields to a signature scheme with expected (rather than strict) polynomial running time.

---

[6] Recall that the RSA problem requires to compute ρ such that $\rho^e = u \mod N$, given (u, N, e).

Nonetheless, the running times of the original version and ours do not differ noticeably. Our experiments have shown, in fact, that the full span of challenge candidates was rarely explored.

We stress that our implementations are not optimized. In particular, we expect faster running times when carrying out the implementation in C/C#. However, a prototype implementation in Python gives still reasonable timings for the purpose of comparing the two transformations.

4.1 Parameter Selection

In order to select reasonable parameters for the two signature schemes we have to assess the hardness of the underlying hardness assumption and take the tightness gap of the reduction into account. The security of the Generalized Okamoto scheme relies on the hardness of the representation problem in a group $\mathbb{G}$ of prime order p, which is equivalent to the discrete logarithm problem [3, Lemma 4.1]. Here, we focus on the case where $\mathbb{G} = GF(p)$ is a Galois fields. More precisely, we take the multiplicative group $\mathbb{Z}_p^*$ where p is a safe prime, i.e. $p = 2q + 1$, and p and q are both prime.

The best known algorithm to solve the discrete logarithm problem in $\mathbb{G} = GF(p)$ is the Number Field Sieve (NFS), with complexity $L_p[1/3, (64/9)^{1/3}]$ for modulus p, where the complexity function $L_p[t, s]$ is defined as $L_p[t, s] = e^{s(1+o(1))(\ln p)^t (\ln \ln p)^{1-t}}$. When estimating security parameters we take previously known attacks and timings into account by saying that if computing discrete logarithms in groups of order p takes time t, than we expect that computing discrete logarithms in groups of order p' takes time roughly $t' \approx t \frac{L_p[1/3, \sqrt[3]{64/9}]}{L_{p'}[1/3, \sqrt[3]{64/9}]}$.

If the difference between p' and p is not too large, the term $o(1)$ goes to zero. A similar strategy was recommended in [36]. We take as reference the 2009 factorization of a 768-bit modulus, which offers roughly 66-bit security ($t \approx 2^{66}$).

Let us now consider both schemes with their respective security reduction. If an adversary $\mathcal{A}$ (t', ε')-breaks the signature scheme obtained via Fiat-Shamir (see [3,32]), there exists an adversary $\mathcal{B}$ with runtime $\approx t'$ that solves the ℓ-representation problem in $\mathbb{G}$ with probability $(\varepsilon')^2 / q_h$, where q_h is the number of queries to the random oracle H. Thus, we need to run $\mathcal{B}$ around $O(q_h/(\varepsilon')^2)$, yielding a runtime $t \approx t' \cdot q_h/(\varepsilon')^2$. For ε' large enough (say $\varepsilon' > 0.1$) and for $q_h \approx t'$, we have $t \approx 2^{165}$ for 80-bit security. Thus, the parameters must be chosen such that computing the discrete logarithm in $\mathbb{G}$ with NFS takes time roughly 2^{165}. This holds for a prime p of roughly 5400 bits. Analogously, for 128-bits of security one needs a prime p of roughly 15000 bits.

We compare these results to Fischlin's scheme. Since the reduction is tight ($\varepsilon \approx \varepsilon' - \varepsilon_{ext} - 2^{-k}$), an adversary $\mathcal{B}$ solves the ℓ-representation problem in $\mathbb{G}$ in time $t \approx t' \cdot (\varepsilon' - \varepsilon_{ext} - 2^{-k})^{-1}$. Recall that ε_{ext} is the extractor's success probability to extract the witness in the security reduction. We have to set parameters r and b such that $\varepsilon_{ext} \leq q_h \cdot 2^{-br}$ is smaller than the advantage of

Table 1. Comparison between Fiat-Shamir (FS) and Fischlin for the Generalized Okamoto signature scheme. The table shows performance and sizes for $\ell = 2$.

	80-bit security				128-bit security			
	FS	Fischlin $r = 7$ $b = 12$	Fischlin $r = 14$ $b = 6$	Fischlin $r = 6$ $b = 14$	FS	Fischlin $r = 7$ $b = 19$	Fischlin $r = 19$ $b = 7$	Fischlin $r = 11$ $b = 12$
Signing time (in sec)	0.463	1.037	**0.103**	3.531	5.3	290.262	**1.889**	4.715
Verification (in sec)	1.16	**0.060**	0.117	0.062	30.89	**0.993**	2.552	1.451
Signature size (in kB)	1.98	1.94	3.87	**1.67**	5.49	**5.22**	14.15	8.2
Public-key size (in kB)	1.98	**0.41**	**0.41**	**0.41**	5.49	**1.12**	**1.12**	**1.12**
Secret-key size (in kB)	1.32	**0.28**	**0.28**	**0.28**	3.66	**0.37**	**0.37**	**0.37**

solving the representation problem ε.[7] We can choose a 1130-bit prime p for $\mathbb{G} = GF(p)$ for 80-bit security and 3048-bit prime for 128-bit security, respectively.

In the following we compare both signature schemes in terms of performance and signature size. We stress that for some signature schemes obtained via Fiat-Shamir or Fischlin (e.g., the Schnorr signature scheme [43]), signatures can be shortened by removing the commitment(s) from the signature because the commitment is re-computable from the challenge and response alone. (This holds in particular for the signature derived from the Generalized Okamoto scheme.) Given the above system parameters, a signature computed via Fiat-Shamir consists of one hash value (the challenge, from $\mathbb{Z}_p^*$), and ℓ elements from $\mathbb{Z}_p^*$ (the response), yielding a signature of size 16200 bits for 80-bit security (resp. 45000 bits for 128-bit security) and $\ell = 2$. On the other hand, a signature computed via Fischlin consists of r challenge values of expected size b bits, and $r \cdot \ell$ elements from $\mathbb{Z}_p^*$. We obtain comparable signature sizes with $r = 7$ for both 80 and 128 bits of security.

In Table 1 and Figures 2 and 3 we illustrate the performances of both schemes in terms of key generation, signing time, verification time, and leakage resilience. The result of the comparison are discussed in the following subsections.

4.2 On Key Generation

The key generation algorithm is exactly the same for both schemes. However, due to the loose reduction of the Fiat-Shamir transform, the resulting signature scheme requires much larger groups than the one derived via Fischlin's transformation.

Recall that, in order to resist special discrete logarithm solvers, we have to instantiate the groups in $\mathbb{Z}_p^*$ where p is a safe prime, i.e., $p = 2q + 1$ with p, q prime. Finding safe primes is an expensive task. Especially, signatures derived via Fiat-Shamir require to sample a safe prime p of size 15000 which may take

---

[7] We derive the bound $\varepsilon_{ext} \leq q_h \cdot 2^{-br}$ by adapting the proof of Corollary 1 to the proposed modifications of the scheme, as described above.

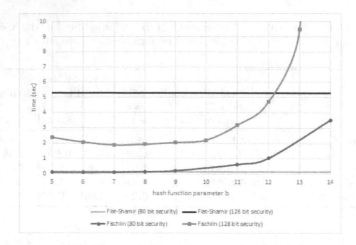

Fig. 2. Signature running time for the Fiat-Shamir and Fischlin transformation

several days or weeks.[8] Even finding a safe prime of 5400 bits (for 80-bit security) took us roughly 12 hours using the safe prime generator of OpenSSL.

As a consequence Fischlin's transformation is preferable when it comes to key generation in both performance and size. Both public and secret-key size of Fischlin's scheme are roughly 80% shorter than the ones in Fiat-Shamir.

4.3   On Signature Generation

For what concerns the signing time, we paid attention to perform the comparison between the two schemes as fair as possible. In particular group operations were implemented in the same way, and the experiments were run on the same machine using the same crypto libraries. As recommended by Fischlin [24], we enhanced the computation of the hash function in Fischlin's scheme by pre-computing and saving part of the hash values $H(m, x, \mathbf{com}, \cdot, \cdot, \cdot)$ (since this part is fixed throughout all loops).[9] We also note that a clever implementation of Fischlin's scheme, requires just additions in the group $\mathbb{G}$ when searching for an appropriate challenge (see step 2. in Fig. 1). Here, one can re-use the previously computed response resp instead of computing it from scratch each time in the loop.

Fischlin's transformation offers a trade-off between performance and signature size. In Fig. 2 we illustrate the runtime of the signature algorithm for different choices of the parameters r and b, and compare it with the case of signatures obtained via Fiat-Shamir. We observe that for similar signature sizes and 80-bit security, singing via Fiat-Shamir is twice as fast as generating signatures via Fischlin. For 128-bit security, the situation is even worse (Fischlin is roughly

---

[8] For this reason we took a slightly larger safe prime p of size 16384. More precisely we took the publicly available prime $p = 2^{16384} - 364486013$ [5].

[9] We note that doing so, we distance our implementation from the random-oracle proof.

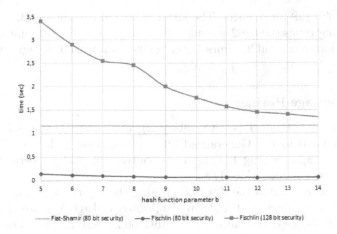

Fig. 3. Runtime of the verification algorithm for the Fiat-Shamir and Fischlin transformation

58 times slower). Nonetheless, we see that if one is flexible with respect to the signature size, Fischlin's transformation yields a signing time which is up to 4.5 times faster than the one for Fiat-Shamir, while the signature size only doubles.

The signing time for Fischlin's transformation increases rapidly from a certain threshold $b = b^*$.[10] For instance, if $b = 14$ and we stick to 80-bit security, the signing algorithm takes time roughly 3.5 seconds. Taking 128-bit security, any $b > 13$ leads to huge signing time. For $b = 13$, a signature requires 9.47 seconds. The signing time for Fischlin and Fiat-Shamir becomes of the same magnitude for $r = 11$ and $b = 12$ (128-bit security).

4.4   On Verification

Recall that, as mentioned in Section 4.1, we implemented the compressed form of the Generalized Okamoto scheme. That is, signatures consist of $(\mathsf{ch}, \mathbf{resp})$ for the Fiat-Shamir transform and $(\mathsf{ch}_i, \mathbf{resp}_i)_{1 \leq i \leq r}$ for Fischlin's transformation. Consequently, the verification algorithm first has to reconstruct the commitment (vector) and then check for the validity of the response(s).

Fig. 3 shows the running time of the verification algorithm for the two schemes. Interestingly, verifying signatures obtained via Fischlin's transformation is significantly faster than verifying signatures obtained via Fiat-Shamir. For comparable signature sizes, Fischlin's transformation yields a verification time which is roughly 19 times faster than Fiat-Shamir for 80-bit security, and roughly 30

---

[10] We believe that the value for the threshold b^* is dependent on the implementation and programming language. For example, for certain parameters, the time needed to compute all hash values exceeds the time needed to perform all necessary group operations. If the gap between the former and the latter changes (e.g., due to a different programming language), one might expect that the threshold b^* shifts, eventually leading to a different outcome for the comparison of signing time.

times faster for 128-bit security. The latter choice of parameters, however, leads to high signing times (209.262 seconds). For comparable signing times, Fischlin's transformation offers still 21 times faster verification with 37% signature size increase.

4.5 On Leakage Resilience

As discussed in Section 3 (see Corollary 2), our analysis of Fischlin's transformation applied to the Generalized Okamoto scheme yields relative leakage asymptotically approaching 1. (This is in contrast to [3,32], where the relative leakage is always smaller than $1/2$.)

In Table 2 one can find the time for running the signing algorithm when considering different values for the desired tolerated leakage λ. Note that to allow higher relative leakage, the parameter ℓ must be increased. This generates a linear blow-up in both the running times and signature size. However, if we set parameters b, r, ℓ such that signatures resulting by applying Fischlin's transformation match Fiat-Shamir in terms of signature size and amount of tolerated leakage, the signing time for Fischlin's transform decreases considerably. This is because the parameter ℓ can be fixed to 2 if we want to tolerate relative leakage less than $1/2$. Thus, one can choose larger r—and smaller b—to meet the signature size of Fiat-Shamir, and end-up with an efficiency gain.

In conclusion, when we take leakage resilience into account, Fischlin's transformation outperforms the Fiat-Shamir transform in signing time, verification time and signature size.

Table 2. Comparison of Fischlin's transformation and the Fiat-Shamir transform for the Generalized Okamoto signature scheme, with different leakage parameter λ. Fischlin is instantiated with r and b such that the resulting signature size is comparable in both schemes. For the timing (*) we selected the fastest parameters r, b where the resulting signature size is even smaller.

Signature running time (in sec)	80-bit security			
	$\lambda \leq 1/4$	$\lambda \leq 3/8$	$\lambda \leq 7/16$	$\lambda \leq 3/4$
	$\ell = 2$	$\ell = 4$	$\ell = 8$	—
	$\ell' = 2$	$\ell' = 2$	$\ell' = 2$	$\ell' = 4$
	$|\sigma| \approx 1.98$	$|\sigma| \approx 3.3$	$|\sigma| \approx 5.93$	$|\sigma| \approx 5.52$
Fiat-Shamir (with ℓ)	0.463	0.951	1.858	—
Fischlin (with ℓ')	1.037	0.114	0.103*	0.287

5   Discussion

We have shown that Fischlin's transformation is a viable approach to get fully leakage-resilient signatures with a tight reduction to the underlying hard problem, and asymptotically optimal relative leakage (in the random oracle model).

This is in contrast to the situation for signatures obtained via Fiat-Shamir (having a non-tight reduction and non-optimal relative leakage). When one takes into account this gap (as demonstrated by our implementation), it becomes evident that, in many important scenarios, Fischlin's transformation might be preferable (or at least comparable) to the Fiat-Shamir transformation in terms of verification time, signing time and key generation.

We conclude with a number of remarks on our results.

- Our model of leakage resilience assumes that the overall amount of leakage is bounded by some a-priori fixed parameter. Using the techniques of [3,1], our results extend to continuous leakage resilience in the so-called "Floppy model" (see also [12]), where one assumes the existence of a (leakage free) floppy that can be used to update the secret key (leaving the corresponding public key unchanged).
- Our analysis can be extended to the context of memory tampering (see, e.g., [18,21,22] and references therein). In particular Theorem 2 can be generalized to the setting of bounded leakage and tamper resilience [12], where the adversary is not only allowed to leak from the state of the signer but can also inject (an a-priori bounded number of) faults into the secret key and then obtain access to a "faulty" signing oracle.
- We remark that in general Fischlin's transformation can be preferable to Fiat-Shamir in all settings where simulation extractability is needed [20].

Furthermore, we ask ourselves how secure signature schemes are against quantum adversaries if the underlying identification scheme is quantum immune. More precisely, does the Fischlin's transformation yield secure signature schemes in the quantum random oracle model [6]. Dagdelen et al. [10] have shown that this does not always hold for the Fiat-Shamir transformation.

Acknowledgments. Özgür Dagdelen is supported by the German Federal Ministry of Education and Research (BMBF) within EC-SPRIDE.

References

1. Agrawal, S., Dodis, Y., Vaikuntanathan, V., Wichs, D.: On continual leakage of discrete log representations. In: Sako, K., Sarkar, P. (eds.) ASIACRYPT 2013, Part II. LNCS, vol. 8270, pp. 401–420. Springer, Heidelberg (2013)
2. Akinyele, J.A., Garman, C., Miers, I., Pagano, M.W., Rushanan, M., Green, M., Rubin, A.D.: Charm: A framework for rapidly prototyping cryptosystems. Journal of Cryptographic Engineering 3(2), 111–128 (2013), http://dx.doi.org/10.1007/s13389-013-0057-3
3. Alwen, J., Dodis, Y., Wichs, D.: Leakage-resilient public-key cryptography in the bounded-retrieval model. In: Halevi, S. (ed.) CRYPTO 2009. LNCS, vol. 5677, pp. 36–54. Springer, Heidelberg (2009)
4. Bellare, M., Rogaway, P.: Random oracles are practical: A paradigm for designing efficient protocols. In: ACM Conference on Computer and Communications Security, pp. 62–73 (1993)

5. blog, K.: Mostly on computers and mathematics. Website Blog,
   http://kenta.blogspot.de/2011/01/cvogqzhd-some-large-safe-primes.html
   (last acess: Januaury 31, 2013)
6. Boneh, D., Dagdelen, Ö., Fischlin, M., Lehmann, A., Schaffner, C., Zhandry, M.:
   Random oracles in a quantum world. In: Lee, D.H., Wang, X. (eds.) ASIACRYPT
   2011. LNCS, vol. 7073, pp. 41–69. Springer, Heidelberg (2011)
7. Boyle, E., Segev, G., Wichs, D.: Fully leakage-resilient signatures. J. Cryptology 26(3), 513–558 (2013)
8. Brakerski, Z., Kalai, Y.T., Katz, J., Vaikuntanathan, V.: Overcoming the hole
   in the bucket: Public-key cryptography resilient to continual memory leakage. In:
   FOCS, pp. 501–510 (2010)
9. Chatterjee, S., Menezes, A., Sarkar, P.: Another look at tightness. IACR Cryptology ePrint Archive 2011, 442 (2011)
10. Dagdelen, Ö., Fischlin, M., Gagliardoni, T.: The fiat–shamir transformation in a
    quantum world. In: Sako, K., Sarkar, P. (eds.) ASIACRYPT 2013, Part II. LNCS,
    vol. 8270, pp. 62–81. Springer, Heidelberg (2013)
11. Damgård, I.: On Σ-protocols. Tech. rep., Aarhus University (2013),
    http://www.daimi.au.dk/~ivan/Sigma.pdf
12. Damgård, I., Faust, S., Mukherjee, P., Venturi, D.: Bounded tamper resilience: How
    to go beyond the algebraic barrier. In: Sako, K., Sarkar, P. (eds.) ASIACRYPT
    2013, Part II. LNCS, vol. 8270, pp. 140–160. Springer, Heidelberg (2013)
13. Davì, F., Dziembowski, S., Venturi, D.: Leakage-resilient storage. In: Garay, J.A.,
    De Prisco, R. (eds.) SCN 2010. LNCS, vol. 6280, pp. 121–137. Springer, Heidelberg
    (2010)
14. Dodis, Y., Haralambiev, K., López-Alt, A., Wichs, D.: Cryptography against con-
    tinuous memory attacks. In: FOCS, pp. 511–520 (2010)
15. Dodis, Y., Haralambiev, K., López-Alt, A., Wichs, D.: Efficient public-key cryptog-
    raphy in the presence of key leakage. In: Abe, M. (ed.) ASIACRYPT 2010. LNCS,
    vol. 6477, pp. 613–631. Springer, Heidelberg (2010)
16. Dodis, Y., Ostrovsky, R., Reyzin, L., Smith, A.: Fuzzy extractors: How to generate
    strong keys from biometrics and other noisy data. SIAM J. Comput. 38(1), 97–139
    (2008)
17. Dziembowski, S., Pietrzak, K.: Leakage-resilient cryptography. In: FOCS, pp.
    293–302 (2008)
18. Dziembowski, S., Pietrzak, K., Wichs, D.: Non-malleable codes. In: ICS, pp.
    434–452 (2010)
19. Faust, S., Kiltz, E., Pietrzak, K., Rothblum, G.N.: Leakage-resilient signatures. In:
    Micciancio, D. (ed.) TCC 2010. LNCS, vol. 5978, pp. 343–360. Springer, Heidelberg
    (2010)
20. Faust, S., Kohlweiss, M., Marson, G.A., Venturi, D.: On the non-malleability of
    the fiat-shamir transform. In: Galbraith, S., Nandi, M. (eds.) INDOCRYPT 2012.
    LNCS, vol. 7668, pp. 60–79. Springer, Heidelberg (2012)
21. Faust, S., Mukherjee, P., Nielsen, J.B., Venturi, D.: Continuous non-malleable
    codes. In: Lindell, Y. (ed.) TCC 2014. LNCS, vol. 8349, pp. 465–488. Springer,
    Heidelberg (2014)
22. Faust, S., Mukherjee, P., Venturi, D., Wichs, D.: Efficient non-malleable codes and
    key-derivation for poly-size tampering circuits. In: Oswald, E. (ed.) EUROCRYPT
    2014. LNCS, vol. 8441, pp. 111–128. Springer, Heidelberg (2014)
23. Fiat, A., Shamir, A.: How to prove yourself: Practical solutions to identification and
    signature problems. In: Odlyzko, A.M. (ed.) Advances in Cryptology - CRYPTO
    1986. LNCS, vol. 263, pp. 186–194. Springer, Heidelberg (1986)

24. Fischlin, M.: Communication-efficient non-interactive proofs of knowledge with online extractors. In: Shoup, V. (ed.) CRYPTO 2005. LNCS, vol. 3621, pp. 152–168. Springer, Heidelberg (2005)

25. Fischlin, M., Fischlin, R.: The representation problem based on factoring. In: Preneel, B. (ed.) CT-RSA 2002. LNCS, vol. 2271, pp. 96–113. Springer, Heidelberg (2002)

26. Garg, S., Bhaskar, R., Lokam, S.V.: Improved bounds on security reductions for discrete log based signatures. In: Wagner, D. (ed.) CRYPTO 2008. LNCS, vol. 5157, pp. 93–107. Springer, Heidelberg (2008)

27. Goh, E.J., Jarecki, S., Katz, J., Wang, N.: Efficient signature schemes with tight reductions to the Diffie-Hellman problems. J. Cryptology 20(4), 493–514 (2007)

28. Goldreich, O.: On the foundations of modern cryptography. In: Kaliski Jr., B.S. (ed.) CRYPTO 1997. LNCS, vol. 1294, pp. 46–74. Springer, Heidelberg (1997)

29. Goldwasser, S., Micali, S., Rivest, R.L.: A digital signature scheme secure against adaptive chosen-message attacks. SIAM J. Comput. 17(2), 281–308 (1988)

30. Guillou, L.C., Quisquater, J.-J.: A "Paradoxical" identity-based signature scheme resulting from zero-knowledge. In: Goldwasser, S. (ed.) CRYPTO 1988. LNCS, vol. 403, pp. 216–231. Springer, Heidelberg (1990)

31. Ishai, Y., Sahai, A., Wagner, D.: Private circuits: Securing hardware against probing attacks. In: Boneh, D. (ed.) CRYPTO 2003. LNCS, vol. 2729, pp. 463–481. Springer, Heidelberg (2003)

32. Katz, J., Vaikuntanathan, V.: Signature schemes with bounded leakage resilience. In: Matsui, M. (ed.) ASIACRYPT 2009. LNCS, vol. 5912, pp. 703–720. Springer, Heidelberg (2009)

33. Koblitz, N., Menezes, A.: Another look at "provable security". J. Cryptology 20(1), 3–37 (2007)

34. Kocher, P.C.: Timing attacks on implementations of Diffie-Hellman, RSA, DSS, and other systems. In: Koblitz, N. (ed.) Advances in Cryptology - CRYPTO 1996. LNCS, vol. 1109, pp. 104–113. Springer, Heidelberg (1996)

35. Kocher, P.C., Jaffe, J., Jun, B.: Differential power analysis. In: Wiener, M. (ed.) CRYPTO 1999. LNCS, vol. 1666, pp. 388–397. Springer, Heidelberg (1999)

36. Lenstra, A.K., Verheul, E.R.: Selecting cryptographic key sizes. In: Imai, H., Zheng, Y. (eds.) PKC 2000. LNCS, vol. 1751, pp. 446–465. Springer, Heidelberg (2000)

37. Micali, S., Reyzin, L.: Physically observable cryptography. In: Naor, M. (ed.) TCC 2004. LNCS, vol. 2951, pp. 278–296. Springer, Heidelberg (2004)

38. Nielsen, J.B., Venturi, D., Zottarel, A.: Leakage-resilient signatures with graceful degradation. In: Krawczyk, H. (ed.) PKC 2014. LNCS, vol. 8383, pp. 362–379. Springer, Heidelberg (2014)

39. Okamoto, T.: Provably secure and practical identification schemes and corresponding signature schemes. In: Brickell, E.F. (ed.) CRYPTO 1992. LNCS, vol. 740, pp. 31–53. Springer, Heidelberg (1993)

40. Paillier, P., Vergnaud, D.: Discrete-log-based signatures may not be equivalent to discrete log. In: Roy, B. (ed.) ASIACRYPT 2005. LNCS, vol. 3788, pp. 1–20. Springer, Heidelberg (2005)

41. Quisquater, J.-J., Samyde, D.: ElectroMagnetic analysis (EMA): Measures and counter-measures for smart cards. In: Attali, S., Jensen, T. (eds.) E-smart 2001. LNCS, vol. 2140, pp. 200–210. Springer, Heidelberg (2001)

42. Schäge, S.: Tight proofs for signature schemes without random oracles. In: Paterson, K.G. (ed.) EUROCRYPT 2011. LNCS, vol. 6632, pp. 189–206. Springer, Heidelberg (2011)

43. Schnorr, C.P.: Efficient identification and signatures for smart cards. In: Quisquater, J.-J., Vandewalle, J. (eds.) Advances in Cryptology - EUROCRYPT 1989. LNCS, vol. 434, pp. 688–689. Springer, Heidelberg (1990)
44. Schütze, T.: Automotive security: Cryptography for car2x communication. In: Embedded World Conference (2011)
45. Seurin, Y.: On the exact security of schnorr-type signatures in the random oracle model. In: Pointcheval, D., Johansson, T. (eds.) EUROCRYPT 2012. LNCS, vol. 7237, pp. 554–571. Springer, Heidelberg (2012)

Anonymous IBE from Quadratic Residuosity with Improved Performance

Michael Clear*, Hitesh Tewari, and Ciarán McGoldrick

School of Computer Science and Statistics,
Trinity College Dublin

Abstract. Identity Based Encryption (IBE) has been constructed from bilinear pairings, lattices and quadratic residuosity. The latter is an attractive basis for an IBE owing to the fact that it is a well-understood hard problem from number theory. Cocks constructed the first such scheme, and subsequent improvements have been made to achieve anonymity and improve space efficiency. However, the anonymous variants of Cocks' scheme thus far are all less efficient than the original. In this paper, we present a new universally-anonymous IBE scheme based on the quadratic residuosity problem. Our scheme has better performance than the universally anonymous scheme from Ateniese and Gasti (CT-RSA 2009) at the expense of more ciphertext expansion.

Keywords: Identity Based Encryption, Anonymous IBE, Cocks Scheme, Quadratic Residuosity.

1 Introduction

Identity-Based Encryption (IBE) is centered around the notion that a user's public key can be efficiently derived from an identity string and system-wide public parameters. The public parameters are chosen by a Trusted Authority (TA) along with a master secret key, which is used to extract secret keys for user identities. IBE was first proposed by Shamir [1]. The first secure IBE schemes were presented by Cocks [2] (based on the quadratic residuosity problem), and Boneh and Franklin [3] (based on bilinear pairings). More recently, there have been IBE constructions based on worst-case lattice problems [4, 5]. Ciphertext expansion in Cocks' scheme is large, which has hindered its practicality. Nevertheless, it is notable as being one of the few known IBE constructions based on number-theoretic assumptions. The quadratic residuosity problem on which it is based has been well studied, and is held to be a hard problem. Since it relies on such a standard assumption, Cocks' scheme has been subject to research efforts to derive more powerful primitives such as anonymous IBE or Public-key Encryption with Keyword Search (PEKS) [6]. It is known that Cocks' scheme is not anonymous.

The notion of anonymity stems from that of key privacy put forward by Bellare et al. [7]. An IBE scheme is said to be anonymous if an adversary cannot

---

* The author's work is funded by the Irish Research Council EMBARK Initiative.

D. Pointcheval and D. Vergnaud (Eds.): AFRICACRYPT 2014, LNCS 8469, pp. 377–397, 2014.

distinguish which identity was used to create a ciphertext, even if the adversary gets to choose a pair of identities to distinguish between. Anonymous IBE is a useful primitive because it can be used to facilitate searching on encrypted data, to allow anonymous broadcasts to be made in a network, and to act as a countermeasure against traffic analysis. A multitude of anonymous IBEs have been constructed based on both pairings and lattices including [3,5,6,8].

Anonymous variants of Cocks' IBE scheme whose security relies on the quadratic residuosity assumption have already been proposed in the literature [9–11]. The most efficient in terms of ciphertext size is due to Boneh, Gentry and Hamburg [10]. However, encryption time in their scheme is quartic in the security parameter, and thus has poor performance. The PEKS scheme in [9] performs better but still requires many Jacobi symbol computations when used as an anonymous IBE. The most time-efficient anonymous IBE to date was presented at CT-RSA 2009 by Ateniese and Gasti [11]. Their construction has similarly-sized ciphertexts to Cocks' original scheme while there is a drop of approximately 30% in performance compared to Cocks according to our experimental results (for a 1024-bit modulus used to encrypt a 128-bit symmetric key; note that IBE is typically used as part of a KEM-DEM). While this is still practical, it is desirable to obtain an anonymous IBE from quadratic residuosity whose performance is on par with the original Cocks scheme, especially for time-critical applications.

1.1   Universal Anonymity

Ateniese and Gasti's scheme also enjoys the property of universal anonymization, first introduced at Asiacrypt 2005 by Hayashi and Tanaka [12]. This property allows any party to anonymize a ciphertext without access to the secret key of the recipient. An illustrative application involves disparate systems distinguished by whether they need to know the intended recipient of encrypted data. Regulations may stipulate that some systems learn the recipient's identity. At some suitable point prior to sending the encrypted data to less trusted systems, the encrypted data can be anonymized by any party without knowledge of the secret key.

1.2   Contributions

We present a new universally anonymous IBE from quadratic residuosity whose performance closely matches that of the original Cocks scheme. Our work builds upon techniques presented in [13], especially the homomorphic property identified therein, to construct a universally anonymous variant of Cocks' scheme that achieves better performance than [11]. Unfortunately, the size of ciphertexts in our scheme is double that of Cocks, and almost double that of [11]. However, we obtain anonymity using a different approach which we believe to be conceptually simpler. We prove this system ANON-IND-ID-CPA secure in the random oracle model and provide both an analytical and experimental comparison between our approach and that of [11].

Another contribution of this paper is a security assessment of a scheme by Jhanwar and Barua [14], which in turn is a variant of the non-anonymous IBE system from [10]. We consider an alternative parameter setting that ensures IND-ID-CPA security. The resulting scheme outperforms the original Cocks scheme, and is slightly more space-efficient. Although the same ideas do not readily allow us to construct an anonymous IBE, the performance benefits provide good motivation for pursuing this in future work. Performance measurements from this scheme are reported along with the others in Section 4. However, due to space constraints, the details of this scheme are deferred to Appendix B.

1.3 Overview of Main Construction

As pointed out in previous works, the main obstacle to achieving anonymity for variants of Cocks' scheme is a property that is unconditionally satisfied for ciphertexts produced under a certain identity id. This property holds with probability negligibly close to $1/2$ with respect to any other identity id′. Thus, it is possible for an adversary to readily distinguish the recipient's identity by checking whether this property holds.

We provide an informal description here to highlight the intuition behind our approach. Let $N = pq$ be an integer where p and q are prime. Let $x \in \mathbb{Z}$. We write $\left(\frac{x}{N}\right)$ to denote the Jacobi symbol of $x \mod N$.

As in [11], we let $H : \{0,1\}^* \to \mathbb{Z}_N^*[+1]$ be a full-domain hash. A message bit is mapped to an element of $\{-1,1\}$ via a mapping $\nu : \{0,1\} \to \{-1,1\}$ with $\nu(0) = 1$ and $\nu(1) = -1$.

An overview of the Cocks scheme is as follows. The Trusted Authority (TA) generates two large primes p and q, which constitute the master secret key. It outputs the public parameters $N = pq$. For any identity id, the public key corresponding to that identity is computed as $a = H(\mathsf{id})$. It will be shown later that given p and q, it is easy to derive an integer $r \in \mathbb{Z}_N$ with

$$r^2 \equiv a \mod N \quad \text{or} \quad r^2 \equiv -a \mod N.$$

Such an r is a secret key for identity id. Now encryption of a message $m \in \{0,1\}$ under identity id is straightforward: an encryptor samples two integers $t_1, t_2 \in \mathbb{Z}_N^*$ uniformly at random subject to the condition that

$$\left(\frac{t_1}{N}\right) = \left(\frac{t_2}{N}\right) = \nu(m).$$

It then computes a ciphertext $(c := t_1 + at_1^{-1}, d := t_2 - at_2^{-1})$. Decryption is also simple: set $e := c$ if $r^2 \equiv a \mod N$; otherwise set $e := d$. Then we decrypt by computing $\nu^{-1}\left(\left(\frac{e + 2r}{N}\right)\right)$. However, to simplify the description, we will focus our attention on the first component of a ciphertext, namely c. In fact, the properties that we will consider concerning such c with respect to a hold analogously for d with respect to $-a$.

It was observed by Galbraith[1] that for any integer c generated as above, it is an invariant that

$$\left(\frac{c^2 - 4a}{N}\right) = 1.$$

We expect this to hold with probability negligibly close to $1/2$ for random a. Hence, an adversary has a non-negligible advantage attacking anonymity. In the XOR-homomorphic variant from [13], the integer c is replaced by a polynomial $c(x) = c_1 x + c_0$ in the quotient ring $R_a = \mathbb{Z}_N[x]/(x^2 - a)$. We can generalize the above test for polynomials in R_a. Define

$$\mathsf{GT}(a, c(x), N) = \left(\frac{c_0^2 - c_1^2 a}{N}\right).$$

Now we define two subsets $G_a = \{c(x) \in R_a : \mathsf{GT}(a, c(x), N) = 1\}$ and $\bar{G}_a = \{c(x) \in R_a : \mathsf{GT}(a, c(x), N) = -1\}$ of R_a. In addition, the set of legally generated ciphertext polynomials (i.e. those in the image of the encryption algorithm) is denoted by the set S_a. It is shown in [13] that $S_a \underset{C}{\approx} G_a$ (computationally indistinguishable) even given access to the secret key r. It is also shown that G_a is a multiplicative group in R_a and S_a is a subgroup of G_a.

The main idea behind our construction is to allow anonymized ciphertexts to be elements of $\bar{G}_a$ half of the time and G_a the other half. Therefore, the adversary cannot use Galbraith's test to distinguish identities. The main problem however is that we don't know what a "ciphertext" in $\bar{G}_a$ decrypts to without knowing the secret key. We can show that a random element in $\bar{G}_a$ can be sampled by multiplying any fixed element in $g(x) \in \bar{G}_a$ by a uniformly random element of G_a. Our idea is to derive this fixed element $g(x)$ from the user's identity using a hash function (modelled as a random oracle in the security proofs), and then multiply it by an encryption of the desired message, which lies in S_a. Since S_a and G_a are computationally indistinguishable, the resultant element $c'(x)$ is also computationally indistinguishable from a random element in $\bar{G}_a$. It can also be shown that the homomorphic property holds even between polynomials in $\bar{G}_a$ and G_a. Therefore, $c'(x)$ is an encryption of the desired message XORed with whatever $g(x)$ decrypts to. Since the decryptor can determine what $g(x)$ decrypts to, she can recover the message.

1.4   Related Work

Di Crescenzo and Saraswat [9] constructed an anonymous variant of Cocks' scheme. In fact their construction is an instance of Public-Key Encryption with Keyword Search (PEKS), a primitive introduced in [6] which allows a sender to encrypt a message with a set of hidden keywords such that a decryptor can only determine whether a specific keyword W appears in the ciphertext if she holds a secret key for W (the secret keys are computed by the TA). The scheme from [9] requires $4k$ elements of $\mathbb{Z}_N$ where k is the length of keywords represented as

---

[1] Reported as emerging from personal communication in [10].

binary strings. Also, encryption requires $4k$ Jacobi symbol evaluations. PEKS captures anonymous IBE as a special case. Two keywords $W_{id}^{(0)}$ and $W_{id}^{(1)}$ representing the messages 0 and 1 respectively are associated with each identity id. Accordingly, secret keys for $W_{id}^{(0)}$ and $W_{id}^{(1)}$ constitute a secret key for identity id.

Boneh, Gentry and Hamburg (BGH) [10] constructed the first space-efficient variant of the Cocks scheme. The size of ciphertexts using their anonymous scheme is quite practical; an ℓ-bit message requires a ciphertext whose size is $\log_2 N + \ell + 1$ bits, which contrasts with $2\ell \cdot \log_2 N$ bits in Cocks. However, encryption in their scheme is time-consuming. Encryption time is dominated by the generation of $\ell + 1$ primes which are needed to help satisfy $\ell + 1$ equations of the form $Rx^2 + Sy^2 \equiv 1 \mod N$. It is reported in [10] that a 1024-bit prime generation takes 123.6 ms on a 2.015 GHz AMD dual-core Athlon64. To encrypt a 128 bit key, one would expect the total time to be on the order of 16 seconds on the same machine since $128 + 1$ primes must be generated. However, the authors give a variant that instead requires primes of length $\log_2 \sqrt{N}$ bits at the expense of an increase in ciphertext length. On the same benchmark machine, a time of 11 ms is reported for a 512-bit prime generation, which brings the total time down to ≈ 1.4 seconds. However, this variant is not anonymous.

While we have not implemented the constructions in [10], we believe they are significantly slower than the scheme in [11] and the one presented in this work. Encryption time is quartic in the security parameter as opposed to cubic for standard number-theoretic schemes. A variant of the non-anonymous BGH construction appeared in [14]. The authors of that work claim their variant achieves higher performance for both encryption and decryption as a trade-off for increased ciphertext size, which is $2 \cdot \log_2 \lceil \sqrt{\ell} \rceil + 2\ell$ bits for an ℓ-bit plaintext. We describe in Appendix B why their proof of security only goes through if a sender encrypts $\log \lambda^{\omega(1)}$ bits where λ is the security parameter. While this fact hinders the space efficiency of the scheme, our experiments show that its performance is on par with Cocks for a similar level of security.

2   Preliminaries

2.1   Notation

A quantity is said to be negligible with respect to some parameter λ, written $\mathsf{negl}(\lambda)$, if it is asymptotically bounded from above by the reciprocal of all polynomials in λ.

For a probability distribution D, we denote by $x \xleftarrow{\$} D$ that x is sampled according to D. If S is a set, $y \xleftarrow{\$} S$ denotes that y is sampled from x according to the uniform distribution on S.

The set of contiguous integers $\{1, \ldots, k\}$ for some $k > 1$ is denoted by $[k]$. Let D_1 and D_2 be distributions. We write $D_1 \approx D_2$ to denote the fact that D_1 and D_2 are statistically indistinguishable. In addition, we write $D_1 \underset{C}{\approx} D_2$ to denote the fact that both distributions are computationally indistinguishable.

2.2   Security Definition for Anonymous IBE (ANON-IND-ID-CPA)

An IBE scheme is said to be anonymous if any PPT adversary has only a negligible advantage in the following game. This is referred to as ANON-IND-ID-CPA security. At the beginning of the game, the adversary $\mathcal{A}$ is handed the public parameters. It then proceeds to make queries for secret keys corresponding to identities $\mathsf{id}_1, \ldots, \mathsf{id}_{q_1}$ for some integer q_1 that is polynomial in the security parameter. Then it sends to the challenger two identities id_0^* and id_1^* such that $\mathsf{id}_0^* \neq \mathsf{id}_1^* \neq \mathsf{id}_i$ for $1 \leq i \leq q_1$. It also sends two messages m_0 and m_1. The challenger samples a bit b uniformly, and sends the encryption of m_b under id_b^* to $\mathcal{A}$. In the final phase, $\mathcal{A}$ is allowed to query secret keys for further identities $\mathsf{id}_{q_1+1}, \ldots, \mathsf{id}_{q_1+q_2}$ where q_2 is polynomial in the security parameter, and $\mathsf{id}_0^* \neq \mathsf{id}_1^* \neq \mathsf{id}_{q_1+i}$ for $1 \leq i \leq q_2$. Finally, $\mathcal{A}$ outputs a guess b' and is said to win if $b' = b$.

2.3   Quadratic Residues and Jacobi Symbols

Let m be an integer. A quadratic residue in the residue ring $\mathbb{Z}_m$ is an integer x such that $x \equiv y^2 \mod m$ for some $y \in \mathbb{Z}_m$. The set of quadratic residues in $\mathbb{Z}_m$ is denoted $\mathbb{QR}(m)$. If m is prime, it is easy to determine whether any $x \in \mathbb{Z}_m$ is a quadratic residue. If m is an odd prime number, we can define the Legendre symbol as a function of any integer $x \in \mathbb{Z}$ with respect to m as

$$\left(\frac{x}{m}\right) = \begin{cases} 1 & \text{if } x \in \mathbb{QR}(m) \\ -1 & \text{if } x \not\equiv 0 \mod m \text{ and } x \notin \mathbb{QR}(m) \\ 0 & \text{if } x \equiv 0 \mod m \end{cases}.$$

The above function can be generalized to positive odd moduli $M = m_1^{\alpha_1} \ldots m_k^{\alpha_k}$ where $m_1, \ldots, m_k$ are prime, and $\alpha_1, \ldots, \alpha_k$ are positive integers. The generalization is called a Jacobi symbol and is defined as

$$\left(\frac{x}{M}\right) = \left(\frac{x}{m_1}\right)^{\alpha_1} \cdots \left(\frac{x}{m_k}\right)^{\alpha_k}.$$

where $\left(\dfrac{x}{m_i}\right)$ denotes the Legendre symbol of x with respect to m_i for $1 \leq i \leq k$. The subset of $\mathbb{Z}_M$ with Jacobi symbol $+1$ is denoted by $\mathbb{J}(M)$; that is, $\mathbb{J}(M) = \{x \in \mathbb{Z} : \left(\dfrac{x}{M}\right) = 1\}$. Naturally, $\mathbb{QR}(M) \subseteq \mathbb{J}(M)$.

2.4   Quadratic Residuosity Problem

Let N be a product of two odd primes p and q. The quadratic residuosity problem is to determine, given input $(N, x) \in \mathbb{Z}_N^2$ where $x \in \mathbb{J}(N)$, whether $x \in \mathbb{QR}(N)$, and it is believed to be intractable.

2.5 Blum Integers

Finally, the schemes in this paper make use of Blum integers. A Blum integer is a product of two primes that are both congruent to 3 modulo 4. As a result, we define $\mathsf{BlumGen}(1^\lambda)$ as a PPT algorithm which takes as input a security parameter λ and outputs two equally-sized primes p and q, whose lengths depend on λ, such that

$$p \equiv q \equiv 3 \pmod 4.$$

2.6 Cocks Scheme

Let $H : \{0,1\}^* \to \mathbb{J}(N)$ be a full-domain hash that sends an identity string $\mathsf{id} \in \{0,1\}^*$ to an integer in $\mathbb{Z}_N$ whose Jacobi symbol is $+1$. A secret key in Cocks' system is a Rabin signature for id. Therefore, to guarantee existential unforgeability of such signatures, the random oracle model is needed.

- Cocks.$\mathbf{Setup}(1^\lambda)$:
    1. Repeat: $(p, q) \leftarrow \mathsf{BlumGen}(1^\lambda)$.

       Note that by definition of $\mathsf{BlumGen}$, we have $p \equiv q \equiv 3 \pmod 4$.
    2. $N \leftarrow pq$
    3. Output $(\mathsf{PP} := N, \mathsf{MSK} := (N, p, q))$
- Cocks.$\mathbf{KeyGen}(\mathsf{MSK}, \mathsf{id})$:
    1. Parse MSK as (N, p, q).
    2. $a \leftarrow H(\mathsf{id})$.
    3. $r \leftarrow a^{\frac{N+5-p-q}{8}} \pmod N$.

       Therefore, either $r^2 \equiv a \pmod N$ or $r^2 \equiv -a \pmod N$.
    4. Output $\mathsf{sk}_{\mathsf{id}} := (N, \mathsf{id}, r)$

Remark 1. It is important that this algorithm always output the same square root, since otherwise N can be factored. To achieve this, one may store the root or calculate it deterministically as done so above.

- Cocks.$\mathbf{Encrypt}(\mathsf{PP}, \mathsf{id}, m)$:
    1. Parse PP as N.
    2. $a \leftarrow H(\mathsf{id})$
    3. Generate $t_1, t_2 \xleftarrow{\$} \mathbb{Z}_N^*$ such that $\left(\dfrac{t_1}{N}\right) = \left(\dfrac{t_2}{N}\right) = \nu(m)$ (Recall that $\nu(m)$ maps $m \in \{0,1\}$ into $\{-1,1\}$).
    4. Output $\psi := (t_1 + at_1^{-1}, t_2 - at_2^{-1})$
- Cocks.$\mathbf{Decrypt}(\mathsf{sk}_{\mathsf{id}}, \psi)$:
    1. Parse ψ as (ψ_1, ψ_2)
    2. Parse $\mathsf{sk}_{\mathsf{id}}$ as (N, id, r)

3. $a \leftarrow H(\mathrm{id})$
4. If $r^2 \equiv a \pmod{N}$, set $d \leftarrow \psi_1$. Else if $r^2 \equiv -a \pmod{N}$, set $d \leftarrow \psi_2$. Else output $\perp$ and abort.
5. Output $\nu^{-1}((\frac{d+2r}{N}))$

3 Time-Efficient Universally Anonymous IBE

3.1 Overview of Our Construction

In order to explain our construction, it is necessary to first describe the XOR-homomorphic variant of Cocks' scheme from [13]. Let $R = \mathbb{Z}_N[x]$ be a polynomial ring over $\mathbb{Z}_N$. Let a be an integer in $\mathbb{J}(N)$. Then let R_a be the quotient ring $R/(x^2 - a)$. Recall the generalization of Galbraith's test to the ring R as follows.

Definition 1 (Galbraith's Test over R). *Define Galbraith's Test for the ring R as the function $\mathsf{GT} : \mathbb{Z}_N \times R \to \{-1, 0, +1\}$ given by*

$$\mathsf{GT}(a, c(x), N) = \left(\frac{c_0^2 - c_1^2 a}{N}\right).$$

Define the subset $G_a \subset R_a$ as follows:

$$G_a = \{c(x) \in R_a : \mathsf{GT}(a, c(x), N) = 1\}.$$

Therefore, this is the subset of R_a that passes Galbraith's test. Define the subset $\bar{G}_a \subset R_a$ as follows:

$$\bar{G}_a = \{c(x) \in R_a : \mathsf{GT}(a, c(x), N) = -1\}.$$

Correspondingly, this is the subset of R_a that fails Galbraith's test. Now define the subset $S_a \subset G_a$:

$$S_a = \{2hx + (t + ah^2 t^{-1}) \in G_a \mid h \in \mathbb{Z}_N, t, (t + ah^2 t^{-1}) \in \mathbb{Z}_N^*\}.$$

The subset S_a is precisely the image of the following algorithm $\mathcal{E}$ which takes as input an integer $a \in \mathbb{J}(N)$ (i.e. $\left(\frac{a}{N}\right) = 1$) along with a message bit $m \in \{0, 1\}$ and produces an element of S_a that encrypts m. This is central to the XOR-homomorphic variant of the Cocks scheme presented in [13], which is referred to as xhIBE in that paper. Like Cocks' original scheme, xhIBE requires a ciphertext to have two components. As such, $\mathcal{E}$ can be viewed as the encryption algorithm for a single component. Accordingly, to encrypt a message m in xhIBE, the sender runs $\mathcal{E}(a, m)$ and $\mathcal{E}(-a, m)$ to produce the first and second component of a ciphertext respectively.

Algorithm $\mathcal{E}(a, m)$:

1. Choose an integer $t \xleftarrow{\$} \mathbb{Z}_N^*$ uniformly such that

$$\left(\frac{t}{N}\right) = \nu(m).$$

2. Choose an integer $h \xleftarrow{\$} \mathbb{Z}_N$ uniformly.

3. Compute $c(x) \leftarrow 2hx + (t + ah^2t^{-1}) \in R$

4. Repeat steps 1-4 until $(t + ah^2t^{-1}) \in \mathbb{Z}_N^*$.

5. Output $c(x)$.

With overwhelming probability, $(t + ah^2t^{-1})$ will be invertible in $\mathbb{Z}_N$.

In addition, we define a decryption algorithm $\mathcal{D}$ which takes an integer $r \in \mathbb{Z}_N$ and a polynomial in R as input, and outputs a bit $m \in \{0, 1\}$. This is defined as follows:

Algorithm $\mathcal{D}(r, c(x))$:

1. Compute $j = \left(\dfrac{c(r)}{N}\right) \in \{-1, 0, +1\}$.

2. If $j = 0$, output $\perp$.

3. Else output $\nu^{-1}(j) \in \{0, 1\}$.

Note that for the sake of notational convenience, it is assumed that N is an implicit input in $\mathcal{E}$ and $\mathcal{D}$. Suppose $a \in \mathbb{QR}(N)$. Then let $r \in \mathbb{Z}_N$ such that $r^2 \equiv a \mod N$. It can be shown that $\mathcal{D}(r, \cdot)$ whose domain is restricted to $S_a = \mathsf{image}(\mathcal{E}(a, \cdot))$ is a group homomorphism $(S_a, *) \rightarrow (\mathbb{Z}_2, +)$. Therefore for $m_1, m_2 \in \{0, 1\}$:

$$\mathcal{D}(r, \mathcal{E}(a, m_1) * \mathcal{E}(a, m_2)) = m_1 \oplus m_2.$$

In fact, for any $c(x), d(x) \in R$ with $\mathcal{D}(r, c(x)), \mathcal{D}(r, d(x)) \in \{0, 1\}$, it holds that

$$\mathcal{D}(r, c(x)d(x)) = \mathcal{D}(r, c(x)) \oplus \mathcal{D}(r, d(x)).$$

Naturally this means that an XOR homomorphism exists even between elements of G_a and $\bar{G}_a$.

Let $g(x) \in \bar{G}_a$. Below are some basic facts which we prove in Section 3.3.

1. $g(x)G_a = \bar{G}_a$.

2. $\{h(x) \xleftarrow{\$} \bar{G}_a\} \approx \{g(x)h'(x) \mid h'(x) \xleftarrow{\$} G_a\}$.

3. $\{h(x) \xleftarrow{\$} \bar{G}_a\} \underset{C}{\approx} \{g(x)h'(x) \mid h'(x) \xleftarrow{\$} S_a\}$.

Property 3 states that the uniform distribution defined over $\bar{G}_a$ and the distribution of multiplying $g(x)$ by uniformly random elements from S_a are computationally indistinguishable (without access to p and q).

We need two hash functions. Like Cocks' scheme, a full-domain hash $H : \{0, 1\}^* \rightarrow \mathbb{J}(N)$ is employed that maps identity strings to elements of $\mathbb{Z}_N$ whose

Jacobi symbol is $+1$. Another hash function $H' : \{0,1\}^* \to R$ is needed that maps an identity string id to an element $g(x) \in R$ such that $\mathsf{GT}(H(\mathsf{id}), g(x), N) = \mathsf{GT}(-H(\mathsf{id}), g(x), N) = -1$ i.e. the $g(x)$ is taken to pass Galbraith's test for both $a = H(\mathsf{id})$ and $-a$. Roughly speaking, an example of constructing such as hash function using H is via a form of rejection sampling i.e. to sample $g'(x)_i \overset{\$}{\leftarrow} H(\mathsf{id} \,\|\, i)$ for consecutive integers $i > 0$ until $\mathsf{GT}(a, g'(x)_i, N) = \mathsf{GT}(-a, g'(x)_i, N) = -1$. In the security proofs, H is modelled as a random oracle on $\mathbb{J}(N)$ and H' is modelled as a random oracle whose response when queried on id is distributed according to the uniform distribution on $\bar{G}_{H(\mathsf{id})} \cap \bar{G}_{-H(\mathsf{id})}$. To anonymize a ciphertext *component* (recall that this discussion is simplified to deal with a single component of a ciphertext corresponding to $a = H(\mathsf{id})$, the steps are repeated for the case of $-a$) $c(x)$ associated with an identity id, the following steps are performed:

1. $a \leftarrow H(\mathsf{id})$

2. $c'(x) \leftarrow \mathcal{E}(a, 0)$.

3. Uniformly sample a bit $b \overset{\$}{\leftarrow} \{0,1\}$.

4. If $b = 0$, output $c'(x)c(x)$.

5. Else compute $g(x) \leftarrow H'(\mathsf{id})$, and output $g(x)c(x)c'(x)$.

Note that the construction is universally anonymous in that anyone can anonymize a ciphertext without having the secret key for the target identity and without access to the random coins used by the encryptor.

The decryption function $\mathcal{D}'$ for our construction is defined in terms of $\mathcal{D}$.

$$\mathcal{D}'(r, c(x)) = \begin{cases} \mathcal{D}(r, c(x)) \oplus \mathcal{D}(r, g(x)) & \text{if } c(x) \in \bar{G}_a \\ \mathcal{D}(r, c(x)) & \text{if } c(x) \in G_a \\ \bot & \text{otherwise} \end{cases}$$

3.2 Formal Description

Our scheme is referred to as UAIBE for the remainder of the paper; a formal description is as follows.

Setup(1^λ) : On input a security parameter 1^λ in unary, generate $(p, q) \leftarrow \mathsf{BlumGen}(1^\lambda)$. Compute $N = pq$. Output public parameters $\mathsf{PP} = (N, H, H')$ and master secret key $\mathsf{MSK} = (N, p, q)$, where H is a hash function $H : \{0,1\}^* \to \mathbb{J}(N)$, and H' is a hash function $H' : \{0,1\}^* \to R$ with the property that for any identity $\mathsf{id} \in \{0,1\}^*$, $a \leftarrow H(\mathsf{id})$ and $g(x) \leftarrow H'(\mathsf{id})$, it holds that

$$\mathsf{GT}(a, g(x), N) = \mathsf{GT}(-a, g(x), N) = -1.$$

KeyGen(MSK, id) : On input master secret key $\mathsf{MSK} = (N, p, q)$ and identity $\mathsf{id} \in \{0,1\}^*$, perform the following steps:

1. Compute $a \leftarrow H(\mathsf{id}) \in \mathbb{J}(N)$.

2. If $r \in \mathbb{QR}(N)$, compute the square root $r = a^{1/2}$;

3. Else compute $r = (-a)^{1/2}$.

4. Output (N, id, r) as the secret key for identity id.

See the description of Cocks' scheme in Section 2.6 for a convenient way to compute a square root in $\mathbb{Z}_N$ deterministically.

Encrypt(PP, id, m): On input public parameters $\text{PP} = (N, H, H')$, an identity id $\in \{0,1\}^*$, and message $m \in \{0,1\}$ run:

1. Compute $a \leftarrow H(\text{id}) \in \mathbb{J}(N)$.

2. Compute $g(x) \leftarrow H'(\text{id}) \in R$.

3. Compute $c(x) \leftarrow \mathcal{E}(a, m)$.

4. Compute $d(x) \leftarrow \mathcal{E}(-a, m)$.

5. Uniformly sample two bits $v_1, v_2 \overset{\$}{\leftarrow} \{0,1\}$.

6. If $v_1 = 1$, then set $c(x) \leftarrow c(x) * g(x)$.

7. If $v_2 = 1$, then set $d(x) \leftarrow d(x) * g(x)$.

8. Output $\mathbf{c} := (c(x), d(x))$.

Decrypt($\text{sk}_{\text{id}}, \mathbf{c}$): On input a secret key $\text{sk}_{\text{id}} = (N, \text{id}, r)$ and a ciphertext $\mathbf{c} = (c(x), d(x))$, do:

1. Compute $a \leftarrow H(\text{id}) \in \mathbb{J}(N)$.

2. Compute $g(x) \leftarrow H'(\text{id}) \in R$.

3. If $r^2 \equiv a \mod N$, set $e(x) \leftarrow c(x)$. Else if $r^2 \equiv -a \mod N$, set $e(x) \leftarrow d(x)$. Else output $\perp$ and abort.

4. If $\text{GT}(r^2 \mod N, e(x)) = -1$, set $e(x) \leftarrow e(x) * g(x)$.

5. Output $\mathcal{D}(r, e(x))$.

3.3 Security

Lemma 1. *Let* $f(x), g(x) \in R_a$. *Then* $\text{GT}(a, f(x)g(x), N) = \text{GT}(a, f(x), N) \cdot \text{GT}(a, g(x), N)$.

Proof. Consider the product $v(x) = f(x)g(x) \in R_a$. We have that $v_0 = f_0 g_0 + f_1 g_1 a$ and $v_1 = f_0 g_1 + f_1 g_0$. It is easy to verify that

$$\left(\frac{(f_0 g_0 + f_1 g_1 a)^2 - (f_0 g_1 + f_1 g_0)^2 a}{N} \right) = \left(\frac{(f_0^2 - a f_1^2)(g_0^2 - a g_1^2)}{N} \right) =$$

$$\text{GT}(a, f(x), N) \cdot \text{GT}(a, g(x), N).$$

$\square$

Lemma 2. *Let* $g(x) \in \bar{G}_a$. *Then* $g(x) \cdot G_a = \bar{G}_a$.

Proof. By Lemma 1, $g(x)h(x) \in \bar{G}_a$ for any $h(x) \in G_a$.

By Lemma 1 in [13], G_a is a multiplicative group in R_a. Hence, $|g(x) \cdot G_a| = |G_a|$. We claim that every $t(x) \in \bar{G}_a$ can be expressed as $g(x)t'(x)$ for some

$t'(x) \in G_a$. Assume the contrary for the purpose of contradiction i.e. there exists a $t(x) \notin g(x) \cdot G_a$. It follows that $t(x) \cdot G_a \cap g(x) \cdot G_a = \emptyset$. But by Lemma 1, $t(x)^2 \in G_a$ and $g(x)t(x) \in G_a$. From the commutativity of R_a, we have $g(x) \cdot t(x)^2 = t(x) \cdot (t(x)g(x))$, which implies that $t(x) \cdot G_a \cap g(x) \cdot G_a \neq \emptyset$, a contradiction. The lemma follows.   □

We include the following result from [13] that is used in the proofs below.

Corollary 1 (Corollary 2, [13]). *The distributions* $\{(N, a, t + ah^2t^{-1}, 2h) : N \leftarrow \mathsf{Setup}(1^\lambda), a \xleftarrow{\$} \mathbb{J}, t, h \xleftarrow{\$} \mathbb{Z}_N^*)\}$ *and* $\{(N, a, z_0, z_1) : N \leftarrow \mathsf{Setup}(1^\lambda), a \xleftarrow{\$} \mathbb{J}, z_0 + z_1 x \xleftarrow{\$} G_a \setminus S_a\}$ *are indistinguishable assuming the hardness of the quadratic residuosity problem.*

Corollary 2. *Let* $g(x) \in \bar{G}_a$. *Then*

1. $\{h(x) \xleftarrow{\$} \bar{G}_a\} \approx \{g(x)h'(x) \mid h'(x) \xleftarrow{\$} G_a\}$.

2. $\{h(x) \xleftarrow{\$} \bar{G}_a\} \underset{C}{\approx} \{g(x)h'(x) \mid h'(x) \xleftarrow{\$} S_a\}$.

Proof. (1). From Lemma 2, each element in $\bar{G}_a$ can be represented as $g(x)h'(x)$ for a unique $h'(x) \in G_a$. Therefore, if $h'(x)$ is sampled uniformly from G_a, then $h'(x)g(x)$ is uniformly distributed in $\bar{G}_a$.

(2). By Corollary 1, $G_a \underset{C}{\approx} S_a$ without knowledge of the prime factors of N, and thus this property follows from (1).   □

Theorem 1. *UAIBE is ANON-IND-ID-CPA-secure in the random oracle model assuming the hardness of the quadratic residuosity problem.*

Proof. We prove the theorem by showing that a poly-bounded adversary has a negligible advantage distinguishing between the following series of games.

Game 0. This is the ANON-IND-ID-CPA game between the challenger and an adversary $\mathcal{A}$ with the scheme UAIBE as described in Section 3.2.

Game 1. The only change in this game from Game 0 is as follows. Let b denote the bit chosen by the challenger to choose either between the tuples (id_0, m_0) or (id_1, m_1) supplied by the adversary. Let $a = H(\mathsf{id}_b)$. Instead of encrypting m_b, we instead encrypt a random bit $b' \in \{0, 1\}$ i.e. we have $c(x) \leftarrow \mathcal{E}(a, b')$ and $d(x) \leftarrow \mathcal{E}(-a, b')$.

We argue that if there is an efficient distinguisher $\mathcal{A}$ that can distinguish between Game 0 and Game 1, then there is efficient adversary $\mathcal{B}$ that can use $\mathcal{A}$ to attack the IND-ID-CPA security of xhIBE. Secret key queries from $\mathcal{A}$ are relayed to $\mathcal{B}$'s oracle. When $\mathcal{A}$ chooses its challenge tuples (id_0, m_0) and (id_1, m_1), perform the following:

1. If $b' = m_b$, output a random bit and abort.

2. Else choose challenge identity $\mathsf{id}^* = \mathsf{id}_b$.

3. When $\mathcal{B}$'s IND-ID-CPA challenger responds with a challenge ciphertext $(c(x)^*, d(x)^*)$, choose two random bits $u_0, u_1 \xleftarrow{\$} \{0, 1\}$: if $u_0 = 1$, set $c(x)^* \leftarrow c(x)^* g(x)$; if $u_1 = 1$, set $d(x)^* \leftarrow d(x)^* g(x)$ where $g(x) \leftarrow H'(\mathsf{id}^*)$ (this oracle can be provided by $\mathcal{B}$).

4. Give $(c(x)^*, d(x)^*)$ to $\mathcal{A}$, and output $\mathcal{A}$'s guess.

If $\mathcal{A}$ has advantage ϵ distinguishing games Game 0 and Game 1, then $\mathcal{B}$ has an advantage of $\frac{1}{2}\epsilon$.

Game 2. To recap, note that the challenge ciphertexts in Game 1 have the distribution $\{(c(x), d(x)) \xleftarrow{\$} S_a \times S_{-a} : a = H(\mathsf{id}_b), b \xleftarrow{\$} \{0, 1\}\}$. This is because by definition for any $a \in \mathbb{J}(N)$, we have $S_a = \mathsf{image}(\mathcal{E}(a, \cdot))$ and $S_{-a} = \mathsf{image}(\mathcal{E}(-a, \cdot))$. The next step is to replace S_a with G_a. Instead of setting $c(x) \leftarrow \mathcal{E}(a, b')$ where $a = H(\mathsf{id}_b)$, we choose $c(x) \xleftarrow{\$} G_a$.

Corollary 1 1 shows that $S_a \underset{C}{\approx} G_a$ for any $a \in \mathbb{J}(N)$ without access to the factorization of N. We follow a similar argument to the above to "embed" the challenge element from either S_a or G_a. We handle secret key queries without the factors of N by programming the oracle responses from H. Suppose the adversary queries the secret key for an identity id'. Assume without loss of generality that it first queries the random oracle H on id'. On the first such query, we uniformly sample a secret key $r' \xleftarrow{\$} \mathbb{Z}_N^*$, set $a' \leftarrow r'^2 \bmod N \in \mathbb{J}(N)$, store the tuple (id', r', a') and return a'. This has the correct distribution and secret keys can easily be extracted. A non-negligible advantage distinguishing Game 1 and Game 2 translates to a non-negligible advantage distinguishing the distributions S_a and G_a, which contradicts Corollary 2 in [13].

Game 3. The change from Game 2 to Game 3 is similar to that from Game 1 to Game 2, namely the second ciphertext component $d(x)$ is sampled from G_{-a} instead of S_{-a} where $a = H(\mathsf{id}_b)$. The argument for indistuinguishability is analogous to that of the last game.

Game 4. This game is identical to Game 3 except that instead of setting $a \leftarrow H(\mathsf{id}_b)$, we instead set $a \xleftarrow{\$} \mathbb{J}(N)$. Furthermore, step 2 of Encrypt is replaced with $g(x) \leftarrow \bar{G}_a \cup \bar{G}_{-a} \in R$.

Clearly, the adversary has a zero advantage in this game since a ciphertext reveals nothing about the challenger's bit b. We now show that a ciphertext in Game 4 is indistinguishable from a ciphertext in Game 3. Observe that *each component* of the latter is computationally indistinguishable from a uniformly random element of the set of units in R. The units in R are precisely those elements $u(x)$ satisfying

$$\mathsf{GT}(a', u(x), N) \in \{-1, 1\}$$

with respect to any $a' \in \mathbb{J}(N)$; that is, the set of units is $G_{a'} \cup \bar{G}_{a'}$.

In Game 3, half of the time the ciphertext component $c(x)$ (resp. $d(x)$) is uniformly distributed in $\bar{G}_a$ (resp. $\bar{G}_{-a}$) according to Corollary 2, and the other half it is uniformly distributed in G_a (resp. G_{-a}), by definition of Game 3.

Thus, each component is a uniformly random element of the set of units in R. But similarly, we have that each component of a ciphertext in Game 4 is also uniformly distributed in the set of units in R. Therefore, both games are indistinguishable to a poly-bounded adversary.

We can conclude that an adversary's advantage is negligible distinguishing between Game 0 and Game 4, which implies that its advantage attacking the ANON-IND-ID-CPA security of UAIBE is also negligible.   □

3.4   Comparison with Ateniese and Gasti's Construction

Our proposed construction has several advantages. Firstly, it is arguably conceptually simpler than existing anonymous variants of Cocks' scheme. Furthermore, like the construction put forward in [11], it is universally anonymous, which may be useful in settings where messages pass through multiple systems, some of which need to know the recipient's identity whereas others should not be privy to this information. Hence, a trusted proxy can be tasked with anonymizing ciphertexts without access to the secret key. The scheme is also group-homomorphic for the XOR operation; this is useful in some settings as discussed in [13], although anonymity must be sacrificed for homomorphic operations to be performed. Another advantage of our scheme is that it faster run-time performance than other anonymous IBEs based on quadratic residuosity. We elaborate more on its performance in this section by comparing it to its nearest rival (in terms of run-tie performance), namely the Ateniese and Gasti (AG) scheme from [11]. However, the most significant downside of the scheme is its poor space efficiency; ciphertext expansion is double that of Cocks, and almost double that of AG.

3.5   Analysis of Ateniese and Gasti's Construction (AG)

Encryption in the AG scheme requires a number of Galbraith test computations per bit of plaintext. Recall that evaluating a Galbraith test entails a costly Jacobi symbol computation. The main intuition behind AG is to "embed" a Cocks ciphertext within a sequence of integers T_i. Its position, k, in such a sequence is distributed according to a geometric distribution with parameter $p = 1/2$. Furthermore, the terms $T_1, \ldots, T_{k-1}$ are chosen such that $\mathsf{GT}(a, T_i, N) = -1$ for $i \in [k-1]$. The intuition behind this approach is grounded in the fact that Galbraith's test can be shown (see Section 2.3 in [11]) to be the "best test" possible in attacking the anonymity of Cocks' scheme. Since the probability of a random element in $\mathbb{Z}_N^*$ passing Galbraith's test is $1/2$, the position of the first element in a random sequence to pass Galbraith's test is distributed according to a geometric distribution with parameter $p = 1/2$. A hash function is used to generate the sequence of integers based on short binary strings incorporated in an AG ciphertext. We defer the details to Appendix A, but it sufficient here to note that ℓ is a global parameter in AG that determines the number of such binary strings (this is closely related to the number of Galbraith tests that must be performed on average during encryption).

Let Y be a random variable representing the number of Galbraith tests evaluated in AG per bit of plaintext. A lower bound for the expected value $E[Y]$ of Y can be derived as

$$E[Y] \geq 4(1 + (\log \kappa - 1) \cdot 2^{-\ell})$$

where κ is the security parameter. A rough lower bound on the variance $\mathsf{Var}(Y)$ is

$$\mathsf{Var}(Y) \geq 2^{2-2\ell}(-8 + 7 \cdot 2^{2\ell} + 2^{1+\ell} - 3 \cdot 2^{2+\ell}\ell).$$

See Appendix A for the derivations of these inequalities. Ateniese and Gasti found $\ell = 6$ to be a good compromise between ciphertext size and performance. See Appendix A for supporting analysis. Setting $\ell = 6$ results in a mean number of Galbraith tests per bit of plaintext of ≈ 4.22 with a standard deviation of ≈ 6.92. Our scheme on the other hand does not require any Galbraith test to be performed during encryption.

4 Experimental Results

To perform an empirical comparison between our scheme and AG, both schemes were implemented in C using the OpenSSL library. Our implementation was based on code provided by the authors of [11]. The following experiment was run for each of the four schemes: Cocks, AG, UAIBE and JB. The latter is a shorthand for our modification to the construction of Jhanwar and Barua described in Appendix B. Note that JB is not anonymous and its inclusion here is to demonstrate the fact that it achieves comparable efficiency to Cocks. Hence, AG and UAIBE are the two anonymous schemes being compared.

1. For each t in the set $\{1024, 2048, 3072, 4096\}$:

   (a) A modulus N of t bits is generated along with primes p and q that constitute the master secret key.

   (b) The public key a and secret key r are derived for some predefined identity string id. A random 128-bit message m is generated.

   (c) The following is repeated 50 times:

      i. Encrypt m under identity id to produce ciphertext c.

      ii. Decrypt c with secret key r and verify the decrypted message matches m.

      iii. The time elapsed performing step 3.(a) and 3.(b) is calculated.

   (d) An average over the times calculated in step 3.(c) is obtained.

The code was compiled with optimization flag '-02' using GCC version 4.4.5-8 with OpenSSL version 0.9.8o. The benchmarks were executed on a machine with 4 GB of RAM and an Intel Core i5-3340M CPU clocked at 2.70 GHz. The benchmark machine was running GNU/Linux 3.2.41 (x86-64). Our implementation however was unoptimized and did not exploit parallelization. For the interested reader, the implementation of encryption in Cocks, AG and UAIBE

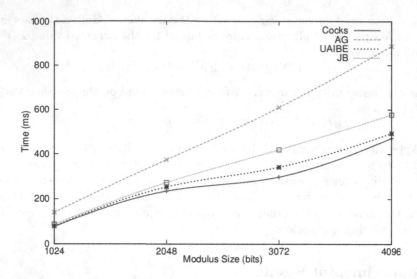

Fig. 1. Average times to encrypt a 128-bit message for Cocks, AG and UAIBE

involved precomputation of random integers with Jacobi symbol -1 and $+1$. This is not needed for JB.

The results of the experiment (average encryption times) are shown in Figure 1. Note that UAIBE and Cocks exhibit similar performance whereas JB is only marginally less efficient than Cocks. On the other hand, AG performs notably worse than UAIBE on average. To illustrate the comparison, encryption and decryption times for all four schemes for the case of a 1024-bit modulus are presented in Table 1.

Table 1. Encryption and decryption times in milliseconds for a 128-bit message with a key size of 1024 bits, averaged over 50 runs

Scheme	Encryption -Mean (Std Dev)	Decryption - Mean (Std Dev)
Cocks	77.39 (3.05)	13.32 (0.14)
AG	140.35 (19.22)	40.79 (1.68)
UAIBE	79.02 (3.14)	27.52 (0.41)
JB	86.78 (0.93)	21.97 (0.42)

Acknowledgments. The authors would like to thank the anonymous reviewers for their many helpful comments.

References

1. Shamir, A.: Identity-based cryptosystems and signature schemes. In: Blakely, G.R., Chaum, D. (eds.) Advances in Cryptology CRYPTO - 1984. LNCS, vol. 196, pp. 47–53. Springer, Heidelberg (1985)
2. Cocks, C.: An identity based encryption scheme based on quadratic residues. In: Honary, B. (ed.) Cryptography and Coding 2001. LNCS, vol. 2260, pp. 360–363. Springer, Heidelberg (2001)
3. Boneh, D., Franklin, M.: Identity-based encryption from the weil pairing. In: Kilian, J. (ed.) CRYPTO 2001. LNCS, vol. 2139, pp. 213–229. Springer, Heidelberg (2001)
4. Gentry, C., Peikert, C., Vaikuntanathan, V.: Trapdoors for hard lattices and new cryptographic constructions. In: STOC 2008: Proceedings of the 40th Annual ACM Symposium on Theory of Computing, pp. 197–206. ACM, New York (2008)
5. Agrawal, S., Boneh, D., Boyen, X.: Efficient lattice (H)IBE in the standard model. In: Gilbert, H. (ed.) EUROCRYPT 2010. LNCS, vol. 6110, pp. 553–572. Springer, Heidelberg (2010)
6. Boneh, D., Crescenzo, G.D., Ostrovsky, R., Persiano, G.: Public key encryption with keyword search. In: Cachin, C., Camenisch, J.L. (eds.) EUROCRYPT 2004. LNCS, vol. 3027, pp. 506–522. Springer, Heidelberg (2004)
7. Bellare, M., Boldyreva, A., Desai, A., Pointcheval, D.: Key-privacy in public-key encryption. In: Boyd, C. (ed.) ASIACRYPT 2001. LNCS, vol. 2248, pp. 566–582. Springer, Heidelberg (2001)
8. Boyen, X., Waters, B.: Anonymous hierarchical identity-based encryption (without random oracles). In: Dwork, C. (ed.) CRYPTO 2006. LNCS, vol. 4117, pp. 290–307. Springer, Heidelberg (2006)
9. Di Crescenzo, G., Saraswat, V.: Public key encryption with searchable keywords based on jacobi symbols. In: Srinathan, K., Rangan, C.P., Yung, M. (eds.) INDOCRYPT 2007. LNCS, vol. 4859, pp. 282–296. Springer, Heidelberg (2007)
10. Boneh, D., Gentry, C., Hamburg, M.: Space-efficient identity based encryption without pairings. In: FOCS, pp. 647–657. IEEE Computer Society (2007)
11. Ateniese, G., Gasti, P.: Universally anonymous IBE based on the quadratic residuosity assumption. In: Fischlin, M. (ed.) CT-RSA 2009. LNCS, vol. 5473, pp. 32–47. Springer, Heidelberg (2009)
12. Hayashi, R., Tanaka, K.: Universally anonymizable public-key encryption. In: Roy, B. (ed.) ASIACRYPT 2005. LNCS, vol. 3788, pp. 293–312. Springer, Heidelberg (2005)
13. Clear, M., Hughes, A., Tewari, H.: Homomorphic encryption with access policies: Characterization and new constructions. In: Youssef, A., Nitaj, A., Hassanien, A.E. (eds.) AFRICACRYPT 2013. LNCS, vol. 7918, pp. 61–87. Springer, Heidelberg (2013)
14. Jhanwar, M.P., Barua, R.: A variant of boneh-gentry-hamburg's pairing-free identity based encryption scheme. In: Yung, M., Liu, P., Lin, D. (eds.) Inscrypt 2008. LNCS, vol. 5487, pp. 314–331. Springer, Heidelberg (2009)
15. Barua, R., Jhanwar, M.: On the number of solutions of the equation $rx^2 + sy^2 = 1$ (mod n). Indian Journal of Statistical 2010 72-A (pt. 1), 226–236 (2010)

A Expected Number of Galbraith Tests in the Ateniese and Gasti Scheme

Ateniese and Gasti proposed the following approach to anonymize a Cocks ciphertext $(c, d) \in \mathbb{Z}_N^*$ which has been computed with public key $a = H(\mathsf{id})$, . Two integers k_1 and k_2 are independently sampled according to a geometric distribution with parameter $1/2$. Two sequences of integers $T_1, \ldots, T_m \in \mathbb{Z}_N^*$ and $V_1, \ldots, V_m \in \mathbb{Z}_N^*$ are randomly generated subject to the condition that for $1 \leq i < k_1$ and $1 \leq j < k_2$

$$\mathsf{GT}(a, Z_1 - T_i, N) = -1 \text{ and } \mathsf{GT}(-a, Z_2 - V_j, N) = -1 \qquad (\text{A.1})$$

where $Z_1 = c + T_{k_1}$ and $Z_2 = d + T_{k_2}$. Note that since $\mathsf{GT}(a, c, N) = 1$ and $\mathsf{GT}(-a, d, N) = 1$ by virtue of (c, d) being a Cocks ciphertext, it obviously holds that $\mathsf{GT}(a, Z_1 - T_{k_1}, N) = \mathsf{GT}(-1, Z_2 - T_{k_2}, N) = 1$. The anonymized ciphertext is outputted as $(Z_1, T_1, \ldots, T_m) \in (\mathbb{Z}_N^*)^{m+1}$ and $(Z_2, V_1, \ldots, V_m) \in (\mathbb{Z}_N^*)^{m+1}$. If m is large enough, i.e. polynomial in the security parameter, it can be shown that this construction is ANON-IND-ID-CPA-secure.

A significant disadvantage of this construction is the fact that $2(m + 1)$ elements of $\mathbb{Z}_N^*$ are needed per bit of plaintext in comparison to the 2 elements required by Cocks. To address this, Ateniese and Gasti present a more space-efficient variant.

The main difference in the space-efficient variant is in how the T_i and V_i are generated. A new global parameter $\ell \in \mathbb{N}$ is fixed. Also, the existence of a hash function $G : \{0, 1\}^* \rightarrow \mathbb{Z}_N$ is assumed. Let X be a multi-bit message. Alice chooses a random identifier MID_X when encrypting X. Now to encrypt the j-th bit of X, she computes a ciphertext

$$(Z_1, \alpha_1, \ldots, \alpha_\ell) \text{ and } (Z_2, \beta_1, \ldots, \beta_\ell)$$

where $\alpha_i, \beta_i \in \{0, 1\}^e$ for $i < \ell$, and $\alpha_\ell, \beta_\ell \in \{0, 1\}^{e'}$. Note that e and $e' > e$ are fixed global parameters. The sequences T_i and V_i are generated as follows:

$$T_i = G(\mathsf{MID} \| 0 \| \alpha_i \| j) \text{ and } V_i = G(\mathsf{MID} \| 1 \| \beta_i \| j) \qquad (\text{A.2})$$

for $1 \leq i < \ell$ and

$$T_i = G(\mathsf{MID} \| 0 \| \alpha_\ell \| j) \text{ and } V_i = G(\mathsf{MID} \| 1 \| \beta_\ell \| j) \qquad (\text{A.3})$$

for $i \geq \ell$. Alice must choose appropriate α_i and β_i in order to satisfy A.1. When $k_1 \leq \ell$ and $k_2 \leq \ell$, this is not too costly because each selection affects only one member of the respective sequence. Moreover, this will be the case with high probability for sufficiently large ℓ, However, as pointed out in [11], in the case when either $k_1 \geq \ell$ or $k_2 \geq \ell$, the cost is exponential in $k_1 - \ell$ or $k_2 - \ell$ respectively.

We now compute the average number of Galbraith tests per bit of plaintext. In fact, it suffices to restrict our attention to a single ciphertext component because we can double the result to obtain the total number of Galbraith tests.

Now the expected number of Galbraith tests is computed as follows. Let X be random variable following a geometric distribution with parameter $1/2$ over the space $\{0, 1, 2, \ldots\}$. Denote by Y' the random variable that determines the number of Galbraith tests performed. There are always at least k Galbraith tests performed, where $k \overset{\$}{\leftarrow} X$. Thus,

$$E[Y'] \geq E[X] = 1.$$

Consider a random variable Z giving the number of tests performed when selecting $\alpha_1, \ldots, \alpha_{\ell-1}$. It holds that $E[Z] = 2 \cdot E[\min(X, \ell - 1)]$, since there are 2 expected trials per α_i for $i \leq k$ subject to the constraint that $k \leq \ell - 1$. We calculate $E[\min(X, \ell - 1)]$ as follows:

$$\sum_{k=0}^{\ell-1} \frac{k}{2^{k+1}} + (\ell - 1) \sum_{k=0}^{\ell-1} \frac{1}{2^{k+1}} = 1 - 2^{1-\ell}.$$

It is necessary to subtract $E[\min(X, \ell - 1)]$ from $E[Z]$ because these particular tests are already incorporated into $E[X]$. Therefore, we now have

$$E[Y'] \geq E[X] + E[Z] - E[\min(X, \ell - 1)] = 2(1 - 2^{-\ell}).$$

There is a $1/2^\ell$ chance that $k > \ell$. In this case, a single binary string, namely $\alpha_\ell \in \{0, 1\}^{e'}$ must be selected that satisfies $k - \ell$ Galbraith tests. Conditioned on $k \geq \ell$, the expected value of k is $\ell + 1$, and the expected number of trials per selection of α_ℓ is therefore $2((\ell + 1) - \ell) = 2$. Now it remains to compute the expected number of selections of α_ℓ. It turns out that this is equivalent to the St. Petersburg lottery. Thus, the expected value is infinite if no bound is set on k and equal to the bound otherwise, all conditioned on $k \geq \ell$. To preserve security, this bound cannot be polylogarithmic in the security parameter κ. However, setting it as such allows us to derive a (loose) lower bound on the number of selections. As a consequence, we formulate a lower bound on the number of selections as

$$\frac{\log \kappa}{2^\ell}.$$

A lower bound on the expected number of tests induced by $k \geq \ell$ is

$$\frac{\log \kappa}{2^{\ell-1}}.$$

Putting all components together yields

$$E[Y'] \geq 2(1 + (\log \kappa - 1) \cdot 2^{-\ell}).$$

A rough lower bound on the variance $\mathsf{Var}(Y')$ can be calculated in a similar manner as

$$\mathsf{Var}(Y') \geq \mathsf{Var}(Z) = 2^{1-2\ell}(-8 + 7 \cdot 2^{2\ell} + 2^{1+\ell} - 3 \cdot 2^{2+\ell}\ell).$$

Figure 2 shows approximations for the mean and standard deviation based on these lower bounds by taking the security parameter κ to be 80 (the value used in [11]). The figure supports the empirical findings of [11] from which $\ell = 6$ was found to be a good compromise between ciphertext size and performance.

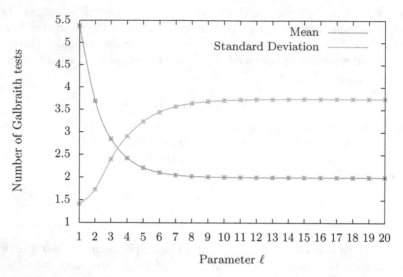

Fig. 2. Dependence on number of Galbraith tests on parameter ℓ

B   Variant of the Boneh, Gentry and Hamburg (BGH) Construction with Improved Performance

Due to space constraints, we do not describe the Boneh, Gentry and Hamburg (BGH) [10] construction here. We refer the reader to [10] and the full version of this work for more details. It is sufficient to point out that their approach hinges on solving equations of the form $Rx^2 + Sy^2 = 1$ modulo N to yield $(x, y) \in \mathbb{Z}_N^2$ and outputting polynomials $f(r) \leftarrow xr + 1$ and $g(s) \leftarrow 2ys + 2$. The method they propose to solve such equations involves the generation of primes, which is the main expense.

B.1   Jhanwar and Burua (JB) Variant

An alternative approach was explored by Jhanwar and Burua in [14], which we refer to as JB. Their approach is based on finding a random point $(x, y) \in \mathbb{Z}_N^2$ on the curve $Rx^2 + Sy^2 = 1$ by making use of the following lemma.

Lemma 3 (Lemma 2.1, [15]). *Let N be prime. Let $R, S \in \mathbb{Z}_N$ where $S \in \mathbb{QR}(N)$. Let s be a square root of S modulo N. Then any solution $(x_0, y_0) \in \mathbb{Z}_N^2$ to the equation $Rx^2 + Sy^2 = 1$ is of the form*

$$\left(\frac{-2st}{R + St^2}, \frac{R - St^2}{s(R + St^2)} \right) \in \mathbb{Z}_N^2$$

for some $t \in \mathbb{Z}_N^*$ such that $R + St^2 \in \mathbb{Z}_N^*$.

In [14], the authors exploit Lemma 3 to generate a random solution to $Rx^2 + Sy^2 = 1$ by choosing a $t \in \mathbb{Z}_N^*$ uniformly at random. However, only the sender, who has access to s, can generate such a solution. Therefore, it is necessary to incorporate the x-coordinate in the ciphertext per bit of plaintext so that the receiver can form the polynomial $f(r) \leftarrow xr + 1$. This leads to considerable ciphertext expansion compared to BGH since 2ℓ elements $x_1, \ldots, x_\ell, \bar{x}_1, \ldots, \bar{x}_\ell$ must be incorporated in the ciphertext. To counteract this considerable blowup in ciphertext size, an optimization is employed in JB, based on the product formula from [10], whereby only 2κ elements need to be sent for some parameter κ.

JB is claimed to be IND-ID-CPA secure by Theorem 2 in [14]. We make an observation here concerning this theorem. Jhanwar and Barua propose setting $\kappa = \lceil \sqrt{\ell} \rceil$ to ensure the ciphertext size is kept "small". However, their argument that Game 5 and Game 6 in the proof of Theorem 2 are indistinguishable in the view of an adversary bounds the probability of an attacker guessing correctly by $\frac{1}{2^\kappa}$. Hence if ℓ is polylogarithmic in the security parameter, it follows that an adversary has a non-negligible advantage distinguishing both games, which invalidates the proof of security. As a result, to guarantee μ bits of security, it becomes necessary to ensure that plaintexts consist of at least μ^2 bits. Concretely, a plaintext of 800 bytes would have to be encrypted to guarantee 80 bits of security if the parameter setting proposed in [14] is employed. A more sensible setting is

$$\kappa = \min(\max(\mu, \sqrt{\ell}), \ell) \tag{B.1}$$

where μ is the desired security level and ℓ is the length of a plaintext in bits. Even with this change, the scheme still provides excellent performance. In concrete terms, we see that to encrypt a 128-bit symmetric key using a 1024-bit modulus, the ciphertext size is 20,512 bytes (note that $\kappa = 80$) in comparison to 32,768 bytes for Cocks. Furthermore, the scheme outperforms Cocks. The modified scheme with κ chosen according to Equation B.1 achieves comparable efficiency to Cocks, but with lower ciphertext expansion. Our experimental results in Section 4 provide a performance comparison.

Anonymity. Given the performance benefits of this scheme, a natural question is whether an anonymous variant can be constructed. Unfortunately, attempts to exploit the same techniques to construct an anonymous IBE have not been successful. It may be tempting to start from the anonymous IBE presented in [10] and incorporate the solutions to the relevant equations in the ciphertext. However, it then becomes easy for an attacker to tell whose identity was used to create a ciphertext.

Expressive Attribute Based Signcryption with Constant-Size Ciphertext

Y. Sreenivasa Rao and Ratna Dutta

Department of Mathematics
Indian Institute of Technology Kharagpur
Kharagpur-721302, India
ysrao,ratna@maths.iitkgp.ernet.in

Abstract. In this paper, we propose a new attribute-based signcryption (ABSC) scheme for linear secret-sharing scheme (LSSS)-realizable monotone access structures that is significantly more efficient than existing ABSC schemes in terms of computation cost and ciphertext size. This new scheme utilizes only 6 pairing operations and the size of ciphertext is *constant*, i.e., independent of the number of attributes used to signcrypt a message. While the secret key size increases by a factor of number of attributes used in the system, the number of pairing evaluations is reduced to constant. Our protocol is proven to provide ciphertext indistinguishability under adaptive chosen ciphertext attacks assuming the hardness of decisional Bilinear Diffie-Hellman Exponent problem and achieves existential unforgeability under adaptive chosen message attack assuming the hardness of computational Diffie-Hellman Exponent problem. The proposed scheme achieves public verifiability of the ciphertext, enabling any party to verify the integrity and validity of the ciphertext.

Keywords: attribute based signcryption, constant-size ciphertext, public ciphertext verifiability, linear secret-sharing scheme.

1 Introduction

In some applications, both encryption and signing are needed to ensure confidentiality and authenticity of the transmitted data. For instance, electronic voting and cloud technology, where the user wants to encrypt his data to guarantee privacy, and at the same time, the user should also be able to prove his genuineness at the destination center (voting center or cloud). One possible approach for such scenarios is to perform both encryption and signing sequentially. However, this technique is inefficient due to the fact that the resulting complexity is sum of the complexities of both the primitives. In order to deploy this approach widely in practice, Zheng [10] devised a novel cryptographic primitive called *signcryption* that is a proper mixture of both encryption and signing in a single primitive. And, the cost of signcryption is significantly smaller than the cumulative cost of both the primitives. With the development of identity based encryption [12], Malone-Lee [11] proposed the first identity based signcryption by combining

D. Pointcheval and D. Vergnaud (Eds.): AFRICACRYPT 2014, LNCS 8469, pp. 398–419, 2014.

encryption and signature schemes in the identity based setting. Later, Sahai and Waters [1] introduced a versatile cryptographic primitive called Attribute Based Encryption (ABE) by assigning a set of abstract credentials (attributes) to each user in the system. ABE is classified as Key-Policy ABE (KP-ABE) [2] or Ciphertext-Policy ABE (CP-ABE) [3]. In KP-ABE, each user is issued a secret key computed according to an access structure $\mathbb{T}$ over his attributes and ciphertext is created with an attribute set L. On the other hand, in CP-ABE, user secret key is generated according to his attribute set L and ciphertext is associated with an access structure $\mathbb{T}$ over receivers' attributes. In both the cases, decryption is admissible only when $\mathbb{T}$ accepts L. A recent direction is to combine the functionalities of ABE [1,2,3,21,8,9,19] and Attribute Based Signature (ABS) [13,14,15,16] to design *Attribute-Based Signcryption* (ABSC) schemes.

Our Contribution. The main focus of this article is to design efficient *constant-size*[1] ciphertext ABSC scheme with *constant* number of pairings. To this end, we use the KP-ABE framework of [19]. We present the *first* signcryption scheme in the attribute based key-policy setting wherein the user secret key is computed according to Linear Secret Sharing Scheme (LSSS)-realizable Monotone Access Structure (MAS) [9]. We give formal *selective* security proofs in the random oracle model for message confidentiality and ciphertext unforgeability based on decisional n-Bilinear Diffie-Hellman Exponent (n-dBDHE) and computational n-Diffie-Hellman Exponent (n-cDHE) assumptions, respectively. More interestingly, our scheme exhibits the *public ciphertext verifiability* similar to [7] which allows any intermediate party, e.g., firewalls, to check the ciphertext's validity before sending to actual recipient. This in turn reduces unnecessary burden on the receiver for unsigncrypting invalid ciphertexts. Our construction also provides insider security with respect to both confidentiality and unforgeability which ensures that the scheme is secure even when either the signcryptor or the unsigncryptor colludes with the adversary against the other.

Table 1 exhibits the efficiency of our scheme against existing attribute based signcryption schemes [4,7,6,5,18,17] in terms of ciphertext size, and computation cost for exponentiations and pairing operations. The functionality comparison of our scheme against [4,7,6,5,18,17] is presented in Table 2.

The ciphertext consists of only 6 group elements in our design, whereas the existing signcryption schemes have ciphertext size that depends on the number of attributes involved to create a ciphertext. On a more positive note, our scheme is pairing efficient as it requires only 6 pairings to unsigncrypt any ciphertext. On the other hand, the number of pairings used in previous constructions are linear to the number of required attributes. To the best of our knowledge, our proposal is the *first* signcryption scheme in the attribute based key-policy setup with constant ciphertext size and constant bilinear pairings. While the signcryption cost of the existing schemes depend on both signing and encryption attributes,

---

[1] According to ABE literature (see [8]), we do not consider the description of the access policy (or the attribute set) as being part of the secret key (or the ciphertext), while measuring its size.

Table 1. Comparison of communication and computation costs of ABSC schemes

	Secret Key (SK) Size		CT Size	Signcryption Cost			Unsigncryption Cost		
	Sig. SK	Dec. SK		Exp. in $\mathbb{G}$	Exp. in $\mathbb{G}_T$	Pair.	Exp. in $\mathbb{G}$	Exp. in $\mathbb{G}_T$	Pair.
[4]	$2L_s$	$3L_e$	$\mathcal{O}(\phi_s+w_e)$	$\mathcal{O}(\phi_s+w_e)$	1	-	$\mathcal{O}(\phi_e)$	$\mathcal{O}(\phi_e)$	$\mathcal{O}(\phi_s+\phi_e)$
[5]	$2L_s$	$2L_e$	$\mathcal{O}(\phi_s+w_e)$	$\mathcal{O}(\phi_s+w_e)$	1	-	-	$\mathcal{O}(\phi_e)$	$\mathcal{O}(\phi_e)$
[6]	$2L_s+1$	$2L_e+1$	$\mathcal{O}(\phi_s+w_e)$	$\mathcal{O}(\phi_s+w_e)$	2	1	-	$\mathcal{O}(\phi_s\log\phi_s+\phi_e\log\phi_e)$	$\mathcal{O}(\phi_s+\phi_e)$
[7]	$2L_s$	$2L_e+2vk+1$	$\mathcal{O}(L_s+u_e)$	$\mathcal{O}(L_s+u_e)$	1	-	-	-	$\mathcal{O}(L_s+u_e)$
[18]	L_s+d-1	L_e+d-1	$\mathcal{O}(\phi_s+w_e)$	$\mathcal{O}(\phi_s+w_e)$	1	-	$\mathcal{O}(\phi_e+w_e)$	$\mathcal{O}(\phi_e)$	$\mathcal{O}(\phi_e)$
[17]	$(L_s+d)^2$	$2L_e$	$\mathcal{O}(\mathsf{id}+w_e)$	$\mathcal{O}(\mathsf{id}+w_e)$	1	-	-	$\mathcal{O}(\phi_e)$	$\mathcal{O}(\mathsf{id}+\phi_e)$
Our	$u_s L_s$	$u_e L_e$	6	$\mathcal{O}(\phi_s)$	1	-	$\mathcal{O}(\phi_e)$	-	6

Note that by size, we mean the number of involved group elements. We exclude the message size from the ciphertext (CT) size. L_s = number of signature attributes annotated to a user's signing secret key, L_e = number of decryption attributes annotated to a user's decryption secret key, ϕ_s = number of signature attributes involved in the signcryption, ϕ_e = minimum number of decryption attributes required to recover a message, w_e = number of encryption attributes used to encrypt a message, u_s = number of signature attributes in the signature attribute space $\mathcal{U}_s$, u_e = number of decryption attributes in the decryption attribute space $\mathcal{U}_e$, vk = bit length of verification key, d = threshold value of the system, id = length of user's identity, CT = ciphertext.

Table 2. Functionality comparison of ABSC schemes

	CP/ KP	Access Structure (AS)		Security		Hardness Assumption		PV	ROM
		Signature AS	Decryption AS	MC	CU	MC	CU		
[4]	KP	Threshold policy	Threshold policy	IND-CCA	EUF	dHmBDH	cmDH	No	No
[5]	KP	Threshold policy	Threshold policy	IND-CCA	-	dBDH	-	No	No
[6]	CP	Monotone tree	Monotone tree	IND-CPA	EUF	Generic group	n-DHI	No	Yes
[7]	CP	Monotone tree	AND-gate policy	IND-CCA	sEUF	dBDH	cDH	Yes	No
[18]	KP	Threshold policy	Threshold policy	IND-CCA	EUF	dBDH	cDH	No	Yes
[17]	KP	Threshold policy	Threshold policy	IND-CCA	EUF	dBDH	n-cDHE	No	Yes
Our	KP	LSSS-realizable	LSSS-realizable	IND-CCA	EUF	n-dBDHE	n-cDHE	Yes	Yes

Note that all the schemes listed in the table are selectively secure. MC = message confidentiality, CU = ciphertext unforgeability, d(Hm)BDH = decisional (hashed modified) bilinear Diffie-Hellman, c(m)DH = computational (modified) Diffie-Hellman, n-DHI = computational n-Diffie-Hellman inversion, IND-CCA, CPA = indistinguishability of ciphertexts under chosen ciphertext, plaintext attack, (s)EUF = (strongly) existential unforgeability, PV = public verifiability, ROM = random oracle model, CP (or KP) = ciphertext (or key) policy.

the same for our construction depends only on signing attributes[2]. The secret key size is increased by a factor of attribute space size in our construction. However, storage is much cheaper nowadays even for a large amount (e.g., smart phones), while the main concerns lie with low bandwidth and computation overhead.

The unsigncryption process in [CCA secure KP-ABE of [19]] + [ABS of [15]] requires 9 pairing operations (in this case, the signing access structure is limited to threshold policy only), while that for our ABSC is only 6. All other complexities are asymptotically same for both the approaches. This in turn implies that cost[our ABSC] < cost[CCA secure KP-ABE of [19]] + cost[ABS of [15]]. In addition, the new ABSC outperforms all the existing ABSC schemes in terms

---

[2] This cost in our construction is related to general LSSS-realizable MAS. If the access structure is a boolean formula as in existing schemes, our signcryption and unsigncryption processes require only 10 and 1 exponentiations respectively. Thus, we achieve *constant* computation cost during signcryption and unsigncryption.

of communication and computation cost while realizing more expressive access policies, namely LSSS-realizable access structures.

Related Work. The first ABSC was introduced by Gagné et al. [4] with formal security definitions of *message confidentiality* and *ciphertext unforgeability* for signcryption in attribute based setting. In order for users to provide different rights for signature and decryption, signing attributes are separated from encryption/decryption attributes. Later, Emura et al. [7] designed an ABSC scheme with dynamic property allowing updation of signing access structures without re-issuing secret keys of users. The signature part makes use of access trees whereas AND-gate policies are used in encryption/decryption process. Wang and Huang [6] proposed another ABSC by employing access trees for both signature and encryption parts. While the security of [6] is argued in the generic group model and using random oracles, the schemes [4,7] are secure in standard model. Hu et al. [5] suggested a fuzzy ABSC in order to introduce authenticated access control in body area networks, although no formal security proof for ciphertext unforgeability is provided in existing security models. Ciphertext size in all these schemes are linear to the sum of required signing and encryption attribute set sizes. Moreover, the number of essential bilinear pairing computations are also linear to the required attribute set sizes. Recently proposed attribute-based ring signcryption [18] and traceable attribute-based signcryption [17] are also suffering from linear-size ciphertexts and hence bilinear pairings as well. This in turn lowers the communication and computation efficiency of [4,7,6,5,18,17] in the sense of ciphertext size and pairing evaluations, respectively.

2 Preliminaries

Notation. Let $s \xleftarrow{\$} S$ denote the operation of picking an element s uniformly at random from the set S. We denote the set $\{1, 2, \ldots, n\}$ as $[n]$. By op $\leftarrow$ Alg(ip), we denote that algorithm Alg($\cdot$) takes as input ip and outputs op. By $f : \mathbb{A} \to \mathbb{B}$, we mean that f is a mapping from $\mathbb{A}$ to $\mathbb{B}$.

Definition 1 (Access Structure). *Let U be the universe of attributes and $\mathcal{P}(U)$ be the collection of all subsets of U. Every non-empty subset $\mathbb{A}$ of $\mathcal{P}(U) \backslash \{\emptyset\}$ is called an access structure. The sets in $\mathbb{A}$ are called authorized sets and the sets not in $\mathbb{A}$ are called unauthorized sets with respect to $\mathbb{A}$. An access structure $\mathbb{A}$ is said to be monotone access structure (MAS) if every superset of an authorized set is again authorized in $\mathbb{A}$, i.e., for any $C \in \mathcal{P}(U)$, with $C \supseteq B$ where $B \in \mathbb{A}$ implies $C \in \mathbb{A}$. An attribute set L satisfies $\mathbb{A}$ if and only if $L \in \mathbb{A}$.*

Definition 2 (Linear Secret-Sharing Scheme (LSSS)). *Let U be the universe of attributes. A secret-sharing scheme $\Pi_{\mathbb{A}}$ for the access structure $\mathbb{A}$ over U is called linear (in $\mathbb{Z}_p$) if $\Pi_{\mathbb{A}}$ consists of the following two polynomial time algorithms, where $\mathbb{M}$ is a matrix of size $\ell \times k$, called the share-generating matrix for $\Pi_{\mathbb{A}}$ and $\rho : [\ell] \to I_U$ is a row labeling function that maps each row $i \in [\ell]$ of the matrix $\mathbb{M}$ to an attribute $\mathsf{att}_{\rho(i)}$ in $\mathbb{A}$, I_U being the index set of the attribute universe U.*

- Distribute$(\mathbb{M}, \rho, \alpha)$: *This algorithm takes as input the share-generating matrix* $\mathbb{M}$, *row labeling function* ρ *and a secret* $\alpha \in \mathbb{Z}_p$ *which is to be shared. It selects* $z_2, z_3, \ldots, z_k \xleftarrow{\$} \mathbb{Z}_p$ *and sets* $\boldsymbol{v} = (\alpha, z_2, z_3, \ldots, z_k) \in \mathbb{Z}_p^k$. *It outputs a set* $\{\boldsymbol{M_i} \cdot \boldsymbol{v} : i \in [\ell]\}$ *of* ℓ *shares, where* $\boldsymbol{M_i} \in \mathbb{Z}_p^k$ *is the* i-th *row of* $\mathbb{M}$. *The share* $\lambda_{\rho(i)} = \boldsymbol{M_i} \cdot \boldsymbol{v}$ *belongs to an attribute* $\rho(i)$.
- Reconstruct$(\mathbb{M}, \rho, W)$: *This algorithm will accept as input* $\mathbb{M}, \rho$ *and a set of attributes* $W \in \mathbb{A}$. *Let* $I = \{i \in [\ell] : \rho(i) \in I_W\}$, *where* I_W *is the index set of the attribute set* W. *It returns a set* $\{\omega_i : i \in I\}$ *of secret reconstruction constants such that* $\sum_{i \in I} \omega_i \lambda_{\rho(i)} = \alpha$ *if* $\{\lambda_{\rho(i)} : i \in I\}$ *is a valid set of shares of the secret* α *according to* $\Pi_{\mathbb{A}}$.

The target vector which is used to characterize access structures is $(1, 0, \ldots, 0)$, i.e., a set $W \in \mathbb{A}$ iff $(1, 0, \ldots, 0)$ is in the linear span of the rows of $\mathbb{M}$ that are indexed by W.

Lemma 1. *[9] Let* $(\mathbb{M}, \rho)$ *be a LSSS access structure realizing an access structure* $\mathbb{A}$ *over the universe* U *of attributes, where* $\mathbb{M}$ *is share-generating matrix of size* $\ell \times k$. *For any* $W \subset U$ *such that* $W \notin \mathbb{A}$, *there exists a polynomial time algorithm that outputs a vector* $\boldsymbol{w} = (-1, w_2, \ldots, w_k) \in \mathbb{Z}_p^k$ *such that* $\boldsymbol{M_i} \cdot \boldsymbol{w} = 0$, *for each row* i *of* $\mathbb{M}$ *for which* $\rho(i) \in I_W$, *here* I_W *is index set of attribute set* W.

2.1 Bilinear Maps and Complexity Assumptions

Bilinear Map. We use multiplicative cyclic groups $\mathbb{G}, \mathbb{G}_T$ of prime order p with an efficiently computable mapping $e : \mathbb{G} \times \mathbb{G} \to \mathbb{G}_T$ such that $e(u^a, v^b) = e(u, v)^{ab}, \forall u, v \in \mathbb{G}, a, b \in \mathbb{Z}_p$ and $e(u, v) \neq 1_{\mathbb{G}_T}$ whenever $u, v \neq 1_{\mathbb{G}}$.

Computational n-DHE Assumption. An algorithm (or an adversary) $\mathcal{A}$ for solving the computational n-DHE (Diffie-Hellman Exponent) problem in $\mathbb{G}$ takes as input a tuple of the form $(g, g^a, \ldots, g^{a^n}, g^{a^{n+2}}, \ldots, g^{a^{2n}}) \in \mathbb{G}^{2n}$, where $a \xleftarrow{\$} \mathbb{Z}_p, g \xleftarrow{\$} \mathbb{G}$ and outputs $g^{a^{n+1}}$. The advantage of $\mathcal{A}$ in solving the computational n-DHE problem is defined as

$$Adv_{\mathcal{A}}^{n\text{-cDHE}} = \Pr\left[\mathcal{A}(g, g^a, \ldots, g^{a^n}, g^{a^{n+2}}, \ldots, g^{a^{2n}}) = g^{a^{n+1}}\right].$$

Definition 3. *The computational n-DHE problem in* $\mathbb{G}$ *is said to be* $(\mathcal{T}, \epsilon)$-hard *if the advantage* $Adv_{\mathcal{A}}^{n\text{-cDHE}} \leq \epsilon$, *for any probabilistic polynomial time (PPT) algorithm* $\mathcal{A}$ *running in time at most* $\mathcal{T}$.

Decisional n-BDHE Assumption. An algorithm $\mathcal{A}$ for solving the decisional n-BDHE (Bilinear Diffie-Hellman Exponent) problem in $(\mathbb{G}, \mathbb{G}_T)$ takes as input a tuple $(\mathbf{y}_{a,\theta}, Z) \in \mathbb{G}^{2n+1} \times \mathbb{G}_T$, where $a, \theta \xleftarrow{\$} \mathbb{Z}_p, g \xleftarrow{\$} \mathbb{G}, g_i = g^{a^i}, \forall i \in [2n] \setminus \{n+1\}, \mathbf{y}_{a,\theta} = (g, g^\theta, g_1, \ldots, g_n, g_{n+2}, \ldots, g_{2n})$ and determines whether $Z = e(g_{n+1}, g^\theta)$ or a random element in $\mathbb{G}_T$. The advantage of a 0/1-valued algorithm $\mathcal{A}$ in solving the decisional n-BDHE problem in $(\mathbb{G}, \mathbb{G}_T)$ is defined to be

$$Adv_{\mathcal{A}}^{n\text{-dBDHE}} = \left|\Pr[\mathcal{A}(\mathbf{y}_{a,\theta}, Z) = 1 | Z = e(g_{n+1}, g^\theta)] - \Pr[\mathcal{A}(\mathbf{y}_{a,\theta}, Z) = 1 | Z \xleftarrow{\$} \mathbb{G}_T]\right|$$

where the probability is over randomly chosen $a, \theta \xleftarrow{\$} \mathbb{Z}_p, g \xleftarrow{\$} \mathbb{G}, Z \xleftarrow{\$} \mathbb{G}_T$.

Definition 4. *The decisional n-BDHE problem in* $(\mathbb{G}, \mathbb{G}_T)$ *is said to be* $(\mathcal{T}, \epsilon)$-*hard if the advantage* $Adv_{\mathcal{A}}^{n\text{-dBDHE}} \leq \epsilon$, *for any PPT algorithm* $\mathcal{A}$ *running in time at most* $\mathcal{T}$.

2.2 Attribute Based Signcryption (ABSC)

In this section, we define attribute based signcryption as a set of five algorithms following [4] wherein $\mathcal{U}_e$ and $\mathcal{U}_s$ respectively are disjoint universe of encryption/decryption attributes and signature attributes. A Central Authority (CA) manages all the (encryption and signature) attributes and their public-secret key pairs by executing the **Setup** algorithm. When a decryptor requests a decryption secret key, the CA creates a decryption access structure $\mathbb{A}_d$ over $\mathcal{U}_e$ according to her role in the system and then computes the decryption secret key $\mathsf{SK}_{\mathbb{A}_d}$ by running **dExtract** algorithm, and finally sends it to the decryptor. Similarly, the CA computes the signing secret key $\mathsf{SK}_{\mathbb{A}_s}$ by executing **sExtract** algorithm with the input a signing access structure $\mathbb{A}_s$ over $\mathcal{U}_s$ and sends to the signcryptor. While the decryption access structure enables what type of ciphertexts the user can decrypt, the signing access structure is used to signcrypt a message.

When a signcryptor wants to signcrypt a message M, it selects a set W_e of encryption attributes that decides a group of legitimate recipients and an authorized signing attribute set W_s of its signing access structure $\mathbb{A}_s$ (i.e., $W_s \in \mathbb{A}_s$), and then executes the **Signcrypt** algorithm with the input $M, \mathsf{SK}_{\mathbb{A}_s}, W_s, W_e$. Here, W_e is used to encrypt a message and W_s is used to sign a message. On receiving the ciphertext $\mathsf{CT}_{(W_s, W_e)}$ of some message M, the decryptor/recipient performs the **Unsigncrypt** algorithm with the input $\mathsf{CT}_{(W_s, W_e)}, \mathsf{SK}_{\mathbb{A}_d}$. The unsigncryption will correctly return M only if $W_e \in \mathbb{A}_d$ and the ciphertext $\mathsf{CT}_{(W_s, W_e)}$ contains a valid signature with signing attributes W_s used in the ciphertext. We denote this ABSC system as follows.

$$\Sigma_{\mathsf{ABSC}} = \begin{bmatrix} (\mathsf{PK}, \mathsf{MK}) & \leftarrow \mathbf{Setup}(\kappa, \mathcal{U}_e, \mathcal{U}_s) \\ \mathsf{SK}_{\mathbb{A}_s} & \leftarrow \mathbf{sExtract}(\mathsf{PK}, \mathsf{MK}, \mathbb{A}_s) \\ \mathsf{SK}_{\mathbb{A}_d} & \leftarrow \mathbf{dExtract}(\mathsf{PK}, \mathsf{MK}, \mathbb{A}_d) \\ \mathsf{CT}_{(W_s, W_e)} & \leftarrow \mathbf{Signcrypt}(\mathsf{PK}, M, \mathsf{SK}_{\mathbb{A}_s}, W_s, W_e) \\ M \text{ or } \perp & \leftarrow \mathbf{Unsigncrypt}(\mathsf{PK}, \mathsf{CT}_{(W_s, W_e)}, \mathsf{SK}_{\mathbb{A}_d}) \end{bmatrix}$$

2.3 Security Definitions for ABSC

Following [4] , we describe the security definitions of *message confidentiality* and *ciphertext unforgeability* for ABSC as follows:

Message Confidentiality. This security notion is defined on indistinguishability of ciphertexts under adaptive chosen ciphertext attack in the selective attribute set model (IND-ABSC-sCCA) through the following game between a challenger $\mathfrak{C}$ and an adversary $\mathcal{A}$.

Init. The adversary $\mathcal{A}$ outputs the target set W_e^* of encryption attributes that will be used to create the challenge ciphertext during Challenge phase.

Setup. The challenger $\mathfrak{C}$ executes $\textbf{Setup}(\kappa, \mathcal{U}_e, \mathcal{U}_s)$, gives the public key PK to the adversary $\mathcal{A}$ and keeps the master secret key MK to itself.

Query Phase 1. The adversary $\mathcal{A}$ is given access to the following oracles which are simulated by the challenger $\mathfrak{C}$.

- *sExtract oracle* $\mathcal{O}_{\mathsf{sE}}(\mathbb{A}_s)$: on input any signing access structure $\mathbb{A}_s$ over signature attributes, the challenger $\mathfrak{C}$ returns $\mathsf{SK}_{\mathbb{A}_s} \leftarrow \textbf{sExtract}(\mathsf{PK}, \mathsf{MK}, \mathbb{A}_s)$ to the adversary $\mathcal{A}$.
- *dExtract oracle* $\mathcal{O}_{\mathsf{dE}}(\mathbb{A}_d)$: on input a decryption access structure $\mathbb{A}_d$ over encryption attributes such that $W_e^* \notin \mathbb{A}_d$, the challenger $\mathfrak{C}$ sends $\mathsf{SK}_{\mathbb{A}_d} \leftarrow \textbf{dExtract}(\mathsf{PK}, \mathsf{MK}, \mathbb{A}_d)$ to $\mathcal{A}$.
- *Signcrypt oracle* $\mathcal{O}_{\mathsf{SC}}(M, W_s, W_e)$: on input a message M, a signing attribute set W_s and an encryption attribute set W_e, $\mathfrak{C}$ samples a signing access structure $\mathbb{A}_s$ such that $W_s \in \mathbb{A}_s$ and returns the ciphertext $\mathsf{CT}_{(W_s, W_e)} \leftarrow \textbf{Signcrypt}(\mathsf{PK}, M, \mathsf{SK}_{\mathbb{A}_s}, W_s, W_e)$ to $\mathcal{A}$, here $\mathsf{SK}_{\mathbb{A}_s} \leftarrow \textbf{sExtract}(\mathsf{PK}, \mathsf{MK}, \mathbb{A}_s)$.
- *Unsigncrypt oracle* $\mathcal{O}_{\mathsf{US}}(\mathsf{CT}_{(W_s, W_e)}, \mathbb{A}_d)$: on input a ciphertext $\mathsf{CT}_{(W_s, W_e)}$ and a decryption access structure $\mathbb{A}_d$ used to decrypt, the challenger $\mathfrak{C}$ first obtains the secret decryption key $\mathsf{SK}_{\mathbb{A}_d} \leftarrow \textbf{dExtract}(\mathsf{PK}, \mathsf{MK}, \mathbb{A}_d)$ and gives the output of $\textbf{Unsigncrypt}(\mathsf{PK}, \mathsf{CT}_{(W_s, W_e)}, \mathsf{SK}_{\mathbb{A}_d})$ to the adversary.

Challenge. The adversary $\mathcal{A}$ outputs two equal length messages M_0^*, M_1^* and a signing attribute set W_s^*. The challenger $\mathfrak{C}$ selects a signing access structure $\mathbb{A}_s^*$ such that $W_s^* \in \mathbb{A}_s^*$ and returns the challenge ciphertext $\mathsf{CT}_{(W_s^*, W_e^*)}^* \leftarrow \textbf{Signcrypt}(\mathsf{PK}, M_b^*, \textbf{sExtract}(\mathsf{PK}, \mathsf{MK}, \mathbb{A}_s^*), W_s^*, W_e^*)$ to the adversary $\mathcal{A}$, where $b \xleftarrow{\$} \{0, 1\}$.

Query Phase 2. The adversary can continue adaptively to make queries as in Query Phase 1 except the queries: $\mathcal{O}_{\mathsf{US}}(\mathsf{CT}_{(W_s^*, W_e^*)}^*, \mathbb{A}_d^*)$, for any $\mathbb{A}_d^*$ with $W_e^* \in \mathbb{A}_d^*$.

Guess. The adversary $\mathcal{A}$ outputs a guess bit $b' \in \{0, 1\}$ and wins the game if $b' = b$.

The advantage of $\mathcal{A}$ in the above game is defined to be $Adv_{\mathcal{A}}^{\mathsf{IND-ABSC-sCCA}} = |\Pr[b' = b] - 1/2|$, where the probability is taken over all random coin tosses.

Remark 1. The adversary $\mathcal{A}$ is allowed to issue the queries $\mathcal{O}_{\mathsf{sE}}(\mathbb{A}_s)$, for any signing access structure $\mathbb{A}_s$ with $W_s^* \in \mathbb{A}_s$, during Query Phase 2. This provides *insider* security, which means that $\mathcal{A}$ cannot get any additional advantage in the foregoing game even though the signing secret key corresponding to the challenge signing attribute set W_s^* is revealed.

Definition 5. *An ABSC scheme is said to be* $(\mathcal{T}, q_{\mathsf{sE}}, q_{\mathsf{dE}}, q_{\mathsf{SC}}, q_{\mathsf{US}}, \epsilon)$-*IND-ABSC-sCCA secure if the advantage* $Adv_{\mathcal{A}}^{\mathsf{IND-ABSC-sCCA}} \leq \epsilon$, *for any PPT adversary* $\mathcal{A}$ *running in time at most* $\mathcal{T}$ *that makes at most* q_{sE} *sExtract queries,* q_{dE} *dExtract queries,* q_{SC} *Signcrypt queries and* q_{US} *Unsigncrypt queries in the above game.*

Ciphertext Unforgeability. This notion of security is defined on existential unforgeability under adaptive chosen message attack in the selective attribute set model (EUF-ABSC-sCMA) through the following game between a challenger $\mathfrak{C}$ and an adversary $\mathcal{A}$.

Init. $\mathcal{A}$ outputs a set of signature attributes W_s^* to $\mathfrak{C}$ that will be used to forge a signature.

Setup. The challenger $\mathfrak{C}$ runs $\mathbf{Setup}(\kappa, \mathcal{U}_e, \mathcal{U}_s)$ and sends the public key PK to the adversary $\mathcal{A}$.

Query Phase. The adversary $\mathcal{A}$ is given access to the following oracles.

- *sExtract oracle* $\mathcal{O}'_{\mathsf{sE}}(\mathbb{A}_s)$: on input a signing access structure $\mathbb{A}_s$ over signature attributes such that $W_s^* \notin \mathbb{A}_s$, the challenger $\mathfrak{C}$ returns $\mathsf{SK}_{\mathbb{A}_s} \leftarrow \mathbf{sExtract}(\mathsf{PK}, \mathsf{MK}, \mathbb{A}_s)$ to $\mathcal{A}$.

- *dExtract oracle* $\mathcal{O}'_{\mathsf{dE}}(\mathbb{A}_d)$: on input any decryption access structure $\mathbb{A}_d$ over encryption attributes, the challenger $\mathfrak{C}$ gives $\mathsf{SK}_{\mathbb{A}_d} \leftarrow \mathbf{dExtract}(\mathsf{PK}, \mathsf{MK}, \mathbb{A}_d)$ to $\mathcal{A}$.

- *Signcrypt oracle* $\mathcal{O}'_{\mathsf{SC}}(M, W_s, W_e)$: on input a message M, a signing attribute set $W_s (\neq W_s^*)$ and an encryption attribute set W_e, the challenger $\mathfrak{C}$ samples a signing access structure $\mathbb{A}_s$ such that $W_s \in \mathbb{A}_s$ and returns the ciphertext $\mathsf{CT}_{(W_s, W_e)} \leftarrow \mathbf{Signcrypt}(\mathsf{PK}, M, \mathsf{SK}_{\mathbb{A}_s}, W_s, W_e)$ to the adversary $\mathcal{A}$, where $\mathsf{SK}_{\mathbb{A}_s} \leftarrow \mathbf{sExtract}(\mathsf{PK}, \mathsf{MK}, \mathbb{A}_s)$.

- *Unsigncrypt oracle* $\mathcal{O}'_{\mathsf{US}}(\mathsf{CT}_{(W_s, W_e)}, \mathbb{A}_d)$: on input a ciphertext $\mathsf{CT}_{(W_s, W_e)}$ and a decryption access structure $\mathbb{A}_d$ used to decrypt, the challenger $\mathfrak{C}$ obtains the secret decryption key $\mathsf{SK}_{\mathbb{A}_d} \leftarrow \mathbf{dExtract}(\mathsf{PK}, \mathsf{MK}, \mathbb{A}_d)$ and sends the output of $\mathbf{Unsigncrypt}(\mathsf{PK}, \mathsf{CT}_{(W_s, W_e)}, \mathsf{SK}_{\mathbb{A}_d})$ to $\mathcal{A}$.

Forgery Phase. The adversary $\mathcal{A}$ outputs a forgery $\mathsf{CT}^*_{(W_s^*, W_e^*)}$ for some message M^* with a decryption access structure $\mathbb{A}_d^*$.

$\mathcal{A}$ wins if the ciphertext $\mathsf{CT}^*_{(W_s^*, W_e^*)}$ is valid and is not obtained from *Signcrypt* oracle, i.e., $M^* \leftarrow \mathbf{Unsigncrypt}(\mathsf{PK}, \mathsf{CT}^*_{(W_s^*, W_e^*)}, \mathbf{dExtract}(\mathsf{PK}, \mathsf{MK}, \mathbb{A}_d^*))$ and $M^* \neq \perp$, and $\mathcal{A}$ did not issue $\mathcal{O}'_{\mathsf{SC}}(M^*, W_s^*, W_e^*)$.

The advantage of $\mathcal{A}$ in the above game is defined as $Adv_{\mathcal{A}}^{\mathsf{EUF-ABSC-sCMA}} = \Pr[\mathcal{A} \text{ wins}]$.

Remark 2. In this security model, $\mathcal{A}$ can query *dExtract* oracle for the receiver's decryption access structure to whom the forgery is created in the foregoing game which captures the *insider* security model for signature unforgeability.

Definition 6. *An ABSC scheme is said to be* $(\mathcal{T}, q_{\mathsf{sE}}, q_{\mathsf{dE}}, q_{\mathsf{SC}}, q_{\mathsf{US}}, \epsilon)$-*EUF-ABSC-sCMA secure if the advantage* $Adv_{\mathcal{A}}^{\mathsf{EUF-ABSC-sCMA}} \leq \epsilon$, *for any PPT adversary $\mathcal{A}$ running in time at most $\mathcal{T}$ that makes at most q_{sE} sExtract queries, q_{dE} dExtract queries, q_{SC} Signcrypt queries and q_{US} Unsigncrypt queries in the above game.*

3 Our ABSC Construction

Let $\mathcal{U}_e = \{\mathsf{att}_y\}$ and $\mathcal{U}_s = \{\mathsf{att}'_x\}$ be the universes of encryption and signature attributes, respectively. In our construction, both signing and decryption access structures are LSSS-realizable. We denote a *signing* LSSS access structure by $(\mathbb{S}, \rho)$ and a *decryption* LSSS access structure by $(\mathbb{D}, \phi)$. We describe now our attribute based signcryption as a set of the following five algorithms.

$\mathbf{Setup}(\kappa, \mathcal{U}_e, \mathcal{U}_s)$. To initialize the system the CA performs the following steps.

- Generate cyclic groups $\mathbb{G}$ and $\mathbb{G}_T$ of prime order p whose size is determined by the security parameter κ. Let g be a generator of the group $\mathbb{G}$ and let $e : \mathbb{G} \times \mathbb{G} \to \mathbb{G}_T$ be an efficiently computable bilinear mapping.
- Choose $\alpha \xleftarrow{\$} \mathbb{Z}_p, K_0, T_0, \delta_1, \delta_2 \xleftarrow{\$} \mathbb{G}$ and set $Y = e(g, g)^\alpha$.
- For each attribute $\mathsf{att}'_x \in \mathcal{U}_s$ (resp., $\mathsf{att}_y \in \mathcal{U}_e$), select $T_x \xleftarrow{\$} \mathbb{G}$ (resp., $K_y \xleftarrow{\$} \mathbb{G}$).
- Let $\{0,1\}^{\ell_m}$ be the message space, i.e., ℓ_m is the length of each message sent. Choose a secure one-time symmetric-key cipher $\Sigma = (\mathsf{Enc}, \mathsf{Dec})$ which takes a plaintext and ciphertext of length $\ell_m + \ell_2$ with key space $\mathbb{G}_T$.
- Sample three one-way, collision resistant cryptographic hash functions $\mathcal{H}_1 : \mathbb{G} \to \mathbb{Z}_p, \mathcal{H}_2 : \{0,1\}^* \to \{0,1\}^{\ell_2}, \mathcal{H}_3 : \{0,1\}^* \to \mathbb{G}$.
- The public key and master secret key are $\mathsf{PK} = [p, g, e, Y, T_0, K_0, \{T_x : \mathsf{att}'_x \in \mathcal{U}_s\}, \{K_y : \mathsf{att}_y \in \mathcal{U}_e\}, \delta_1, \delta_2, \Sigma, \mathcal{H}_1, \mathcal{H}_2, \mathcal{H}_3]$ and $\mathsf{MK} = \alpha$, respectively.

sExtract($\mathsf{PK}, \mathsf{MK}, (\mathbb{S}, \rho)$). Each row i of the signing share-generating matrix $\mathbb{S}$ of size $\ell_s \times k_s$ is associated with an attribute $\mathsf{att}'_{\rho(i)}$. The CA carries out the following steps and returns the signing secret key to a legitimate signcryptor.

- Obtain a set $\{\lambda_{\rho(i)} = \boldsymbol{S}_i \cdot \boldsymbol{v}_s : i \in [\ell_s]\} \leftarrow \mathsf{Distribute}(\mathbb{S}, \rho, \alpha)$ of ℓ_s shares, where $\boldsymbol{S}_i$ is i-th row of $\mathbb{S}$, $\boldsymbol{v}_s \xleftarrow{\$} \mathbb{Z}_p^{k_s}$ such that $\boldsymbol{v}_s \cdot \boldsymbol{1} = \alpha$, $\boldsymbol{1} = (1, 0, \ldots, 0)$ being a vector of length k_s.
- For each row $i \in [\ell_s]$, choose $r_i \xleftarrow{\$} \mathbb{Z}_p$ and compute $D_{s,i} = g^{\lambda_{\rho(i)}}(T_0 T_{\rho(i)})^{r_i}$, $D'_{s,i} = g^{r_i}, D''_{s,i} = \{D''_{s,i,x} : D''_{s,i,x} = T_x^{r_i}, \forall\, \mathsf{att}'_x \in \mathcal{U}_s \setminus \{\mathsf{att}'_{\rho(i)}\}\}$.
- Return the signing secret key as $\mathsf{SK}_{(\mathbb{S}, \rho)} = [(\mathbb{S}, \rho), \{D_{s,i}, D'_{s,i}, D''_{s,i} : i \in [\ell_s]\}]$.

dExtract($\mathsf{PK}, \mathsf{MK}, (\mathbb{D}, \phi)$). Each row i of the decryption share-generating matrix $\mathbb{D}$ of size $\ell_e \times k_e$ is associated with an attribute $\mathsf{att}_{\phi(i)}$. In order to issue the decryption secret key to a legitimate decryptor, the CA executes as follows.

- Compute a set $\{\lambda_{\phi(i)} = \boldsymbol{D}_i \cdot \boldsymbol{v}_e : i \in [\ell_e]\} \leftarrow \mathsf{Distribute}(\mathbb{D}, \phi, \alpha)$ of ℓ_e shares, where $\boldsymbol{D}_i$ is i-th row of $\mathbb{D}$, $\boldsymbol{v}_e \xleftarrow{\$} \mathbb{Z}_p^{k_e}$ with $\boldsymbol{v}_e \cdot \boldsymbol{1} = \alpha$, here $\boldsymbol{1} = (1, 0, \ldots, 0)$ is a vector of length k_e.
- For each row $i \in [\ell_e]$, choose $\tau_i \xleftarrow{\$} \mathbb{Z}_p$ and compute $D_{e,i} = g^{\lambda_{\phi(i)}}(K_0 K_{\phi(i)})^{\tau_i}$, $D'_{e,i} = g^{\tau_i}, D''_{e,i} = \{D''_{e,i,y} : D''_{e,i,y} = K_y^{\tau_i}, \forall\, \mathsf{att}_y \in \mathcal{U}_e \setminus \{\mathsf{att}_{\phi(i)}\}\}$.
- Return the secret decryption key as $\mathsf{SK}_{(\mathbb{D}, \phi)} = [(\mathbb{D}, \phi), \{D_{e,i}, D'_{e,i}, D''_{e,i} : i \in [\ell_e]\}]$.

Signcrypt($\mathsf{PK}, M, \mathsf{SK}_{(\mathbb{S}, \rho)}, W_s, W_e$). To signcrypt a message $M \in \{0,1\}^{\ell_m}$, the signcryptor executes the following steps. Where $\mathsf{SK}_{(\mathbb{S}, \rho)} = [(\mathbb{S}, \rho), \{D_{s,i}, D'_{s,i}, D''_{s,i} : i \in [\ell_s]\}]$.

- Select an authorized signature attribute set W_s of the signing LSSS access structure $(\mathbb{S}, \rho)$ hold by the signcryptor and choose a set of encryption attributes W_e which describes the intended recipients.
- Obtains a set $\{\omega_i : i \in I_s\} \leftarrow \mathsf{Reconstruct}(\mathbb{S}, \rho, W_s)$ of reconstruction constants, where $I_s = \{i \in [\ell_s] : \mathsf{att}'_{\rho(i)} \in W_s\}$.

- Sample $\theta, \vartheta \xleftarrow{\$} \mathbb{Z}_p$ and compute
  key $= Y^\theta, C_1 = g^\theta, C_2 = (K_0 \prod_{\text{att}_y \in W_e} K_y)^\theta, \sigma_1 = g^{\theta\vartheta}$.
- Compute $\mu = \mathcal{H}_1(C_1)$ and set $C_3 = (\delta_1^\mu \delta_2)^\theta$.
- Choose $\xi \xleftarrow{\$} \mathbb{Z}_p$ and set $\sigma_2 = g^\xi \prod_{i \in I_s} (D'_{s,i})^{\omega_i}$.
- Compute $r = \mathcal{H}_2(M, \text{key}, C_1, C_2, C_3, \sigma_1, \sigma_2, W_s, W_e)$ and encrypt the message as $C = \text{Enc}_{\text{key}}(M\|r)$.
- Set $Q = \mathcal{H}_3(C, C_1, \sigma_1, W_s, W_e)$ and generate
  $\sigma_3 = \prod_{i \in I_s} \left(D_{s,i} \cdot \prod_{\text{att}'_x \in W_s, x \neq \rho(i)} D''_{s,i,x}\right)^{\omega_i} \times (T_0 \prod_{\text{att}'_x \in W_s} T_x)^\xi Q^\theta C_3^\vartheta$.

The signcryption of M is $\text{CT}_{(W_s, W_e)} = [W_s, W_e, C, C_1, C_2, C_3, \sigma_1, \sigma_2, \sigma_3]$.

Note that since W_s satisfies $\mathbb{S}$, $\sum_{i \in I_s} \omega_i \lambda_{\rho(i)} = \alpha$ (this implicitly holds and we use this fact in Lemma 2 below), although the secret shares $\{\lambda_{\rho(i)}\}_{i \in I_s}$ are not explicitly known to the signcryptor and hence so is α. However, the secret α can correctly be embedded in the exponent as $g^{\sum_{i \in I_s} \omega_i \lambda_{\rho(i)}} = g^\alpha$ in the ciphertext component σ_3 by using the secret key components $D_{s,i}, D''_{s,i}$ (see Lemma 2 below).
Unsigncrypt$(\text{PK}, \text{CT}_{(W_s, W_e)}, \text{SK}_{(\mathbb{D}, \phi)})$. The decryptor performs as follows.

1. Compute $Q' = \mathcal{H}_3(C, C_1, \sigma_1, W_s, W_e), \mu' = \mathcal{H}_1(C_1)$ and check the validity of the ciphertext $\text{CT}_{(W_s, W_e)}$ as

$$e(\sigma_3, g) \stackrel{?}{=} Y \cdot e\left(T_0 \prod_{\text{att}'_x \in W_s} T_x, \sigma_2\right) \cdot e(Q', C_1) \cdot e(\delta_1^{\mu'} \delta_2, \sigma_1), \quad (1)$$

  if it is invalid, output $\perp$; otherwise, proceed as follows.
2. Obtain $\{\nu_i : i \in I_e\} \leftarrow \text{Reconstruct}(\mathbb{D}, \phi, W_e)$, where $I_e = \{i \in [\ell_e] : \text{att}_{\phi(i)} \in W_e\}$.
3. Compute $E_1 = \prod_{i \in I_e} \left(D_{e,i} \cdot \prod_{\text{att}_y \in W_e, y \neq \phi(i)} D''_{e,i,y}\right)^{\nu_i}, E_2 = \prod_{i \in I_e} (D'_{e,i})^{\nu_i}$ and recover the symmetric decryption key as key$' = e(C_1, E_1)/e(C_2, E_2)$.
4. Decrypt the message $M'\|r' = \text{Dec}_{\text{key}'}(C)$ and compute
   $\bar{r} = \mathcal{H}_2(M', \text{key}', C_1, C_2, C_3, \sigma_1, \sigma_2, W_s, W_e)$.
5. If $r' = \bar{r}$, accept the message as $M = M'$; else, output $\perp$.

Note that the exponents $\{\omega_i\}$ and $\{\nu_i\}$ are 1 for boolean formulas [9]. Hence, Signcrypt and Unsigncrypt require only 10 and 1 exponentiations, respectively.

Remark 3. We note here that the verification process stated in Eq. (1) is formulated based on the public key parameters and the ciphertext components, thereby any user who has access to the ciphertext can verify the integrity and validity of the sender and the ciphertext. This provides the property of *Public Ciphertext Verifiability* to our scheme.

In addition, the receiver can convince a third party TP that the sender has signed the message concealed in the ciphertext without exposing his secret key. Precisely, the receiver computes key$'$ using his secret key and sends the ciphertext $\text{CT}_{(W_s, W_e)} = [W_s, W_e, C, C_1, C_2, C_3, \sigma_1, \sigma_2, \sigma_3]$ along with key$'$ to TP. Then, TP can perform the steps 1, 4 and 5 of the above Unsigncrypt algorithm serially and can conclude that whether $\text{CT}_{(W_s, W_e)}$ is valid. This algorithm is known as

TP-Verify. The above signcryption with the TP-Verify is called *Third Party Verifiable Attribute Based Signcryption.* For ease of presentation, we consider security models without TP-Verify; however, our security proofs will work for the security models with TP-Verify similar to [20].

The correctness of the unsigncryption is proved in the following two lemmas.

Lemma 2. *The* **Unsigncrypt** *algorithm can verify whether the received ciphertext* $\mathsf{CT}_{(W_s, W_e)}$ *has been altered or not according to Eq. (1).*

Proof. Assume the ciphertext $\mathsf{CT}_{(W_s, W_e)} = [W_s, W_e, C, C_1, C_2, C_3, \sigma_1, \sigma_2, \sigma_3]$ is not altered. Then $Q' = \mathcal{H}_3(C, C_1, \sigma_1, W_s, W_e) = Q, \mu' = \mathcal{H}_1(C_1) = \mu$ and hence Eq. (1) is established as follows. Since W_s satisfies $(\mathbb{S}, \rho)$, we have that $\sum_{i \in I_s} \omega_i \lambda_{\rho(i)} = \alpha$. Then,

$$\prod_{i \in I_s} \left(D_{s,i} \cdot \prod_{\mathrm{att}'_x \in W_s, x \neq \rho(i)} D''_{s,i,x} \right)^{\omega_i} = \prod_{i \in I_s} \left(g^{\lambda_{\rho(i)}} (T_0 T_{\rho(i)})^{r_i} \cdot \prod_{\mathrm{att}'_x \in W_s, x \neq \rho(i)} T_x^{r_i} \right)^{\omega_i}$$

$$= g^{\sum_{i \in I_s} \omega_i \lambda_{\rho(i)}} \prod_{i \in I_s} \left(T_0^{r_i} \prod_{\mathrm{att}'_x \in W_s} T_x^{r_i} \right)^{\omega_i}$$

$$= g^{\alpha} \left(T_0 \prod_{\mathrm{att}'_x \in W_s} T_x \right)^{\sum_{i \in I_s} r_i \omega_i},$$

$$\sigma_3 = \prod_{i \in I_s} \left(D_{s,i} \cdot \prod_{\mathrm{att}'_x \in W_s, x \neq \rho(i)} D''_{s,i,x} \right)^{\omega_i} \times \left(T_0 \prod_{\mathrm{att}'_x \in W_s} T_x \right)^{\xi} Q^{\theta} C_3^{\vartheta}$$

$$= g^{\alpha} \left(T_0 \prod_{\mathrm{att}'_x \in W_s} T_x \right)^{\sum_{i \in I_s} r_i \omega_i} \times \left(T_0 \prod_{\mathrm{att}'_x \in W_s} T_x \right)^{\xi} Q^{\theta} (\delta_1^{\mu} \delta_2)^{\theta \vartheta}$$

$$= g^{\alpha} \left(T_0 \prod_{\mathrm{att}'_x \in W_s} T_x \right)^{\xi + \sum_{i \in I_s} r_i \omega_i} Q^{\theta} (\delta_1^{\mu} \delta_2)^{\theta \vartheta}.$$

Now, $\sigma_2 = g^{\xi} \prod_{i \in I_s} (D'_{s,i})^{\omega_i} = g^{\xi} g^{\sum_{i \in I_s} r_i \omega_i} = g^{\xi + \sum_{i \in I_s} r_i \omega_i}$. Hence,

$$e(\sigma_3, g) = e \left(g^{\alpha} \left(T_0 \prod_{\mathrm{att}'_x \in W_s} T_x \right)^{\xi + \sum_{i \in I_s} r_i \omega_i} Q^{\theta} (\delta_1^{\mu} \delta_2)^{\theta \vartheta}, g \right)$$

$$= e(g, g)^{\alpha} \cdot e \left(T_0 \prod_{\mathrm{att}'_x \in W_s} T_x, g^{\xi + \sum_{i \in I_s} r_i \omega_i} \right) \cdot e(Q, g^{\theta}) \cdot e(\delta_1^{\mu} \delta_2, g^{\theta \vartheta})$$

$$= Y \cdot e \left(T_0 \prod_{\mathrm{att}'_x \in W_s} T_x, \sigma_2 \right) \cdot e(Q, C_1) \cdot e(\delta_1^{\mu} \delta_2, \sigma_1)$$

This proves the lemma. $\qquad \square$

Lemma 3. *The* **Unsigncrypt** *$(\mathsf{PK}, \mathsf{CT}_{(W_s, W_e)}, \mathsf{SK}_{(\mathbb{D}, \phi)})$ algorithm can correctly unsigncrypt the ciphertext $\mathsf{CT}_{(W_s, W_e)}$ if the Eq. (1) is valid and W_e satisfies the decryption LSSS access structure $(\mathbb{D}, \phi)$.*

Proof. Since the Eq. (1) is valid, all the ciphertext components are consistent. The identity $\sum_{i \in I_e} \nu_i \lambda_{\phi(i)} = \alpha$ implicitly holds since W_e satisfies $(\mathbb{D}, \phi)$. Now, compute

$$E_1 = \prod_{i \in I_e} \left(g^{\lambda_{\phi(i)}} (K_0 K_{\phi(i)})^{\tau_i} \cdot \prod_{\text{att}_y \in W_e, y \neq \phi(i)} K_y^{\tau_i} \right)^{\nu_i}$$

$$= g^{\sum_{i \in I_e} \nu_i \lambda_{\phi(i)}} \prod_{i \in I_e} \left(K_0^{\tau_i} \prod_{\text{att}_y \in W_e} K_y^{\tau_i} \right)^{\nu_i} = g^{\alpha} \left(K_0 \prod_{\text{att}_y \in W_e} K_y \right)^{\sum_{i \in I_e} \tau_i \nu_i},$$

$$E_2 = \prod_{i \in I_e} (D'_{e,i})^{\nu_i} = \prod_{i \in I_e} g^{\tau_i \nu_i} = g^{\sum_{i \in I_e} \tau_i \nu_i}.$$

Therefore,

$$\mathsf{key}' = \frac{e(C_1, E_1)}{e(C_2, E_2)} = \frac{e(g^{\theta}, g^{\alpha}(K_0 \prod_{\text{att}_y \in W_e} K_y)^{\sum_{i \in I_e} \tau_i \nu_i})}{e((K_0 \prod_{\text{att}_y \in W_e} K_y)^{\theta}, g^{\sum_{i \in I_e} \tau_i \nu_i})} = Y^{\theta} = \mathsf{key}.$$

Thus, $M \| r = \mathsf{Dec}_{\mathsf{key}}(C)$ and $r = \mathcal{H}_2(M, \mathsf{key}, C_1, C_2, C_3, \sigma_1, \sigma_2, W_s, W_e)$. This completes the proof. $\qquad \square$

4   Security Proof

Theorem 1 (Indistinguishability). *Assume that the encryption attribute universe $\mathcal{U}_e$ has n attributes and collision-resistant hash functions exist. Then our attribute based signcryption scheme is $(\mathcal{T}, q_{\mathsf{sE}}, q_{\mathsf{dE}}, q_{\mathsf{SC}}, q_{\mathsf{US}}, \epsilon)$-IND-ABSC-sCCA secure in the random oracle model, assuming the decisional n-BDHE problem in $(\mathbb{G}, \mathbb{G}_T)$ is $(\mathcal{T}', \epsilon')$-hard, where $\epsilon' = \epsilon - (q_{\mathsf{US}}/p)$ and $\mathcal{T}' = \mathcal{T} + q_{\mathsf{SC}} \cdot \mathcal{T}_{enc} + q_{\mathsf{US}} \cdot \mathcal{T}_{dec} + \mathcal{O}(|\mathcal{U}_s|^2 \cdot (q_{\mathsf{sE}} + q_{\mathsf{SC}}) + n^2 \cdot (q_{\mathsf{dE}} + q_{\mathsf{US}})) \cdot \mathcal{T}_{exp} + \mathcal{O}(q_{\mathsf{US}}) \cdot \mathcal{T}_{pair}$. Here, $\mathcal{T}_{enc}$ and $\mathcal{T}_{dec}$ denote the running time of symmetric-key encryption and decryption, respectively. $\mathcal{T}_{exp}$ and $\mathcal{T}_{pair}$ denote the running time of one exponentiation and one pairing computation, respectively.*

Proof. Suppose there is an adversary $\mathcal{A}$ which can $(\mathcal{T}, q_{\mathsf{sE}}, q_{\mathsf{dE}}, q_{\mathsf{SC}}, q_{\mathsf{US}}, \epsilon)$-*break* our signcryption scheme in the IND-ABSC-sCCA security model. We will construct a distinguisher $\mathfrak{D}$ to show that the decisional n-BDHE problem in $(\mathbb{G}, \mathbb{G}_T)$ is *not* $(\mathcal{T}', \epsilon')$-hard. On input the decisional n-BDHE challenge $(\mathbf{y}_{a,\theta}, Z)$, where $\mathbf{y}_{a,\theta} = (g, g^{\theta}, g_1, \dots, g_n, g_{n+2}, \dots, g_{2n})$, $g_i = g^{a^i}$, and $Z = e(g_{n+1}, g^{\theta})$ or a random element of $\mathbb{G}_T$, the distinguisher $\mathfrak{D}$ attempts to output 1 if $Z = e(g_{n+1}, g^{\theta})$ and 0 otherwise. Now, $\mathfrak{D}$ plays the role of a challenger in IND-ABSC-sCCA game and interacts with $\mathcal{A}$ as follows. Throughout the simulation, we use three one-way, collision resistant cryptographic hash functions $\mathcal{H}_1 : \mathbb{G} \to \mathbb{Z}_p$, $\mathcal{H}_2 : \{0,1\}^* \to \{0,1\}^{\ell_2}$, $\mathcal{H}_3 : \{0,1\}^* \to \mathbb{G}$, where $\mathcal{H}_3$ is modeled as a random oracle. In addition, we use a secure one-time symmetric-key cipher $\Sigma = (\mathsf{Enc}, \mathsf{Dec})$ which takes a plaintext and ciphertext of length $\ell_m + \ell_2$ with key space $\mathbb{G}_T$. Let $\mathcal{U}_e = \{\text{att}_1, \text{att}_2, \dots, \text{att}_n\}$ be the encryption attribute universe.

Init. The adversary submits the target encryption attribute set W_e^*.

Setup. The distinguisher $\mathfrak{D}$ programs the public parameters as follows.

- Sample $\alpha' \xleftarrow{\$} \mathbb{Z}_p$ and implicitly set $\alpha = \alpha' + a^{n+1}$ by letting
  $Y = e(g,g)^{\alpha'} e(g_1, g_n) = e(g,g)^{\alpha}$.
- For each $\text{att}_y \in \mathcal{U}_e$ (i.e., $y \in [n]$), sample $\gamma_y \xleftarrow{\$} \mathbb{Z}_p$ and compute
  $K_y = g^{\gamma_y} g_{n+1-y}$.
- Sample $\gamma_0 \xleftarrow{\$} \mathbb{Z}_p$ and set $K_0 = g^{\gamma_0} \prod_{\text{att}_y \in W_e^*} K_y^{-1}$.
- Sample $t_0 \xleftarrow{\$} \mathbb{Z}_p$, $t_x \xleftarrow{\$} \mathbb{Z}_p$, for each $\text{att}'_x \in \mathcal{U}_s$ and set $T_0 = g_1 g^{t_0}, T_x = g^{t_x}$.
- Set $C_1^* = g^\theta$ and $\mu^* = \mathcal{H}_1(C_1^*)$. Note that g^θ is taken from n-BDHE challenge
  $(g, g^\theta, g_1, \ldots, g_n, g_{n+2}, \ldots, g_{2n}, Z) \in \mathbb{G}^{2n+1} \times \mathbb{G}_T$.
- Compute $\delta_1 = g_n^{1/\mu^*}$, $\delta_2 = g_n^{-1} g^d$, where $d \xleftarrow{\$} \mathbb{Z}_p$.

Note that from $\mathcal{A}$'s point of view, all public parameters have the same distribution as in the original construction.

Query Phase 1. $\mathcal{A}$ issues a series of queries to which $\mathcal{D}$ responds as follows.

sExtract oracle $\mathcal{O}_{\mathsf{sE}}(\mathbb{S}, \rho)$: When $\mathcal{A}$ asks for signing secret key corresponding to the signing LSSS access structure $(\mathbb{S}, \rho)$, the distinguisher $\mathcal{D}$ constructs the signing secret key $\mathsf{SK}_{(\mathbb{S}, \rho)}$ as follows. Note that $\mathcal{D}$ does not know a^{n+1} and the master secret key $\alpha = \alpha' + a^{n+1}$, but has the knowledge of α'. Let $\mathbb{S} = (S_{i,j})_{\ell_s \times k_s}$ and $\boldsymbol{S}_i$ is the i-th row of the matrix $\mathbb{S}$.

The distinguisher $\mathcal{D}$ samples $z_2, \ldots, z_{k_s} \xleftarrow{\$} \mathbb{Z}_p$ and implicitly sets $\boldsymbol{v}_s = (\alpha' + a^{n+1}, z_2, \ldots, z_{k_s})$, which will be used for generating secret shares of α as in the original scheme. The vector $\boldsymbol{v}_s$ can be written as $\boldsymbol{v}_s = \boldsymbol{w}_s + (a^{n+1}, 0, \ldots, 0)$, where $\boldsymbol{w}_s = (\alpha', z_2, \ldots, z_{k_s})$ which is known to $\mathcal{D}$. Observe that $\lambda_{\rho(i)} = \boldsymbol{S}_i \cdot \boldsymbol{v}_s = \boldsymbol{S}_i \cdot \boldsymbol{w}_s + a^{n+1} S_{i,1}$ contains the term a^{n+1} and hence $g^{\lambda_{\rho(i)}}$ contains terms of the form $g^{a^{n+1}} = g_{n+1}$ which is also unknown to $\mathcal{D}$. Therefore, $\mathcal{D}$ must make sure that there are no terms of the form g_{n+1} involved in secret key components. To this end, the distinguisher implicitly creates suitable r_i values in such a way that the unknown terms are eliminated automatically. Now, the signing secret key corresponding to each row $\boldsymbol{S}_i, i \in [\ell_s]$, of $\mathbb{S}$ is computed by $\mathcal{D}$ as follows. Sample $r'_i \xleftarrow{\$} \mathbb{Z}_p$ and implicitly set $r_i = r'_i - a^n S_{i,1}$. Then compute

$$D_{s,i} = g^{\boldsymbol{S}_i \cdot \boldsymbol{w}_s} \left(T_0 T_{\rho(i)}\right)^{r'_i} g_n^{-(t_0 + t_{\rho(i)}) S_{i,1}}, \quad D'_{s,i} = g^{r'_i} g_n^{-S_{i,1}}$$

$$D''_{s,i} = \{D''_{s,i,x} : D''_{s,i,x} = T_x^{r'_i} g_n^{-t_x S_{i,1}}, \forall \, \text{att}'_x \in \mathcal{U}_s \setminus \{\text{att}_{\rho(i)}\}\}$$

Return the secret key $\mathsf{SK}_{(\mathbb{S}, \rho)} = [(\mathbb{S}, \rho), \{D_{s,i}, D'_{s,i}, D''_{s,i} : i \in [\ell_s]\}]$ to $\mathcal{A}$.

Claim 1. The adversary's view to the above values of $D_{s,i}, D'_{s,i}, D''_{s,i}, \forall i \in [\ell_s]$ simulated by $\mathcal{D}$ are identical to that of the original construction.

Proof. Note that $\lambda_{\rho(i)} = \boldsymbol{S}_i \cdot \boldsymbol{w}_s + a^{n+1} S_{i,1}$ and $r_i = r'_i - a^n S_{i,1}$.

$$D_{s,i} = g^{\boldsymbol{S}_i \cdot \boldsymbol{w}_s} \boxed{g_{n+1}^{S_{i,1}}} (T_0 T_{\rho(i)})^{r'_i} \boxed{g_{n+1}^{-S_{i,1}}} g_n^{-(t_0 + t_{\rho(i)}) S_{i,1}}$$

$$= g^{\boldsymbol{S}_i \cdot \boldsymbol{w}_s} g_{n+1}^{S_{i,1}} (T_0 T_{\rho(i)})^{r'_i} (g_1 g^{t_0} g^{t_{\rho(i)}})^{-a^n S_{i,1}} = g^{\lambda_{\rho(i)}} (T_0 T_{\rho(i)})^{r_i},$$

$$D'_{s,i} = g^{r'_i} g_n^{-S_{i,1}} = g^{r'_i - a^n S_{i,1}} = g^{r_i},$$

$$D''_{s,i,x} = T_x^{r'_i} g_n^{-t_x S_{i,1}} = T_x^{r'_i - a^n S_{i,1}} = T_x^{r_i}. \qquad \square \; (\text{of Claim 1})$$

dExtract oracle $\mathcal{O}_{dE}(\mathbb{D}, \phi)$: The distinguisher $\mathcal{D}$ constructs secret decryption key $\mathsf{SK}_{(\mathbb{D}, \phi)}$ as follows for a decryption LSSS access structure $(\mathbb{D}, \phi)$ such that W_e^* does not satisfy $\mathbb{D}$.

Let $\ell_e \times k_e$ be the size of a share-generating matrix $\mathbb{D}$. Since W_e^* does not satisfy $\mathbb{D}$, by Lemma 1, there exists a vector $\boldsymbol{w} = (-1, w_2, \ldots, w_{k_e}) \in \mathbb{Z}_p^{k_e}$ such that $\boldsymbol{D}_i \cdot \boldsymbol{w} = 0$, for all rows i where $\mathsf{att}_{\phi(i)} \in W_e^*$, where $\boldsymbol{D}_i$ is i-th row of the matrix $\mathbb{D}$. The distinguisher selects $z_2', z_3', \ldots, z_{k_e}' \xleftarrow{\$} \mathbb{Z}_p$ and implicitly sets $\boldsymbol{v}_e = (\alpha' + a^{n+1}, -(\alpha' + a^{n+1})w_2 + z_2', \ldots, -(\alpha' + a^{n+1})w_{k_e} + z_{k_e}') \in \mathbb{Z}_p^{k_e}$. Note that $\boldsymbol{v}_e$ can be written as $\boldsymbol{v}_e = -(\alpha' + a^{n+1})\boldsymbol{w} + \boldsymbol{v}'$, where $\boldsymbol{v}' = (0, z_2', \ldots, z_{k_e}')$. Now, the secret decryption key corresponding to each row $\boldsymbol{D}_i, i \in [\ell_e]$, of $\mathbb{D}$ is computed as one of the following two cases.

Case (1): For i where $\mathsf{att}_{\phi(i)} \in W_e^*$.

In this case, $\lambda_{\phi(i)} = \boldsymbol{D}_i \cdot \boldsymbol{v}_e = -(\alpha' + a^{n+1})\boldsymbol{D}_i \cdot \boldsymbol{w} + \boldsymbol{D}_i \cdot \boldsymbol{v}' = \boldsymbol{D}_i \cdot \boldsymbol{v}'$. Then, $\mathcal{D}$ chooses $\tau_i \xleftarrow{\$} \mathbb{Z}_p$ and computes $D_{e,i} = g^{\boldsymbol{D}_i \cdot \boldsymbol{v}'}(K_0 K_{\phi(i)})^{\tau_i}, D_{e,i}' = g^{\tau_i}$, $D_{e,i}'' = \{D_{e,i,y}'' : D_{e,i,y}'' = K_y^{\tau_i}, \forall\, y \in [n] \setminus \{\phi(i)\}\}$

It is easy to see that the values of $D_{e,i}, D_{e,i}', D_{e,i}''$ are identical to that of original scheme from $\mathcal{A}$'s point of view.

Case (2): For i where $\mathsf{att}_{\phi(i)} \notin W_e^*$.

In this case, $\phi(i) \neq y, \forall \mathsf{att}_y \in W_e^*$. Note that $\lambda_{\phi(i)} = \boldsymbol{D}_i \cdot \boldsymbol{v}_e = \boldsymbol{D}_i \cdot (\boldsymbol{v}' - \alpha'\boldsymbol{w}) - (\boldsymbol{D}_i \cdot \boldsymbol{w})a^{n+1}$. The distinguisher samples $\tau_i' \xleftarrow{\$} \mathbb{Z}_p$ and implicitly sets $\tau_i = \tau_i' + (\boldsymbol{D}_i \cdot \boldsymbol{w})a^{\phi(i)}$. Then computes

$$D_{e,i} = g^{\boldsymbol{D}_i \cdot (\boldsymbol{v}' - \alpha'\boldsymbol{w})}(K_0 K_{\phi(i)})^{\tau_i'} \left(g_{\phi(i)}^{\gamma_0 + \gamma_{\phi(i)}} \cdot \prod_{\mathsf{att}_y \in W_e^*} (g_{\phi(i)}^{-\gamma_y} \cdot g_{n+1-y+\phi(i)}^{-1}) \right)^{\boldsymbol{D}_i \cdot \boldsymbol{w}},$$

$$D_{e,i}' = g^{\tau_i'} g_{\phi(i)}^{\boldsymbol{D}_i \cdot \boldsymbol{w}},$$

$$D_{e,i}'' = \{D_{e,i,y}'' : D_{e,i,y}'' = K_y^{\tau_i'}(g_{\phi(i)}^{\gamma_y} \cdot g_{n+1-y+\phi(i)})^{\boldsymbol{D}_i \cdot \boldsymbol{w}}, \forall\, y \in [n] \setminus \{\phi(i)\}\}.$$

Claim 2. The simulated values of $D_{e,i}, D_{e,i}', D_{e,i}''$ by $\mathcal{D}$ are identical to that of original scheme from the $\mathcal{A}$'s point of view.

Proof.

$$D_{e,i} = g^{\boldsymbol{D}_i \cdot (\boldsymbol{v}' - \alpha'\boldsymbol{w})}(K_0 K_{\phi(i)})^{\tau_i'} \left(g^{\gamma_0} \prod_{\mathsf{att}_y \in W_e^*} K_y^{-1} \right)^{(\boldsymbol{D}_i \cdot \boldsymbol{w})a^{\phi(i)}} g_{\phi(i)}^{(\boldsymbol{D}_i \cdot \boldsymbol{w})\gamma_{\phi(i)}}$$

$$= g^{\boldsymbol{D}_i \cdot (\boldsymbol{v}' - \alpha'\boldsymbol{w})} \boxed{g_{n+1}^{-(\boldsymbol{D}_i \cdot \boldsymbol{w})}} (K_0 K_{\phi(i)})^{\tau_i'} K_0^{(\boldsymbol{D}_i \cdot \boldsymbol{w})a^{\phi(i)}} g_{\phi(i)}^{(\boldsymbol{D}_i \cdot \boldsymbol{w})\gamma_{\phi(i)}} \boxed{g_{n+1}^{(\boldsymbol{D}_i \cdot \boldsymbol{w})}}$$

$$= g^{\boldsymbol{D}_i \cdot (\boldsymbol{v}' - \alpha'\boldsymbol{w})} g_{n+1}^{-(\boldsymbol{D}_i \cdot \boldsymbol{w})}(K_0 K_{\phi(i)})^{\tau_i'} \left(K_0 \cdot g^{\gamma_{\phi(i)}} g_{n+1-\phi(i)} \right)^{(\boldsymbol{D}_i \cdot \boldsymbol{w})a^{\phi(i)}}$$

$$= g^{\boldsymbol{D}_i \cdot (\boldsymbol{v}' - \alpha'\boldsymbol{w}) - (\boldsymbol{D}_i \cdot \boldsymbol{w})a^{n+1}}(K_0 K_{\phi(i)})^{\tau_i' + (\boldsymbol{D}_i \cdot \boldsymbol{w})a^{\phi(i)}}$$

$$= g^{\lambda_{\phi(i)}}(K_0 K_{\phi(i)})^{\tau_i},$$

$$D_{e,i}' = g^{\tau_i'} g_{\phi(i)}^{\boldsymbol{D}_i \cdot \boldsymbol{w}} = g^{\tau_i' + (\boldsymbol{D}_i \cdot \boldsymbol{w})a^{\phi(i)}} = g^{\tau_i},$$

$$D_{e,i,y}'' = K_y^{\tau_i'}\left(g^{\gamma_y} g_{n+1-y} \right)^{(\boldsymbol{D}_i \cdot \boldsymbol{w})a^{\phi(i)}} = K_y^{\tau_i}, \forall\, y \in [n] \setminus \{\phi(i)\}. \qquad \square \text{ (of Claim 2)}$$

$\mathcal{H}_3$ *hash oracle* $\mathcal{O}_{\mathcal{H}_3}(C, C_1, \sigma_1, W_s, W_e)$: In order to answer $\mathcal{H}_3$ queries, the distinguisher $\mathfrak{D}$ maintains a list HList_3 of records $[\mathsf{IP}, \eta, \mathcal{H}_3(\mathsf{IP})]$ as described below.

When the adversary $\mathcal{A}$ queries the $\mathcal{O}_{\mathcal{H}_3}$ hash oracle with input of the form $\mathsf{IP} = (C, C_1, \sigma_1, W_s, W_e)$, the distinguisher first checks if the list HList_3 contains the record $[\mathsf{IP}, \eta, \mathcal{H}_3(\mathsf{IP})]$. If yes, $\mathfrak{D}$ returns $\mathcal{H}_3(\mathsf{IP})$. Otherwise, $\mathfrak{D}$ samples $\eta \xleftarrow{\$} \mathbb{Z}_p$ and sets $\mathcal{H}_3(\mathsf{IP}) = g^\eta$. The distinguisher stores the new record $[\mathsf{IP}, \eta, \mathcal{H}_3(\mathsf{IP})]$ in the list and returns $\mathcal{H}_3(\mathsf{IP})$ to the adversary.

Signcrypt oracle $\mathcal{O}_{\mathsf{SC}}(M, W_s, W_e)$: The adversary $\mathcal{A}$ queries the distinguisher $\mathfrak{D}$ on a tuple (M, W_s, W_e) consisting of a message, signing and encryption attribute sets, respectively. The distinguisher then selects a signing access structure $(\mathbb{S}, \rho)$ such that W_s satisfies $(\mathbb{S}, \rho)$ and generates the secret key $\mathsf{SK}_{(\mathbb{S},\rho)}$ for $(\mathbb{S}, \rho)$ by running the *sExtract oracle* $\mathcal{O}_{\mathsf{sE}}(\mathbb{S}, \rho)$ described above. Finally, $\mathfrak{D}$ sends the ciphertext $\mathsf{CT}_{(W_s, W_e)} \leftarrow \mathbf{Signcrypt}(\mathsf{PK}, M, \mathsf{SK}_{(\mathbb{S},\rho)}, W_s, W_e)$ to $\mathcal{A}$.

Unsigncrypt oracle $\mathcal{O}_{\mathsf{US}}(\mathsf{CT}_{(W_s, W_e)}, (\mathbb{D}, \phi))$: The ciphertext is $\mathsf{CT}_{(W_s, W_e)} = [W_s, W_e, C, C_1, C_2, C_3, \sigma_1, \sigma_2, \sigma_3]$. The distinguisher $\mathfrak{D}$ checks whether $C_1 = C_1^*$. If yes, the simulation aborts (since $C_1 = g^\theta$ is random in $\mathcal{A}$'s view, the probability of this type of ciphertext submitted by the adversary is at most $1/p$). Otherwise, $\mathfrak{D}$ proceeds as follows.

If W_e^* does not satisfy the decryption access structure $(\mathbb{D}, \phi)$, the distinguisher $\mathfrak{D}$ computes the secret decryption key $\mathsf{SK}_{(\mathbb{D},\phi)} \leftarrow \mathcal{O}_{\mathsf{dE}}(\mathbb{D}, \phi)$ and returns the output of $\mathbf{Unsigncrypt}(\mathsf{PK}, \mathsf{CT}_{(W_s, W_e)}, \mathsf{SK}_{(\mathbb{D},\phi)})$ to the adversary. On the other hand, if W_e^* satisfies $(\mathbb{D}, \phi)$, the distinguisher performs as follows.

The distinguisher $\mathfrak{D}$ first checks the validity of the ciphertext $\mathsf{CT}_{(W_s, W_e)}$ according to Eq. (1). If it is invalid, outputs $\perp$. Otherwise, set $\mu = \mathcal{H}_1(C_1)$ and compute

$$\mathsf{key}' = e(C_3/C_1^d, g_1)^{(\frac{\mu}{\mu^*}-1)^{-1}} \cdot e(C_1, g^{\alpha'}). \tag{2}$$

Note that $\mu = \mathcal{H}_1(C_1) \neq \mathcal{H}_1(C_1^*) = \mu^*$ since $C_1 \neq C^*$ and $\mathcal{H}_1$ is collision-resistant. Finally, the message is calculated as $M' \| r' = \mathsf{Dec}_{\mathsf{key}'}(C)$ and return M' if $r' = \mathcal{H}_2(M', \mathsf{key}', C_1, C_2, C_3, \sigma_1, \sigma_2, W_s, W_e)$. Otherwise, output $\perp$.

Claim 3. The value of key' *in Eq. (2) simulated by* $\mathfrak{D}$ *is identical to that of real construction from* $\mathcal{A}$*'s point of view.*

Proof. $e(C_3/C_1^d, g_1) = e((\delta_1^\mu \delta_2)^\theta / g^{\theta d}, g_1) = e((g_n^{\mu/\mu^*} g_n^{-1} g^d)^\theta / g^{\theta d}, g_1)$
$= e(g_n^\theta, g_1)^{(\frac{\mu}{\mu^*}-1)} = e(C_1, g^{a^{n+1}})^{(\frac{\mu}{\mu^*}-1)}$ and hence
$e(C_3/C_1^d, g_1)^{(\frac{\mu}{\mu^*}-1)^{-1}} \cdot e(C_1, g^{\alpha'}) = e(C_1, g^{\alpha' + a^{n+1}}) = e(C_1, g^\alpha)$.
Since Eq. (1) is valid, all the ciphertext components are consistent and hence $e(C_1, K_0 \prod_{\mathsf{att}_y \in W_e} K_y) = e(C_2, g)$. Therefore,

$$e(C_3/C_1^d, g_1)^{(\frac{\mu}{\mu^*}-1)^{-1}} \cdot e(C_1, g^{\alpha'}) = e(C_1, g^\alpha) \cdot \frac{e(C_1, K_0 \prod_{\mathsf{att}_y \in W_e} K_y)^{\sum_{i \in I_e} \tau_i \nu_i}}{e(C_2, g)^{\sum_{i \in I_e} \tau_i \nu_i}}$$

$$= \frac{e(C_1, g^\alpha (K_0 \prod_{\mathsf{att}_y \in W_e} K_y)^{\sum_{i \in I_e} \tau_i \nu_i})}{e(C_2, g^{\sum_{i \in I_e} \tau_i \nu_i})}$$

$$= \frac{e(C_1, E_1)}{e(C_2, E_2)} = \mathsf{key}'.$$

We refer to the values of E_1 and E_2 in Lemma 3. $\square$ *(of Claim 3)*

Challenge. The adversary $\mathcal{A}$ outputs two equal length messages M_0^*, M_1^* along with a signing attribute set W_s^*. The distinguisher $\mathfrak{D}$ then chooses $b \xleftarrow{\$} \{0,1\}$ and signcrypts M_b^* under the challenge encryption attribute set W_e^* and signing attribute set W_s^*. The components of challenge ciphertext $\mathsf{CT}_{(W_s^*,W_e^*)}^*$ are simulated as follows:

- $C_1^* = g^\theta$ (which is programmed during Setup phase),
- $C_2^* = (g^\theta)^{\gamma_0}$, $\mathsf{key}^* = Z \cdot e(g^\theta, g^{\alpha'})$, $\sigma_1^* = (g^\theta)^\vartheta$, where $\vartheta \xleftarrow{\$} \mathbb{Z}_p$,
- $C_3^* = (g^\theta)^d$, where $\mu^* = \mathcal{H}_1(C_1^*)$,
- $\sigma_2^* = g^{\xi' + \Omega_{W_s^*}} g_n^{-1}$,

    where $\xi', \Omega_{W_s^*} \xleftarrow{\$} \mathbb{Z}_p$ and implicitly sets $\xi = \xi' - a^n, \sum_{i \in I_s} r_i \omega_i = \Omega_{W_s^*}$,
- $C^* = \mathsf{Enc}_{\mathsf{key}^*}(M_b \| r^*)$, where $r^* = \mathcal{H}_2(M_b, \mathsf{key}^*, C_1^*, C_2^*, C_3^*, \sigma_1^*, \sigma_2^*, W_s^*, W_e^*)$,
- $\sigma_3^* = g^{\alpha'} \left(T_0 \prod_{\mathsf{att'}_x \in W_s^*} T_x\right)^{\xi' + \Omega_{W_s^*}} \left(g_n^{-t_0} \prod_{\mathsf{att'}_x \in W_s^*} g_n^{-t_x}\right) (g^\theta)^{\eta^* + d\vartheta}$, where η^*
    is retrieved from the record $[\mathsf{IP}, \eta^*, \mathcal{H}_3(\mathsf{IP})]$ with $\mathsf{IP} = (C^*, C_1^*, \sigma_1^*, W_s^*, W_e^*)$
    added to the list HList_3.

Claim 4. If $Z = e(g_{n+1}, g^\theta)$, then the simulated challenge ciphertext $\mathsf{CT}_{(W_s^*,W_e^*)}^*$ is same as in the original ciphertext construction of the message M_b^* under the encryption attribute set W_e^* and signature attribute set W_s^*.

Proof. Suppose $Z = e(g_{n+1}, g^\theta)$.

- $C_1^* = g^\theta$, $\mu^* = \mathcal{H}_1(C_1^*) = \mathcal{H}_1(g^\theta)$,
- $C_2^* = (g^\theta)^{\gamma_0} = (g^{\gamma_0})^\theta = \left(K_0 \prod_{\mathsf{att}_y \in W_e^*} K_y\right)^\theta$,
- $\sigma_1^* = (g^\theta)^\vartheta = g^{\theta\vartheta}$,
- $C_3^* = (g^\theta)^d = (g^d)^\theta = ((g_n^{1/\mu^*})^{\mu^*} g_n^{-1} g^d)^\theta = (\delta_1^{\mu^*} \delta_2)^\theta$,
- $\sigma_2^* = g^{\xi' + \Omega_{W_s^*}} g_n^{-1} = g^{\xi' - a^n} g^{\sum_{i \in I_s} r_i \omega_i} = g^\xi \prod_{i \in I_s} (D_{s,i}')^{\omega_i}$, where implicitly
    $D_{s,i}' = g^{r_i}, \forall i \in I_s$,
- $\mathsf{key}^* = Z \cdot e(g^\theta, g^{\alpha'}) = e(g_{n+1}, g^\theta) \cdot e(g^\theta, g^{\alpha'}) = e(g,g)^{\theta(\alpha' + a^{n+1})} = Y^\theta$,
- $C^* = \mathsf{Enc}_{\mathsf{key}^*}(M_b \| r^*)$, where $r^* = \mathcal{H}_2(M_b, \mathsf{key}^*, C_1^*, C_2^*, C_3^*, \sigma_1^*, \sigma_2^*, W_s^*, W_e^*)$,
- $\mathcal{H}_3(C^*, C_1^*, \sigma_1^*, W_s^*, W_e^*) = g^{\eta^*} = Q^*$ and

$$\sigma_3^* = g^{\alpha'} \boxed{g_{n+1}} \left(T_0 \prod_{\mathsf{att'}_x \in W_s^*} T_x\right)^{\xi' + \Omega_{W_s^*}} \left(\boxed{g_{n+1}^{-1}} g_n^{-t_0} \prod_{\mathsf{att'}_x \in W_s^*} g_n^{-t_x}\right) (g^\theta)^{\eta^* + d\vartheta}$$

$$= g^{\alpha' + a^{n+1}} \left(T_0 \prod_{\mathsf{att'}_x \in W_s^*} T_x\right)^{\xi' + \Omega_{W_s^*}} \left(g_1 g^{t_0} \prod_{\mathsf{att'}_x \in W_s^*} g^{t_x}\right)^{-a^n} (g^{\eta^*})^\theta (g^{\theta d})^\vartheta$$

$$= g^\alpha \left(T_0 \prod_{\mathsf{att'}_x \in W_s^*} T_x\right)^{\sum_{i \in I_s} r_i \omega_i} \left(T_0 \prod_{\mathsf{att'}_x \in W_s^*} T_x\right)^{\xi' - a^n} (Q^*)^\theta (C_3^*)^\vartheta$$

$$= \prod_{i \in I_s} \left(D_{s,i} \cdot \prod_{\mathsf{att'}_x \in W_s^*, x \neq \rho(i)} D_{s,i,x}''\right)^{\omega_i} \times \left(T_0 \prod_{\mathsf{att'}_x \in W_s^*} T_x\right)^\xi (Q^*)^\theta (C_3^*)^\vartheta,$$

since, from Lemma 2, we have that

$$\prod_{i \in I_s} \left(D_{s,i} \prod_{\text{att}'_x \in W_s, x \neq \rho(i)} D''_{s,i,x} \right)^{\omega_i} = g^\alpha \left(T_0 \prod_{\text{att}'_x \in W_s} T_x \right)^{\sum_{i \in I_s} r_i \omega_i}.$$

This completes the proof. □ (*of Claim 4*)

From the claim 4, we can say that if $Z = e(g_{n+1}, g^\theta)$, then the challenge ciphertext $\mathsf{CT}^*_{(W_s^*, W_e^*)}$ is a valid signcryption of the message M_b^* under the encryption attribute set W_e^* and signature attribute set W_s^*. If Z is a random element in $\mathbb{G}_T$, then the challenge ciphertext is independent of b in the adversary's view.

Query Phase 2. The adversary $\mathcal{A}$ performs a second series of queries and the distinguisher $\mathfrak{D}$ answers to these queries in the same way as it did in the Query Phase 1. Here, $\mathcal{A}$ cannot ask the *Unsigncrypt oracle* for the challenge ciphertext $\mathsf{CT}^*_{(W_s^*, W_e^*)}$ received during Challenge Phase. Note that $\mathcal{A}$ can query the *sExtract oracle* for any signing access structure $\mathbb{A}_s$ with $W_s^* \in \mathbb{A}_s$, which makes our scheme *insider* secure.

Guess. The adversary $\mathcal{A}$ outputs its guess $b' \in \{0, 1\}$ on b. If $b' = b$, then $\mathfrak{D}$ outputs 1 in the decisional n-BDHE game to guess that $Z = e(g_{n+1}, g^\theta)$; otherwise it outputs 0 to indicate that Z is a random element in $\mathbb{G}_T$.

Probability Analysis. The event in which $\mathfrak{D}$ aborts the game is when $\mathcal{A}$ queries the *Unsigncrypt oracle* $\mathcal{O}_{\mathsf{US}}(\cdot)$ with the ciphertext satisfying $C_1 = C_1^*$. The probability of this event happened is at most q_{US}/p, where q_{US} is the maximum number of Unsigncrypt queries made by the adversary. If $\mathfrak{D}$ does not abort and $Z = e(g_{n+1}, g^\theta)$, then $\mathfrak{D}$ provides a perfect simulation and hence

$$\Pr[\mathfrak{D}(\mathbf{y}_{a,\theta}, e(g_{n+1}, g^\theta)) = 1] > \frac{1}{2} + \epsilon - \frac{q_{\mathsf{US}}}{p}.$$

If Z is a random element $X \in \mathbb{G}_T$, then $\mathcal{A}$ cannot obtain any information about M_b^* and hence $\Pr[\mathfrak{D}(\mathbf{y}_{a,\theta}, X) = 1] = 1/2$. Therefore, the distinguisher $\mathfrak{D}$ can solve the n-dBDHE problem with advantage at least $\epsilon - (q_{\mathsf{US}}/p)$, i.e., $Adv_{\mathfrak{D}}^{n\text{-dBDHE}} > \epsilon - (q_{\mathsf{US}}/p)$. Thus, the decisional n-BDHE problem in $(\mathbb{G}, \mathbb{G}_T)$ is not $(\mathcal{T}', \epsilon')$-hard, where $\epsilon' = \epsilon - (q_{\mathsf{US}}/p)$. □

Theorem 2 (Unforgeability). *Assume the signing attribute universe $\mathcal{U}_s$ has n attributes and collision-resistant hash functions exist. Then, our attribute based signcryption scheme is $(\mathcal{T}, q_{\mathsf{sE}}, q_{\mathsf{dE}}, q_{\mathsf{SC}}, q_{\mathsf{US}}, \epsilon)$-EUF-ABSC-sCMA secure in the random oracle model, assuming that the computational n-DHE problem in $\mathbb{G}$ is $(\mathcal{T}', \epsilon)$-hard, where $\mathcal{T}' = \mathcal{T} + q_{\mathsf{SC}} \cdot \mathcal{T}_{enc} + q_{\mathsf{US}} \cdot \mathcal{T}_{dec} + \mathcal{O}(n^2 \cdot (q_{\mathsf{sE}} + q_{\mathsf{SC}}) + |\mathcal{U}_e|^2 \cdot (q_{\mathsf{dE}} + q_{\mathsf{US}})) \cdot \mathcal{T}_{exp} + \mathcal{O}(q_{\mathsf{SC}} + q_{\mathsf{US}}) \cdot \mathcal{T}_{pair}$. Here, $\mathcal{T}_{enc}$ and $\mathcal{T}_{dec}$ denote the running time of symmetric-key encryption and decryption, respectively. $\mathcal{T}_{exp}$ and $\mathcal{T}_{pair}$ denote the running time of one exponentiation and one pairing computation, respectively.*

Proof. Assume that there is an adversary (or an algorithm) $\mathcal{A}$ which can $(\mathcal{T}, q_{\mathsf{sE}}, q_{\mathsf{dE}}, q_{\mathsf{SC}}, q_{\mathsf{US}}, \epsilon)$-*break* our signcryption scheme in the EUF-ABSC-sCMA security model. We will construct another algorithm $\mathcal{C}$ that can solve the computational n-DHE problem with advantage at least ϵ' in time at most $\mathcal{T}'$. Hereafter, we refer $\mathcal{C}$ as a challenger.

The challenger $\mathcal{C}$ is given the computational n-DHE instance $(g, g^a, \ldots, g^{a^n}, g^{a^{n+2}}, \ldots, g^{a^{2n}}) \in \mathbb{G}^{2n}$, where $a \xleftarrow{\$} \mathbb{Z}_p$ and a generator $g \xleftarrow{\$} \mathbb{G}$. In order to calculate $g^{a^{n+1}}$ from the tuple given, $\mathcal{C}$ runs the adversary $\mathcal{A}$ answering its queries in each phase of the EUF-ABSC-sCMA security game as follows. As in Theorem 1, we use, throughout the simulation, three one-way, collision resistant cryptographic hash functions $\mathcal{H}_1 : \mathbb{G} \to \mathbb{Z}_p$, $\mathcal{H}_2 : \{0,1\}^* \to \{0,1\}^{\ell_2}$, $\mathcal{H}_3 : \{0,1\}^* \to \mathbb{G}$, where $\mathcal{H}_3$ is modeled as a random oracle. In addition, we use a secure one-time symmetric-key cipher $\Sigma = (\mathsf{Enc}, \mathsf{Dec})$ which takes a plaintext and ciphertext of length $\ell_m + \ell_2$ with key space $\mathbb{G}_T$. Let $\mathcal{U}_s = \{\mathsf{att'}_1, \mathsf{att'}_2, \ldots, \mathsf{att'}_n\}$ be the signature attribute universe.

Init. The adversary $\mathcal{A}$ sends the challenge signing attribute set W_s^* to the challenger $\mathcal{C}$.

Setup. The challenger $\mathcal{C}$ programs the public parameters as follows.

- Set $Y = e(g,g)^{\alpha'} \cdot e(g_1, g_n) = e(g,g)^{\alpha}$ by implicitly setting $\alpha = \alpha' + a^{n+1}$, where $\alpha' \xleftarrow{\$} \mathbb{Z}_p$ is sampled by $\mathcal{C}$.
- For each $\mathsf{att'}_x \in \mathcal{U}_s$ (i.e., $x \in [n]$), choose $t_x \xleftarrow{\$} \mathbb{Z}_p$ and compute $T_x = g^{t_x} g_{n+1-x}$.
- Select $t_0 \xleftarrow{\$} \mathbb{Z}_p$ and set $T_0 = g^{t_0} \prod_{\mathsf{att'}_x \in W_s^*} T_x^{-1}$.
- Sample $\gamma_0 \xleftarrow{\$} \mathbb{Z}_p$, $\gamma_y \xleftarrow{\$} \mathbb{Z}_p$, for each $\mathsf{att}_y \in \mathcal{U}_e$ and set $K_0 = g_1 g^{\gamma_0}, K_y = g^{\gamma_y}$.
- Set $\delta_1 = g^d, \delta_2 = g^{d'}$, where $d, d' \xleftarrow{\$} \mathbb{Z}_p$.

All public parameters have the same distribution as in the original construction in the adversary's view.

Query Phase. The adversary $\mathcal{A}$ issues a series of queries to which $\mathcal{C}$ responds as follows.

sExtract oracle $\mathcal{O}'_{\mathsf{sE}}(\mathbb{S}, \rho)$: When $\mathcal{A}$ queries $\mathcal{C}$ for signing secret key on a signing access structure $(\mathbb{S}, \rho)$ such that W_s^* does not satisfy $(\mathbb{S}, \rho)$, the challenger $\mathcal{C}$ proceeds exactly as it did in *dExtract oracle* $\mathcal{O}_{\mathsf{dE}}(\cdot)$ of Theorem 1[3] and returns the obtained secret key $\mathsf{SK}_{(\mathbb{S}, \rho)} = [(\mathbb{S}, \rho), \{D_{s,i}, D'_{s,i}, D''_{s,i} : i \in [\ell_s]\}]$ to $\mathcal{A}$.

dExtract oracle $\mathcal{O}'_{\mathsf{dE}}(\mathbb{D}, \phi)$: The adversary $\mathcal{A}$ can request the secret decryption key for any decryption access structure $(\mathbb{D}, \phi)$ over encryption attributes. The challenger computes the corresponding secret decryption key $\mathsf{SK}_{(\mathbb{D}, \phi)} = [(\mathbb{D}, \phi), \{D_{e,i}, D'_{e,i}, D''_{e,i} : i \in [\ell_e]\}]$ similar to the *sExtract oracle* $\mathcal{O}_{\mathsf{sE}}(\cdot)$ of Theorem 1 and sends it to $\mathcal{A}$.

$\mathcal{H}_3$ *hash oracle* $\mathcal{O}'_{\mathcal{H}_3}(C, C_1, \sigma_1, W_s, W_e)$: The challenger $\mathcal{C}$ maintains a list HList'_3 of records $[\mathsf{IP}, \eta, \mathcal{H}_3(\mathsf{IP})]$ to answer $\mathcal{H}_3$ queries. When the adversary $\mathcal{A}$ queries the $\mathcal{O}'_{\mathcal{H}_3}$ hash oracle with input of the form $\mathsf{IP} = (C, C_1, \sigma_1, W_s, W_e)$, the challenger checks if the list HList'_3 contains the record $[\mathsf{IP}, \eta, \mathcal{H}_3(\mathsf{IP})]$. If yes, $\mathcal{C}$ returns $\mathcal{H}_3(\mathsf{IP})$. Otherwise, $\mathcal{C}$ samples $\eta \xleftarrow{\$} \mathbb{Z}_p$ and sets

---

[3] This is possible due to the fact that the computation procedure of public parameters of signing (resp. decryption) attributes is same as that of decryption (resp. signing) attributes in Theorem 1.

$$\mathcal{H}_3(\mathsf{IP}) = \mathcal{H}_3(C, C_1, \sigma_1, W_s, W_e) = \begin{cases} g_n^{-1} g^\eta, & \text{if } W_s \neq W_s^*, \\ g^\eta, & \text{if } W_s = W_s^*. \end{cases}$$

The challenger stores the new record $[\mathsf{IP}, \eta, \mathcal{H}_3(\mathsf{IP})]$ in the list and returns $\mathcal{H}_3(\mathsf{IP})$ to the adversary.

Signcrypt oracle $\mathcal{O}'_{\mathsf{SC}}(M, W_s, W_e)$: The challenger $\mathcal{C}$ formulates a signing access structure $(\mathbb{S}, \rho)$ with W_s an authorized set. If the challenge signing attribute set W_s^* does not satisfy $(\mathbb{S}, \rho)$, the challenger $\mathcal{C}$ can obtain the secret signing key $\mathsf{SK}_{(\mathbb{S},\rho)} \leftarrow \mathcal{O}'_{\mathsf{sE}}(\mathbb{S}, \rho)$ and returns the ciphertext $\mathsf{CT}_{(W_s,W_e)} \leftarrow$ **Signcrypt**$(\mathsf{PK}, M, \mathsf{SK}_{(\mathbb{S},\rho)}, W_s, W_e)$ to $\mathcal{A}$. Suppose W_s^* satisfies $(\mathbb{S}, \rho)$. In this case, $\mathcal{C}$ performs as follows.

- Sample $\theta' \xleftarrow{\$} \mathbb{Z}_p$ and implicitly set $\theta = \theta' + a$. Compute $C_1 = g^{\theta'} g_1$.

- $C_2 = \left(K_0 \prod_{\mathsf{att}_y \in W_e} K_y \right)^{\theta'} \left(g_2 g_1^{\gamma_0} \prod_{\mathsf{att}_y \in W_e} g_1^{\gamma_y} \right)$.

- Select $\vartheta \xleftarrow{\$} \mathbb{Z}_p$ and set $\sigma_1 = C_1^\vartheta$.

- $C_3 = (\delta_1^\mu \delta_2)^{\theta'} (g_1^{\mu d + d'})$, where $\mu = \mathcal{H}_3(C_1)$.

- Choose $\xi' \xleftarrow{\$} \mathbb{Z}_p$ and set $\sigma_2 = g^{\xi'}$. Note that ξ' implicitly contains the term $\sum_{i \in I_s} r_i \omega_i$, i.e., we can assume that $\xi' = \xi + \sum_{i \in I_s} r_i \omega_i$.

- $\mathsf{key} = Y^{\theta'} \cdot e(g, g_1)^{\alpha'} \cdot e(g_2, g_n)$.

- Compute $r = \mathcal{H}_2(M, \mathsf{key}, C_1, C_2, C_3, \sigma_1, \sigma_2, W_s, W_e)$ and $C = \mathsf{Enc}_{\mathsf{key}}(M \| r)$.

- Process the list HList'_3 and set $Q = g_n^{-1} g^\eta$ since $W_s \neq W_s^*$.

- Compute $\sigma_3 = g^{\alpha'} \left(T_0 \prod_{\mathsf{att}'_x \in W_s} T_x \right)^{\xi'} (g_n^{-1} g^\eta)^{\theta'} g_1^\eta C_3^\vartheta$.

The ciphertext $\mathsf{CT}_{(W_s,W_e)} = [W_s, W_e, C, C_1, C_2, C_3, \sigma_1, \sigma_2, \sigma_3]$ is sent to $\mathcal{A}$.

Unsigncrypt oracle $\mathcal{O}'_{\mathsf{US}}(\mathsf{CT}_{(W_s,W_e)}, (\mathbb{D}, \phi))$: $\mathcal{A}$ can perform an unsigncryption query on a ciphertext $\mathsf{CT}_{(W_s,W_e)}$ and a decryption access structure $(\mathbb{D}, \phi)$. The challenger $\mathcal{C}$ computes the secret decryption key $\mathsf{SK}_{(\mathbb{D},\phi)} \leftarrow \mathcal{O}'_{\mathsf{dE}}(\mathbb{D}, \phi)$ and returns the output of **Unsigncrypt**$(\mathsf{PK}, \mathsf{CT}_{(W_s,W_e)}, \mathsf{SK}_{(\mathbb{D},\phi)})$ to the adversary.

Forgery Phase. $\mathcal{A}$ produces a *valid* forgery $\mathsf{CT}^*_{(W_s^*,W_e^*)} = [W_s^*, W_e^*, C^*, C_1^*, C_2^*, C_3^*, \sigma_1^*, \sigma_2^*, \sigma_3^*]$ for some message M^* and an encryption attribute set W_e^*. Then, the challenger solves the computational n-DHE problem as follows.

Since $[W_s^*, W_e^*, C^*, C_1^*, C_2^*, C_3^*, \sigma_1^*, \sigma_2^*, \sigma_3^*]$ is a valid signcryption of M^*, it must pass the verification test stated in Eq. (1), which means that

$\sigma_3^* = g^\alpha \left(T_0 \prod_{\mathsf{att}'_x \in W_s^*} T_x \right)^{\xi'} Q^\theta (C_3^*)^\vartheta$, $\sigma_2^* = g^{\xi'}$, $C_1^* = g^\theta$, $\sigma_1^* = g^{\theta\vartheta}$, $C_3^* = (\delta_1^{\mu^*} \delta_2)^\theta = g^{(d\mu^* + d')\theta}$, where $\alpha = \alpha' + a^{n+1}$, $\xi' = \xi + \sum_{i \in I_s} r_i \omega_i, \mu^* = \mathcal{H}_1(C_1^*)$, and $Q = \mathcal{H}_3(C^*, C_1^*, \sigma_1^*, W_s^*, W_e^*)$.

The challenger $\mathcal{C}$ retrieves η from HList'_3 as $Q = \mathcal{H}_3(C^*, C_1^*, \sigma_1^*, W_s^*, W_e^*) = g^\eta$ and computes

$$\frac{\sigma_3^*}{g^{\alpha'} \cdot (\sigma_2^*)^{t_0} \cdot (C_1^*)^{\eta} \cdot (\sigma_1^*)^{(d\mu^* + d')}} = \frac{g^{\alpha} (T_0 \prod_{\text{att}'_x \in W_s^*} T_x)^{\xi'} Q^{\theta} (C_3^*)^{\vartheta}}{g^{\alpha'} \cdot (g^{\xi'})^{t_0} \cdot (g^{\theta})^{\eta} \cdot (g^{\theta\vartheta})^{(d\mu^* + d')}}$$

$$= \frac{g^{\alpha' + a^{n+1}} (T_0 \prod_{\text{att}'_x \in W_s^*} T_x)^{\xi'} Q^{\theta} (C_3^*)^{\vartheta}}{g^{\alpha'} \cdot (g^{t_0})^{\xi'} \cdot (g^{\eta})^{\theta} \cdot (g^{(d\mu^* + d')\theta})^{\vartheta}}$$

$$= \frac{g^{\alpha'} g^{a^{n+1}} (T_0 \prod_{\text{att}'_x \in W_s^*} T_x)^{\xi'} Q^{\theta} (C_3^*)^{\vartheta}}{g^{\alpha'} (T_0 \prod_{\text{att}'_x \in W_s^*} T_x)^{\xi'} Q^{\theta} (C_3^*)^{\vartheta}} = g^{a^{n+1}}$$

Success Probability. The advantage of C in solving the computational n-DHE problem is

$$Adv_C^{n\text{-cDHE}} = \Pr\left[C(g, g^a, \ldots, g^{a^n}, g^{a^{n+2}}, \ldots, g^{a^{2n}}) = g^{a^{n+1}}\right]$$
$$= \Pr\left[A \text{ wins the EUF-ABSC-sCMA game}\right]$$
$$= Adv_A^{\text{EUF-ABSC-sCMA}} > \epsilon.$$

Therefore, the challenger C can solve the computational n-DHE problem with advantage at least ϵ, if A creates valid forgery with advantage ϵ.   □

5 Discussion

Reducing the number of secret keys. In our construction, the secret keys used for decryption and signature are essentially the same. As pointed out in [4], it would be possible to use the same attribute universe and the same key generation algorithm for both encryption and signature by defining appropriate security model. This reduces the number of secret keys by exactly half.

Non-monotone access structure realization. We can build an ABSC for Non-Monotone Access Structure (nonMAS) with constant-size ciphertext and constant number of pairings by employing the *moving from MAS to nonMAS* technique [21] that represents non-monotone access structures in terms of monotone access structures with *negative* attributes.

6 Conclusion

We present the *first* constant-size ciphertext attribute based key-policy signcryption scheme with constant number of pairings. Both ciphertext confidentiality and unfogeability against selective adversary have been proven under decisional n-BDHE and computational n-DHE assumptions, respectively, in the random oracle model. Additionally, it provides public ciphertext verifiability property which allows any third party to check the integrity and validity of the ciphertext. The secret key size in our scheme increases by a factor of number of attributes used in the system.

Acknowledgement. The authors would like to thank the anonymous reviewers of this paper for their valuable comments and suggestions.

References

1. Sahai, A., Waters, B.: Fuzzy Identity-Based Encryption. In: Cramer, R. (ed.) EUROCRYPT 2005. LNCS, vol. 3494, pp. 457–473. Springer, Heidelberg (2005)
2. Goyal, V., Pandey, O., Sahai, A., Waters, B.: Attribute Based Encryption for Fine-Grained Access Control of Encrypted Data. In: ACM Conference on Computer and Communications Security, pp. 89–98 (2006)
3. Bethencourt, J., Sahai, A., Waters, B.: Ciphertext-Policy Attribute-Based Encryption. In: IEEE Symposium on Security and Privacy, pp. 321–334 (2007)
4. Gagné, M., Narayan, S., Safavi-Naini, R.: Threshold Attribute-Based Signcryption. In: Garay, J.A., De Prisco, R. (eds.) SCN 2010. LNCS, vol. 6280, pp. 154–171. Springer, Heidelberg (2010)
5. Hu, C., Zhang, N., Li, H., Cheng, X., Liao, X.: Body Area Network Security: A Fuzzy Attribute-based Signcryption Scheme. IEEE Journal on Selected Areas in Communications 31(9), 37–46 (2013)
6. Wang, C., Huang, J.: Attribute-based Signcryption with Ciphertext-policy and Claim-predicate Mechanism. In: CIS 2011, pp. 905–909 (2011)
7. Emura, K., Miyaji, A., Rahman, M.S.: Dynamic Attribute-Based Signcryption without Random Oracles. Int. J. Applied Cryptography 2(3), 199–211
8. Attrapadung, N., Herranz, J., Laguillaumie, F., Libert, B., de Panafieu, E., Ràfols, C.: Attribute-Based Encryption Schemes with Constant-Size Ciphertexts. Theor. Comput. Sci. 422, 15–38 (2012)
9. Waters, B.: Ciphertext-Policy Attribute-Based Encryption: An Expressive, Efficient, and Provably Secure Realization. Cryptology ePrint report 2008/290 (2008)
10. Zheng, Y.: Digital Signcryption or How to Achieve Cost (Signature & Encryption) << Cost(Signature) + Cost(Encryption). In: Kaliski Jr., B.S. (ed.) CRYPTO 1997. LNCS, vol. 1294, pp. 165–179. Springer, Heidelberg (1997)
11. Malone-Lee, J.: Identity-based signcryption. Cryptology ePrint Archive, Report 2002/098 (2002)
12. Shamir, A.: Identity-Based Cryptosystems and Signature Schemes. In: Blakely, G.R., Chaum, D. (eds.) Advances in Cryptology CRYPTO -1984. LNCS, vol. 196, pp. 47–53. Springer, Heidelberg (1985)
13. Maji, H.K., Prabhakaran, M., Rosulek, M.: Attribute-Based signatures. In: Kiayias, A. (ed.) CT-RSA 2011. LNCS, vol. 6558, pp. 376–392. Springer, Heidelberg (2011); The first version available as Cryptology ePrint report 2008/328
14. Ge, A., Ma, C., Zhang, Z.: Attribute-Based Signature Scheme with Constant Size Signature in the Standard Model. IET Information Security 6(2), 1–8 (2012)
15. Herranz, J., Laguillaumie, F., Libert, B., Ràfols, C.: Short Attribute-Based Signatures for Threshold Predicates. In: Dunkelman, O. (ed.) CT-RSA 2012. LNCS, vol. 7178, pp. 51–67. Springer, Heidelberg (2012)
16. Gagné, M., Narayan, S., Safavi-Naini, R.: Short Pairing-Efficient Threshold-Attribute-Based Signature. In: Abdalla, M., Lange, T. (eds.) Pairing 2012. LNCS, vol. 7708, pp. 295–313. Springer, Heidelberg (2013)
17. Wei, J., Hu, X., Liu, W.: Traceable attribute-based signcryption. Security Comm. Networks, doi: 10.1002/sec.940 (2013)
18. Guo, Z., Li, M., Fan, X.: Attribute-based ring signcryption scheme. Security Comm. Networks 6, 790–796 (2013), doi:10.1002/sec.614
19. Rao, Y.S., Dutta, R.: Computationally Efficient Expressive Key-Policy Attribute Based Encryption Schemes with Constant-Size Ciphertext. In: Qing, S., Zhou, J., Liu, D. (eds.) ICICS 2013. LNCS, vol. 8233, pp. 346–362. Springer, Heidelberg (2013)

20. Selvi, S.S.D., Sree Vivek, S., Pandu Rangan, C.: Identity Based Public Verifiable Signcryption Scheme. In: Heng, S.-H., Kurosawa, K. (eds.) ProvSec 2010. LNCS, vol. 6402, pp. 244–260. Springer, Heidelberg (2010)
21. Ostrovksy, R., Sahai, A., Waters, B.: Attribute Based Encryption with Non-Monotonic Access Structures. In: ACM Conference on Computer and Communications Security, pp. 195–203 (2007)

DRECON: DPA Resistant Encryption by Construction

Suvadeep Hajra[1], Chester Rebeiro[1], Shivam Bhasin[2], Gaurav Bajaj[1],
Sahil Sharma[1], Sylvain Guilley[2,3], and Debdeep Mukhopadhyay[1]

[1] Dept. of Computer Science and Engineering,
Indian Institute of Technology Kharagpur, India
{suvadeep.hajra,chetrebeiro,bajaj.gaurav92,shlshrm000,
debdeep.mukhopadhyay}@gmail.com
[2] Institut MINES-TELECOM, TELECOM ParisTech,
Department COMELEC, 46 rue Barrault,
75634 PARIS Cedex 13, France
{bhasin,guilley}@telecom-paristech.fr
[3] Secure-IC S.A.S., 80 avenue des Buttes de Coësmes,
35700 Rennes, France

Abstract. Side-channel attacks are considered as one of the biggest threats against modern crypto-systems. This motivates the design of ciphers which are naturally resistant against side-channel attacks. The present paper proposes a scheme called DRECON to construct a block cipher with innate protection against differential power attacks (DPA). The scheme is motivated by tweakable block ciphers and is shown to be secure against first-order DPA using information theoretic metrics. DRECON is shown to be less expensive than masking and re-keying countermeasures from the implementation perspective and can be efficiently realized in both hardware and software platforms. On FPGAs especially, DRECON can optimally utilize the abundant block RAMs available and therefore have minimal overheads. We estimate the cost overhead of DRECON in micro-controllers and FPGAs, two common targets for cryptographic applications. Finally we demonstrate practical side-channel resistance of a DRECON implementation on a Xilinx Virtex-5 FPGA (SASEBO GII board).

1 Introduction

In 1998, Paul Kocher demonstrated a new class of cryptographic attacks known as differential power analysis (DPA) [13], which utilize information leakages from power or electro-magnetic radiation of the cipher's implementation. Since then, several DPA attacks have been demonstrated on almost every crypto-system in use. Today DPA has become one of the biggest threats to modern security systems. Over the years there have been several attempts to prevent these attacks. A current trend is to either eliminate [37,38] or randomize [2,7] side-channel leakage. An alternate trend is a modification of the protocols, for example,

D. Pointcheval and D. Vergnaud (Eds.): AFRICACRYPT 2014, LNCS 8469, pp. 420–439, 2014.

by changing a secret in every encryption [8,19,28]. However in both these prevention methodologies, the underlying cryptographic algorithm is unchanged and by itself remains weak against the attack. Further, many of the countermeasures are ad-hoc, platform dependent, or require customized development processes. It has also been seen that several countermeasures can only make the attack more difficult and not fully protected.

Better protection can be achieved if the cryptographic algorithm itself is designed with DPA in the hindsight in addition to the conventional cryptanalytic attacks as the primitives used can be chosen with better side-channel attack resistance. Due to this reason, research is being carried out in developing cryptographic primitives that are easily protectable by masking [29,6], or that can inherently tolerate these attacks. In this paper, we show how cryptographically good sboxes can be arranged in such a way that would result in a cipher with increased resistance against DPA. This strategy ensures DPA resistance without compromising on classical cryptographic properties.

The scheme we present is called DRECON (DPA Resistant Encryption by CONstruction), which attempts to design a complete block cipher with DPA prevention as a pre-requisite. The scheme currently guarantees first-order security, and can be used as a starting point to build ciphers with higher order DPA resistance. The construction is inspired from tweakable block ciphers [14], where in addition to the plaintext and key, the cipher takes a *tweak*. However, unlike the tweakable block ciphers in [14], the construction requires the tweak to be kept secret. The tweak is used to choose an sbox from a given pool of cryptographically strong sboxes, thus modifying the mapping between the plaintext and the ciphertext. Protection against DPA is obtained based on the assumption that the tweak is exclusively shared between the sender and the receiver and modified in every encryption. Besides the fact that the primitives used in DRECON have higher resistance against DPA attacks, there are several advantages over contemporary DPA countermeasures. Compared to randomization techniques such as masking, we show that encryptions in software can be done faster. In hardware, the area and performance overheads are considerably less compared to masking. Further no custom libraries or design flows are required as compared to hiding countermeasures such as [37,38] and unlike protocol countermeasures such as [8,19,28], key expansion needs to be done just once. The construction is supported by information theoretic proofs of security. We show that the DRECON is resistant against the first-order (1O) DPA attack in the presence of glitches also. We have experimentally validated the result on a version of DRECON using the powerful correlation-collision attacks.

The organization of the paper is as follows: in Section 2, the necessary background for DPA is presented along with an introduction to commonly used countermeasures. Section 3 presents DRECON and evaluates its security against DPA. In Section 4, implementation aspects of the scheme are presented for both hardware and software platforms. We also validate its resistance to correlation-collision attack on the SASEBO-GII side-channel evaluation board [33]. The final section has the conclusion of the paper.

2   Preliminaries

2.1   Notations

We denote random variables by capital letters (e.g. X) and their realization by small letters (e.g. x). The universe for the variable is represented as a calligraphic letter (e.g. $\mathcal{X}$).

Let x be part of the plaintext which gets ex-ored with part of a sub-key, k, in an encrypting block cipher. Assume that x and k are chosen from $\mathcal{X}$ and $\mathcal{K}$ respectively and its choice is represented by the random variable X and K respectively. Generally, X is ex-ored with a unknown but fixed key k and then undergoes a non-linear transformation with an sbox. We denote this operation by $S(X \oplus k)$.

2.2   Differential Power Attacks

The aim of a DPA adversary is to use the side-channel leakage from either $X \oplus k$ or $S(X \oplus k)$ to reveal the secret data k. The steps involved is to rank each possible candidate $k^* \in \mathcal{K}$ for the key k by statistically comparing the actual leakage with a model of the leakage. The candidate ranked first is the most likely and the one ranked last is the least likely candidate. The o−th order average success rate of the attack is the probability with which the correct value of k has a rank between 1 and o [36].

The success of the attack depends on how much information gets leaked. In 2004, Micali and Reyzin used leakage functions (denoted ϕ) to encapsulate the information leaked through the side-channels [21]. The *ideal leakage function* is one which leaks the entire internal state of the cipher. It can be defined as:

$$\mathsf{Id} : y \in \mathbb{F}_2^n \mapsto y \in \mathbb{F}_2^n \ , \tag{1}$$

where $y = S(x \oplus k)$ is the intermediate state. A more realistic leakage function is the Hamming weight leakage, which leaks the Hamming weight of y. It is defined as

$$\mathsf{HW} : y \in \mathbb{F}_2^n \mapsto \sum_{i=0}^{n-1} y_i \in \mathbb{N} \ , \tag{2}$$

where y_i is the ith bit of the intermediate state y.

2.3   Countermeasures for DPA

Countermeasures for DPA are applied at the implementation level or at the protocol level. The most common countermeasures used in the implementation level are masking, shuffling, and hiding. The advantage of this is that they can be applied on any cipher algorithm. On the other hand, they are affected by the platform of implementation and do not always provide provable security. Several *hiding* schemes have been proposed, which essentially use side-channel resistant logic styles in order to prevent information leakage through the power consumption. Examples of this can be found in [37,38]. These countermeasures may require specific CMOS libraries or full custom designs. Masking and shuffling do not have these limitations and will be discussed here in greater detail.

Masking: *Masking* is the most frequently used countermeasure [2,7] applied to implementations. A p-order masking scheme involves spreading each sensitive variable Z into $p + 1$ shares $Z_0, \ldots, Z_p$ maintaining the invariant $Z = g(Z_0, \cdots, Z_p)$. Each of the Z_i's are uniformly random and the joint distribution of any p variables are independent of Z. Thus, any collection of variables less than or equal to p contains no information about the sensitive variable Z. The most commonly used masking is the *first order masking* (denoted 1O masking) where a single uniformly random mask is used.

Let M be a random variable M with entropy $h_m \leq n$. In the 1O Boolean masking scheme, it gets added to the sensitive variable $X \oplus k$ resulting in two shares: $X \oplus k \oplus M$ and M. Each sbox S is also replaced by a masked sbox S_M such that $S_M(X \oplus k \oplus M, M, M') = S(X \oplus k) \oplus M'$. In other words, the masked sbox S_M first unmasks the randomized variable $X \oplus k \oplus M$, passes it through the sbox S and then re-masks the output $S(X \oplus k)$ by the output mask M'. The mask M is also replaced by the new mask M', thus the invariant $S(X \oplus k) = S_M(X \oplus k \oplus M, M, M') \oplus M'$ is maintained. Now, the 1O side-channel leakage has the form $\phi(S(X \oplus k) \oplus M')$. In the case where the leakage function $\phi = \mathsf{Id}$ (Equation 1), the entire output $S(X \oplus k) \oplus M'$ is revealed to the adversary. Information leakage is measured by the mutual information (abridged I) between what can be observed by the attacker and the sensitive variable. Renaming the variable M' as M:

$$\mathsf{I}[S(X \oplus K) \oplus M, X; K] = n - h_m \ . \tag{3}$$

This means that the countermeasure is perfect at 1O if and only if M has entropy $h_m = n$, *i.e.* M is uniformly distributed over $\mathbb{F}_2^n$.

In the case where $\phi = \mathsf{HW}$ (defined in Equation 2), only the Hamming weight of $S(X \oplus k) \oplus M$ is revealed. The information leakage is equal to:

$$\mathsf{I}[\mathsf{HW}(S(X \oplus K) \oplus M), X; K] = \mathsf{H}[\mathsf{HW}(K)]$$
$$- \sum_{x,k} \mathsf{P}[K = k]\mathsf{P}[X = x] \cdot \mathsf{H}[\mathsf{HW}(S(x \oplus k) \oplus M)] \ . \tag{4}$$

If M is independent of X, the second term of the difference is equals to $\mathsf{H}[\mathsf{HW}(M)]$. The value of Equation (4) is lower than that of Equation (3), but a priori hard to make null if $h_m < n$.

If $h_m = n$, the single mask can perfectly shield against first-order attacks but not against attacks of higher order such as [20]. This is because, in a second-order attack the adversary is capable of obtaining the leakage $\phi(M)$ in addition to $\phi(S(X \oplus k) \oplus M)$. However, the complexity of the attack increases. The complexity is reduced significantly when the computations of M and $S(X \oplus k) \oplus M$ overlap. The leakage then takes the form $\phi(M) + \phi(S(X \oplus k) \oplus M)$ and follows a distribution whose higher order moment depends on $X \oplus k$. These attacks are known as *univariate higher order attacks* (the one which uses variance is called univariate second order attack [39]).

A 1O masking scheme is secure against a 1O attack in an idealistic model. Most of the model is based on the assumption that the output of a circuit switches only once in a clock cycle. However due to asymmetric path delay, output of a CMOS gate may switch more than once in a clock cycle. This phenomenon is referred to as 'glitch' [16]. Since most of the masking schemes combine the masked value $X \oplus k \oplus M$ and the masks M, M' within the same combinatorial circuit, the leakage takes the form $\phi(X \oplus k \oplus M, M, M')$. Due to the glitches, this leakage $\phi(X \oplus k \oplus M, M, M')$ becomes strongly correlated to the unmasked sensitive variable $X \oplus k$ [27]. Consequently, the circuit becomes vulnerable to first order DPA attacks [17,4,22,23]. Secure implementation of non-linear function in the presence of glitches has been proposed in [27]. However, implementation of such a scheme increases the hardware cost drastically [25,31].

Shuffling: An alternate randomization technique is *shuffling* [10,34]. Here instead of a random mask being added, executions of several sensitive operations are shuffled in time. If the execution of an operation is spread over m different signals, then the information per signal is reduced m times. This works well because DPA can target a single signal at a time. However 1O DPA attacks can defeat shuffling using m^2 times traces [3].

Protocol Level: DPA requires several power traces in order to successfully retrieve the secret key due to the noise present in the target device and due to the non-injective nature of the leakage function. Protocol level countermeasures prevent the adversary from collecting the required number of traces. In [28], Kocher suggests to update the key on a regular basis. The rate of updation should be fast enough to prevent an adversary from collecting the necessary traces. The updation rate should be evaluated for each device and cipher implementation. In the strongest form, every encryption is done with a new key.

There are various ways in which key updation (or *re-keying*) can be done. Abdalla and Bellare in [1] classify them into two schemes: parallel and serial. In the *parallel re-keying scheme*, a key update is derived directly from the master key using a suitable function f. For example the ith key update (denoted K_i) can be obtained from $K_i = f(K, i)$, where K is a master key. Methods of key updation using this scheme have been suggested in [19] and [8]. To obtain provable security using the scheme, it is required that the key updates are precomputed and stored in memory. Thus the number of encryptions is limited by the size of memory. In the serial re-keying scheme, a new key is obtained from the previous key using a suitable function f. For example, the ith key can be obtained from the previous key as follows: $K_i = f(K_{i-1})$, while the first key used is derived from the master key (*i.e.* $K_1 = f(K)$). Re-keying mechanisms using this technique were suggested in [28] and [18].

The drawback of the re-keying mechanisms is that with each key update, new round keys have to be computed. Thus, the overhead is not only in the generation of the new key, but also the computation of the key expansion algorithm. This can add significant overheads in the performance, especially in software implementations. Our proposal does not suffer from this drawback. In DRECON, the round keys are fixed for all encryptions. Instead, only the tweak is updated by a function similar to f used in the previous schemes.

3   The DRECON Scheme

In 2002, Liskov, Rivest, and Wagner introduced *tweakable block ciphers* to add more variability to the functionality [14]. Here an additional secret input called the *tweak* is present, which if changed alters the map between the plaintext and ciphertext thereby obtaining more variations in the mapping. Both the sender and the receiver need to know the tweak in addition to the secret key. The proposal in this paper is inspired by tweakable block ciphers, and uses a regularly changing tweak to stymie differential power attacks. In this section, we present the proposal and then compare its security with that of 1O masking.

DRECON: The secret in DRECON comprises of the tuple (t, k), where t is called the tweak and k the key used in the block cipher. The key k is held constant for all encryptions, while the tweak t changes for each encryption, using a **tweak generation algorithm**. The tweak is used to select a function from the set $\mathcal{F}\{F_1, F_2, \cdots, F_r\}$, where $F_j : \{0,1\}^n \mapsto \{0,1\}^n$ and $(1 \leq j \leq r)$, are cryptographically strong sbox functions. For every application of the sbox on X, a function from $\mathcal{F}$ is selected depending on the value of the tweak (t) and applied to X. This sbox, known as the *tweaked-sbox*, is represented by $\boldsymbol{S}(\cdot, \cdot)$ and defined as follows:

$$S(t, X) \leftarrow F_t(X) \qquad \text{where } t \xleftarrow{R} \{1, 2, \cdots, r\}.$$

In a typical iterative block cipher, the first round key is added to plaintext before the sbox operation and the sbox operation has the form $S(x \oplus k)$. However, in DRECON, we choose to omit the whitening at the beginning of the encryption. Thus, each round except the last round consists of substitution layer, diffusion layer and key addition layer. The last round consists of only substitution layer. The sboxes of the substitution layers is replaced by the tweaked-sbox. For all round, the same tweaks are used though two different tweaked-sboxes of the same round use two different tweaks independently. The first round of DRECON is shown in Figure 1. It may be noted that DRECON requires no key whitening at the beginning and end of the block cipher since the tweaked-sboxes provide the required randomization of the input and output respectively.

3.1   Information Theoretic Analysis

First we will analyse the security of the above scheme for glitch-free circuit. Then we will analyse its resistance in the presence of glitches. For all the analysis, we consider the known plaintext attacks where its distribution is uniformly random.

In the Absence of Glitch: Let us assume there is no diffusion layer, since its presence does not dilute the side-channel security of DRECON. Let T be the random variable representing the tweak and having entropy h_t. In the worst case, the entire state gets leaked (that is $\phi = \mathsf{Id}$). Hence, one can get the 1O leakage $\boldsymbol{S}(T, X) \oplus k$. The information leakage can be shown to be

$$\mathsf{I}[\boldsymbol{S}(T, X) \oplus K, X; K] = \mathsf{H}[X] - \sum_x \mathsf{P}[X = x] \cdot \mathsf{H}[\boldsymbol{S}(T, x)] . \qquad (5)$$

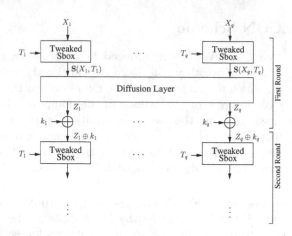

Fig. 1. First round of DRECON. The same structure is repeated for all rounds except the last round which is consisted of only substitution layer.

This quantity is greater than or equal to $n - h_t$ if $h_t \leq n$. When $h_t > n$, it is greater than or equal to 0.

Comparing Equations (5) and (3), we note that DRECON performs as good as 10 masking scheme for the Id leakage functions (the worst case), provided the following propositions are respected.

Proposition 1. $\mathsf{I}[S(T, X) \oplus K, X; K] = n - h_t$ if and only if $\nexists(x, t_0, t_1) \in \mathbb{F}_2^n \times \mathcal{T} \times \mathcal{T}$, such that $S(t_0, x) = S(t_1, x)$.

This means that for all $x \in \mathbb{F}_2^n$, the values taken by the random variable $S(T, x)$ are of cardinality 2^{h_t}.

Proposition 2. $\mathsf{I}[S(T, X) \oplus K, X; K] = n - h_t$ if and only if $\forall x, t \in \mathcal{T} \mapsto S(t, x) \in \mathbb{F}_2^n$ is balanced.

Thus, to make the information leakage null, h_t should be atleast n. In other words, the tweak should be atleast n bit long.

In the Presence of Glitches: Let us first assume that there is no glitch in the key addition layer. This assumption is aligned with the existing observations in the literature [16] and much of the effort has been directed to make a nonlinear circuit resistant in the presence of glitches [27]. We also assume that tweak T is following an uniformly random distribution with entropy $h_t \geq n$ and the tweaked-sbox satisfies the balancedness property of Proposition 2.

Under the above assumptions, the output of the substitution layer in the first round (see Figure 1) is a uniformly random unknown variable. Hence the leakages of the linear layer of the first round take the form $\phi(R \oplus k)$ for some uniformly distributed unknown random variable R. Thus, univariate attack targeting that layer is not feasible. However, the leakage of an sbox $S(R \oplus k, T)$ in an intermediate round takes the form $\phi(R \oplus k, T)$ which may leak some information of the secret k if there exists a dependency between R and T. To analyse the case further, we assume that the diffusion layer consists of q/m maximum distance separable (MDS) mapping $\mathsf{M} : (\{0,1\}^n)^m \mapsto (\{0,1\}^n)^m$ where q is the number of sboxes present in a single round, $m \geq 2$ and q is divisible by m. Lemma 1 provides the following result.

Lemma 1. *Let $X_1, \ldots, X_q$ be the random variables representing the plaintext inputs of DRECON. Let $Z \oplus k$ and T be the inputs to an sbox $S(Z \oplus k, T)$ of an intermediate round where T is the random variable representing the tweak input to the sbox. Then the random variable Z is independent to the joint distribution of $X_1, \ldots, X_q$ and T.*

The proof of the above lemma is given in Appendix A. Before computing the bound of the information leakage, we state without proof two well known results of information theory [5]:

Lemma 2. *Let $U_1, \ldots, U_r$ be r mutually independent variables. Then*

$$\mathsf{H}[U_1, \ldots, U_r] = \sum_{i=1}^{r} \mathsf{H}[U_i]$$

Lemma 3. *Let U_1 and U_2 be two random variables. Then*

$$\mathsf{I}[U_1; U_2] = \mathsf{H}[U_1] + \mathsf{H}[U_2] - \mathsf{H}[U_1, U_2]$$

Applying the above three lemmas, the information leakage $\phi(Z \oplus k, T)$ due to the sbox $S(Z \oplus k, T)$ can be computed as

$$\mathsf{I}[\phi(Z_1 \oplus K, T), X_1, \ldots, X_q; K] \leq \mathsf{I}[Z_1 \oplus K, T, X_1, \ldots, X_q; K]$$
$$= 0 .$$

Hence, DRECON is resistant against 1O DPA in the presence of glitches also.

DRECON and Shuffling: DRECON is not a shuffling countermeasure. A typical shuffling countermeasure would have computed the correct result sometimes and at others something which is totally uncorrelated from the data. Thus shuffling is not perfect at first order. On the other hand, DRECON is sound at 1O using the mutual information metric.

DRECON and Masking: As discussed in Section 2.3, a simple masking like Boolean masking provides perfect secrecy against 1O DPA only in glitch free circuits. By custom design of circuits, glitches can be reduced, but can never be eliminated totally. On the other hand, masking schemes that of [27] provides high resistance against 1O DPA in the presence of glitches, but are very costly to implement. Thus, DRECON provides a cost-effective alternative to those masking schemes against 1O DPA in the presence of glitches.

DRECON and Re-keying: DRECON, in some sense, is similar to re-keying mechanisms. However unlike re-keying, the key is held constant for all encryptions while it is the tweak that changes and needs to be synchronized between the sender and receiver. If the key is changed such as in [28,8,18,19], then in addition to the generation of the new key, the key expansion algorithm has to be executed in order to generate the new round keys. DRECON doesn't suffer from this drawback since it is only the tweak that needs to be generated.

4   An Application of the DRECON Scheme

DRECON is implementation friendly for both software and hardware platforms and can be easily derived from any legacy block cipher, preferably one which has small sboxes of dimensions for example 4×4. When implemented with DRECON, each of the sboxes of the legacy cipher is chosen from a pool of cryptographically equal strong sboxes for each encryption based on the unknown tweak such that Proposition 2 is satisfied. Thus, the classical blackbox cryptanalytic attacks are no more applicable under the assumption that the tweak is a uniformly random variable parameter.

In this article, we consider two implementations of DRECON which are based on AES algorithm. The first implementation is more resource friendly and based on a simplified AES algorithm. The simplified AES algorithm, which we name as 4×4 AES, follows the standard AES specification, except that the 8×8 sbox is replaced by a pair of 4×4 sboxes. Thus, there are 32 sbox access per round instead of 16 for the regular AES algorithm. The second implementation is based on the standard AES which we refer as 8×8 AES.

The adapted $n \times n$ AES algorithm with DRECON is called $n \times n$ DRECON-AES where possible values of n is 4 and 8. The DRECON-AES has the following properties. Each round of DRECON-AES has the same structure as that of $n \times n$ AES except the *AddRoundKeys* of the first and the last round are omitted. The *ShiftRow* operation of the last round is also omitted, and thus last round is left with only the *SubBytes* operation (Figure 2). Further, each $n \times n$ bit sbox is replaced by a $n \times n$ bit tweaked-sbox. Each tweaked-sbox is a set of 2^n ($r = 2^n, n = 4$ or 8) non-linear functions having the equal cryptographic strength, which put together satisfies Proposition 2. The criteria for selection of these sboxes is specified in Section 4.3.

4.1   Operation of DRECON-AES

Using DRECON-AES to secure communication between a sender and receiver has three phases as shown in Figure 3. The phases are explained below.

- **Bootstrapping:** To bootstrap, both parties need to agree on a secret key as well as a secret master tweak. Standard key exchange protocols can be used for the purpose.
- **Key Expansion:** The next step is to generate the round keys at both ends using a key scheduling algorithm. The sboxes in AES key scheduling algorithm are replaced by the tweaked-sbox. The tweak bits are generated by a

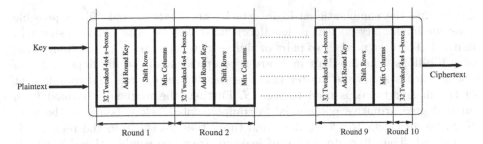

Fig. 2. 4 × 4 AES Adapted for DRECON

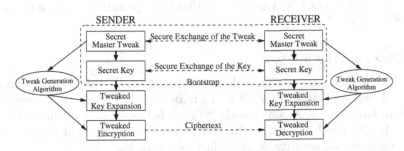

Fig. 3. Application of DRECON

tweak generation algorithm discussed in Section 4.2. The same tweak is used for all accesses during the key generation. The round keys are thus generated once and stored which are then used for every encryption until there is a change in the session key.

- **Encryption:** is then performed. Each encryption requires 128 bits of tweak to be generated, since each sbox takes a tweak input of equal size of its original input and all rounds use the same tweak.

The entire operation of the 4 × 4 DRECON-AES is summarized in Algorithm 1 of Appendix B. The operations of the 8 × 8 DRECON-AES are similar to those of 4 × 4 DRECON-AES except every pair of consecutive 4 × 4 tweaked-sboxes is replaced by a 8 × 8 tweaked-sbox.

4.2 Tweak Generation Algorithm

From the master tweak agreed upon by the sender and receiver, tweaks need to be generated for each encryption. The tweak generation needs to produce uniformly random tweaks in the range of 1 to r in order to select one of the r sboxes (for DRECON-AES $r = 16$ or 256). Further, the algorithm needs to be secure against power attacks as is discussed in detail in [19].

Any mask generation function (MGF) or stream cipher implemented in a secure manner can be used as a tweak generator. However, given the fact that the adversary has no control or knowledge of the input and output of the tweak generator, lightweight solutions can be developed by balancing registers and minimizing the combinational logic, which can otherwise leak [9]. A possible construction for a tweak generation algorithm makes use of an LFSR as shown in figure 4. The design uses two pairs of shift registers (S and $\overline{S}$), each comprising of n flip-flops. The flip-flops in $\overline{S}$ are a complement of the flip-flops in S. To obtain such a state, the master tweak is used to seed S and the complement of the master tweak is used to seed $\overline{S}$. Further, the feedback obtained from an n degree primitive polynomial is complemented before being fed back to $\overline{S}$. Since all clocks toggle at the same time, the leakage from the registers is minimised. The alternate source of leakage, from the combinational paths, is also kept minimum by choosing a primitive polynomial with small number of coefficients. For DRECON-AES, $n = 128$ and the primitive polynomial chosen was $\alpha^{128} \oplus \alpha^7 \oplus \alpha^2 \oplus \alpha \oplus 1$.

4.3 Choosing the S-boxes

Proposition 2 mandates that in order to make $\mathsf{I}[\boldsymbol{S}(T, X) \oplus K, X; K]$ minimum, $\boldsymbol{S}(T, x)$ should be balanced for all x. To make $\mathsf{I}[\boldsymbol{S}(T, X) \oplus K, X; K]$ zero, it is enough to have $h_t = n$. Again to make $\boldsymbol{S}(T, x)$ balanced, we can choose the size of $\mathcal{T}$ to be 2^n. Further, each of the sboxes needs to have good cryptographic properties to ensure security against black box attacks.

Exhaustive search can be used to find such sboxes. However, when the size of the sbox is large, it becomes infeasible. We choose the set of sboxes which are obtained using an affine transformations of a cryptographically strong sbox. That is, if $\mathsf{S}(\cdot)$ is a cryptographically strong sbox, we find a set of 2^n strong sboxes by setting $\mathsf{F}_i(x) = \alpha\mathsf{S}(x) \oplus i$ for all $i = 0, \cdots, 2^n - 1$ where α is an invertible matrix of dimensions $n \times n$. Since affine transformation does not changes the cryptographic properties of sboxes, all the sboxes of the set possess equal cryptographic straingth of the original sbox.

4.4 Software Implementation of DRECON-AES

DRECON-AES can be efficiently implemented on a micro-controller. We define a micro-controller model to compare the cost of DRECON-AES with the first-order masking of AES. We use an 8-bit micro-controller model [11] which takes:

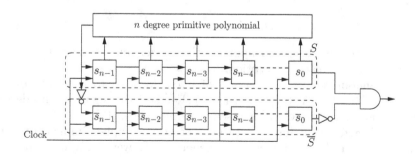

Fig. 4. Tweak Generation for DRECON

- load/store from/to RAM takes one clock cycle,
- load/store from/to ROM takes two clock cycles,
- XOR two registers takes one clock cycle,
- shift by one bit takes one clock cycle.
- swapping two nibbles of a byte takes one clock cycle.

We assume that the tweaked-sbox is stored in ROM. The masked implementations are assumed to use *GLUT* (global look-up table) with the same input and the output mask (as it is first order protection). The GLUT is also stored in ROM. Key Expansion is considered precomputed and thus omitted.

8 × 8 DRECON-AES: In the given scenario, the cost of each AES sub-operation for a standard 8 × 8 AES in terms of number of cycles are [11]:

- *SubBytes* (SB): 80
- *ShiftRows* (SR): 24
- *MixColumns* (MC): 256
- *AddRoundKey* (ARK): 64
- Memory Size: 256 Bytes

In DRECON-AES, the linear operations are exactly same as AES. The *SubBytes* is the only component which has changed. For 8 × 8 DRECON-AES, the number of clock cycles required for the *SubBytes* operation is $SB^* = 96$. It also needs one SR and two ARK operations less than the standard AES. Thus, the full 8 × 8 DRECON-AES (without Key Expansion) would need $Encryption = 10 \times SB^* + 9 \times SR + 9 \times MC + 9 \times ARK$ i.e. 4056 clock cycles and 256×256 bytes or 64Kbytes of ROM.

For masking using GLUT approach, the total number of 8 × 8 sboxes are 256 which is same as 8 × 8 DRECON-AES. The number of cycles to compute *SubBytes* is also same as 8 × 8 DRECON-AES i.e. $SB^* = 96$ clock cycles. Apart from *SubBytes*, there is one initial masking and a final demasking. This is called Extra Mask Addition (MA) and takes as many clock cycles as ARK. At each round, the *MixColumns* and *ShiftRows* operations need to be performed for the mask also. Thus a total masked AES (without Key Expansion), would need $Encryption = 10 \times SB^* + 20 \times SR + 18 \times MC + 11 \times ARK + 2 \times MA$ i.e. 6880 clock cycles and 256×256 bytes or 64Kbytes of ROM. Total cost estimation of 8 × 8 DRECON-AES and masked AES is shown in Table 1.

Table 1. Cost Comparison for 8-bit Micro-controller in terms of number of clock cycles taken for various operations

Architecture	SB	SR	MC	ARK	ROM (bytes)	MA	Encryption
AES	80	24	256	64	256	0	4048
8 × 8 DRECON-AES	96	24	256	64	64K	0	4056
Masked AES	96	24	256	64	64K	64	6880

Table 2. Cost Comparison for 8-bit Micro-controller in terms of number of clock cycles taken for various operations

Architecture	SB	SR	MC	ARK	ROM (bytes)	MA	Encryption
4 × 4 DRECON-AES	256	24	256	64	128	0	5656
Masked Alternative	224	24	256	64	128	64	8160

4 × 4 DRECON-AES: Implementation of 4 × 4 DRECON-AES in 8-bit micro-controller is a bit more tricky. The tweaked-sbox comprising of 16 non-linear functions can be stored in 128-bytes of ROM. The tweak determines the correct sbox for the current operation by additional XOR (Figure 5). Since the ROM in micro-controllers are often organised in bytes, a conditional swap operation followed by AND operation with 0x0F determines the correct output nibble.

With the defined model in the beginning of Section 4.4 into consideration, we have derived the number of clock cycles for each encryption and compare with its masking counterpart. Table 2 gives the number of cycles for *SubBytes* (SB), *ShiftRows* (SR), *MixColumns* (MC), *AddRoundKeys* (ARK). The table also lists the ROM required for the implementations.

The 8 × 8 DRECON-AES has almost null performance overhead, while its masking counterpart has a significant performance overhead, both having a 256x memory overhead. The memory overhead of the 8 × 8 DRECON-AES can be further reduced to as low as 16x for 4 × 4 DRECON-AES only at the cost of slightly higher performance overhead. For both the cases, DRECON-AES has an advantage over its masking counterpart in terms of the performance while having same memory overhead.

4.5   Hardware Implementation DRECON-AES

In hardware, two implementation styles are followed for DRECON-AES. The first one is *parallel implementation* which is preferred when sbox is small as in

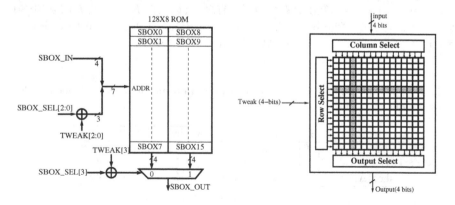

Fig. 5. 4 × 4 tweaked-sbox in Software   **Fig. 6.** 4 × 4 tweaked-sbox with in Hardware

Table 3. Comparing Resource Requirements for 8 × 8 DRECON-AES with Masking on an FPGA (XC5VLX50-2FF324)

Implementation	Slices	LUTs	Registers	Clock Cycles	Clock Period (ns)
4 × 4 AES[1]	1120	3472	1270	11	11.14
Masked 4 × 4 AES	3427	10589	1765	11	23.83
4 × 4 DRECON-AES	1379	3868	1583	11	10.3

[1] 4 × 4 AES is an implementation of the AES-128 algorithm with the 8 × 8 bit sbox replaced by a pair of 4 × 4 cryptographically strong sboxes.

the case of 4 × 4 DRECON-AES. The second is the *serialized implementation* used when sboxes are larger or when small area implementations are required. We adopted serialized implementation for 8 × 8 DRECON-AES.

4 × 4 DRECON-AES: Much like the RSM countermeasure proposed in [26], DRECON makes excessive use of tables. This especially suits FPGA platforms which possess large memory blocks (BRAM) to implement arrays of sboxes. The BRAM are used to store the pool of sboxes ($\mathcal{F}$) efficiently. BRAM based un protected implementation of ciphers have been shown to offer higher resistance against DPA as compared to other unprotected implementations [35]. The memory is addressed by a 4 bit tweak as shown in Figure 6. For DRECON-AES, we replicated this memory 32 times; once for each tweaked-sbox in the round. The value of the tweak is used to select a row while the input data selects a column in order to obtain the result. There are 32 such structures, one for each of the 32 substitution functions, present in the design. We use distributed RAM instead of BRAM to accelerate the attack. Resource requirements for the DRECON-AES implementation is compared with masked implementations of an equivalent AES with 4 × 4 sboxes in Table 3. The estimation for the masked implementation is computed from [32].

8 × 8 DRECON-AES: To implement 8 × 8 DRECON-AES, we followed the design of [23]. Serialised architecture with a rotating shift register and a single sbox is used. Each sbox access is split into two cycles. In the first cycle, two bytes - one from the state register and other from the tweak register - are applied to the input of the tweaked-sbox and its output is saved into the state byte. In the next cycle, both the state register and the tweak register are rotated by one byte. Thus, the *SubBytes* operation requires $2 \times 16 = 32$ clock cycles. The purpose of performing the state-byte update and rotation of state register in two different clock cycles is to be able to perform a sound security analysis of the implementation [23]. After the *SubBytes* operation, *ShiftRows*, *MixColumns* and *AddRoundKey* operations are performed in one clock cycle making a single round consisting of 33 clock cycles. The 8 × 8 tweaked-sbox consists of 2^8 8 × 8 sboxes. Each of the 2^8 sboxes was generated from AES sbox using the second strategy of Section 4.3. The resource requirement of 8 × 8 DRECON-AES implemented

Table 4. Comparing Resource Requirements for 8×8 DRECON-AES with Masking on an FPGA (XC5VLX50-2FF324)

Implementation	Slices	LUTs	Registers	Clock Period (ns)
Masked AES[1]	3948	13278	1592	14.955
8×8 DRECON-AES	1355	3716	1568	10.789

[1]The implementation exclude the PRNG used to generate mask.

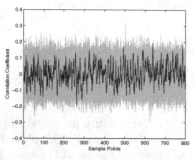
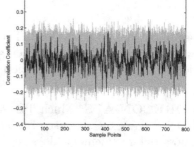

(a) 10 correlation-collision attack   (b) 20 correlation-collision attack

Fig. 7. Results of correlation-collision attack on 8×8 DRECON-AES using about $10,00,000$ traces. The correlation coefficients for wrong key differences are shown in grey and for the correct key difference (in this case $k_1 \oplus k_2 = 108$) in black.

in SASEBO-GII FPGA platform is shown in Table 4. The table also shows the resource requirement of the second scheme of [15] implemented in the same platform.

4.6   Attack on the Hardware Implementation

Recently correlation-collision attacks [24,22,23] have become a very effective tool to expose the vulnerabilities of 1O masking scheme in the presence of glitches. In [23], the authors have discussed the security of several ROM-based masking scheme of AES against correlation-collision attacks. In this section, we provide a similar security analysis of 8×8 DRECON-AES.

In *correlation-collision* attack [24], two sets of leakages of the same sbox instant during two different clock cycles with two different inputs, $x_i \oplus k_i$ and $x_j \oplus k_j$, are compared using a statistical test. If the two sets of leakages are similar in some statistical sense, a collision between the two sets are detected by the statistical test. The collision assures the relation $x_i \oplus k_i = x_j \oplus k_j$ or $k_i \oplus k_j = x_i \oplus x_j$. Correlation-collision attack was originally proposed to detect collision using the correlation between the mean values of the two sets of leakages. However in [22,23], it has been used to consider the higher order moments of the leakages also.

To verify the resistance of 8×8 DRECON-AES against DPA attack in the presence of glitches, we performed correlation-collision attack using both 1O and the 2O moments as described in [22]. For this evaluation also, SASEBO-GII board was used. The algorithm was implemented in Virtex 5 XC5VLX50 FPGA of SASEBO-GII which is driven by a clock frequency of 2 MHz. The power traces were acquired using Tektronix MSO 4034B Oscilloscope at the rate of 2.5 GS/s i.e. 1, 250 samples per clock period. Figure 7 shows the result of both 1O and the 2O correlation-collision attack on the leakages of the first and the second sbox access of the second round 8×8 DRECON-AES using about 10, 00, 000 traces. It may be noted that similar implementations of several masking schemes have been reported to be vulnerable to correlation-collision attack in [23].

5   Conclusion

DRECON provides a simple and efficient method of constructing block ciphers with inherent and provable security against DPA. The use of off-the-shelf sboxes ensures that the cipher is secure against classical cryptanalysis. In a glitch-free scenario, the security against DPA is proved to be equal to first-order Boolean masking from an information theoretic perspective. Its resistance against univariate DPA is also proved in the presence of glitches. Additionally, the first-order and second-order univariate DPA security is validated empirically with implementations on the SASEBO GII side-channel evaluation board. From the implementation perspective, DRECON has several advantages over standard countermeasures such as masking, hiding, and re-keying. In future, we hope to extend DRECON to provide security against higher-order DPA attacks as well.

References

1. Abdalla, M., Bellare, M.: Increasing the Lifetime of a Key: A Comparative Analysis of the Security of Re-keying Techniques. In: Okamoto, T. (ed.) ASIACRYPT 2000. LNCS, vol. 1976, pp. 546–559. Springer, Heidelberg (2000)
2. Chari, S., Jutla, C.S., Rao, J.R., Rohatgi, P.: Towards Sound Approaches to Counteract Power-Analysis Attacks. In: Wiener (ed.) [40], pp. 398–412
3. Clavier, C., Coron, J.S., Dabbous, N.: Differential Power Analysis in the Presence of Hardware Countermeasures. In: Koç, Ç.K., Paar (eds.) [12], pp. 252–263
4. Clavier, C., Feix, B., Gagnerot, G., Roussellet, M., Verneuil, V.: Improved Collision-Correlation Power Analysis on First Order Protected AES. In: Preneel, B., Takagi, T. (eds.) [30], pp. 49–62
5. Cover, T.M., Thomas, J.A.: Elements of Information Theory, 2nd edn. Series in Telecommunications and Signal Processing. Wiley-Interscience (July 2006)
6. Gérard, B., Grosso, V., Naya-Plasencia, M., Standaert, F.-X.: Block Ciphers That Are Easier to Mask: How Far Can We Go? In: Bertoni, G., Coron, J.-S. (eds.) CHES 2013. LNCS, vol. 8086, pp. 383–399. Springer, Heidelberg (2013)
7. Goubin, L., Patarin, J.: DES and Differential Power Analysis (The "Duplication" Method). In: Koç, Ç.K., Paar, C. (eds.) CHES 1999. LNCS, vol. 1717, pp. 158–172. Springer, Heidelberg (1999)

8. Guajardo, J., Mennink, B.: On side-channel resistant block cipher usage. In: Burmester, M., Tsudik, G., Magliveras, S., Ilić, I. (eds.) ISC 2010. LNCS, vol. 6531, pp. 254–268. Springer, Heidelberg (2011)

9. Guilley, S., Sauvage, L., Flament, F., Vong, V.N., Hoogvorst, P., Pacalet, R.: Evaluation of Power Constant Dual-Rail Logics Countermeasures against DPA with Design Time Security Metrics. IEEE Trans. Computers 59(9), 1250–1263 (2010)

10. Herbst, C., Oswald, E., Mangard, S.: An AES Smart Card Implementation Resistant to Power Analysis Attacks. In: Zhou, J., Yung, M., Bao, F. (eds.) ACNS 2006. LNCS, vol. 3989, pp. 239–252. Springer, Heidelberg (2006)

11. Hoheisel, A.: Side-Channel Analysis Resistant Implementation of AES on Automotive Processors. Master's thesis, Ruhr-University Bochum, Germany (June 2009)

12. Paar, C., Koç, Ç.K. (eds.): CHES 2000. LNCS, vol. 1965. Springer, Heidelberg (2000)

13. Kocher, P.C., Jaffe, J., Jun, B.: Differential Power Analysis. In: Wiener (ed.) [40], pp. 388–397

14. Liskov, M., Rivest, R.L., Wagner, D.: Tweakable block ciphers. In: Yung, M. (ed.) CRYPTO 2002. LNCS, vol. 2442, pp. 31–46. Springer, Heidelberg (2002)

15. Maghrebi, H., Prouff, E., Guilley, S., Danger, J.-L.: A first-order leak-free masking countermeasure. In: Dunkelman, O. (ed.) CT-RSA 2012. LNCS, vol. 7178, pp. 156–170. Springer, Heidelberg (2012)

16. Mangard, S., Popp, T., Gammel, B.M.: Side-Channel Leakage of Masked CMOS Gates. In: Menezes, A. (ed.) CT-RSA 2005. LNCS, vol. 3376, pp. 351–365. Springer, Heidelberg (2005)

17. Mangard, S., Pramstaller, N., Oswald, E.: Successfully Attacking Masked AES Hardware Implementations. In: Rao, J.R., Sunar, B. (eds.) CHES 2005. LNCS, vol. 3659, pp. 157–171. Springer, Heidelberg (2005)

18. McEvoy, R.P., Tunstall, M., Whelan, C., Murphy, C.C., Marnane, W.P.: All-or-Nothing Transforms as a Countermeasure to Differential Side-Channel Analysis. IACR Cryptology ePrint Archive 2009, 185 (2009)

19. Medwed, M., Standaert, F.X., Großschädl, J., Regazzoni, F.: Fresh Re-keying: Security against Side-Channel and Fault Attacks for Low-Cost Devices. In: Bernstein, D.J., Lange, T. (eds.) AFRICACRYPT 2010. LNCS, vol. 6055, pp. 279–296. Springer, Heidelberg (2010)

20. Messerges, T.S.: Using Second-Order Power Analysis to Attack DPA Resistant Software. In: Koç, Ç.K., Paar (eds.) [12], pp. 238–251

21. Micali, S., Reyzin, L.: Physically Observable Cryptography (Extended Abstract). In: Naor, M. (ed.) TCC 2004. LNCS, vol. 2951, pp. 278–296. Springer, Heidelberg (2004)

22. Moradi, A.: Statistical Tools Flavor Side-Channel Collision Attacks. In: Pointcheval, D., Johansson, T. (eds.) EUROCRYPT 2012. LNCS, vol. 7237, pp. 428–445. Springer, Heidelberg (2012)

23. Moradi, A., Mischke, O.: How Far Should Theory Be from Practice? - Evaluation of a Countermeasure. In: Prouff, E., Schaumont, P. (eds.) CHES 2012. LNCS, vol. 7428, pp. 92–106. Springer, Heidelberg (2012)

24. Moradi, A., Mischke, O., Eisenbarth, T.: Correlation-Enhanced Power Analysis Collision Attack. In: Mangard, S., Standaert, F.-X. (eds.) CHES 2010. LNCS, vol. 6225, pp. 125–139. Springer, Heidelberg (2010)

25. Moradi, A., Poschmann, A., Ling, S., Paar, C., Wang, H.: Pushing the Limits: A Very Compact and a Threshold Implementation of AES. In: Paterson, K.G. (ed.) EUROCRYPT 2011. LNCS, vol. 6632, pp. 69–88. Springer, Heidelberg (2011)

26. Nassar, M., Souissi, Y., Guilley, S., Danger, J.L.: RSM: A small and fast counter-measure for AES, secure against 1st and 2nd-order zero-offset SCAs. In: Rosenstiel, W., Thiele, L. (eds.) DATE, pp. 1173–1178. IEEE (2012)

27. Nikova, S., Rijmen, V., Schläffer, M.: Secure Hardware Implementation of Nonlinear Functions in the Presence of Glitches. J. Cryptology 24(2), 292–321 (2011)

28. Kocher, P.C.: Leak-Resistant Cryptograhic Indexed Key Update, US Patent 6539092 (2003)

29. Piret, G., Roche, T., Carlet, C.: PICARO – A Block Cipher Allowing Efficient Higher-Order Side-Channel Resistance. In: Bao, F., Samarati, P., Zhou, J. (eds.) ACNS 2012. LNCS, vol. 7341, pp. 311–328. Springer, Heidelberg (2012)

30. Preneel, B., Takagi, T. (eds.): CHES 2011. LNCS, vol. 6917, pp. 2011–2013. Springer, Heidelberg (2011)

31. Prouff, E., Roche, T.: Higher-Order Glitches Free Implementation of the AES Using Secure Multi-party Computation Protocols. In: Preneel, B., Takagi, T. (eds.) [30], pp. 63–78

32. Regazzoni, F., Yi, W., Standaert, F.X.: FPGA Implementations of the AES Masked Against Power Analysis Attacks. In: Proceedings of 2nd International Workshop on Constructive Side-Channel Analysis and Secure Design (COSADE) (February 2011)

33. Research Center for Information Security National Institute of Advanced Industrial Science and Technology: Side-channel Attack Standard Evaluation Board SASEBO-GII Specification, Version 1.01 (2009)

34. Rivain, M., Prouff, E., Doget, J.: Higher-Order Masking and Shuffling for Software Implementations of Block Ciphers. In: Clavier, C., Gaj, K. (eds.) CHES 2009. LNCS, vol. 5747, pp. 171–188. Springer, Heidelberg (2009)

35. Shah, S., Velegalati, R., Kaps, J.P., Hwang, D.: Investigation of DPA Resistance of Block RAMs in Cryptographic Implementations on FPGAs. In: Prasanna, V.K., Becker, J., Cumplido, R. (eds.) ReConFig, pp. 274–279. IEEE Computer Society (2010)

36. Standaert, F.X., Pereira, O., Yu, Y., Quisquater, J.J., Yung, M., Oswald, E.: Leakage Resilient Cryptography in Practice. Cryptology ePrint Archive, Report 2009/341 (2009), http://eprint.iacr.org/

37. Tiri, K., Akmal, M., Verbauwhede, I.: A Dynamic and Differential CMOS Logic with Signal Independent Power Consumption to Withstand Differential Power Analysis on Smart Cards. In: ESSCIRC 2002, pp. 403–406 (2002)

38. Tiri, K., Verbauwhede, I.: A Logic Level Design Methodology for a Secure DPA Resistant ASIC or FPGA Implementation. In: DATE, pp. 246–251. IEEE Computer Society (2004)

39. Waddle, J., Wagner, D.: Towards Efficient Second-Order Power Analysis. In: Joye, M., Quisquater, J.-J. (eds.) CHES 2004. LNCS, vol. 3156, pp. 1–15. Springer, Heidelberg (2004)

40. Wiener, M. (ed.): CRYPTO 1999. LNCS, vol. 1666. Springer, Heidelberg (1999)

Appendix A: Proof of Lemma 1

Without loss of generality, we assume that the sbox in the statement of Lemma 1 belongs to the second round. Thus, the random variable Z is the output of a MDS mapping M of the diffusion layer in the first round (see Figure 1) which takes the outputs of m sboxes of the first round as inputs. Let $S(X_{i_1}, T_{i_1})$, $S(X_{i_2}, T_{i_2})$, ..., $S(X_{i_m}, T_{i_m})$ be the inputs to the the MDS mapping. Since the operation is a MDS mapping, it can be realised by a $q \times q$ matrix whose elements are essentially non-zero. Thus Z can be represented as $a_1 \cdot S(X_{i_1}, T_{i_1}) \oplus \cdots \oplus a_m \cdot S(X_{i_m}, T_{i_m})$ where $a_j \in \{0, 1\}^n \setminus \{0\}^n$. We can now compute the posterior probability of $Z = z$ given $X_1, \ldots, X_q, T$ as

$$
\begin{aligned}
P[Z = z | X_1 = x_1, \ldots, X_q = x_q, T = t] &= P[Z = z | X_{i_1} = x_{i_1}, \ldots, X_{i_m} = x_{i_m}, T = t] \\
&= P[a_1 \cdot S(X_{i_1}, T_{i_1}) \oplus \cdots \oplus a_m \cdot S(X_{i_m}, T_{i_m}) = z \\
&\qquad | X_{i_1} = x_{i_1}, \ldots, X_{i_m} = x_{i_m}, T = t] \\
&= P[a_1 \cdot S(x_{i_1}, T_{i_1}) \oplus \cdots \oplus a_m \cdot S(x_{i_m}, T_{i_m}) = z \\
&\qquad | T = t]
\end{aligned}
$$

The variable T may or may not belong to $\{T_{i_1}, \cdots, T_{i_m}\}$. Let us first assume that T does not belong to $\{T_{i_1}, \cdots, T_{i_m}\}$. In that case, the above probability can be given by

$$
P[Z = z | X_1 = x_1, \ldots, X_q = x_q, T = t] = P[a_1 \cdot S(x_{i_1}, T_{i_1}) \oplus \cdots \oplus a_m \cdot S(x_{i_m}, T_{i_m}) = z]
$$

Since, the sbox $S(\cdot, \cdot)$ satisfies Proposition 2 and T_is are uniformly random, the variable $a_1 \cdot S(x_{i_1}, T_{i_1}) \oplus \cdots \oplus a_m \cdot S(x_{i_m}, T_{i_m})$ is also uniformly random and consequently $P[Z = z | X_1 = x_1, \ldots, X_q = x_q, T = t] = 1/2^n$.

On the other hand, if T belongs to $\{T_{i_1}, \cdots, T_{i_q}\}$, let say $T = T_{i_1}$, the posterior probability of $Z = z$ can be given by

$$
\begin{aligned}
P[Z = z | X_1 = x_1, \ldots, X_q = x_q, T = t] &= P[a_1 \cdot S(x_{i_1}, T_{i_1}) \oplus \cdots \oplus a_m \cdot S(x_{i_m}, T_{i_m}) = z \\
&\qquad | T_{i_1} = t] \\
&= P[a_1 \cdot S(x_{i_1}, t) \oplus \cdots \oplus a_m \cdot S(x_{i_m}, T_{i_m}) = z] \\
&= P[a_2 \cdot S(x_{i_2}, T_{i_2}) \oplus \cdots \oplus a_m \cdot S(x_{i_m}, T_{i_m}) = \\
&\qquad z \oplus a_1 \cdot S(x_{i_1}, t)]
\end{aligned}
$$

Since $m \geq 2$, this probability is also equates to $1/2^n$. Thus, in both the cases the posterior probabilities of $Z = z$ is $1/2^n$ which is equals to its a priori probability $P[Z = z]$. Thus we conclude that Z is independent of the joint distribution of the variables $X_1, \ldots, X_q$ and T.

Appendix B: Algorithm of 4×4 DRECON-AES

Algorithm 1: Compute 4×4 DRECON-AES

Input: T: Master Tweak, ntr: number of encryption, P: Array of ntr plaintext
Output: C: Array of ntr ciphertext

```
 1 begin
 2 │    Generate Key Expansion Tweak T_K ∈ [0, 63] from Master Tweak T
 3 │    Generate 16 Tweaked Sboxes
 4 │    KeyExpansion k[1] ... k[9]
 5 │
 6 │    for i = 1 to ntr do
 7 │    │    Generate: 128 − bit Session Tweak T_S
 8 │    │    state ← P[i]
 9 │    │    for round r = 1 to 9 do
10 │    │    │    for nibble n = 1 to 32 do
11 │    │    │    │    Sbox ← Sboxes[T_S[(4 ∗ n : 4 ∗ n + 3)mod16]]
12 │    │    │    │    state[n] ← Sbox(state[n])          (SubBytes)
13 │    │    │    end
14 │    │    │    state ← ShiftRows(state)
15 │    │    │    state ← MixColumns(state)
16 │    │    │    state ← AddRoundKey(state, k[r])
17 │    │    end
18 │    │    for nibble n = 1 to 32 do
19 │    │    │    Sbox ← Sboxes[T_S[(4 ∗ n : 4 ∗ n + 3)mod16]]
20 │    │    │    state[n] ← Sbox(state[n])          (Final SubBytes)
21 │    │    end
22 │    │    C[i] ← state
23 │    end
24 │    return C
25 end
```

Counter-bDM: A Provably Secure Family of Multi-Block-Length Compression Functions

Farzaneh Abed, Christian Forler, Eik List, Stefan Lucks, and Jakob Wenzel

Bauhaus-Universität Weimar, Germany
{farzaneh.abed,christian.forler,eik.list,stefan.lucks,
jakob.wenzel}@uni-weimar.de

Abstract. Block-cipher-based compression functions serve an important purpose in cryptography since they allow to turn a given block cipher into a one-way hash function. While there are a number of secure double-block-length compression functions, there is little research on generalized constructions. This paper introduces the COUNTER-bDM family of multi-block-length compression functions, which, to the best of our knowledge, is the first provably secure block-cipher-based compression function with freely scalable output size. We present generic collision- and preimage-security proofs for it, and compare our results with those of existing double-block-length constructions. Our security bounds show that our construction is competitive with the best collision- and equal to the best preimage-security bound of existing double-block-length constructions.

Keywords: block cipher, compression function, hash function, provable security.

1 Introduction

While the SHA-3 competition has encouraged many new interesting ideas for designing hash and compression functions (e.g., the sponge framework [3]), one of the most popular approaches is to use a given block cipher and turn it into a one-way function. While the roots to this simple principle can be tracked back to Rabin [33] at the end of the 70s, the knowledge about it is still highly relevant today. For instance, the standardized SHA-1 and SHA-2 hash function families base on the SHACAL-1/2 ciphers. But also many submissions for the SHA-3 contest, such as – Blake [2], Skein [37], or SHAvite-3 [4] – are built on block ciphers. The advantages are obvious: not only can compression-function designers profit from the pseudo-randomness of an IND-CCA-secure cipher, but also do they require only a single primitive to obtain both encryption and hashing – an important matter when designing hardware for resource-constrained devices.

The best understood principle for block-cipher-based compression functions are so-called *single-block-length* constructions, which compress a $2n$-bit input to an n-bit output, where n is the state size of the cipher. However, the state size of the AES is 128 bits, which yields a 64-bit collision security, which is

D. Pointcheval and D. Vergnaud (Eds.): AFRICACRYPT 2014, LNCS 8469, pp. 440–458, 2014.

insufficient for many applications. As a consequence, one is usually interested in double-block-length or, more generally, multi-block-length block-cipher-based hash functions, which take an (an)-bit input and produce a (bn)-bit output, for $a > b \geq 2$.

Related Work. The idea of double-block-length hashing can be attributed to Meyer and Schilling and their proposal of the rate-1/2 and rate-1/4 hash functions MDC-2 and MDC-4 [6] in 1988. Together with the Davies-Meyer-like schemes ABREAST-DM and TANDEM-DM from Lai and Massey [24], these four are commonly known as *classical* constructions. A number of further double-block-length functions have been proposed recently. According to Mennink [34], these can be ordered into the classes DBL^{2n} – which employ a cipher with a $2n$-bit key – and DBL^n – which use a cipher with an n-bit key (see [41] for example). The former class contains ABREAST-DM, the variants by Lee and Kwon [27], TANDEM-DM, HIROSE-DM [17], Stam's supercharged Type-I compression function [30,43,44], as well as the generalizations by Özen and Stam [38] and by Hirose [16].

Moreover, Fleischmann et al. generalized several classes of Davies-Meyer designs and proposed a class of cyclic constructions that contains the compression functions WEIMAR-DM, ADD-k-DM, and CUBE-DM [12,14]. A more detailed review of related work is provided in Appendix A. All of the mentioned provide a birthday- type collision security; in addition, there are security proofs for WEIMAR-DM, HIROSE-DM, TANDEM-DM, and ABREAST-DM are given in [12,17,26,27,29].

While double-block-length hashing can offer an acceptable collision security, a variety of applications demand secure multi-block-length functions with a freely scalable output of the compression function. For instance, public-key signature schemes expect inputs of the exact length of the signing key. Moreover, in the era of SHA-3, hash values with a length of ≥ 256 bits are standard. But it is still an open research question how to create provably secure b-block-length compression functions for $b > 2$.

Contribution. First, we define the class MBL^{bn} for multi-block-length compression functions that employ a (bn, n)-bit keyed block cipher $E : \{0,1\}^{bn} \times \{0,1\}^n \to \{0,1\}^n$, and produce a bn-bit chaining value. Then, we present a freely scalable multi-block-length compression function, called COUNTER-bDM, which, to the best of our knowledge, is the first provably secure multi-block-length compression function for $b > 2$. It is a generalization of the double-block-length compression function HIROSE-DM [18]. For the generic COUNTER-bDM, we present a detailed security analysis for proofs of collision and preimage security, which employs the idea of super queries by Armknecht et al. [1]. Similar approaches were presented by Mennink [34] and Lee [25].

For $b = 2$ our resulting collision-security bound shows that every adversary that wants to find a collision with advantage 1/2 requires $2^{125.18}$ queries, which is comparable to the currently best collision- security bound of WEIMAR-DM [12]. Concerning preimage security, we obtain a near-optimal bound of 2^{251} queries,

Table 1. Comparison of security results on double-block-length compression functions, evaluated for $n = 128$ bits and a success probability of $1/2$. For CYCLIC-DM, $k > 1$; for ADD-K-DM $k' \geq 2$.

Compression function	Collision bound		Preimage bound	
ABREAST-DM [24]	$2^{124.42}$	[14,26]	2^{246}	[1]
ADD-K-DM [14]	$2^{127-k'}$	[14]	$\approx 2^{128}$	[14,26]
Counter-2DM [Sec. 3]	$2^{125.18}$	[Sec. 5]	2^{251}	[Sec. 6]
CUBE-DM [14]	$2^{125.41}$	[14]	$\approx 2^{128}$	[14,26]
CYCLIC-DM (cycle length > 2) [14]	2^{127-k}	[14]	$\approx 2^{128}$	[14,26]
CYCLIC-DM (cycle length 2) [14]	$2^{124.55}$	[14]	$\approx 2^{128}$	[14,26]
HIROSE-DM [17]	$2^{125.23}$	[13]	2^{251}	[1]
LEE/KWON [27]	$2^{125.0}$	[26]	$\approx 2^{128}$	[14,26]
TANDEM-DM [24]	$2^{120.87}$	[29]	2^{246}	[1]
WEIMAR-DM [12]	$2^{126.73}$	[9]	2^{251}	[12]

which is equivalent to the currently best bound of WEIMAR-DM. Table 1 compares our bounds with that of previously published double-block-length compression functions.

Outline. In what remains, Section 2 revisits the basic notions concerning block-cipher-based compression functions. Section 3 introduces COUNTER-bDM. Section 4 summarizes the formal security definitions that are essential for our analysis. In Section 5 we present the proof for the collision security of COUNTER-bDM. Section 6 then derives the preimage-security bound. Finally, Section 7 concludes the paper.

2   Basic Notions

This section recaps the relevant basic notions. We borrow the description of block-cipher-based compression functions from [12]:

Definition 1 (Block Cipher). *Let $k, n \geq 1$ be integers. We define a (k, n)-bit block cipher as a keyed family of permutations, which consists of an encryption function $E : \{0,1\}^k \times \{0,1\}^n \rightarrow \{0,1\}^n$, and its inverse (decryption) function $D = E^{-1} : \{0,1\}^k \times \{0,1\}^n \rightarrow \{0,1\}^n$. Both take a k-bit key K and an n-bit input block X, and produce an n-bit output Y, where $D_K(E_K(X)) = X$, for all $X \in \{0,1\}^n, K \in \{0,1\}^k$. We denote by $\texttt{Block}(k, n)$ the set of all (k, n)-bit block ciphers.*

Definition 2 (Single-Block-Length Compression Function). *Let $n \geq 1$ be an integer. A single-block-length (SBL) block-cipher-based compression function is a function $H^{SBL} : \{0,1\}^n \times \{0,1\}^n \rightarrow \{0,1\}^n$ which uses a block cipher from $\texttt{Block}(n, n)$.*

The idea was discussed in the literature first by Rabin [33]. Most SBL functions use a block cipher from $\texttt{Block}(n, n)$ and compress a $2n$-bit string to an n-bit string. A popular example is the Davies-Meyer (DM) [46] mode:

$$H^{DM}(M, U) = E_M(U) \oplus U,$$

which is essentially used twice inside HIROSE-DM and b times, in slightly modified fashion, inside COUNTER-bDM.

Definition 3 (Multi-Block-Length Compression Function). *Let $b, n \geq 1$ be integers. A multi-block-length (MBL) block-cipher-based compression function is a function $H^{MBL} : \{0,1\}^{bn} \times \{0,1\}^n \to \{0,1\}^{bn}$, which takes an n-bit message and a bn-bit chaining value, and outputs a new bn-bit chaining value.*

Independent Ciphers. The sophisticated task of proving the security for a multi-block-length compression function simplifies greatly if one can ensure that the b outputs of the individual block-cipher calls in one invocation of the compression function are independent and distinct from each other. Previous double-block-length constructions achieve this requirement by either...

Distinct Permutations: ... using b independent permutations in the compression function. This approach is used, e.g., by the early construction of Hirose [16] or those by Rogaway and Steinberger [41].

Distinct Keys: ... guaranteeing that all key inputs K_i used for the block-cipher calls inside one compression-function call are different: $K_i \neq K_j$, $1 \leq i < j \leq b$, which results in having de facto different permutations. This approach is used, e.g., by WEIMAR-DM [12].

Distinct Plaintexts: ... guaranteeing that all b plaintext inputs X_i used as inputs to the block cipher in one compression-function call are different: $X_i \neq X_j$, $1 \leq i < j \leq b$. This approach is used, e.g., by CUBE-DM [14] or HIROSE-DM [18].

The first approach renders unpractical in practice since it requires multiple permutation implementations of the class MBL^{bn}. The further two approaches are similar. However, using a different key in every block-cipher call implies the potential need of running the key schedule of the underlying block cipher multiple times. Therefore, we employ the latter strategy function for COUNTER-bDM, i.e., we ensure that all plaintext inputs to the block-cipher calls are different.

3 Counter-bDM

This section defines the COUNTER-bDM family of multi-block-length compression functions. Note that we use H^{CbDM} as short notion of COUNTER-bDM.

Definition 4 (Counter-bDM). *Let E be a block cipher from $\texttt{Block}(bn, n)$. The compression function $H^{CbDM} : \{0,1\}^{bn} \times \{0,1\}^n \to \{0,1\}^{bn}$ is defined by*

$$H^{CbDM}(M, U_1, \ldots, U_b) = (V_1, \ldots, V_b),$$

where the outputs V_i are given by $V_i = E_K(U_1 \oplus (i-1)) \oplus U_1$, with $K = U_2 \mid\mid \ldots \mid\mid U_b \mid\mid M$.

Two concrete examples of our multi-block-length compression-function family, COUNTER-3DM (left) and COUNTER-4DM (right), are illustrated in Figure 1. However, in our security analysis in Sections 5 and 6 we consider the generic version COUNTER-bDM.

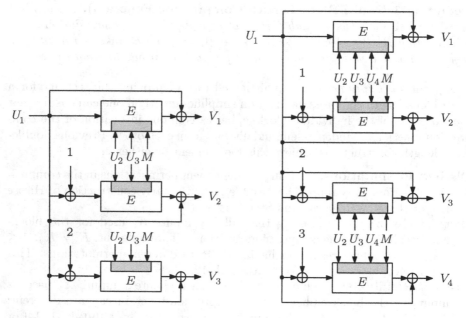

Fig. 1. Two examplary compression functions H^{C3DM} (left) and H^{C4DM} (right) from the family of compression functions H^{CbDM}

It is easy to see that, due to the XOR with the counter $i-1$, all plaintext inputs X_i to the block-cipher calls are pair-wise distinct. Additionally, since all values $i-1$ are in the range of $[0, \ldots, b-1]$, the counter values affect only the least significant $\lceil log_2(b) \rceil$ bits of the plaintexts. We call the most significant $n - \lceil log_2(b) \rceil$ bits of the plaintexts a *common prefix*.

Definition 5 (Common-Prefix Property). *Let $X = X_{pre} \mid\mid X_{post}$, $X \in \{0,1\}^n$ be an n-bit integer, where X_{pre} denotes the $n - \lceil log_2(b) \rceil$ most significant bits, and X_{post} the $\lceil log_2(b) \rceil$ least significant bits of X. Further, let $X_i = X \oplus (i-1)$ (with $1 \le i \le b$) denote the values which are used as plaintext inputs to the block-cipher calls in one invocation of H^{CbDM}. Then, all values X_i share the same common prefix $X_{pre} \in \{0,1\}^{n-\lceil log_2(b) \rceil}$.*

Remark 1. For the remainder of this paper, we denote by $c = 2^{\lceil log_2(b) \rceil} \ge b$ the maximal number of plaintexts $X = X_{pre} \mid\mid X_{post}$ which can share the same prefix X_{pre}.

We will see later that both the pair-wise distinct plaintexts and the common-prefix property will be beneficial for an easy-to-grasp security analysis of COUNTER-bDM.

4   Proof Preliminaries

This section formally describes the notions and properties that are relevant for our security analysis of COUNTER-bDM.

4.1   Proof Model

The security of a block-cipher-based compression function should depend only on the security of the construction, and not on that of the (potentially insecure) chosen block cipher inside. Thus, one usually considers the *ideal-cipher model*, wherein a block cipher is modeled as a family of random n-bit random permutations $\{E_K\}$. The permutation E that is used in the compression function is chosen at random from $\texttt{Block}(k, n)$: $E \xleftarrow{\$} \texttt{Block}(k, n)$. Thus, we follow the notions by Black et al. [5].

An adversary $\mathcal{A}$ is defined as a probabilistic, computationally unbounded algorithm that is limited only by a number of q queries it can ask to an oracle E. For any of its queries, the adversary is allowed to ask either a forward (encryption) query $E_K(X) = Y$, or a backward (decryption) query $X = D_K(Y)$, where $X, Y \in \{0,1\}^n$ and $\forall X : D_K(E_K(X)) = X$. Each query Q^i is stored as a 3-tuple (X_i, Y_i, K_i) in a query history $\mathcal{Q}$, where we denote by $\mathcal{Q}_i$ the state of the query history after i queries have been asked by the adversary, for $1 \leq i \leq q$. We further borrow two usual assumptions about $\mathcal{A}$ from [12]:

1. If $\mathcal{A}$ has successfully found a collision or a preimage for H^{CbDM}, it has obtained the necessary encryption or decryption results only by making queries to the oracle E.
2. $\mathcal{A}$ does not ask queries to which it already knows the answer, e.g., if $\mathcal{A}$ already knows the answer to a forward query $Y = E_K(X)$, it will not ask $D_K(Y)$ – which must return X – and vice versa.

4.2   Collision-Security

We define the collision security of our compression function H^{CbDM} by the advantage of an adversary $\mathcal{A}$ to win Experiment 1.

Experiment 1 (Collision-Finding Experiment $\texttt{Exp-Coll}_{\mathcal{A}, H^{CbDM}}(bn)$)

1. *An adversary $\mathcal{A}$ is given oracle access to a block cipher $E \in \texttt{Block}(bn, n)$.*
2. *After asking at most q queries (X_i, Y_i, K_i) for $1 \leq i \leq q$, it outputs a pair $(M, U_1, \ldots, U_b), (M', U'_1, \ldots, U'_b) \in \{0,1\}^{(b+1)n} \times \{0,1\}^{(b+1)n}$.*

3. *The adversary wins the experiment iff its output is a valid collision for* H^{CbDM}, *i.e.,*

$$H^{CbDM}(M, U_1, \ldots, U_b) = H^{CbDM}(M', U'_1, \ldots, U'_b) \text{ and}$$
$$(M, U_1, \ldots, U_b) \neq (M', U'_1, \ldots, U'_b).$$

Otherwise, $\mathcal{A}$ *loses the experiment.*

The advantage of an adversary $\mathcal{A}$ to find such a collision for H^{CbDM} is given by the probability that $\mathcal{A}$ can win Experiment 1, or formally written, by

$$\mathbf{Adv}^{COLL}_{H^{CbDM}}(\mathcal{A}) = \Pr\left[\texttt{Exp-Coll}_{\mathcal{A}, H^{CbDM}}(bn) = 1\right]$$

Since we only limit the adversary by the number of queries, it is allows to ask to E, we write

$$\mathbf{Adv}^{COLL}_{H^{CbDM}}(q) := \max_{\mathcal{A}}\left\{\mathbf{Adv}^{COLL}_{H^{CbDM}}(\mathcal{A})\right\},$$

where the maximum is taken over all adversaries that ask at most q oracle queries in total.

4.3   Preimage Security

There are various notions considering preimage security (see [40] for example). We adapt that of *everywhere preimage security* (EPRE), which was introduced by Rogaway and Shrimpton in [40]. There, the adversary commits to a hash value before it makes any queries to the oracle. The preimage security of our compression function H^{CbDM} is therefore defined by the advantage that an adversary $\mathcal{A}$ wins Experiment 2.

Experiment 2 (Preimage-Finding Experiment $\texttt{Exp-ePre}_{\mathcal{A}, H^{CbDM}}(bn)$**)**

1. *An adversary* $\mathcal{A}$ *is given oracle access to a block cipher* $E \in \texttt{Block}(bn, n)$. *Before it makes any queries, it announces a hash value* $(V_1, \ldots, V_b) \in \{0, 1\}^{bn}$.
2. *After asking at most* q *queries* (X_i, Y_i, K_i) *for* $1 \leq i \leq q$, *it outputs a* $(b+1)$-*tuple* $(M, U_1, \ldots, U_b) \in \{0, 1\}^{(b+1)n}$.
3. *The adversary wins the experiment iff its output is a valid preimage for* $(V_1, \ldots, V_b)$ *and* H^{CbDM}, *i.e.,*

$$H^{CbDM}(M, U_1, \ldots, U_b) = (V_1, \ldots, V_b).$$

Otherwise, $\mathcal{A}$ *loses the experiment.*

We let $\mathbf{Adv}^{EPRE}_{H^{CbDM}}(\mathcal{A})$ be true iff $\texttt{Exp-ePre}_{\mathcal{A}, H^{CbDM}}(bn)$ returns 1. The pre-committed hash value $(V_1, \ldots, V_b)$ is an omitted parameter of $\mathbf{Adv}^{EPRE}_{H^{CbDM}}(\mathcal{A})$. We define

$$\mathbf{Adv}^{EPRE}_{H^{CbDM}}(q) := \max_{\mathcal{A}}\left\{\mathbf{Adv}^{EPRE}_{H^{CbDM}}(\mathcal{A})\right\},$$

where the maximum is taken over all adversaries that ask at most q oracle queries in total.

5   Collision-Security Analysis of Counter-bDM

Let $\mathcal{A}$ be a collision-finding adversary for H^{CbDM} that can ask queries to an oracle E. In between $\mathcal{A}$ and E, we construct another adversary $\mathcal{A}'$ which simulates $\mathcal{A}$, but sometimes is allowed to make additional queries to E that are not taken into account. Since $\mathcal{A}'$ is more powerful than $\mathcal{A}$, it is easy to see that it suffices for us to upper bound the success probability of $\mathcal{A}'$. Thereby, we say that an adversary $\mathcal{A}$ (or $\mathcal{A}'$, respectively) is *successful* if its query history contains the means of computing a collision for H^{CbDM}.

Attack Setting. During the attack, $\mathcal{A}$ maintains a query history $\mathcal{Q}$ wherein it stores all queries it poses to E. An entry in the query history of $\mathcal{A}$ is a tuple (K, X, Y), where $Y = E_K(X)$. Simultaneously, $\mathcal{A}'$ maintains a query list $\mathcal{L}$ which contains all input/output pairs to the compression function H^{CbDM} that can be computed by $\mathcal{A}$. An entry $L \in \mathcal{L}$ is a tuple $(K, X, Y_1, \ldots, Y_c) \in \{0,1\}^{(b+1+c)n}$, where $K \in \{0,1\}^{bn}$, $X \in \{0,1\}^n$ is the input to the compression function H^{CbDM}, and $c = 2^{\lceil log_2(b) \rceil}$ (see Remark 1). The values $Y_i \in \{0,1\}^n$ are given as the results of the forward queries $Y_i = E_K(X \oplus (i-1))$, for $1 \leq i \leq c$. Moreover, we define $\mathcal{L}_j$ to denote the state of $\mathcal{L}$, which contains the first j queries of $\mathcal{A}'$, with $j \geq 1$.

Collision Events. When E is modeled as an ideal cipher, we run into problems when $\mathcal{A}$ asks close to or even more than $q = 2^n$ queries. In the case when $\mathcal{A}$ asks q queries under the same key to E and q reaches $2^n - 1$, E loses its randomness. As a remedy to this problem, Armknecht et al. proposed the idea of *super queries* [1]; given some key K, $\mathcal{A}'$ can pose regular queries to E or D until $N/2$ queries with the same key K have been added to its query list $\mathcal{L}$, where $N = 2^n$.

   If $\mathcal{L}$ contains $N/2$ queries for a key K and $\mathcal{A}$ requests another query for the key K from $\mathcal{A}'$, then, $\mathcal{A}'$ poses all remaining queries $(K, *, *)$ under this key to E at once. In this case, we say that a *super query* occurred. All queries that are part of a super query *are not taken into account*, i.e., they do not add to q, the number of queries $\mathcal{A}$ is allowed to ask. Since these free queries are asked at once, one no longer has to consider the success probability of a single query; instead, one can consider the event that $\mathcal{A}'$ is successful with any of the contained queries. Thus, E does not lose its randomness. In the following, we define three mutually exclusive events which cover all case when $\mathcal{A}'$ can be successful.

NormalQueryWin($\mathcal{L}$). This describes the case when $\mathcal{A}'$ finds a collision with its current query L^j and a query $L^r \in \mathcal{L}_{j-1}$, where L^j was a normal query.
SuperQueryWin($\mathcal{L}$). This describes the case when $\mathcal{A}'$ finds a collision with its current query L^j and a query $L^r \in \mathcal{L}_{j-1}$, where L^j was part of a super query.
SameQueryWin($\mathcal{L}$). This describes the case when $\mathcal{A}'$ finds a collision within the same entry $L^j \in \mathcal{L}$.

   Since the adversary can only win if it finds a collision using either one of the mentioned events, it is sufficient for us to upper bound the sum of the probabilities. Thus, it holds that

$$\mathbf{Adv}_{H^{CbDM}}^{COLL}(q) \leq \Pr[\text{NormalQueryWin}(\mathcal{L})] + \Pr[\text{SuperQueryWin}(\mathcal{L})] \quad (1)$$
$$+ \Pr[\text{SameQueryWin}(\mathcal{L})].$$

Remark 2. Note that a tuple $L \in \mathcal{L}$ consists of $c = 2^{\lceil log_2(b) \rceil}$ query results. Since c always divides $N/2$, i.e., $c \mid N/2$, each tuple L is either part of a normal query or a super query, but never both.

Before we present our bound, we describe more precisely what we mean by $\mathcal{A}'$ has found a collision for H^{CbDM}. Let $L^r = (K^r, X^r, Y_1^r, \ldots, Y_c^r)$ represent the r-th entry in $\mathcal{L}$, and $L^j = (K^j, X^j, Y_1^j, \ldots, Y_c^j)$ the j-th entry in $\mathcal{L}$, where $1 \leq r < j \leq q$. We say that L^r and L^j provide the means for computing a collision if $\exists \ell, m \in \{0, \ldots, c-1\}$ so that b equations of the following form hold:

$$E_{K^r}(X^r \oplus \ell \oplus 0) \oplus X^r = E_{K^j}(X^j \oplus m \oplus 0) \oplus X^j,$$
$$E_{K^r}(X^r \oplus \ell \oplus 1) \oplus X^r = E_{K^j}(X^j \oplus m \oplus 1) \oplus X^j,$$
$$\vdots$$
$$E_{K^r}(X^r \oplus \ell \oplus (b-1)) \oplus X^r = E_{K^j}(X^j \oplus m \oplus (b-1)) \oplus X^j.$$

Theorem 3. *Let $N = 2^n$. Then, it applies that*

$$\mathbf{Adv}_{H^{CbDM}}^{COLL}(q) \leq \frac{c^2 \cdot 2^b \cdot q^2}{N^b} + \frac{c^3 \cdot 2^{b+2} \cdot q^2}{N^{b+1}}.$$

Proof. After $\mathcal{A}$ has asked a (normal) forward query $Y^j = E_{K^j}(X^j)$ or a (normal) backward query $X^j = D_{K^j}(Y^j)$, $\mathcal{A}'$ checks if $\mathcal{L}_{j-1}$ already contains an entry $L^r = (K^j, X_{pre}^j \| *, *, \ldots, *)$, where X_{pre}^j denotes the prefix of X^j (see Definition 5) and $*$ denotes arbitrary values. In the following, we analyze two possible cases.

Case 1: L^r is not in $\mathcal{L}_{j-1}$. In this case, $\mathcal{A}'$ labels Y^j as Y_1^j and asks $(c-1)$ further queries to E that are not taken into account:

$$\forall i \in \{2, \ldots, c\}: \quad Y_i^j = E_{K^j}(X^j \oplus (i-1)).$$

$\mathcal{A}'$ creates the tuple $L^j = (K^j, X^j, Y^j, \ldots, Y_c^j)$ and appends it to its query list, i.e., $\mathcal{L}_j = \mathcal{L}_{j-1} \cup \{L^j\}$. Now, we have to upper bound the success probability of $\mathcal{A}'$ to find a collision for H^{CbDM}, i.e., the success probabilities for the events mentioned above.

Subcase 1.1: **NormalQueryWin($\mathcal{L}$).** In this case, the adversary finds a collision using a normal query L^j and a query L^r that was already contained in $\mathcal{L}$. While super queries may have occurred for different keys before, the query history of $\mathcal{A}'$ may contain at most $N/2 - c$ plaintext-ciphertext pairs for the current key K^j. So, our random permutation E samples the query responses $Y_1^j, \ldots, Y_c^j$ for the current query at random from a set of size of at least $N/2 + c \geq N/2$ elements.

Hence, the probability that one equation from above holds for some fixed ℓ and m can be upper bounded by $1/(N/2)$; and the probability for b equations to hold is then given by

$$\frac{1}{(N/2)^b} = \frac{2^b}{N^b}.$$

There are c^2 possible combinations for ℓ and m, s.t. b values V_i^j can form a valid collision with b values V_i^r, with $i \in \{0, \ldots, b-1\}$. Thus, $\mathcal{A}'$ has a success probability for finding a collision for H^{CbDM} for two fixed queries L^j and L^r is at most

$$\frac{c^2}{(N/2)^b} = \frac{c^2 \cdot 2^b}{N^b}.$$

Since the j-th query can form a collision with any of the previous entries in $\mathcal{L}_{j-1}$, we have to determine the maximum number of queries in $\mathcal{L}_{j-1}$. If $\mathcal{A}'$ obtained a super query for each key it queried before, $\mathcal{L}_{j-1}$ may contain up to $2(j-1)$ entries. Since the winning query has to be a normal query in this case, $\mathcal{L}$ can contain at most q normal queries and up to $(q-1)$ queries (without the current one) resulting from super queries in the history. This would imply that one had to sum up the probabilities up to $2q - 1$:

$$\sum_{j=1}^{2q-1} \frac{2(j-1) \cdot c^2 \cdot 2^b}{N^b}.$$

However, we can do better. In the NormalQueryWin($\mathcal{L}$) case, $\mathcal{A}'$ will not win if its last (winning) query was part of a super query. Hence, we do not need to test if any of the super queries will produce a collision with any of their respective previous queries, and we have to test only possible collisions with the (at most q) normal queries. Nevertheless, $\mathcal{A}'$ still has to test each of the q normal queries if they collide with any of the at most $2q$ previous queries (including those which were part of a super query). Therefore, the success probability of $\mathcal{A}'$ to find a collision for H^{CbDM} can be upper bounded by

$$\Pr[\mathsf{NormalQueryWin}(\mathcal{L})] \leq \sum_{j=1}^{q} \frac{2(j-1) \cdot c^2 \cdot 2^b}{N^b} \leq \frac{q^2 \cdot c^2 \cdot 2^b}{N^b}. \tag{2}$$

Subcase 1.2: **SuperQueryWin($\mathcal{L}$).** In this case, $\mathcal{A}'$ wins with a super query, i.e., it has asked the $(N/2 + 1)$-th query for K^j, triggering a super query to occur. We can reuse the argument from Subcase 1.1 that the success probability of $\mathcal{A}'$ to obtain b colliding equations for two fixed queries L^r, L^j can be upper bounded by

$$\frac{c^2}{(N/2)^b}.$$

Here, the query history $\mathcal{L}_q$ contains at most $2q$ queries. But this time, we do not have to test if any of the q normal queries produces a collision with any of their respective predecessors. Hence, we can upper bound the success probability of $\mathcal{A}'$ to find a collision for H^{CbDM} with one super query by

$$\frac{2q \cdot c^2 \cdot 2^b}{N^b}.$$

For a super query to occur, $\mathcal{A}$ has to ask at least $N/(2c)$ regular queries. Thus, there can be at most $q/(N/2c)$ super queries in $\mathcal{L}$ and we obtain

$$\Pr[\mathsf{SuperQueryWin}(\mathcal{L})] \leq \frac{2q \cdot c^2 \cdot 2^b}{N^b} \cdot \frac{q}{N/2c} = \frac{c^3 \cdot 2^{b+2} \cdot q^2}{N^{b+1}}. \qquad (3)$$

Subcase 1.3: **SameQueryWin($\mathcal{L}$).** In this case, $\mathcal{A}'$ wins if it finds two integers $\ell, m \in \{0, \ldots, c-1\}$ with $\ell \neq m$ s.t.:

$$E_{K^j}(X^j \oplus \ell \oplus 0) \oplus X^j = E_{K^j}(X^j \oplus m \oplus 0) \oplus X^j,$$
$$E_{K^j}(X^j \oplus \ell \oplus 1) \oplus X^j = E_{K^j}(X^j \oplus m \oplus 1) \oplus X^j,$$
$$\vdots$$
$$E_{K^j}(X^r \oplus \ell \oplus (b-1)) \oplus X^j = E_{K^j}(X^j \oplus m \oplus (b-1)) \oplus X^j.$$

However, due to the XOR with the distinct values $i-1$, all plaintext inputs $X^j \oplus (i-1)$ in one compression-function call differ from each other. Furthermore, since all plaintext inputs are encrypted under the same key K^j and E is an ideal block cipher, their corresponding outputs Y_i^j are all different and uniformly distributed, and so are the values $Y_i^j \oplus X^j$ after the feed-forward operation. Hence, it is not possible for $\mathcal{A}'$ to find a collision for H^{CbDM} among the values $Y_i^j \oplus X^j$:

$$\Pr[\mathsf{SameQueryWin}(\mathcal{L})] = 0. \qquad (4)$$

Case 2: L^r **is in** $\mathcal{L}_{j-1}$. In this case, the key K^j and the plaintext prefix X_{pre}^j of $\mathcal{A}$'s current query $(K^j, X_{pre}^j \,\|\, X_{post'}^j)$ are already stored in some entry $L^r \in \mathcal{L}_{j-1}$, where $L^r = (K^r, X_{pre}^r \,\|\, X_{post}^r, Y_1^r, \ldots, Y_c^r)$. $\mathcal{A}'$ just extracts $Y_{(X_{post}^r \oplus X_{post'}^j)+1}^r$ from L^r, and passes it to $\mathcal{A}$. This implies that $\mathcal{A}$ can learn only information which $\mathcal{A}'$ already possesses. Thus,

$$\mathbf{Adv}_{H^{CbDM}}^{COLL}(\mathcal{A}) \leq \mathbf{Adv}_{H^{CbDM}}^{COLL}(\mathcal{A}').$$

Our claim is given by summing up equations (2), (3), and (4).   □

Table 2 shows the minimal number of queries q an adversary has to ask in order to obtain an advantage of $\mathbf{Adv}_{H^{CbDM}}^{COLL}(q) = 1/2$ for the most practical block lengths $n \in \{64, 128\}$ and depending on b.

Table 2. Minimum number of block-cipher queries q that an adversary must ask in order to find a collision for H^{CbDM} with advantage $1/2$

	n = 64			n = 128	
#blocks b	#queries q	optimal bound $2^{bn/2}$	#blocks b	#queries q	optimal bound $2^{bn/2}$
2	$2^{61.50}$	2^{64}	2	$2^{125.50}$	2^{128}
4	$2^{123.50}$	2^{128}	4	$2^{251.50}$	2^{256}
8	$2^{248.50}$	2^{256}	8	$2^{504.50}$	2^{512}

6 Preimage-Security Analysis of Counter-bDM

Attack Setting. Let $(V_1, \ldots, V_b) \in \{0,1\}^{bn}$ be the point to invert (see Definition 4), chosen by an adversary $\mathcal{A}$ before it makes any query to E. We define that $\mathcal{A}$ has the goal to find a preimage for $(V_1, \ldots, V_b)$ as described in Experiment 2. For our preimage-security analysis, we adapt the procedure from our collision analysis, i.e., we construct another adversary $\mathcal{A}'$, which simulates $\mathcal{A}$, but sometimes is allowed to make additional queries to E that are not taken into account. Again, since $\mathcal{A}'$ is more powerful than $\mathcal{A}$, it suffices to upper bound the success probability of $\mathcal{A}'$. Here, we say that $\mathcal{A}'$ is *successful* if its query history $\mathcal{Q}$ contains the means of computing a preimage for $(V_1, \ldots, V_b)$.

The procedures of $\mathcal{A}$ and $\mathcal{A}'$ asking queries to the oracle E and building the query histories $\mathcal{Q}$ and $\mathcal{L}$ are the same as that described in our collision-security proof. Furthermore, we adopt the events NormalQueryWin($\mathcal{L}$) and SuperQueryWin($\mathcal{L}$) from there, which in this context, cover all possible winning events for $\mathcal{A}'$. Thus, it holds that

$$\mathbf{Adv}^{EPRE}_{H^{CbDM}}(q) \leq \Pr[\mathsf{NormalQueryWin}(\mathcal{L})] + \Pr[\mathsf{SuperQueryWin}(\mathcal{L})]. \quad (5)$$

Before we present our bound, we describe more precisely what is meant by $\mathcal{A}'$ has found a preimage for H^{CbDM}. Let $L^j = (K^j, X^j, Y_1^j, \ldots, Y_c^j)$ represent the j-th entry in $\mathcal{L}$. We say that L^j contains the means of computing a preimage if $\exists \ell \in \{0, \ldots, c-1\}$, so that the following b equations hold:

$$E_{K^j}(X^j \oplus \ell) \oplus X^j = V_1$$
$$E_{K^j}(X^j \oplus \ell \oplus 1) \oplus X^j = V_2$$
$$\vdots$$
$$E_{K^j}(X^j \oplus \ell \oplus (b-1)) \oplus X^j = V_b.$$

Theorem 4. *Let $N = 2^n$. Then, it applies that*

$$\mathbf{Adv}^{EPRE}_{H^{CbDM}}(q) \leq \frac{c \cdot 2^{b+1} \cdot q}{N^b}.$$

Proof. After $\mathcal{A}$ has asked a (normal) forward query $Y^j = E_{K^j}(X^j)$ or a (normal) backward query $X^j = D_{K^j}(Y^j)$, $\mathcal{A}'$ checks if $\mathcal{L}_{j-1}$ already contains an entry $L^r = (K^j, X^j_{pre} \| *, *, \ldots, *)$, where X^j_{pre} denotes the prefix of X^j. In the following, we analyze the possible cases and upper bound their success probabilities separately.

Case 1: L^r **is not in** $\mathcal{L}_{j-1}$. In this case, $\mathcal{A}'$ labels Y as Y^j_1 and asks $c - 1$ further queries to E that are not taken into account:

$$\forall i \in \{2, \ldots, c\} : \quad Y^j_i = E_{K^j}(X^j \oplus (i - 1)).$$

Then, $\mathcal{A}'$ creates the tuple $L^j = (K^j, X^j, Y^j_1, \ldots, Y^j_c)$ and appends it to its query list, i.e., $\mathcal{L}_j = \mathcal{L}_{j-1} \cup \{L^j\}$. Note that due to the XOR with $i - 1$, all plaintexts X^j_i, with $i \leq i \leq c$, are pair-wise distinct. Thus, all ciphertexts Y^j_i, and the results of all feed-forward operations $(Y^j_i \oplus X^j)$ are always uniformly distributed.

In the following, we have to upper bound the success probability of $\mathcal{A}'$ to find a preimage for H^{CbDM} using either a normal query or a super query.

Subcase 1.1: **NormalQueryWin($\mathcal{L}$).** Since we assume that the winning query is a normal one, $\mathcal{A}'$ can have collected at most $N/2 - c$ queries for the current key K^j. Thus, E samples the query responses $Y^j_1, \ldots, Y^j_c$ at random from a set of size of at least $N/2 + c \geq N/2$ elements. From the c values Y_i of L^j, the probability that one equation $E_{K^j}(X^j \oplus \ell) \oplus (X^j \oplus \ell) = V_i$ from above holds for some fixed value of ℓ, can be upper bounded by $1/(N/2)$. The probability that b equations from above hold for a fixed ℓ can be upper bounded by $1/(N/2)^b$. Since there are c possible values for ℓ, the probability to obtain a preimage with the j-th query is given by

$$\frac{c}{(N/2)^b} = \frac{c \cdot 2^b}{N^b}.$$

Since $\mathcal{A}'$ is allowed to ask at most q queries, it applies that

$$\Pr[\mathsf{NormalQueryWin}(\mathcal{L})] \leq \frac{c \cdot 2^b \cdot q}{N^b}. \tag{6}$$

Subcase 1.2: **SuperQueryWin($\mathcal{L}$).** In this case, $\mathcal{A}'$ has already posed and stored $N/2c$ queries for the key K^j of its winning query. From the super query, it obtains the remaining $N/2c$ queries for K^j. We denote the latter set of queries by $\mathcal{SQ}$. From above, we already know that the probability that one point $L^j \in \mathcal{SQ}$ satisfies the preimage property can be upper bounded by

$$\frac{c}{(N/2)^b} = \frac{c \cdot 2^b}{N^b}.$$

Since the adversary obtains $N/2c$ points from the super query, the success probability that one of them yields a preimage for the given point is given by

$$\frac{N}{2c} \cdot \frac{c \cdot 2^b}{N^b} = \frac{2^{b-1}}{N^{b-1}}.$$

Table 3. Minimum number of block-cipher queries q that an adversary must ask in order to find a preimage for H^{CbDM} with advantage $1/2$

	n = 64			n = 128	
#blocks	#queries	optimal bound	#blocks	#queries	optimal bound
b	q	2^{bn}	b	q	2^{bn}
2	2^{123}	2^{128}	2	2^{251}	2^{256}
4	2^{248}	2^{256}	4	2^{504}	2^{512}
8	2^{499}	2^{512}	8	2^{1011}	2^{1024}

For every super query to occur, $\mathcal{A}'$ has to collect $N/2c$ queries in advance. Thus, there are at most $q/(N/2c)$ super queries and we obtain

$$\Pr[\mathsf{SuperQueryWin}(\mathcal{L})] \leq \frac{q}{N/2c} \cdot \frac{2^{b-1}}{N^{b-1}} = \frac{c \cdot 2^b \cdot q}{N^b}. \qquad (7)$$

Case 2: L^r is in $\mathcal{L}_{j-1}$. Like in the Case 2 of our collision-security proof, the key K^j and the plaintext prefix X_{pre}^j of $\mathcal{A}$'s current query $(K^j, X_{pre}^j \parallel X_{post'}^j)$ are already stored in some entry $L^r \in \mathcal{L}_{j-1}$, where $L^r = (K^j, X_{pre}^j \parallel X_{post}^j, Y_1^r, \ldots, Y_c^r)$. Again, $\mathcal{A}'$ extracts $Y_{(X_{post}^r \oplus X_{post'}^j)+1}^r$ from L^r, and passes it to $\mathcal{A}$. This implies that $\mathcal{A}$ can learn only information that $\mathcal{A}'$ already possesses and

$$\mathbf{Adv}_{HCbDM}^{COLL}(\mathcal{A}) \leq \mathbf{Adv}_{HCbDM}^{COLL}(\mathcal{A}').$$

Our claim is given by summing up equations (6) and (7).   □

For $n = 128$ and $\mathbf{Adv}_{HCbDM}^{EPRE}(q) = 1/2$, we list in Table 3 the amounts of queries q an adversary has to make, depending on the value of b.

7   Conclusion and Outlook

This paper introduced COUNTER-bDM – the first provably secure family of multi-block-length compression functions, that maps $(b+1)n$-bit inputs to bn-bit outputs for arbitrary $b \geq 2$. With COUNTER-bDM, we propose a simple, though, very neat design, that not only avoids costly requirements such as the need of having independent ciphers, or having to run the key schedule multiple times, but also simplifies the analysis greatly. In our collision- and preimage-security analysis we provided proofs for arbitrary block lengths $b > 2$. It remains an open research topic to find a multi-block-length hash function with arbitrary output size employing an n-bit or at most $2n$-bit keyed block cipher.

References

1. Armknecht, F., Fleischmann, E., Krause, M., Lee, J., Stam, M., Steinberger, J.: The Preimage Security of Double-Block-Length Compression Functions. In: Lee, D.H., Wang, X. (eds.) ASIACRYPT 2011. LNCS, vol. 7073, pp. 233–251. Springer, Heidelberg (2011)
2. Aumasson, J.-P., Henzen, L., Meier, W., Phan, R.C.-W.: SHA-3 proposal BLAKE. Submission to NIST, Round 3 (2010)
3. Bertoni, G., Daemen, J., Peeters, M., Van Assche, G.: Sponge functions. Ecrypt Hash Workshop (May 2007)
4. Biham, E., Dunkelman, O.: The SHAvite-3 Hash Function. Submission to NIST, Round 2 (2009)
5. Black, J.A., Rogaway, P., Shrimpton, T.: Black-Box Analysis of the Block-Cipher-Based Hash-Function Constructions from PGV. In: Yung, M. (ed.) CRYPTO 2002. LNCS, vol. 2442, pp. 320–335. Springer, Heidelberg (2002)
6. Meyer, C., Matyas, S.: Secure Program Load With Manipulation Detection Code (1988)
7. Chang, D., Nandi, M., Lee, J., Sung, J., Hong, S., Lim, J., Park, H., Chun, K.: Compression Function Design Principles Supporting Variable Output Lengths from a Single Small Function. IEICE Transactions 91-A(9), 2607–2614 (2008)
8. Coppersmith, D., Pilpel, S., Meyer, C.H., Matyas, S.M., Hyden, M.M., Oseas, J., Brachtl, B., Schilling, M.: Data Authentication Using Modification Dectection Codes Based on a Public One-Way Encryption Function. U.S. Patent No. 4,908,861 (March 13, 1990)
9. Ewan Fleischmann. Analysis and Design of Blockcipher Based Cryptographic Algorithms. PhD thesis, Bauhaus-Universität Weimar (2013)
10. Fleischmann, E., Forler, C., Gorski, M., Lucks, S.: Collision-Resistant Double-Length Hashing. In: Heng, S.-H., Kurosawa, K. (eds.) ProvSec 2010. LNCS, vol. 6402, pp. 102–118. Springer, Heidelberg (2010)
11. Fleischmann, E., Forler, C., Lucks, S.: The Collision Security of MDC-4. In: Mitrokotsa, A., Vaudenay, S. (eds.) AFRICACRYPT 2012. LNCS, vol. 7374, pp. 252–269. Springer, Heidelberg (2012)
12. Fleischmann, E., Forler, C., Lucks, S., Wenzel, J.: Weimar-DM: A Highly Secure Double-Length Compression Function. In: Susilo, W., Mu, Y., Seberry, J. (eds.) ACISP 2012. LNCS, vol. 7372, pp. 152–165. Springer, Heidelberg (2012)
13. Fleischmann, E., Gorski, M., Lucks, S.: On the Security of Tandem-DM. In: Dunkelman, O. (ed.) FSE 2009. LNCS, vol. 5665, pp. 84–103. Springer, Heidelberg (2009)
14. Fleischmann, E., Gorski, M., Lucks, S.: Security of Cyclic Double Block Length Hash Functions. In: Parker, M.G. (ed.) Cryptography and Coding 2009. LNCS, vol. 5921, pp. 153–175. Springer, Heidelberg (2009)
15. Hattori, M., Hirose, S., Yoshida, S.: Analysis of Double Block Length Hash Functions. In: Paterson, K.G. (ed.) Cryptography and Coding 2003. LNCS, vol. 2898, pp. 290–302. Springer, Heidelberg (2003)
16. Hirose, S.: Provably Secure Double-Block-Length Hash Functions in a Black-Box Model. In: Park, C., Chee, S. (eds.) ICISC 2004. LNCS, vol. 3506, pp. 330–342. Springer, Heidelberg (2005)
17. Hirose, S.: Some Plausible Constructions of Double-Block-Length Hash Functions. In: Robshaw, M. (ed.) FSE 2006. LNCS, vol. 4047, pp. 210–225. Springer, Heidelberg (2006)

18. Hirose, S.: Some Plausible Constructions of Double-Block-Length Hash Functions. In: Robshaw, M. (ed.) FSE 2006. LNCS, vol. 4047, pp. 210–225. Springer, Heidelberg (2006)
19. Hohl, W., Lai, X., Meier, T., Waldvogel, C.: Security of Iterated Hash Functions Based on Block Ciphers. In: Stinson, D.R. (ed.) Advances in Cryptology - CRYPTO 1993. LNCS, vol. 773, pp. 379–390. Springer, Heidelberg (1994)
20. ISO/IEC. ISO DIS 10118-2: Information technology - Security techniques - Hash-functions, Part 2: Hash-functions using an n-bit block cipher algorithm. First released in 1992 (2000)
21. Knudsen, L.R., Lai, X., Preneel, B.: Attacks on Fast Double Block Length Hash Functions. J. Cryptology 11(1), 59–72 (1998)
22. Knudsen, L.R., Muller, F.: Some Attacks Against a Double Length Hash Proposal. In: Roy, B. (ed.) ASIACRYPT 2005. LNCS, vol. 3788, pp. 462–473. Springer, Heidelberg (2005)
23. Krause, M., Armknecht, F., Fleischmann, E.: Preimage Resistance Beyond the Birthday Bound: Double-Length Hashing Revisited. IACR Cryptology ePrint Archive 2010, 519 (2010)
24. Lai, X., Massey, J.L.: Hash Functions Based on Block Ciphers. In: Rueppel, R.A. (ed.) Advances in Cryptology - EUROCRYPT1992. LNCS, vol. 658, pp. 55–70. Springer, Heidelberg (1993)
25. Lee, J.: Provable Security of the Knudsen-Preneel Compression Functions In: Wang, X., Sako, K. (eds.) ASIACRYPT 2012. LNCS, vol. 7658, pp. 504–525. Springer, Heidelberg (2012)
26. Lee, J., Kwon, D.: The Security of Abreast-DM in the Ideal Cipher Model. Cryptology ePrint Archive, Report 2009/225 (2009), http://eprint.iacr.org/
27. Lee, J., Kwon, D.: The Security of Abreast-DM in the Ideal Cipher Model. IEICE Transactions 94-A(1), 104–109 (2011)
28. Lee, J., Stam, M.: MJH: A Faster Alternative to MDC-2. In: Kiayias, A. (ed.) CT-RSA 2011. LNCS, vol. 6558, pp. 213–236. Springer, Heidelberg (2011)
29. Lee, J., Stam, M., Steinberger, J.: The Collision Security of Tandem-DM in the Ideal Cipher Model. In: Rogaway, P. (ed.) CRYPTO 2011. LNCS, vol. 6841, pp. 561–577. Springer, Heidelberg (2011)
30. Lee, J., Steinberger, J.P.: Multiproperty-Preserving Domain Extension Using Polynomial-Based Modes of Operation. IEEE Transactions on Information Theory 58(9), 6165–6182 (2012)
31. Lucks, S.: A Collision-Resistant Rate-1 Double-Block-Length Hash Function. In: Symmetric Cryptography (2007)
32. Luo, Y., Lai, X.: Attacks On a Double Length Blockcipher-based Hash Proposal. IACR Cryptology ePrint Archive 2011, 238 (2011)
33. Rabin, M.: Digitalized Signatures. In: De Millo, R., Dobkin, D., Jones, A., Lipton, R. (eds.) Foundations of Secure Computation, pp. 155–168. Academic Press (1978)
34. Mennink, B.: Optimal Collision Security in Double Block Length Hashing with Single Length Key. In: Wang, X., Sako, K. (eds.) ASIACRYPT 2012. LNCS, vol. 7658, pp. 526–543. Springer, Heidelberg (2012)
35. Merkle, R.C.: One Way Hash Functions and DES. In: Brassard, G. (ed.) Advances in Cryptology - CRYPT0 1989. LNCS, vol. 435, pp. 428–446. Springer, Heidelberg (1990)
36. Nandi, M., Lee, W.I., Sakurai, K., Lee, S.-J.: Security Analysis of a 2/3-Rate Double Length Compression Function in the Black-Box Model. In: Gilbert, H., Handschuh, H. (eds.) FSE 2005. LNCS, vol. 3557, pp. 243–254. Springer, Heidelberg (2005)

37. Ferguson, N., Lucks, S., Schneier, B., Whiting, D., Bellare, M., Kohno, T., Callas, J., Walker, J.: Skein Source Code and Test Vectors, http://www.skein-hash.info/downloads
38. Özen, O., Stam, M.: Another Glance at Double-Length Hashing. In: Parker, M.G. (ed.) Cryptography and Coding 2009. LNCS, vol. 5921, pp. 176–201. Springer, Heidelberg (2009)
39. Peyrin, T., Gilbert, H., Muller, F., Robshaw, M.J.B.: Combining Compression Functions and Block Cipher-Based Hash Functions. In: Lai, X., Chen, K. (eds.) ASIACRYPT 2006. LNCS, vol. 4284, pp. 315–331. Springer, Heidelberg (2006)
40. Rogaway, P., Shrimpton, T.: Cryptographic Hash-Function Basics: Definitions, Implications, and Separations for Preimage Resistance, Second-Preimage Resistance, and Collision Resistance. In: Roy, B., Meier, W. (eds.) FSE 2004. LNCS, vol. 3017, pp. 371–388. Springer, Heidelberg (2004)
41. Rogaway, P., Steinberger, J.P.: Constructing Cryptographic Hash Functions from Fixed-Key Blockciphers. In: Wagner, D. (ed.) CRYPTO 2008. LNCS, vol. 5157, pp. 433–450. Springer, Heidelberg (2008)
42. Satoh, T., Haga, M., Kurosawa, K.: Towards Secure and Fast Hash Functions. TIE-ICE: IEICE Transactions on Communications/Electronics/Information and Systems (1999)
43. Stam, M.: Beyond Uniformity: Better Security/Efficiency Tradeoffs for Compression Functions. In: Wagner, D. (ed.) CRYPTO 2008. LNCS, vol. 5157, pp. 397–412. Springer, Heidelberg (2008)
44. Stam, M.: Blockcipher-Based Hashing Revisited. In: Dunkelman, O. (ed.) FSE 2009. LNCS, vol. 5665, pp. 67–83. Springer, Heidelberg (2009)
45. Steinberger, J.P.: The Collision Intractability of MDC-2 in the Ideal Cipher Model. In: Naor, M. (ed.) EUROCRYPT 2007. LNCS, vol. 4515, pp. 34–51. Springer, Heidelberg (2007)
46. Robert, S., Winternitz: A Secure One-Way Hash Function Built from DES. In: IEEE Symposium on Security and Privacy, pp. 88–90 (1984)

A   Related Work

This part summarizes related work regarding to single- and double-block-length hash functions.

Double-Block-Length Schemes. The essentially first double-block-length hash functions were presented by Merkle [35], who proposed three constructions on the basis of DES. Today, there are four so-called "classical" double-block-length constructions, which were introduced in the early 1990s: MDC-2, MDC-4, ABREAST-DM, and TANDEM-DM. MDC-2 and MDC-4 [8,20] are (n,n)-bit double-block-length hash functions with rates $1/2$ and $1/4$, respectively. For MDC-2, Steinberger [45] proved in 2006 that no adversary asking less than $2^{74.9}$ queries will obtain a significant advantage at finding a collision. In a sophisticated proof, it was shown by Fleischmann, Forler, and Lucks [11] in 2012, that for MDC-4 an adversary requires at least $2^{74.7}$ queries to find a collision with an advantage of $1/2$.

Concerning rate-1 double-block-length hash functions, Lucks [31] presented a first construction at Dagstuhl'07. Stam [44] also proposed a rate-1 single-call double-block-length function, for which he showed an almost-optimal collision-resistance, up to a logarithmic factor. However, while Lucks and Stam claimed a rate-1 property for their constructions, those are actually much slower, as pointed out by Luo and Lai [32]. At CRYPTO'93, Hohl et al. [19] analyzed the security of compression functions of rate-1/2 double-block-length hash functions. In 1998, Knudsen, Lai, and Preneel [21] discussed the security of rate-1 double-block-length hash functions. In 1999, Satoh, Haga, and Kurosawa [42] as well as Hattori, Hirose, and Yoshida [15] in 2003 attacked rate-1 double-block-length hash functions. At FSE'05, Nandi et al. [36] presented a rate-2/3 compression function, which was later analyzed by Knudsen and Muller at ASIACRYPT'05 [22]. At CT-RSA'11, Lee and Stam [28] presented a faster alternative to MDC-2, called MJH.

Double-Block-Length Schemes with Birthday-Type Collision Security.

ABREAST-DM and TANDEM-DM base on the famous Davies-Meyer scheme, and have been presented by Lai and Massey [24] at EUROCRYPT'92. In 2004, Hirose added a large class of rate-1/2 double-block-length hash functions, composed of two independent $(2n, n)$-bit block ciphers, with $2n$ being the key and n the block size [16] . At FSE'06, he proposed a new scheme called HIROSE-DM [17], which dropped the requirement of independent ciphers, and for which he provided a collision-security proof in the ideal-cipher model, stating that no adversary asking less than $2^{124.55}$ queries can find a collision with probability $\geq 1/2$.

In [39], Peyrin et al. analyzed techniques to construct larger compression functions by combining smaller ones. The authors proposed $3n$-to-$2n$-bit and $4n$-to-$2n$-bit constructions composed of five public functions, yet they did not show proofs for their concepts.

In 2008, Chang et al. introduced a generic framework for purf-based multi block length constructions [7], where purf denotes a public random function.

Considering TANDEM-DM, Fleischmann, Gorski, and Lucks [13] gave a collision-security proof at FSE'09, showing that no adversary can obtain a significant advantage without making at least $2^{120.4}$ queries. In 2010, Lee, Stam, and Steinberger [29] have shown that the proof of Fleischmann et al. has several non-trivial flaws. Further, they provided a bound of $2^{120.87}$ queries for a collision adversary.

For ABREAST-DM, Fleischmann, Gorski, and Lucks [14] as well as Lee and Kwon presented, independent from each other, collision-security bound of $2^{124.42}$ queries. More general, [14] introduced the class notion of CYCLIC-DL, which included the constructions ABREAST-DM, CYCLIC-DM, ADD-k-DM, and CUBE-DM, and applied similar proofs for these. At IMA'09, Özen and Stam [38] proposed a framework for double-block-length hash functions by extending the generalized framework by Stam at FSE'09 for single-call hash functions. Still, their framework based on the usage of two independent block ciphers. At ProvSec'10, Fleischmann et al. [10] extended their general classification of double-block-length hash functions by the classes GENERIC-DL, SERIAL-DL, and PARALLEL-DL. For the framework by Özen and Stam, they relaxed the requirement of distinct independent block ciphers and gave collision bounds for TANDEM-DM and CYCLIC-DM. In [23], Krause, Armknecht, and Fleischmann provided techniques for proving asymptotically-optimal preimage-resistance bounds for block-cipher-based double-length, double-call hash functions. They introduced a new Davies-Meyer double-block-length hash function for which they proved that no adversary asking less than 2^{2n-5} queries can find a preimage with probability $\geq 1/2$. At ACISP'12, Fleischmann et al. [12] showed a very similar Davies-Meyer construction – called WEIMAR-DM– for which they could prove the currently best collision-security bound of $2^{126.23}$ queries, and the currently best preimage-security bound among the previously known double-block-length hash function.

Universal Hash-Function Families:
From Hashing to Authentication

Basel Alomair

National Center for Cybersecurity Technology (C4C)
King Abdulaziz City for Science and Technology (KACST)
alomair@kacst.edu.sa

Abstract. Due to their potential use as building blocks for constructing highly efficient message authentication codes (MACs), universal hash-function families have been attracting increasing research attention, both from the design and analysis points of view. In universal hash-function families based MACs, the message to be authenticated is first compressed using a universal hash function and, then, the compressed image is encrypted to produce the authentication tag. Many definitions of universal hash families have appeared in the literature. The main focus of earlier definitions is to classify universal hash functions based on their message collision properties. In this paper, we introduce a different classification of universal hash families. As opposed to classifying universal hash families based on message collision probabilities, our classification aims to give direct relation between universal hash families used as building blocks to design MACs and the encryption algorithm used to process their hashed images. We give two examples of universal hash families with equivalent collision resiliency. We show that, while one constructs secure MACs, the other can lead to insecure MAC construction even when coupled with an encryption algorithm that provides perfect secrecy (in Shannon's sense). We formally define two classes of universal hash families: *independent* and *dependent* universal hash families. We show that, while independent universal hash families provide the desired unforgeability independently of the used encryption algorithm, the security of MACs based on dependent universal hash families is not guaranteed for all choices of encryption algorithms. We conclude by giving a sufficient condition on the encryption algorithm that guarantees the construction of secure MACs, even when combined with a dependent hash family.

1 Introduction

With todays technology, huge amount of data is being transmitted over insecure channels, such as wireless communications, Internet communications, etc. In most scenarios, users have no control over the route transmitted messages take in their way to the destination. Since some, if not all, links that messages take in their route can be insecure, it is desirable, even necessary in many applications, to protect exchanged messages against malicious users. Message integrity, in particular, is one of the most concerning problems when communicating through

D. Pointcheval and D. Vergnaud (Eds.): AFRICACRYPT 2014, LNCS 8469, pp. 459–474, 2014.

insecure channels. Fortunately, however, the literature of cryptography is rich with techniques to protect the integrity of messages transmitted over insecure channels, and message authentication codes (MACs) are amongst the most extensively used primitives for preserving message integrity.

Universal hash-function families based MACs belong to a class of MACs that has been increasingly popular due its fast implementation. In MACs based on universal hash functions, the message to be authenticated is first compressed using a universal hash function and then the compressed image is encrypted to produce the authentication tag.[1] Processing messages using hash functions is much faster than processing them block by block using block ciphers. Since the compressed image is usually much shorter than the message itself, applying a cryptographic function on the compressed image can be accomplished efficiently. To date, this construction of MACs is the fastest technique for message authentication [1–3, 8, 9, 11, 15].

A typical example of a universal hash function works as follows. The message is broken into multiple blocks of a predefined length, e.g., $M = m_1 || \cdots || m_B$, where B is the number of blocks. Let each block be of size N bits and choose an N-bit long prime integer p. Define a universal hash function by the secret key $K = k_1 || \cdots || k_B$, where each k_i is chosen randomly from the set $\{1, 2, \cdots, p-1\}$, along with the operation

$$h(M) = \sum_{i=1}^{B} k_i m_i \mod p. \tag{1}$$

(It will be shown in Section 3 that this is indeed a universal hash function.)

As can be seen from equation (1), the compressed image of an NB-bit long message is only N-bit long. The problem, however, with such universal hash functions is that they are not cryptographic functions. That is, the observation of multiple message-image pairs can reveal information about the secret key K (for example, by constructing a system of linear equations of the form $MK = H$ mod p). This implies that the output of the universal hash function must be encrypted before transmission (to maintain the secrecy of the universal hash function's key).

When two distinct messages collide (i.e., hash to the same image), however, their authentication tag will be the same. Therefore, if given a message-tag pair, call it (M, τ), one can come up with a different message M' that hashes to the same value, the pair (M', τ) will be accepted as authentic. That is, in universal hashing based MACs, an adversary able to come up with distinct colliding messages can forge valid authentication tags with high probabilities, even if the hashing key remains secret. Consequently, a critical property of universal hash-function families is their resilience to message collision. Since Carter

---

[1] Since the purpose of the authentication tag is preserving message integrity, encryption algorithm can be replaced with other noninvertible cryptographic primitives. For instance, cryptographic hash functions such as the SHA family have been used in the construction of universal hashing based MACs [3].

and Wegman introduced universal hash families [4,5] and their potential use for message authentication [16,17], many definitions that classify universal hash-function families based on their message collision resiliency have appeared (see, e.g., [5,9,10,13]).[2]

In this work, we investigate the relation between the universal hash-function family used for message compression and the encryption algorithm used to process the compressed image. We give two examples of universal hash families with equivalent message collision probabilities. Although the two universal hash families are equivalent when characterized based on their resiliency against message collision, they possess completely different properties when used to construct message authentication codes. We give two examples in which we couple the two universal hash families with an encryption algorithm that provides perfect secrecy (in Shannon's information theoretic sense) and show that, while one universal hash family results in a secure MAC algorithm, the other universal hash family results in a totally insecure MAC algorithm. Motivated by these examples, we propose a new classification of universal hash-function families that is not based on their message collision resiliency. That is, based on their security implications on the constructed MAC, we classify universal hash-function families into *independent* and *dependent* universal hash-function families. We provide a formal definition for each of the two categories. For independent universal hash-function families, we show that any encryption algorithm that maintains the secrecy of the universal hash function's key suffices to provide message integrity. On the other hand, we show that the integrity of messages compressed with dependent universal hash-function families is not guaranteed, even if the encryption algorithm is perfectly secret. On the positive side, we derive a sufficient condition on the encryption algorithm that guarantees the secure construction of message authentication codes, even when coupled with dependent universal hash-function families.

The rest of the paper is organized as follows. In Section 2 we give a list of the used notations, current classification of universal hash families, and some preliminaries. Section 3 provides the two examples of universal hash families. In Section 4, we introduce a hypothetical scenario that will help in our classification of universal hash families. In Section 5 we formally define independent and dependent universal hash families. We analyze the two classes of universal hash families in Section 6, and conclude our paper in Section 7.

2   Notations and Definitions

In this section we list the notations and definitions that will be used for the remainder of the paper, and state the algebraic preliminaries that will be used for our security analysis.

---

[2] Although some definitions reach beyond the collision probabilities, such as the ϵ-AXU and the ϵ-AΔU [9], they are directed to specific constructions and do not generalize to all possible MAC constructions.

2.1 Notations

The following notations will be used throughout the rest of the paper.

- For two sets A and B, the set $C = A \backslash B$ contains all elements in A that are not in B.
- For the finite integer ring $\mathbb{Z}_n$, the notation $\mathbb{Z}_n^*$ denotes the multiplicative group modulo n; that is, the set of integers relatively prime (co-prime) to n.
- For any non-empty set I, the cardinality of the set is denoted as $|I|$.
- For any two strings a and b, $(a \parallel b)$ denotes the concatenation operation.
- For the rest of the paper, addition and multiplication are performed over elements in the ring $\mathbb{Z}_n$, even if the "mod n" part is dropped for ease of notations.
- For any two integers a and b, $\gcd(a, b)$ is the greatest common divisor of a and b.
- For an element a in a ring R, the element a^{-1} denotes the multiplicative inverse of a in R, if it exists.

2.2 Definitions

A family of hash functions $\mathcal{H}$ is specified by a finite set of keys $\mathcal{K}$. Each key $k \in \mathcal{K}$ defines a member of the family $\mathcal{H}_k \in \mathcal{H}$. As opposed to thinking of $\mathcal{H}$ as a set of functions from $\mathcal{D}$ to $\mathcal{R}$, it can be viewed as a single function $\mathcal{H} : \mathcal{K} \times \mathcal{D} \to \mathcal{R}$, whose first argument is usually written as a subscript. A random element $h \in \mathcal{H}$ is determined by selecting a $k \in \mathcal{K}$ uniformly at random and setting $h = \mathcal{H}_k$. The following are previously defined classes of universal hash families [4, 8–10, 13, 16]

Definition 1 (Universal Hash Families). *Let $\mathcal{H} = \{h : \mathcal{D} \to \mathcal{R}\}$ be a family of hash functions.*

- $\mathcal{H}$ *is said to be universal if for all distinct $M, M' \in \mathcal{D}$, we have that*

$$\Pr_{h \leftarrow \mathcal{H}}[h(M) = h(M')] = 1/|\mathcal{R}|.$$

- $\mathcal{H}$ *is said to be ϵ-almost universal, denoted ϵ-AU, if for all distinct $M, M' \in \mathcal{D}$, we have that $\Pr_{h \leftarrow \mathcal{H}}[h(M) = h(M')] \leq \epsilon$.*
- $\mathcal{H}$ *is said to be strongly universal if for all distinct $M, M' \in \mathcal{D}$ and all $a, b \in \mathcal{R}$, we have that $\Pr_{h \leftarrow \mathcal{H}}[h(M) = a, h(M') = b] = 1/|\mathcal{R}|^2$.*
- $\mathcal{H}$ *is said to be ϵ-almost-strongly universal, denoted ϵ-ASU, if for all distinct $M, M' \in \mathcal{D}$ and all $a, b \in \mathcal{R}$, we have that $\Pr_{h \leftarrow \mathcal{H}}[h(M) = a, h(M') = b] = \epsilon/|\mathcal{R}|$.*
- $\mathcal{H}$ *is said to be ϵ-almost XOR universal, denoted ϵ-AXU, if for all distinct $M, M' \in \mathcal{D}$, and any $a \in \mathcal{R}$ we have that $\Pr_{h \leftarrow \mathcal{H}}[h(M) \oplus h(M') = a] \leq \epsilon$.*
- $\mathcal{H}$ *is said to be Δ-universal if for all distinct $M, M' \in \mathcal{D}$, and any $a \in \mathcal{R}$ we have that $\Pr_{h \leftarrow \mathcal{H}}[h(M) - h(M') = a] = 1/|\mathcal{R}|$, where $\mathcal{R}$ is an Abelian group and '-' denotes the subtraction operation over $\mathcal{R}$.*

– $\mathcal{H}$ is said to be ϵ-almost-Δ-universal, denoted ϵ-$A\Delta U$, if for all distinct $M, M' \in \mathcal{D}$, and any $a \in \mathcal{R}$ we have that $\mathrm{Pr}_{h \leftarrow \mathcal{H}}[h(M) - h(M') = a] \leq \epsilon$, where $\mathcal{R}$ is an Abelian group and '-' denotes the subtraction operation over $\mathcal{R}$.

A message authentication scheme consists of a signing algorithm $\mathcal{S}$ and a verifying algorithm $\mathcal{V}$. The signing algorithm might be probabilistic, while the verifying one is usually not. On input a key K and a message M, algorithm $\mathcal{S}$ outputs a string τ called the authentication tag, or simply the "tag" of M. On input a key K, a message M, and a tag τ, algorithm $\mathcal{V}$ outputs a bit, with 1 standing for accept and 0 for reject. Authentic tags must be accepted with probability one. That is, if $\tau = \mathcal{S}(K, M)$, it must be the case that $\mathcal{V}(K, M, \tau) = 1$ for any key K, message M, and tag τ.

A message authentication code (MAC) algorithm based on universal hash-function families is composed of two primitives: a universal hash function and an encryption algorithm. Given a message M to be authenticated, the message is first compressed into a short string of convenient length. Then, the compressed image is encrypted with an encryption algorithm. The output of the encryption algorithm is the authentication tag of the message.

Another definition that will be used in the rest of the paper is the notion of negligible functions.

Definition 2. *[6] [Negligible Functions] A function $\gamma : \mathbb{N} \to \mathbb{R}$ is said to be negligible if for any nonzero polynomial p, there exists N_0 such that for all $N > N_0$, $|\gamma(N)| < \frac{1}{|p(N)|}$. That is, the function is said to be negligible if it converges to zero faster than the reciprocal of any polynomial function.*

2.3   Preliminaries

We list, and prove, below two facts about integer rings that will be used in our analysis.

Lemma 1. *Let $\mathbb{Z}_n$ be any finite integer ring, and let α and β be nonzero elements of $\mathbb{Z}_n$. Then, $\alpha\beta \equiv 0 \bmod n$ only if both α and β are non invertible elements of $\mathbb{Z}_n$. That is, for any nonzero elements α and β in $\mathbb{Z}_n$, the following one-way implication holds*

$$\alpha\beta \equiv 0 \quad \bmod n \Rightarrow \{\alpha, \beta \in \mathbb{Z}_n \backslash \mathbb{Z}_n^*\} \tag{2}$$

Proof. Let α and β be nonzero elements of $\mathbb{Z}_n$. Without loss of generality, assume that $\alpha \in \mathbb{Z}_n^*$; that is, there exists an element $\alpha^{-1} \in \mathbb{Z}_n$ so that $\alpha\alpha^{-1} \equiv 1 \bmod n$. Then,

$$\alpha\beta \equiv 0 \quad \bmod n \Rightarrow \alpha^{-1}\alpha\beta \equiv 0 \quad \bmod n \Rightarrow \beta \equiv 0 \quad \bmod n,$$

a contradiction to the hypothesis that β is not the zero element; and the lemma follows.   □

Lemma 2. *Given an integer $k \in \mathbb{Z}_n^*$, the following must hold:*

1. *for an r_1 uniformly distributed over $\mathbb{Z}_n^*$, the value ϵ_1 given by*

$$\epsilon_1 \equiv r_1 k \quad \mod n \tag{3}$$

   is uniformly distributed over $\mathbb{Z}_n^*$.
2. *for an r_2 uniformly distributed over $\mathbb{Z}_n \backslash \mathbb{Z}_n^*$, the value ϵ_2 given by*

$$\epsilon_2 \equiv r_2 k \quad \mod n \tag{4}$$

   is uniformly distributed over $\mathbb{Z}_n \backslash \mathbb{Z}_n^*$.

Proof. To prove the first part, it suffices to show that for every $\epsilon_1 \in \mathbb{Z}_n^*$, there exists an $r_1 \in \mathbb{Z}_n^*$ that satisfies equation (3) and that this r_1 is unique.

Fix any $\epsilon_1 \in \mathbb{Z}_n^*$ and any $k \in \mathbb{Z}_n^*$. Since $k \in \mathbb{Z}_n^*$, by Bézout's lemma [14], k^{-1} does exist. That is, there exists $k^{-1} \in \mathbb{Z}_n^*$ so that

$$k^{-1}k \equiv 1 \quad \mod n, \tag{5}$$

and multiplying both sides of equation (5) by ϵ_1 gives:

$$(\epsilon_1 k^{-1})k \equiv \epsilon_1 \quad \mod n. \tag{6}$$

Hence, $r_1 = \epsilon_1 k^{-1} \mod n$ satisfies equation (3). Further, $r_1 \in \mathbb{Z}_n^*$ since $r_1^{-1} = \epsilon_1^{-1}k$ does exist. Therefore, there exists an $r_1 \in \mathbb{Z}_n^*$ that satisfies equation (3).

To show that this r_1 is unique, let $r_1' \neq r_1$ also satisfies equation (3). Then,

$$r_1' k \equiv \epsilon_1 \quad \mod n. \tag{7}$$

Multiplying both equations (3) and (7) by k^{-1} gives:

$$r_1 \equiv \epsilon_1 k^{-1} \quad \mod n, \tag{8}$$

and

$$r_1' \equiv \epsilon_1 k^{-1} \quad \mod n. \tag{9}$$

Therefore, $r_1 \equiv r_1' \mod n$ and, hence, the r_1 in $\mathbb{Z}_n^*$ that satisfies equation (3) for any fixed $\epsilon_1 \in \mathbb{Z}_n^*$ is unique.

The proof of the second part of the lemma is similar to the proof of the first part and, thus, it is omitted. □

3   Two Universal Hash Families

In this section, we give two examples of universal hash-function families that are equivalent in their message collision resiliency.

3.1   Description of the Universal Hash Families

Fix a security parameter N. Without loss of generality, assume the message can be divided into B blocks of length N-bits, that is $M = m_1||m_2||\ldots||m_B$, where $m_i \in \mathbb{Z}_{2^N}$ denotes the ith message block. Let $K = k_1||k_2||\ldots||k_B$ for k_i's drawn uniformly and independently at random from $\mathbb{Z}_{2^N}$. The key K is the shared secret key that will be used for message compression. (We overload m_i and k_i to denote both the N-bit strings and their integer representation as elements of $\mathbb{Z}_{2^N}$, depending on the context.) Note that the hashing key K can be used to hash messages of arbitrary lengths, not only NB-bit messages or shorter. The extension to hash arbitrary-length messages with the same key can be achieved with a variety of methods, e.g., [3, 8, 17]; the details of such extensions are out of the scope of this paper.

The First Example: Universal Hash Family 1 (UHF1). Choose an N-bit prime integer p. For every message M to be authenticated, the compressed image of message M is computed as:

$$h(M) = \sum_{i=1}^{B} k_i m_i \mod p, \tag{10}$$

assuming each k_i is chosen uniformly at random from the multiplicative group $\mathbb{Z}_p^*$.

The Second Example: Universal Hash Family 2 (UHF2). Let k_1 be drawn uniformly at random from the multiplicative group $\mathbb{Z}_{2^N}^*$, and let $K = k_2||k_3||\ldots||k_B$ for k_i's drawn uniformly and independently at random from $\mathbb{Z}_{2^N}$. For every message M to be authenticated, the compressed image of message M is computed as:

$$h(M) = k_1 m_1 \mod 2^N + \sum_{i=2}^{B} k_i m_i \mod 2^{2N}. \tag{11}$$

We will show in the next section that both UHF1 and UHF2 lead to a universal hash family. In either universal hash family, the authentication tag, τ, is computed by applying an encryption algorithm on the compressed image. That is,

$$\tau = \mathcal{E}\Big(h(M)\Big), \tag{12}$$

where $\mathcal{E}$ is the used encryption algorithm.

3.2   Message Collision Analysis

We show in this section that both UHF1 and UHF2 are universal hash families with equivalent resiliency to message collision.

UHF1. Recall that two messages, $M = m_1 \parallel \cdots \parallel m_B$ and $M' = m'_1 \parallel \cdots \parallel m'_B$, will have the same image if and only if the following holds:

$$h(M) \equiv \sum_{i=1}^{B} k_i m_i \equiv \sum_{i=1}^{B} k_i m'_i \equiv h(M') \mod p. \tag{13}$$

We consider now the three possible scenarios: single-block difference, two-block difference, and more than two-block difference.

a) Single block: without loss of generality, assume that only the first message block is different. That is, $m_i = m'_i$ for all $i \neq 1$ but $m_1 \not\equiv m'_1 \mod p$. Let $m'_1 \equiv m_1 + \delta \mod p$ for some $\delta \in \mathbb{Z}_p \backslash \{0\}$. Since only the first message block is different, equation (13) is equivalent to

$$k_1 m_1 \equiv k_1 m'_1 \equiv k_1 (m_1 + \delta) \equiv k_1 m_1 + k_1 \delta \mod p. \tag{14}$$

That is, M and M' will have the same image iff $k_1 \delta \equiv 0 \mod p$. Since neither k_1 nor δ is the zero element in the field $\mathbb{Z}_p$, M and M' can never have the same image if they are different by only a single block.

b) Two blocks: without loss of generality, assume that only the first two message blocks are different. That is, $m_i = m'_i$ for all $i \neq 1, 2$ but $m_1 \not\equiv m'_1 \mod p$ and $m_2 \not\equiv m'_2 \mod p$. Let $m'_1 \equiv m_1 + \delta_1 \mod p$ and $m'_2 \equiv m_2 + \delta_2 \mod p$ for some $\delta_1, \delta_2 \in \mathbb{Z}_p \backslash \{0\}$. Since only the first two message blocks are different, equation (13) is equivalent to

$$k_1 m_1 + k_2 m_2 \equiv k_1 m'_1 + k_2 m'_2 \equiv (k_1 m_1 + k_1 \delta_1) + (k_2 m_2 + k_2 \delta_2) \mod p. \tag{15}$$

That is, M and M' will have the same image iff

$$k_1 \delta_1 + k_2 \delta_2 \equiv 0 \mod p. \tag{16}$$

By Lemma 2, the values $k_1 \delta_1$ and $k_2 \delta_2$ are uniformly distributed over $\mathbb{Z}_p \backslash \{0\}$. Consequently, the probability of satisfying equation (16) is $1/(p-1)$. Given that p is an N-bit prime, the probability of satisfying equation (16) can be bounded by $1/2^{N-1}$.

c) More than two blocks: let $m'_i \equiv m_i + \delta_i \not\equiv m_i \mod p; \forall i \in I \subseteq \{1, 2, \cdots, B\}$; $|I| \geq 3$. Then, equation (13) is equivalent to

$$k_i \delta_i + \sum_{\substack{j \in I \\ j \neq i}} k_j \delta_j \equiv 0 \mod p, \tag{17}$$

for some $i \in I$. The only difference between this case and the case in which only two blocks are different is that $\sum_{j \in I, j \neq i} k_j \delta_j$ in equation (17) can still be congruent to zero modulo p, while $k_2 \delta_2$ in equation (16) can be chosen not to be congruent to zero in the two block case. Consequently, similar to the two block analysis, the probability of satisfying equation (17) is $1/p$.

UHF2. Recall that two messages, $M = m_1 \| \cdots \| m_B$ and $M' = m'_1 \| \cdots \| m'_B$, will have the same image if and only if the following holds:

$$h(M) \equiv k_1 m_1 \mod 2^N + \sum_{i=2}^{B} k_i m_i \mod 2^{2N}$$

$$\equiv k_1 m'_1 \mod 2^N + \sum_{i=2}^{B} k_i m'_i \mod 2^{2N} \equiv h(M'). \tag{18}$$

When the k_i's, for $i = 2, \cdots, B$, are chosen uniformly at random from $\mathbb{Z}_{2^N}$, the probability that $\sum_{i=2}^{B} k_i m_i \equiv \sum_{i=2}^{B} k_i m'_i \mod 2^{2N}$ is known to be $1/2^N$ for any distinct messages $M \neq M'$ (see [3] for a proof). For an odd $k_1 \in \mathbb{Z}_{2^N}^*$, the probability that $k_1 m_1 \equiv k_1 m'_1 \mod 2^N$ is zero. This is a direct consequence of Lemma 1 and the fact that any odd integer is invertible in $\mathbb{Z}_{2^N}$. Therefore, the probability of satisfying equation (18) is $1/2^N$.

Corollary 1. *Both UHF1 and UHF2 are* $\frac{1}{2^{N-1}}$-*AU.*

In the following sections, we will show that, although both UHF1 and UHF2 are $\frac{1}{2^{N-1}}$-AU, the resulting MAC is secure when UHF1 is used while a man-in-the-middle can forge a valid MAC with probability one when UHF2 is used. We start by describing a hypothetical scenario to help formalizing our results.

4 Hashed-Image Attack

As it is typically assumed that the adversary has the ability to modify the transmitted message and its tag, it is not assumed that the adversary can modify the compressed image before it is passed to the encryption algorithm, and there is a good reason for this assumption: only the message and the corresponding tag are transmitted in the clear, which makes them vulnerable to modification (on the other hand, an adversary will need a physical access to the MAC hardware in order to be able to modify the hashed image before it is passed to the cryptographic function). However, we assume the possibility of hashed-image attacks in our analysis because it leads to our formalization and helps understanding other practical attacks (as will be detailed later).

Assume now that both the message M and its compressed image $h(M)$ have been modified to M' and $h'(M)$, respectively. Then, if the compressed image of the modified message is equal to the modified compressed image of the original message, the modified message will pass the integrity check. That is, if an adversary can modify M to M' and $h(M)$ to $h'(M)$ such that $h'(M) = h(M')$, by receiving M' the intended user will compute $h(M')$. Since $h'(M) = h(M')$, the computed authentication tag at both the sender and the receiver ends will be the same. Therefore, the modified M' will be accepted as an authentic message.

We will now analyze the probability of successfully launching such an attack on the universal hash families of Section 3.

4.1   UHF1

Assume that $h'(M) \equiv h(M) + \gamma \mod p$, for some $\gamma \in \mathbb{Z}_p \backslash \{0\}$. Recall that a modified message-image pair $(M', h'(M))$ will be accepted if the following condition holds:

$$h(M') \equiv \sum_{i=1}^{B} k_i m_i' \overset{?}{\equiv} \sum_{i=1}^{B} k_i m_i + \gamma \equiv h'(M) \quad \mod p. \tag{19}$$

To analyze equation (19), we will break the problem into two cases: modifying a single message block and modifying more than one message block.

1. Without loss of generality, assume that only the first message block m_1 has been modified to $m_1' \equiv m_1 + \delta_1 \not\equiv m_1 \mod p$. Then, equation (19) is equivalent to

$$k_1 \delta_1 \equiv \gamma \quad \mod p, \tag{20}$$

   for an unknown $k_1 \in \mathbb{Z}_p^*$, and some $\delta_1, \gamma \in \mathbb{Z}_p \backslash \{0\}$ of the adversary's choice.
2. Assume that two or more message blocks have been modified, i.e., $m_i' \equiv m_i + \delta_i \not\equiv m_i \mod p$; $\forall i \in I \subseteq \{1, 2, \cdots, B\}$; $|I| \geq 2$. Then, equation (19) is equivalent to

$$\sum_{i \in I} k_i \delta_i \equiv \gamma \quad \mod p, \tag{21}$$

   for unknown k_i's in $\mathbb{Z}_p^*$, and some δ_i's and γ in $\mathbb{Z}_p \backslash \{0\}$ of the adversary's choice.

Lemma 3. *In UHF1, an adversary modifying both the message M and its compressed image $h(M)$ to M' and $h'(M)$ so that $h'(M) = h(M')$ will be successful with a negligible probability.*

Proof. Observe that, by Lemma 2, the value of $k_1 \delta_1$ in equation (20) is uniformly distributed over $\mathbb{Z}_p^*$. Therefore, the probability of choosing a pair (δ_1, γ) that satisfies equation (20) is $1/(p-1)$. Similarly, by Lemma 2 and the fact that $\sum_{i \in I} k_i \delta_i$ can be congruent to *zero* modulo p, the probability of satisfying equation (21) is at most $1/p$. Therefore, the probability of satisfying equation (19) is at most $1/2^{N-1}$, a negligible function in the security parameter N.   □

4.2   UHF2

Assume that $h'(M) = h(M) + \gamma$, for some $\gamma \in \mathbb{Z}_{2^{2N}} \backslash \{0\}$. Recall that a modified message-image pair $(M', h'(M))$ will be accepted if the following condition holds:

$$h(M') \equiv k_1 m_1' \mod 2^N + \sum_{i=2}^{B} k_i m_i' \mod 2^{2N}$$

$$\overset{?}{\equiv} k_1 m_1 \mod 2^N + \sum_{i=2}^{B} k_i m_i \mod 2^{2N} + \gamma \equiv h'(M). \tag{22}$$

Lemma 4. *In UHF2, an adversary modifying both the message M and its compressed image $h(M)$ to M' and $h'(M)$ so that $h'(M) = h(M')$ can be successful with probability one.*

Proof. Recall that the key k_1 is chosen from $\mathbb{Z}_{2^N}^*$. Therefore, k_1 must be an odd integer. Let $m_1' = m_1 + \delta_1$, for $\delta_1 = 2^{N-1}$, and write $k_1 = 2r + 1$, for some integer r. Then,

$$k_1\delta_1 \equiv (2r+1)2^{N-1} \equiv 2^{N-1} \mod 2^N, \tag{23}$$

for any integer r. That is, by replacing m_1 with $m_1' = m_1 + 2^{N-1}$ and setting $\gamma = 2^{N-1}$, equation (22) will be satisfied with probability one, and the lemma follows. $\square$

5 Classification of Universal Hash-Function Families

In this section we define two general classes of universal hash families. As mentioned earlier, this classification will have security implications on the encryption algorithm that can be combined with the universal hash function to construct the MAC. We start with the notion of independent universal hash-function families.

Definition 3 (Independent universal hash-function families). *Let $\mathcal{H} = \{h : \mathcal{D} \to \mathcal{R}\}$ be a family of hash functions. We say that $\mathcal{H}$ is independent if for any message $M \in \mathcal{D}$, for all messages $M' \neq M$, for any function f, the probability that $\Pr[h(M') = f(h(M))]$ is negligible for any $h \in \mathcal{H}$.*

As can be inferred from their name, independent universal hash families can be used to construct a secure MAC independently of the combined encryption algorithm (as will be shown later). Observe that this definition can be viewed as a generalization of the ϵ-AXU and the ϵ-AΔU in which the function f is not restricted to be an XOR or a linear function. The second class of universal hash families is the dependent one defined as follows.

Definition 4 (Dependent universal hash-function families). *Let $\mathcal{H} = \{h : \mathcal{D} \to \mathcal{R}\}$ be a family of hash functions. We say that $\mathcal{H}$ is dependent if for a message $M \in \mathcal{D}$, there exists a message $M' \neq M$ and a function f so that the probability that $\Pr[h(M') = f(h(M))]$ is non-negligible for at least one $h \in \mathcal{H}$.*

Intuitively, the security of MACs based on dependent universal hash-function families will depend on the combined encryption primitive. By Lemma 3, the UHF1 of Section 3 is independent while, by Lemma 4, the UHF2 is dependent.

Observe the difference between our classification of universal hash families and the previous classifications in Definition 1. Our definitions deal with the ability to come up with a message M' different than the original message M and predict its hashed image *as a function of the image of* M (not strictly the absolute value of $h(M')$ or a linear or XOR function of $h(M)$). This classification gives a general and direct relation between the used universal hash function and the used encryption algorithm.

6   Analysis of the Two Classes of Universal Hash Families

We will show in this section the effect of the used universal hash family on the security of the constructed MAC. We will restrict our analysis to man-in-the-middle (MITM) attacks; the generalization to other attacks is straightforward.

Recall that there are two cases to be considered in MITM attacks: modifying plaintext only and modifying both the plaintext and its corresponding tag. When only the plaintext message M is modified to M', the fact that the used hash function is ϵ-AU guarantees that the compressed image $h(M')$ will be different than $h(M)$ with probability $1 - \epsilon$. Now, if the compressed image is processed with any one-to-one encryption algorithm, the tag $\tau = \mathcal{E}(h(M))$ will be accepted only if $\mathcal{E}(h(M')) = \tau$. Since $\mathcal{E}$ is one-to-one, this implies that $h(M') = h(M)$, which will occur with the negligible probability ϵ.

Consider now the use of many-to-one primitives, such as cryptographic hash functions. Of course, if $h(M') = h(M)$ the tag will be accepted. This, however, will occur with a negligible probability ϵ. Since the cryptographic primitive is many-to-one, however, there is a possibility that $\tau = \mathcal{E}(h(M)) = \mathcal{E}(h(M'))$ for $h(M') \neq h(M)$. Typical cryptographic hash functions, however, are pseudorandom functions; hence, it is computationally infeasible to predict their outputs on two different inputs $h(M)$ and $h(M')$ (see the proof Theorem 1).

Therefore, the fact that the used hash family is ϵ-AU guarantees that it is computationally infeasible to modify only the plaintext message in a way undetected by its unmodified tag (provided that the secret keys are protected against exposure). This is regardless of whether the used universal hash family is independent or not. In what follows, we show how this is not the case when both the message and its tag are modified.

6.1   Independent Universal Hash Families

Here we will show that any encryption algorithm can be used to construct a secure MAC when combined with an independent universal hash family, as long as the primitive provides the necessary protection against key exposure of the universal hash family. Observe that a MITM modifying a message-tag pair to (M', τ') will be successful if the following holds.

$$\tau' = \tau + \delta = \mathcal{E}\Big(h(M)\Big) + \delta \overset{?}{=} \mathcal{E}\Big(h(M')\Big), \tag{24}$$

Assume now that the compressed image is not processed by any encryption algorithm. Then, equation (24) can be reduced to

$$\tau' = \tau + \delta = h(M) + \delta \stackrel{?}{=} h(M'). \tag{25}$$

That is, the problem is reduced to modifying the compressed image by some value δ so that $f(h(M)) := h(M) + \delta = h(M')$. For independent universal hash families such as UHF1, however, the probability of satisfying equation (25) is negligible (by Definition 3 of independent universal hash families).

Now, consider the use of any semantically secure encryption scheme[3] to process the hashed image. The adversary's inability to satisfy equation (24) follows directly from her inability to satisfy equation (25) when the universal hash family is independent, and her inability to predict the ciphertext of any given plaintext.

Therefore, the use of independent universal hash families can lead to the secure construction of universal hash functions based MACs regardless of the choice of the cryptographic primitive, given that it provides the required security against key exposure. Thus the name "independent universal hash-function families".

6.2 Dependent Universal Hash Families

In this section, we will show that dependent universal hash families, although can lead to secure constructions of MACs, can also lead to totally insecure MACs, depending on the used encryption algorithm. Consider the use of a perfectly secret additive one-time pad cipher as the cryptographic primitive. (An additive one-time pad cipher is an encryption algorithm that adds the message to be encrypted to a key of equal length. When they key is used for only a single encryption, it is known that such encryption is perfectly secret in Shannon's information theoretic sense [12].)

Consider now a MITM modifying both the transmitted message and its corresponding tag to $M' \neq M$ and $\tau' = \tau + \delta$, respectively. Then, the modified message-tag pair will be accepted as valid if the following holds

$$\tau' = \tau + \delta = \left(h(M) + k_e\right) + \delta \stackrel{?}{=} \left(h(M') + k_e\right), \tag{26}$$

where k_e is the one-time key.

Consider now the use of UHF2 for message compression, the dependent universal hash family of Section 5. Then, equation (26) can be written as

$$\tau' \equiv \tau + \delta \equiv \left(k_1 m_1 \mod 2^N + \sum_{i=2}^{B} k_i m_i \mod 2^{2N} + k_e\right) + \delta$$

$$\stackrel{?}{\equiv} \left(k_1 m_1' \mod 2^N + \sum_{i=2}^{B} k_i m_i' \mod 2^{2N} + k_e\right). \tag{27}$$

---

[3] Semantic security is shown to be equivalent to indistinguishability under chosen plaintext in [7]. Which means that given two plaintexts of the adversary's choice and a ciphertext corresponding to one of them, the adversary cannot determine, with probability significantly higher than 1/2, to which plaintext the ciphertext corresponds.

Assume that the first message block has been modified to $m'_1 := m_1 + 2^{N-1}$ and recall that k_1 is an odd integer; then, similar to the proof of Lemma 4, equation (27) can be written as

$$\tau' \equiv \tau + \delta \equiv \left(k_1 m_1 \mod 2^N + \sum_{i=2}^{B} k_i m_i \mod 2^{2N} + k_e \right) + \delta$$

$$\overset{?}{\equiv} \left(k_1 m_1 \mod 2^N + \sum_{i=2}^{B} k_i m_i \mod 2^{2N} + k_e \right) + 2^{N-1}. \qquad (28)$$

Hence, choosing $\delta = 2^{N-1}$ will satisfy equation (28). That is, modifying the first message block by 2^{N-1} and modifying the authentication tag by 2^{N-1} will go undetected. In other words, forgery can succeed with probability one. This shows that dependent universal hash-function families can lead to totally insecure MAC construction, even when coupled with an encryption algorithm that provides the highest degree of secrecy. (The use of one-time pad ciphers to encrypt the compressed image has been proposed in, e.g., [8,16].)

However, this does not imply that dependent universal hash families cannot be used for the construction of secure MACs. In fact, Theorem 1 states that a sufficient condition for the construction of a MAC based on an ϵ-AU family to be secure is to be combined with a secure pseudorandom function.

Theorem 1. *Let $\epsilon \geq 0$ be a real number and let $\mathcal{H} = \{h : \{0,1\}^* \to \{0,1\}^*\}$ be an ϵ-AU family of hash functions. Let $\mathcal{F} = \{f : \{0,1\}^* \to \{0,1\}^\beta\}$, where β is a positive integer, be a pseudorandom function family that can be distinguished from a true random function family with a probability at most $\delta \geq 0$. Then the probability of forging a valid tag for MAC:=$f(h(M))$, where f and h are any members of $\mathcal{F}$ and $\mathcal{H}$ respectively, is at most $\epsilon + \delta + 2^{-\beta}$.*

Proof. Let $\epsilon \geq 0$ be a real number and let $\mathcal{H} = \{h : \{0,1\}^* \to \{0,1\}^*\}$ be an ϵ-AU family of hash functions. Let $\mathcal{F} = \{f : \{0,1\}^* \to \{0,1\}^\beta\}$, where β is a positive integer, be a pseudorandom function family that can be distinguished from a true random function family with a probability at most $\delta \geq 0$. Let $\tau = f(h(M))$ be the authentication tag for a message M. Assume an adversary is given the (M, τ) pair and wants to authenticate a different message M'. Define C to be the event that a collision in the hashing phase occurred, and let $\overline{C}$ denote the complement of C.

Observe that if $h(M) \neq h(M')$ then the adversary chance of predicting the correct tag is bounded by the adversary's chance of predicting the output of the pseudorandom function. If f is truly random and $h(M) \neq h(M')$ then the adversary's probability of successfully predicting the valid τ is $2^{-\beta}$. However, since f is not truly random, the adversary's probability of predicting its output is bounded by $\delta + 2^{-\beta}$. Therefore, we have that $\Pr(\text{forgery}|\overline{C}) \leq \delta + 2^{-\beta}$.

Consequently,

$$\Pr(\text{forgery}) = \Pr(\text{forgery}|C) \cdot \Pr(C) + \Pr(\text{forgery}|\overline{C}) \cdot \Pr(\overline{C}) \qquad (29)$$

$$\leq \Pr(C) + \Pr(\text{forgery}|\overline{C}) \qquad (30)$$

$$\leq \epsilon + \delta + 2^{-\beta}, \qquad (31)$$

and the theorem follows.   □

The use of pseudorandom functions to process the compressed image has appeared in, e.g., [3].

7 Conclusion

In this work, we studied the use of universal hash-function families in the construction of message authentication codes (MACs). We showed that driving the probability of message collision as small as possible does not guarantee a secure MAC construction. We gave an example in which coupling a universal hash-function family with a perfectly secret encryption algorithm can lead to completely insecure MAC. We gave another examples of a universal hash-function family with the same message collision resiliency that can lead to secure MAC construction when coupled with any secure encryption algorithm. Based on these two examples, we classified universal hash families into two classes: dependent and independent universal hash families. We give a sufficient condition on the encryption algorithm so that MACs based on dependent universal hash families are secure; namely, the cryptographic primitive is a pseudorandom function (PRF).

References

1. Alomair, B.: Authenticated Encryption: How Reordering Can Impact Performance. In: Bao, F., Samarati, P., Zhou, J. (eds.) ACNS 2012. LNCS, vol. 7341, pp. 84–99. Springer, Heidelberg (2012)
2. Alomair, B., Clark, A., Poovendran, R.: The Power of Primes: Security of Authentication Based on a Universal Hash-Function Family. Journal of Mathematical Cryptology 4(2) (2010)
3. Black, J., Halevi, S., Krawczyk, H., Krovetz, T., Rogaway, P.: UMAC: Fast and Secure Message Authentication. In: Wiener, M. (ed.) CRYPTO 1999. LNCS, vol. 1666, pp. 216–233. Springer, Heidelberg (1999)
4. Carter, J., Wegman, M.: Universal classes of hash functions. In: Proceedings of the Ninth Annual ACM Symposium on Theory of Computing-STOC 1977, pp. 106–112. ACM, New York (1977)
5. Carter, L., Wegman, M.: Universal hash functions. Journal of Computer and System Sciences, JCSS 18(2), 143–154 (1979)
6. Goldreich, O.: Foundations of Cryptography. Cambridge University Press (2001)
7. Goldwasser, S., Micali, S.: Probabilistic encryption. Journal of Computer and System Sciences 28(2), 270–299 (1984)

8. Halevi, S., Krawczyk, H.: MMH: Software message authentication in the gbit/Second rates. In: Biham, E. (ed.) FSE 1997. LNCS, vol. 1267, pp. 172–189. Springer, Heidelberg (1997)
9. Krawczyk, H.: LFSR-based hashing and authentication. In: Desmedt, Y.G. (ed.) Advances in Cryptology - CRYPTO 1994. LNCS, vol. 839, pp. 129–139. Springer, Heidelberg (1994)
10. Krawczyk, H.: New hash functions for message authentication. In: Guillou, L.C., Quisquater, J.-J. (eds.) Advances in Cryptology - EUROCRYPT 1995. LNCS, vol. 921, pp. 301–310. Springer, Heidelberg (1995)
11. Rogaway, P.: Bucket hashing and its application to fast message authentication. Journal of Cryptology 12(2), 91–115 (1999)
12. Shannon, C.: Communication Theory and Secrecy Systems. Bell Telephone Laboratories (1949)
13. Stinson, D.: Universal hashing and authentication codes. Designs, Codes and Cryptography 4(3), 369–380 (1994)
14. Tignol, J.: Galois' Theory of Algebraic Equations. World Scientific (2001)
15. van Tilborg, H.: Encyclopedia of cryptography and security. Springer (2005)
16. Wegman, M., Carter, J.: New classes and applications of hash functions. In: 20th Annual Symposium on Foundations of Computer Science-FOCS 1979, pp. 175–182 (1979)
17. Wegman, M., Carter, L.: New hash functions and their use in authentication and set equality. Journal of Computer and System Sciences, JCSS 22(3), 265–279 (1981)

Author Index